The Role of the Oceans as a Waste Disposal Option

NATO ASI Series

Advanced Science Institutes Series

A series presenting the results of activities sponsored by the NATO Science Committee, which aims at the dissemination of advanced scientific and technological knowledge, with a view to strengthening links between scientific communities.

The series is published by an international board of publishers in conjunction with the NATO Scientific Affairs Division

A Life Sciences	Plenum Publishing Corporation
B Physics	London and New York
C Mathematical and Physical Sciences	D. Reidel Publishing Company Dordrecht, Boston, Lancaster and Tokyo
D Behavioural and Social Sciences	Martinus Nijhoff Publishers
E Engineering and Materials Sciences	The Hague, Boston and Lancaster
F Computer and Systems Sciences	Springer-Verlag
G Ecological Sciences	Berlin, Heidelberg, New York and Tokyo

Series C: Mathematical and Physical Sciences Vol. 172

The Role of the Oceans as a Waste Disposal Option

edited by

G. Kullenberg

Department of Physical Oceanography,
University of Copenhagen, Denmark

D. Reidel Publishing Company

Dordrecht / Boston / Lancaster / Tokyo

Published in cooperation with NATO Scientific Affairs Division

Proceedings of the NATO Advanced Research Workshop on
Scientific Basis for the Role of the Oceans as a Waste Disposal Option
Vilamoura, Portugal
April 24-30, 1985

Library of Congress Cataloging in Publication Data

NATO Advanced Research Workshop on Scientific Basis for the Role of the Oceans as a Waste
    Disposal Option (1985 : Vila Moura, Portugal)
    The role of the oceans as a waste disposal option.

    (NATO ASI series. Series C, Mathematical and physical sciences; vol. 172)
    "Sponsored by the NATO Science Committee"—Ser. t.p.
    "Published in cooperation with NATO Scientific Affairs Division."
    Includes index.
    1. Waste disposal in the ocean—Congresses.   I.   Kullenberg, Gunnar.
II. NATO Science Committee.   III.   Title.   IV.   Series: NATO ASI series. Series C,
Mathematical and physical sciences; vol. 172.
TD763.N38   1985   363.7'28   86–475
ISBN-13:978-94-010-8557-1   e-ISBN-13:978-94-009-4628-6
DOI:10.1007/978-94-009-4628-6

Published by D. Reidel Publishing Company
P.O. Box 17, 3300 AA Dordrecht, Holland

Sold and distributed in the U.S.A. and Canada
by Kluwer Academic Publishers,
190 Old Derby Street, Hingham, MA 02043, U.S.A.

In all other countries, sold and distributed
by Kluwer Academic Publishers Group,
P.O. Box 322, 3300 AH Dordrecht, Holland

D. Reidel Publishing Company is a member of the Kluwer Academic Publishers Group

TABLE OF CONTENTS

FOREWORD AND ACKNOWLEDGEMENTS

This publication is a result of the meeting on scientific basis for the role of the oceans as a waste disposal option, held in Vilamoura, Algarve, Portugal, April 24 - 30, 1985. It was sponsored by the Special Panel on Marine Sciences of the NATO Science Committee. The organizing committee consisted of E.D. Goldberg (La Jolla), M.W. Holdgate (London), G. Kullenberg (Copenhagen), A.D. McIntyre (Aberdeen), J.H. Steele (Woods Hole), and J.C.J. Nihoul (Liège), the latter representing the Marine Sciences Panel. The group, except M.W. Holdgate, met in Hamburg in August 1983 to formulate the programme. This meeting, also sponsored by the Marine Sciences Panel, resulted in a proposal to the Panel, which then recommended the proposal for funding by the Panel.

The Director-General, Professor Remy Freire agreed that the National Institute for Fisheries Research, Lisbon, would co-sponsor the meeting and Ing. M. de Barros from this Institute became a member of the organizing committee and was responsible for the local organization. The Institute also provided secretarial assistance at the meeting. The support and very valuable assistance of the National Institute for Fisheries Research is hereby acknowledged with great appreciation. We also acknowledge the extra support received for travelling and expenses for the U.S. participants from the German Marshall Fund of the United States and the Alcoa Foundation. The generous support of the Marine Sciences Panel, without which the meeting would not have been made possible, is of course also gratefully acknowledged. The meeting was of a novel and different character from those usually sponsored.

G. Kullenberg

THE OCEANS AS A WASTE DISPOSAL OPTION - MANAGEMENT, DECISION MAKING
AND POLICY

M W Holdgate and P T McIntosh
Department of the Environment, London*

1. Introduction

The central thesis of this paper is not whether it is right for some of
the wastes of modern industrial civilisations to be disposed of in the
oceans - because the natural flux of elements in the atmosphere and the
rivers makes this process inevitable. It is rather about the choices we
can and should make about the pathways, quantities, rates and locations
through and at which wastes enter the seas. It contends that these
choices should depend upon objective and scientific evaluations as to
the likely effects of particular substances in the sea, and
accompanying social and economic evaluations of the acceptability of
those effects compared with the consequences of adopting other options
for disposal.

There is a primary dichotomy in the choice of policy. It can be argued
that because the oceans are regarded by many as the common heritage of
mankind, and a resource shared among all nations, it is wrong in
principle for particular industrialised states to contaminate oceanic
waters by discharging their wastes. This argument is often applied with
particular vigour to categories of wastes which excite strong social
emotions - among which radioactive substances are paramount. Holders of
this philosophy would argue that no wastes should be deposited in, or
released to, the seas if it can be avoided, and where unavoidable, all
waste discharges should be reduced to that minimum which is achievable
with the best available modern technology. Such an approach leads
logically to the adoption of uniform, maximum emission standards which
should not be exceeded anywhere - regardless of the state or dilution
capacity of the receiving waters, and of the costs of using the
technology - and to the progessive tightening of the standards as new
technologies are developed.

---

* The views expressed in this paper are those of the authors and not
necessarily those of the United Kingdom Department of the Environment.

G. Kullenberg (ed.), The Role of the Oceans as a Waste Disposal Option, 1–18.

The alternative approach is to seek a specified high standard of
environmental quality for the waters concerned and to ensure that
discharges and emissions do not lead to concentrations which exceed
that standard, and consequently to an unacceptable environmental
impact. Such a policy implies a judgement for each particular discharge
- taking other discharges, the desired uses or objectives for the
affected waters, and the geographical and ecological variability into
account - because dispersion capacities and the sensitivity of living
organisms vary with situations. It may also require a substance by
substance approach because toxicity thresholds depend on the flora and
fauna present. Possible synergistic effects must also be taken into
consideration.

Both philosophies are reflected in national waste disposal management
policies for the acquate environment today. The former is evidenced by
the stated aim of certain European and other governments to ensure that
at all times the best available technical means is used to minimise the
contamination of the sea with potentially harmful substances.The second
is reflected in the approach which the United Kingdom Government in
particular has adopted, namely to establish Environmental Quality
Objectives and Standards which permit the use of the capacity of the
sea to absorb wastes and render them harmless but, at the same time,
seek to avoid unacceptable environmental degradation. What is
'acceptable' is of course a matter of value judgement and further
argument. Both approaches are permitted by relevant international
agreements (eg by the European Community water quality Directives).

Recently, however, both of these policies have been qualified by the
recognition of two basic considerations. The first is a lack of
scientific knowledge. After many decades of oceanographic and
biological research we still remain uncertain about the pathways of
many substances in the oceans and the thresholds at which they are
likely to have effect on marine ecosystems - especially as a result of
slow accumulation over long periods of time. Moreover, because of the
vastness of the ocean, that process of accumulation is likely to be
only slowly reversible should society decide that it must act to modify
a trend. Accordingly many governments, environmental interests and
individuals are increasingly stressing the need for a precautionary and
preventative approach to environmental pollution. This is in turn
leading to a demand for environmental quality objectives and standards
with a substantial safety margin built in, or for the adoption of
minimal practicable discharges as a further component of that
preventative action.

The second development has been the mounting recognition of the wisdom
of the doctrine of 'best practicable environmental option' (originally
set out by the United Kingdom Royal Commission on Environmental
Pollution in their fifth report -1976). 'BPEO' is an approach seeking
to make the best use of various environmental sectors, to minimise
pollution damage overall. It accepts that wastes have to be disposed of
and that this process brings them into the environment somewhere, in

one form or another. It recognises that many kinds of waste could be
deposited in the oceans, passed to landfill, or incinerated and
dispersed through the atmosphere, or recycled or generated in reduced
amounts through the adoption of cleaner technology. It calls for a
judgement as to which of these processes is least damaging to the
environment – although in deciding which option should be pursued,
economic factors will also need to be taken into account – hence the
use of the word 'practicable' which denotes a blend of what is
technically feasible and economically sustainable. Taken in its widest
form, BPEO sets decisions relating to the environment in the arena of
the full range of social and economic judgements.

Clearly, if the philosophy that wastes should not be disposed of at sea
at all was adopted this would represent a judgement that this outlet
was not to be treated as the BPEO, and alternative routes should always
be pursued regardless of the relative economic costs and impacts of,
for example, land as against marine disposal. There are those who argue
that this is the right course, but this paper does not take that
argument further, except to repeat that it seems inevitable that the
oceans must receive somepart of the world's wastes. If, on the other
hand, it is accepted that some release of wastes to the sea is
acceptable so long as damage is not done – and this is the guiding
philosophy implicit in modern definitions of pollution like that
adopted in the Paris Convention and in the approach of the London and
Oslo Conventions and the Regional Seas Conventions of UNEP – then
information and analysis is needed to guide the decisions.

In choosing an outlet we need in principle to be informed about a wide
range of topics. These include:

    (a) the concentrations a discharged waste is likely to attain in
the sea (exposure or 'dose');

    (b) the sensitivity of targets liable to be affected ('response');

    (c) the consequential ecological impacts, not only on the target
species but on others interacting with and dependant upon those
species;

    (d) the likely timescale on which the discharged material will
accumulate in the marine environment or marine organisms, the rate of
removal to permanent 'sinks' outside the ecosystems involved, and other
parameters governing the likely duration of exposure;

    (e) the geographical scale of the effects of the discharges and
their variability within the affected areas;

    (f) the acceptability of those effects to the human community;

(g) the consequences of rejecting the ocean disposal option. These include the feasibility, costs and impacts of alternative ways of disposing of, or reducing, the wastes.

Clearly these components of a decision blend scientific, economic, social and political considerations. There is nothing new in this; most important decisions which affect many people and wide areas should be informed in a similar way.

Such a process of analysis does not, of course, require that the same policy will be appropriate for all substances. For example, it would not be unreasonable to decide that a minimalist approach was preferable for substances about whose long term effects we were most ignorant or concerning which the public was most apprehensive. Some communities are effectively conceding this point by reducing releases of radioactive materials to the ocean to a minimum, while continuing to permit the discharge of heavy metals and organochlorines via rivers and atmospheric pathways, and the dumping of metal-contaminated spoils in continental seas and deliberately using inland and marine waters for the controlled disposal of relatively benign substances or those creating only local effects.

There are therefore two major scientific questions we need to be able to answer in guiding a decision on policy:

(a) Is scientific knowledge adequate to allow the prediction of the consequences of a particular disposal route of a particular substance – and within what margins of error?

(b) Is it then possible to estimate the safety margins that we should seek?

There are then three social or economic questions:

(c) On objective grounds, are the predicted effects acceptable – economically or ecologically?

(d) What are the costs and effects of the alternative pathways for disposal?

(e) How far do public preference and unquantifiable aesthetic judgements and other uncertainties tilt the scales?

There must be uncertainty in the answers to all these questions and often rather arbitrary judgements must be made – but that is no reason for not considering each so far as practicable.

2. Scientific considerations influencing decision

Of 1,380,000 x 10^{12} m^3 of water in the hydrosphere, 1,350,000 x 10^{12} m^3 (98%) is in the oceans. Annually, only some 33 x 10^{12} m^3 (0.002%) water passes from land to ocean through river flow. It is self-evident that were this inflow uniformly mixed, the impact on the oceans would be negligible. But problems can arise because of localisation and the inpact of other sources.

Pollutants reach the sea by three pathways: rivers and coastal outfalls; atmospheric transport; and direct input from ships, oil rigs and off-shore structures. Table 4 compares some of these inputs to the North Sea from the United Kingdom, and may be set alongside Table 5 which gives somewhat less recent data for the Oslo Commission Area, which represents the north east Atlantic as a whole. It is clear from these tables that the atmospheric pathway is the dominant one for many of these substances. Fossil fuel combustion is apparently the main means by which iron and copper enter the sea, mainly through the wide dispersion of fine particulate material (Global 2000, 1980).

So far as the open oceans are concerned, many of the inputs of substances by man are trivial compared with natural fluxes. For example, the input of some 200 x 10^6 tonnes per year of carbon via rivers and a comparable quantity via the atmosphere amounts to only some 2% of the net primary production of carbon in the oceans, 90% of which is due to the phytoplankton. The greater proportion of nitrogen and phosphorus nutrients originating from the land is used up by plant life in the inshore waters, where local eutrophication may occur, rather than dispersing to stimulate plant growth over the oceans as a whole.

In contrast, Table 1 compares the estimated input of various metals by natural river erosion and as a result of mining activities and Table 2 estimates of oil releases to the world ocean from various sources; these are clearly both areas in which human activity has accelerated natural fluxes perceptibly. Various international syntheses (eg Holdgate, Kassas and White, 1982) treat petroleum derived hydrocarbons, metals, synthetic organochlorine pesticides and radionuclides as the only four classes of contaminant liable to accumulate on an oceanic scale; it is noted that discharges of some of them, such as DDT and PCBs, are likely to fall as a consequence of control policies, while DDT in particular is fairly efficiently scavenged by settling particles and deposited in the sea bed (where it may nonetheless accumulate in benthic organisms). Lead has demonstrably increased in concentration in the surface waters of the sea especially because of its input as an aerosol, and this trend is likely to be reversed as a result of the phasing out of lead as a petroleum additive in North America and Western Europe.

Local 'hot spots' or more widely, but still restricted, affected areas generally occur in coastal waters, especially at mouths of major rivers, and in the coastal or landlocked seas. This is because, compared with the oceans, coastal waters are often shallow and receive the direct inpact of river flows and because the volume of such seas is small. The flushing times of such waters may be relatively long – in the North Sea between perhaps 50 and 150 days, according to geographical location and season; the Baltic has an even longer flushing time and interchange between the Mediterranean and the open Atlantic proceeds on a very restricted scale.

A recent review of the state of the North Sea, which can be taken as typical of the more polluted shelf seas of the world (FRG, 1984) indicated that the main pollution problems occurred in the mouths of the major rivers, particularly those flowing from the Federal Republic of Germany, the Netherlands, Belgium and the United Kingdom: in landlocked inshore waters like the Waddensee, and in the immediate vicinity of sewage sludge dumping grounds like that in the Thames estuary. There was also some concern over oil pollution from ships, particularly off the German, Dutch and Belgian coasts, and over discharges from oil rigs. The general conclusion was however that the marine resources of the North Sea as a whole were not suffering from pollution and that priority action needed to be given to reducing the riverine inputs; some concern was also expressed over the amount of sewage sludge dumped in the North Sea and calls were made for it to be reviewed.

This example illustrates clearly some of the choices to be made. Most of the sewage sludge in the North Sea is dumped from the United Kingdom. The amount – approximately 0.32×10^6 tonnes of dry solid per annum – is equivalent to approximately 25% of sewage sludge generated in the UK. Some 70% is disposed of on land (60% of it either on farm land or in landfill sites) and some 5% is burned (no doubt giving rise to some aerosols and particulates which find their way back into the sea). Metal contamination in sludges disposed of on agricultural land is of increasing concern, because of the tendency of these contaminants to persist in the soil, and the result of antipathy to both marine and farmland disposal could be increasing pressure for the dumping of these sludges in landfill sites, since incineration is both wasteful of energy and expensive and gives rise to its own environmental problems.

This discussion is important because it demonstrates that the community already make certain involuntary choices between the air, rivers, direct outfalls and dumping as pathways for the disposal of wastes at sea. It is interesting to see how dominant the atmospheric pathways are for many substances – without, of course, creating the same high local concentrations that a direct outfall or a dumping ground is liable to produce. In many circumstances the policy options which we are now able to consider would result only in altering the ratio of these inputs rather than withholding material from the seas altogether. The main exception is where it is practicable and socially acceptable to retain materials on land, in excavated landfill sites, deep depositories or engineered long term stores.

But disposal to land is not problem free. If aquifers are prone to
contamination, nitrate and phosphate may leach from the wastes and
cause eutrophication and problems for supply, while the processes of
decomposition in the sludges can generate methane (although the latter
can be tapped off and utilised under some circumstances). Co-disposal
of sewage sludges with domestic refuse is a possibility, but clearly
the diversion to land of the substantial volumes now going to sea would
bring social costs, particularly in those areas where there are not
very many suitable on-land sites. This option is being explored
thoroughly for radioactive substances, where the alternative to sea
disposal is a mixture of land based options ranging from encapsulation
(vitrification) and long term dry storage for high level wastes,
followed by disposal in specially constructed deep underground shafts
or shafts; similar storage followed by deep disposal for the more
active category of intermediate level wastes; land surface disposal in
encapsulated form for the less intractable intermediate level wastes;
and land burial for low-level wastes. The best practicable
environmental option for the category of wastes hitherto dumped at sea
is currently being reviewed in the UK, and also by the Commission of
the London Dumping Convention and by the IAEA and NEA.

Most reviews suggest that marine pollution on an oceanic scale is
unlikely to have a significant environmental impact (GESAMP:1981).
However, some individual scientists have argued that the coverage of
considerable oceanic water areas with oil films, with entry of
polycyclic aromatic hydrocarbons into marine ecosystems and with
physical changes at the ocean atmosphere interface, may have subtle but
serious ecological effects. It is also argued that the mean
concentration of certain widespread pollutants is now only one or two
orders of magnitude below the mean critical value where primary
production of phytoplankton is reduced by 50% (Izrael, 1984). Such
analysis leads to the argument that we should not be complacent about
the resilence of the oceans as a whole in the face of pollution, and
should certainly move ahead with coordinated research in this area.

Nonetheless the general consensus is that it is in the continental and
landlocked seas, particularly estuaries and inshore waters, that the
impact of marine pollution are most acute. These are the most shallow
and land-locked parts of the marine environment, are closest to urban
and industrial settlements, and are most likely to require discharges
of waste. Over the world as a whole, sewage, petroleum hydrocarbons,
agricultural wastes, and effluents from chemical, food, metal and pulp
and paper industries are among their most serious polluters
(Table 3). They are therefore the areas most in need of protection.

3. Management, decisions and policies.

3.1 The generation of Wastes

Certain of the kinds of wastes now disposed in the ocean seem likely to
continue to be deposited there on approximately the present scale. For

example there seems no alternative to most of the marine dumping of
dredged spoils, often arising within the marine or estuarine
environment. While it is conceivable that changes in sewage treatment
technology will alter the composition of sewage sludges (for example
reclaiming a higher proportion of metals and other intractable
contaminants of industrial origin, or preventing their arising because
of improved industrial technology) it is likely that in the foreseeable
future sewage sludges will require disposal in approximately their
present quantities. Advances in industrial technology may also reduce
the emission of some metals, but measures purely aimed at pollution
abatement may be costly. Mention has already been made of the
reductions in airborne lead likely to arise from the widespread use of
unleaded petrol, and the amount of cadmium discharged to the sea
through rivers over the past twenty years has fallen as a result of
tighter controls on the use of this metal. Inputs of persistent
organochlorine pesticides and mercury are also likely to fall as a
result of changes in use. Increased political emphasis on prevention of
pollution rather than cure is likely to stimulate the development of
low-waste and non-waste industrial processes. Changes in ship design
and more especially the more rigorous enforcement of agreed
international measures, under the MARPOL Convention may also reduce
hydrocarbon inputs. But it is prudent to regard all these trends as at
best marginal adjustments, offset in many resumed industrial growth and
the emergence of new industries, some of which (eg the electronics
industry) are by no means free from environmentally intractable
byproducts. Biotechnology itself brings with it the possibility of new
forms of contamination in the world environment.

It is therefore unwise to think of the trend towards low-waste or
non-waste processes as solving all the problems of marine
contamination, even though there are some outstanding recent examples
of reduced water pollution resulting from such industrial changes
(Table 6). Similarly, the evolution of sewage treatment processes will
not provide an instant panacea. Technical means exist to reduce the
BOD, nitrate and phosphate levels of sewage effluents through tertiary
treatment, but the costs are high and the practice is likely to be
introduced only where it is considered essential in order to maintain
the quality of waters for supply or to avoid extreme risks of
eutrophication.

One approach, which has wide support, is to emphasise the use of later
and clearer technolgies as new and replacement plants are constructed.
This is not a minimalist strategy - it is rather taking advantage of
the often lower incremental costs of pollution abatement that can arise
with new processes compared with add-on controls at existing plant.

3.2. Scientific judgements concerning waste disposal at sea

As discussed above, the scientific evidence suggests that the current
levels of use of the open oceans as a sink for the world's waste are
not leading to widespread measurable damage. Even so more research is

clearly desirable, particularly in order to explore the possible long
term ecological impacts of oil, petroleum hydrocarbons, persistent
pesticides and metals. The disposal of radioactive materials in the
deep ocean or in the deep ocean bed is to be the subject of continuing
scrutiny at international level.

Where the scientific evidence does justify pressure for tightened
controls is in estuaries, and parts of inshore and landlocked seas. An
increasing number of reviews is drawing attention to these as critical
areas. For example, the most recent Worldwatch Institute Report (Brown
et al, 1985) emphasises the importance of estuaries as nursery grounds
for many fish that are caught offshore as well as being the site of
productive shellfish and fin-fish industries, and points out that these
are also the regions of the sea most at risk from damaging pollution.
National legislation and Regional Conventions including those
negotiated within the Regional Seas Programme of UNEP already apply to
many of these waters, but Brown et al. illustrates by reference to
Chesapeake Bay in the USA the magnitude of the effort required to clean
up a major "hot spot" ($1 billion in that case). The general approach
is likely to follow that suggested recently to be appropriate for the
North Sea, namely to pay particular attention to inputs through direct
discharges to estuaries and coastal waters and via in-flowing rivers,
and to improve the quality of dumped wastes. Land-use planning and the
adoption of measures to control soil erosion will be important in some
areas. There is also a need for continuing monitoring of the state of
these seas, especially to ensure that potential new problem areas are
detected and remedied before they become acute.

3.3. The impact of public opinion

However, we can predict that there will continue to be increasing
public and political demands for improvements in environmental quality,
and that some forms of waste disposal in the ocean may be regarded as
increasingly unacceptable because they are perceived as creating long
term risks, because the uncertainties are regarded as too great, or
because it is considered they should be phased out as part of the
precautionary process. Scientists, economists and those whose principal
concern is finance are often prone to criticise media and environmental
groups for demanding more stringent (and expensive) action on
insufficient evidence. If this is to be avoided the scientific
community will need to expound the results of research and monitoring
clearly and convincingly, so developing a broader consensus. Public
opinion needs to be guided towards recognizing that the best
practicable environmental option for waste disposal should be checked
in order to make certain that substitute policies are not even less
acceptable environmentally than disposal at sea.

A methodology for conducting BPEO analysis, relating scientific impacts
and economic costs, together with public perceptions of risks and with
appropriate allowances for uncertainty, should be developed as a matter
of urgency; this process could take into account the adaptive

environmental assessment approach of Holling (1978). If, as seems
likely, public opinion continues to favour a reduction of inputs to the
sea where this can conveniently be done, pressure to find economically
and socially acceptable alternatives will clearly grow.

3.4. The state of current controls

The seas are a subject to a wide and interlocking series of
internationally agreed control measures. The dumping of wastes at sea
generally is controlled by the London Convention, while the dumping of
wastes in the North East Atlantic is regulated by the Oslo Convention
and pollution from landbased sources by the Paris Convention. There are
some ten UNEP Regional Seas Conventions in existence or in draft which
cover the coastal waters of most of the industralised and some
considerable parts of the developing world, and general guidelines
concerning land based sources have been agreed. Shipping is the subject
of control by IMO Conventions, and there are also international
measures dealing with oil rigs and platforms, and specifically with the
transport of hazardous goods. Within Europe, a series of EEC Directives
covers the discharge to the marine and freshwater environments of
particularly intractable substances. Fine tuning to improve the
efficacy of marine pollution prevention is clearly possible within
these agreements, and we would argue that the recent North Sea
Conference demonstrated that there is no need, in that case, for
special additional supplementary organisations or legislation; this may
well apply with equal validity to other seas. But it may be that a
greater willingness to use or adapt existing mechanisms would bring
about important benefits.

Most developed countries have national policies and control systems
capable of regulating disposals of material to the oceans. In developed
countries the trend in recent decades has been towards improvement in
the quality of rivers and estuaries, and increasing constraints on the
release of substances to the sea. The overall trend is therefore likely
to be towards progressive improvement including the elimination of the
most hazardous releases. As an aid in this process, cost effective new
technology, which allows the reduction of discharges of persistent and
intractable substances, would be of great social value.

3.5. Prospects and Policy

It is clear that the oceans are an inevitable sink for a substantial
part of the wastes of modern industrial civilisations. Damage from
this process is generally and fortunately localised, although there are
some severe problems and others will develop. But as attention turns
increasingly to a preventative approach to pollution, we should pay
greater attention to means of excluding from pathways to the sea those
substances whose toxicity, persistence, or other properties makes them
especially undesirable. We should encourage a more critical scrutiny of
what constitutes the 'best practicable environmental option' for
disposing of those wastes that we are bound to go on generating and the
avoidance of generating wastes by taking opportunities for recycling

using cleaner technolgies: A general policy could be, so far as practicable, one of protecting undamaged areas and of restoring, or at least, preventing further deterioration of, damaged areas – although the costs can be great and the timescales long.

If such measures and policies are followed then the prognosis for the seas is a good one.

TABLE 1

Estimates from Annual River Discharges of Amounts of Metals Injected into the Oceans Annually by Geological Processes and by Man		
	By Geological Processes (in rivers)	By Man (in mining)
	(in thousands of metric tons)	
Iron	25,000	319,000
Manganese	440	1,600
Copper	375	4,460
Zinc	370	3,930
Nickel	300	358
Lead	180	2,330
Molybdenum	13	57
Silver	5	7
Mercury	3	7
Tin	1.5	166
Antimony	1.3	40

Source: Michael Waldichuk, Global Marine Pollution: An Overview, Paris; UNESCO, 1977, p.20

TABLE 2

Best Estimates of Petroleum Hydrocarbons Intro- duced into the Oceans Annually		
Source	Best Estimate	Probable Range
	(millions of metric tons)	
Natural seeps	0.6	0.2-1.0
Offshore production	0.08	0.08-0.15
Transportation		
LOT tankers	0.31	0.15-0.4
Non-LOT tankers	0.77	0.65-1.0
Dry docking	0.25	0.2 -0.3
Terminal operations	0.003	0.0015-0.005
Bilges bunkering	0.5	0.4-0.7
Tanker accidents	0.2	0.12-0.25
Nontanker accidents	0.1	0.02-0.15
Coastal refineries	0.2	0.2-0.3
Atmosphere	0.6	0.4-0.8
Coastal municipal wastes	0.3	–
Coastal nonrefining		
industrial wastes	0.3	–
Urban runoff	0.3	0.1-0.5
River runoff	1.6	–
Total	6.113	

Source: National Academy of Sciences, Petroleum in the Marine
Environment Washington, 1975. p.6

TABLE 3
Marine Pollution in various Regional Seas

Water Discharge or other Process of Activity Potentially Causing Contamination	Baltic Sea	North Sea	Mediterranean Sea	Persian Gulf	West African Areas	South African Areas	Indian Ocean Region	South-east Asian Areas	Japanese Coastal Waters	North American Areas	Caribbean Sea	South-west Atlantic Region	South-east Pacific Region	Australian Areas	New Zealand Coastal Waters
Sewage	x	x	x	x	x	x	x	x	x	x	x	x	x	x	x
Petroleum Hydrocarbon (Maritime Transport)	x	x	x	x	x	x	x	x	x	x	x	x			
Petroleum Hydrocarbon (Exploration and Exploitation)		x		x	x			x		x	x	x	x		
Petrochemical Industry		x	x	x					x	x	x				
Mining		x					x		x				x	x	
Radioactive Wastes	x	x	x				x		x	x		x			
Food and Beverage Processing	x	x	x		x					x	x	x	x	x	x
Metal Industries		x	x		x				x	x		x			x
Chemical Industries	x	x	x						x	x					
Pulp and Paper Manufacture	x				x					x			x	x	x
Agriculture runoff (Pesticides and Fertilizer)			x		x		x	x		x					x
Siltation from Agriculture and Coastal Development						x	x	x			x				
Sea-salt Extraction							x				x				
Thermal Effluents							x	x		x	x	x	x		
Dumping of Sewage Sludge and Dredge Spoils		x								x	x				

Source: GESAMP, repeated in Holdgate, Kassas and White (1982)

Table 4. Inputs to the North Sea from the United Kingdom

Substance	Atmosphere	Dumping	Rivers & Direct Discharge
Cadmium	530	12–18.3	37 – 41
Mercury	5.6	10.1 – 12.5	10 – 16
Copper	4.9×10^3	1192 – 1281	600 – 700
Lead	5.6×10^3	1384 – 1394	400 – 650
Zinc	1.45×10^4	5690 – 6040	2600
Chromium	720	1358 – 1363	500 – 700
Nickel	1.65×10^3	632 – 644	450 – 550
Arsenic	420	less than 0.35	250

All figures metric tonnes per annum.

(Source, Heriot-Watt University report to DOE, unpublished, May 1984)

Table 5.   Inputs of Pollution to the Oslo Convention Area

	Domestic sewerage	Industrial waste	Domestic + Industrial	Rivers	Dumping	Atmospheric deposition
Total flow (million m³/y)	5,664	3,432	9,393	316,514	-	-
Contaminant (tonnes/y):						
Nitrogen	109,999	70,255	202,481	973,010	22,202	(400,000)a
Phosphorus	29,759	25,042	56,249	94,794	13,048	(~2,000)b
Suspended solids	388,000	9,354,100	9,893,100	5,188,000	-	-
BOD	452,000	395,000	1,125,000	938,000	-	-
Iron	4,958	16,051	28,336	246,588	-	(105,000)c
Manganese	310	-	-	30,207d	-	(4,100)c
Cadmium	43d	38	80d	421d	89	(530)c
Copper	598d	891	1,492	2,786d	2,426	(4,900)c
Chromium	176d	170	381d	2,678d	2,712	(720)c
Nickel	219d	173	391d	2,417d	527	(1,650)c
Lead	246d	785	1,726	3,831d	4,248	(5,600)c
Zinc	1,279d	11,053	13,719	19,275	9,131	(14,500)c
Mercury	17	6.3	23.3	36.4	35	(5.6)c

Source: Holdgate, White and Kassas (1982), who quote detailed origin of data.
Most information from ICES.

Note   a, b, for the Baltic alone
       c, for the North Sea alone
       d, estimated by compiler of ICES tables

Data are in most cases for 1974 or 1975. Input from France, Spain, Portugal
and North America not available.

TABLE 6 Cost-effectiveness of some new technologies reducing pollution in France

Process	Company	Cost of operating conventional or destructive pollution control process (French Francs)	Profit of alternative recovery process (French Francs)
Recovery of hydrocarbon in an oil refinery	Raffinerie Flf Feyzin (Rhone)	Investment: Nil Operating costs: 2,438,000	Investment: 11,000,000 Operating costs: 2,644,000 Sales of recovered product: 8,000,000 Gross operating profit: 5,356,000
Recovery of Methionine mother liquor by evaporation	Société Alimentaire Equilibree de Commentry (Allier)	Investment: 9,600,000 Operating costs: 960,000	Investment: 7,000,000 Operating cost: 10,500,000 Sale of recovered product: 13,000,000
Recovery of protein and potassium from a yeast factory	Société Industrielle de la Levure Fala (SILF), Usine de Strasbourg (Bas-Rhin)	Investment: 10,800,000 Operating costs: 1,080,000	Investment: 5,200,000 Operating costs: 860,000 Sale of recovered product: 1,015,500 Gross operating profit: 155,500
Recovery of lead and tin from furnace fumes	Société des Alliages d'Etain et Derives Montreuil (Seine-Saint Denis)	Investment: Nil Operating costs: Nil Sale of recovered product 4,400 Profit: 4,400	Investment: 300,000 Operating costs: 200 Sales of recovered product: 8,930 Gross operating profit: 8,730
Conversion of phosphoric acid waste into plasterboard	Rhone Progil, Les Roches deCondrieu (Isère), Rouen (Seine-Maritime)	Investment: 9,000,000 Operating costs: 5,000,000	Investment: 35,000,000 Operating costs: 73,000,000 Sales of recovered product: 73,500,000
Water recycle in fiberboard plant	Isorel, Castel Jaloux (Tarn-et-Garonne)	Investment: 5,000,000 Operating costs: 500,000	Investment: 2,500,000 Operating costs: 100,000 Sales of recovered product: 350,000 Gross operating profit 250,000

TABLE 6 CONTINUED

Process	Company	Cost of operating conventional or destructive pollution control process (French Francs)	Profit of alternative recovery process (French Francs)
Recycle of effluents in glue and galatine manufacture	Societe des Establissements Georges Alquier Bout-du Pont-de-l'Alin, Mazamet Tarn)	Investment: 534,000 Operating costs: 53,000	Investment 248,000 Operating costs: — Reduced consumption of chemicals and sale of recovered product: 18,000 Gross operating profit: 18,000
Recovery of iron dust in steel works	Sacilor, Gandrange, (Moselle)	Investment: 3,700,000 Operating costs: 1,850,000	Investment: 9,800,000 Operating costs: 3,250,000
Recovery of plum juice	Establissements Laparee Castelnaud de Gratecombe (Lot-et-Garonne)	Investment: 768,000 Operating Costs: 77,000	Investment: 235,000 Operating costs: 140,000 Sale of recovered product: 247,500 Gross Operating profit: 107,500
Recovery of glycerine in a soap factory	Savonnerie du Lutterbach (Haut-Rhin)	Investment: 600,000 Operating costs: 60,000	Investment: 400,000 Operating costs: 101,700 Sale of recovered product: 280,000 Gross operating profit: 178,300
Recovery of quarry washings	Societe d'Exploitation de l'Entreprise Hirsaint-Lary (Hautes-Pyranees)		Investment: 188,000 Operating costs: 3,200 Sale of recovered product: 11,000 Gross operating profit: 7,800

Source: French government statistics cited by Holdgate, Kassas and White (1982).

REFERENCES

Brown, Lester R, et. al. (1985). State of the World, 1985 A Worldwatch
Institute Report on Progress towards a Sustainable Society.
W W Norton & Co, New York and London

FRG (1984). Background paper on the state of the North Sea, and other
papers for the Conference on the North Sea and organised by the Federal
Republic of Germany, November 1984.

GESAMP (1981). The Health of the Oceans. Report of the Group of Experts
on Scientific Aspects of Marine Pollution. Food and Agriculture
Organization, Rome.

Global 2000 (1980). Entering the Twenty-First Century. The Global 2000
Report to the President. G D Barney, Study Director. Council on
Environmental Quality and Department of State, Washington DC. Holdgate
M W, Kassas M and White G F (Eds) 1982. The World Environment,
1972-1982. A Report by the United Nations Environment Programme.
Dublin: Tycooly International.

Holling C S (Ed) (1978). Adaptive Environmental Assessment and
Management. J Wiley & Sons, Chichester.

Israel Yu. A (1984). Paper presented to Inter-Parliamentary Union:
United National Environment Programme Conference on the Environment,
November-December 1981, Nairobi.

Royal Commission on Environmental Pollution (1976. Air Pollution
Control - an Integrated Approach. Fith Report, (Cmnd 6371). HMSO,
London.

ACCEPTABLE ENVIRONMENTAL CHANGE FROM WASTE DISPOSAL

Edward D. Goldberg
Scripps Institution of Oceanography
Ocean Research Division, A-020
La Jolla, California 92093

ABSTRACT. Environmental alterations result from the disposition of
waste materials. The identification of acceptable changes is a major
criterion of effective waste husbandry. The protection of public
health and of ecosystem integrity constitute the primary guidelines for
appropriate waste management. Still, subjective judgements are often
necessary and can supplement the assessments based on scientific and
technological data. Three examples of possible changes in the marine
environment as the consequence of the entry of societal discards are
considered: (1) the disposal of sewage sludge; (2) the release of tin
butyls through their use in anti-fouling paints on boats and ships; and
(3) artificial radionuclide releases. Once the maximum acceptable
levels or fluxes of polluting substances are determined, the translation of these numbers into regulatory and monitoring activity is essential. Public perception of environmental problems can conflict with
rational disposal tactics.

1.   INTRODUCTION

The introduction of waste materials to any environment will result in
changes of its physical, chemical and biological properties. Sometimes
the change will be beneficial to societal interests; sometimes it will
be detrimental. For over thirty years marine scientists have recognized that the nature of seawaters can be altered by human activities
(Goldberg, 1976). Initially, there was the concern about the widespread loss of resources through the improper discard of radioactive
debris from nuclear power generating facilities. As a consequence the
goals for the management of radioactive wastes were formulated to minimize any deleterious effects of the artificial radionuclides upon human
health.
    Subsequent impacts of pesticides upon non-target organisms, first
recognized in the early 1960s, directed attention to the protection of
ecosystem integrity. The potential destruction of entire groups of
organisms by biocides brought about use restrictions in many northern
hemispheric countries. At the present time there are fears that the

19

G. Kullenberg (ed.), The Role of the Oceans as a Waste Disposal Option, 19–26.

excessive use of halogenated hydrocarbon pesticides in the tropics and southern hemisphere may cause a repetition of the ecodisasters of the 1960s in the northern hemisphere (Goldberg, 1983).

Once the polluting material or collective of substances is identified, cause/effects relationships are sought with regard to the impact upon the receptor. Environmental levels are then established which define acceptable alterations to the makeup of the environment. The British in their husbandry of radioactive wastes discharged to their coastal waters developed the "critical pathways" approach which sought to regulate the release of those radionuclides which might jeopardize the health of the most exposed individuals (Hunt, 1982). In the United States a similar model was developed which was applied not only to public health but also to the maintenance of ecosystem vitality, "the assimilative capacity" approach (NOAA, 1979). It is emphasized that in both concepts concentration levels, which could be monitored, were sought such that unacceptable alterations of the environment could be detected.

Of crucial importance to those responsible for managing our environment is an ability to regulate inputs of materials that can threaten resources. Scientific wisdom must be translated into simple numerical concepts which can be written into our laws. But the formulation of such concepts often may require subjective inputs. Herein, I will consider several waste disposal problems whose resolution may depend both upon judgement and upon hard scientific data. Finally, it must be recognized that public perceptions on the resolution of waste management problems can be in conflict with scientific evaluations. As a consequence, the management strategy finally adopted may not be the more rational one but one based upon public acceptance.

2.   SEWAGE SLUDGE

An increasing world population with an increasing material usage is producing an increasing amount of waste which ends up in domestic sewage systems. Further, with increasing treatment of the sewage, an increasing volume of sewage sludge is being produced. Where will it go? The sludge does have a restricted land usage, especially as a soil additive, inasmuch as it may contain high concentrations of metals, some of which may be deleterious to plant growth. With large populations of the world living near sea coasts, marine disposal appears especially attractive. However, the materials do contain large amounts of organic matter whose dissolved oxygen demand can potentially interfere with life processes of animals. What amounts of alteration are allowable from marine discard of sewage sludge and how can disposals be managed through appropriate monitoring procedures.

A rather unique approach to the problem has been proposed by Jackson (1982) who adopted the concept that an unacceptable change involved the reduction of the oxygen levels of the water to less than 4 micromolar. Below this level, some organisms cannot survive. For any given site location and depending upon the flux of sewage sludge, models can be formulated to predict whether this level will be maintained.

Using accepted parameters on mixing processes in the San Pedro-Santa Barbara basins of Southern California, a model was constructed. Pipe disposal at 800 meters of varying amounts of digested sewage sludge appeared to have crucial impacts on the oxygen concentrations but shallower disposal, say at 400 meters, would not. The simple and easily measurable dissolved oxygen gas in principle can flag an unacceptable alteration to the environment. Other parameters can also be considered, such as the biostimulants nitrogen and phosphorus that could possibly lead to eutrophication, perhaps an unacceptable alteration.

Possibly a more subjective approach merits consideration. For example, in the Southern California Bight region about five percent of the fish, the Dover Sole, have surface abnormalities such as lesions, discolorations or fin-rot (SCCWRP, 1984). These alterations are associated with the inputs of domestic wastes from the eleven million citizens inhabiting the adjacent lands. Is this five percent figure an acceptable trade off for waste accommodation? Should the number be one percent or ten percent? Most important is the basis for the designation of a number. Perhaps, the assimilative capacity of the area for domestic wastes might be based upon the area whose community structure is measurably altered by the discharge. Would a five percent change be acceptable? Possibly, marine ecologists can develop substantial and usable schemes in which to designate the acceptability of change due to waste disposal.

3. THE PROBLEM OF ORGANOTINS

The entry of organotins to coastal environments provides a significant problem in the management of toxic chemicals. Although not technically a waste, they do provide an interesting "for instance" for the control of a societal discard that disturbs marine ecosystems. Many countries of the world are now reviewing the problems associated with the release of the organotins from boats and ships to harbor waters. Can we ascertain with present knowledge the flux of organotins to a given part of the coastal ocean that will result in acceptable change?

The biocidal properties of organotin compounds were initially recognized in the early 1950s by Dutch scientists. These substances have subsequently been used as fungicides, bactericides, and preservatives for woods, textiles, paper and electrical equipment (a review is given by Bennett, 1982). The major compounds are tributyl tin oxide, tributyl tin fluoride, triphenyltin chloride, triphenyltin hydroxide and tricyclohexyltin hydroxide. Two of these compounds, tributyl tin oxide and tributyl tin chloride are incorporated into marine paints as antifouling agents. There are a number of attributes of these compounds that make them especially attractive and preferable to conventionally used anti-fouling compounds such as cuprous oxides. First of all, they provide more effective protection against fouling organisms for long periods. Secondly, they do not promote corrosion. Finally, they degrade to relatively harmless compounds with time.

The release of the organotins from the paint, usually given in weight per unit area per unit time, is the significant parameter in

their use. The minimum value, consistent with anti-fouling activity, is
sought. A greater release is non-economic. A smaller release is ineffective. There are three types of formulations: a direct mixing of the
organotins with the paint constituents; incorporation into a chloroprene
rubber; or incorporation into acrylate or methacrylate polymers.

There are other entries of organotin biocides to the marine environment. For example, toxic impacts of organotin effluents from the
manufacture of "odor-free" socks where the biocide is applied as an
anti-bacterial agent have been uncovered in North Carolina (Cardwell et
al., 1984). The wastes from their manufacture have been implicated in
fish kills through discharge into natural waterways which enter the
oceans. This impact is known as the "toxic sock syndrome".

Both the military and civilian operators of boats and ships find
the use of organotin anti-fouling coatings especially desirable. They
have a five to seven year service life which reduces the need for more
closely spaced underwater cleanings of ship bottoms. Their use results
in a fifteen percent lower fuel consumption, as a consequence of a
reduction in frictional effects due to biofouling. These applications
translate into savings of hundreds of millions of dollars per year for
large fleets. Presently used copper oxide based anti-fouling paints
last for periods of only two years.

The butyl tins are introduced to the marine environment both from
leakage directly from the applied paints and from painting and blasting
operations. A typical release rate from ship hulls is of the order of
0.1 micrograms per square centimeter per day. The most common fouling
species are barnacles and seagrasses although molluscs, tubeworms,
hydroids and sponges are other known offenders. Established communities
reduce the smoothness of the hull and consequently increase friction and
drag through the water. Typical paints contain up to ten percent tributyl tin compounds.

There has been far more extensive work upon the impact of butyl
tins on organisms rather than on humans. Health effects and acceptable
daily intake rates are yet to be identified through public health
studies. There is a limited amount of field data on the persistence of
the compounds in the marine environment. The most sensitive organism so
far studied is the juvenile mysid shrimp which has a toxicity value of
0.5 micrograms/liter for a 96 hour LC50. Acute (short term responses)
toxicities have been measured for many organisms. Long term (chronic)
toxicities have not as yet been investigated. They can be approached
by applying a safety factor of ten to the acute levels, i.e., on the
basis of present wisdom, a level of 0.05 micrograms per liter or less
would protect marine organisms from adverse effects.

The persistence of butyl tins in the marine environment is poorly
known. Photolysis may be the most important mode of abiotic degradation
with environmental half-lives estimated to fall between 18 days and somewhat greater than 90 days. Another pathway from the aqueous system is
adsorption on particulates and subsequent sedimentation. Degradation by
microorganisms can limit the persistence of the tributyl tins in the
marine environment. Laboratory studies indicate that butyl tins in
fungal and bacterial cultures can have biological half-lives of weeks or
fractions of a week. In abiotic sediments where sulfides are present,

the tributyl tin sulfide forms irreversibly and is removed from any
toxic activity. In sediments biodegradation is the principal removal
mechanism and the half-lives in aerobic and anaerobic sediments are
reported to be of the order of 116 and 815 days, respectively.

Although present information is fragmentary, we have, I submit,
adequate information to make a judgement as to whether to allow ships
treated with tin butyls to utilize a specific harbor. If the yearly
average area of ship bottoms painted with tributyl tins in a given
harbor is known, the leakage of tin butyls into the waters can be estimated. Knowing the flushing times of harbor waters, coupled with the
particulate loading of the waters and with persistences based upon biological and abiotic degradations, estimates of tributyl tin levels can
be ascertained. The results can be compared with the upper limit of
0.05 microgram/liter proposed on the basis of the acute dose to shrimp
larvae. Even though this approach is somewhat primitive, it can be
checked by monitoring programs for the tributyl tin levels in the water.

An optional procedure to manage the use of tin butyls might be
based upon a trade off. Clearly there are operational and economic
benefits from the use of tin tributyls. Also the biocides will impact
upon the more sensitive organisms, especially in areas where the tin
tributyls have a short path to the organisms from the ship bottom, such
as in dock areas. There are reports that arthropods and bivalves have
disappeared from the dock areas in some marinas. The question then
becomes how much of a harbor area can be sacrificed in order to minimize
ship and boat maintenance and fuel costs. This is the same type of
question that can be posed for the disposal of domestic wastes. A
possible solution is then to accept changes in the flora and fauna of
the harbor area. Is the displacement of some indigenous organisms by
those that are more tolerant of butyl tins a reasonable exchange for
the economies resulting to the ship and boat owners?

4. THE RELEASE OF ARTIFICIAL RADIONUCLIDES

There are many instances in which public perception, as opposed to
scholarly assessment, has defined acceptable environmental alterations.
Several recent instances involving low level artificial radioactive
waste disposal address this point. The planned marked reduction in the
release of low level radioactive wastes into the coastal waters of the
United Kingdom and the decision to dispose of nuclear submarine hulks
on land as opposed to the deep sea are responses in the main to public
opinion rather than to conventional scientific and engineering wisdom
and to economic considerations. It must be emphasized that the informed
public can be put into a state of confusion by exposure to conflicting
views of knowledgeable scientists.

Obtaining public acceptance of a rational discharge strategy based
upon scholarly assessments will be difficult with the prevailing mood
that the oceans should be kept inviolate with respect to societal
wastes. This became especially evident with the problem of the disposal
of decommissioned, defueled naval submarines. The United States Department of the Navy prepared an environmental impact statement in which

land versus sea disposal options were considered (NAVY, 1984). The former involved storage of the radioactive parts of the submarines at the Savannah River Plant in South Carolina or at Hanford, Washington, both presently existing nuclear waste disposal sites. An alternative would be the sinking of the entire submarine, without the reactor core, to the seafloor in waters deeper than 4.0 km.

The assessment of the options involved potential impacts upon the environment, the uses of resources such as land or materials, the impact upon ecosystems, the effects on public health, especially during protective custody of the vessels, and the relative costs. Sea disposal turned out to be the least costly option. There were no evident impacts upon public health through exposure to radiation or to the environment in the land or sea disposal options. All alternative disposal techniques could not be examined inasmuch as such information gathering would have added additional costs to what appears to be an already expensive undertaking. Disposal at sea would have cost two million dollars, while burying the radioactive reactor compartments and the disposal of the non-radioactive parts of the ship would cost a little over seven million dollars. With one hundred ships slated for disposal, the sea option would result in a savings of around a half a billion dollars.

Environmental groups, citizens and the U.S. Environmental Protection Agency challenged the validity of the assessment. For example a fundamental concern of one environmental group, The Ocean Society, involved the absence of detailed knowledge about some aspects of deep sea ecology. Yet there were neither identified problems in the deep sea involving the transfer back to society of the radionuclides nor serious impacts upon ecosystems. Some scientists were concerned about the marine disposition of these low level radioactivities as providing a precedent for more extensive ocean dumping activities.

As a consequence of the public furor, the Navy abandoned its plan to sink the retired nuclear submarine hulks to the seafloor and instead to bury them in disposal yards. Eight ships are presently awaiting this fate. This example illustrates the urgency for marine scientists to formulate more persuasive arguments to convince the public of economically and scientifically sound courses of action in cases where the sea is proposed as waste receptable. In this case, the U.S. citizenry will pay more for a land disposal tactic that is not better than the sea alternative.

For nearly a quarter of a century, the United Kingdom nuclear establishments have discharged wastes to the coastal waters. The Fisheries Radiobiological Laboratory of the Ministry of Agriculture, Fisheries and Food have monitored these discharges to establish whether the resulting public radiation exposure is within nationally acceptable limits. Annually, reports are issued to identify the most exposed individuals and the radiation doses they received, either from the consumption of seafoods or from exposure on beaches or in boats (Hunt, 1982). The monitoring programs indicated that at no times were any individuals receiving amounts of radiation greater than those recommended by the International Commission on Radiological Protection.

However, in 1983 a television program suggested that there was an

increased incidence of childhood luekemia in the neighborhood of the
Sellafield Nuclear Facilities perhaps as a consequence of the artificially produced radioactivity entering the environment. A Commission
was formed to inquire into the problem and headed by Sir Douglas Black
(Black, 1984). The Commission concluded that the hypothesis can neither
be categorically dismissed nor can it be readily proven. Mortalities
from childhood cancer, particularly from those other than leukemia,
appeared to be near the national average; but the possibility of local
pockets of high incidence could not be excluded. The Commission pin-
pointed some great difficulties in relating measured environmental
levels to actual exposure. Population exposures are determined on
instrumental measurements of radionuclides in various parts of the
environment. Perhaps, there are unidentified sites of concentration
which act as a path back to human society. Also, unplanned discharges,
not detected by the monitoring programs, could have delivered a significant dose via an unsuspected route. Still, using models with most conservative assumptions, the Commission found no evidence of a general
risk to children or adults living near the nuclear facilities compared
to their near neighbors.

     The report had critics both among the lay public and within the
scientific community. Shortly after its issuance in 1984, there was a
series of articles in Nature and in the popular press contesting the
conclusions of the Black Commission (see for example Pomiankowski,
1984). Sophisticated statistical analyses of the data, only transferable with difficulty, if at all, to the lay public formed the bases of
the disagreement. Both the report and its critics call for further
research on the possible health consequences of the discharges. However, the Black Commission Report, suggesting that there is no demonstrable relationship between the discharges and the incidence of childhood cancer, will not overcome the momentum to greatly decrease the
disposition of the radwastes to the sea. Present intents of the British
Government are to markedly reduce the marine discharges over the next
decade. Can we more effectively and in an economically reasonable way
manage these wastes on land?

5.   OVERVIEW

The first step in pinpointing an unacceptable alteration to the marine
environment through waste discharge is the identification of what
resource is being protected--human health, ecosystem integrity, recreational use, transportation or aesthetics. Once the acceptable level
of a given waste in the ocean system is agreed upon on a scientific basis
(or the rate of introduction of the waste), the translation of numerical
levels or rates into public law is possible. However, sometimes objectively determined criteria cannot be formulated and recourse to subjective judgements takes place. For example, in the case of the entry
of tin butyls to the environment through use in ship and boat paints,
the level to protect the most sensitive organism so far investigated,
the mysid shrimp, is known. However, it may be worthwhile to consider
an alteration in the communities structures of organisms in the harbor

as a rational trade off for the thousands upon thousands of dollars
savings in fossil fuel and maintenance costs to operate the ships. The
determination of the area of the harbor that can be sacrificed is a
subjective judgement. Similar trade offs can be formulated for the
disposition of sewage sludge to a coastal area where the waste problems
of large populations can be rationally handled.

The identification of the minimum amount of information necessary
to evaluate whether or not to put a given waste into the ocean is also
a very important consideration (Goldberg, 1984). In general the essential scientific and technological information to consider the marine
disposal of wastes can be divided into three parts: the source term;
the impact upon marine resources; and the mixing or dilution with the
receiving waters. These three sets of data are site specific and are
interrelated. With such information, monitoring programs can be devised
to ensure that unacceptable impacts have not occurred or are not possible of attainment in the near future.

6.   REFERENCES

Bennett, R.F. (1983). 'Industrial development of organotin chemicals.'
     Ind. Chem. Bull. 2, 171–176.
Black, D. (1984).   Investigation of the possible increased incidence
     of cancer in West Cumbria. Her Majesty's Stationary Office,
     London, 104 pp.
Cardwell, R.D., Pavlou, S.P. and Kadeg, R.D. (1984). 'An assessment of
     the environmental chemistry and aquatic toxicology of trialkyltin
     compounds.' Prepared for the Organotin Environmental Programme
     Association. September, 1984. 68 pp.
Goldberg, E.D. (1976). The Health of the Oceans. UNESCO Press, Paris,
     170 pp.
Goldberg, E.D. (1983). 'Can the oceans be protected?' Can. J. Fish.
     Aquat. Sci. 40, Suppl. 2, 349–353.
Goldberg, E.D. (1984). 'Information needs for ocean waste disposal.'
     La Mer 22, 327–333.
Hunt, C.J. (1982).   Radioactivity in surface and coastal waters of
     the British Isles, 1980. Ministry of Agriculture, Fisheries and
     Food. Aquat. Environ. Monitoring Report No. 8. 35 pp.
Jackson, G.A. (1982). 'Sludge disposal in the Southern California
     Basins.' Environ. Sci. Technol. 16, 746–757.
NAVY (1984). Final environmental impact statement on the disposal of
     decommissioned, defueled naval submarine reactor plants. United
     States Department of the Navy. 3 volumes.
Pomiankowski, A. (1984). 'Cancer incidence in Sellafield.' Nature
     311, 100.
NOAA (1979). 'The assimilative capacity of U.S. coastal waters for
     pollutants.' Proceedings of a Workshop at Crystal Mountain
     Washington, July 29–August 4, 1979.
SCCWRP (1984). Southern California Coastal Water Research Project
     Biennial Report 1983 – 1984. Long Beach California. 332 pp.

NATURAL VARIABILITY AND WASTE DISPOSAL OPTIONS

John H. Steele
Woods Hole Oceanographic Institution
Woods Hole, MA   02543
U.S.A.

ABSTRACT. On the land and in the sea there is considerable natural variability in the physical and chemical environment and in the communities that inhabit these environments. These variations occur at all space and time scales and affect our choice of areas which are suitable or unsuitable for disposal of wastes. The natural variability also places major constraints on our capability to assess the consequences of disposal. Without adequate understanding of the underlying processes causing these changes we cannot determine what are appropriate options, particularly if we are concerned with effects on species other than man.

This paper focusses on questions of effects at the population or community level and on the ocean where these problems appear to be most acute. The sensitivity of these issues for marine populations is, possibly, the single most critical factor in decisions on land versus sea disposal.

BACKGROUND

We now have an extensive and fairly reliable description of the rates of addition of many chemicals to the ocean from the enhanced industrial activities of the last century. Certainly there are always new potential toxins but the present capabilities of analytic marine chemistry do not appear inadequate to handle identified problems.

Our knowledge of ocean physics at all scales from the local to the global has increased tremendously in the last decade and is probably more than adequate to deal with specific, or with general problems of advection or dispersal of contaminants (Csanady, this volume). The role of sediment transport, particularly in sporadic events, appears to be a tractable problem.

The early short-term LC-50 techniques for study of effects on individual organisms have now been replaced by a wide range of approaches from detailed biochemical analyses to longer-term and larger-scale bio-assays. Many of these provide extremely sensitive indicators of effects of low levels of contaminants on individual organisms. Especially they can be used to study accumulation and transformation of the toxins in the organisms.

27

G. Kullenberg (ed.), The Role of the Oceans as a Waste Disposal Option, 27–38.

The major accomplishments of these developments have been in the
chemical monitoring of the pathways by which toxins can return to the
producer, man. Although, as stated, there are the continuing concerns
about discharge of new chemical forms; and also the need to decrease
accidental discharges, the critical questions relate to acceptable
exposure or intake by humans. We accept that small groups or even
individuals should not be at risk. In this context the intractable
questions arise in defining acceptable risk for individuals or,
especially social groups.

Is this chemical route to man the critical path? If so, then it
is possible to plan relatively coherent research and monitoring
strategies in the sea and many such programs are already underway
internationally. There is a further major benefit. If man is the
critical target, then comparisons of the land and sea options have a
common focus. This does not eliminate the many problems in making such
comparisons, particularly those involving amenity and aesthetics but it .
brings nearer the potential for joint technical analyses. Thus the
longest-term question about accumulation of toxins in groundwater or in
nearshore sediments will be relevant if we are comparing only the
question of return to man as body burdens.

It is apparent, however, that there are other concerns. These
relate to other species considered as reproducing populations and the
potential effects on the size or even survival of these populations.
This is a very different kind of question. For certain shellfish
populations, closure of beds to harvesting because of contamination may
increase the population densities. There are, however, sufficient
documented cases of reduction in local communities such as certain bird
species to raise more general questions about long-term and
large-scale, chronic effects on species. These can occur in ways that
do not affect us in a direct toxicological manner or even in terms of
aesthetics or amenity.

There are inherent questions about the values we place on such
changes but I want to deal now with the problems of identifying the
possible effects at the population or community level. These problems
exist on land but, in the context of disposal options, it is in the
marine environment that they are most contentious.

We should admit that we have been essentially unable even to
define the methods which could be used, let alone supply answers. The
conceptual problem is to determine the critical population processes.
The technical difficulty is the inherently great variability in most
marine populations in space and time; particularly those on the
continental shelf which are most exposed. The conceptual and technical
aspects are intimately related since the variability is not just a
statistical or sampling problem, but reflects the ways in which species
within communities interact and, especially, how they deal with
external perturbations in their environment. Thus Gray (1979) states,
"the presence of a species in a polluted area may be more a question of
life history strategy than tolerance of adverse environmental
conditions. If this hypothesis is correct, considerable doubt must be
placed on the ecological relevance of data from toxicity tests."

It is a fact that we have not detected any effect of pollution on

the abundance of open-sea fish populations (Waldichuck, 1979) mainly
because of the large fluctuations in year-to-year recruitment to these
stocks. This lack of evidence can result in two opposite conclusions.
Because there is no observable effect, there is no effect; or because
the variability is so great, the effects are not observable but may
still be there.

Underlying these questions is our need to understand the nature of
the response of populations to the natural variations which occur in
their environments. I have argued elsewhere (Steele, 1985) that there
are significant differences between the space and time variability of
terrestrial and marine environments and these have produced very
different types of response from their respective communities.
Additional perturbations which we add to the system need to be placed
in the context of these natural variations and the community responses
to them.

The ocean has two physical properties not found on land - great
dispersive capability and large heat capacity. Taken together, ocean
physics displays relative uniformity at small scales but great
variations at larger scales - particularly long-time scales. Thus
ocean temperatures, excluding predictable cycles such as the diurnal or
annual, show relatively small variation at short, interannual time
scales, whereas atmospheric temperature variability has about the same
amplitude at interannual and decadal frequencies. We are now conscious
of these factors in relation to climate trends. Marine populations
have adapted to these aspects in various ways which differ
significantly from terrestrial species. In simple terms I have
suggested (Steele, 1985) that terrestrial species attempt to eliminate
as far as possible the effects of the smaller scales of variability,
whereas marine species accept and utilize the physical processes. This
is most evident in the evolution of reproductive strategies. The
consequences are seen in the very different approaches of marine and
terrestrial ecologists to the study of longer-term population changes.
The former look mainly at the internal community structure to explain
persistence or quasi-cyclical change; whereas the latter attempt to
relate variations to climate factors. These real differences can be
critical to the understanding of the consequences of additional
environmental changes.

For terrestrial animals there is a more clearly perceived
distinction between "external" and "internal" factors. Thus for sea
birds the immediate and local population effects of DDT through
egg-shell thinning can be appreciated, even if we still differ on the
larger-scale community aspects. For marine organisms, even when we
know the biochemical pathways and consequences for individual eggs or
larvae, we are essentially unable to relate these results to population
criteria. To repeat, this is not merely a statistical problem but
arises because the narrow limits on clutch size in birds is a very
different adaptation to natural environmental stresses compared with
the large egg production and high recruitment variability in nearly all
marine fish.

Thus a comparison of marine and terrestrial options has, as one
critical component, the need for methods to go from effects on

individual marine organisms to population consequences. I suggest that
the route, at present, is circumstantial and depends on understanding
the responses to natural variability. Let me give two specific
examples.

CASE STUDIES

Consider the waste disposal at Dumpsite 106 off New York. Capuzzo
(1985) has shown significant effects of 96-hour exposures on
physiological indices and especially on reproductive rate for three
species of copepod. For two of these species there are decreases in
reproduction at concentrations of 10^{-5} and 10^{-4} but no significant
effects at 10^{-6}. The third species (Temora) seemed minimally effected
by the higher concentrations.

Materials dumped on each occasion are of the order of $500m^3$ and
the initial dilution is about 10^{-4} (Wiesenburg & Brooks, 1983).
Assuming conservatively, a depth distribution of 20m then the area
which would be covered uniformly is $0.25m^2$ or less. Csanady (1981) has
investigated the expected dilution rates in this environment and
proposes that the combined effects of diffusion and shear dispersion
would decrease the concentration to 10^{-6} in a few days. Thus the time
scales of Capuzzo's experiments are appropriate to the expected
exposures at DS106. What are the appropriate space and time scales for
the variability of zooplankton populations?

One method, used by Mackas (1984) in a study on the west coast of
Canada, is to compute the spatial coherence scales for zooplankton
population parameters. In terms of community structure, these length
scales are of the order of 30-50km. This agrees well with the expected
scales for animals with life cycles of 20-40 days (Frost, 1980) and
Mackas attributes the observed coherence scales to reproducing cohorts.
In areal units these scales are of the order of 5.10^3km^2. Thus for the
pollutants at DS106 to have a significant effect on population
structure, the affected area would need to be about 200 times larger,
requiring presumably an equivalent increase in quantity of material
dumped.

The studies by Mackas also show that total zooplankton biomass has
much shorter coherence scales of the order of 10km. These can be
associated with day-to-day variability in feeding aggregations and
correspond to the short coherence scales in their phytoplankton food
which are of the order of 5km or less. These very short scales
emphasize the sampling problems inherent in attempting to observe any
significant changes in general estimates of plankton populations.

This specific example allows one to make several points. The
essential idea is that the variability of the system which is normally
supposed to act against estimation of effects is used here as the
measure to define the critical population scale. Secondly, this scale
is taken from data from a quite different region. The assumption is
that the order of magnitude of the coherence may be relatively
constant. This in turn is based on assumptions about the general
nature of physical dispersion processes and lifetimes of the animals
involved. A further conclusion is that the possible effect (or lack of

it) at the population level must be is inferred rather than observed. Because of the shorter coherence scales for biomass, it would be difficult if not impossible to observe a direct effect on community structure by routine sampling and the density of observations required to deduce coherence scales on each occasion would preclude general use.

The underlying concept is that variability as a function of scale contains more useful information than the mean. The critical factors are; (1) that there is a regularity in these spectra of variance, and (2) that we can relate these spectra to population parameters.

There have been many studies of the variance spectra of temperature and phytoplankton distributions from near-surface transits (Platt and Denman, 1975). These studies show general regularity in distribution with a slope of (wave number)$^{-2}$ which derives from the consequences of scale dependent dispersion. Thus the inherent mixing processes in the ocean impose patterns on the variance which can be used as one, partial, description of the variability of the system. There are similar spectra in time (Whitledge, 1983). Fewer data are available for zooplankton and pelagic fish. They show similar distributions but with shallower slopes. There are no comparable data for benthos due to sampling problems.

Many of our concerns, however, involve questions about long-term effects from low-level chronic exposure; particularly in relation to fish populations. How do such possible effects interact with natural variability, especially the longer-term trends? There is considerable discussion about the causes of these quasi-cyclical changes observed both in terrestrial and marine populations (Cushing, 1982). For pelagic fish stocks I have suggested (Steele & Henderson, 1984) that the large variability can be simulated with a simple stochastic model of a system with two equilibrium states. The problem can be typified by the changes in North Sea herring stocks after 1960 (Saville & Bailey, 1980). Recruitment expressed as year class strength, Figure 1, is highly variable but the data suggest a relatively sudden switch in two to three years from a high to a low population level. This kind of jump can be attributed to overfishing, to natural environmental factors, or to pollution. Mathematically this pattern of change can be produced by the simple equation

$$dP/dt = aP(1 - P/b) - P^2/(1 + P^2) \tag{1}$$

where the "natural" variability in the population P can be introduced through stochastic variation in the population growth rate $\underline{a}$. For our purpose here we can consider manmade effects as changing the carrying capacity $\underline{b}$; for example by decreasing spawning grounds or food supply. If the system were deterministic with a steady trend in $\underline{a}$ (or in $\underline{b}$) then at some point the population would jump from one (low/high) state to the other, Figure 2a. This simulates the concern inherent in the idea that gradual environmental change could have relatively sudden and dramatic consequences. When, however, the natural year-to-year variations in recruitment are very large so that $\underline{a}$ can vary annually across the range of the steady state regimes then a very different type of response is found. High frequency variability can be absorbed but

the lower frequency components of the variance in _a_ can induce jumps
between states. The period between these jumps depends on _b_ so that a
decrease in the carrying capacity will increase the frequency of these
jumps rather than their amplitude, Figure 2B. This very simple model
may help to elucidate the kinds of response that may occur. But its
simplicty precludes any specific statements about the particular casual
factórs which produce these effects. These examples are intended to
show that the variability in space and time is not merely "noise" but
an important parameter. Further, there is no simple additive relation
between natural (noise) variability and manmade (signal) changes
whereby the latter could be distinguished from the former by a large
enough sampling program.

DISCUSSION

     I began by considering the relative success of chemical monitoring
programs to determine pathways back to man. By comparison what may be
termed population effects monitoring has been essentially unsuccessful
for open-sea pelagic systems (MacIntyre, 1984).
     Effects on benthic fauna are only noticeable when there is also
definite evidence of physical alteration in quiescent area where
drilling muds or sewage sludge are dumped. Possibly the best
documented example (Topping, 1976) is sludge from the city of Glasgow
(population one million) in the Clyde Sea area where an area of about
$15km^2$ at 100m depth is significantly altered physically, in terms of
organic matter in the sediment and in the fauna. Television surveys
show a clearly demarked area of contamination so that the costs and
benefits can be evaluated relatively well. (The annual loss to
fishermen was put at $185,000.) The value judgements about sludge
dumping, however, are very different in different countries
(Dethlefsen, this volume).
     This example relates to long-term but very small-scale
perturbation. At the opposite extreme are our concerns with massive
open-sea oil spills where the time scale of the cause is relatively
short, but the spatial scale may be large. One authoritative report
(Royal Commission on Enrivonmental Pollution, 1981) states, "oil
spill...causes transient local effects on populations". Another
equally authoritative report (NRC, 1985) said, "Petroleum can have a
seriously adverse effect on local environments persisting, in some
cases unaltered, for decades". Part of this divergenece in opinion
depends on value judgements about "adversity"; part depends on the
technical definitions of such terms as transience and persistence. For
open sea effects the "worst case" example is a spill at a time and
place which might cause virtual elimination of a larval year class.
Figure 1 displays the great natural variability but also indicates that
the consequences of a very poor year class may be different for a large
and then for a small spawning stock. Johnston (1977) has attempted to
assess the economic loss for the North Sea haddock stocks of such an
oil spill and estimated for the worst case a loss of about $2m from a
total, yearly fishery revenue of $600m (in 1977). This is close to the
value assessed for the probable loss for fisheries from the AMOCO CADIZ

(U.S. Department of Commerce/NOAA, 1985). It is clear, however, that such estimates are exceedingly circumstantial and we have not developed any generally acceptable means for direct monitoring.

Given this lack of success we may ask whether such direct monitoring can answer questions about the effects of contaminants. Such monitoring depends on the precept that we can separate natural and manmade variations in population changes and regard them as additive and this may apply to some terrestrial systems. The examples given here are intended to show that this view does not translate into the marine environment.

As an overgeneralization, terrestrial reproductive systems are adapted to eliminate, as far as possible, the effects of environmental variability whereas marine systems accept and utilize the physical and chemical environment. Thus it is easier to evaluate the consequences of effects such as eggshell thinning on bird populations compared with, say, mortality in fish eggs. However, for marine populations, one can distinguish possible responses to different space and time scales of perturbation in relation to the inherent scales of the populations.

I have attempted to generalize from the examples given here, Figure 3. For some part of the system, S, I assume it is possible to give characteristic space (km) and time (day) scales to this population or community, such as those for zooplankton or pelagic fish. Then the space/time domain can be divided into four regions depending upon whether the scales are greater or less than those under consideration. The time scales will normally correspond to the life span with the assumption that the population can absorb fluctuations at much shorter periods but will respond to longer period events. The shorter and longer space scales can correspond, potentially, to coherence lengths (e.g., plankton) or ambits (e.g., fish stocks).

In each box examples of natural variability are given together with possible effects derived from man's activities. The plankton example shows that, with relatively discrete and localized introduction of contaminants, some knowledge of the space and time scales of variability in community structure could permit indirect inferences to be made. These suggest that if the space and time scales are both less than the corresponding population scales, then the population is unlikely to be affected.

For larger scale and longer term or "chronic" exposure we do not have any real understanding of the factors determining the natural variability although there is general acceptance that these depend indirectly on "ocean climate". My second example was intended to show that there may be complicated interactions between this variability and any manmade effects. The fact that we cannot readily distinguish the consequences of natural change from those of heavy fishing indicate that at present we are unlikely to separate out the role of contaminants at these scales. The major concern in this context is nutrient addition - for example, in the southern North Sea or the New York Bight where increased phytoplankton growth can be expected but where the spatial scales are still small but not negligible in relation to most fish stocks.

When one or other scale is larger, there is the traditional choice

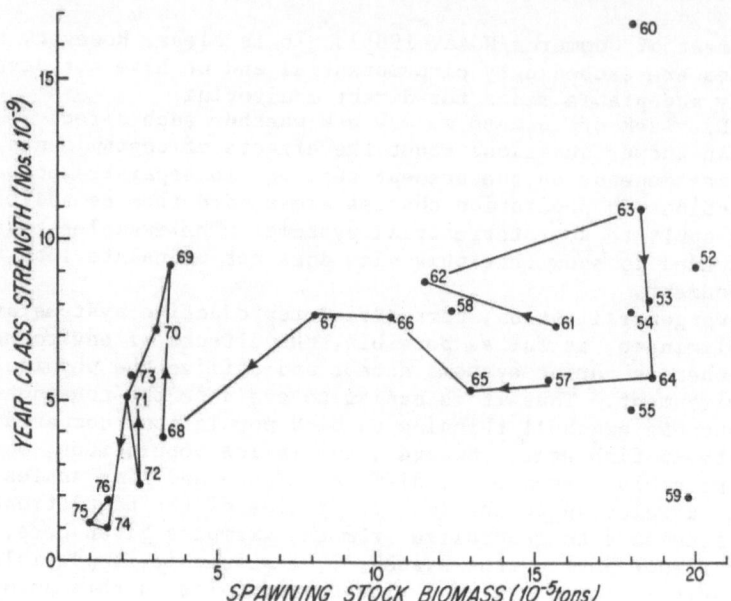

Figure 1.   The relation of recruitment to spawning stock for North Sea herring, 1952-1976 (From Saville & Bailey, 1980).

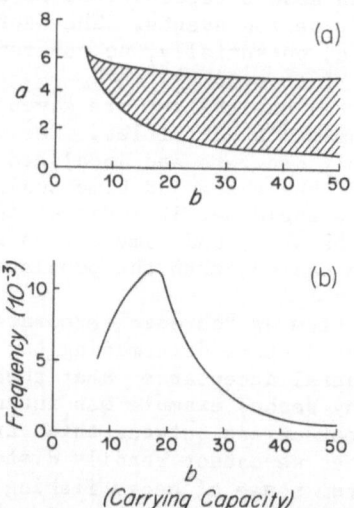

(Carrying Capacity)

Figure 2.   (a) Solutions of equation (1) have simple or triple values depending on a and b (see May, 1974 and Steele and Henderson, 1984 for details).   Variation in a across the tripled valued shaded region induces jumps between high and low values of p.   (b) The frequency of jumps between high and low values of p depends on b.   The jumps become much more rapid as b decreases.

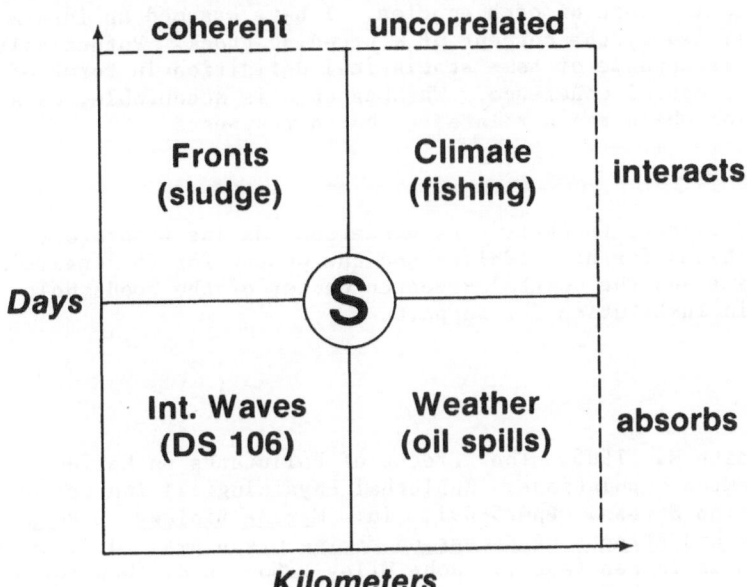

Figure 3.   A summary of the discussion of scale-related responses of a
population, s, to natural and manmade perturbation. The scales, in log
units would be, typically, 1 - 10 km and 1 - 10 days (See Steele,
1981).

between containment and dispersal. This is exemplified by the problems
of sludge disposal. The consequences of containment are illustrated by
the clearly defined and circumscribed effects of dumping in the Clyde
Sea area. But this containment is not always possible. At the other
extreme, oil spills in the open sea, although relatively short term,
can involve large areas - and large volumes if artifical dispersants
are used. The assumption, implicit in Figure 3, is that there will be
no observable effects on longer lived populations and this is confirmed
by available reports. Yet there is still significant social and
political distaste for this disposal option. It is ironic that the
unique capability of the ocean compared with the land - large-scale
dispersal - appears to be unacceptable.
     This analysis, summarized in Figure 3, depends on the assumption
that, for a given problem we can define a target population or
community and associate with it, characteristic space and time scales.
This neglects the links within the whole food web where the
interactions are across the space and time dimensions (Steele, 1978).
Nutrient additions can have relatively large spatial scales for
phytoplankton but be small for fish populations.
     Finally, there are both technical questions and social concerns
about the definition of a "population". Do we mean a patch of

zooplankton or benthos; or a shoal of pelagic fish; or do we take the
total geographic range of each species. I have assumed an intermediate
level exemplified by the concept of a breeding stock. Potentially, at
least, this is capable of some statistical definition in terms of
spatial and temporal coherence. Whether this is acceptable, is a
social problem which again relates to human responses.

ACKNOWLEDGEMENTS

I would like to thank Eric W. Henderson, Marine Laboratory,
Aberdeen Scotland for his modeling and the Center for the Analysis of
Marine Systems and the Coastal Research Center of the Woods Hole
Oceanographic Institution for support.

REFERENCES

Capuzzo, Judith M. 1985. The Effects of Pollutants on Marine
    Zooplankton Populations: Sublethal Physiological Indices of
    Population Stress. Pp475-491. In: Marine Biology of Polar
    Regions and Effects of Stress on Marine Organisms. J.S. Gray and
    M.E. Christiansen (eds.). John Wiley & Sons Ltd., New York.

Csanady, G.T. 1981. An analysis of dumpsite diffusion experiements.
    Pp109-129. In: Ocean Dumping of Industrial Wastes, B.H. Ketchum,
    D.R. Kester and P.K. Park (eds.). Plenum Press, New York and
    London.

Csanady, G.T. (this volume). Dispersal of particulate waste on an
    open continental shelf. NATO Advanced Research Workshop on
    Scientific Basis for the Role of the Oceans as a Waste Disposal
    Option, 24-30 April 1985, Algarve, Portugal.

Cushing, D.H. 1982. Climate and Fisheries. Academic Press, London.
    Pp373.

Dethlefsen, Volkert. (this volume). Sewage Sludge Disposal -
    Experiences from F.R.G. NATO Advanced Research Workshop on
    Scientific Basis for the Role of the Oceans as a Waste Disposal
    Option, 24-30 April 1985, Algarve, Portugal.

Frost, B.W. 1980. The Inadequacy of Body Size as an Indicator of
    Niches in the Zooplankton. Pp742-753. In: The evolution and
    ecology of zooplankton populations, W.C. Kerfoot (ed.). Amer.
    Soc. Limnol. Oceanogr. Spec. Symposium III.

Gray, J.S. 1979. Pollution-induced changes in populations. Phil.
    Trans. R. Soc. Lond. B. 286: 545-561.

Johnston, R. 1977. What North Sea Oil Might Cost Fisheries.
    Rapp. P.-v Reun. Cons. int. Explor. Mer. 171: 212-223.

MacIntyre, A.D. 1984. What Happened to Biological Effects Monitoring. (Editorial). Marine Pollution Bulletin. **15**(11): 391-392.

Mackas, David L. 1984. Spatial autocorrelation of plankton community composition in a continental shelf ecosystem. Limnol. Oceanogr. **29**(3): 451-471.

May, R.M. 1974. Stability and Complexity in Model Ecosystems. 2nd edn. Pp265. Princeton University Press.

National Research Council. 1985. Oil in the Sea - Inputs, Fates, and Effects. Pp601. National Academy Press, Washington, DC.

Platt, T. and K.L. Denman. 1975. Spectral analysis in ecology. Ann. Rev. Ecol. Syst. **6**: 189-210.

Royal Commission on Environmental Pollution. Sir Hans Kornberg (Chairman). Eighth Report - Oil Pollution of the Sea. Pp307. Presented to Parliament, October 1981. Her Majesty's Stationery Office.

Saville, A. and R.S. Bailey. 1980. The assessment and management of the herring stocks in the North Sea and to the west of Scotland. Rapp. P.-v. Reun. cons. int. Explor. Mer. **40**(1): 67-75.

Steele, John .H. 1978. Some comments on plankton patches. Pp1-20. In: Spatial Pattern in Plankton Communities. J.H. Steele (ed.). Plenum Press, New York.

Steele, John H. 1981. Some Varieties of Biological Oceanography. 376-383. In: Evolution of Physical Oceanography. Bruce A. Warren and Carl Wunsch (eds.). MIT Press, Cambridge and London.

Steele, John H. and Eric W. Henderson. 1984. Modeling long term fluctuations in fish stocks. Science. **224**(4652): 985-987.

Steele, John H. 1985. A comparison of terrestrial and marine ecological systems. Nature. **313**(6001): 355-358.

Topping, G. 1976. Sewage and the sea. Pp303-351. In: Marine Pollution. R. Johnston (ed.). Academic Press, London, New York, San Francisco.

U.S. Department of Commerce/National Oceanic and Atmospheric Administration. Assessing the Social Costs of Oil Spills: The AMOCO CADIZ Case Study. July 1983.

Waldichuck, M. 1979. Review of the problems. Phil. Trans. R. Soc. Lond. B. **286**: 399-424.

Whitledge, Terry E. and Creighton D. Wirick. 1983. Observations of chlorophyll concentrations off Long Island from a moored in situ fluorometer. Deep-Sea Res. 30(3A): 397-309.

Wiesenburg, Denis, A. and James A. Brooks. 1983. Eddy-enhanced dispersion of Ocean-Dumped Organic Waste at Deep Water Dumpsite - 106. Can. J. Fish. Aquat. Sci. 40(suppl. 2): 248-261.

BASIC FACTORS AFFECTING THE LAND/FRESHWATER VERSUS THE SEA OPTION
FOR WASTE DISPOSAL

Charles Osterberg
Office of Health and Environment
U.S. Department of Energy
Washington, D.C., USA 20545

ABSTRACT: The air, land, and water are tied together in one
system, so that actions taken toward one will impact on the others.
But not all three are equally productive, nor are all three as easily
polluted. The vast ocean does not provide much food to man, compared
with the land, nor does it appear that its yield can be greatly increased
in the near future. The land not only provides most of our food, fibers,
and building materials, but it covers, shields and stores the most precious resource of all, the freshwater. Because of the bounty of the land
and its role in the freshwater cycle, I argue for more protection for
the land and less for the sea. Gravity cleanses the air and the sea,
and most natural processes carry contaminants into the ocean. The ocean
is, and always has been, Nature's trash basket. Its special properties
may help man solve his disposal problems, with little environmental
costs.

1. INTRODUCTION

        Our planet is an integrated system of air, land and water.
Of these, only the land has been considered to be "retentive," i.e.,
able to hold contaminants firmly. The air and sea, on the other hand,
are dispersive media, characterized by largely horizontal motions that
move quickly around the circumference of the globe.
        Because the earth was believed to be a good sponge, most wastes
have been placed on land. The more toxic the wastes, the better chance
they will end up on land, for containment there has been considered
reliable. Fears are expressed that pollutants in the ocean will quickly
find their way back to man, while wastes in the land were thought to
be effectively isolated from further contact with human beings.   These
concepts are changing with increasing evidence that, although sometimes
resembling a giant sponge able to absorb and hold pollutants, soil is
a very leaky sponge with many unknown characteristics. As a result,
its carrying capacity, or ability to retain materials, is easily overloaded so that contaminants once thought to be safely fixed in the soil
are showing up in ground waters, some distance from the source.

G. Kullenberg (ed.), The Role of the Oceans as a Waste Disposal Option, 39–53.

Gaseous materials released into the air quickly circle the globe, as did fallout from nuclear testing, and as the materials responsible for acid rain now do. But larger particles drop to the ground and smaller particles usually coalesce and fall, or are washed out by rain, so the troposphere, at least, is self cleansing. So too is the ocean. While the currents carry water and neutrally buoyant materials thousands of miles, particulate materials gradually fall in response to gravity. Pollutants in the water bind to the particles and are swept to the bottom. The eventual resting place of materials disposed of into the ocean is in the sediments (Broecker, 1974).

Air, land and ocean are the three media where wastes from civilization must go. Wastes can be expected to increase as the world's population grows. Perhaps the amounts of waste will rise even faster than the population as resources dwindle and become more dilute, thus requiring larger amounts of energy for recovery, to overcome the 2nd law of thermodynamics. Yet most wastes cannot be converted to gases or vapors prior to disposal, and even if they could, it is not clear that the atmosphere is the best place to release pollutants. For example, methane levels in the atmosphere, according to cores taken in polar ice, have doubled in the last 400 years, CO_2 levels will double by early in the next century, ozone levels are affected by current releases, and airborne acid rain is a present day problem.

Realistically, for most wastes, we have only two disposal options, the land and the ocean. However, the latter option has been essentially foreclosed in the U.S. because of the restrictive laws favored by enviromental lobby groups which make the ocean the medium of last resort, leaving only the land easily available for waste disposal.

As scientists concerned with the welfare of planet Earth, we should make sure that the strengths and weaknesses of the environment are understood so that impending decisions, which are urgently needed, can be made intelligently. This paper examines some strengths and weaknesses of the land and sea in serving mankind, and concludes that the land yields far more food and materials than does the sea; while the ocean, although relatively unproductive of food, is better designed to absorb many of the wastes of civilization. The weak link in the chain of life support systems on planet Earth is clearly the freshwater, which must be protected at all costs if mankind is to survive.

2. FOOD FROM THE LAND AND FROM THE SEA

2.1. The Sea

Life is found in the deepest trenches, and throughout the 1370 million km^3 of seawater. Let us call this the ocean biosphere, using the terminology of Vallentyne (1971), who defines the biosphere as that part of the world inhabited by living things. On land, life exists only near the surface; I estimate that on land, almost all living organisms, except for some roots of certain trees, are confined to the top 10 meters of soil. Since the area of the land portion of the globe is 154 million km^2, the volume of land to a depth of 10 m, which I call the land biosphere, would be 1.54 million km^3. Thus, the ratio of the ocean biosphere to the land biosphere is 1370 to 1.54, or nearly 900 to 1.

What is important to life is not the volume of the living space so much as it is the food available, all of which comes from photosynthesis. Only plants use sunlight, water, carbon dioxide, and trace materials to build the organic compounds animals need for food. Here again the ocean appears to have the better of it. Yentch (1977) estimates that the ocean absorbs seven-eighths of the sunlight falling on the globe, and phytoplankton throughout the top 100 m or so receive enough light for photosynthesis, so surely, one would think, the ocean must outproduce the land. Especially since the ocean, which covers 71% of the earth's surface, has an area nearly 2.5 times that of land.

And yet Lieth, 1972, as quoted by Larcher (1975), estimates that of the world's total production of 155.2 x 10^9 tons of net primary productivity (dry matter) per year, 100.2 x 10^9 tons derive from the land and its associated freshwaters. The remaining 55.0 x 10^9 tons per year are found diluted in those 1370 million km^3 of seawater.

Oceanographers know why the ocean is such a poor producer of food. Sunlight is confined to the upper 2 or 3% of the ocean's waters in the top 100 meters or so, while the nutrients vitally needed for photosynthesis—the phosphates, nitrates, and some trace metals—are trapped beneath the thermocline, too far from the sunlight to be of any value in photosynthesis. Since the average depth of the ocean is about 3800 meters, most of the water column is not producing any food at all. Using bathometric data from Sverdrup, et al. (1942) and fishery data from Warner (1983), I estimate that 85% of our seafood comes from only 2% of the ocean's volume, those shelf waters of 1000 m depth or less. Nearly all of the remaining 15% of fish are taken from the top 1000 meters overlying deeper water.

Having marine life distributed throughout the huge seawater biosphere, even in the deepest trenches, is a major problem, for food from the sea is a very dilute resource. von Arx (1979/1980) reports that edible biomass in the ocean exists at just one part per million, which makes it only the 16th most abundant material in seawater, even more dilute than boron (4.5 ppm.) and fluorine (1.3 ppm.). Such a dilute resource usually requires considerable energy for recovery.

A second problem is that food chains in the ocean are longer and leakier than those on land. This is a natural result of the fact that phytoplankton, the grass of the sea, are microscopic and must pass through several trophic levels before achieving enough size to be used on our tables. La Fond and La Fond (1971) estimate that it takes 1000 pounds of phytoplankton to put one pound of tuna, salmon, swordfish, halibut, or any other of the predaceous fishes man seems to prefer, on our dinner plates. Since Isaacs (1969) estimates that the maximum efficiency of transfer of material (or energy) from one trophic level to another is 15%, this would represent a food chain of 3.64 steps. Or if it were 5 steps, as shown in Figure 1, 1000 pounds of phytoplankton at 15% efficiency would yield less than an ounce of fish on our plates. Put another way, it would take an average efficiency of 25.1% for the 5 step food chain to yield one pound of flesh from 1000

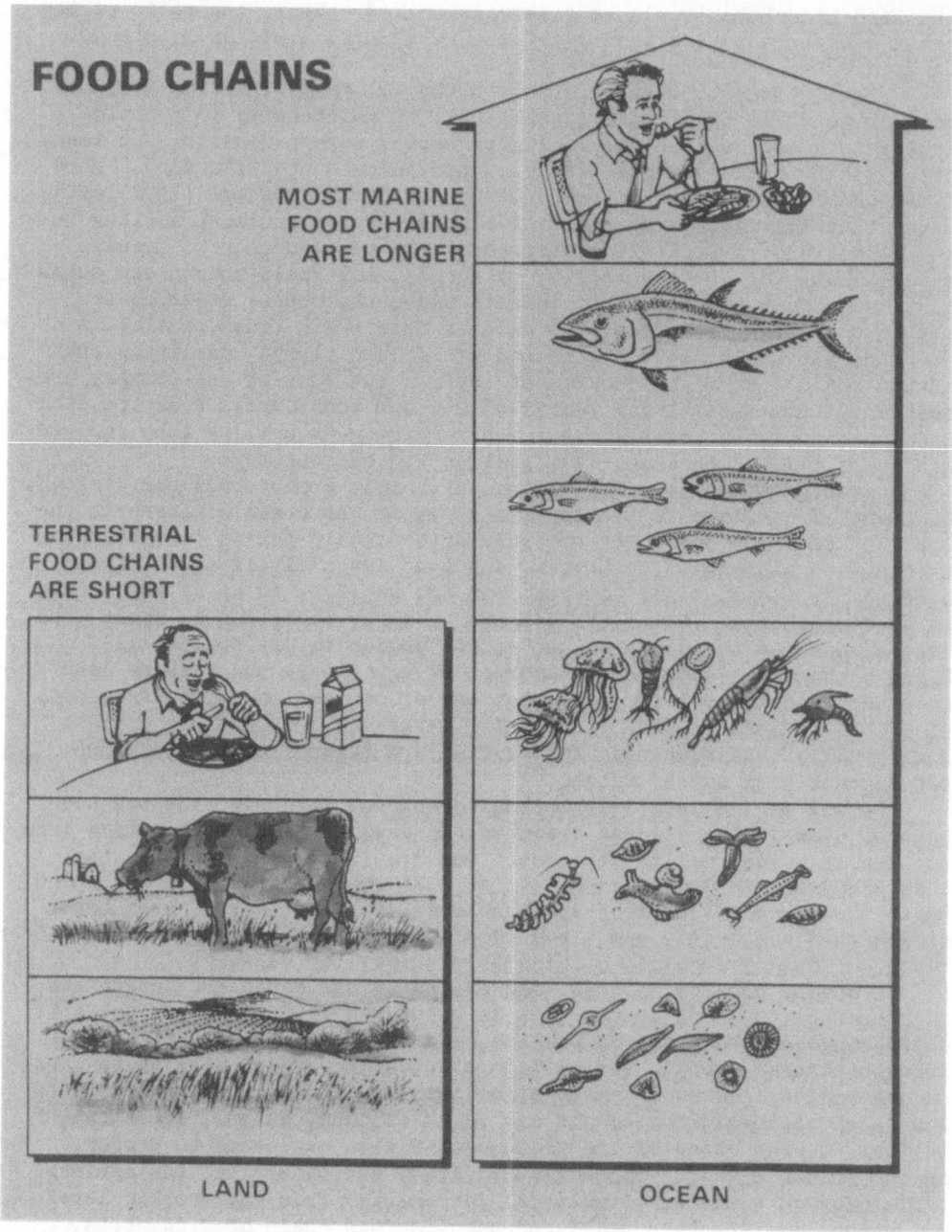

Figure 1. Typical food chains on land and at sea. Marine food chains are usually much longer, and thus less efficient than those on land in transferring energy from the sun to man.

pounds of phytoplankton.

Leaks occur because predators take at least as many of the fish as does man. Birds, seals, and larger fish get their share before the fisherman gets his. Life in the ocean is a struggle, a battle to eat and avoid being eaten. This chaotic natural system is designed for survival of the fittest but not to efficiently provide food for man, and it does not. Unlike the managed systems on land where the rapidly growing heifers and lambs get ample feed and protection, in the sea the young and fastest growing are chased hither and yon, rarely get enough to eat, and only a few survive to grow large enough to be caught by fishermen.

Finally, as the easier, more bountiful stocks of fish are fished as fast as they can reproduce, increases in seafood must come from either deeper waters or by searching farther afield. Since the ocean has a third dimension, depth, and volume increases as the cube root, fishing in deeper water means much more water has to be searched and fished.

Deep water fishing is inefficient. I remember as a young scientist that research trawls down to 3000 m took up to 13 hours to complete, with meagre yields. While modern equipment might speed the process, fishing at great depths requires more time and energy. Sooner or later, the cost of energy will limit the profitablity of deep fishing, and probably has already, for the Food and Agriculture Organization (1983) statistics show that the catch of rat-tails (Grenadiers), the most abundant edible fish in the depths, is falling (Figure 2). Warner (1983) writes that a typical trawl for rat-tails takes about twice the time, and much larger and powerful gear than is required for cod and other popular fish, and the yields are usually much smaller.

And yet any major increases in the fisheries of northern oceans will probably have to come from deeper water. Isaacs (1971) writes, "All of our observations (with his 'monster' camera, which photographed bottom fishes gathering around a bait) except those of the Antarctic to date have shown the grenadier to be the only fish of any quantity at depths below 3000 meters." He also said, "bathypelagic populations are sparce, and in this region living creatures are less than one hundredmillionth of the water volume. Nevertheless, the zone is of immense dimensions and the total populations may be large" (Isaacs, 1969). Analysis of FAO data show that 53% of the world catch of Grenadiers are roundnose grenadiers (<u>Coryphaenoides rupestris</u>), 37% blue grenadiers (<u>Macruronus novaezealandiae</u>), 6% are Grenadiers (<u>Macrourus spp.</u>), and 5% Patagonian grenadiers (<u>Macruronus magellanicus</u>).

Warner, who observed the Russians fishing for grenadiers, said that only "Soviets and other Communist-bloc nations, desperate for animal protein and equally desperate to avoid sudden layoffs of ships and fishermen, may fish for grenadiers for some years to come. But who else?" The data show that the U.S.S.R. takes 78% of the grenadiers caught. Poland (6%), Korean Republic (5%), Germany FR (9%) and Germany DM RP (2%) essentially take the rest. The time and energy required make the grenadier a very expensive source of protein.

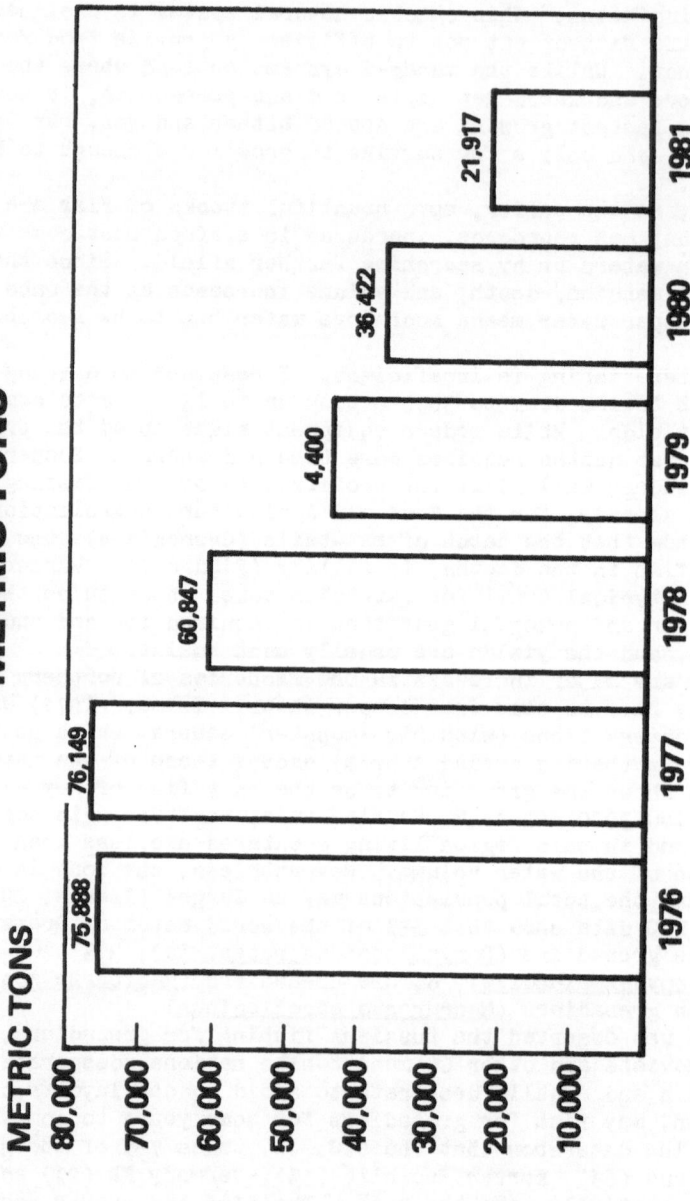

Figure 2.   The catch of Grenadiers has declined sharply in the last 5 or 6 years, perhaps indicating either overfishing or the increased cost of energy and the difficulty in fishing at great depths, or all of these. Note that the vertical scale on this figure is 10^{-3} that of Figure 4.

2.2. The Land

Seen from space, planet earth is dominated by the oceans, with pitifully little land. We are nearly inundated with seawater. Not only is 71% of the surface under water, but 55% of the planet's surface is covered with at least 3000 m of water. This deep seawater provides almost none of man's food yet has fewer anthropogenic contaminants than most of the streams and underground aquifers which provide our drinking water, because the latter are in more intimate contact with civilization's wastes.

The world's 4 plus billion people, are, of course, land animals. Not all of the land is suitable for man, so we are clustered in valleys, along seacoasts, and near major rivers. While 99.4 percent of the liquid water on the globe is seawater, it is the relatively few drops of freshwater on and under the land that sustain human life.

Unlike the ocean, the land is wildly productive of food and fiber for mankind's use. Plants are concentrated on the surface of the earth, where sunshine is available. Nutrients are in the surface soil, or can be added, and water from rains falls directly on the plants, or can be brought as needed.

Having plants, light, nutrients and water concentrated together is the recipe for plant growth, and the land does this much better than the ocean. According to Lieth, quoted earlier, the land outproduces the ocean (net primary productivity, dry matter) by more than 4 to 1 per unit area.

The biggest gain to man's food supply, however, probably comes about because of the shorter food chains on land (Figure 1). We can eat many plants, especially the seeds and fruit, directly, thus escaping the losses that occur in the ocean when the materials in plants (phytoplankton) must pass up several steps of the food chain before entering our diet. And, since we commonly feed on predacious marine fishes, we do so only inefficiently; we almost never eat predacious land animals. For, as Figure 1 shows, land food chains have only 1 or 2 steps, while 5 trophic level steps are often seen in the sea.

If efficiency is only 10%, 1000 pounds of phytoplankton would drop to .01 pounds after 5 trophic level steps, while 1000 pounds of wheat on land could be eaten directly, or, using the same transfer efficiency above, fed to animals to get 100 pounds of meat. As we shall see later, terrestrial food chains are generally much more efficient than those in the sea.

While the relative length of marine and terrestrial food chains causes a great difference in the relative yields to man from the land and sea, also subtracting from the yield of the ocean is the fact that the ocean is a wild, natural system, with billions of young fishes and eggs disappearing into the gullets of predators. On land, because of husbandry, animals are bred for maximum milk, egg, or meat production, fed ideal diets, and protected from predators and disease. Only a few mature animals, which don't gain much weight and thus use food inefficiently, are kept mainly as breeding stock for producing the rapid growing young which yield the maximum amount of meat per unit of feed.

Because of management, chickens can provide 481 pounds of meat

Land Food Chains

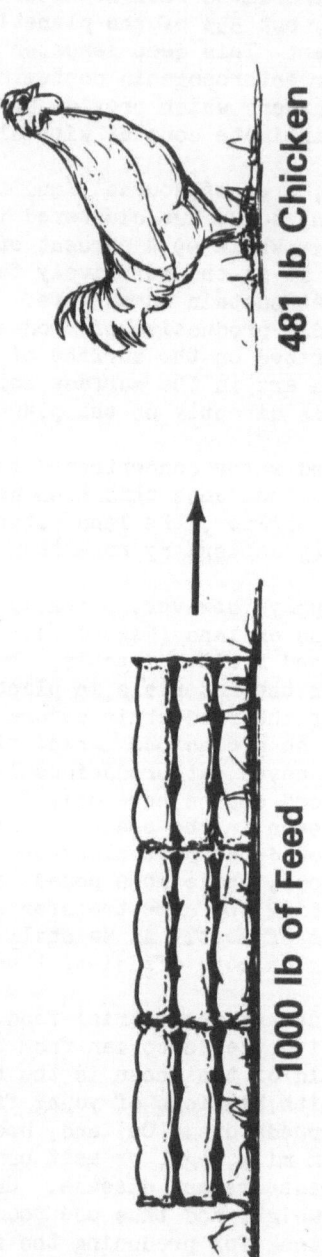

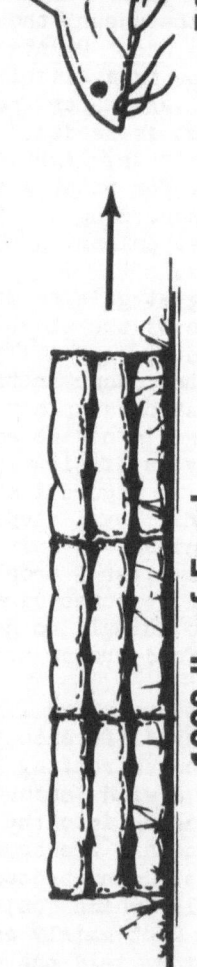

481 lb Chicken

510 lb Catfish

1000 lb of Feed

1000 lb of Feed

Figure 3. The efficiency of managed land food chains reaches an apex in cultured catfish and chickens raised for meat. Special care and diets, with few losses to disease (and none to predation), are required to achieve these efficiencies.

for every 1000 pounds of food consumed. Catfish can do even better, gaining 510 pounds on 1000 pounds of food (Figure 3). These transfer efficiencies are 48 and 51 percent, respectively, far above anything predicted in the ocean, but circumstances are not comparable. Instead of running free and competing, the chickens and catfish are confined and pampered, but their lives are short since they are killed to take advantage of a maximum return on the investment. For catfish, that means when they weigh about 1 pound, while chickens can look forward to about 6 weeks of life (U.S. Department of Agriculture).

Kleiber (1975) writes that a steer gains 240 pounds from 1300 pounds of hay, and so do 300 rabbits, for an efficiency of over 18%. But another factor becomes important, because the rabbits can convert 1300 pounds of hay into 240 pounds of meat in only 30 days, while it takes the steer 120 days for the same weight gain. This is an important consideration that I haven't tried to evaluate, but phytoplankton and zooplankton reproduce and grow very quickly compared with large land animals. Thus the materials they contain are quickly recycled. However, the average age of fishes such as salmon, tuna and rat-tails is greater than the average age of chickens, catfish, and cattle when slaughtered.

Cattle and sheep grazing in the fields are much less efficient producers of protein than are chickens, but, protected by man, there are few losses to predators, disease or starvation. And, being ruminants, these animals can utilize cellulose, the most abundant organic molecule on earth, transforming it into edible food for man. Since man cannot digest cellulose, without ruminents he would be unable to use any of the vast store of energy bound up in grazing land. Permanent pasture, range, meadows, forest and woodland comprise 55% of the world's land area and have the potential of producing 5.8×10^{12} megacalories of metabolizing energy in forage annually, so the valuable role of ruminents in converting cellulose to protein for human consumption is obvious (Baldwin, 1984). There is very little cellulose in the sea, where it has largely been replaced by chitin.

The bottom line on food, then, is that most of it comes from the land, either directly or through short, very efficient food chains. Food from the sea is produced inefficiently and little of the sun's energy reaching the sea returns to man in his food, and these losses are largely beyond our control.

Whatever and wherever the losses, the ocean is outproduced in organic carbon by the land about 4 to 1 per unit area. But it is in the production of food for mankind that the ocean is badly outclassed. For the ocean only produces about 1% of man's food (Roels, 1982), or about 3% of his protein, which is about 10% of his animal protein (Holt, 1969). According to the Food and Agriculture Organization Food Balance Sheets, "the percentage of fish consumption (not production) for human food on total food consumed (not produced) is less than one percent in terms of calories, and more than five percent in terms of proteins (Becker, 1984). Let us say that 1% of our food comes from the sea. Since the ratio of sea to land is 71%/29%, or 2.45, and the land produces 100 times more food for humans than the sea, so the land outproduces the sea in food for man by 245 to 1; about 80 to 1 in protein; and 10 to 1 in animal protein. These facts should give pause to ad-

Figure 4. Food and Agriculture Organization total for seafood catches. Not all of these organisms are used to feed man, since many go to cattle and chickens.

vocates of waste disposal on land. The land's bounty is so overwhelming that it should be protected for that reason alone.

There are other reasons. Since land is solid (about 5.5 times
as dense as seawater), it doesn't turnover and renew itself as do the
sea and atmosphere. Particles of pollutants settle on the surface,
near the root zone of plants, or on their leaves, where uptake is often
rapid. Food chains on land are short, and the pollutants are quickly
passed on to man.

But the greatest reason for protecting the land is that it spawns,
shields and stores the freshwaters on which life on land depends. Current U.S. practices fail to protect these waters. "The groundwater
has suffered the most," according to Hopkins (1984). "The Environmental Protection Agency has identified 15,000 'uncontrolled sites'--sites
that are clearly contributing to contamination of water or air. Of
these, at least 347 pose direct threats to drinking water supplies and
could cause birth defects, cancer, and other diseases, according to
the E.P.A. (Ibid).

3. FOOD IN THE FUTURE

While the numbers clearly indicate that the land produces most
of our food, should we not anticipate that in the future, those 1370
million km^3 of seawater may represent the solution to the world's food
supply problems?

The most optomistic estimate comes from Wise (1984), who writes
that so far, we have only heavily fished the shelves of the northern
hemisphere, and that southern shores have about 3 times as much shelf
area with waters of 1000 m or less, so should outproduce the northern
hemisphere. Ironically, these coastal shelves that produce seafood
so abundantly are geologically not of the sea, but are actually submerged extensions of the continents. How barren the sea would be without the land!

Wise (1984) also believes our fish supplies would double if none
were used for animal feed, and a 10 to 15% gain could be achieved if
fish currently discarded were eaten. Best guesses are that the fisheries can eventually yield 150-250 million tons per year, about three
times the present catch.

Perhaps confirmation that new shelf fisheries are being exploited comes from the FAO, (1982), which reports that the tonnage of fish
caught by third world countries is growing at about 4.5% per year, while
catches of the better established fishery nations are increasing only
1% per year.

However, it seems unlikely that the deep ocean can add more than
a few percent of the total catch, which in 1983 was about 77 million
tons of mixed seafood (Figure 4). Rat-tails have already been discussed, but that still leaves the pelagic fishes in the water column.
However, data from Pearcy and Laurs (1966) show that the numbers of
organisms in the water column drop sharply with depth, while Vinogradov
(1962) pointed out that below 1000 m net plankton biomass decreases
exponentially.

Even the fishery for Antarctic krill doesn't offer much hope, and it is a very difficult, dangerous, and energy intensive fishery that will suffer most from increased energy costs.

Largest gains in food must come from the land. A few comparisons will show how this might happen. The U.S. has a fairly advanced farm industry. With only 5% of the world's population and 7% of the world's land but 13% of the arable land or permanent pasture, the U.S. produces:

1) 20% of the grain
2) 6.7% of the fruit and vegetables
3) 16.9% of the meat
4) 13.7% of the eggs
5) 14% of the milk
6) 19.3% of the cheese
7) 5% of the seafood

It is in the last 5 categories, animal protein, that the U.S. is both a large supplier and consumer; and, of course, it is animal protein that is needed around the world because of its essential nutrients.

Even though the U.S. produces a fifth of the world's grain, "the average yield of the six major U.S. grain crops is only about 20 to 35 percent of record yields," suggesting that substantial improvements lie ahead (Gifford, et. al, 1984).

However, Pimental (1978) cautions that already 15% of our fossil-fuel energy goes into food production. If a U.S. diet is provided to all 4 billion members of the world by U.S. methods, the known reserves of petroleum would be consumed for this purpose alone in 13 years. The U.S. system uses too much energy to be duplicated everywhere.

But each kilogram of fertilizer produces an average yield increase of 10 kg of grain in the developing world, and rice plots protected by insecticides yield almost twice as much as unprotected plots. According to D. Gale Johnson, University of Chicago, "---irrigated land yields from two to four times as much food as the same land before it was irrigated. Thus, irrigating one hectare of land is the same as finding 1 to 3 additional hectares of cropland." (Mellor and Adams, 1984).

Certainly the land has greater potential to provide more food than does the sea, and even a few percent increase would greatly exceed the present yield from the ocean.

4. WASTES ON LAND AND IN THE SEA

4.1. The Sea

Nearly all rivers run to the sea, and the rains carry many contaminants off the land into the coastal ocean. While this often pollutes our estuaries with materials disposed of on the watershed, it also brings nutrients into the coastal zone and perhaps accounts for about 10% of the increased productivity there.

However, since as indicated earlier, 85% of our seafood comes from coastal waters of 1000 m or less, the coastal zone is a critical area which should be protected from pollution. After all, 98% of the sea-

water is elsewhere, in the deep ocean, and it is far more able to absorb
wastes without ill effects. In the first place, the deep water regime
is the biggest biome (55% of the planet's surface is covered with 3000
m or more seawater) in the world. Thus it is less vulnerable than
smaller biomes, which could be more easily contaminated. It has less
animals per cubic meter of water, or square meter of bottom, and almost
none of these are in the food chains of man.

4.1.1. Cleansing Processes. Pollutants in seawater are swept out by
particles, which settle to the bottom and tend to isolate materials
in the sediments. Turekian (1983) calls it a "strong particle veil,
typically present at ocean margins, that scavenges the nuclides (and
other things) from the water column."
     Spencer and Bacon (1981) report that removal from the water column
occurs preferentially near the continental margins, and the rate of
this boundry removal is governed largely by the redox conditions within
the sediment column, which are ultimately controlled by the biological
productivity in the overlying surface waters. The SEEP Project (Shelf
Edge Exchange Processes) of the Department of Energy, which involves
several oceanographic groups, has shown how coastal waters cleanse themselves, transporting materials to sea, and eventually to the bottom.
     Perhaps most convincing are the real-world experiments carried
out that show how quickly the ocean recovers from contamination. In
the Marshall Islands in the Pacific Ocean, Enewetak, the scene of 46
atomic bomb explosions, was thoroughly contaminated with radioactivity.
Most of it ended up in the lagoon. Even so, marine food chains were
soon nearly free of radioactivity, and could be consumed in quantity.
Not so the terrestrial plants and animals. Even now, over thirty years
later, coconuts, breadfruit, and other plants, and land crabs, a native
favorite, are too radioactive from strontium-90 and cesium-137 to be
eaten regularly by the natives. However, those radionuclides are diluted by the massive amounts of stable strontium, calcium, and cesium
and potassium in seawater, and pose almost no threat in marine food
chains, despite their great potency in terrestrial food chains.
     The Mediterranean Sea is another example. The edges are clearly
polluted from sewage and contaminants from the great populations along
shore. However, only a few miles out at sea, the water is clean.
Osterberg and Keckes (1977) reported that the open Mediterranean was
hardly any more polluted than the open Atlantic and Pacific Oceans,
which is to say, practically not at all. Clearly cleansing mechanisms
are active.

4.2. The Land

     Winds and rain scour and clean the land. Heavy rains in particular
flush contaminants down streams into the ocean. Lighter rains percolate
through the soil, dissolving contaminants in or under the soil, taking
them into the groundwater. Levels of groundwater change with season
and rainfall, often saturating waste dumps that were thought to be above
the water level.
     Efforts are made to protect the groundwater by using plastic or
clay liners, but the life of these safeguards is limited. Hazardous

wastes have been placed in rocks of low permeability, such as shale
and glacial till. Unfortunately, these are rocks most apt to fracture.
Professor Pettyjohn, Oklahoma State University, in a report to the President's Council of Envirnomental Quality in 1984, concluded, "Consequently, instead of retaining the leachate, the fractures permit rapid
transfer of fluids along paths not readily traceable."

There is bacterial degradation of wastes in and on land, just as
there is in the ocean, but the eventual resting place of most materials
released on land and the residual of those burned in the air is the
ocean, Nature's trash basket. Thus it has always been, and ever shall
be.

The views expressed are those of the author, not the Department of
Energy.

5. REFERENCES

Baldwin, R. L. 1984. 'Digestion and metabolism of ruminants, Bio-
    science, 34 (9):244-249.
Becker, K., 1984. Chief, Basic Data Unit, Statistics Division, Food
    and Agriculture Organization, Rome, in a letter dated December
    21, 1984.
Broecker, Wallace, 1974. Chemical Oceanography, Harcourt Brace
    Ivanovich, Inc., New York. pp. 214.
Food and Agriculture Organization, 1982. The State of Food and Agri-
    culture, 'World review livestock production: a world perspective'
    Rome, pp. 203.
Food and Agriculture Organization, 1983. 1981 Yearbook of Fishery
    Statistics 52, FAO Fishery Series, Rome. pp.79.
Gifford, Roger, J.H. Thorne, W.D. Hitz, and Robert Giaquinta.
    1984. 'Crop productivity and photoassimilate partioning,'
    Science 225:801-808.
Holt, S.J. 1969. 'The food resources of the ocean.' In: The Ocean
    A Scientific American Book. W.H. Freeman and Company. San Fran-
    cisco, Calif. pp. 94-106.
Hopkins, Ed. 1984. Groundwater contamination; out of sight, out of
    mind?' OutdoorAmerica, Winter, 40 (1):6-34.
Isaacs, John. 1969. 'The nature of oceanic life,' In: The Ocean,
    A Scientific American Book. W.H. Freeman and Co., San Francisco,
    Calif. pp. 65-79.
Isaacs, John, and Richard Swartzlose. 1971. 'Near bottom animal pop-
    ulation and related deep circulation in the ocean,' Report to the
    U.S. Atomic Energy Commission, 27, July, 1971, Washington, D.C.
    12 numbered pages, unpublished.
Kleiber, Max. 1975. The Fire of Life, Robert Krieger Publishing Co.
    Huntingdon, New York, pp. 453.
Larcher, W. 1975. Physiological Plant Ecology, Springer-Verlag,
    Berlin, pp. 340.

LaFond, E.C., and K.G. LaFond. 1971. 'Oceanography and its relation to marine organic production.' In: Fertility of the Sea. John Costlow (Ed.), Gordon and Breach Science Publishers, New York. pp.241-265.

Mellor, John W., and Richard H. Adams. 1984. 'Feeding the under-developed world,' Chemical and Engineering News, 62 (17):32-39.

Osterberg, Charles, and S. Keckes. 1977. 'The state of pollution in the Mediterranean Sea,' Ambio 6:321-326.

Pearcy, William G., and R.M. Laurs. 1966. 'Vertical migration and distribution of mesopelagic fishes off Oregon.' Deep-Sea Research, 13, pp. 153-165.

Pettyjohn, Wayne A. 1984. 'Improving the knowledge base on surface water and groundwater processes,' unpublished papar prepared for Geochemical and Hydrologic Processes, Panel Meeting (September 24-25) in support of the Council of Environmental Quality Conference on Long-term Environmental Research and Development, Washington, D.C., pp. 19.

Pimentel, D., and Marcia Pimentel 1978. 'Dimensions of the world food problem and losses to pests,' World Food, Pest Losses and the Environment. AAAS Selected Symposium Series, Westview Press, Inc. Boulder, Colorado. pp. 1-6.

Roels, Oswald. 1982. 'Mariculture fertilized by upwelling.' Sea Technology, 23 (8):63-67.

Spencer, Derek, Michael Bacon, and Peter Brewer. 1981. 'Models of the distribution of ^{210}Pb in a section across the North Equatorial Atlantic Ocean,' Journal of Marine Research 39(1):119-138.

Sverdrup, H.U., Martin Johnson, and R.H. Fleming. 1942. The Oceans. Prentice-Hall, Inc. Englewood Cliffs, New Jersey, pp. 1087.

Turekian, Carl. 1983. 'The fate of nuclides in natural water systems,' Report to the U.S. Department of Energy, Washington, D.C., November 30, 1983, unpublished.

U.S. Department of Agriculture, Statistical Reporting Service, Washington, D.C., 1984. Personal communication from T. Stucker and L. Van Meer.

Vallentyne, J.R. 1971. 'Biosphere.' In: Encyclopedia of Science and Technology, 2, McGraw Hill Book Co., N,Y. pg. 251,

Vinogradov, M.E. 1962. 'Feeding of the deep-sea zooplankton,' Intern. Cong. Explor. Sea Rap. et Pro-verbaux des Reunions, 153:114-120.

von Arx, William. 1978/1979, Winter. 'Prospects for a psychozoic era,' Oceanus, pp. 3-11.

Warner, W. Warner. 1981. Distant Water. Atlantic Monthly Press Book, Little, Brown and Company, Boston, pp. 338.

Wise, John. 1984. 'The future of food from the sea,' In: The Resourceful Earth, J. Simon and H. Kahn (Eds.) Basil Blackwell Publisher, Ltd., Oxford, England, pp. 113-127.

Yentsch, Charles. 1977. 'Phytoplankton growth in the sea; a coalescence of disciplines.' In: Primary Productivity in the Sea, P. Falkowski (Ed.). pp. 17-32.

SEWAGE TREATMENT AND DISPOSAL – CONSTRAINTS AND OPPORTUNITIES

Dr. A. L. Downing
Binnie & Partners
Artillery House
Artillery Row
London SW1P 1RX
United Kingdom

ABSTRACT. The legislative, technical, land use and economic considerations influencing design of schemes for disposal of sewage and sewage
effluents to the sea are briefly reviewed. Three case studies are
outlined with particular reference to the comparison of costs of
alternative disposal options. Finally, opportunities for improvement
of present disposal practice are suggested featuring the scope for
wider application of new technology.

1. INTRODUCTION

The problem of sewage disposal, which stemmed originally principally
from concern about public health, began to assume serious proportions
once people started to congregate together in dense urban communities
and it was rapidly intensified by the introduction of the water closet
towards the end of the eighteenth century. The resulting waste waters were
commonly discharged to drains intended originally for surface water and
this led rapidly to a marked deterioration in the condition of waterways
to which the drainage was released. The problem was compounded in many
countries by the development of industries, by the general natural increases in population, and, with rising standards of living and productive capacity, by the increased use of water in the home and in
industry.

Construction of sewers to remove the waste waters from the centres
of population led to some improvement in the urban environment, though
collecting waste waters together for discharge at a relatively small
number of points inevitably increased the potential for severely adverse
effects of pollution around the outfalls of these discharges. However,
it was at the same time a vital pre-cursor to the introduction of engineering works to bring about the orderly disposal of the wastes in
such a way as to reduce impacts to acceptable levels.

Inherent in the practices that have been subsequently evolved has
been the belief justified by experience that the most convenient method
of disposal, that of releasing waste waters either directly to the sea
or to rivers from which they would eventually pass to the sea, was quite
satisfactory if properly engineered. Even with modern technology which

55

G. Kullenberg (ed.), *The Role of the Oceans as a Waste Disposal Option*, 55–71.

would render elimination of virtually all impurities from waste waters
technically feasible, thus permitting their recycle for domestic and
industrial reuse, disposal directly or indirectly to the sea is still
almost always the preferred option, essentially because the costs and
other disbenefits of recycling are normally unacceptably large. The
current issue therefore is not whether the sea should be used as a disposal medium but how best it should be used for this purpose. Nowadays,
there are many options ranging at one extreme from the extensive purification to permit recycle, as already mentioned, to discharge with
minimal treatment in such a way that the pollutants are dispersed and
diluted to an extent which does not cause harm. The task of the design
engineer is to choose the best option for the local circumstances, which
however must take account of potential effects outside the immediate
locality.

2. SCALE OF THE PROBLEM

It is appropriate before considering the choice of method of disposal to
review briefly the constituents of sewage which for one reason or another
are undesirable and must therefore be removed or dispersed to levels that
do not cause harm; and also to consider the quantities of sewage and its
constituents that require disposal.

2.1. Composition

The concentrations of such constituents will depend on the volume of
water used per head of population and the proportion of industrial waste
in the sewage. The figures quoted below relate to wholly domestic sewage
from people using 150 l of water per head per day.

2.1.1. Pathogens. Pathogens can almost always be found in sewage from
urban communities. The numbers per unit volume depend considerably on
the health of the population. They are liable to be especially numerous
in countries where water-borne diseases are endemic but even in developed
countries, where cases of such disease are rare, some can usually be
detected. In the case of some diseases the latter occurrence is due to a
fraction of the population being carriers who do not suffer the symptoms
of the disease. In UK for example one in every 50,000 people is said to
be a carrier of typhoid but the incidence of the disease in UK attributable to water-borne transmission is minute. Generally, however, the
likely presence of pathogens in sewage demands disposal in such a way as
to prevent exposure of the population to any associated health risks.
Because pathogens often occur in low numbers and their enumeration is
frequently laborious, a widespread practice is to judge the potential
for transmission of disease in terms of numbers of normally non-pathogenic
coliform bacteria, which are present in relatively large numbers and can
be more readily enumerated. They thus serve as useful indicator organisms.
Numbers in sewage are usually around 10 to 50 million per 100 ml and in
well purified effluents one to two orders of magnitude lower.

2.1.2. "Visible" pollutants. In developed countries the population increasingly values its amenities and does not wish to be reminded of the necessity of disposing of its waste waters to rivers and the sea. The minimum requirement of any disposal scheme therefore is to ensure that no objects obviously derived from sewage are visible at the water's edge.

Constituents with the potential for rendering discharges visible include any gross solids especially if these are buoyant enough to float, suspended matter conferring turbidity, surface-active materials (soaps, detergents, fats, oils) causing the water surface to assume a greasy or glassy appearance ("slick"), and coloured substances especially industrial dyestuffs. Typically, concentrations of suspended matter in crude sewages are of the order of 500 mg/l, of grease about 100 to 150 mg/l and of dtergents 10 to 20 mg/l. Well over 90 per cent of such such constituents are normally removed during treatment of sewage by conventional means to render the effluent suitable for release to a river.

2.1.3. Oxygen-consuming substances. Most of the organic matter in sewage is biologically degradable and the natural processes of oxidation in natural waters will ensure the eventual conversion of such substances into carbon dioxide, and bacterial cells. In this process oxygen is consumed and this may in some circumstances lead to the complete depletion of the dissolved oxygen content of a natural water. This may occur, for example, where incompletely treated sewage is discharged to a small stream; in the absence of dissolved oxygen fish will be killed and offensive conditions will develop. Deoxygenation is not often a problem in marine waters. Concentrations of oxygen consuming substances are usually assessed in terms of their biochemical oxygen demand (BOD) which in crude sewage is usually around 500 mg/l.

All sewage contains ammonia which also can be oxidised by appropriate micro-organisms either in the sewage works or in natural waters provided that there is an adequate supply of dissolved oxygen. Concentrations in crude sewage are around 50 mg/l. Ammonia is extremely toxic to fish, the toxicity being greater at low dissolved oxygen concentrations and high pH values. Ammonia is also undesirable because it, or its oxidation products, may promote growth of algae (see below) and particularly when the water is a source for potable supply as it complicates the processes of disinfection by chlorination.

Sewage treatment works are therefore increasingly designed to ensure complete oxidation of ammonia to nitrate.

2.1.4. Phosphates. In parts of the world where sewage effluents are discharged to lakes and tideless seas there is concern over the incidence of eutrophication - the enhanced growth of algae and larger plants. Some forms of algal growth look unsightly and can cause deoxygenation, difficulties in water treatment, and unpleasant tastes in drinking waters. Phosphates derived from sewage are often a limiting factor and in some countries (particularly Scandinavia and South Africa) removal of phosphate forms an essential stage in sewage treatment. Concentration of phosphorus (as P) is usually about 10 mg/l.

In some cases nitrogen may also be an important factor and must then be removed usually by reduction of nitrate to nitrogen gas as the final stage of sewage treatment.

2.1.5. <u>Toxic metals.</u> Sewages, particularly those containing effluents
from industrial processes, may contain elements such as cadmium, mercury
and lead which are toxic to some forms of life and undesirable in drinking
water even at low concentrations. Sewage treatment processes are not
designed to remove toxic metals though some removal does occur, often
creating difficulties if the sludges are to be disposed of to agricultural land. Such metals are best removed from industrial effluents
before discharge to the sewer, and drainage authorities impose increasingly severe restrictions to ensure that this is done.

2.1.6. <u>"Micro-organics".</u> Many waterworks intakes on rivers are necessarily sited below points of discharge of sewage effluents and it is possible, using modern sensitive methods of analysis, to detect in drinking
water supplies many substances derived from sewage effluents. Attempts
have been made to determine whether such substances have any harmful
effect and in particular to determine whether they are carcinogenic,
mutagenic, or teratogenic. To date such attempts have failed to reveal
any significant effect. For example in a study commissioned by the UK
Department of the Environment the relative degree of reuse of water in
the London area was compared with the mortality from a variety of
causes - principally various forms of cancer. When socio-economic
factors were taken into account no correlation could be found. Nevertheless in view of the increasing variety of new substances released to
the water environment each year the position must be kept continually
under review. Concentration of the non-biodegradable dissolved fraction
in sewage is about 50 mg/l expressed as COD.

2.2. Volumes and Dilution

The volume of sewage per head varies from place to place but the average
world-wide is probably around 200 l/day. If the world's population is
taken to be about 4,000 million and about 25% of this population reside
in towns the flow of urban sewage or sewage effluent which eventually
would pass to the sea would be about 1.5×10^8 m^3/day.
      The flow of water entering the sea from rivers and other external
sources has been reported to be about 10^{12} m^3/day. The volume of the
world's oceans is about 1.4×10^{18} m^3 so if the oceans were uniformly
mixed the retention time for a particle of water and any dissolved or
suspended matter that travelled with it would thus be about 3,000 years. It
follows that if steady conditions were developed and the sea were uniformly mixed the dilution of sewage or sewage derived water with other water
would be about 7,000 to 1. If any truly conservative substance were
present in "sewage water" and not in the diluting water or in any other
input then ultimately its concentration in the sea would attain a concentration one-seven-thousandth of that in the sewage water. The time
taken for this concentration in the sea to rise to within one per cent
of the equilibrium value if none were initially present would be about
5 residence-times, that is, about 15,000 years.
      If one supposed that release of sewage water to the sea in appre-
ciable quantity was confined to say the last 300 years and the content
of the conservative non-biodegradable organics were say 50 mg/l then

the increased concentration of these materials in the sea up to the
present time would be about 0.7 µg/l. This is 3 to 4 orders of magnitude below levels in rivers which support healthy aquatic communities
and in water that has been drunk without harm by the public for many
decades, so there would appear to be no immediate cause for concern over
accumulation of such materials in the oceans as a whole. Similar conclusions appear to apply to other components especially when account is taken
of the fact that the majority are subject to removal processes resulting
in half-lives much smaller than the average residence-time of water in
the sea.

On a local scale, however, the situation is very different because
mixing of waste waters released to the sea with the ambient sea water
takes place at a finite rate and in the immediate vicinity of outfalls
concentrations of pollutants inevitably approach those in the discharge
itself. The essential problem in direct disposal of sewage to the sea is
therefore to prevent concentrations in coastal waters in the neighbourhood
of outfalls from attaining unsatisfactory levels.

3. DESIGN OF SCHEMES

3.1. Administrative Consideration

In devising a scheme of treatment and disposal which will deal with the
large volumes of waste water produced by a developed community and which
will overcome the detrimental effects previously mentioned many factors
must be evaluated so that a satisfactory compromise may be reached. In
general the compromise will reconcile the requirements of regulatory
authorities with the results which can be achieved by available technical methods at a cost which is acceptable to the community.

The bases of such regulations are commonly relationships (often
termed water quality criteria) between concentration (or sometimes mass
loading) of a pollutant in water and its effect on the suitability of
the water for a given use. There are substantial compilations of such
criteria in the literature, most of the data stemming from the work of
organisations such as authoritative national research laboratories and
environmental protection agencies and international bodies such as the
World Health Organisation.

Armed with such criteria a judgement can be made by regulatory
bodies as to the maximum concentration of pollutants tolerable in water
bodies. In some countries regulatory bodies prefer to ensure that such
concentrations are not exceeded by imposition of what are termed Uniform
Emission Standards which specify limits to the quality of all discharges,
irrespective of the degree of dilution afforded in receiving waters and
other local factors. Other countries prefer to set Environmental Quality
Objectives (EQOs) and to prescribe the conditions for achieving these
in terms of the maximum permissible concentrations of pollutants, often
termed Environmental Quality Standards (EQSs), in the receiving waters.
Effluent quality standards can then be varied according to local circumstances providing the EQSs are met. In the modern practice of many regulatory bodies adopting the EQO approach, the standards promulgated are

related to the proposed uses of the water. For example, in Western
Australia the regulations identify 16 possible uses of marine waters and
standards for up to several dozen quality characteristics are prescribed
for each use. Usually such regulations permit the limits to be exceeded
in so-called mixing zones, to avoid the necessity of the limits having
to be complied with in the effluent itself. However, the size and location
of such zones is usually defined, having regard to such features as
bathing areas or the travel paths of migratory marine organisms.

The majority of countries do not yet lay down regulations as
detailed as those for Western Australia but in relation to sewage disposal
an increasingly common requirement is a restriction on the number of
coliform bacteria or of the particular faecal species <u>Escherichia</u> <u>Coli</u>
in bathing waters.

National standards will, of course, normally be consistent with any
international agreements to which the country concerned is party. Thus
member countries of the European Economic Community are bound to enforce
by national legislation the Directives issued by the Community. Many
countries are signatories of broader international agreements such as
the Oslo Convention and the Paris Convention, which among other things
could affect disposal of sludges derived from treatment of sewage.

Despite the mass of information on water quality criteria such data
are by no means comprehensive and many uncertainties exist about the safe
levels of pollutants in the sea. To cover these uncertainties, regulatory
bodies often include safety factors frequently of one or two orders of
magnitude. Others prefer to require adoption of the Best Practicable
Environmental Option, as described in an earlier paper in this Workshop;
and some pressure groups argue that the uncertainties are so great that
the only satisfactory policy in the case of some pollutants is total
prohibition of their release.

3.2. Technical Considerations

The technical problem of the designer of treatment and disposal facilities
is to meet in the most cost-effective manner, such requirements as may
be laid down by regulatory authorities and by the community that the
facilities are intended to serve. Such requirements and thus the choice
of plant vary with local circumstances.

3.2.1. <u>Treatment options</u>. In developed countries sewage is usually
treated before discharge to rivers by a sequence of which the most common
components are screening, grit removal, primary sedimentation, biological
oxidation and secondary settlement, though there are many variants of
this basic scheme. Such treatment is usually designed to reduce the
concentration of suspended solids and BOD to respectively not more than
30 and 20 mg/l and increasingly frequently nowadays to oxidise the bulk
of the ammonia to nitrate. In the process most of the material that would
give rise to visible evidence of pollution (gross solids, surface active
matter, many coloured materials derived from industry), a proportion of
heavy metals and many other industrial pollutants are removed and bacterial
content is reduced by one or two orders of magnitude. Certain industrial
pollutants could penetrate these conventional processes, and result in
 ailure of the effluent to comply with standards limiting their

concentration. Also some would interfere with the treatment processes, particularly the biological stages, causing them to fail. To prevent such occurrences, control authorities normally impose limits on the concentrations of such substances in discharges of industrial waste waters to sewers.

Where discharges of sewage effluent are to water bodies drawn on for potable supply additional treatment (often termed "polishing") is often provided to reduce the concentration of suspended matter and associated BOD in the final effluent to not more than 10 mg/l. This can be achieved in various ways, of which one of the most popular is sand filtration.

In the case of some discharges, usually to inland water bodies, to prevent excessive growths of algae in the receiving waters treatment is given to reduce the content of assimilable forms of phosphorus (P) or nitrogen (N) or both. Again there are many ways in which this can be done. Usually the most effective methods of removing P depends on chemical precipitation though biological methods are used. Removal of nitrogen is often most economically achieved by biological methods, though physical or physico-chemical techniques such as counter current air stripping of ammonia or its removal by ion-exchange are also effective.

In some schemes polished sewage effluent is used as a water supply for those industrial purposes for which a particularly high quality is not required. In such cases it is usually disinfected by addition of chlorine. It is being increasingly frequently used for irrigation usually after disinfection by chlorination or by passage through matura-tion ponds having a sufficiently long detention period to ensure that few micro organisms survive to reach the final effluent.

In just a few situations, because of local water shortages sewage is treated to yield water of potable quality, either for direct recycle to consumers after blending with supplies from more conventional sources, or to serve as a barrier against the intrusion of sea water into aquifers drawn on for public supply. In such schemes additional processes to those already mentioned are brought into play including adsorption on activated-carbon to reduce the residual concentration of dissolved organics and reverse-osmosis which removes the majority of the residual content of both organic and inorganic pollutants.

3.2.2. Dispersion in coastal waters. At a coastal site there is the further option of discharging to sea through a submerged outfall. In broad terms the further offshore a particular waste water is discharged the smaller, usually, is its environmental effect. Thus the stronger the waste, the further offshore it must be discharged to prevent unacceptable environment-al impact. However, the cost of conveyance by pipeline or tunnel to the point of release increases with the increase in the distance of this point from the shore.

The strength of waste waters can be reduced by treatment but the cost of treatment increases with the extent of the reduction required. Thus a principal task of the designer of a disposal scheme is to establish the economic balance between the extent of treatment before release and the distance of the point of release from the shore.

The balance of advantage depends on many factors but particularly the hydrography and hydrodynamics of the coastal waters, the topography and stability of the ground both above and below the water, the

availability of land, the nature of the waste waters, the type of sewerage systems used to collect them, the water quality that must be maintained
in the receiving waters and the uses made of them, the treatment processes
available, and the costs of all the relevant engineering structures.

When sewage is released into markedly saline coastal water having
no major vertical gradient in salinity, it rises to the surface under
the influence of buoyancy, sometimes to form a characteristic disturbance
termed a "boil". During this process initial dilution takes place, the
extent of which is dependent on several factors including the ratio of
the diameter of the orifice to its depth of submergence, the velocity
of the issuing jet and the ambient current velocity. The greater the
dilution achieved the lower the frequency of formation and visibility
of slick. The relationship tends to be site specific but broadly, if
dilutions exceed 100, slick would be rarely visible, except perhaps to
a close observer in a boat or aircraft.

If the initial dilution achievable at an otherwise appropriate outfall location but discharging from a single orifice is insufficient,
satisfactory dilutions can usually be attained by distributing the sewage
through an array of orifices, termed a multi-port diffuser.

After initial dilution the sewage is further dispersed by the combined
action of tidal drift and eddy currents. The intensity of these effects
also tends to be site specific and local investigation of the hydrodynamics
of the water movements must be made to provide firm estimates of the
dilution that will be obtained at given distances from the. boil. Usually,
however, dilutions by at least a factor of 10 beyond the initial dilution
are achieved within a few hundred metres of the boil along the direct
path of the "plume" drifting from the boil; and much higher dilution at
similar distances at right angles to the direction of tidal drift.

In the case of certain pollutants such as non-biodegradable dissolved
organics, the concentrations produced in coastal waters by the release
of sewage are determined almost solely by initial dilution and subsequent
dispersion. In the case of many others concentrations in the water are
also reduced by other processes of self-purification including settlement, biological oxidation, utilisation as nutrients by marine organisms,
and particularly in the case of bacteria, decay due to various properties
of the coastal environment.

Using modern methods, the combined influences of these various processes can be described with adequate accuracy using mathematical models,
though the magnitudes of the parameters in these models are site specific
and have to be determined by local investigation.

The stability of the sea bed is a vital factor in outfall design
and in the choice of method of construction. If the bed is highly mobile,
lengths of pipe may become unsupported with consequent danger of fracture.
Geophysical surveys using, for example, side-scan sonar and sparker devices
can be used to assess stability and to decide the needs for protection
which may be achieved by burying the pipe below the mobile layers, or
by "mattressing" techniques. A common method of installation consists
of laying a pipe in a trench by one of three techniques, comprising
bottom pull, lay barge and float and sink methods. However, in areas of
strong persistent currents it may not be possible to dredge a trench and
keep it open during installation.

In locations where the superficial bed strata are too unstable, the bed is very uneven or rock strewn or changes sharply in profile, a tunnelled outfall may be appropriate. However, such outfalls are seldom economic for flows requiring diameters of less than 1.5 m and often present considerable difficulties in the installation and operation of diffusers.

Many other factors have to be considered in arriving at an appropriate design. Suffice it to note, however, that nowadays in virtually all coastal outfall schemes designed in developed countries, sewage receives screening and usually grit removal before discharge. Frequently the cheapest satisfactory scheme is one in which these are the only pretreatments, as is indicated in some case studies briefly summarised in a later section. However, discharge of sewage which has been partially treated (for example by sedimentation or high rate biofiltration) is not uncommon and in some circumstances discharge after a higher degree of treatment through a shorter outfall may be economic.

Usually, for a town of say 10,000 inhabitants, when only preliminary treatment is given, the point of discharge would have to be at least 400 m seaward of the low water mark to meet bacterial standards imposed in Europe and satisfy minimum aesthetic requirements. The current trend of practice, particularly for larger populations, has been for the points of discharge to be located at greater distances offshore, not infrequently more than 1 km for small populations and several kilometres for larger ones.

3.2.3. Optimization of degree of treatment and outfall length.

White and Agg[1] have suggested a simple approximate approach relating the required initial dilution at an outfall (Co/Cm) with the length of the outfall (L) and a constant (Ψ) whose value is selected according to the level of amenity required at the beach in question. For a beach of high amenity value $\Psi = 10^8$ m^2. According to White and Agg:

$$\frac{C_o}{C_m} - 1 = \frac{\Psi}{L^2}$$

and hence for a dilution factor of, say 50, the length L is 1,430 m.

For a partially purified sewage a shorter outfall is needed to maintain the same level of pollution at the beach and for simplicity the same dilution factor is assumed (that is same depth of water and discharge arrangements). The required level of purification may then be plotted against length as shown in Figure 1.

On the same figure the percentages removed by various treatment schemes are shown. These suggest that sedimentation alone would reduce the required length of outfall to about 1,100 m and full conventional treatment to, say 300 m.

This treatment is a gross over-simplification in terms of the mathematics and also in terms of local conditions but nevertheless indicates the degree of flexibility which may be achieved by a trade-off between length of outfall and degree of treatment.

3.2.4. Bacterial contamination from rivers.

In passing, the point may also be made that in many countries including the United Kingdom sewage effluents are not disinfected, the normal processes of primary and secondary treatment removing say 99% of the initial concentration of coliform bacteria leaving a concentration in the effluent of say 10^6 organisms per 100 ml. This effluent would normally be discharged to a river or stream.

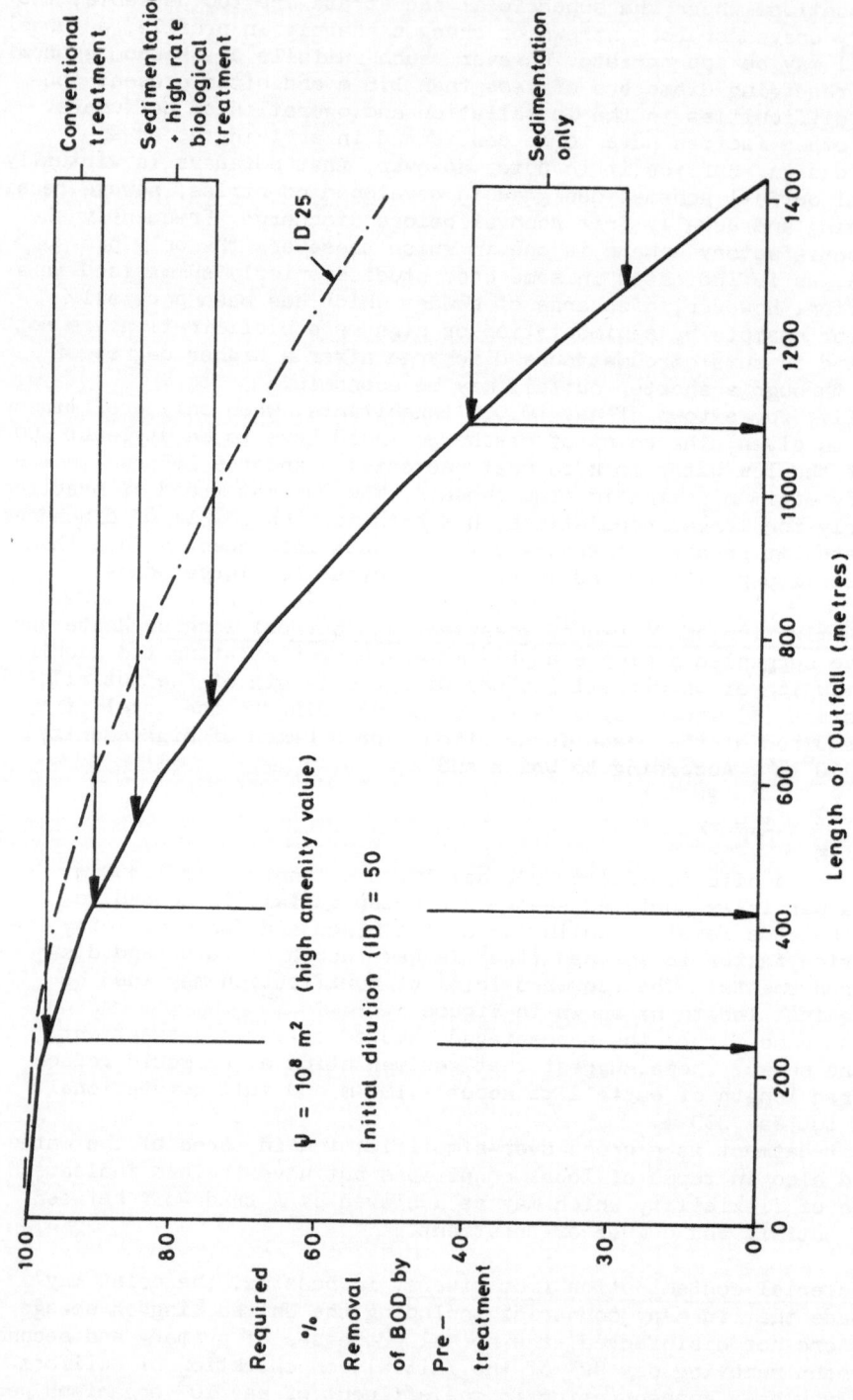

Figure 1.   Required degree of pre-treatment related to outfall length (after White and Agg)

Where such streams enter the sea close to a bathing beach the level of contamination of the inshore waters would be comparable with that to be expected from discharge of screened sewage from a relatively short outfall.

An outfall at Bridport in the south-west of England was the subject of detailed studies by the former Water Pollution Research Laboratory. This outfall was so constructed that sewage could be discharged at various distances from the shore. It was found that the level of bacterial pollution at the beach from the river Brit (which received treated sewage from inland works) was more significant than that from the outfall discharging at 680 metres offshore, and may have been greater even when the outfall discharged at 430 metres.

It cannot be assumed therefore that bacterial pollution of a beach can be eliminated by choice of a sewage treatment works on land rather than a sea outfall. Where it is essential to reduce numbers of pathogens at a beach the choice is between a works provided with facilities for disinfection of effluent and a properly designed long sea outfall. In some cases, as a temporary measure, sewage discharged through a short outfall may be disinfected by chlorination.

3.3. Land Use Considerations

Despite efforts to intensify sewage treatment and thus to reduce the size of works the necessary area of land may not be easy to find in a district which has already been developed or where geographical conditions are unsuitable. This is particularly the case in coastal towns where sewerage systems tend to slope towards the sea. One solution is to pump the sewage out of the town to a works discharging to the local river and this method has been adopted in some towns in the United Kingdom. Care is needed in designing the pumping station to ensure that it is inconspicuous and that it does not emit noise or odours. The danger of beach pollution even from treated sewage effluent has already been mentioned.

3.4. Sludge Disposal

The organic pollutants removed from sewage are partially oxidised to carbon dioxide but a large proportion is either precipitated from the sewage or converted into new bacterial cells and must be disposed of as a sludge. The proportion converted to sludge is approximately one half but will be higher in high rate processes. Some newer treatment processes using oxygen or other intensive methods of aeration are claimed to produce less sludge but these claims on the whole have not been sustainable.

Disposal of sludges therefore represents a major problem at sewage works and represents around half the total cost of operation. Sludge has a high water content (95% or more) and many methods are available to reduce this and so facilitate disposal. Some of the organic matter in the sludge may be converted to methane by anaerobic fermentation thus reducing the weight of solids in the sludge and yielding a less objectionable product. The method and degree of treatment of the sludge will depend on the method employed for its disposal. These options are to be discussed by another contributor but in the context of the present paper it should be remarked that problems of treatment and disposal of sewage sludge may play a major role in deciding on the situation and design of a sewage treatment plant. In some cases expected difficulties in sludge disposal

could well lead to a decision to dispose of sewage through a suitably-
designed sea outfall rather than by treatment at an inland works.

4.   CASE STUDIES

To illustrate typical differences in the costs of alternatives, three
schemes are briefly considered below. The alternative chosen appears
to have proved satisfactory in each case.

4.1. North Wirral

Calvert(2) quotes the case of a scheme for disposal of sewage from two
adjacent towns in the north-west of England where one of the local authorities was reluctant to permit discharge to sea, while the other was
unwilling to permit treatment on land. Ultimately the difficulty was
resolved by construction of an outfall 5 km in length. Despite its length,
the outfall proved less expensive than the alternatives costed (Table 1).

4.2. Ingoldmells

This scheme, and the other which follows, was among those developed by
the author's firm. For Ingoldmells, a holiday area on the east coast
of England, three alternatives were examined:
- extension of the existing works to give complete primary and
  secondary treatment with discharge to a drainage channel
  leading to the sea;
- discharge of screened and degritted sewage through an
  outfall 0.6 metres in diameter and 4.8 kilometres long;
- use of the existing works to give partial treatment by
  high-rate filtration with discharge of effluent through
  a shorter sea outfall 2.4 kilometres long.
    Two rates of flow were examined, 0.8 million gallons (3,636 m³) per
day representing the provision to be made initially and 1.3 million gallons
(5,909 m³) per day representing a further stage of development.
    The capital costs and the total annual costs obtained by adding
the loan repayments to operating costs, are shown in Table 1. At the
higher planned flow either of the systems employing a sea outfall was
considerably less expensive than treatment on land, though in the short
term extension of the existing works was marginally the least expensive.

4.3. Cape Peron, Perth, Western Australia

In 1980 alternatives were examined for disposal of an ultimate flow of
250,000 m³/day of sewage from the city of Perth. The principal options
investigated were:
- land treatment and disposal;
- advanced treatment with nitrogen removal and disposal
  to a partially enclosed bay (Owen Anchorage) adjacent
  to the terminal point of the existing sewerage system;
- primary treatment and discharge to deep ocean waters.
    Superficially land disposal initially appeared attractive since
it offered the possibility of recharging groundwater and thus increasing

COMPARISONS OF THE COSTS OF ALTERNATIVES FOR DISPOSAL OF THE SEWAGE FROM THREE LOCATIONS

Location

North Wirral, UK*

		Full treatment at outfall	Primary treatment	'Semi-full' treatment at outfall	Long sea outfall	Works and lagoon treatment outfall to river	Outfall to Dee Estuary
1974 prices	C	2.68	2.28	2.51	1.93	2.60	2.18
(£ millions)	0	0.26	0.22	0.25	0.16	0.26	0.17

Ingoldmells, UK

		Extension of Works to give full treatment		Long sea outfall		Partial treatment and shorter sea outfall	
		(a)	(b)	(a)	(b)	(a)	(b)
1966 prices	C	0.32	0.20	0.46	0.008	0.37	0.038
(£ millions; 6% discount rate)	0	0.032	0.050	0.035	0.037	0.033	0.038

Perth, Western Australia +

		Advanced secondary treatment	Land disposal	Sea outfall 4km	8.8km
1980 prices	C	47	100+	36	70
(Australian $ millions)	A	1.6	uncertain	0.6	0.6

C: Capital cost
0: Annual operating cost plus loan repayment
A: Annual operating cost only

(a) DWF:3636m³/d
(b) For further DWF of 2273m³/d

* After Calvert (2)
+ After Metropolitan Water Board(3)

resources for public supply. Three main options were studied but all presented major technical, social and environmental difficulties; also they
were all much more expensive than ocean disposal schemes (Table 1).

Studies of the current patterns in Owen Anchorage indicated that
its waters would be poorly flushed during periods of several weeks, especially during the autumn. It was concluded that, even with nitrogen
removal, the possibilities of unsightly algal blooms developing was too
great to make disposal to the Anchorage a desirable option.

In investigating the third major option, four pipeline routes were
examined involving discharges at distances offshore ranging from 4 to
17 km.

The preferred option proved to be one involving conveyance of the
sewage by 23 km of land pipeline to Cape Peron, and then discharge by
submarine pipeline, 4.2 m long, terminating in a 316 m diffuser section
at a depth of 20 m.

The conclusions of the study were embodied in an Environmental Impact
Statement submitted to the Western Australian State Government and exposed
for public comment. The recommended scheme was adopted and brought into
operation in mid-1984(3).

5. OPPORTUNITIES

It has often been noted that if the wastes in sewage could be conveyed
by some means other than a water-carriage system the problem of water
pollution could be greatly reduced. Certain schemes based on this idea
have been adopted on a small scale but none have proved successful enough
to prompt their wider adoption or indeed to encourage any hope that worthwhile development of this type can be expected in the foreseeable future.

In the absence of any such radical solution there are, nevertheless,
manifold opportunities for improvement of disposal practice using existing
technology.

Undoubtedly in some countries there are still many outfalls that
produce unsightly evidence of pollution because they have become damaged
or overloaded or were located too close to shore at a time when dispersion
mechanisms were not so well understood or the public's requirements were
less stringent. Obviously, there are straightforward opportunities for
rectifying such situations either by giving greater treatment of sewage
before discharge or lengthening the outfall pipelines. In the latter
context there is a need for designers to give greater recognition to
a natural concern, especially among bathers about health risks and interference with aesthetic enjoyment. The sea is a mobile medium and in the
absence of any barrier between those enjoying body contact recreation
and an outfall that can be seen, no amount of reassurance, however soundly
based, will convince everyone that they are not at risk. There is thus a
particular need to ensure that outfalls are not visible from the shore
and certainly that there is no evidence of pollution from them.

Also apparent is the scope for the wider use of a range of newer
processes by industry, particularly flotation, electrolysis using fluid-
ised bed electrodes, ultrafiltration and reverse osmosis, to give effect
to the opportunities of recovering valuable materials for reuse or sale
as by-products and at the same time reduce release of the more pernicious
pollutants to sewers.

Looking further ahead there is undoubtedly scope for environmental scientists to refine and extend water quality criteria so that standards can be based on sounder premises; and for engineers to find more cost-effective ways of meeting the standards by improving both pipeline technology and treatment processes.

A particularly intriguing prospect potentially having wide application is a new development in the field of cross-flow microfiltration. The principle of this process, which has been known for about 15 years, relies on a fast tangential flow of liquid across the surface of a filtering medium preventing accumulation of layer of filtered solids to an extent which would cause clogging and reduction in flux (Figure 2). A variety of pre-coats can be used which in the simplest form of the process enable suspended matter down to about 0.1 micron to be removed. One obvious application of the process is in polishing effluents from conventional biological treatment of waste waters. The most popular current method of improving such effluents is by sand filtration, a process producing final effluents containing average concentrations of suspended solids of about 5 to 6 mg/l. Associated with each mg/l of these suspended solids is a BOD usually of around 0.6 mg/l and coliforms bacteria in numbers around 10^5/l. Crossflow filters used in place of sand filters could remove these impurities completely at costs lower than those of sand filtration. Chemicals have to be used to pre-coat the medium but only in small quantities at intervals which can be as long as several weeks.

Another application demonstrated to be technically feasible in the laboratory is to use the process in place of sedimentation tanks for the retention of biomass in bio-reactors. The ability to prevent thick cakes forming on the membrane allows concentrations of suspended solids in the reactor to be increased beyond that normally maintainable and thus the size and cost of the reactors to be substantially reduced. Clearly such uses of the process, if demonstrated to be practicable at full scale, would permit effluents discharged to rivers and thus indirectly to the sea to be better purified at lower cost than at present.

An even more intriguing possibility is that the pre-coat can consist of materials, such as zicronium salts which for reasons not yet clear will retain dissolved impurities.

Finally in the longer term one must hope that advances in genetic engineering will permit the efficiency and versatility of biological processes to be substantially improved.

(a) BASIC PRINCIPLE

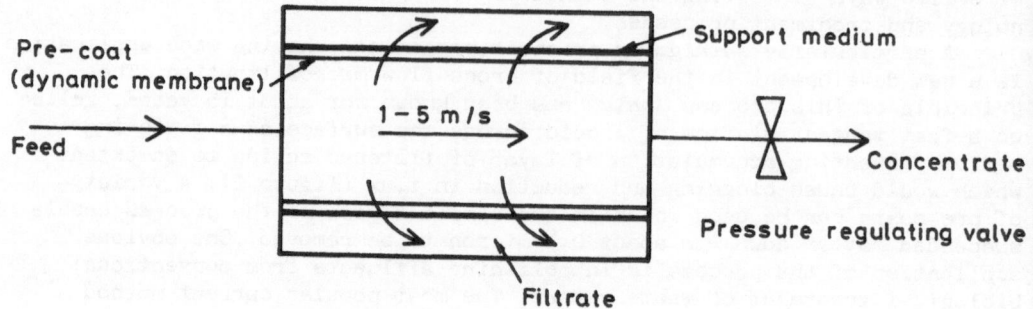

(b) TAPERED MULTISTAGE OPERATION FOR FILTRATION

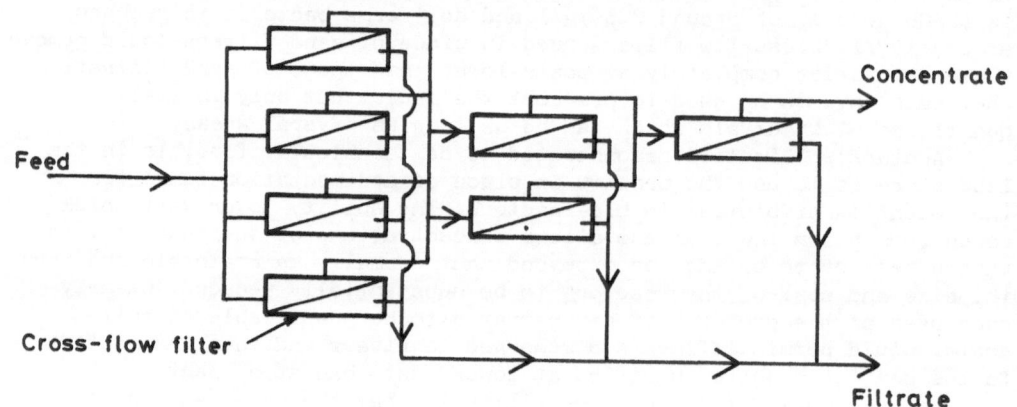

(c) SINGLE STAGE OPERATION FOR RECYCLING

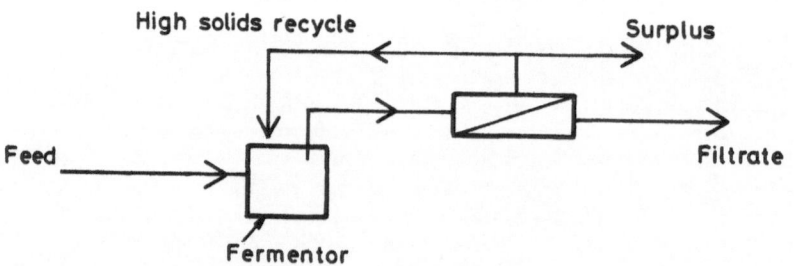

Figure 2. Basic principle and two modes of operation of cross-flow microfilters.

REFERENCES

1. White, W.R. and Agg, A.R. Outlet Design. In 'Discharge of Sewage
    from Sea Outfalls' (Ed. A.L.H. Gameson) Pergamon Press, Oxford,
    1975, pp. 265-274.

2. Calvert, J.T. The Case against Treatment. In 'Discharge of Sewage
    from Sea Outfalls' (Ed. A.L.H. Gameson) Pergamon Press, Oxford,
    1975, pp. 173-177.

3. Metropolitan Water Board, Perth Western Australia. Cape Peron
    Ocean Outlet - Environmental Progress Report, Sept., 1981.

ENGINEERING OF OCEAN OUTFALLS

Philip J.W. Roberts
School of Civil Engineering
Georgia Institute of Technology
Atlanta, Georgia 30332
U.S.A.

ABSTRACT. The criteria and design philosophy for ocean outfalls are
reviewed with particular emphasis on hydraulic and sanitary aspects.
It is argued that the outfall and treatment plant must be regarded as
an interrelated system whose characteristics depend on what receiving
standards are to be maintained. Three case studies of ocean and
estuarine outfalls for disposal of combined sewage, urban runoff, and
industrial waste are discussed with particular emphasis on the design
criteria used for each, the options considered, and reasons for final
choices made.

1. INTRODUCTION

The objective of an ocean outfall system is to rapidly disperse
wastewater so as to minimize any harmful effects of the discharge on
the receiving water. The purpose of this paper is to discuss some of
the engineering aspects of ocean outfalls and to present case studies
to show how this objective has been achieved in various situations. As
the philosophy and criteria used in the design of ocean outfalls have
been extensively discussed in the writings of Brooks and co-workers
(4,5,9,14,17) they will not be discussed in detail here. Rather, in
the next section we will briefly summarize these criteria. The primary
purpose of this paper is to discuss three case studies of ocean
outfalls with which the author has been involved. These three cases
form a disparate range of conditions; they are: municipal sewage
outfalls from the City of San Francisco into the Pacific Ocean and from
the City of Seattle into Puget Sound; and a combined industrial and
municipal wastewater outfall into the Indian Ocean off Richards Bay,
South Africa. The design criteria for each project, the options
considered, and the reasons for the final choices made will be
discussed. Emphasis is placed on hydraulic and environmental aspects;
construction aspects will be only briefly discussed. It should be kept
in mind, however, that constructability can often be the overriding
factor at a particular site.

G. Kullenberg (ed.), The Role of the Oceans as a Waste Disposal Option, 73–109.
© *1986 by D. Reidel Publishing Company.*

2. DESIGN CRITERIA AND PHILOSOPHY

The outfall and onshore treatment facilities should be regarded as an interrelated system whose characteristics depend on what receiving water standards must be maintained. This is because a tradeoff often exists between increasing levels of treatment and increased outfall performance, achieved by lengthening the outfall and diffuser and by discharging into deeper waters. The outfall should disperse the effluent sufficiently that ecosystem products of the effluent, for example organic carbon and nutrients, can be kept within allowable limits. The ocean itself provides a high assimilative capacity for this type of effluent. To accomplish these requirements, modern outfall design practice consists of outfall lengths of several kilometers terminating in long multiport diffusers which may themselves be the order of a kilometer in length. The diffuser discharges into water depths of up to 70 m, or in extreme cases deeper, resulting in sewage dilutions of the order of several hundred parts to one within a short distance of the outfall. Discharges should be into non-critical areas, or areas of limited ecological significance, and should avoid regions of poor flushing, such as enclosed bays. The outfall should be long enough that adequate mortality of bacteria and viruses occurs for any sewage which reaches shore. This can enable shoreline bacterial standards to be met without the need for costly disinfection. The avoidance of shoreline contamination can be further enhanced by utilizing the natural density stratification present in the ocean, due to salinity and/or temperature variations with depth, to prevent the diluted wastefield from reaching the surface. This also reduces any aesthetic problems which may arise from a surfacing wastefield. A further advantage of a submerged field may be the accelerated diffusion of the field by shear dispersion caused by trapping near the strongest stratification where velocity shear is strongest in the ocean.

Specific receiving water standards will often be set by government agencies. In California, for example, the State Water Resources Control Board has published a plan (26) for the protection of the ocean waters of California which specifies water quality criteria to be met by ocean outfalls. The plan specifies shoreline bacteriological standards, and physical, chemical, and biological requirements to be met in the receiving waters near the discharge. Minimum treatment levels are specified in the plan's "Table A," reproduced in Table I, and water quality objectives for toxic materials to be achieved after initial dilution are specified in "Table B," reproduced as Table II. This type of regulation allows a systems approach to the design of the outfall and degree of treatment. Specific effluent requirements, for example, treatment to secondary level do not provide this flexibility and may require unnecessary expense due to overtreatment. This may waste the ability of the ocean to assimilate BOD, for example, as dissolved oxygen depletion is rarely a problem in coastal waters. Frequently, only preliminary treatment will be necessary to remove most particulate matter and floatables which may come to shore. Of course, no toxic substances should be discharged, for example DDT, and PCB's, which should be eliminated by source control.

TABLE I. "Table A" effluent quality requirements from
California Ocean Water Plan (26)

	Unit of measurement	Limiting Concentrations		Maximum at any time
		Monthly (30 day Average)	Weekly (7 day Average)	
Grease and Oil	mg/l	25	40	75
Suspended Solids	mg/l	75 Percent Removal		
Settleable Solids	ml/l	1.0	1.5	3.0
Turbidity	JTU	75	100	225
pH	units	within limits of 6.0 to 9.0 at all times		
Toxicity Concentration	tu	1.5	2.0	2.5

The details of the hydraulic design have also been dealt with in detail elsewhere (9,17). Briefly, the hydraulic design should result in roughly uniform division of flow between the discharge ports at all flows. The velocity in the pipe should be high enough to maintain scour and prevent deposition of particles; this will be accomplished by peak flow velocities between about 0.5 and 1.0 m/s. Once the outfall is flowing seawater intrusion into the ports should be prevented, this can be achieved by maintaining port densimetric Froude numbers greater than 1 and a total port area less than the outfall pipe cross-sectional area. The total headloss in the outfall and diffuser should be kept small to minimize pumping (if used) costs. The hydraulic capacity is usually chosen so that the system can discharge peak flows against the highest tide ever observed. This may result in an overly conservative design, however, and can sometimes be relaxed by considering the joint probability of very high tides and peak flows. The ports should discharge horizontally for optimum dilution, and can be simple circular holes in the pipe wall or can be risers if the diffuser is submerged. The port design should be simple; nozzle gadgets which may require high maintenance should be avoided.

TABLE II. "Table B" receiving water quality requirements to be met after completion of initial dilution in the California Ocean Waters Plan (26).

	Unit of Measurement	Limiting Concentrations		
		6-Month Median	Daily Maximum	Instantaneous Maximum
Arsenic	mg/l	0.008	0.032	0.08
Cadmium	mg/l	0.003	0.012	0.03
Total Chromium	mg/l	0.002	0.008	0.02
Copper	mg/l	0.005	0.020	0.05
Lead	mg/l	0.008	0.032	0.08
Mercury	mg/l	0.00014	0.00056	0.0014
Nickel	mg/l	0.02	0.08	0.2
Silver	mg/l	0.00045	0.0018	0.0045
Zinc	mg/l	0.020	0.08	0.2
Cyanide	mg/l	0.005	0.02	0.05
Phenolic Compounds	mg/l	0.03	0.12	0.3
Total Chlorine Residual	mg/l	0.002		
Ammonia (expressed as nitrogen)	mg/l	0.6	2.4	6.0
Toxicity Concentration	tu	0.05	-	-
Total Chlorinated Pesticides and PCB's	mg/l	0.002	0.004	0.006
Radioactivity	Not to exceed limits specified in Section 30269 of the California Administrative Code.			

3. CASE STUDIES

3.1 San Francisco

3.1.1 Project Description. The City of San Francisco has a combined sewerage system that transports domestic and industrial sewage and, when rainfall occurs, stormwater runoff. The continuous flow of domestic and industrial sewage, which we shall call "dry weather flow," is currently treated and discharged through three outfalls, one into the Pacific Ocean, and two into the San Francisco Bay. Even a modest amount of rainfall, however, causes the hydraulic capacity of the sewerage system to be exceeded. When this occurs the mixture of sewage and stormwater runoff, which we shall call "wet weather flow," is discharged at 41 locations into the Ocean and Bay. The beaches are contaminated for several days following each of these events.

The original master plan to alleviate this problem called for the transport of all flows by a system of tunnels to a location on the southern, Pacific Ocean, side of the city near Lake Merced as shown in Figure 1. After treatment, the flows would be discharged to the

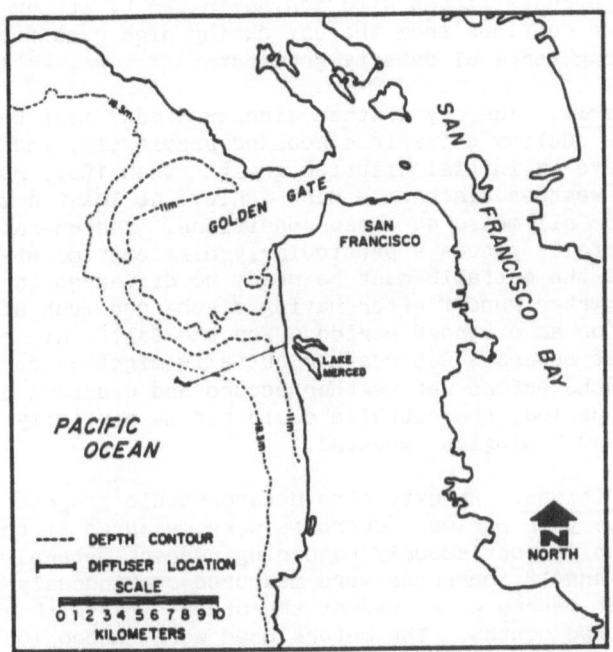

Figure 1. Location map for ocean outfall system proposed for the City of San Francisco (from Isaacson, et al. (12)).

Pacific Ocean through a system of outfalls. The peak flow to be conveyed by the outfalls is dependent on rainfall and on the storage capacity provided in the collection system. That is, a large quantity of stormwater could be stored and then released at a relatively slow rate for treatment and discharge. Conversely, a small storage capacity would require high rates of discharge through the outfall system to avoid overflows. Following economic and engineering analyses, a peak capacity for the outfall system of 43.8 m^3/s (1000 mgd) was chosen. Of this, an average of 4.38 m^3/s (100 mgd) would be the dry weather flow and the remainder stormwater runoff. The dry weather flow is continuous, although varying diurnally, and the wet weather flow is expected to occur for an average of 350 hours per year, primarily during the winter months of December through February. This system depends on stormwater flows from the whole city being brought to the Lake Merced discharge site and would require construction of a cross-town tunnel to transport sewage and stormwater runoff from the

eastern half of the city. Without this tunnel, the peak wet weather
flows are reduced to 19.7 m^3/s (450 mgd).

The proposed discharge site has a number of unusual features. The
sea floor slope is quite gentle, resulting in discharge depths which
are less than usual for typical major west coast United States and
Hawaii outfalls. Currents at the site are dominated by strong tides,
and large freshwater outflows from the bay during high runoff periods
can cause very strong vertical density gradients.

3.1.2 <u>Design Criteria</u>. The dry weather discharge must meet the usual
hydraulic and water quality criteria discussed previously, and in
addition must achieve an initial dilution greater than 100:1 most of
the time. The wet weather discharges must achieve at least a 25:1
initial dilution for all measured ocean conditions. The operation of
the wet weather outfalls places a particularly difficult maintenance
requirement in that the outfalls must be ready to discharge large
quantities of stormwater runoff after having discharged much smaller
dry weather flows for an extended period. For the California climate,
this could mean that separate wet weather outfalls might be dormant for
as long as nine months before wet weather occurs and causes a discharge.
During the dormant period, the outfalls could become partially blocked
by sand intrusion and biological growths.

3.1.3 <u>Oceanic Conditions</u>. An extensive oceanographic program was
conducted over a one year period. Currents were measured at the sites
shown in Figure 2 using continuously recording current meters. At
stations 7, 8, 11, and 12, currents were measured continuously for one
year; roving current meters were used at the other sites with each site
being occupied for two months. The meters used were Endeco 105's which
measured speed and direction averaged over half-hour intervals.
Density stratifications were measured at many sites, including the
current meter stations shown in Figure 2, at least once per month. In
addition, stratifications at some sites were measured at hourly
intervals over several complete tidal cycles. Other observations
included measurements of the movements of surface and subsurface
drogues and drifters.

The currents were analyzed by time-series methods. First, the
directions of the principal axes at all stations were computed. These
are the orthogonal axes which maximize and minimize, respectively, the
variance of the currents projected on to them. The directions of the
principal axes are shown in Figure 2 for a typical month. The
components of the currents projected along the principal axes are known
as the principal components, and a typical time series of the
principal components (for station 7) are shown in Figure 3. The major
principal axes at all stations point approximately to the Golden Gate.
The average kinetic energy, or variance, of the currents decreases with
distance from the Golden Gate, as illustrated by the lengths of the
principal axes shown in Figure 2. The average kinetic energy of the
first principal components are generally much greater than of the
second, and inspection of Figure 3 suggests that the first principal
components are strongly tidal. This is confirmed by the energy spectra

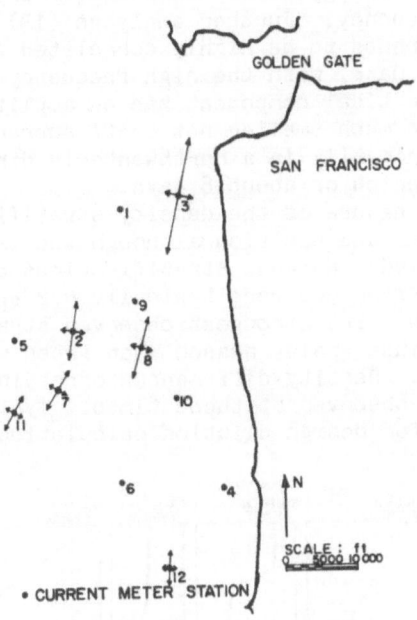

Figure 2. Current meter locations and principal axis directions of the
currents measured in a typical month (from Roberts (23)).

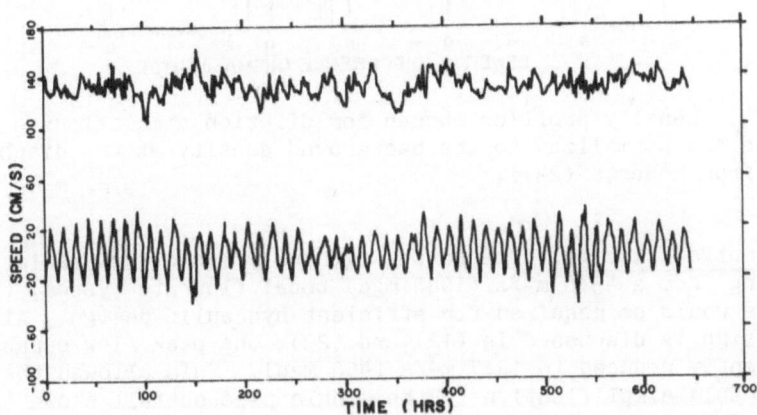

Figure 3. Principal components of currents measured at Station 7 for a
typical month. The second principal component (top) is displaced 140
cm/s above the first principal component (bottom) (from Roberts (23)).

(13,23) which show the energy to be concentrated primarily at the
semidiurnal tidal frequency. Further analyses (13) showed the tidal
components of the currents to be highly correlated with the tidal
current at the Golden Gate, with the high frequency content being
basically random. The tidal component has an amplitude of about 25
cm/s superimposed on a much smaller net drift current of about 2.5 cm/s.
The net drift moves primarily in a northwesterly direction and has
fluctuations with a period of about 8 days.

The strength and nature of the density stratification varied
substantially with time and position although the water column was
almost always stratified. Weakest stratifications occurring during
fall when density differences were typically 0.2 σ_t units (0.0002
g/cm^3) over 25 m depth. The strongest observed stratifications
occurred during the winter rainy season when fresh water runoff into
the Bay was very high. Density differences exceeding 4 σ_t units over
the water column were observed at these times. Typical density
profiles were chosen for design dilution calculations; they are shown
in Figure 4.

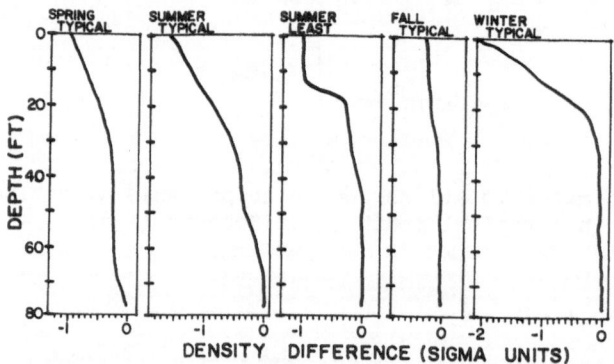

Figure 4. Density profiles chosen for dilution simulations. The
profiles are normalized to the background density at the discharge
depth (from Roberts (23)).

3.1.4 <u>Outfall Design</u>. Due to the extremely large flow variation
resulting from a 43.8 m^3/s (1000 mgd) total flowrate system, three
outfalls would be required for efficient hydraulic design. Although
this design is discussed in (12) and (23), the peak flow capacity was
subsequently reduced to 19.7 m^3/s (450 mgd). This allowed the
considerable simplification to the single pipe outfall shown in Figure
1. The dry weather design flows are: 1.75 m^3/s (40 mgd) minimum; 4.69
m^3/s (107 mgd) average; and 6.22 m^3/s (142 mgd) maximum.

The outfall pipe is concrete, 3.66 m (12 ft) in diameter, and
terminates in a diffuser 922 m (3024 ft) long. The diffuser has 680
ports with an average spacing of 1.36 m (4.5 ft) whose diameters vary
between 10.97 cm (3.60 in) and 13.1 cm (4.31 in). This design was
based on mathematical and physical modeling at the California Institute

of Technology (10,11,12) using measured oceanic conditions. The diffuser loading of about 30 mgd/ft is small compared to other typical diffusers shown in Table 10.1 (9). The small loading is necessitated by the requirement of high dilution under slack water for the occasionally strong measured stratifications. As diffusers oriented perpendicular to a current will result in higher dilutions than at other angles (20,21), the diffuser was oriented perpendicular to the first principal axis of the currents at the discharge site (see Figures 2 and 3) to gain the maximum beneficial effect of the currents on dilution.

Roberts (23) applied a mathematical model of initial dilution to investigate the effects of varying current speed and direction on this diffuser. Typical predicted temporal variations in dilution are shown in Figure 5. This prediction was made using measured currents, the

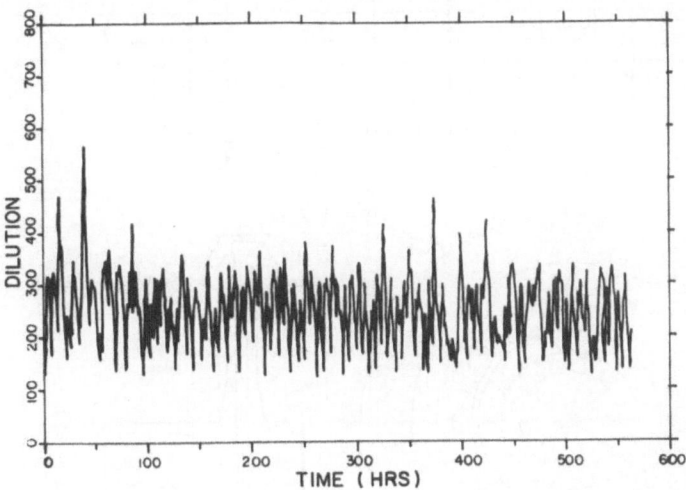

Figure 5. Predicted variation in initial dilution for dry weather discharge conditions caused by varying current speed and direction (from Roberts (23)).

spring density stratification (Figure 4), and expected diurnal variation of dry weather flow. The initial dilution varies widely; for all conditions tested, the dilution varied from about 100 at strongest stratification, highest flowrate, and weakest current to 1900 at weakest stratification, lowest flowrate, and strongest current. The wastefield rise height was also found to vary considerably, but was almost always submerged. Thus, although the relative shallowness and strong density stratification of the discharge site might suggest a poor location for the discharge, the strong currents result in high dilutions and considerable variation in rise height, enhancing the subsequent diffusion of the wastefield. The probability of onshore

transport is very low; if it were higher, the diffuser might be
oriented to be more perpendicular to the shoreline to obtain optimum
dilution during onshore currents. The City of San Francisco has
applied for a waiver from the requirement for secondary treatment of
effluent for this pipeline.

The shallowness of the water necessitates burying the diffuser to
protect it from wave action during storms. Thus, riser pipes are
necessary to connect the diffuser with discharge ports located above
the sea floor as shown in Figure 6. As these risers are relatively

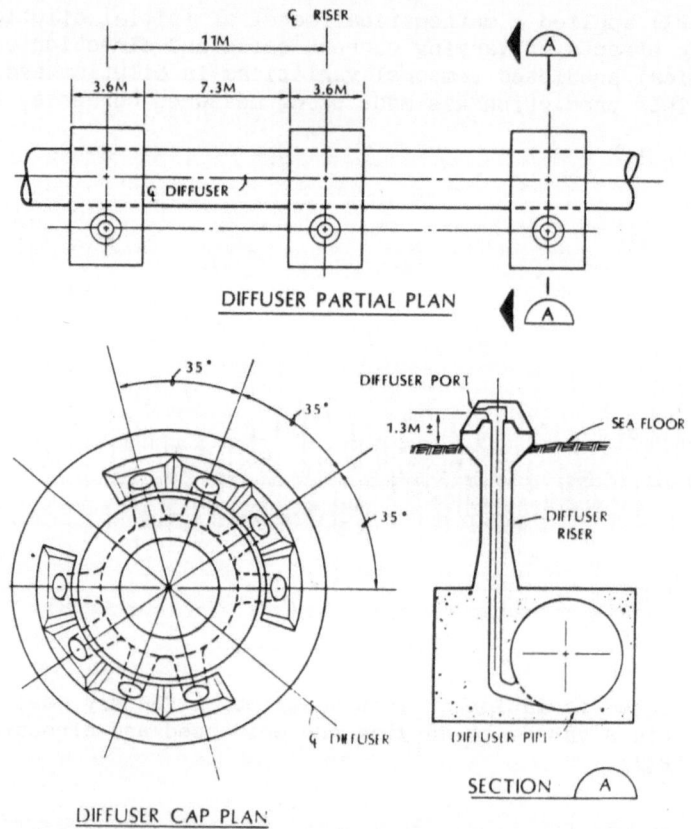

Figure 6. Diffuser riser and port design (courtesy of CH2M Hill, Inc.)

massive structures, it is clearly advantageous to maximize the number
of ports per riser, and hence minimize the number of risers. The final
design consists of 8 ports per riser, with the risers spaced at 11.0 m
(36 ft) intervals. Hydraulic model tests performed at Caltech
(10,11,12) verified that the initial dilution for this design was not
strongly dependent on the number of ports per riser, provided the total
diffuser length remained constant. Outfalls with four ports per riser

have been built in England (12) and the Philippines (6). The riser shafts take off from the diffuser pipe invert in order to reduce trapping of sand pumped into the ports by waves during low flows, and to enhance flushing of sand and particulates from the diffuser.

3.2 Richards Bay, South Africa

3.2.1 Project Description. A major industrial facility planned for Richards Bay, South Africa, will generate large quantities of industrial and municipal wastewaters. It has been decided to discharge these wastewaters via an outfall into the Indian Ocean, Figure 7. The projected wastewater composition is shown in Table III, and is

TABLE III. Projected Wastewater Flows and Characteristics for Richards Bay outfalls

Source	Description	Density kg/m^3	Flowrate m^3/day
Mondi	Pulp mill effluent Temperature approximately 50°C	990	90,000
Triomf	Fertilizer factory Gypsum slurry. Approximately 45% solids by weight, 26% by volume	2,300	2,800
Triomf	Fluoride, 12.3 kg/m^3	1,000	7,500
Others		997	23,450
Combined effluent		1,020	132,750

particularly noticeable for the large quantities of gypsum and fluoride produced by the fertilizer factory. If the wastewaters were all mixed together, the combined flowrate would be 132,750 m^3/d (35 mgd), they would have a density of 1.020 g/cc, slightly less than that of the receiving seawater, and they would contain a fluoride concentration of 693 ppm. In addition, the combined wastewaters would contain the other normal constituents of domestic sewage. The unusual wastewater characteristics constitute a challenging outfall design project.

3.2.2 Design Criteria. In addition to the usual hydraulic and sanitary design criteria, the gypsum and fluoride place special requirements on the design. The water quality standards specify maintenance of a fluoride concentration less than 5 ppm after initial

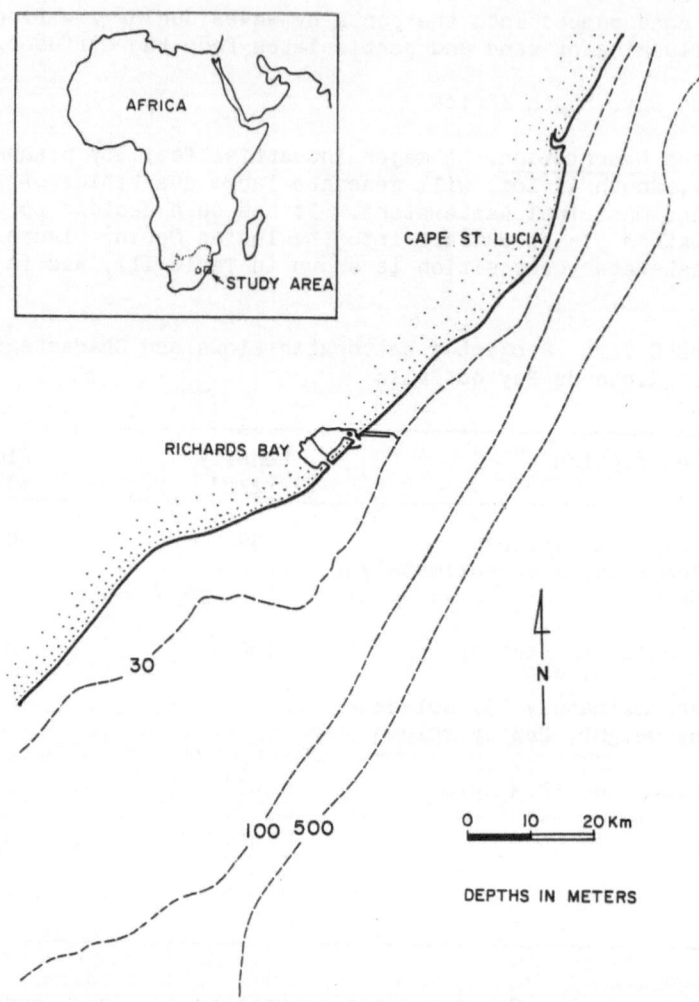

Figure 7. Location map for ocean outfall proposed for Richards Bay,
South Africa.

dilution; this standard is based on the work of (7). The background
concentration is 1.5 ppm, so an increment of 3.5 ppm is allowed. The
gypsum, $CaSO_4 2H_2O$, is composed of needle shaped particles with a median
diameter of 140 μ; a large fraction of the particles are very fine
grains with a diameter less than 25 μ. The median fall velocity of the
particles is about 8 mm/s. This gypsum is soluble in seawater, and
tests were conducted at the National Research Institute for Oceanology
(NRIO) to investigate the degree and rate of solubility. It was found

that gypsum concentrations of 1.5, 2.5, and 3.5 g/ℓ all resulted in about 90% ultimate solubility, with about 85% dissolution after 5 minutes. The solubility limit was about 5 g/ℓ. Hydraulic model tests of the behavior of gypsum deposited directly on the seafloor under the action of waves showed the gypsum to form a hard, elastic, layer. It was decided that the discharge criteria for the gypsum should be that the diffuser results in sufficient dilution to cause most of the gypsum to go into solution before reaching the ocean floor. This will result in dispersion of the gypsum over a wide area under the action of ocean currents and turbulence.

It was decided to use high density polyethyline pipe (HDPE) for the ocean outfall. The advantages of this type of pipe have been discussed elsewhere (15). They can be extruded in very long sections of up to 400 m and at Richards Bay these lengths were joined onshore and towed offshore in lengths exceeding 1 km with anchor weights attached, then sunk into place by controlled air venting. When lying on the seabed they can yield slightly to wave and current forces and conform to underscour of the seabed without the failure that can occur in rigid pipelines. Extensive erosion protection is also unnecessary. For a pumped outfall, failure of the pump or variations in pumping rates can cause transients to be transmitted down the pipeline. The large wall pressures resulting from these transients are considerably reduced in HDPE pipes by the flexibility of the pipe wall. Despite the advantages of HDPE pipe, considerable care is needed in their design and construction. Quality control is essential to ensure that the raw materials used, the extrusion process, and installation procedures are satisfactory. Scratch marks should be avoided, welding must be carefully controlled, and the pipe should not be bent through small radii of curvature during sinking. As the pipe and weights are buoyant in seawater, air must be prevented from entering the outfall once in operation. Sufficient weighting must be applied so that movement under waves is not too large since this can lead to structural damage or excessive abrasion resulting from movement over a rock bottom. Nevertheless, when used and designed with care, plastic outfalls can be an economical alternative to rigid materials.

A series of tests was conducted at the University of Cape Town to investigate the hydraulic characteristics of gypsum slurry flow in HDPE pipes. Headloss was measured as a function of gypsum concentration, pipe diameter, and flow velocity, and the critical velocity to avoid particulate deposition was investigated. Tests with clear water showed the pipe to have an equivalent sand grain roughness height, ε, of 0.03 mm. The slurry behaved as a slow settling pseudo homogeneous, Newtonian fluid, and the friction headloss, h_{LS}, of the slurry was found to be given by:

$$h_{LS} = S \ h_{LW} \tag{1}$$

where S is the specific gravity of the slurry, and h_{LW} the friction
headloss of clear water flowing at the same conditions. The velocity
at which deposition occurred, U_{dep}, was found to be related to the pipe
diameter D by the empirical equation:

$$U_{dep} = 1.29 \; D^{0.34} \hspace{6cm} (2)$$

where U_{dep} is in m/s, and D is in m. The pipe velocity criteria was
chosen to ensure that the pipe velocity exceeded U_{dep}, and a minimum
port diameter of 75 mm was chosen to avoid clogging.

3.2.3 <u>Oceanic Conditions</u>. Detailed studies of nearshore bathymetry,
sediment and beach profiles, and oceanographic conditions were
undertaken by NRIO and as these studies are presented in detail
elsewhere (18,19), they will only be summarized here. A perspective
view of the local bathymetry is shown in Figure 8. Following a fairly

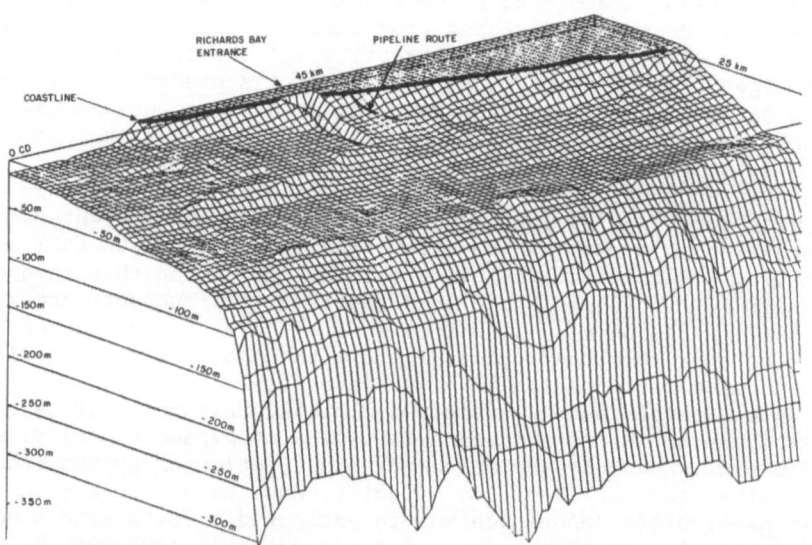

Figure 8. Bathymetry in vicinity of Richards Bay

steep beach slope, the bottom slopes gently to the edge of the
continental shelf which lies about 25 km offshore. At the edge of the
continental shelf the water depth is about 50 m, and further offshore
the bottom plunges steeply. The bottom is fairly rocky but is overlain
by sand, mud, or gravel in layers of varying thickness up to about 6 m.
Occasionally there is no overlay and the rock lies exposed.

The principal oceanic feature of the area is the Agulhas current,
Figure 9. This warm current flows swiftly southwards with typical

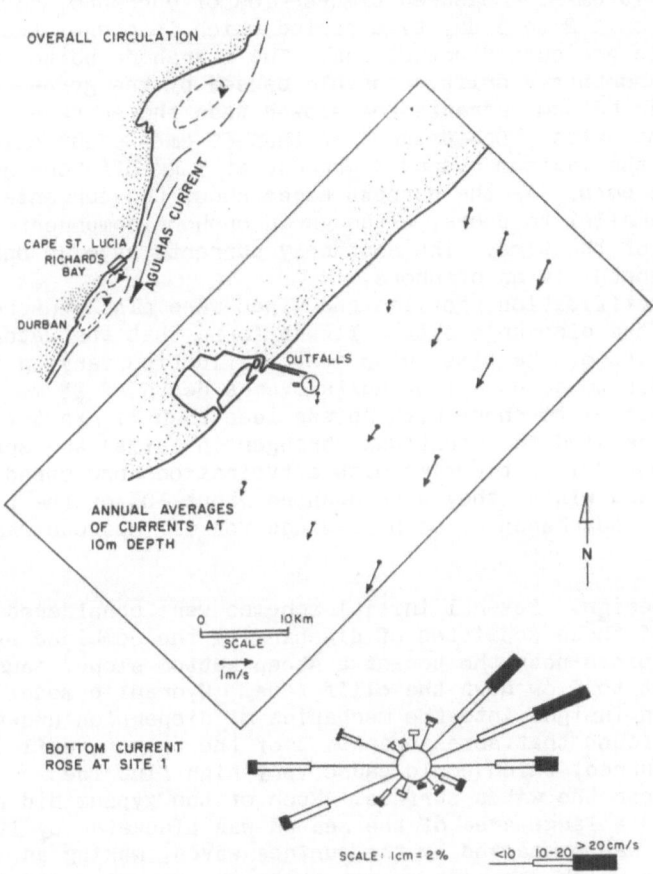

Figure 9. Current characteristics near Richards Bay

surface speeds of 1 m/s. The inshore boundary of the current is near
the edge of the continental shelf. At the discharge site, this is
about 25 km offshore, but meanders can cause it to vary between 10 and
30 km offshore. The Agulhas current separates at Cape St. Lucia,
causing the probable inshore circulation pattern shown in Figure 9.
The gyres shown are speculative (but plausible) as it is difficult to
measure features of this type on this scale. The water inshore of the
Agulhas is drawn up from about 50 to 100 m deep in the Agulhas current.
This could be water replenishing the inshore surface water entrained by
the fast moving Agulhas current. Local wind stress also exerts an

important influence in the shallow coastal waters. The resulting
surface currents at the 10 m isobath are southgoing, possibly due to
local wind stress, with the average speed decreasing as the shore is
approached. Near the diffusers there is an average southerly current
of the order of 10 cm/s. Measured time-series of currents, however,
show a reversal on a 2 to 5 day time period which is correlated with
local atmospheric pressure fluctuations. The nearshore bottom currents
have an average northerly drift, possibly caused by the gyres, of about
8 cm/s. The near bottom currents are slower than the surface currents
and show frequent calms. Speeds greater than 50 cm/s occur only about
3% of time, and the maximum observed current at 5 km offshore near the
diffusers was 65 cm/s. As the current roses show, the currents are
predominantly parallel to shore, although an onshore component exists
about 10 to 40% of the time. The southerly currents have an onshore
component; the northerly an offshore.

Density stratification profiles and winds were also measured in
the vicinity of the discharge site. It was found that the water column
was stratified most of the time, with a stratification varying from
about 0.3 σ_t units to about 1.1 σ_t units over a depth of 28 m. Winds
tended to be parallel to shore with speeds less than 11 m/s 95% of the
time. The onshore wind component was stronger in summer and spring,
and occurred about 20% of the time with a typical onshore speed of 5
m/s. In autumn and winter they were onshore about 10% of the time.
The frequency of occurrence of onshore winds for all seasons was about
13%.

3.2.4 Outfall Design. Several initial schemes were considered and
rejected. One of these consisted of discharging the combined effluent
from open-ended pipes near the top of a steep bottom slope, causing a
turbidity current to flow down the cliff face. Hydraulic model tests
were done to gain insight into the mechanics of dispersion under this
scheme. It was found that some separation of the buoyant effluent
constituents occurred, which would cause very high fluoride
concentrations near the water surface. Much of the gypsum did not go
into solution and a large area of the seabed was blanketed by it. The
blanketed gypsum was compacted by the surface waves, making an
"elastic" surface which was very resistant to resuspension and further
transport. Due to these poor dispersion conditions, this scheme was
rejected.

As previously discussed, HDPE pipe was used for the outfall. The
maximum pipe size currently available is 1200 mm OD, which means that
two pipes will be required to carry the flows. Initially, it was
proposed to combine the effluents and discharge them through two
similar outfall and multiport diffuser combinations. Although the
combined effluent density (Table III) is very close to that of seawater
(1025 kg/m³), changes in future discharges could result in the effluent
being either nonbuoyant, or slightly buoyant, or slightly negatively
buoyant. The overriding consideration for such a scheme is the
fluoride concentration after dilution. The fluoride concentrations in
the combined effluent is 693 ppm (Table III), so an initial dilution of
198 is required to reduce the fluoride concentration to the allowable

increment of 3.5 ppm. Assuming the effluent to behave like a neutrally buoyant jet, we can approximate the average dilution by (Fischer, et al. (9) Eq. 9.18):

$$S_{av} = 0.28 \frac{x}{d} \tag{3}$$

Thus, a dilution of 198 times occurs about 700 diameters from the nozzle, which, for a port diameter of 7.5 cm, is 53 m from the nozzle. To achieve this at the surface in the 25 m depth anticipated would require a fairly shallow trajectory of about 30° upwards. Such a jet, however, would probably be trapped by the ambient density stratification, and be subject to considerably reduced dilutions.

Preliminary considerations showed the efficacy of a separated effluent scheme, whereby the gypsum would be discharged through one pipe, and all the other buoyant effluents, including the fluoride, discharged through another. The diffuser for the gypsum line could be designed to achieve sufficient dilution to cause the gypsum to go into solution, and the diffuser for the buoyant line could be designed to achieve sufficient dilution to meet the fluoride requirement. Discussion with the onshore facilities design engineers showed it was possible to separate the effluents in this way. Because of the better effluent density control and predictability resulting from the separated effluent scheme, and its better ability to handle future unexpected flow variations, it was chosen over the combined effluent scheme. Some of the effluent from the Mondi plant was needed to transport the gypsum, and some seawater is necessary, as discussed below, to maintain sufficient dilution under reasonable power requirements. The flows are pumped to the shore through ultra high molecular weight high density polyethylene pipes of 450 and 1200 mm OD to another pump and surge tank and then to the ocean outfalls. The outfall alignment was chosen as shown in Figure 10 to avoid rocky areas and to lie on sand. The hydraulic design of the two outfalls is discussed below.

Discharge of a dense effluent, such as gypsum slurry, by an inclined jet results in a trajectory like that shown in Figure 11. The dilution criteria adopted specified that the concentration of gypsum on the jet centerline be below the solubility limit at the jet apex, causing most of the gypsum to go into solution there. This requires a concentration less than about 3.5 g/ℓ based on the previously discussed solubility tests. In addition, the rise height should be such that the jet does not reach the water surface. Zeitoun, et al. (27) investigated experimentally the discharge of dense effluents from nozzles of various angles and found a 60° jet inclination to provide the longest path length and therefore highest dilution. As a vertical jet falls back on itself under stagnant ambient conditions, causing severely reduced dilution, the 60° inclination was chosen for the nozzles. As there was little data available on the effects of currents on inclined jets, an experimental program was conducted as part of the design study to investigate this phenomena (24). It was found that the severe dilution reduction which can occur for vertical jets did not

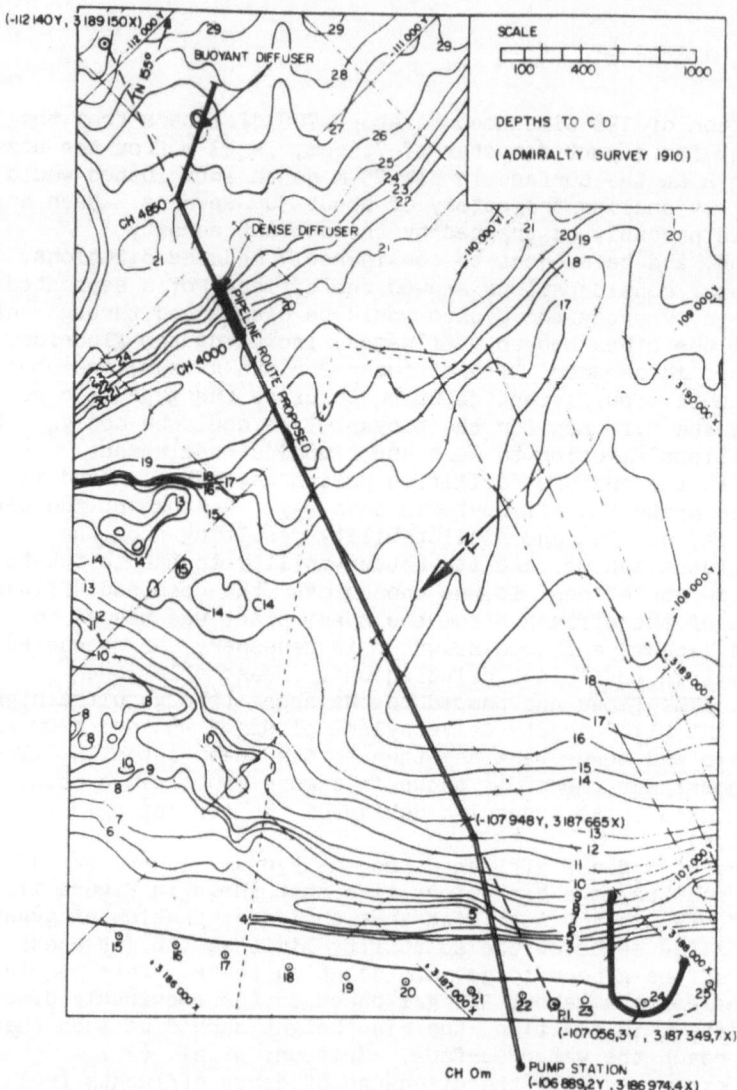

Figure 10.   Pipeline routes for separated effluent scheme (from 18)).

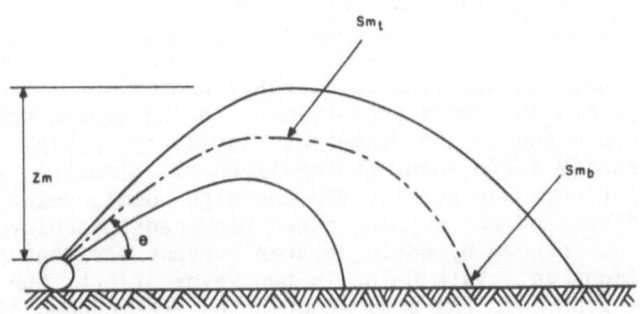

Figure 11. Characteristics of a dense jet (from 18)).

occur for the 60° jets, even for a jet discharging into a current whose speed is such that the jet fell back on itself. It was recommended that the inclined jets be pointed in a generally offshore direction to impart a horizontal momentum component to aid in clearing the effluent from the discharge site. This capability is a further advantage of inclined jets over vertical.

The characteristics of an inclined dense jet are shown in Figure 11. Denoting Z_m as the terminal rise height, and S_{mt} and S_{mb} the dilutions at the terminal rise height and impact point respectively, Roberts and Toms (24) show that for source Froude number, F > 20:

$$\frac{S_{mt}}{F} = 0.38 \tag{4}$$

$$\frac{S_{mb}}{F} = 1.03 \tag{5}$$

$$\text{and } \frac{Z_m}{dF} = 2.08 \tag{6}$$

where $F = u_j / \sqrt{g \frac{\Delta\rho}{\rho} d}$, and u_j is the jet exit velocity, g the acceleration due to gravity, $\Delta\rho$ the density difference between effluent and

ambient water, and d the nozzle diameter. These equations were derived
originally by Zeitoun, et al. (27), but the coefficients given in Eqs.
4 to 6 are those derived from Roberts and Toms (24). It is instructive
to write Eq. 4 in terms of the quantities which can be varied by the
designer:

$$S_{mt} \alpha \, u_j \, \Delta\rho^{-1/2} \, d^{-1/2} \tag{7}$$

from which we see that the dilution is directly proportional to the
discharge velocity, and inversely proportional to the square roots of
the density difference and nozzle diameter. Obviously, a high
velocity, small density difference, and small nozzle diameter, will
produce a high dilution. The penalty for the high nozzle velocity is
the large jet headloss and the pumping power required to achieve it.
This headloss can be reduced by adding either buoyant effluent or
seawater to the discharge. This helps in two ways: first, the
predilution of the gypsum in the pipe reduces the jet-induced dilution
required for dissolution; and second, the density difference, $\Delta\rho$, is
reduced thereby reducing the jet velocity required to reach this
dilution. There is a limit to the amount of make-up water, however, as
if this becomes too large the pipe velocity and friction headloss in
the pipe become very large.

To investigate the trade-off between pipe friction loss and jet
headloss, hydraulic calculations were performed. The total headloss,
h_L, in the pipe can be approximated by:

$$h_L \approx f \, \frac{L}{D} \, \frac{u_p^2}{2g} \; + \; \frac{1}{c_D^2} \, \frac{u_j^2}{2g} \tag{8}$$

$$\underbrace{\text{pipe friction}}_{\text{loss}} \quad \underbrace{\text{jet headloss}}$$

where f is the Darcy friction factor, L the total outfall length
including the diffuser, D the outfall diameter, U_p the pipe velocity,
and C_D the coefficient of discharge of the ports. f was calculated
from the Colebrook formula:

$$\frac{1}{f^{1/2}} = -2.0 \log \left(\frac{\varepsilon/D}{3.7} + \frac{2.51}{\text{Re} \, f^{1/2}} \right) \tag{9}$$

where ε is the pipe roughness height, assumed to be 0.03 mm, and Re =
$U_p D/\nu$ is the pipe Reynolds number, where ν is the fluid kinematic
viscosity. The headloss was corrected for the slurry characteristics
by Eq. 1. A discharge coefficient, C_D, of 0.9 was assumed for the
nozzles. The power P required to pump the effluent is given by

$$P = \rho_e \, g \, Q \, h_L \tag{10}$$

where ρ_e is the effluent density, and Q the total flowrate.

The question now is: How many ports and what flowrate are required to produce sufficient dilution at the terminal rise height while minimizing the pumping power? An outfall pipe length of 4.5 km, corresponding to discharge in about 22m water depth was chosen; the pipe OD was 900 mm, and ID was 798 mm. From Eq. 7, it can be seen that it is desirable to use the smallest nozzles possible which prevent clogging, this was taken as 75 mm diameter. An arbitrary quantity of seawater of density 1.024 g/cm³ was then added to the gypsum flow, and the minimum number of ports required to achieve concentrations of 1.5, 2.0, or 2.5 g/ℓ at the jet apex, according to Eq. 4, were calculated. The corresponding headloss and power required were then computed from Eqs. 8, 9, 10, and 1. This procedure was repeated for different seawater flow rates to obtain the variation of pumping power and outfall headloss with seawater flowrate; the results are shown in Figure 12. As can be seen, a flowrate of about 88,000 m³/day, with 15

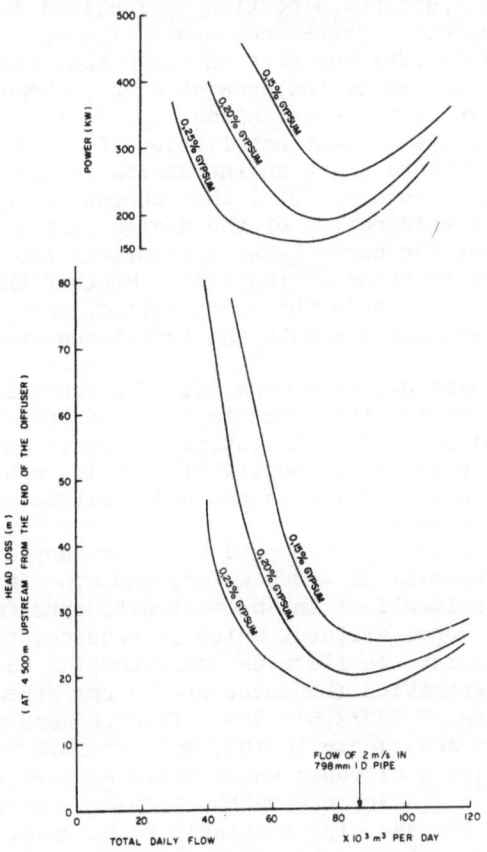

Figure 12. Effect of seawater added to the gypsum flow on the pumping power and outfall headloss required to achieve the specified gypsum concentrations at the jet apex (from (18)).

ports minimizes the pumping power required and was chosen for final
design. For lower total flowrates, the number of ports must be reduced
to increase the jet induced dilution, causing the jet velocity head to
dominate; for higher flowrates, the number of ports is increased and
the jet velocity reduced, but then the friction headloss becomes very
large. Eq. 2 suggests a pipe velocity of 1.2 m/s to prevent gypsum
deposition in the pipeline. As this equation is based on tests in much
smaller pipes, it was decided to use a more conservative value of 2 m/s
for design. This is close to that resulting from the optimum flow in
the 798 mm ID pipeline. The ports were spaced to avoid interference
between adjacent jets, and originally one port per riser was proposed,
but this was later changed to two ports per riser to avoid transverse
forces on the pipeline. The terminal and impact dilutions calculated
from Eqs. 4 and 5, are 40 and 80.

The dilutions for which the dense effluent diffuser was designed
are for stagnant water. Because there was little data available on the
effects of varying current speed and direction on inclined dense jets,
an extensive series of laboratory tests were done. The results are
presented in Roberts and Toms (24) where it is shown that the current
speed effect can be characterized by the parameter u_rF, where $u_r = u_a/u_j$ is the ratio of the ambient current speed, u_a, to the jet
velocity u_j. For $u_rF \ll 1$, the current has little effect, but for
larger values of u_rF the currents exert an increasing influence.
According to the experimental results (24), the current begins to exert
an effect for $u_rF > 0.2$. Consideration of the design variables and
current statistics show that the currents will result in increased
dilutions for a substantial fraction of the time. Most of the gypsum
will go into solution before reaching the ocean bottom, and will be
dispersed and diffused by ambient currents and turbulence over a wide
area.

Final hydraulics calculations were done with a program similar to
that described in Fischer, et al. (9). Due to the high pressures and
low velocities in the pipeline, the contribution of the velocity head
to the total energy in the pipeline is negligible. This results in a
uniform distribution of C_D and velocity in the ports without having to
resort to varying the port size.

The flow rate remaining to be discharged from the buoyant outfall
is 109,950 m³/day with a density of 1,000 kg/m³, and containing 92
ton/day of fluoride. (This is all of the buoyant effluents from Table
III, less 20,000 m³/day of Mondi effluent which is required to
transport the gypsum.) The in-pipe fluoride concentration is thus 837
ppm, and to achieve a concentration increment of 3.5 ppm after initial
dilution requires a dilution of 837/3.5 = 239. This is somewhat higher
than typical designs, which are closer to 100, but feasible.

It was decided to design a diffuser which would achieve the
maximum possible dilution in a discharge depth of 25 m. Hydraulic
considerations require the largest pipe available to be used, which is
923 mm ID. The dilution is maximized by using as many ports as
possible, within hydraulic constraints, and spacing them far enough
apart that the individual jets do not merge. The maximum number of
ports is dictated by the requirement that the ports flow full, which

requires that the total jet area be less than the pipe cross-sectional
area. Experience suggested by Fischer, et al. (9) indicates the best
area ratio to lie between 1/3 and 2/3, depending on the bottom slope.
We chose an area factor of 0.7, and again taking the nominal nozzle
size to be the minimum permitted of 75 mm, yields:

$$\frac{\text{Total nozzle area}}{\text{Pipe area}} = \frac{n \frac{\pi}{4} d^2}{\frac{\pi}{4} D^2} = 0.7$$

or n = 106 ports. To prevent merging, the experimental results of
Liseth(15) suggest keeping $y/\ell > 5$, where y is the water depth, and ℓ
the average port spacing = diffuser length ÷ total number of ports.
Thus, for a 25 m depth, ℓ = 5 m, and the total diffuser length = 106 x
5 = 530 m.

The dilution for such a diffuser can now be estimated. Assuming
uniform flow distribution, the flow per port is 109,950 m³/day ÷ 106 =
1037 m³/day = 0.0120 m³/s. The jet velocity = $4Q/\pi d^2$ = 2.72 m/s, and
the densimetric Froude number, F = 20.1. In this case, y/dF is 16.6 so
the effect of source momentum flux on surface dilution can be neglected
(see Roberts (20), Brooks (4)). The minimum dilution can then be
calculated from the plume formula for a non-merging jet (20):

$$S_m = 0.091 \ g_o'^{1/3} \ y^{5/3} \ (Q_t/n)^{-2/3} \tag{11}$$

where $g_o' = g \ \Delta\rho/\rho$, and Q_t is the total flowrate. The average
dilution, S_a, is about 1.77 times S_m, yielding $S_m \sim 235$, and $S_a \sim 417$
for present conditions. The surface concentration of fluoride is thus
837/417 ÷ 1.5 = 3.5 ppm, which meets the water quality requirement of
5.0 ppm. More detailed calculations incorporating the effects of
stratification and source momentum flux (18) show the plume to be
submerged over 50 percent of the time and to meet the fluoride
requirement most of the time.

3.3 Alki Point, Seattle

3.3.1 Project Description. The Municipality of Metropolitan Seattle
(Metro), Washington operates a number of treatment plants and outfalls
which discharge into Puget Sound (3), as shown in Figures 13 and 14.
The Alki Point system is a partially separated sewer which receives
substantial volumes of primarily roof drainage during wet weather. Due
to the limited hydraulic capacity of the Alki Point treatment plant and
outfall, overflows of untreated combined sewage occur around the
shoreline   during wet weather, resulting in poorly diluted effluent
reaching adjacent beaches. As these waters support a wide range of
beneficial uses including fishing, shell fishing, swimming, scuba
diving, beachcombing, and boating, it has been recommended that the
existing Alki Point outfall be improved. The studies and resulting

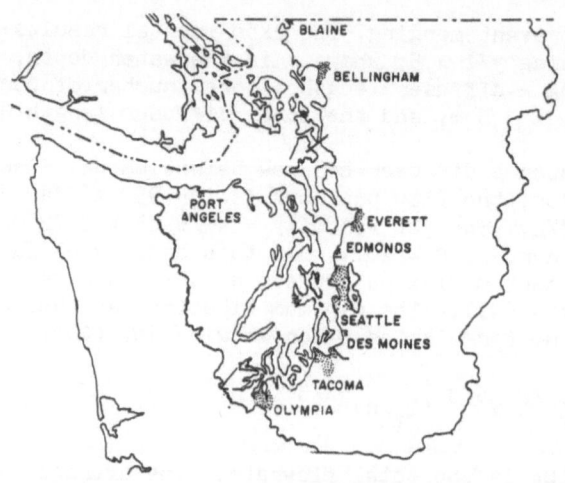

Figure 13. Location map for City of Seattle and Puget Sound (from Birke, et al. (3)).

recommendations for outfall modification are discussed here.
     Puget Sound is the third largest sound in the United States. Its water surface area is about 6300 km² (2433 mi²), composed of a large number of bays and inlets and more than 300 islands; the average width is about 50 km (30 mi). It is a fjord-like basin with an average depth greater than 100 m (330 ft), and a total volume of about 169 km³ (40.5 mi³). From both a depth and volume standpoint, Puget Sound is nearly three times greater than any estuarine system in the Continental United States. Off Alki Point, the bottom slopes steeply, reaching depths of about 61 m (200 ft) at only 610 m (2000 ft) offshore. The existing outfall from the Alki treatment plant is 1.07 m (42 inches) in diameter, extends 366 m (1200 ft) offshore, and terminates in an open-ended pipe discharging horizontally at a depth of 24 m (79 ft). The average daily dry-weather flow is about 0.22 m³/s (5 mgd), but this increases considerably during wet weather, up to about 1.31 m³/s (30

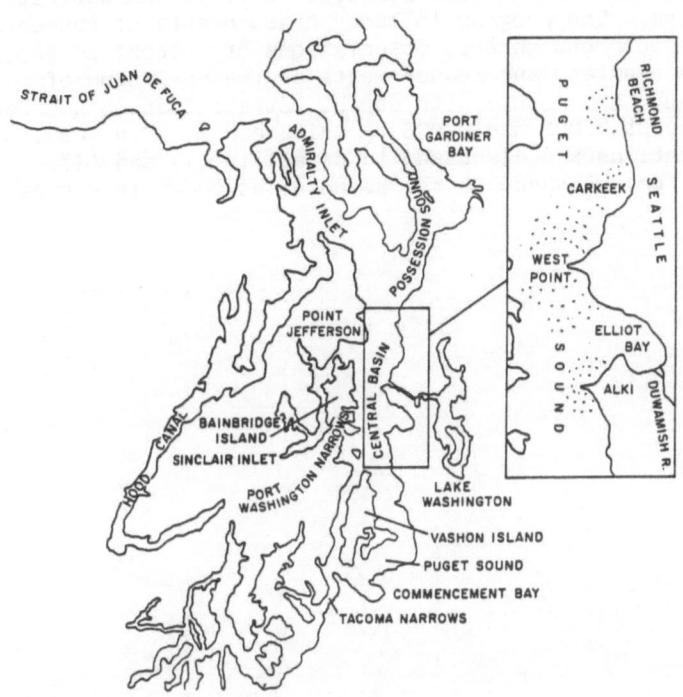

Figure 14.   Seattle combined overflow sewers (from Birke, et al. (3)).

mgd). The capacities of the pump stations owned by Metro have been
increased to pass all flows resulting from the once-per-year storm
event. As a result, future storm flows into the treatment plant will
be considerably higher than storm flows recorded prior to the upgrade.

3.3.2 Design Criteria. Flow capacities were chosen based on measured
flows, future population projections, and historical rainfall records.
The average design flows were chosen to be 0.23 m³/s (5.3 mgd) for the
summer and 0.37 m³/s (8.5 mgd) for the winter, with a summer peak flow
of 1.31 m³/s (30 mgd). A peak wet weather capacity of 3.94 m³/s (90
mgd) was chosen to pass the flow resulting from the once-per-year
rainfall. As the combination of the once-per-year storm and annual
peak tide is highly unlikely, the head requirement for discharge of the
peak wet weather flow against the historical highest tide observed was
relaxed to a combination resulting in less than one overflow per year.

3.3.3 Oceanic Conditions. An extensive measurement program was
undertaken by Northern Technical Services, Inc. to characterize the
receiving waters. The program included measurements of currents by
moored Aanderaa current meters, observations of subsurface drogue
movements, dye studies, and measurements of the variation of
conductivity and temperature with depth. Stratification was found to
vary between about 0.002 and 0.025 σ_t units per m. The results of
these investigations are discussed in detail in (1) and (2).
     Currents were measured at the seven sites shown in Figure 15 using

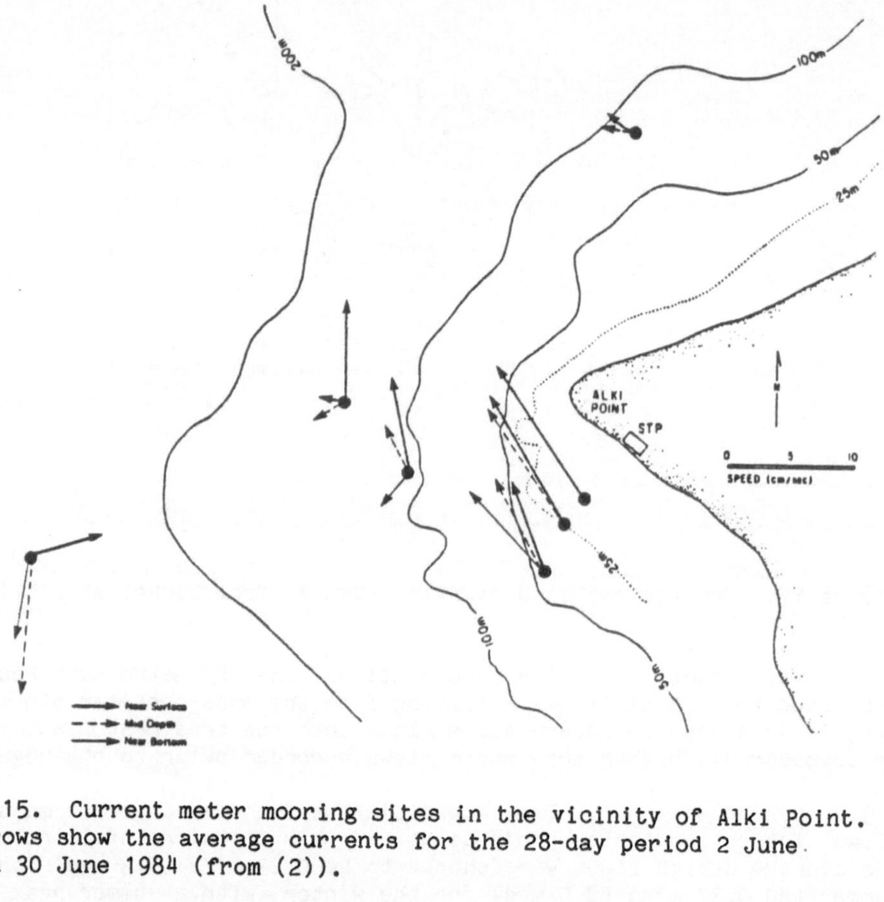

Figure 15. Current meter mooring sites in the vicinity of Alki Point.
The arrows show the average currents for the 28-day period 2 June.
through 30 June 1984 (from (2)).

1, 2, or 3 meters at each site for a total of 18 meters. The resulting
speeds and direction measurements were averaged over 15 minutes
intervals. Two deployments were used: a spring deployment from 16 May
to 3 July; and a fall deployment from 13 September to 17 October.
Average current speeds and directions for the spring deployment are
shown in Figure 15; the principal axis directions and principal
components of the near surface meters are shown in Figures 16, 17, and

18.   The currents in the vicinity of Alki Point are dominated by tide

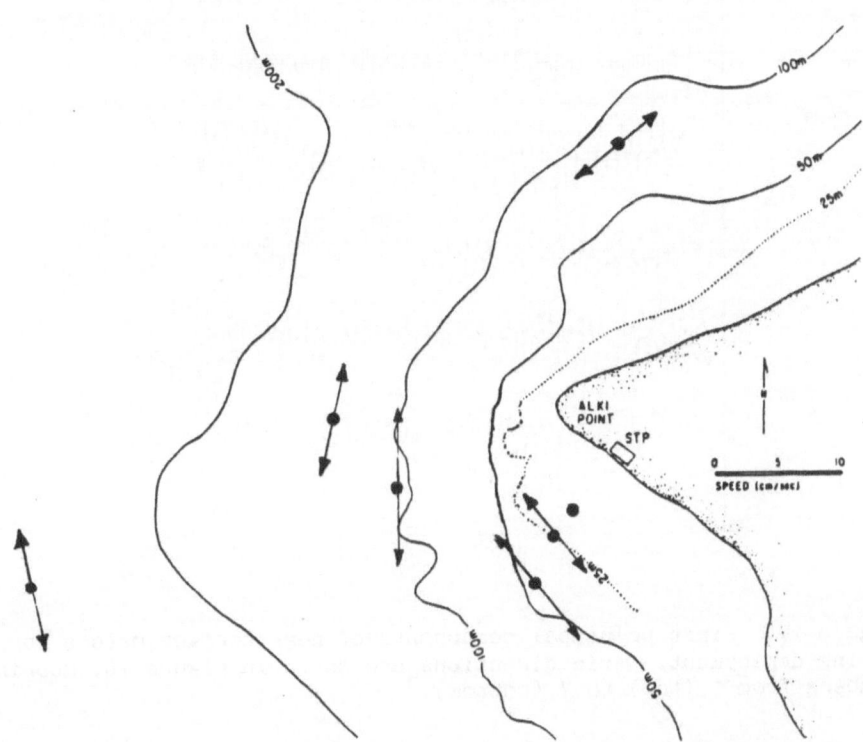

Figure 16. Directions of principal axes near surface meters during
spring deployment.

and topography, and the principal axes are generally aligned with the
local depth contours. The currents are strong, with peak velocities of
about 50 cm/s, and it is apparent that they have a considerable low
frequency content. The average currents, Figure 15, show a strong
surface drift generally in a seaward direction at speeds of the order
of 10 cm/s. In the deep waters the average current directions are
consistent with the classical two-layer gravitational estuarine
circulation. Near to shore, however, the mean currents are apparently
driven by residual circulations caused by the interaction of the tide
and local bathymetry. The low frequency currents are variable, with
speeds up to about 20 cm/s (2).

3.3.4 Outfall Design. A very large number of possible outfall
configurations, involving different discharge sites and depths,

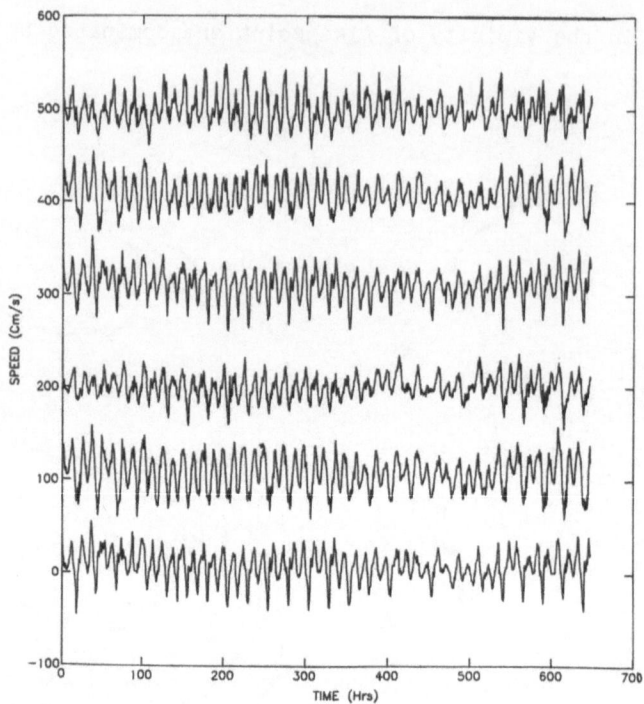

Figure 17. First principal components of near surface meters for spring deployment. Axis directions are shown in Figure 16, mooring numbers from 1 (top) to 7 (bottom).

diffuser designs, and flow configurations were considered. The design flowrate is from 0.11 to 3.94 m³/s (2.5 to 90 mgd) which is too large a range for efficient hydraulic operation of one outfall. Two outfalls are therefore required, and as a result of analyses based on hydraulic considerations and preliminary screening, three alternative scenarios were chosen for detailed evaluation. All three scenarios involve modification of the existing outfall plus either construction of a new outfall at Alki Point or transport of wastewaters for discharge at Duwamish Head. The potential Alki Point discharge sites are shown in Figure 19 and are classified as moderate or deep-water discharges. The moderate depth outfalls discharge into about 50 m (164 ft); these are alternatives A-1, A-2, and A-3 on Figure 19. The deep-water outfalls discharge into 100 m (325 ft); these are alternatives B, F, and G. Of these alternatives, A-2 is the best moderate-depth, and B the best deep discharge, due to the more gentle bottom slopes at their diffuser locations. The final new outfall locationalternatives are therefore A-2, B, and Duwamish Head.

With any of the alternatives, the existing outfall will be used only about 10 times per year as an overflow. The use of a diffuser is

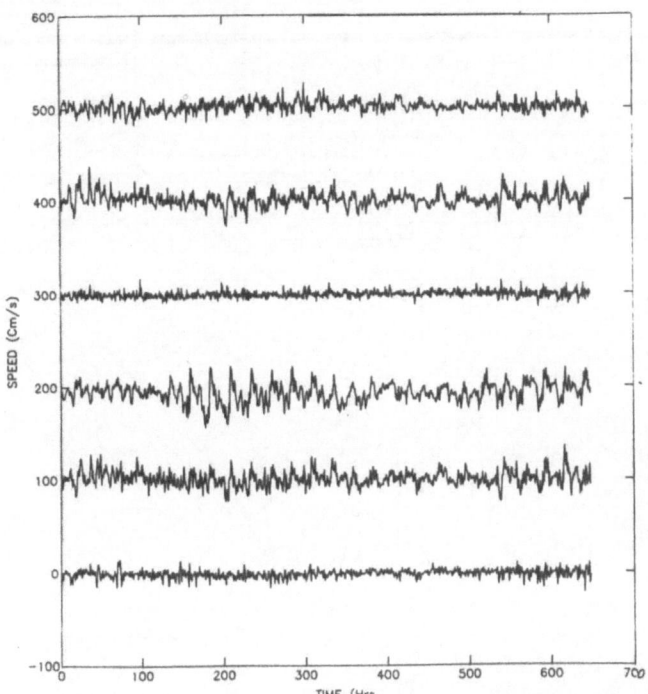

Figure 18.   Second principal components of near surface meters for
spring deployment.   Axis directions are shown in Figure 16, mooring
numbers from 1 (top) to 7 (bottom).

therefore inappropriate, as it would be subject to sediment intrusion
and biofouling which would be difficult to flush when called for.   As
illustrated by Equation 11, dilution in unstratified water is very
sensitive to water depth, y, and to the number of discharge ports, n.
Increasing the number of ports to 2 by adding a Y end, and lengthening
the outfall to discharge into 50 m (164 ft) rather than the present 24
m (79 ft) should increase the dilution of a surfacing plume, according
to Equation 11, by a factor of 5.4.   More detailed calculations (1)
show a somewhat reduced effect, probably caused by jet merging and
source momentum flux effects.   The effect on theminimum dilution is
still large, however, showing an increase from 19 to 64.   This
modification causes very little reduction in hydraulic capacity
compared to the addition of a diffuser, and was therefore considered to
be the best upgrade to the existing outfall.   This upgrade would be
used intermittently for combined, partially treated sewage.
        The new outfall alternatives were subjected to detailed
comparisons of initial dilution, subsequent transport, and hydraulics
(1,2).   Initial dilutions were calculated using the U.S. EPA program
MERGE to evaluate the effects of currents and plume merging.   The

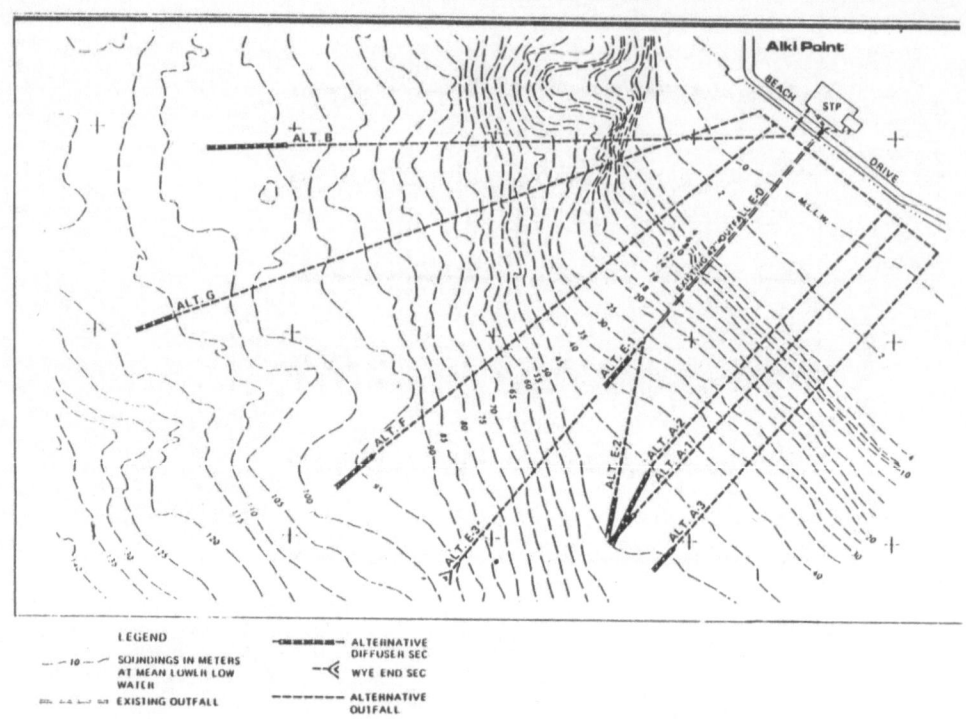

Figure 19. Alternative Alki Point outfalls (from (1)).

results suggest that the Duwamish Head site and the Alki Point deep
discharge (site B) give minimum dilutions of 178 and 120, respectively.
Both outfalls would give submerged fields all the time, with the
Duwamish field being more deeply submerged. The new moderate depth
outfall (A-2), would give minimum dilutions comparable to B, but the
wastefield would surface occasionally.

     The advective transport of the wastefield due to currents were
also evaluated for all sites (2). As the currents are continuously
varying, the location of the wastefield at any time should be
considered to be a stochastic variable. To quantify this effect,
streaklines were computed from the current meter data to estimate the
shape of the wastefield centerline for particles released from the
discharge site within a specified travel time after release. The
current meter records were interpolated spatially by an inverse
distance squared weighting scheme to predict currents between mooring
sites. The computed centerline was then overlaid by a grid of 250 m by
250 m squares and each grid square containing the centerline was noted.
The procedure was then repeated for each specified travel time within
the total current meter record. The number of times the centerline lay

within each grid square was summed and divided by the total number of releases. These numbers represent the fraction of time that each grid square was occupied by the wastefield within the specified travel time after release. This is the same as the probability of the wastefield lying in a grid square within that time and is closely related to the "visitation frequency" used by Csanady (8). Finally, a contouring program was used to obtain smooth probability contours. This type of calculation is often more useful than the rather uncertain subsequent diffusion calculations. That is, it is more important to know where the wastefield goes, rather than whether the subsequent dilution is 3 or 5.

Simulations were done using current meter data taken at depths closest to the predicted trapping depths. For example, Figure 20 shows the transport results for 3, 6, 12 and 24 hours after release from the existing outfall, using the currents shown in Figures 17 and 18. The effect of the tidal currents is to spread the wastefield a distance roughly equal to the tidal excursion. For longer times of release, say 24 hours, the contours are elongated to the Northwest as a result of the low frequency currents in that direction (Figure 15). The predicted onshore probabilities are conservatively high, as the currents should become more parallel to the shoreline as it is approached. Nevertheless, the results suggest that with the existing outfall a substantial length of shoreline south of Alki Point will be impacted by the wastefield for a substantial fraction of time. It is not realistic to extrapolate the results for very long travel times, as the computed particle trajectories then pass outside the area covered by the current meters. Also, little onshore transport will occur beyond the depth contour corresponding to the wastefield trapping depth due to the limiting of vertical mixing by the ambient stratification. For these reasons, the travel times were reduced to six hours for comparison, and shoreward transport beyond the trapping depth contour was curtailed. The results for the four alternative discharge sites are shown in Figure 21. The results indicate that a surfacing plume at site A2 would have a nonzero onshore transport probability, although the usual submergence at this site will reduce this probability substantially. Only sites B and Duwamish Head show zero onshore transport for the surfacing field. These sites can probably meet shoreline bacterial standards without chlorination.

The costs of the alternative outfalls are: Extend the existing outfall to 50 m depth and add a Y end, $2.7 million; construct the moderate depth outfall (A2), $7.4 million; construct the deep water outfall (B), $12.7 million. The Duwamish Head outfall is projected to discharge into 183 m (600 ft) depths, considerably deeper than any existing sewage outfall; its cost for an additional outfall for the Alki Point wastewater, including the new force mains and pumping systems, is $24.4 million.

Metro is currently considering upgrading its whole sewage treatment system to secondary level. Because of this, the final outfall configurations are uncertain. Nevertheless, since it is apparent that considerable improvements in water quality can be made in the interim by the recommended modifications to the existing outfall,

Metro has decided to extend the existing outfall to a depth of 50 m.

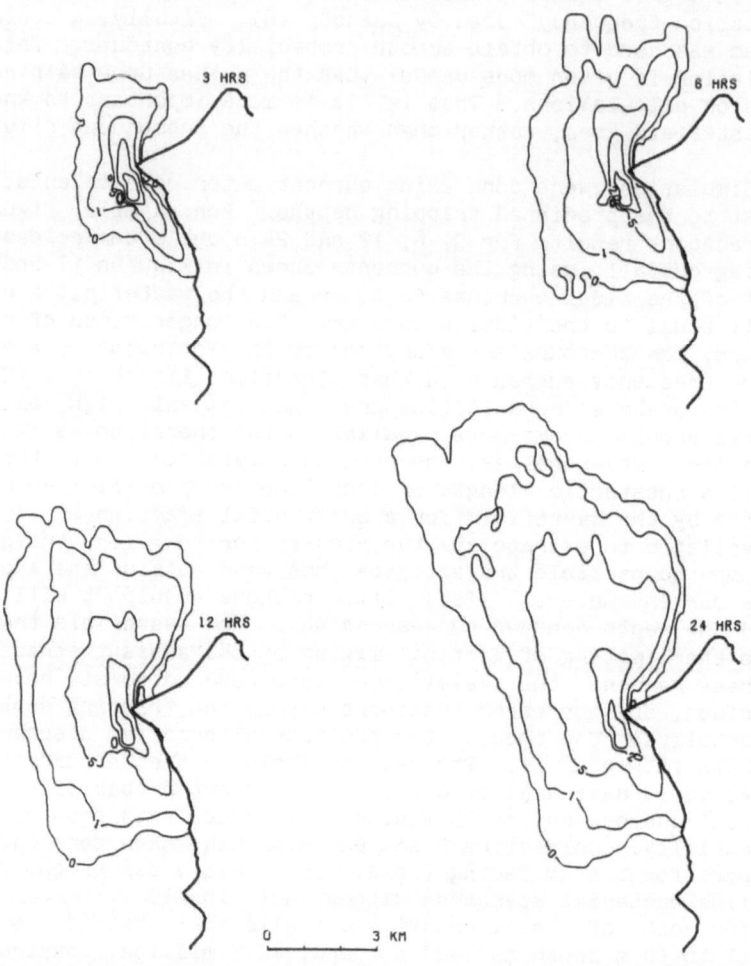

Figure 20. Contours of percent probability that wastefield centerline lies in a gird square 250 m by 250 m within specified times after release from outfall alternative A2.

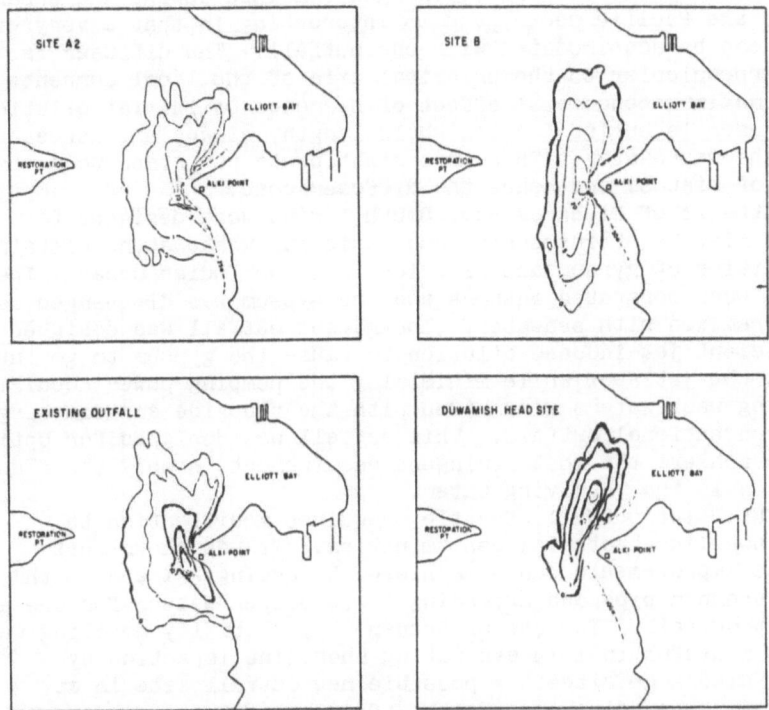

Figure 21. Transport probability for six hour travel times for four alternative discharge sites (from (2)).

4. SUMMARY AND CONCLUSIONS

In this paper, we have discussed some of the criteria and philosophy for the design of ocean outfalls and have given three case studies. We have tried to emphasize that the outfall and treatment plant design should constitute an interrelated system whose characteristics depend on what receiving water standards are to be maintained. Examples of receiving water standards were discussed and it was suggested that often only preliminary treatment may be needed to remove particulates and floatables. Specific effluent requirements, for example, treatment to secondary level, may require unnecessary expense due to overtreatment. Secondary treatment, which primarily reduces BOD, is probably unnecessary for many open coastal sites where BOD can be readily assimilated with little reduction in dissolved oxygen.

The San Francisco outfall discharges combined sewage and urban runoff into the Pacific Ocean, and is interesting in that a very wide flow range can be accommodated with one outfall. The diffuser is oriented perpendicular to the principal axis of the local currents to obtain the maximum beneficial effect of currents on initial dilution. As the diffuser is buried for its whole length, risers are necessary to discharge the wastewater. There are eight ports per riser to minimize the number of risers, and hence the diffuser cost.

The outfalls of Richards Bay, South Africa were designed to discharge a mixture of industrial and municipal wastewaters containing large quantities of gypsum and fluoride into the Indian Ocean. The wastewaters were separated onshore and the gypsum was discharged as a dense jet premixed with seawater. The gypsum outfall was designed to cause sufficient jet-induced dilution to cause the gypsum to go into solution at the jet apex while minimizing the pumping power requiring. The remaining wastewaters were mixed with the fluoride and discharged through a conventional outfall. This outfall was designed for optimum dilution to achieve the most stringent requirement, namely the fluoride concentration in the receiving water.

The Alki Point outfall, Seattle, requires modification to alleviate shoreline bacterial contamination. It was shown that considerable improvements can be achieved by adding a Y end to the present open-ended pipe and extending it to deeper waters for use as an intermittent outfall. The use of transport probability modeling was shown to be a useful tool in evaluating shoreline impaction by alternative discharge sites. A possible new outfall site is at Duwamish Head, where discharge depths of 183 m (600 ft) are currently being contemplated, far deeper than any presently in use.

ACKNOWLEDGEMENTS

Many people contributed to this paper. In particular, I would like to acknowledge Dr. Noel Williams, of CH2M Hill, San Francisco, and Drs. N.H. Brooks and R.C.Y. Koh of the California Institute of Technology, for supplying information on the San Francisco outfall study. I thank the American Society of Civil Engineers for permission to reproduce Figures 1 through 5. The Mhlatuzi Water Board at Richards Bay, South Africa, the operators of the Richards Bay pipeline, are thanked for their permission to publish the information and Figures describing the pipeline. The engineering consultants to the Water Board were Campbell, Bernstein and Irving of Durban while the National Research Institute for Oceanology (NRIO) of the South Africa Council for Scientific & Industrial Research (CSIR) were responsible for most of the oceanographic studies, dilution calculations, and pipeline planning and design aspects. The author acted as a specialist consultant to NRIO on the dilution and hydraulic aspects of the design. The Alki Point studies were conducted under the overall direction of Mr. George Capestany of Parametrix, Seattle. Dr. John Downing of Northern Technical Services, Inc., supervised the oceanographic studies; Mr. Bill Fox of Parametrix performed the initial dilution calculations; and Mr. Gary Graham of CH2M Hill performed the hydraulic analyses. Thanks are particularly extended to John Downing of Nortec and Douglas Houck of the Municipality of Metropolitan Seattle for their timely supply of the reports describing this work and for permission to reproduce Figures from these reports. Elizabeth T. Harding of the MIT Sea Grant College Program is thanked for her permission to reproduce Figures 13 and 14. Finally, Noel Williams of CH2M Hill, Geoff Toms of NRIO, John Downing of Nortec, and Douglas Houck of Metro are thanked for their helpful comments on an earlier draft of this paper.

REFERENCES

1. "Alki Wastewater Treatment Plant Outfall Improvements Predesign Study. Technical Report 8.1. Initial Dilution," Municipality of Metropolitan Seattle, Water Quality Division, April 1985.

2. "Alki Wastewater Treatment Plant Outfall Improvements Predesign Study. Technical Report 8.2. Circulation and Effluent Transport," Municipality of Metropolitan Seattle, Water Quality Division, April 1985.

3. Birke, L.E., Jr., et al. (1983), "Puget Sound Case Study," Ocean Disposal of Municipal Wastewater: Impacts on the Coastal Environment, MIT Sea Grant College Program, 83-33.

4. Brooks, N.H. (1980), "Synthesis of Stratified Flow Phenomena for Design of Ocean Outfalls," 2nd International Symposium on Stratified Flows, Trondheim, Norway, 24-27 June, pp. 809-831.

5. Brooks, N.H. (1983), "Evaluation of Key Issues and Alternative Strategies," Ch. 12 of Ocean Disposal of Municipal Wastewaters: Impacts on the Coastal Environment, Sea Grant College Program, MIT, Cambridge, Mass., pp. 707-759.

6. Chao, J. (1984), "Discussion of "Plume Dilution for Diffusers with Multiport Risers," J. Hydraulic Engineering, ASCE, 110, No. 1, pp. 94-95.

7. Connell, A.D., and Airey, D.D., (1982), "The Chronic Effects of Fluoride on the Estuarine Amphipods Grandidierella lutosa and G. Lignorum," Water Research, 16, No. 8, pp. 1313-1317.

8. Csanady, G.T. (1983), "Dispersal by Randomly Varying Currents," J. Fluid Mechanics, 132, pp. 375-394.

9. Fisher, H.B., et al. (1979), Mixing in Inland and Coastal Waters, Academic Press.

10. Isaacson, M.S., et al. (1978), "Sectional Hydraulic Modeling of
     Study of Plume Behavior: San Francisco Southwest Ocean Outfall
     Project. Progress Report," W.M. Keck Laboratory of Hydraulics and
     Water Resources, California Institute of Technology, Tech. Memo.
     78-2, January 1978.

11. Isaacson, M.S., et al. (1979), "Sectional Hydraulic Modeling of
     Study of Plume Behavior: San Francisco Southwest Ocean Outfall
     Project. Final Report," W.M. Keck Laboratory of Hydraulics and
     Water Resources, California Institute of Technology, Tech. Memo.
     79-4, October 1979.

12. Isaacson, M.S., et al. (1983), "Plume Dilution for Diffusers with
     Multiport Risers," J. Hydraulic Engineering, ASCE, 109, No. 2, pp.
     199-220.

13. Koh, R.C.Y. (1977), "Analysis of Multiple Time Series by Principal
     Components with Application to Ocean Currents off San Francisco,"
     W.M. Keck Laboratory of Hydraulics and Water Resources, California
     Institute of Technology, Tech. Memo. 77-4.

14. Koh, R.C.Y., and Brooks, N.H. (1975), "Fluid Mechanics of
     Wastewater Disposal in the Ocean," Ann. Rev. of Fluid Mech., 1,
     pp. 187-211.

15. Larsen, I. (1979), "Emmissarios Submarinos Flexíveis," 10th
     Congress o Brasileiro de Engenhario Sanitaria, Maraus, Brasil,
     21-26 January, 1979.

16. Liseth, P. (1976), "Wastewater Disposal by Submerged Manifolds,"
     J. Hydraulics Division, ASCE, 102, No. HY1, pp. 1-13.

17. Rawn, A.M., et al. (1960), "Diffusers for Disposal of Sewage in
     Seawater," J. San. Engg. Div., ASCE, 86, SA2, pp. 65-105.

18. "Richards Bay Ocean Outfalls. Engineering Design Aspects. Vol.
     III," Coastal Engineering and Hydraulics Divisions, National
     Research Institute for Oceanology, Stellenbosch, South Africa,
     CSIR Report C/SEA 8231/3, October 1982.

19. "Richards Bay Ocean Outfall. Final Report," Coastal Engineering
     and Hydraulics Divisions, National Research Institute for
     Oceanology, Stellenbosch, South Africa, CSIR Report C/SEA 8116,
     May 1981.

20. Roberts, P.J.W. (1977), "Dispersion of Buoyant Wastewater
     Discharged from Outfall Diffusers of Finite Length," W.M. Keck
     Laboratory of Hydraulics and Water Resources, California Institute
     of Technology, Report No. KH-R-35.

21. Roberts, P.J.W. (1979), "Line Plume and Ocean Outfall Dispersion,"
     J. Hydraulics Division, ASCE, 105, No. HY4, pp. 313-331.

22. Roberts, P.J.W. (1979), "A Mathematical Model of Initial Dilution
     for Deepwater Ocean Outfalls," Proceeding of Conference:
     Conservation and Utilization of Water and Energy Resources, San
     Francisco, California, pp. 318-225.

23. Roberts, P.J.W. (1980), "Ocean Outfall Dilution: Effects of
     Currents," J. Hydraulics Division, ASCE, 106, No. HY5, pp.
     769-782.

24. Roberts, P.J.W., and Toms, G. (1985), "Inclined Dense Jets in a
     Flowing Ambient," to be submitted to J. of Hydraulic Engineering,
     ASCE.

25. Roberts, P.J.W., and Toms, G. (1985), "Design of an Ocean Outfall
     System for Dense and Buoyant Effluents," to be submitted to J. of
     Hydraulic Engineering, ASCE.

26. "Water Quality Control Plan for Ocean Waters of California," State
     Water Resources Control Board, Sacramento, California, reprinted
     February 1981.

27. Zeitoun, M.A., et al. (1970), "Conceptual Designs of Ocean Outfall
     Systems for Desalination Plants," Res. and Develop. Progress
     Report No. 550, Office of Saline Water, U.S. Dept. of Interior.

SEWAGE SLUDGE DISPOSAL OPTIONS

P. C. Wood,
Ministry of Agriculture, Fisheries and Food,
Fisheries Laboratory,
Burnham-on-Crouch, Essex, CMO 8EJ, UK

ABSTRACT. A review is made of the nature of sewage sludge, and of the advantages and disadvantages of disposing of it to land, by incineration, and to sea. The major factors which influence the impact of sludge disposal, and the effects of sludge disposal on the marine environment are considered. Balancing the economic costs of the various methods of sludge disposal and their environmental effects are discussed.

1. INTRODUCTION

Sewage sludge is an inevitable by-product of sewage treatment, and hence is associated mainly with developed countries which have extensive sewage treatment facilities. All countries aspire to increase the extent of sewage treatment and the amounts of sewage sludge which arise are therefore also expected to increase. Thus, sewage sludge production cannot be halted, and it falls on nations to find methods of disposal which are environmentally and economically acceptable. There are essentially three ways in which sewage sludge can be disposed of: to land, by incineration, and by disposal at sea via a pipeline or by dumping vessel. This paper will briefly describe the range of major options available, their advantages and disadvantages, and the major factors which influence selection of a particular option. As this meeting is concerned mainly with the use of the sea as an option for waste disposal, particular attention will be given to marine disposal of sewage sludge.

2. SEWAGE SLUDGE - ORIGIN AND NATURE

It has been estimated that in European countries, the proportion of the population served by sewerage varies from 25% in Greece to 95% in the UK, the proportion of the total European population served being 72% (Vincent and Critchley, 1982). In addition, 37% of European industrial load enters domestic sewerage systems. In the USA and Canada, about 75%

111

G. Kullenberg (ed.), The Role of the Oceans as a Waste Disposal Option, 111–124.

of the domestic population is sewered.

Current European sludge production is 7 million tds (tonnes dry solids)/yr, the principal producers being Germany (2.2 million tds/yr), the UK (1.5 million tds/yr), Italy (1.2 million tds/yr) and France (0.84 million tds/yr). Thus, very substantial amounts of a somewhat objectionable material are produced, whose safe disposal plays an important part in environmental protection and will do so increasingly, for sludge production in Europe is expected to rise to 9.5 tds/yr by 1990.

Sewage sludge is a collective term applied to the solids derived from domestic sewage following various levels of treatment. Raw sludge containing between 2 to 7% dry solids (DOE, 1981) may arise at the primary stage of treatment (settlement) but further arisings occur following secondary biological treatment. These sludges may be subjected to a range of treatments designed to reduce the water content, odours, and pathogens, or to improve their condition so making them a more acceptable material for application in various ways. Dewatering may be achieved by thickening in gravity tanks, by solids floation or by centrifuge, with or without the addition of chemicals. The dry solids content of sludge so treated may range from 3 to 12%. Storage of raw sludge in open lagoons for a period of months to years reduces the water content and often kills some of the pathogens. Cold anaerobic digestion in open tanks for 4 to 6 months may lead to a 30%-40% reduction of organic matter. Warm anaerobic digestion in tanks takes about half the time and is more controllable. Both processes lead to a reduction in the numbers of pathogens. Further dewatering may be achieved by air drying or by mechanical dewatering to produce a sludge cake usually containing between 20 and 50% of dry solids. Air drying may also lead to a reduction of pathogens and of nitrogen. Sludge which has been biologically treated may contain fewer pathogens than raw sludge, but viable eggs of parasites can still be present.

There is thus a wide range of processes available for the treatment and disposal of sludge, but the final choice depends on the outlets available, and the environmental and economic factors which apply to each particular arising.

3. COMPOSITION OF SLUDGE

Because of variations in the quality of sewage entering the sewage treatment works and the subsequent treatment of the sludge, wide variations occur in sludge composition. Domestic sewage collecting systems often receive industrial water containing metals and organic wastes; in some areas up to 50% of the flow may be of industrial origin. On the other hand, small domestic sewage treatment works in wholly rural areas receive virtually no industrial waste, but still contain small amounts of various metals.

Nutrient content may be enhanced by effluents from food processing establishments (abattoirs, canneries), tanneries etc, but losses of nitrogen may occur during the digestion of sludge (Table I). Despite

this, nitrogen, phosphorus and potassium are usually present in sewage
sludge in significant amounts.

TABLE I   Components of sewage sludges disposed
to land in the UK (mg/kg dry solids)
1977-1978 (DOE, 1981)

Solids	Mean	Range
Nutrients#		
Nitrogen	2.1-5.1	–
Phosphorus	1.0-1.8	–
Potassium	0.2-0.5	
Potentially toxic metals		
Zinc	1820	199 -19000
Copper	613	36 - 2889
Nickel	188	5 - 3036
Chromium	744	7 -10356
Cadmium	29	0.4- 183
Lead	550	19 - 3538
Organohalogen compounds (1978)		
Dieldrin	–	< 0.05 -17.0
γBHC	–	0.1 - 0.5
DDT	–	0.02 - 0.8
PCB	–	< 0.004- 5.0

Range of means depends on method of treat-
ment of sludge.

Contaminants consisting of metals and persistent organic compounds
mainly from industrial effluents, but also from domestic water distribution systems and from urban run off, enter sewage collecting systems,
and a high proportion of them become attached to the solids present in
sewage sludge. From an examination of the levels of metals found in
sewage sludge applied to land in the UK it can be seen that sludges
arising in industrialised areas may contain over 10 times the average
levels.

Less is known about the presence of persistent organic compounds in
sewage sludge. Although levels are generally low, dieldrin, γBHC, DDT
and PCBs have all been detected in sludges, some at milligram levels on
a dry solids basis. Other organic compounds will reflect their input
from industrial sources.

An important feature affecting the choice of a disposal route for
sewage sludge is the pathogen content. Sewage received at a treatment
works reflects the infectious and carrier status of the population in
the catchment area. The numbers and types of pathogens in sewage sludge
are not accurately known because of uncertainties associated with

variability of treatment, the incidence in the human population, the
season, and the lack of precision and standardisation in the methods
used for isolating and enumerating pathogenic organisms. Four groups of
pathogenic organisms have been found in sewage sludges in the UK, but it
seems likely that the range and type of organism will be greater in
developing countries where the incidence of disease in the population is
greater. Salmonellae responsible for enteric diseases are the commonest
group of bacterial pathogens (Table II), but the eggs of parasitic worms,
pathogenic viruses and protozoan cysts (Davis, 1980) are also present.

TABLE II. Pathogens present in sewage sludges

1.   Bacteria, particularly Salmonella species.

2.   Eggs of parasitic worms, particularly the
      tape worms (Taenia saginata, and T. solium)
      but also Ascaris and Trichuris.

3.   Viruses, particularly the Enteroviruses.

4.   Protozoan cysts such as Giardia
      intestinalis.

Clearly the numbers and types of pathogens present in sludge, and the
risk of human infection associated with each form of sludge disposal are
a major factor influencing the choice of disposal method.

4. DISPOSAL ROUTES FOR SEWAGE SLUDGE

In Europe, of the 7 million tds of sludge produced annually, 43% is
tipped at landfill sites or used in land reclamation, 37% is used in
agricutlure, horticulture, parks etc, 8% is incinerated and 7% is
disposed at sea (Vincent and Critchley, 1982) (Table III). In the USA,
proportionally more is disposed of via incineration (21%) and to sea
(18%) with the result that only about half the European figure is tipped
to land (24%).

5. LAND DISPOSAL OF SEWAGE SLUDGE

Disposal to agricultural land, including other outlets where sludge is
used as a fertilizer, accounts for about a third of the sludge produced
in Europe and the USA (Table III). National dependence varies from
insignificant in Greece and Ireland to 80% and 90% in Switzerland and
Luxembourg. Agricultural disposal by the large European producers
(France, Germany and the UK) accounts for between 30 and 40% of
production.

TABLE III. Sludge disposal in Europe

Country	Current sludge production ('000 tds/yr)	Use of disposal routes (% of sludge produced)					
		Agricultural land*	Other land+	Incineration	Sea Vessels	Pipelines	Unspecified
UK‡	1500	41	26	4	27	2	0
Belgium	70	15	83	2	0	0	0
Denmark	130	45	45	10	0	0	0
France	840	30	50	20	0	0	0
Germany	2200	39	49	8	2	0	2
Greece	3	0	100	0	←------45---→		0
Ireland	20	4	51	0	←------45---→		20
Italy	1200	20	55	5	0	0	20
Luxembourg	11	90	10	0	0	0	0
Netherlands‡	230	60#	27	2	0	11	0
Austria	140	small	large	30	0	0	0
Finland	130	40	45	0	0	0	15
Norway	55	18	82	0	0	0	0
Spain	45	60	←----20---→		←------20---→		10
Sweden	210	60	←----30---→		0	0	10
Switzerland	150	80	10	10	0	0	0
Europe	6934	37	43	8	←-------7---→		5
USA	4500	31	24	21	←------18---→		6

* Including horticulture, allotments and gardens
+ Mainly land-fill but including land reclamation and forest
‡ Values given are % of sludge as disposed
# Including 10% used to produce commercial soil conditioners

Source: (Vincent and Critchley, 1982)

The potential beneficial effects of land disposal are due to the presence of nutrients, potassium, calcium and trace metals, and of particulate organic matter (DOE, 1981). A major influence on the selection of this option is the financial saving to the community, particularly where disposal areas are close to the sewage treatment works.

The potential for adverse effects arising from land disposal gives considerable cause for concern, and requires disposal authorities in areas of high population density to examine the use of this outlet with great care. Adverse effects arise from direct physical contamination of buildings, road and growing crops during transport and application. Odour and fly problems can be considerable unless the sludge is adequately treated. Of particular concern is the risk of direct contamination of surface and underground water supplies, particularly where land is sloping or fissured. Contamination with nitrates can be a particular problem. Soil may also be contaminated by excessive applications of nitrate, which may be transmitted to man and animals via crops. Contamination of soil by toxic metals, may affect the texture of the soil, which may become phytotoxic for sensitive species. Because of the complexity of the mechanism of phytotoxicity, some restraint on the rate of application of high metal-containing sludges is necessary. Uptake of metals, particularly cadmium, by fast-growing crops such as lettuces, can lead to the excessive intake by consumers. In the UK, guidelines for the rates of application of sewage sludge to various types of soil have been laid down (DOE, 1981), and within the European Commission, similar standards are the subject of discussion. As a result, careful monitoring of levels of metals in soil and crops is essential where sludges of industrial origin are applied repeatedly to the same land.

Persistent organic compounds may find their way into the human food chain by direct contamination of crops, ingestion by grazing animals and their excretion in milk. Because oil-producing crops (rape, linseed) may absorb organohalogen compounds, a limit of 10 mg/kg (dry solids) of PCB's in sewage sludge applied to agricultural land has been set in the United States.

Risks of infection by pathogenic organisms limit the areas and times in which sewage sludge may be applied to agricultural land. Grazing animals may become infected directly, and man may become infected by direct contact, or by consumption of milk, crops, etc. For these reasons, strict guidelines must be observed to ensure that the risk of disease transmission is kept to a minimum.

In Europe, tipping of sludge to land accounts for 43% of arisings, national dependence varying between 10% and 100% (Table III). Sludge may be applied to tips in liquid form or in a partially dewatered state, on its own or with other waste materials in order to avoid odour or leachate problems (DOE, 1981). Care is required to ensure that the tipping of liquid sludge does not contaminate surface or ground water supplies. Land reclamation using sludge may be employed where sites are to be recovered from industrial excavations and top-soil is not available. However, application rates may be limited if the sludge is contaminated by toxic metals.

6. INCINERATION

In Europe the incineration of sludge accounts for between 2% and 30% of national arisings, although several countries do not incinerate any sludge (Table III). In the USA, about 20% of sludge is incinerated. The main reason for the limited used of this method of disposal lies in the cost, for equipment is sophisticated, and fuel costs are high. In addition, there are the costs of dewatering which is usually required before the material can be efficiently incinerated.

The advantages of incineration are that except for the conservative contaminants which may be present, the aqueous and organic component is volatilised or destroyed, and only limited space is required for ultimate disposal of the small amount of ash that arises.

The major disadvantage of this form of disposal is the very substantial cost both of dewatering and of burning. Incineration sites close to urban areas are often difficult to find, and can bring about unacceptable amenity problems. Environmental disadvantages resulting from the incineration of sludges containing industrial wastes include possible discharge to the atmosphere of volatile and particulate metals, and liquid and gaseous organic compounds.

7. SEA DISPOSAL

According to statistics supplied by the London Dumping Convention (LDC, 1984), between 1976 and 1981 the total amount of sewage sludge disposed to sea annually was between 12 and 15 million tonnes (wet weight), equivalent to about 500,000 tds/yr. In 1981, about 8 million tonnes was dumped in Europe, and 6 million tonnes in North America. Only four European countries disposed of sewage sludge to sea (Table III). The United Kingdom disposed of 27% (272,800 tds/yr) of its arisings to sea, Ireland 45% (9000 tds/yr), Spain 20% (9,000 tds/yr), the Netherlands 11% (36,000 tds/yr) and the FRG 2%, (34,000 tds/yr). The Netherlands disposal was via a long sea outfall from the coast. It should be noted that disposal by the FRG has since ceased. In the USA, 18% of sludge is disposed at sea. These figures probably underestimate the total amounts as they exclude those disposal operations carried out in internal waters, and sewage solids present in untreated sewage outfalls.

From this brief review of data, it is clear that countries have a widely differing dependence on the sea disposal route. In some instances, particularly in developing countries, only a small proportion of the population is sewered. In other countries, although sewered, whole sewage is discharged to sea, and the need to dispose of sludge does not arise. In some instances, centres of population are so far away from coastal regions that there is no economic or social incentive to dispose of sewage sludge at sea, whereas in other countries, fears of environmental damage have disuaded the authorities from carrying out sea disposal, particularly where the economic advantages of so doing, are marginal.

8. SELECTION OF SITES FOR SEA DISPOSAL

In deciding whether or not sea disposal of sewage sludge is an option which should be used, the authorities must have regard for the environmental consequences of disposal. Clearly, these consequences vary with the quantity of sludge to be disposed of, the hydrographic regime of the area of disposal and the likelihood that other activities adjacent to the dumping site will be affected. Sewage sludge disposal is probably the best researched and monitored dumping activity that takes place, with numerous publications from the USA and the UK which describe the physical, chemical and biological characteristics of dumping sites, as well as the impact of disposal activities. What is apparent from all these studies is the site-specific nature of the impact, and how unwise it is to extrapolate the circumstances of one disposal operation to another.

A study of the influence of site-specific characteristics on the effects of sewage sludge dumping has been made by Norton and Champ (in press). Although only 8 sites were considered, the conclusions are of general application as the sites were selected on the basis of the amounts dumped and the hydrographic regime (Table IV). Thus the amounts

TABLE IV.   Characteristics of dumpsites

Disposal site	Quantity dumped#		Distance from coast (km)	Water depth (m)	Maximum current speeds (cm/s)	
	Wet weight	Solids			Tidal*	Wave induced (max)
Lyme Bay (UK)	62	2.1	20	40	65- 35	30
Off Plymouth (UK)	76	4.1	8	50	50- 30	50
Bristol Channel (UK)	330	8.8	20	36	120- 67	44
German Bight (FRG)	290	14.5	18	20	-	200
Liverpool Bay (UK)	1,665	65.2	20	25	90- 50	56
Firth of Clyde (UK)	1,400	70.0	2	100	25-	very low
Thames estuary	4.620	124.7	15	20	130-120	35
New York Bight	4,371	171.6	12	22	- 25	200

\# Thousands of tonnes in year of report
* Surface-bottom

of sludge dumped varied from 2,100 tds/yr in Lyme Bay to 171,600 tds/yr in New York Bight, the distance from the coast ranged from 2 km in the Firth of Clyde to 20 km at 3 sites, and the water depth varied from 20-50 m at 7 sites to 100 m in the Firth of Clyde. Of particular relevance are the current characteristics at the eight sites. Tidal currents varied widely, there being a nearly five-fold difference in maximum velocities (25 cm/s in New York Bight, and 130 cm/s in the Thames estuary). Similarly, water-induced near-bed orbital velocities varied from 35 cm/s in the Thames estuary to 200 cm/s in the German Bight and in New York Bight. From assessment of the hydrographic data and examination of the results of sea-bed monitoring data, the authors were able to classify the sites according to their dispersive characteristics. They found that two of the sites (Thames estuary and Bristol Channel) were continually dispersive because the tidal currents were erosional at all stages of tide, two sites (Plymouth and Liverpool Bay) were regularly flushed by spring tides or waves and three sites (German Bight, Lyme Bay and New York Bight) were rarely flushed because tidal velocities were not erosional and disturbance caused by waves was infrequent. The Firth of Clyde site was a totally accumulative site because of low tidal and wave-induced currents.

Thus, from this brief review it is evident that there is no uniform basis for the selection of sites according to their hydrographic regime. Other factors also must be considered, e.g. the distance of the disposal site from the port of loading, an operational matter which is closely related to the economics of the disposal operation. The location of the site must also take into account other activities in the area of dumping. These include dredging, navigation, fishing and recreation, and must take into account physical interference with vessels undertaking these activities, as well as the consequences of the dumping action.

Thus, a dumping site must be selected which ensures that sewage sludge does not settle on areas being dredged, does not affect fisheries including the ecosystems which support them, nor make water quality and beaches unacceptable for recreational activities such as boating, swimming, etc. From variations in the degree of perturbation observed between dump sites, it is evident that national authorities vary in their degree of tolerance of environmental change.

9. FACTORS AFFECTING THE IMPACT OF SLUDGE DISPOSAL

As would be expected from the diversity of characteristics at dump sites and major differences between the amounts dumped at each site, the effects of dumping activities vary widely.

Comparison of the effects of disposal at two sites which receive comparable amounts of sewage sludge, showed different levels of impact. Thus, the Lyme Bay and Plymouth sites (Table IV) each receive small amounts of sludge (2,100 and 4,100 tds/yr respectively) into 40-50 m of water. At Lyme Bay (Eagle et al., 1978) dispersal processes were found to be ineffective in preventing accumulations of dumped material in the sediments, levels of carbon, nitrogen and certain metals being significantly higher in the area most commonly used for dumping. Some reduction of species diversity was noted. At the Plymouth site, only minor effects attributable to sewage sludge were evident (Eagle et al., 1979), and levels of carbon and metals were lower than in the adjacent coastal region. No effects on benthic species could be detected. The lack of impact at the latter site was attributed to the more exposed position of the site which was subject to disturbance by winter storms and associated high dispersal characteristics.

A similar comparison can be made between two of the major sites receiving sewage sludge, viz. the Thames estuary and New York Bight which receive annually 124,700 tds and 171,600 tds respectively. Environmental effects can be readily observed in New York Bight (Norton and Champ, in press), with gross elevations of carbon in the sediments and concentrations of metals elevated ten fold. Benthic populations showed major changes of diversity and biomass, compared with non-impacted areas. On the other hand, in the Thames estuary (Norton et al., 1981) the area affected by the dumping activity was restricted in size, and benthic populations were mainly affected by the nature and mobility of the substrate, although a secondary minor effect due to sludge particles was observed. It is evident that the different levels of impact seen at the two sites is due to the differing hydrographic regimes, maximum tidal velocities in the Thames estuary being 5 times greater than those in New York Bight.

It is clear therefore that the effect of each proposed disposal operation needs to be evaluated taking into account the nature and quantity of the sludge, and the hydrographic characteristics of the disposal area.

10. EFFECTS OF SEWAGE SLUDGE DISPOSAL

It is not intended here to give a detailed account of all the effects
which have been observed following the disposal of sludge, but in considering the option of sea disposal, a potential discharger should be
aware of the range of likely effects which might result (Norton and
Rolfe, 1978). Site selection should wherever possible minimise the
nature and scale of these effects. As indicated above, in carefully
selected sites which optimise dispersal, only one or two of these
effects may be evident; in others, the full range of effects may be
observed.

Near to the sea bed, the high organic loading might result in low
dissolved oxygen concentrations, but this is only likely to be observed
in areas with little natural energy, and might be attributed in whole or
in part to terrigenous inputs or natural causes. An increase in faecal
organisms (bacteria, viruses and parasites) may be found in the water
column, and some care will need to be given to the prevention of
contamination of molluscan shellfish or of recreational areas, particularly where statutory limits are placed on water quality. In sediments,
a wide range of changes may take place but these are likely only where
sludge particles accumulate e.g. where mud and fine clays occur.
Changes observed in sediments include an increase in carbon and nitrogen
and of metals such as lead, mercury, chromium, zinc and copper. The
levels of contamination of the sediments will depend to some extent upon
the origin of the sludge, for sludges derived from industrialised areas
often contain substantial amounts of materials used by industries in the
catchment area. Increases in cadmium in the sediments do not usually
occur, as cadmium generally passes from sludge into the water phase soon
after release.

Changes in the sediment, particularly increases in the organic particulate matter, may cause substantial changes in benthic communities.
This may lead to changes in biomass and diversity, but in areas where
sediment composition is patchy, the classical pattern of change is not
generally observed. Major changes in the benthic popoulation occur in
areas of low energy receiving large amounts of sludge (Halcrow et al.,
1973). Most of the evidence related to benthic change has been derived
from examination of sediments using 1-2 mm sieves; there is evidence
however that meiobenthos also undergoes similar changes.

Of particular importance are changes which might be caused to fish
and fisheries. There is a risk of direct contamination of fish by
faecal bacteria and viruses, but this can be reduced or eliminated by
careful selection of the site. In other areas, substrate changes may be
sufficient to make it unsuitable for sessile crustacean species, such as
Nephrops norvegicus (McIntyre and Johnson, 1975). Fishing activities in
which traps are used may be curtailed, and limitations may be placed on
trawling. An important consideration is the effect of industrial components of sludge, such as heavy metals and organohalogen compounds on the
quality of fish and shellfish. These substances may enter the food
chain and lead to levels of accumulation which make consumable species
unacceptable for human consumption. There is evidence that improvement
of sludge quality by limiting the entry of some industrial wastes, such

as mercury, into the sewage system, can lead to significant reductions
in fish concentrations over a relatively short time scale. However,
some consideration must be given to other sources of potentially
polluting materials (coastal discharges and river inputs), particularly
when sludge disposal sites are situated close to the coast (Preston and
Portmann, 1981).

11. SELECTING THE BEST OPTION FOR SLUDGE DISPOSAL

This paper has described the range of disposal options available, and
the advantages and disadvantages of each option. In many developing
countries and in certain areas of developed countries, the need to
dispose of sewage sludge does not arise as sewage treatment works do not
exist and raw untreated sewage is discharged directly into rivers,
estuaries and coastal waters. In some countries, raw or partially
treated sewage, still containing most of the organic material is discharged through long sea outfalls at carefully selected sites. In other
countries, sewage sludge derived from treatment of sewage is discharged
at sea via land-based outfalls. Whilst the discharge of whole or partially treated sewage is only likely to have local effects (a topic outside the scope of this paper), the discharge of substantial amounts of
sewage sludge from land-based pipelines is not an environmentally
attractive option because of limitations placed on selection of the
point of discharge by engineering and economic considerations. Furthermore, disposal from ships allows flexibility of site selection, whereas
land-based outfalls are permanent.

When looking at the remaining options, e.g. whether to dispose
sludge to land, to sea by ships or by incineration, it is necessary to
consider the advantages and disadvantages of each method of disposal.
Table V lists the major factors which have to be taken into account when
deciding which route is likely to be the most satisfactory in any particular circumstances. It is clear that every option has its benefits,
and its restraints, e.g. whilst agricultural use of sewage sludge can
utilise an important natural resource, in some circumstances there may
be associated restraints such as phytotoxicity, or risks of transmitting
bacteria and parasites to cattle and man. Similarly, the advantages of
cheapness and nuisance-free handling of sewage sludges for sea disposal
from certain treatment works must be offset against the risk of damage
to the marine environment.

What is also clear is the need to make a site specific evaluation
of each of the options, for the benefits and restraints at one site may
not be the same at another. Thus, sewage sludges which arise at sewage
works situated at coastal sites can often be readily transported to sea,
whereas their transport for land-fill or agricultural use would impose
major problems arising from the use of road transport, and cause public
nuisance. This is particularly so for sludge arising in large cities
where the nearest available land may be costly and up to 50-100 miles
(80-160 km) away. On the other hand, land disposal is the most suitable
means of disposal of sludges which arise in rural communities or in
towns with close proximity to agricultural land. For cities situated

TABLE V. Benefits and restraints of the major options for the disposal
          of sewage sludge

Option	Use/method	Benefits*	Restraints#
Land disposal	Agricultural and horticultural purposes	Nutrient, trace metal and organic material utilisation	Smells Nuisance Phytotoxicity Metal burden Parasites Bacteria/virus Pollution of ground water Availability of land
	Reclamation and landfill	Material retained on site	Availability of land Cost of dewatering
Incineration	–	Ash disposal volume small	Very costly Smells Gaseous and droplet emissions Smoke
Sea disposal	Land-based outfall	Cheapness No terrestrial nuisance	Limited choice of sea disposal site
	Seagoing vessel	Cheapness No terrestrial nuisance No land requirement	Interference with other activities marine environmental degradation Fish quality

* Benefits do not apply to all sites
# Selection of proper site for these activities can overcome restraints
  or reduce them to acceptable levels

away from the coast, the option of sea disposal is denied because of
heavy transportation costs.

In selecting a disposal route, the economics of the various
alternatives cannot be ignored. In a study made in the UK in 1981, it
was found that the costs of the various forms of disposal varied widely
(Table VI). Not only did the average costs of disposal by each of the

TABLE VI.   Relative costs of sludge treatment and disposal in UK, 1981

	% costs/tonne dry solids‡		Treatment
	Average	Range	
Disposal at sea	100	60- 290	None required
Utilisation on farmland	200	100- 290	Stabilisation/ digestion
Controlled tipping	140	100- 310	Dewatering to produce cake
Incineration	460	270-1140	Dewatering and incineration

‡ Shown as a percentage of the average cost of disposal at sea

(Based on Collinge and Bruce, 1981)

major routes vary (by a factor of four), but the costs of various
similar disposal operations varied widely according to site.   There was
for instance a five-fold difference in the cost of sea-disposal
operations and a three-fold difference in the cost of land-based
operations.   The higher costs were mainly due to the distance of
disposal sites, and the need for additional treatment of some sludges.
      Disposal authorities must clearly take into account both environ-
mental and economic aspects when deciding which method of disposal to
adopt.   They are required to make an assessment which will allow them to
select the disposal route offering the maximum environmental protection
at acceptable economic cost or some justifiable balance between the two.
The selection of a disposal route will be influenced by local site
considerations, and by the national attitude towards waste disposal.
Some countries take the view that providing a disposal activity does not
exceed the assimilative capacity of the area of disposal (whether it be
to land, sea or by incineration), sewage sludge may be disposed of in
any way which is economically available.   Others take the view that dispersal of sewage sludge should be avoided even if there are no adverse
consequences, and that sludge should be used on land or retained in
tips, even if the costs are excessive.   This subject is clearly outside
the scope of this paper.

12.   REFERENCES

COLLINGE, V. K. AND BRUCE, A. M. 1981.   Sewage sludge disposal: a
      strategic review and assessment of research needs.   Medmenham,
      Water Research Centre, Medmenham Laboratory, Technical Report, TR
      166, 31 pp.

DAVIS, R. D. 1980. Control of contamination problems in the treatment and disposal of sewage sludge. Medmenham, Water Research Centre Medmenham Laboratory, Technical Report, TR **156**, 79 pp.

DEPARTMENT OF THE ENVIRONMENT AND NATIONAL WATER COUNCIL, 1981. Report of the Sub-committee on the Disposal of Sewage Sludge to Land. London, HMSO, 83 pp (Standing Technical Committee Reports No. **20**).

EAGLE, R. A., HARDIMAN, P. A., NORTON, M. G. AND NUNNY, R. S. 1978. The field assessment of effects of dumping wastes at sea: 3. A survey of the sewage sludge disposal area in Lyme Bay. Fish Res. Tech. Rep., MAFF Direct. Fish. Res., Lowestoft, No **49**, 22 pp.

EAGLE, R. A., HARDIMAN, P. A., NORTON, M. G. and NUNNY, R. S. 1979. The field assessment of effects of dumping wastes at sea: 4. A survey of the sewage sludge disposal area off Plymouth. Fish. Res. Tech. Rep., MAFF Direct. Fish. Res., Lowestoft, No **50**, 24 pp.

HALCROW, W., MACKAY, D. W. and THORNTON, I. 1973. The distribution of trace metals and fauna in the Firth of Clyde in relation to the disposal of sewage sludge. J. Mar. Biol. Ass. U.K., **53** (3), 721–739.

LONDON DUMPING CONVENTION, 1984. Consideration of reports on dumping. Report on the nature and quantities of wastes dumped at sea. LDC/SG8/INF **4**.

McINTYRE, A. D. and JOHNSTON, R. 1975. Effects of nutrient enrichment from sewage in the sea. pp. 131–141 In: Gameson, A. L. H. ed. Discharge of Sewage from Sea Outfalls. Oxford, Pergamon Press.

NORTON, M. G. and CHAMP, M. A. (In press). The influence of site-specific characteristics on the effects of sewage sludge dumping. In: Duedall, T. W. et al. eds. Wastes in the Ocean. 3. Chichester, John Wiley.

NORTON, M. G., EAGLE, R. A., NUNNY, R. S., ROLFE, M. S., HARDIMAN, P. A. and HAMPSON, B. L. 1981. The field assessment of effects of dumping wastes at sea: 8. Sewage sludge dumping in the outer Thames estuary. Fish. Res. Tech. Rep., MAFF Direct. Fish. Res., Lowestoft, No **62**, 62 pp.

NORTON, M. G. and ROLFE, M. S. 1978. The field assessment of effects of dumping wastes at sea. 1. An introduction. Fish. Res. Tech. Rep., MAFF Direct. Fish. Res., Lowestoft, No **45**, 9 pp.

PRESTON, A. and PORTMANN, J. E. 1981. Critical path analysis applied to the control of mercury input to United Kingdom coastal waters. Environ. Pollut. B, **2** (6), 1981, pp 451–464.

VINCENT, A. J. AND CRITCHLEY, R. F. 1982. A review of sewage sludge treatment and disposal in Europe. Report **442-M** made by WRC Environmental Protection, Medmenham Laboratory, Henley Road, Medmenham, PO Box 16, Marlow, Bucks SL7 2HD.

THE ENVIRONMENTAL IMPACT OF SLUDGE DUMPING AT SEA AND OTHER DISPOSAL
OPTIONS IN THE U.K. - TRACE METAL INPUTS

M. Hutton & C. Symon
Monitoring and Assessment Research Centre
459A Fulham Road
London SW10 0QX
U.K.

1. INTRODUCTION

Coastal waters surrounding industrial nations have for many years been
major recipients of wastes from human activities, either in the form
of direct discharges or indirectly, via river inputs. Recent concern
over such inputs has focused attention on the environmental effects of
contaminant discharges, particularly in those waters susceptible to
such inputs. The North Sea is one such water considered to be
sensitive to contaminant inputs because it receives discharges from
numerous industrialized regions and major rivers, yet is largely
landlocked and relatively shallow with a restricted exchange of water
(ICES 1983; Royal Commission on Environmental Pollution 1984). The
international nature of contaminant inputs to the North Sea and to the
north-east Atlantic as a whole has resulted in an international
approach to control, based on the Oslo and Paris Conventions. The
Oslo Convention regulates the marine dumping of wastes to the north-
east Atlantic, while the Paris Convention controls land-based
discharges in the same region (Oslo and Paris Commissions 1984). Both
Conventions categorize contaminants according to their potential
impact on the marine environment: Annex I substances, such as mercury
(Hg), cadmium (Cd) and PCBs are toxic, persistent and show a
propensity for bioaccumulation. Annex II substances are considered to
be less toxic and include arsenic (As), lead (Pb), zinc (Zn) and
cyanides.
 Despite the international agreements reached by contracting
countries, differences in attitude have arisen in the last five years
with regard to the dumping of wastes which contain Annex I and II
substances in the North Sea (Jenkins 1982). Certain European
countries oppose the dumping of sewage sludge to the North Sea and
have now curtailed these activities (Oslo and Paris Commissions
1984). In contrast, the U.K. currently disposes about a quarter of
its total sewage sludge to coastal waters (DOE/NWC 1983). This
practice is not only controlled by licensing but also observes the
conditions of the Oslo Convention and is supported by regular
monitoring of the disposal sites. Nevertheless, the U.K. is under

G. Kullenberg (ed.), The Role of the Oceans as a Waste Disposal Option, 125–138.
© *1986 by D. Reidel Publishing Company.*

pressure to cease this form of sludge disposal (ENDS 1984). This
pressure is based on the premise that the accumulation of persistent
contaminants in the North Sea from the dumping of sludge poses a
hazard to the indigenous biota (Jenkins 1982). There are, however,
other sources of contaminant discharge to the North Sea and attempts
have been made to place the U.K. practice of sludge dumping in
perspective by comparing the quantities of contaminants discharged via
this activity with those entering from other sources (Norton 1982;
Mance and O'Donnell 1984; Hill, Mance and O'Donnell 1984). However,
an overall assessment of sludge disposal must also examine the
relative size of contaminant inputs to the different environmental
sectors resulting from the other sludge disposal options. This will
allow an insight into the implications of any ban on sludge dumping by
examining the consequent increase in contaminant discharges from the
remaining disposal options.

This paper evaluates the overall significance of U.K. sludge
disposal to sea from the perspective of environmental release of
selected metals listed in Annexes I and II of the Oslo and Paris
Conventions. Current sewage sludge disposal practice and constraints
in the U.K. are described and estimates given for the corresponding
quantities of metals associated with these options. The quantities of
metals discharged by sea disposal of sludge are compared with inputs
from other human sources, both for U.K. coastal waters and the North
Sea as a whole. Finally, the environmental significance of a
hypothetical ban on sludge dumping is considered by examining the size
and implications of increased metal release associated with the
remaining disposal options.

No consideration is given in this paper to the quantities of
metals entering the North Sea and U.K. coastal waters via atmospheric
deposition. This was because close inspection of the most widely used
published values suggested that these may, in some cases, be
overestimates. Currently, we are critically re-evaluating the
relative importance of this route of metal input.

2. SLUDGE DISPOSAL IN THE U.K.

2.1. Disposal options and associated metal burdens

Table I depicts the relative importance of the sludge disposal routes
in the U.K., together with the quantities of metals associated with
these options. A major proportion of U.K. sludge is disposed of to
land, with 40 % applied to agricultural land and 27 % to landfill and
other non-agricultural land. Sludge dumping at sea is another
important disposal route and also accounts for 27 % of the total. The
remaining options are less important, with 4 % being incinerated and
2 % discharged to coastal waters via pipeline. Although application
to agricultural land is the single largest disposal route, similar or
greater quantities of metals are associated with sea dumping and
disposal to landfill. This is because the sludges disposed by these
routes contain higher metal concentrations (DOE/NWC 1983).

Table I   The quantities of metals associated with sewage sludge in the U.K. according to disposal option, 1980.

Disposal route	tds x 10^6	Metal input (tyr^{-1})			
		Cd	Hg	Pb	As
Agricultural land	0.473	7.5	1.3	155	2.0
Landfill and other non-agricultural land	0.319	8.5	1.8	148	2.8
Dumped at sea*	0.321	7.3	2.8	203	1.5
Pipeline to sea	0.019	0.3	0.1	10	0.1
Incineration	0.045	2.3	0.7	28	1.2
U.K. Total	1.177	25.9	6.7	544	7.6

Source: DOE/NWC (1983)
* Differences exist in the metal burdens ascribed to sludge dumped at sea in this table and those given by the Oslo Commission in Table VI. The reason for this disparity is not known.

2.2. Constraints on sludge disposal in the U.K.

Sludge disposal in the U.K. is controlled by both legislative and non-statutory guidelines. Those controls of most relevance to trace metals are shown in Table II. It is apparent that sea disposal is mostly affected by international controls while land disposal is currently covered by national constraints. The guidelines which limit the addition of metals to agricultural soils are designed to protect crops, groundwaters, livestock and humans. These guidelines contain recommended maximum addition rates of 12 potentially toxic elements (DOE/NWC 1981b), but in almost all cases, it is the Cd limit which restricts sludge application to agricultural soils (Davis and Coker 1980; Davis 1984). In this respect, it is of interest to note that the CEC Directive on the agricultural use of sludge proposes a more stringent limit for Cd (Council of the European Communities 1982). This directive could therefore lead to further restrictions on the agricultural use of sludge and increase the demand on the other means of disposal at a time when there is also international pressure to curtail the sea disposal of sludge.

The most important air emission constraint on sludge incineration in the U.K. relates to the release of suspended particulates (DOE/NWC

1978). These standards were proposed over a decade ago but have never
become statutory regulations. No emission standards exist for trace
metals, but the guidelines for particulate emissions will also
restrict metal release. Sludge incinerator ash is disposed to
landfills in accordance with the normal requirements for disposal of
municipal solid waste (DOE/NWC 1981a). These requirements also apply
to the landfill of sewage sludge itself.

Table II Constraints affecting sewage sludge disposal in the U.K.

Disposal route	Constraint
Agricultural land	Non-legislative guidelines limiting the addition of metals to soils
Landfill and other non-agricultural land	Licence and consent granted by waste disposal authority
Dumped at sea	Oslo and London Conventions. U.K. Dumping at Sea Act
Pipeline	Paris Convention
Incineration	Recommended limits of particulate emissions

Sources: DOE/NWC (1981a b)

3. METAL DISCHARGES FROM SLUDGE DUMPING AND OTHER ACTIVITIES

3.1. Metal discharges to the North Sea

Table III presents estimates of the quantities of selected metals
discharged to the North Sea. It is apparent that rivers are the
largest sources of Cd, Hg and As. For Pb, dredged spoils represent
the most important discharge pathway. The size of both these sources
is mainly a consequence of the very large sediment inputs from the
rivers Rhine and Elbe and the large quantity of material dredged from
the Rhine estuary (Hill et al. 1984). However, it must be stressed
that the extent of metal enrichment on the particulate material will
determine the toxicological significance of these discharges. It is
possible that a proportion of the suspended particulate load of rivers
flowing into the North Sea is not enriched in trace metals and these
inputs simply reflect the natural flux of material between land and
ocean resulting from crustal weathering. Further work on the chemical
composition of river particulate material and extent of trace metal
enrichment is needed in order to determine the significance of this
source of input.

Table III Estimated quantities of trace metals entering the North Sea from land-based sources.

Source and country	Metal input (tyr^{-1})			
	Cd	Hg	Pb	As
RIVERS				
U.K.	21.5	5.1	396	57
Other countries	136	21.5	1509	527
DIRECT DISCHARGES				
U.K.				
sewage	13.9	1.5	64	10.2
industrial wastes	6.6	6.2	57	194
Other countries	2.2	0	29	1.1
SEA DISPOSAL				
Sewage sludge				
U.K.	3.6	0.9	104	–
other countries	0.4	0.1	11	–
Industrial wastes				
U.K.	0.4	0.4	244	–
other countries	0.1	0.1	2	–
Dredged spoils				
U.K.	5.4	7.3	495	–
other countries	43.2	18.7	2490	–
TOTALS	233.3	61.8	5401	789.3

River inputs and direct discharges refer to 1980; Source: Mance and O'Donnell (1984). Sea disposal data refer to 1982; Source: Oslo Commission (1984a).

An insight into the metal content of dredged spoils dumped in the North Sea can be gained by examining the relevant data supplied to the Oslo Commission (Oslo Commission 1984a). Table IV presents the published quantities of dredged materials and associated metal burdens dumped in the Oslo Convention area by Belgium, the Netherlands and the U.K. The table also shows the estimated average concentrations of these metals, calculated from the published data. Although these values are in wet weight, they nevertheless indicate that dredged spoils show only modest enrichment of trace metals compared with background values of river-suspended matter (Martin and Meybeck

1979). It would therefore appear that dredged spoils dumped in the
North Sea do not pose a particularly significant environmental hazard,
at least with respect to their metal content.

Table IV Quantities of metals in dredged spoils dumped in the
Oslo Convention area by selected countries in 1982, together
with the estimated metal concentrations of this material

Country	Quantity ($\times 10^6$ tyr^{-1})	Annual input (t)/ Average concn. (mg kg^{-1})		
		Cd	Hg	Pb
Belgium	47.7	–	10.5/0.2	1542/32.3
The Netherlands	21.4	43.2/2.0	8.2/0.38	948/44.3
U.K.	12.9	7.0/0.5	11.2/0.87	789/61

Quantities of dredged spoils and annual metal inputs from
Oslo Commission (1984a)

It is evident from Table III that the U.K. sea disposal of sewage
sludge makes only a small contribution to the total metal inputs of
the North Sea. However, the relevance of this kind of broad
comparison is questionable, given that about 90% of metal inputs from
U.K. sludge dumping takes place at just one site in the Thames Estuary
(Oslo Commission 1984a).

Inspection of Table III reveals that another sector of the U.K.
sewage treatment industry, namely the direct discharge of sewage to
coastal waters, is a larger source of Cd and Hg to the North Sea than
the dumping of sludge. The importance of direct sewage discharges was
first revealed by Mance and O'Donnell (1984) and previous estimates of
metal inputs to the North Sea have not specifically considered this
pathway (e.g. Norton 1982; Critchley 1983).

3.2. Metal discharges to U.K. coastal waters

Table V presents the estimated quantities of metals discharged to U.K.
coastal waters. Inspection of the values in this table reveals a
change in the relative importance of different sources compared with
the overall situation in the North Sea, particularly in the case of
Cd. Thus, Cd inputs from direct sewage discharges are relatively
large and are only exceeded by river inputs. Industrial discharges
are also a significant source of Cd input to U.K. coastal waters.

Table V Estimated quantities of trace metals entering
U.K. coastal waters from land-based sources

Source	Metal input (tyr^{-1})			
	Cd	Hg	Pb	As
RIVERS	38.7	13.9	878	118
DIRECT DISCHARGES				
Sewage	24.4	1.8	146	15
Sludge via pipeline	0.3	0.1	10	0.1
Industrial discharges	22.2	9.5	134	195
SEA DISPOSAL				
Sewage sludge	5.1	1.4	165	1.5
Industrial wastes	0.4	0.4	245	–
Dredging spoils*	7.0	11.2	789	–
TOTALS	98.1	38.3	2367	330

River inputs and direct discharges refer to 1980; Source:
Mance and O'Donnell (1984). Sea disposal data refer to
1982; Source: Oslo Commission (1984a).
* Values refer only to England and Wales.

Further inspection of the published discharge data reveals that
metal inputs via direct sewage discharge are not evenly distributed
around the U.K. coast (Paris Commission 1984). For example, of the
24.4 tyr^{-1} Cd entering coastal waters, 11 tyr^{-1} Cd are discharged
to the Thames Estuary alone. The Thames Estuary also receives
3.2 tyr^{-1} Cd from sludge dumping (Oslo Commission 1984a), resulting
in a total Cd input of 14.2 tyr^{-1} from the sewage sector. In
comparison, direct industrial discharges of Cd to the Thames Estuary
total 3.5 tyr^{-1} while river inputs are estimated to be 1.9–2.5
tyr^{-1} (Paris Commission 1984).
  Finally, Table VI provides an insight into the declining
importance of sludge dumping in U.K. coastal waters from the
perspective of metal inputs. The table shows considerable declines in
the quantities of Cd and Hg associated with this practice over the
period 1976-1982. It should be stressed that the actual quantities of
sludge disposed in this manner have not declined.

Table VI Temporal trends in the quantities of metals
associated with the dumping of sewage sludge in U.K.
coastal waters

	Annual input for each year (t)						
Metal	1976	1977	1978	1979	1980	1981	1982
Cd	10.2	11.4	8.6	8.6	9.3	6.5	5.1
Hg	4.0	3.8	2.8	2.8	3.6	2.4	1.4
Pb	128	156	208	197	182	160	165

Source: Oslo Commission (1984a b)

4. IMPLICATIONS OF ABOLISHING THE MARINE DISPOSAL OF SEWAGE

The abolition of sludge dumping at sea by the U.K. would necessitate
an increase in the importance of the remaining disposal options.
Given that larger metal inputs may arise from direct sewage discharges
(Tables III and IV), it is conceivable that this form of disposal may
also be curtailed in the future. This section examines the
environmental fate of the metal burdens in the sludge and sewage which
is currently discharged to U.K. coastal waters, assuming a
hypothetical ban on their disposal routes. Two contrasting options
have been selected to accommodate this ban on sea disposal: sludge
incineration and application to agricultural land. It must be
stressed that no attempt has been made to investigate the economic
consequences of these changes in sewage disposal. A fuller analysis
should examine this issue, given that economic factors will strongly
influence the practical feasibility of such changes in disposal.

4.1. Sludge incineration

Currently, sludge incineration in the U.K. is only a minor disposal
option and accounts for 3-4% of the sludge produced (Table 1). In
other countries, however, this disposal route is more important; for
example, it accounts for 25% of the sludge produced in the U.S.A. (EPA
1984). Table VII presents estimates of the current atmospheric
emissions of selected trace metals from sludge incineration in the
U.K. For comparison, the corresponding emissions from refuse
incineration in the U.K. are also given. Evidently, refuse
incineration releases considerably greater quantities of these
metals. Table VII shows the extent of increase in atmospheric metal
emissions if all sludge currently dumped at sea were diverted to the
incineration pathway. These values suggest that emissions from this

source will still be relatively modest compared with refuse
incineration. Table VII also depicts the "worst-case situation",
where the sewage currently discharged to coastal waters is treated to
form a sludge in which all the metals are retained and all of this
material subsequently enters the incineration pathway. Even in this
highly unlikely situation, metal emissions are still no larger than
those currently arising from refuse incineration.

Table VII Hypothetical impact of abolishing the
marine disposal of sludge and sewage in the U.K. by
diversion to the incineration pathway -
atmospheric emissions of metals

Source	Metal emission (tyr^{-1})			
	Cd	Hg	Pb	As
SLUDGE INCINERATION				
Present	0.2	0.6	1.2	0.03
+Dumped sludge	0.5	1.4	7.0	0.03
+Sewage discharges	2.4	1.6	6.1	0.4
Total	3.1	3.6	14.3	0.46
REFUSE INCINERATION	4.5	5.4	130	0.4

Estimates of current atmospheric metal emissions for
sludge and refuse incineration taken from Hutton et
al. (1984). Projected emissions for sludge incineration assume present-day emission characteristics

Sludge incineration not only results in atmospheric emissions of
metals but also produces ashes enriched in certain metals (DOE/NWC
1978). These residues are disposed of to landfill and therefore pose
a potential hazard to groundwater quality. Table VIII presents the
estimated metal inputs to landfill arising from sludge incineration,
both at present and for the two hypothetical situations. For
comparison, the corresponding inputs resulting from the disposal of
refuse incinerator ashes are also given. Once again, it is evident
that even in the extreme situation described above, metal inputs from
landfill of sludge incinerator ashes would be smaller than those which
currently arise from refuse incineration.

4.2. Agricultural utilization

About half the sewage sludge generated in the U.K. is applied to
agricultural land, 60% of which enters arable land with the remainder

going to pasture (DOE/NWC 1983). This practice results in
considerable metal inputs, but, as stated earlier, it is the presence
of Cd which restricts the application of sludge to agricultural land.
From the perspective of human exposure, sludge applicaton to arable
land is of greatest concern because the soil-crop plant pathway is
susceptible to increases in soil cadmium, while the soil-livestock
pathway is not (Ryan, Pahren and Lucas 1982).

Table VIII Hypothetical impact of abolishing the
marine disposal of sludge and sewage in the U.K. by
diversion to the incineration pathway - metal inputs
to landfill

	Metal input (tyr^{-1})			
Source	Cd	Hg	Pb	As
SLUDGE INCINERATION				
Present	2.0	0.06	27	1.2
+Dumped sludge	4.9	0.1	168	1.6
+Sewage discharges	22.0	0.2	140	14.6
Total	28.9	0.4	335	17.4
REFUSE INCINERATION	42.0	0.6	2.9×10^3	17.7

Estimates of current metal inputs to landfill for
sludge and refuse incineration taken from Hutton et
al. (1984). Projected inputs for sludge incineration
assume present-day partitioning characteristics

This section first examines the quantities of Cd currently
entering arable land from sludge application in the U.K. The values
obtained are placed in perspective by comparison with the other major
sources of Cd input to U.K. agricultural land - atmospheric deposition
and the application of phosphate fertilizers (Hutton 1982). The
situation is examined at the national level and also for the two water
authorities responsible for the majority of sludge dumped at sea:
Thames Water Authority (TWA) and North West Water Authority (NWWA)
(DOE/NWC 1983). The increase in Cd inputs resulting from a diversion
of the sludge currently dumped at sea to the agricultural pathway is
examined at the national level and for TWA and NWWA. In the latter
cases, it has been assumed that only the sludge which is currently
being dumped by the two water authorities is applied to the
agricultural land in that authority. It is also assumed that the
increased application of sludge is divided between arable land and
pasture in the same proportions as occur at present.

Table IX Hypothetical impact of abolishing sludge dumping at sea by two U.K. water authorities and the diversion to agricultural land - Cd inputs to arable land

Region	Cd input from sewage sludge (tyr^{-1})		Cd inputs from other sources (tyr^{-1})	
	Present	Hypothetical	Atmospheric deposition	Phosphate fertilizers
North West Water Authority	0.1	0.3	0.4	0.6
Thames Water Authority	1.8	4.0	1.3	2.0
U.K. total	4.7	7.9	15.4	23.6

Source of estimates: Hutton et al. (1984)

Table IX shows the current and hypothetical Cd inputs to U.K. arable land, based on the assumptions described above. At both the national level and in NWWA, sludge application is a relatively small source of Cd. A different situation is apparent for TWA, where Cd inputs from sludge application currently rival those from the two other major sources. The difference between the two water authorities reflects a contrasting agricultural utilization of sludge. In TWA, about 75% of the sludge used for this purpose is applied to arable land and 25% to pasture; in NWWA, the relative proportions are reversed (DOE/NWC 1983).

In the scenario shown in Table IX, sludge application would be the largest source of Cd input to arable land in TWA. This may pose problems both with respect to adhering to the DOE guidelines for agricultural sludge application and in relation to increased soil and crop Cd levels. However, it should be stressed that in both TWA and in the U.K. as a whole, only a small proportion of arable land actually receives sludge applications (Davis 1984). This is in contrast to the other two sources which, by their nature, discharge Cd to all arable land. Thus, the marketing and distribution of crops grown on sludge-amended soils will dictate the impact of the situation depicted in Table IX on the dietary Cd exposure in local populations. Currently, work at MARC is being undertaken to quantify the relative importance of this pathway.

5. CONCLUSIONS

This paper, in common with other inventories of metal discharge to the
North Sea, has drawn attention to the small contribution made by
sludge disposal to the total inputs of metals in this area. However,
it is considered that the use of such broad comparisons to evaluate
the significance of sludge dumping can be misleading. This is because
the environmental impact of metal inputs from sources such as dredged
spoils is questionable, given the modest degree of metal enrichment in
this material. Concern should instead be directed to those sources
discharging materials highly enriched in metals, as these can lead to
elevated concentrations in sediments and the water column. It is also
questionable whether discrete "point source" metal discharges such as
sludge dumping and diffuse inputs from rivers are directly comparable.
Indeed, it is not clear at present which of these forms of metal input
is of greater environmental significance, with a high dilution but
large area affected on the one hand, and a restricted dispersal but
high concentration on the other. An analagous problem exists with the
agricultural application of sludge and human exposure to Cd via crop
uptake. This relates to the dilemma of either applying sludge and Cd
at low rates to a large land area or at high rates to a smaller area
(Ryan et al. 1982).

One source of metal inputs to the North Sea highlighted in this
paper was the direct discharge of sewage by the U.K. This source was
responsible for considerable inputs of the Annex I metals, Hg and Cd.
Given the size of these inputs, the corresponding sewage discharge
data from other countries bordering the North Sea should be obtained.
Similarly, information on direct industrial discharges from these
countries is incomplete and needs to be obtained to construct a
complete inventory for the area.

Examination of the sources of metals in U.K. coastal waters again
revealed a large contribution from direct sewage discharges,
particularly for Cd. Closer inspection of the U.K. pattern of sewage
discharge and sludge dumping revealed an irregular distribution of
inputs. The Thames Estuary, in particular, receives about half the
national total of Cd in sewage discharges and about 90% of the Cd in
sludge dumped at sea. This finding underlines the deficiencies of
broad-based inventories in which localized inputs would not be
detected.

The two simple scenarios, developed in response to a hypothetical
ban on the marine disposal of sludge and sewage, ignored several
important economic and technical constraints. Nevertheless, the
results obtained do give an indication of the magnitude of metal
inputs resulting from increases in sludge incineration and
agricultural utilization in the U.K. It was found that even in the
worst-case situation for sludge incineration, atmospheric discharges
of metals and inputs to landfill would be no larger than are currently
experienced from the incineration of refuse in the U.K. It was
predicted that a diversion of the sludge currently dumped at sea to
U.K. agricultural land would result in larger Cd inputs than at
present but these are considerably smaller than inputs from other

sources. Once again, however, examination of inputs at the national
level can conceal local problems, as illustrated by the situation in
TWA. Here, total Cd inputs from sludge application to arable land
currently rival those from other sources. In the scenario developed
in this paper, sludge application would result in larger Cd inputs
than the other sources put together.

It must be stressed that the simple methods used to examine the
impact of abolishing sludge disposal at sea have produced only
preliminary results. Nevertheless, the approach illustrates the need
to explore fully the implications of such a ban for all sectors of the
environment.

6. REFERENCES

Council of the European Community 1982 Proposal for a Council directive
  on the use of sewage sludge in agriculture, Offical Journal of the
  European Communities No.C 264, 13-8.
Critchley, R. F. 1983 An assessment of trace metal inputs and pathways
  to the marine and terrestrial environments. In: Heavy Metals in the
  Environment, Heidelberg, CEP Consultants, Edinburgh, pp.1108-1111.
Davis, R. D. 1984 Cadmium in sludges used as a fertilizer, Experientia
  40, 117-126.
Davis, R. D. and Coker, E. G. 1980 Cadmium in Agriculture, with
  Special Reference to the Utilization of Sewage Sludge on Land, Water
  Research Centre Technical Report TR 139.
DOE/NWC 1978 Report of the Sub-Committee on the Disposal of Sewage
  Sludge by Incineration, Standing Committee on the Disposal of Sewage
  Sludge, Department of the Environment/National Water Council.
DOE/NWC 1981a Report of the Sub-Committee on the Economics of Sewage
  Sludge Disposal, Standing Committee on the Disposal of Sewage
  Sludge, Department of the Environment/National Water Council.
DOE/NWC 1981b Report of the Sub-Committee on the Disposal of Sewage
  Sludge to Land, Standing Committee on the Disposal of Sewage Sludge,
  Department of the Environment/National Water Council.
DOE/NWC 1983 Sewage Sludge Survey 1980 Data, Standing Committee on the
  Disposal of Sewage Sludge, Department of the Environment/National
  Water Council.
EPA 1984 Environmental Regulations and Technology. Use and Disposal of
  Municipal Wastewater Sludge EPA 625/10-84-003, U. S. Environmental
  Protection Agency, Washington, D.C.
ENDS 1984 U.K.'s defensive efforts pay off at North Sea conference,
  Environmental Data Service 118, 23-24.
Hill, J. M., Mance, G. and O'Donnell, A. R. 1984 The Quantities of Some
  Heavy Metals Entering the North Sea, Water Research Centre Technical
  Report TR 205.
Hutton, M. 1982 Cadmium in the European Community. MARC Report No.26,
  Monitoring and Assessment Research Centre, Chelsea College,
  University of London.
Hutton, M., Chilvers, D., Coleman, D. and Symon, C. 1984 Trace Metals

in the U.K.: Sources, Pathways and Human Impact, Water Research
     Centre Contract No.SO 4140RX.
ICES 1983 Flushing Times of the North Sea, ICES Cooperative Research
     Report No.123, International Council for the Exploration of the Sea,
     Copenhagen.
Jenkins, S. H. 1982 Disposing of Sewage Sludge to Sea, Mar. Pollut.
     Bull. 13, 37-39.
Mance, G, and O'Donnell, A. R. 1984 The magniture of land-based
     discharges of heavy metals entering coastal waters of the United
     Kingdom and the North Sea. Environmental Contamination,
     International Conference 1984, CEP Consultants Ltd, Edinburgh, pp.
     542-8.
Martin, J. and Meybeck, M. 1979 Elemental mass-balance of material
     carried by major world rivers, Mar. Chem. 7, 173-206.
Norton, R. L. 1982 Assessment of Pollution Loads to the North Sea,
     Water REsearch Centre Technical Report TR 182.
Oslo Commission 1984a Ninth Annual Report, Oslo Commission, London.
Oslo Commission 1984b Eighth Annual Report, Oslo Commission, London.
Oslo and Paris Commissions 1984 The First Decade, Oslo and Paris
     Commissions, London.
Paris Commission 1984 Sixth Annual Report, Paris Commission, London.
Royal Commission on Environmental Pollution 1984 Tenth Report.
     Tackling Pollution - Experience and Prospects, Her Majesty's
     Stationery Office, London.
Ryan, J. A., Pahren, H. R. and Lucas, J. B. 1982 Controlling cadmium
     in the human food chain: a review and rationale based on health
     effects, Environ. Res. 28, 251-302.

FRESHWATERS AS WASTE DISPOSAL SYSTEMS: AN INTERPRETATION OF THE EXPERIMENTAL LAKES AREA, CANADA WHOLE-ECOSYSTEM EXPERIMENTS

R. H. Hesslein, D. W. Schindler, S. E. Bayley
and G.J. Brunskill
Department of Fisheries and Oceans
Freshwater Institute
501 University Crescent
Winnipeg, Manitoba, Canada   R3T 2N6

ABSTRACT. Over the past 16 years a variety of whole-ecosystem experiments have been carried out at the Experimental Lakes Area, northwestern Ontario, Canada. Although these experiments were designed to study the impact of contamination or perturbation of the systems by pollutants, the results have been interpreted in terms of the effectiveness of these systems for waste disposal. We include discussion of the addition of radiotracers of metals, lake acidification with sulfuric and nitric acids, fertilization with phosphorus, nitrogen and carbon and acid irrigation of a bog area. Many detailed investigations into the processes controlling the ability of watersheds and lakes to "dispose" of pollutants are discussed along with some successful modelling of these processes. Important interactions between contamination by metals and concurrent eutrophication or acidification are demonstrated by our experiments. Long-term whole ecosystem studies are essential for resolving pollution questions in view of the natural variability observed. We suggest regulations require background data and strict bookkeeping of actual waste disposal so that the impacts may be assessed in a scientific manner.

1. INTRODUCTION

Disposal of wastes in freshwater ecosystems poses a problem not shared by terrestrial and marine disposal: Humans rely on freshwaters for drinking water. In most areas of the world, freshwaters are so scarce that the same water bodies must be used for drinking and waste disposal.

Finding solutions to this dilemma has been a high research priority with the Canadian Department of Fisheries and Oceans for many years. In addition to the above problems, contamination of fishes, which are an important economic resource and often form the primary source of protein for native people, has been subject of concern. One method of researching this problem has been to deliberately add pollutants in known amounts and at known rates to

139

G. Kullenberg (ed.), The Role of the Oceans as a Waste Disposal Option, 139–163.

small freshwater lakes in a remote area, carefully documenting
rates, pathways, and mechanisms of removal, effects in key species
in the food web, and bioconcentration of toxic pollutants in species
which are used by man. This work has been done in the Experimental
Lakes Area of northwestern Ontario for sixteen years.

Several categories of pollutants have been studied in a variety
of experiments. These include sulfuric and nitric acids,
phosphorus, nitrogen and carbon at a variety of rates and ratios,
and reactor-produced and fallout radionuclides. Studies of natural
control systems have been carried out concurrently with the
experiments enabling us to contrast pollutant effects with natural
variations in the chemical and biological components. While the
experimental designs of these studies were not strictly
intercomparable, because they were done with a variety of objectives
in mind, it is possible to compare a number of common factors which
relate to the topic of this conference.

A scientific basis for options of disposal of wastes resulting
from human activities can be considered initially as a choice of the
level of entropy or dispersion. The low entropy option is
characterized by containment of highly concentrated wastes. This
has the advantages of limited space, possibility of relocation, and
highly controlled distribution. The disadvantages are a high
potential for accidental release, highly concentrated release, and
site protection problems. Natural environments used for low entropy
disposal can be considered as sites only, not disposal systems.

The use of natural environments as disposal systems, as we
describe below, is the high entropy option. This approach is
characterized by wide dispersion or small inputs resulting in low
concentrations with lower chance for sudden catastrophic changes
in the status of the system. However, recovery or removal of
materials is very difficult and expensive. Long-term failures in
the disposal system are difficult to reverse, and local decreases in
entropy (i.e. concentration increases) are difficult to control.

Depending on the rate and amount of a pollutant added, it can
perturb an ecosystem by toxifying one or more species in the food
web or by disrupting geochemical cycles, or it can merely
contaminate the system by leaving detectable residues in major
ecosystem compartments without harming ecosystem function. Bormann
(1) considered the contamination phase to be an early or low-level
stage of pollution, assigning it to "stage 0" of a three stage
rating of ecosystem responses to stress. Some of our studies were
of this type, while others were carried out to perturb whole
ecosystems or their key components. For the purposes of this paper,
we shall simply refer to the two types of pollutants or levels of
pollutant addition as contaminants and pollutants. The effects of
pollutants and the pathways of both pollutants and contaminants
cannot be determined with confidence either from laboratory studies,
or from the study of unpolluted systems. Interactions of pollutants
with the biogeochemical cycles of other elements, the effects of
diffusion, oxidation-reduction reactions, particle settling,
food-chain pathways, and interactions between pollutants are

examples of ecosystem-level reactions which are difficult to predict without studying ecosystems before and after they are polluted.

2.   RESULTS

2.1.   Radionuclides as Contaminant Tracers

A series of experiments using reactor-produced radionuclides as tracers in tube enclosures and whole lakes was begun in 1975 using four 10 metre dia tubes attached to the sediments of Lake 223 in 2 metres of water (2). This experiment served as a pilot for subsequent whole lake experiments in Lake 224 in 1976 (3) and Lake 226 in 1977 (4). A large number of experiments were carried out in 1 metre dia. tubes during 1981-1983 (5) to examine in detail some of the processes and rates observed in the whole lake experiments.

The early experiments concentrated on determining the pathways followed by the tracers and the rates of movement. The primary sink for all of the tracers removed from the water was the sediment. The removal, expressed as first-order rate coefficients, ranged from 0.03-0.13 d^{-1} in the Lake 223 tubes (2) to 0.008-0.074 d^{-1} in the whole lakes (3, 4) (Table 1). Particle settling and direct sorption to the sediments were identified as the important pathways to the sediments. The relative importance of each to the individual tracers was found to be determined by the degree of association of the tracers with particulate material which was established rapidly (<2 days) (5) and remained relatively constant throughout the experiments. In general, radionuclides with a high affinity for suspended particles were removed from the lakes by settling. These often appeared in the hypolimnetic sediments, even if additions were made to the surface waters of thermally stratified lakes. Some of these (Co60, Fe59) were remobilized from hypolimnetic sediments, apparently due to redox reactions (4). In contrast, radionuclides which were not associated with particles were confined to the epilimnion, with removal to epilimnetic sediments being the most important sink. Sorption to sediment surfaces, rather than particle settling, appeared to be the most important process. The very limited mixing across the thermocline (6, 7) prevented these tracers from reaching the hypolimnetic sediments until the thermal stratification disappeared with fall mixing of the lake, Cs134 is the most conspicuous example (3, 4).

In the tubes in Lake 223 the effect of decreased pH (four tubes at pH's of 5.1, 5.7, 6.7, 6.8) was to decrease the association with particles for isotopes of cobalt, iron, manganese, and zinc. Even the largest shift from particles to solution, from >80% filterable to <1% in the case of manganese, altered the removal rate by less than a factor of two. By altering the affinity of isotopes for particles, acidification of a whole lake could cause the "sink" for some radionuclides to shift from the hypolimnion to the epilimnion.

Analyses of sediment cores from lakes 224 and 226 (3, 4) showed that the deposition of tracers was quite variable even within

Table 1. First-order removal rates for radioisotopes from the water columns of sediments of enclosures and whole lakes at ELA (all units d^{-1}).

Isotope	Lake 223 tubes[1]				Lake[2] 224	Lake[3] 226N	Lake[3] 226S
	B (pH 5.1)	C (pH 5.7)	D (pH 6.7)	E (pH 6.8)			
^{74}As	0.061	0.053	0.035	0.056			
^{133}Ba	0.047	0.064	0.037	0.046			
^{134}Cs	0.051	0.065	0.047	0.049	0.016	0.042	0.042
^{51}Cr	0.108	0.047	0.062	0.051			
^{60}Co	0.087	0.112	0.102	0.105	0.025	0.056	0.052
^{59}Fe	0.081	0.126	0.064	0.098	0.039	0.074	0.063
^{54}Mn	0.060	0.072	0.105	0.099			
^{203}Hg	0.043	0.062	0.054	0.055	0.040	0.063	0.063
^{75}Se	0.032	0.045	0.028	0.028	0.008	0.032	0.027
^{85}Sr						0.032	0.023
^{228}Th	0.062	0.066	0.072	0.060			
^{48}V	0.128	0.119	0.055	0.088			
^{65}Zn	0.055	0.071	0.082	0.081	0.019	0.053	0.034

[1] (2)
[2] (3)
[3] (4)

a small area but that the ratios of the tracers in the sediments at a single depth horizon was quite consistent.

The tracer additions to Lake 226 (4) provided an opportunity to investigate the effect of elevated trophic status (see section on eutrophication) on the geochemical pathways and rates and to study the behavior of tracers as influenced by the anoxic conditions existing in the hypolimnia of both subbasins. Tracers <60% associated with particles in oligotrophic Lake 224 (Se, Hg, Cs, Zn) showed increased association with particles while the filterable percent of Fe and Co decreased slightly (Table 2). Desorption of Fe, Co, and Hg tracers from sedimenting particles was observed in both hypolimnetic basins, producing hypolimnetic concentrations higher than the initial concentrations of the epilimnia. These concentrations disappeared with the introduction of oxygen during fall mixing but recurred with anoxia during the winter.

2.1.2. Modelling Radionuclide Results. Although simple physical-chemical models had been proposed to explain the fates of the tracers in the early tube and whole lake experiments, a very significant improvement in the understanding of the controlling process was made in the analyses of the data from the 1 metre dia. tubes (5, 8, 9). These studies also made use of laboratory experiments on the kinetics and equilibria of sorption of tracers to suspended and sedimented particles. Because of the complex interactions in the model, numerical finite difference methods were used, and so only the basic features of the models will be presented here. A tracer entering the system is partitioned by the model between the particulate and dissolved phases by a sorption and a desorption rate coefficient (the ratio of which represents the more common distribution coefficient, Kd) (5). Particles are removed according to their settling velocities and dissolved phases by direct sorption to the sediments. The sediment sorption in the model is controlled in part by molecular diffusion through a boundary layer at the sediment water interface (5, 10). The thickness of this layer has been determined by analyses of the tracer data and by the measurement of the dissolution of gypsum plates placed in the lake sediments. The value is in the range of 0.4-4.0 mm in the ELA lakes (5, 4). Direct sorption to sediments is also controlled by the sediment sorption/desorption coefficients. Within the model sediments, processes of diffusion in pore waters and bioturbation of the solid phase are included. This model has a high degree of success in explaining the fates of tracers of sixteen different elements in the tubes and seven in the whole lakes (5, 4).

2.1.3. Fallout and Natural Radionuclides. In addition to the above described short-term studies, investigations of the longer term tracers, Cs137 from atmospheric thermonuclear testing and the radionuclides of the U238 decay series have been carried out since 1979 (11). These studies have been aimed at establishing a mass balance for Lake 239 at ELA involving atmospheric and stream inputs, outflows and estimates of sediment accumulation. For all but U238

Table 2. Percent of radioisotopes associated with suspended particulate material (>0.45 μ).

Isotope	B (pH 5.1)	C (pH 5.7)	Lake 223 tubes[1] D (pH 6.7)	E (pH 6.8)	Lake[2] 224	Lake[3] 226N	Lake[3] 226S
74As	2	15	10	0			
133Ba	0	3	10	5			
134Cs	10	14	12	15	3	7	6
51Cr	54	79	37	67			
60Co	6	35	97	88	90	68	77
59Fe	63	98	99	99	98	88	83
54Mn	1	20	100	68			
203Hg	58	67	57	52	56	74	70
75Se	37	45	39	35	30	41	47
85Sr	-	-	-	-	-	0	1
228Th	30	35	0	41			
48V	61	73	48	44			
65Zn	5	23	75	50	15	67	70

[1] (2)
[2] (3)
[3] (4)

the calculated sedimentation exceeds the total input (Table 3).
This may be due in part to the different time scales of
measurement. The inputs are based on three years of data while the
sediment data represent an average over about ·30 years.
Alternatively, the rate of supply of nuclides in mineral components
to the sediments may not be reliably measured in the inputs, i.e.
suspended particles entering as dust, shore erosion or stream
bedload may escape measurement in our studies. An estimate of lake
retention of the nuclides was made by (Inflow-Outflow)/Inflow.
These values (Table 4N) show that only Pb210 is very effectively
retained by the lake. The rate coefficients for removal to the
sediments, ((Inflow-Outflow) /(Mass in the lake)), is also given in
Table 4).

2.1.4. Comparison of Instantaneous and Continuous Additions of
Radionuclides. There is an apparently large discrepancy between the
rate coefficients for removal to sediments for nuclides in the short
term experiments (5, 4) and those derived from the Lake 239 mass
balance (11) (Table 4). In fact, the results are entirely
consistent, once the differences in the availability of sediments
for sorbtion are taken into account. The availability of sediments
for sorbtion of a single addition of tracer includes the suspended
particles plus hundreds of years of deposits which are accessed by
the processes of diffusion and sediment mixing. The sediment
material available to the continuously added tracer consists only of
the steady state supply of particles to the lake, as all the
previously deposited particles have already equilibrated with the
tracer. Based on calculations from Pb210 sedimentation rates the
supply of particles to Lake 239 is about 100 g m^{-2} yr^{-1} or 0.27 g
m^{-2} d^{-1} (11, 12). ·Because the lake has a mean depth of 10 metres
the mean contact time for particles with water is 2.7 * 10^{-8} g cm^{-3}
d^{-1} (sed. rate/mean depth). Although Kd values for uranium,
thorium, and radium have not been experimentally determined for Lake
239 sediments they can be estimated from the data for filterable
portion of the nuclides (11) and compared to other nuclides for
which Kds have been determined (5, 8). Pb210 is highly
particle-reactive and probably has a Kd = 10^5 cm^3 g^{-1}. Th228 had
intermediate association with particles in the Lake 223 tube
experiments and in Lake 239 which suggests a Kd = 2 * 10^4 cm^3 g^{-1}.
Ra226 and uranium are not highly associated with particles and
probably have Kd = 5 * 10^3 cm^3 g^{-1}. The product of these values and
the contact times gives removal rates to the sediments respectively
at 0.0027 d^{-1}, 0.00054 d^{-1} and 0.00014 d^{-1}. The agreement of these
estimates with the values in Table 4 is quite good, considering the
approximations involved. The response to a change in the input rate
of contaminants to a lake would be expected to be initially
controlled by the equilibration with the accumulated sediments but
in the long term by the sedimentation rate.

2.2 LAKE ACIDIFICATION

Table 3. Estimates of input, output, and masses of natural radionuclides to Lake 239 ELA (all units Bq m^{-2} yr^{-1} except mass in lake Bq m^{-2}).[1]

	^{238}U	^{234}U	^{230}Th	^{226}Ra	^{210}Pb	^{232}Th
Stream inflow (I)	8.1	6.0	0.19	1.3	3.5	0.28
Direct atmospheric input (A)					90	
Outflow (O)	3.7	4.5	0.081	0.74	10	0.24
Sedimentation (S)	7.3	9.4	2.5	2.6	125	3.5
Accumulation (I+A-O)	4.4	1.5	0.109	0.56	83.5	0.04
Mass in lake	20	25	0.44	4.0	64	1.9

1(11)

Table 4. Retention$_1$ ((Inflow-outflow)/Inflow in %) and removal rate coefficients ((Inflow-outflow)/mass in lake in d^{-1}).[1]

	^{238}U	^{234}U	^{230}Th	^{226}Ra	^{210}Pb	^{232}Th
Retention	54	25	57	43	89	14
Removal	0.00060	0.00016	0.00068	0.00038	0.00357	0.00005

1(11)

Whole lake experiments to investigate the responses of the
geochemical cycles and biological communities to acid rain
were initiated in 1976 with the acidification of Lake 223 by
addition of sulfuric acid. The pH of the lake was gradually lowered
over the period of 1976-1983 from a natural value of 6.8 to 5.0
(13) (Table 5). In 1982 two additional acidification experiments
were initiated in the two basins of Lake 302, one with the addition
of sulfuric acid and one with nitric acid.

Table 5. Sulfuric acid additions and pH values for Lake 223 1975 to
1983 (lake volume = 19.51×10^5 m^3).[1]

Year	H_2SO_4 added (Kg)	Mean epilimnion pH
1975	-	6.8
1976	9968	6.49
1977	5239	6.13
1978	6799	5.93
1979	5069	5.64
1980	5576	5.59
1981	6392	5.02
1982	6222	5.09
1983	4014	5.13

[1](14).

The effects of acidification on lakes must be mediated by the
effect on the buffering (alkalinity) systems of the lakes. Prior to
the Lake 223 acidification, two hypotheses prevailed concerning the
ability of lakes to buffer acid inputs. The first was that
alkalinity was a conservative property in the lake, produced in the
terrestrial drainage basin by weathering processes and flowing out
of the lake concentrated primarily by evaporation. The second was
the assumption that built up reserves of buffering capacity in the
lake might be considerable, but that these would eventually be
depleted and only the steady state supply from the inflows would
remain. Both of these hypotheses have been shown to be false,
although our understanding of the processes is not yet complete.

We have shown that significant differences between the annual
inflow and outflow masses of titrated alkalinity exist for Lake 239,
a control basin (14). This requires internal alkalinity generation
of >100 meq m^{-2} yr^{-1} based on 9-13 years of records. Ion exchange
reactions in the sediments yielded Ca, Mg, and K to the lake,
accounting for respectively 50%, 10%, and 8% of the internal
alkalinity generation (14). Biological uptake and possibly
denitrification of nitrate accounted for another 17%. Sulfate
reduction and iron sulfide precipitation in the sediments accounted
for 13% (Table 6) (14). While the precipitation of iron sulfides as
a consequence of microbial sulfate reduction is a contributor to
alkalinity within Lake 239, which is well oxygenated throughout most
of the water column, its influence is greater in the anoxic
hypolimnia of other lakes (15, 16). In addition sulfate reduction
also increases in proportion to sulfate concentrations, so that it

becomes the predominant alkalinity generating mechanism in lakes acidified with sulfuric acid (16).

Table 6. Within-lake ionic processes affecting the alkalinity budgets of Lake 239 (natural) and Lake 223 (acidified to pH 5 with H_2SO_4). Values are either long-term averages, in meq m^{-2} yr^{-1}, or % of total alkalinity production.[1]

Ion	Lake 239 (meq m^{-2} yr^{-1})	% of alkalinity generated	Lake 223 (meq m^{-2} yr^{-1})	% of alkalinity generated
Ca^{2+}	52	45	55	19
Mg^{2+}	10	9	-23	- 8
Na^+	2	2	1	0
K^+	8	7	- 6	- 2
NH_4^+	-13	-11	-14	- 5
Mn^{2+}	- 1	- 1	7	2
Fe^{3+}	-19	-17	- 6	- 2
SO_4^{2-}	15	13	246	85
Cl^-	17	15	12	4
NO_3^-	17	15	16	5
	101		288	

[1](14)

Investigations at ELA have shown that rather than being depleted, the internal sources of alkalinity respond positively to increased loading of sulfuric and nitric acids (14, 16). In both lake 223 and 302S (sulfuric acid) large increases in sulfate reduction have been documented in response to the increased sulfate concentrations (14). In laboratory experiments using incubations of sediment cores Rudd et al. (17) have shown with sulfate and tracer S35 that sulfate reduction is limited by the sulfate concentration in the overlying water. Cook and Schindler (15) have also shown that rates of sulfate reduction are proportional to sulfate concentration in the hypolimnion of Lake 223. The linear relationship between the rate of reduction and sulfate concentration explains the observed increases shown in alkalinity generation following lake acidification. But the microbial response is not 100% efficient and sulfate reduction in Lake 223 and in thousands of lakes in North America and Scandinavia has not prevented acidification by sulfuric acid in precipitation. Furthermore, because the effectiveness of sulfate reduction as a source of alkalinity is dependent on the availability of iron, it is possible that in time this source will decline with the depletion of sulfide reactive iron reserves. The mass of iron in the sediments probably represents several hundreds or thousands of years of accumulation (18). In Lake 239 this accumulation is 3 g m^{-2} yr^{-1} Fe (19, 12).

The addition of nitric acid to lakes has analogous effects on alkalinity generation (20). 'Both denitrification of nitrate to N_2 and assimilitory nitrate reduction are potential buffering reactions (20):

$$(CH_2O)_{106}(NH_3)_{16}(H_3PO_4) + 84.8H^+ + 84.8NO_3^- \rightarrow 106CO_2 +$$

$$42.4N_2 + 16NH_3 + H_3PO_4 + 148.4H_2O$$

$$106CO_2 + 16H^+ + 16NO_3^- + H_3PO_4 + 122H_2O \rightarrow$$

$$(CH_2O)_{106}(NO_3)_{16}(H_3PO_4) + 138O_2$$

The energy gain by bacteria from nitrate reduction is significantly greater than from sulfate reduction (11614 kcal/mole vs. 1605 kcal/mole). As shown in sediment core incubations this leads to very effective utilization of nitrate by microbes (21). In addition, nitrate is in great demand ·by photosynthesizing plants. This effective utilization is apparent in the very large buffering response documented in the nitric acid addition experiment in Lake 302N (Figure 1). The denitrifying bacteria have been shown to be inhibited by low pH, however, and with sufficiently high rates of acidification it is possible that this buffering mechanism may also fail as its active zone is in the top few millimeters of the sediments, the first zone to suffer a decrease in pH.

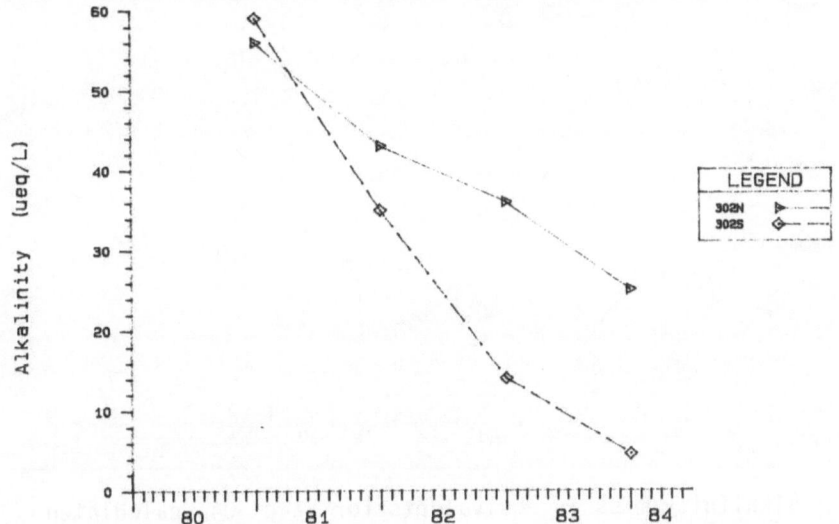

Figure 1. Mean annual alkalinity values for the north and south basins of Lake 302 showing the greater effect of sulfuric acid. The first year, 1981 is prior to addition of nitric acid to Lake 302N and sulfuric acid to Lake 302S.

In Lake 239, as well as in other lakes, there are alkalinity-consuming as well as alkalinity-producing reactions at work, and the net alkalinity generation is a balance between the two. As in the case of sulfate and nitrate reduction, many of these reactions are biologically-mediated.

A graphic illustration of how efficient such processes can be, was provided by the acidification of Lake 304, which decreased from pH values >6 to 4.5 due to addition of ammonium chloride (22). In the first phase of the experiment, lasting two years, phosphate was added with the ammonium, and the primary mechanism consuming alkalinity was the assimilation of NH_4^+ by phytoplankton during eutrophication. Similar mechanisms caused the acidification of Lago d'Orta, Italy due to inputs of silk factory effluents (23).

In a second two-year stage, Lake 304 was fertilized with NH_4Cl but not phosphate. The efficiency of acidification by assimilatory nitrogen uptake was reduced, but after NH_4^+ accumulated to about 2 mg L^{-1}, a strong pulse of acidification accompanied nitrification of the ammonium under winter ice (22).

In a final two year phase, the alkalinity of the lakes was rejuvenated by adding phosphate and nitrate, generating considerable alkalinity, as expected. All three phases of the experiment are shown in Fig. 2.

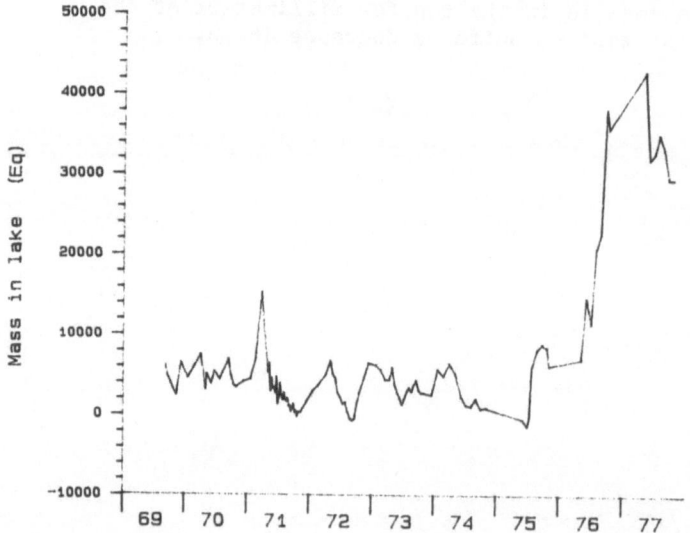

Figure 2. Alkalinity mass in equivalents for Lake 304, calculated from pH and dissolved inorganic carbon with CO_2 - assumed to be in equilibrium with the atmosphere. NH_4Cl was added as fertilizer in 1971-72, none in 1973-74 and $NaNO_3$ in 1975-76. (Data from 22).

2.3   EUTROPHICATION

The ELA eutrophication studies which began in 1969 consisted of a
series of whole-lake experiments supported by many detailed
investigations of specific processes through the use of in situ
enclosures, isotope tracers, and laboratory experiments. The major
goal of this work was to demonstrate the relative importance of
increased loading of phosphorus, nitrogen, and carbon in raising the
trophic status of natural oligotrophic lake systems, forming a
rational and quantitative basis for the control of sewage effluents
to the Laurentian Great Lakes and other Canadian fresh waters. The
single most important conclusion of the experiments was that
controlling phosphorus always controlled the degree of
eutrophication, independent of its rate of supply. This was partly
due to the efficiency with which phosphorus was removed to and
retained in the sediments. It was also because biogeochemical
mechanisms were able to compensate for deficiences in nitrogen and
carbon at any tested level of phosphorus input (24, 25). As a
result, the degree of eutrophication could be predicted quite
accurately by very simple models based on phosphorus input and water
renewal (26). Further goals of the research were to determine
whether the efficiency of sedimentation of phosphorus would remain
constant over years of continued loading, and how the potential for
lake recovery would be affected by prolonged artificial
eutrophication.

Recent assessment of the longest-running whole lake fertilization
experiment, that of Lake 227, has shown that the efficiency of
sedimentation of the added phosphorus has not changed over a 15 year
period (21) (Figure 3). Consistently 90% of the P has gone to the
sediments while the remaining 10% has contributed to a buildup in
the water column three times the pre-fertilization concentration.
The P concentration in the lake is controlled by the P input and the
water renewal of the lake (21, 26).

In an earlier discussion of the P sedimentation in Lake 227 it
had been postulated that the high efficiency was partially due to
large reserves of old sediments being made available for sorption by
bioturbation (27). A recent change in the lake has allowed us to
re-evaluate this possibility. Apparently, because of the increased
extent and duration of anoxic conditions brought on by the high
level of productivity, the bioturbation has ceased for at least the
past 8 years, as indicated by analyses of the paleodiatom record in
the sediments (28). Contrary to our expectation, however, the
cessation of bioturbation did not alter the efficiency of P
sedimentation even though the concentrations of phosphorus in
sediments have increased considerably (21). The steady-state supply
of P-binding capacity to the sediments is therefore sufficient to
maintain the efficient P removal under the current experimental
loading regime (0.5 g P m^{-2} y^{-1}), even though the mud-water
interface in the hypolimnion has been anoxic for several years.
While higher loading rates might cause saturation of P-binding

capacity in the sediments, such loadings would be unusual in Canadian lakes.

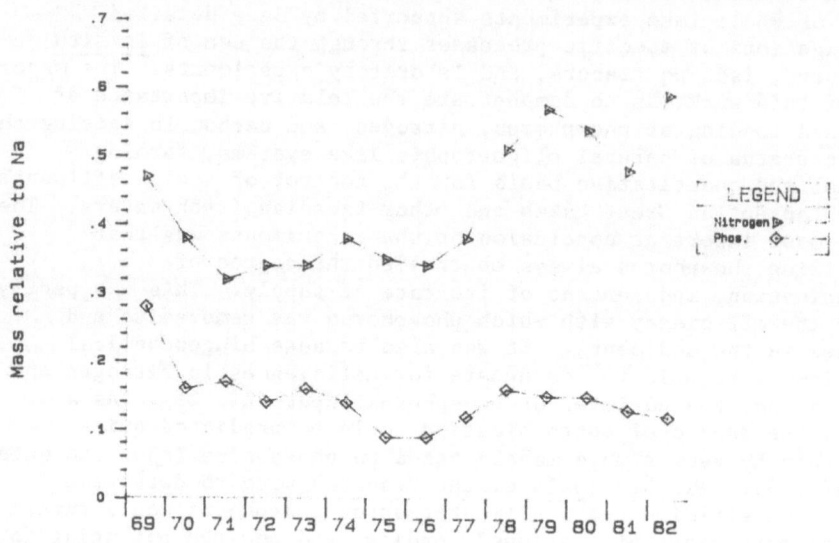

Figure 3. Mean annual masses of nitrogen and phosphorus in the water column of Lake 227 relative to (divided by) their expected masses had they acted like the conservative tracer, sodium (Na would have a value of 1.0). The lake has been fertilized with phosphoric acid and sodium nitrate since 1970. (Data from 21).

In terms of the influence on downstream lakes, the quality of the P in the outflow may be as important as the quantity. The modelling of P dynamics in Lake 227 showed that the lake removed a constant proportion of <u>input</u>, not of <u>mass</u> in the lake as is commonly assumed (21). This result would certainly not be expected if all forms of P in the lake were equally available to the algae which cause sedimentation. We therefore have postulated that the residual P is not in a form available to the algae. This hypothesis is supported by the results of a whole lake addition of the radiotracer P32 to Lake 227 during the summer of 1978 (29). The tracer was rapidly taken up by phytoplankton and bacteria (> 95% after 2 h), because of the great preference for PO$_4$ over other forms of dissolved phosphorus (30). As the experiment progressed, and P was recycled through the various pools, the specific activity of the dissolved fraction (>0.2 μm) never reached more than 50% of the specific activity of the total P. This maximum value was reached within two to three days after addition, indicating that about half of the dissolved P is rapidly turned over and half is essentially inert.

Both basins of Lake 226, separated by a curtain, were fertilized from 1973 until 1981 with P, N and C in the north basin

(226N) and only N and C in the south basin (226S). This experiment
demonstrated the ineffectiveness of N and C alone in causing
eutrophication, and how blooms of bluegreens can appear to help
maintain "Redfield Ratio" proportions of nitrogen to phosphorus
(31, 32). The cessation of fertilization in 1980 has shown how
quickly a lake can recover to pre-fertilization levels once inputs
cease. Within two years the levels of primary productivity and
algal biomass as well as the concentrations of P, N, and C, were
within the envelope of pre-fertilization values (Fig 4).
Phytoplankton which were dominated by Cyanophyceae during the
phosphorus additions, returned to the Chrysophycean-dominated
assemblage which characterizes natural oligotrophic lakes of the
area (33, 31). The water renewal of 0.5 yr^{-1} as well as the
internal nutrient removal mechanisms contributed to the rate of
recovery.

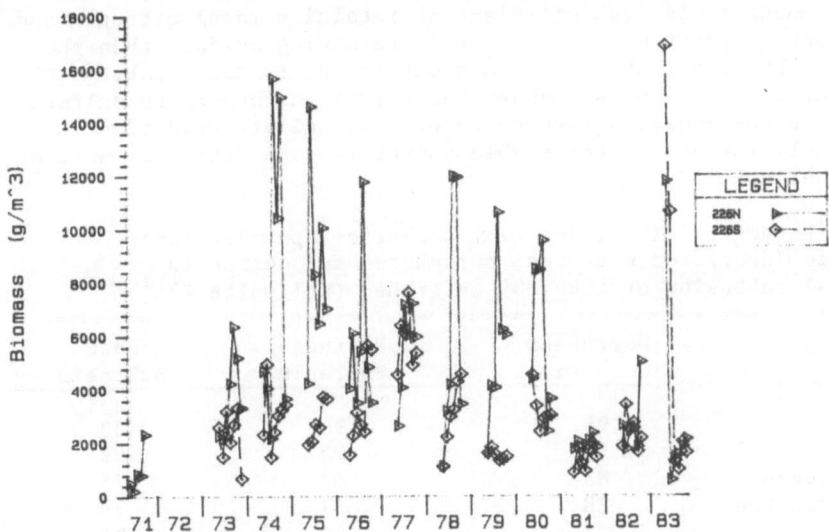

Figure 4. Phytoplankton biomass in Lake 226 north and south basins.
Phosphorus, nitrogen and carbon were added to the north basin and
only nitrogen and carbon to the south basin from 1973 to 1980.
Fertilization ceased in 1980. (Data from 31).

2.4 TERRESTRIAL WATERSHEDS AND WETLANDS

In recent years, some of the effort at the ELA has been shifted to
the study of upland and wetland portions of our watersheds, due to
their influence on the chemical characteristics of downstream
receiving waters. The studies include analyses of the export of all
major cations and anions by the streams and the responses of one
wetland to experimental acidification.

Three watersheds of Lake 239 have been intensively monitored since 1970 for streamflow and output of dissolved chemical constitutents. The sub basins, referred to as East, Northwest, and Northeast contain respectively a higher proportion of wetland area (7%, 8%, 25%) and consequently have more humic character in their outflows. The mean outflow pH's were for the East, 6.1, Northwest, 5.5, and Northeast, 4.9. Precipitation and its chemical composition have also been measured since 1969 (34).

The retention efficiency of the watersheds ((Atmospheric input-Outflow)/(Atmospheric input)) has been very high for the nutrients, nitrogen and phosphorus, and especially for nitrogen as nitrate and ammonia (Table 7). Forest fires in the East and Northeast watersheds in 1974 and in all three watersheds in 1980 caused the release of large amounts of nitrogen and phosphorus immediately after the fires, but recovery of the retention of most forms of N and P was very rapid (35). Based on the 1969-1983 data the Northeast subbasin is less efficient at retaining total nitrogen and total phosphorus, but more efficient at retaining sulfate than the other two subbasins (Table 7). This may be due to the greater wetland influence in the Northeast basin (36) which permits sulfate reduction in the anoxic saturated zone. The sulfate reduction significantly influences the stable sulfur isotope ratios as well as the mass balance (37).

Table 7. Retention efficiency (ΣAtmospheric-Input-Outflow)/ Atmospheric Input) for nitrogen, phosphorus and sulfur in terrestrial subbasins of Lake 239 watershed (all units %).[1]

	Northwest Subbasin	Northeast Subbasin	East Subbasin
Nitrate	95	98	88
Ammonia	98	95	97
Total nitrogen	83	62	75
Total phosphorus	88	57	71
Sulfate	7	66	25

[1](36)

In 1983 an acidification experiment was begun in the wetland portion of the Northeast sub-basin to answer critical questions about the effects of acid precipitation on key chemical and biological processes in wetlands. This area was chosen because of its limited area and because of the high retention of both sulfate and nitrate under natural conditions. A spray irrigation system was constructed to allow the uniform application of water and acid at a rate of 2 mm h^{-1} over 3 ha area. The system intake draws freshwater from the surface of Roddy Lake, an oligotrophic control lake. Acid (54% H_2SO_4, 46% HNO_3 based on H^+ equivalents) is introduced to the main pipe by a metered high pressure pump adjusted to give a pH of 3 in the spray. This low pH reduces the amount of water which must be applied and its application is followed by a rinse of lake water (pH 6-7) to prevent "burning" of the foliage. Starting in mid-1983, acid was added monthly during the ice free season. Each application consisted of one hour of lake water to wet the foliage, 4-5 hours of acid addition, and 30 minutes of lake water rinse.

The nitrate and sulfate concentrations were greatly elevated by the application of acid, but returned to pre-application values in 1-2 days and 1-2 weeks, respectively (Figure 5). Partly because of this removal of strong acid anions and because of other buffering in the bog, the pH depression produced by the acid application is small and of short duration. Most of the nitrate is probably taken up by the bog flora, which normally exists on a nitrogen poor substrate. Preliminary results indicate that denitrification also increases after acidification. In the long term the nitrogen is retained as plant material or is lost to the atmosphere by denitrification. Sulfate from the acidification is either reduced to sulfide and lost

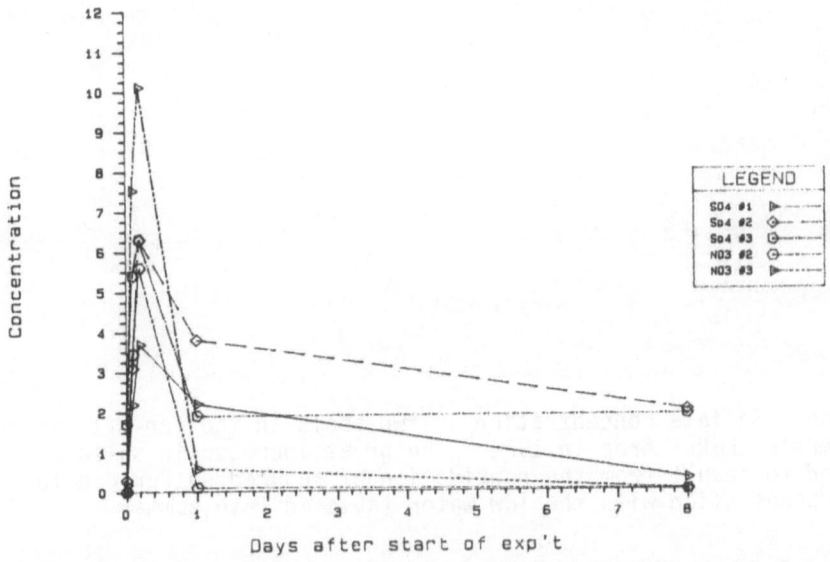

Figure 5. The disappearance of nitrate (mg/10 L) and sulfate (mg/L) in a bog pool during three experimental additions of acid in 1983.

to the atmosphere as H_2S or precipitated with iron, reduced and deposited as elemental sulfur, or fixed as organic sulfur. The sulfur deposited as a sulfide could be reoxidized by penetration of oxygen into the peat as the water table drops during prolonged dry periods. Sulfate reduced in this oxidation could be flushed from the bog with the resumption of high flow resulting in a significant sulfuric acid pulse to the receiving waters.

The oxidation of sulfur compounds to sulfate was documented in 1984 in two fen stations after a prolonged dry period (Fig. 6). The same reoxidation phenomena has occurred in Swedish groundwaters resulting in corrosive concentrations of sulfuric acid in drinking water (37). The change in redox potential resulting from water level fluctuations can also be important in determining the retention by wetlands of P and N in treated sewage (39, 40, 41). As previously mentioned forest fire also affects the retention efficiencies of P and N. This could be an important consideration in the management of terrestrial basins as waste disposal systems.

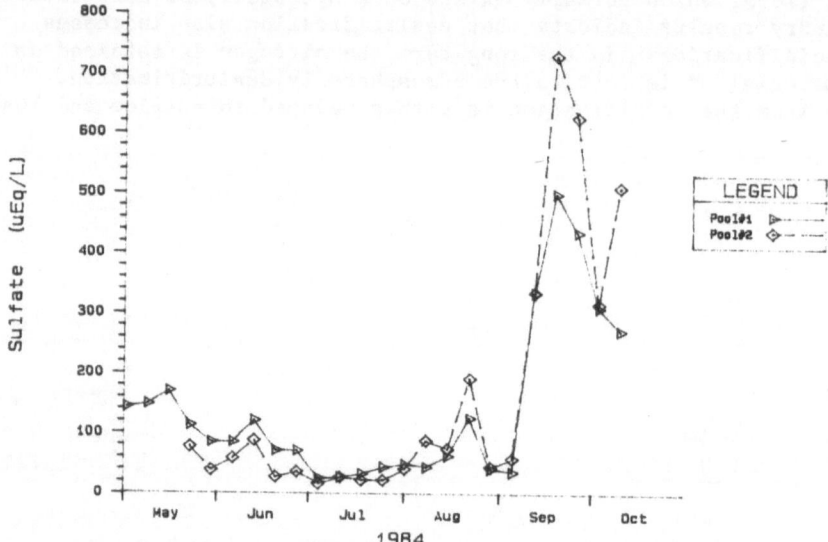

Figure 6. Sulfate concentration in two pools in the fen site at the Experimental Lakes Area in 1984. The great increase in sulfate is believed to result from the reoxidation of reduced sulfur due to oxygen penetration with the low water table in late summer.

3. GENERAL DISCUSSION

In the preceding presentation we have concentrated on the biogeochemical characteristics and processes which control the responses of freshwater and upland watershed areas to waste disposal. We have not dealt with the second part of the waste disposal problem, the consequences with respect to the biological communities and the continued use of the ecosystem by man for recreation, as a food and water source, or, in the cases of some of Human's more widespread waste disposal, as a functioning part of a balanced world ecology. The work at ELA on the consequences of experimental pollutant additions is too extensive to be covered

completely in this presentation, but we must describe a few of the
important biological impacts and some of the synergistic
interactions between the effects of various waste materials.

Although the metal radiotracers added to lakes and enclosures
were in most cases rapidly and efficiently removed to the sediments
this did not prevent significant accumulation in the higher
organisms. ˙Bioconcentration factors ((Concentration in
organism/Concentration in water) at steady state) predicted by
modelling the measured uptake and loss were as high as 30 000 for
some tracers in fish. In most cases, the uptake was rapid and loss
rates very slow. Such high bioconcentration factors are in some
cases due to the low ionic strength of the ELA waters as some of the
tracers were likely taken up as analogues of non-trace ions. This
effect was well documented for the case of Ra226 and calcium.
˙Because of the high efficiencies of sediments for holding metal
contaminants and the low sedimentation rates, the recovery of
contaminated fish stocks can be very slow. This is particularly
evident in the mercury polluted English-Wabigoon river system in
northwestern Ontario. The mercury effluent from a pulp and paper
mill ceased in the early 1970's, but the fish still have
concentrations of 2-4 $\mu g \ g^{-1}$, well above the Canadian limit for sale
of 0.5 $\mu g \ g^{-1}$.

We have shown that the rates of removal of the radiotracers
from the water column, the equilibrium distribution between
particulate and solution phases (Kd), and the importance of reducing
conditions in the lakes are affected by acidification and
eutrophication. Radiotracer affinity for suspended particles was
significantly reduced at lower pH in tubes in 1975 and 1979.
Tracers of Fe, Mn, Co, were most affected. The 1979 experiments
further demonstrated consistent and predictable return to solution
from sediments in acidified tubes. The increased particle abundance
from increased algal growth in artificially eutrophied Lake 226
caused increased association of many radiotracers with suspended
particles relative to oligotrophic Lake 224. Eutrophication also
produces a greater degree of oxygen depletion in the hypolimnia of
lakes which leads to increased release of many metals from particles
and sediments, notably iron, manganese, mercury, cobalt, and radium.

Many physiological and population responses have been
documented in the acidification of Lake 223 including the virtual
disappearance of common species such as opposum shrimp, Mysis
relicta, the fathead minnow, Pimephales promelas, and the crayfish
Orconectes virilis, as well as a decrease in the condition factor of
lake trout (Salvelinus namaycush). These responses have been
ascribed to the direct effects of hydrogen ion or the loss of the
bicarbonate buffer in the lake. Some indirect effects, paralleling
those observed for the radiotracers, such as increased
concentrations of manganese, zinc, and aluminum, may also be
influencing the organisms. These changes could have important
impacts on the biota by enhancing the physiological impact of
hydrogen ion or by contaminating the edible species (13).

The consequences of eutrophication are well known; algal
blooms, odor and poor taste in the water, and oxygen depletion
leading to fish kills. There may be some positive advantages of
controlled eutrophication in the increased production of fish (42),
but the loss of desirable species may also occur.

4. CONCLUSION

Based on our understanding of the functioning of natural ecosystems,
the responses of lakes and watersheds to pollutants, and the
consequences of those responses, we must devise strategies for
choosing which freshwater ecosystems to use for waste disposal. The
susceptibility of lakes to acidification, eutrophication, metal
contamination, etc. must be projected from the relationships between
the physical, chemical, and biological characteristics such as
morphometry, water renewal, sedimentation, thermal regime, major ion
and nutrient chemistry, and species composition. We must further
project the perturbation of these characteristics under the
influence of the pollutant. The fertilization of the hypolimnion of
Lake 302N at ELA is an example of this process of strategical choice
(43). During 1972-1976 we added phosphoric acid, ammonium chloride,
and sucrose to the hypolimnion of Lake 302N at rates similar to
those used in the several other fertilization experiments carried
out in the epilimnia of ELA lakes. We had postulated that the
strong thermal stratification during the period of addition would
prevent the nutrients from reaching the euphotic zone and promoting
algal growth. We also hoped that microbial sedimentation and
sediment sorption of phosphate would deplete the stores of this
critical nutrient in the hypolimnion before the breakdown of
stratification in the fall which would mix the nutrients to the
surface water for the following year's growing season.

The 302N fertilization scheme resulted in increases in primary
productivity and algal biomass for the whole lake of only 10-21% of
those observed in the lakes fertilized in the epilimnia. The
response in the epilimnion of Lake 302N was just 5-8% of that in the
other lakes (Table 8). Oxygen consumption in the hypolimnion of
Lake 302N was increased by the fertilization, but this was probably
due to the addition of reduced nitrogen and carbon compounds. The
use of well oxidized wastes should overcome this problem. This
method of hypolimnetic disposal could also be useful with metals, as
sediment sorption and sedimentation could retard or prevent the
movement into the euphotic food web. Because the Lake 302N
experiment lasted only five years it is unlikely that the system had
reached equilibrium with the fertilization. It is possible that the
lake would not continue to respond as effectively to long term
fertilization.

We believe that whole ecosystem experiments are ideal ways to
assess the potential options for waste disposal. We also must
stress the importance of carrying out these kind of experiments over
long periods of time. Many of the experiments that we have been
conducting may seem long-term to some researchers and certainly to

Table 8. A comparison of nutrients chlorophyll, phytoplankton volume and production in the epilimnia during the ice-free season. Data re 5 yr averages except for production, 3 yr.

$$Eff = \frac{(L302N - L302S)}{(L227 - L302S)} \times 100^1$$

	Lake 302S	Lake 302N	Lake 227	Eff
ΣP, mg·m^{-3}	7.6	13.8	39.8	19
ΣN, mg·m^{-3}	283	358	833	14
ΣC, mg·m^{-3}	1649	1878	5389	6
Chlorophyll a, mg·m^{-3}	4.0	6.1	38.8	6
Phytoplankton volume, mm^3·m^{-3}	1797	2394	9499	8
Phytoplankton production, gC·m^{-3}·yr^{-1}	25.7	39.3	172.1	9

1(43)

agencies interested in fast answers to waste disposal problems. However, in the Canadian Shield lakes it typically takes 20-50 years to bury sediments to a depth below the depth of bioturbation. For many pollutants a new steady state will not be achieved in less time. Many other features of ecosystems cannot be deduced from small-scale or short-duration experiments. For example, laboratory eutrophication bioassays are frequently interpreted as indicating that control of nitrogen inputs to lakes are necessary, ignoring the species changes and atmosphere-water exchanges that help ameliorate nitrogen shortages in natural lakes. Likewise, it would be difficult to develop a laboratory study which would include the complex food-web interactions which led to the decline of lake trout in Lake 223, over a nine year period (44).

We realize that these kinds of experiments cannot be performed for all potential waste disposal sites. In fact, many "experimental" perturbations and contaminations are being carried out all over the world by actual waste disposal. Unfortunately, these "experiments" are very poorly controlled and a great deal of potentially useful information is lost. These could be useful if 1) background information was available and 2) if regulations for waste disposal required that precise records of effluent concentration and flow be kept. This would allow the scientific assessment of the pathways and effects of the pollutants. Accurate mass balances could be calculated and the efficiency of the disposal system continuously reassessed. There is no logic to the present regulatory structure which forces scientists studying polluted ecosystems to spend considerable effort reconstructing, with much uncertainty, input records which could be easily monitored.

5. ACKNOWLEDGEMENTS

We would like to thank M. Turner, D. Findlay and K. Mills for
providing information for figures and tables. L. Wilson was of
great help in organizing the manuscript.

6. REFERENCES

1. Bormann, F. H. 1982. 'The effects of air pollution on the
    England landscape.' Ambio 11: 338-346.
2. Schindler, D.W., R.H. Hesslein, R. Wagemann and W.S. Broecker.
    1980. 'Effects of acidification on mobilization on heavy
    metals and radionuclides from the sediments of a freshwater
    lake.' Can. J. Fish. Aquat. Sci. 37: 373-377.
3. Hesslein, R.H., W.S. Broecker and D.W. Schindler. 1980.
    'Fates of metal radiotracers added to a whole lake.
    Sediment-water interactions.' Can. J. Fish. Aquat. Sci. 37:
    378-386.
4. Hesslein, R.H. 'Whole-lake metal radiotracer movement in
    fertilized lake basins.' Can. J. Fish. Aquat. Sci.
    (Submitted).
5. Santschi, P.H., U.P. Nyfeller, R.F. Anderson, S.L. Schiff, P.
    O'Hara and R.H. Hesslein. 1984. 'Response of radiocactive
    trace metals to acid-base titrations in controlled experimental
    ecosystems: comparison of results from enclosure and whole-lake
    radio-tracer additions.' Can. J. Fish. Aquat. Sci.
    (Submitted).
6. Hesslein, R.H. and P. Quay. 1973. 'Vertical eddy diffusion
    studies in the thermocline of a small stratified lake.' J.
    Fish. Res. Board Can. 30: 1491-1500.
7. Quay, P.D., W.S. Broecker, R.H. Hesslein and D.W. Schindler.
    1980. 'Vertical diffusion rates determined by tritium tracer
    experiments in the thermocline and hypolimnion of two lakes.'
    Limnol. Oceanogr. 21: 357-364.
8. Nyfeller, U.P., Y.H. Li and P.H. Santschi. 1984. 'A kinetic
    approach to describe trace element distribution between
    particles and solution in natural aquatic systems.' Geochim.
    Cosmochim. Acta 48: 1513-1522.
9. Anderson, R.F., P.H. Santschi, U.P. Nyfeller and S.L. Schiff.
    1982. Comparing the geochemical behaviors of stable metals and
    radiotracers. Americ. Geophys. Union and Amer. Assoc. of
    Limnol. and Oceanogr. San Francisco, Dec. 7-15, 1982, EOS
    63(45), 1009.
10. Santschi, P.H., P. Bower, U.P. Nyffeler, A. Azevedo and W.S.
    Broecker. 1983. 'Estimates of the resistance to chemical
    transport posed by the deep-sea boundary layer.' Limnol.
    Oceanogr. 28: 899-912.
11. Brunskill, G.J. and P. Wilkinson. 1985. 'Annual supply of
    ^{238}U, ^{234}U, ^{230}Th, ^{226}Ra, ^{210}Pb, and ^{210}Po to Lake 239
    (Experimental Lakes Area, northwestern Ontario) from
    terrestrial and atmospheric sources.' Can. J. Fish. Aquat.

Sci. (Submitted).
12. Kipphut, G.W. and R.H. Hesslein. 1982. 'Nutrient retention and sediment accumulation in a small lake.' 45th Annual meeting Amer. Soc. Limnol. and Oceanogr. Raleigh, NC, 14-17 June 1982.
13. Schindler, D.W., K.H. Mills, D.F. Malley, D.L. Findlay, J.A. Shearer, I.J. Davies, M.A. Turner, G.A. Linsey and D.R. Cruikshank. 'Chronic ecosystem stress: The effects of eight years of experimental acidification on a small lake.' Science. (Accepted).
14. Cook, R.B., C.A. Kelly, D.W. Schindler and M.A. Turner. 'Mechanisms of hydrogen ion neutralization in an experimentally acidified lake.' Limnol. Oceanogr. (Submitted).
15. Cook, R.B. and D.W. Schindler. 1984. 'Distributions of ferrous iron and sulfide in an anoxic hypolimnion.' Can. J. Fish. Aquat. Sci. 41: 286-293.
16. Kelly, C.A. and J.W.M. Rudd. 1984. 'Epilimnetic sulfate reduction and its relationship to lake acidification.' Biogeochemistry 1: 63-77.
17. Rudd, J.W.M., C.A. Kelly, V. St. Louis, R.H. Hesslein, A. Furutani and M. Holoka. 'Microbial consumption of nitric and sulfuric acids in sediments of acidified lakes in four regions of the world.' Limnol. Oceanogr. (To be submitted).
18. Campbell, P. and T. Torgersen. 1980. 'Maintenance of iron meromixis by iron redeposition in a rapidly flushed monimolimnion.' Can. J. Fish. Aquat. Sci. 37: 1303-1313.
19. Brunskill, G.J., D. Povoledo, B.W. Graham and M.P. Stainton. 1971. 'Chemistry of surface sediments of sixteen lakes in the Experimental Lakes Area, northwestern Ontario.' J. Fish Res. Board Can. 28: 277-294.
20. Kelly, C.A., J.W.M. Rudd, R.B. Cook and D.W. Schindler. 1983. 'The potential importance of bacterial processes in regulating rate of lake acidification.' Limnol. Oceanogr. 27: 868-882.
21. Schindler, D.W., R.H. Hesslein and M.A. Turner. 1985. The exchange of nutrients between sediments and water of Lake 227 after fifteen years of experimental eutrophication.' Can. J. Fish. Aquat. Sci. (Submitted).
22. Schindler, D.W., M.A. Turner and R.H. Hesslein. 1985. 'Acidification and alkalinization of lakes by experimental addition of nitrogen compounds.' Biogeochemistry. (In press).
23. Gerletti, M. and Provini, A. 1978. Effect of nitrification in Orta Lake. Prog. Wat. Tech. 10: p. 839-851.
24. Schindler, D.W. 1977. 'Evolution of phosphorus limitation in lakes: Natural mechanisms compensate for deficiences of nitrogen and carbon in eutrophied lakes.' Science (Wash. DC) 195: 260-262.
25. Schindler, D.W. 1985. 'The coupling of elemental cycles by organisms: evidence from whole lake chemical perturbations. p. 225-250. In W. Stumm (ed.), Chemical processes in lakes. John Wiley & Sons, New York, NY.
26. Schindler, D.W., E.J. Fee and T. Ruszczynski. 1978.

'Phosphorus input and its consequences for phytoplankton standing crop and production in the Experimental Lakes Area and in similar lakes.' J. Fish. Res. Board Can. 35: 190-196.

27. Schindler, D.W., R. Hesslein and G. Kipphut. 1977. 'Interactions between sediments and overlying waters in an experimentally-eutrophied Precambrian Shield Lake.' p. 235-243. In H.L. Golterman (ed.), Interactions between sediments and freshwater. Proc. Symp., Amsterdam, Sept. 1976. Junk, The Hague, PUDOC, Wageningen.

28. Davidson, G.A. 1985. 'A new method for producing permanent qualitative slides for microfossil analysis.' Can. J. Fish. Aquat. Sci. (To be submitted).

29. Levine, S.N., M.P. Stainton and D.W. Schindler. 1985. A radiotracer study of phosphorus cycling in Lake 227, northwestern Ontario. Can. J. Fish. Aquat. Sci. (Submitted).

30. Lean, D.R.S. 1973. 'Movements of phosphorus between its biologically important forms in lake water.' J. Fish Res. Board Can. 3: 1525-1636.

31. Findlay, D.L. and S.E.M. Kasian. 1985. 'Effects of whole lake eutrophication on phytoplankton community in Lake 226.' Can. J. Fish. Aquat. Sci. (Submitted).

32. Mills, K.H. Responses of lake whitefish to fertilization of Lake 226, the Experimental Lakes Area. Can. J. Fish. Aquat. Sci. (Under review).

33. Schindler, D.W. and S.K. Holmgren. 1971. 'Primary production and phytoplankton in the Experimental Lakes Area (ELA) northwestern Ontario and other low-carbonate waters and a liquid scintillation method for determining ^{14}C activity in photosynthesis.' J. Fish. Res. Board Can. 28: 189-201.

34. Linsey, G.A. 1985. 'Atmospheric deposition of nutrients and major ions at the Experimental Lakes Area in northwestern Ontario, 1980-1983.' Can. J. Fish. Aquat. Sci. (To be submitted).

35. Schindler, D.W., R.W. Newbury, K.G. Beaty, J. Prokopowich, T. Ruszczynski and J.A. Dalton. 1980. 'Effects of a windstorm and forest fire on chemical losses from forested watersheds and on the quality of receiving stream.' Can. J. Fish. Aquat. Sci. 37: 328-334.

36. Bayley, S.E. 1985. 'Chemical mass balance budgets of a bog watershed in the Precambrian Shield of Canada. (In prep.).

37. Hesslein, R.H., M.J. Capel and D.E. Fox. 1984. 'Sulfur isotope studies in natural and experimentally acidified Canadian Shield Lakes. Proc. 3rd Int. Symp. on Interactions between Sediments and Water. Geneva. 27-31 Aug. 1984.

38. Hultberg, H. and S. Johansson. 1981. 'Acid groundwater.' Nordic Hydrology 12: 51-64.

39. Bayley, S.E., J. Zoltek, Jr., A.J. Hermann, T.J. Dolan and L. Tortora. 1985. 'Experimental manipulation of nutrients and water in a freshwater marsh: effects on biomass decomposition.' Limnol. Oceanogr. (In press).

40. Dolan, T.J., S.E. Bayley, J. Zoltek, and A.J. Hermann. 1981.

'Phosphorus dynamics of a Florida freshwater marsh receiving treated waste water.' J. Appl. Ecol. 18: 305-219.

41. Boyt, R.L., S.E. Bayley and J. Zoltek. 1977. 'Removal of nutrients from treated municipal wastewater by wetland vegetation.' J. Water Poll. Control Fed. 49: 789-799.

42. Colby, P.J., G.R. Spangler, D.A. Hurley and A.M. McCombie. 1972. Effects of eutrophication on salmonid communities in oligotrophic lakes. J. Fish. Res. Board Can. 29: 975-983.

43. 'Schindler, D.W., T. Ruszczynski and E.J. Fee. 1980. 'Hypolimnion injection of nutrient effluents as a method for reducing eutrophication.' Can. J. Fish. Aquat. Sci. 37: 320-327.

44. Mills, K.H., S.M. Chalanchuk, L.C. Mohr and I.J. Davies. The responses of the fish populations of Lake 223, the Experimental Lakes Area, to eight years of experimental acidification. Can. J. Fish. Aquat. Sci. (Under review).

A COMPARISON OF AQUATIC AND TERRESTRIAL NUTRIENT CYCLING AND PRODUCTION PROCESSES IN NATURAL ECOSYSTEMS, WITH REFERENCE TO ECOLOGICAL CONCEPTS OF RELEVANCE TO SOME WASTE DISPOSAL ISSUES

John R. Kelly and Simon A. Levin
Ecosystems Research Center
Corson Hall
Cornell University
Ithaca, NY 14850

ABSTRACT. Results of a comparative analysis, using data over annual cycles from about 200 different natural areas and a marine mesocosm experiment, suggest general patterns in the importance of both internal and external sources of nutrients to autotrophic production. The range of reported external loading of nitrogen and phosphorus to ecosystems spans several orders of magnitude, with a wide range evident for each of several classes of terrestrial and aquatic ecosystems. The pattern of net primary production across this loading range suggests where internal sources of nutrients contribute significantly to the supply of available nutrients. For aquatic ecosystems in general the external loading/net production relation exhibits a striking pattern; many coastal marine areas seem to be near saturating levels of nutrient loading, and here the efficiency of autotrophic use of available nutrients may be quite low. Our analysis points out the importance of considerations of scale in assessing waste disposal options; this is particularly a key issue for marine ecosystems, which need to be considered across a wide variety of scales.

1. INTRODUCTION

Anthropogenic stresses upon ecosystems carry with them the need to assess possible ecological consequences; and this need is even greater when, as in the case of waste disposal, information on comparative responses of ecosystems can guide choices among available options. Yet, we are severely limited in our ability to predict ecological effects; the extent of environmental change likely to ensue from a proposed human actvity rarely can be suggested with confidence.

Myriad dynamic biotic and abiotic factors interact to define the functional and structural character of any ecosystem. Thus, general principles are rarely sufficient to provide the site-specific information needed for any particular application. Potential alteration of a unique combination of forces in a particular situation becomes the key concern and in-depth understanding of the particular ecosystem is required. When considering an array of systems as possible receiving bodies for inputs, one must take a tiered approach, in which preliminary decisions are made on the basis of coarser, broad-scale information, and in which one moves from initial classifications to more specific ones at later stages. A comparative approach to the identification of common or dissimilar features of ecosystem responses is essential to this task, and may eventually help reduce the dimensionality of the site-specific problem.

165

G. Kullenberg (ed.), The Role of the Oceans as a Waste Disposal Option, 165–203.
© 1986 by D. Reidel Publishing Company.

With the above concerns in mind, we consider the generic problem of the role of oceans, or other arenas, for waste disposal. It is appropriate to begin by looking at the coarse level for major similarities and differences in the factors shaping the responses of different ecosystems to inputs. Comparative studies within selected types of ecosystems have proven useful in identifying variables influencing ecosystem processes and biotic composition. In particular, comparative analyses among a narrowly defined set of physically similar ecosystems have been instrumental in highlighting fundamental factors which help define local ecology. In contrast, cross-system analyses using a spectrum of ecosystems across the ecological landscape, considering both terrestrial and aquatic realms, have not been very common. This situation is not surprising, in part because the separate disciplines have often dealt with different issues, but also because the coarseness of the comparison can be unsatisfying to one filled with appreciation for the complexity of just one individual ecosystem. However, since many waste disposal problems are raising issues which involve multiple media, cross-ecosystem analyses are unavoidable (e.g., Capuzzo et al., this volume).

Among those dynamic ecological features which lend themselves to relatively straightforward comparison across ecosystems are processes, such as nutrient cycling and primary production, that are common to all ecosystems. In this paper we examine general patterns of nutrient inputs to, and autotrophic activity within, various terrestrial and aquatic ecosystems. Our synthesis relies on data from the literature on about 200 field sites (forests, grasslands, agricultural fields, wetlands, estuaries, coastal and open ocean gyres) around the world. A recent experimental mesocosm study of coastal nutrient enrichment conducted at the Marine Ecosystems Research Laboratory (MERL) at the University of Rhode Island (USA) is also discussed, as the results bear specifically upon ocean disposal concerns.

The data summary provides necessary, but not sufficient, information to answer several fundamental ecological questions. Does the relationship of productivity to nutrient supply vary in a general way across ecosystems? Does the importance of recycling to productivity vary across ecosystems in some systematic way? How does the efficiency of biotic use of nutrients vary? While addressing some aspects of these questions, our prime intent in this paper is to use cross-system analysis to raise two general notions, both of which are related to ecological disturbances from waste disposal practices and of current interest in ecological theory. The first notion involves the importance of internal and external controls on biotic functions in ecosystems, and is relevant to issues of monitoring for impacts and recovery from disturbance. The second notion involves scale, and is relevant to determination of the extent of impact, the issue of dispersion vs. containment of waste, and the interpretation and extrapolation of results from individual studies.

2. SOURCES OF INFORMATION, MANIPULATIONS,CAUTIONS

2.1. Comparative analysis of different ecosystems

The search through very disparate literatures can indeed be a most exasperating experience, and for this reason alone it is not surprising that there does not exist a wealth of comparative cross-system studies. Several types of problems emerge. One type is relatively trivial and includes the lengthy and frustrating process of converting to a standard currency for comparison. Terrestrial studies often report nutrient flows in kg ha^{-1} and lacustrine studies generally use g m^{-2} [$= 10$ x (kg ha^{-1})]. Oceanographic studies sometimes use mass, but usually characterize concentrations by atoms, using either ug-at or umoles l^{-1} (note: uM = ug-at where a compound has one atom of the u element of interest); mmoles m^{-2} yr^{-1} is a common reporting unit for flows. The reason for this convention is twofold. The expression clarifies that, for example, the value refers to the N in NO_3^- or NH_4^+ and does not include the weight of the oxygen or hydrogen. Additionally, organisms are sensitive to atoms, not weight. The mixed use of weight or atoms in aquatic studies has contributed to misinterpretations of element ratios because the atomic and weight ratios of N/P, for example, differ by a factor of 2.21 (~ 31/14). In comparisons of freshwater or terrestrial with marine studies, this problem must always be kept in mind. Where we discuss a ratio in this paper, its basis, either atoms or weight, is given. Aquatic studies, with good reason, often develop element budgets using volumetric units such as mg l^{-1} or mmoles m^{-3}. In some cases data are available to convert to an areal basis; yet it is difficult to determine whether the conversion is appropriate given the nature of the data collection, an especially acute problem when dealing with a spatially or temporally heterogeneous process or environment. To include terrestrial sites, we report areal units, but in deference to the perception of organisms include atomic, along with mass scales, on major figure axes.

The difficulty of assessing the quality of data sets is not trivial. Arbitrarily, we limited the comparison to annual budgets. The extent to which interpolations or calculations, rather than direct measurements, have been employed in calculating annual budgets is not always clear in the original citation, and much less so in seminal reviews of a restricted set of ecosystems. Additionally, many reported budgets are truly synthetic, constructed from data collected from different years. We surveyed many more papers than were finally included; careful scrutiny and rejection (usually for confusing documentation or incomplete measurements) have whittled down our data points. In the end, we limited data for this paper to relatively well recognized sources (Appendix 1). Reported values span a significant breadth of "natural" (non-experimental) conditions reported for each different system type. The summary thus constitutes a substantial, but far from comprehensive, review of the literature. We continue efforts in this area. Major sources for sites are given with the data set in Appendix 1; addition of all references for each site and detailed discussion of individual data points, with their noted imperfections, would itself comprise a report of some length. The present summary includes 58 lakes, 25 marine and coastal areas, 26 forested sites, 7 grasslands, and a generic desert (range) for which both a "net" primary production (NPP, as carbon) and nutrient input (either phosphorus or nitrogen) have been reported for one or more annual cycles.

Another non-trivial concern is data comparability. Methods for estimating net primary productivity vary significantly. (For methods and comparative rates of NPP see reviews of Riley (1972), Lieth and Whittaker (1975), and NAS (1975).) For terrestrial systems, studies report annual incremental increase in biomass. Routinely reported as organic matter, values have been converted to carbon (assuming 1 g organic matter = 0.5 g C). For the grasslands data set of the IBP North American syntheses, total net production includes both incremental gain in above and belowground biomass (Sims and Coupland, 1979). Litterfall has not been included in this estimate of NPP. For forested ecosystems around the world, temperate and boreal sites, both deciduous and coniferous are included (Cole and Rapp, 1980). Most studies include annual increments in overstory foliage, branches, and boles; some include understory vegetation (usually a small amount). The estimate of NPP comprises all these increments (when reported) plus litterfall. Litterfall often represents a large fraction of NPP. A few studies report belowground net production, and values are indicated in the data summary. Agricultural systems usually report yield or "net nitrogen output", a measure whose relationship to NPP is unclear, not simply approximated by some standard conversion factor, and not included in these comparisons.

For aquatic systems, a routine method for determining primary production rates is the ^{14}C method. Details of the methodology and the formulations used to calculate annual values vary across laboratories; short term incubations are the usual, and values are reported in gC of organic tissue produced. What this method actually measures is still a matter of some controversy (cf. Peterson, 1981; Valiela, 1984; Oviatt et al., in press), but it is commonly used as an estimate of NPP.

For other reasons, a ^{14}C-derived rate may misrepresent production for the ecosystem. For example, ecosystem NPP may be underestimated substantially unless a separate effort has been made to measure NPP of the benthic plants (and associated epiphytes) not measured in a bottle incubation. For shallow aquatic ecosystems, where data exist for production of submerged aquatic or emergent vegetation, such data have been included. NPP values of our data summary, for the most part, reflect pelagic NPP. Additionally, the open ocean values have been extrapolated to an annual value from a small number of measurements; for intense upwelling (e.g., off Peru), NPP may be high by a factor of at least two-- a range is given in graphs.

We are limited to the productivity data at hand, but one exception to the routine methodology deserves mention. Smith (1984) recently reported "net ecosystem production" (NEP) of three oligotrophic warm water estuarine embayments. NEP is not directly comparable to, but is less than, NPP (cf. Nixon and Pilson, 1984). The exception is singled out in tables and graphs because NEP, although not yet a routine measure of aquatic ecosystems, may eventually offer a better basis for the comparison of terrestrial, agricultural and aquatic productivity-- potential biomass yield is of more direct concern to man.

In addition to sites where both productivity and a nutrient input (nitrogen, phosphorus, or both) were available, the combination of annual nitrogen and phosphorus inputs have been reported. Our summary includes an additional 23 estuarine and marine areas (several for the same area for a different year, or by a different investigator), 14 forested ecosystems, 5 wetlands, and 43 agricultural systems. Like production measures, analytical methods for nutrients differ, but perhaps not enough to cause problems with data comparability. However, the degree to which all sources of input, and all forms of a nutrient, have been characterized varies both within and across ecosystems. Since there are not many estimates for nitrogen fixation for entire ecosystems, this is not included as an N input (in our summaries), although a note on its magnitude is given for some systems (Table 1). Most N and P input values can probably be assumed to be minimal, rather than maximal estimates.

We made two major data manipulations to develop the cross-system analysis. The first concerns lake data. The role of phosphorus inputs in lake eutrophication, convincingly demonstrated by Schindler (1974), helped make P the focus of freshwater studies-- P loading: chlorophyll (or productivity) predictive relationships, mostly based on variations on Vollenweider's model (1975,1976), are published for lakes and reservoirs (cf. Rast and Lee, 1978; Schindler, 1981; Jones and Lee, 1982). In contrast, terrestrial systems often have emphasized nitrogen cycling except for the less common situation where there are P-deficient soils. Marine studies have focused more on nitrogen, especially since Ryther and Dunstan (1971) implicated it as the primary limiting nutrient in coastal waters (Carpenter and Capone, 1983). To allow a direct comparison across these situations and to maximize the number of data points, we used nitrogen as the generic nutrient. In doing so we created a variable, *N input, to serve as a proxy for P input in lake studies. In spite of the ability of some lakes to make up N deficits by withdrawing N from the atmosphere (e.g., Schindler, 1974; 1976; 1977), from the ELA studies it is also apparent that, as a provisional conclusion "elevated supplies of both nitrogen and phosphorus are necessary to get a significant production response" (Fee, 1979). N/P loading in weight ratios from 15:1 to 5:1 (or 33:1 to 11:1, atoms) showed more productivity response than either nutrient alone (Fee,1979). The "Redfield" model for marine plankton tissue needs (Redfield, 1934; 1958; Redfield et al., 1963; Goldman et al., 1979) lies within this range. Values for individual (as contrasted with natural community) freshwater plankton species at low P concentrations may be higher by a factor greater than two (Valiela, 1984). But lacking any better guidelines and as a first approximation across the entire range of loading, we calculated *N input by converting from P input, using Redfield's 16:1 N/P ratio (atoms). The conversion would be inappropriate where macrophytes contribute greatly, either in fresh or salt waters (e.g., Likens et al., 1981; Smith, 1984). Both N inputs (where given) and *N inputs for lakes are given in Appendix 1. *N inputs differ significantly from N inputs in that they are lower, by a factor of about 3 to 4 at low P loading (see below). At intermediate and high loading N and *N are most often within a factor of 2 or 3 (either one may be higher) of each other. The effect of the conversion at the level considered in this paper thus is generally minimal; we note specific exceptions. Why even make the conversion then? *N allows direct cross-system comparisons, not flipping back and forth across elements, yet does not alter, at all, the pattern observed between P inputs and productivity-- in that sense the conversion represents only a scale change (i.e. divide *N (atoms) by 16 to get the actual P value).

A final concern is the calculation of nutrient inputs in the sea. In large bodies of water the "boundaries" are transient, difficult to establish, and the delineation of an "ecosystem" may seem more arbitrary than in terrestrial or limnological studies where a landscape helps define an area. The perfect conceptualization does not exist, and one chooses the scale and defines the system for a given purpose. For example, one can consider the entire ocean as a system, where the "inputs" are land runoff, atmospheric washout and aerosol or particle settling, and (if considered "external") hydrothermal submarine nutrient flows. One can also consider entire gyres, a geographically defined system (e.g. the New York Bight or Apex), or the photic zone above a thermocline as ecosystems. For the photic zone case, inputs include precipitation, horizontal mixing or advection if surrounding systems are distinct, but also include vertical advection and mixing. Nitrogen tracer techniques have been used to distinguish between "new" (from outside the euphotic zone) and "used" (based on autochthonous processes) nitrogen sources in contributing to primary production of particulate matter (Dugdale and Goering, 1967; Dugdale, 1967; Dugdale, 1976; Eppley et al., 1979; Peterson and Eppley, 1979). We calculated the "inputs" to some open pelagic areas using recent data and models (Peterson and Eppley, 1979; also Laws, 1983). The calculation simply converts from "new" production in carbon units to "new" and thus, "externally" supplied nitrogen, by use of the familar Redfield ratio for organic tissue; this is appropriate in this case because it is based on open ocean statistical averages. Our calculated inputs compare reasonably with those of the method of McCarthy and Carpenter (1983). To highlight the fact that we have calculated these values, we also denote inputs by *N.

In all, the cautions are many. The warning accompanying agricultural nutrient budgets (Frissel, 1976, p. 313) is appropriate: "consideration of summarizing graphs and tables only may result in generalizations which are misleading. Verify conclusions by consulting the original nutrient balances." Our intent is not to develop a general predictive model--it would be inappropriate. At this stage of development of quantitative cross-system comparisons, we must remain ever uncomfortable with the validity of our suggestions, and use the studies primarily as a key to macroscopic insight, where detecting anomalous points can be as useful as general patterns in guiding future, more refined efforts.

2.2. An experimental enrichment study for coastal waters

There is a great and diverse literature on fertilization and enrichment experiments in different systems, and this is critical information for addressing the questions posed in this paper. A great many studies are of little comparative use because they did not measure properties at the scale of the ecosystem. However, many are appropriate (e.g., the ELA enrichment studies of Schindler and colleagues). Detailed comparison of fertilization responses is beyond the scope of the present paper; but since there is a focus on ocean waste disposal, we would like to discuss some results emerging from a recent ecosystem study on coastal eutrophication. In contrast to studies of lake eutrophication, the development of coastal nutrient budgets, examination of productivity: nutrient relationships, and experimental studies at the ecosystem level are only recent enterprises (cf. Boynton et al., 1982; Kemp et al., 1982; Nixon, 1983; Nixon and Pilson, 1983; Nixon et al., 1984).

An enrichment gradient study in large ($13 \, m^3$) experimental mesocosms, designed to simulate a shallow coastal ecosystem such as Narragansett Bay (RI, USA), was conducted over several years. The MERL systems and experiments have been described in many publications (cf. Nixon et al., 1984; Oviatt et al., 1984); the cylindrical tanks are 5 m tall, are maintained with a 30 cm deep benthic community taken from the Bay, and have light, temperature, salinity, and flushing regimes similar to the lower Bay. The gradient experiment was conducted to examine the effects of nutrient enrichment across a range of nutrient loading rates typical of a variety of U.S. estuaries, and nutrients (N, P, Si) were added in ratios measured for sewage discharge to Narragansett Bay (Nixon et al., 1984). Great detail on the rationale, design, and some of the numerous results of the multiple year, continuous daily nutrient addition experiment are available or forthcoming-- Nixon (1983), Nixon and Pilson (1983), Nixon et al. (in press), Seitzinger and Nixon (in press), and Berounsky and Nixon (in press). We present some results on nutrient cycling and productivity (Kelly et al. and Oviatt et al., both in press) in comparison to that gathered for the cross-system analysis.

3. PATTERNS ACROSS ECOSYSTEMS

3.1. A variety of input rates

The range of nutrient loading to all types of natural ecosystems is indeed large (Figure 1, Table 1). Since all systems receive some amount of precipitation input, where this is the primary input one can find examples, for each category we reviewed, having nitrogen inputs between about 0.1 and 1.0 $gN \, m^{-2} \, yr^{-1}$. The position of lakes, estuaries, and some types of wetlands in the landscape dictates that their input range should be large. Compared to the relatively flat, sometimes oblique (highly sloped) rainfall collectors represented by many terrestrial areas, or the slightly convex collector represented by an ombrotrophic bog, downstream aquatic ecosystems have a greater potential volume of water input. Gravity causes them act as collectors for not only the rains and dusts falling directly upon their surface area, but also for those falling on an entire watershed, sweeping areas sometimes much larger than their own. These waters, of course, carry both weathered mineral nutrients and unretained rainfall nutrients exported by the upland system in surface or subsurface flows. Because of the variety in configurations of downstream areas to upland collectors a range in nutrient inputs is possible due simply to scale, because the "effective collection area" differs (Kelly and Harwell, 1982;1985; Nixon et al., in press). Of course, humans play a large role in nutrient inputs. Civilization

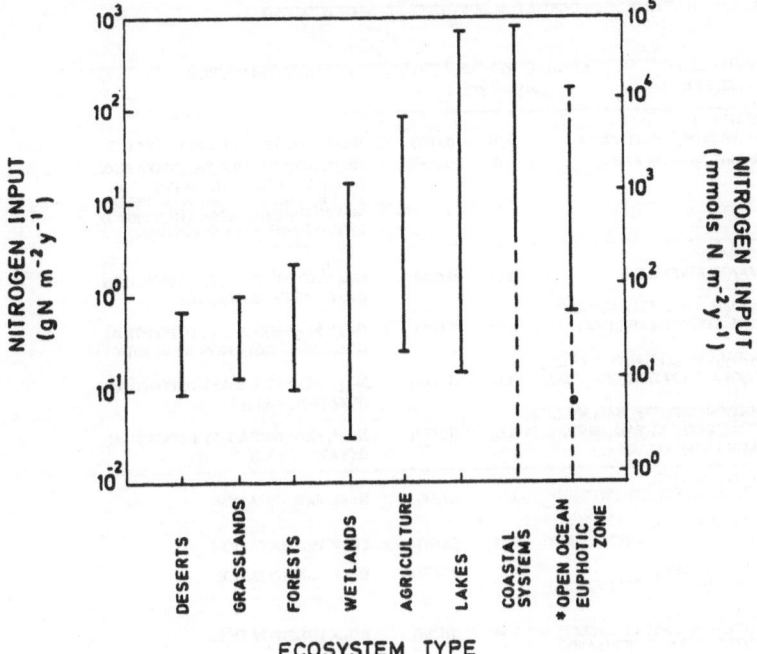

Figure 1: The range of nitrogen inputs to ecosystems.

Data are from values reported in the literature (Table 1, Appendix 1). Values do not include nitrogen fixation. For coastal systems (which here include salt marshes), dotted lines indicate that values will extend lower where land runoff doesn't contribute (e.g. tropical embayments). Net import from adjacent surface systems has been measured in several such instances (Smith 1984), but its relationship to gross influx is uncertain. For open ocean surface water systems, the * denotes that values have been calculated, not measured. The dotted line at the upper end of the range indicates the Peru upwelling based on either 6 mo or 1 yr period of intense upwelling. The solid point represents an average for the oceans on a global basis, considering the inputs to be precipitation and land runoff. The dotted line at the lower end of open ocean range is to suggest the possibility that inputs are lower in areas where there is little precipitation, very weak upwelling, or even downwelling (that imports surface waters with indetectable N). It is likely that lower input terrestrial systems also occur.

TABLE 1: LOW TO HIGH RANGE OF NITROGEN INPUTS FOR DIFFERENT TYPES OF ECOSYSTEMS

TYPE	REFERENCE(S) DESCRIPTION	ANNUAL NITROGEN INPUT gN m^{-2} yr^{-1}		COMMENTS ON SOURCES
DESERT	WEST(1978), CRAWFORD&GOSZ(1982)	0.09	(LOW)	NH$_4^+$, NO$_3^-$ IN BULK PRECIPITATION
	GENERIC SYSTEM RANGE	0.69	(HIGH)	NH$_4^+$, NO$_3^-$ IN BULK PRECIPITATION (PARTICULATE INPUTS & FIXATION MAY BE EQUAL TO OR AN ORDER OF MAGNITUDE GREATER THAN PRECIP, BUT ARE NOT WELL QUANTIFIED)
GRASSLANDS	MEDINA(1982), SAVANNAH(VENEZUELA)	0.13	(LOW)	NH$_4^+$, NO$_3^-$ IN BULK PRECIPITATION (NO FIXATION ESTIMATES)
	WOODMANSEE(1979),'ALE'(IBP), SHRUB-STEPPE(WASH,USA)	0.20	(LOW)	NH$_4^+$, NO$_3^-$ WET & DRY DEPOSITION (FIXATION < 0.05 + NON-SYMBIOTIC)
	WOODMANSEE(1979), OSAGE(IBP) TALLGRASS PRAIRIE (OKLA,USA)	1.0	(HIGH)	NH$_4^+$, NO$_3^-$ WET & DRY DEPOSITION (FIXATION < 0.10)
	WOODMANSEE(1979), SAN JOAQUIN (IBP) MEDITERRANEAN ANNUAL GRASSLAND(CALIF,USA)	1.0	(HIGH)	NH$_4^+$, NO$_3^-$ WET & DRY DEPOSITION (FIXATION < 2.5)
FORESTS	LIKENS et al.(1977), CONIFEROUS FOREST(SWEDEN)	0.1	(LOW)	BULK PRECIPITATION
	COLE&RAPP(1980), SPRUCE FOREST(KARELIA,USSR)	0.11	(LOW)	BULK PRECIPITATION
	LIKENS et al.(1977) DECIDUOUS FOREST(OHIO,USA)	2.0	(HIGH)	BULK PRECIPITATION
	COLE&RAPP(1980) SPRUCE PLANTATIONS, BEECH FORESTS (FED.REP. GERMANY)	2.18	(HIGH)	BULK PRECIPITATION
WETLANDS (FRESHWATER)	VAN CLEVE&ALEXANDER(1981), WET MEADOW TUNDRA(AK,USA)	0.03	(LOW)	BULK PRECIPITATION (FIXATION= 0.07 g m^{-2} yr^{-1})
	ULEVOHA et al.(1976), WET MEADOW(CZECHOSLOVAKIA)	8.8-13.8	(HIGH)	BULK PRECIPITATION &FLOOD WATER RUNOFF(ONLY INORGANIC N)
	DAY&KEMP(1985), SWAMP FOREST(LA,USA)	15.54	(HIGH)	BULK PRECIPITATION&PUMPED RUNOFF FROM LEVEE (TOTAL N MEASURED)
AGRICULTURAL ECOSYSTEMS	[ALL IN FRISSEL(1977)] STEPPE AND SEMI-DESERT OF PATAGONIA(S. AMERICA)	0.25	(LOW)	DRY AND WET DEPOSITION (FIXATION SAME MAGNITUDE)
	SHRUB STEPPE(ARGENTINA)	0.3	(LOW)	"
	HIGH TABLE LAND, MIXED LIVESTOCK&ARABLE(MID-ANDES, S.AMERICA)	0.3	(LOW)	"
	SEMI-ARID WILDLIFE PASTURE(ISRAEL)	0.5	(LOW)	"
	DAIRY FARM (NETHERLANDS)	78.3	(HIGH)	INPUT OF LIVESTOCK FEED, FERTILIZER, DRY & WET DEPOSITION (FIXATION 'NEGLIBLE')
	MIXED FARM, LEYS IN ROTATION WITH ARABLE CROP (FRANCE)	80.0	(HIGH)	FERTILIZER INPUT. (DEPOSITION AND FIXATION CONSIDERED 'TRACE')
	FORAGE FARM, WITHOUT GRAZING (FRANCE)	80.0	(HIGH)	"
LAKES	SCHINDLER et al.(1974) CHAR LAKE(NWT,CANADA)	0.31	(LOW)	BULK PRECIPITATION, INFLOW STREAMS(TOTAL N)
	RAST & LEE(1968) WALDO LAKE(OR,USA)	0.33	(LOW)	ESTIMATED BULK PRECIPITATION& BASIN RUNOFF
	NYHOLM (1978) LAKE OLLERUP(DENMARK)	200.0	(HIGH)	INPUTS FROM RIVERS&STREAMS, EFFLUENT DISCHARGE, RUNOFF FROM AGRICULTURE, PRECIPITATION
	NYHOLM (1978) LAKE BRAS (DENMARK)	713.0	(HIGH)	INPUTS FROM RIVERS&STREAMS, EFFLUENT DISCHARGE, RUNOFF FROM AGRICULTURE, PRECIPITATION

TABLE 1: (CONTINUED)

	NYHOLM(1978) BASIN 1 OF LAKE TANGE (DENMARK) Not included in Fig 1 range for whole lakes	1209.0	(HIGH)	INPUTS FROM RIVERS&STREAMS, EFFLUENT DISCHARGE, RUNOFF FROM AGRICULTURE, PRECIPITATION
COASTAL SYSTEMS (SALTWATER WETLANDS)	CORRELL(1981) INTERTIDAL WETLAND(MD,USA)	10.94	(LOW?)	BULK PRECIPITATION&LAND DRAINAGE(SURFACE&GROUND- WATER)(TOTAL N MEASURED) (GROSS TIDAL INFLUX NOT INCLUDED)
	VALIELA&TEAL(1979) SALT MARSH(MA,USA)	74.4	(HIGH)	BULK PRECIPITATION, GROUND- WATER, FIXATION(6.8), & GROSS TIDAL INFLUX(54.2)(TOTAL N MEASURED)
(ESTUARIES, BAYS,LAGOONS, SOUNDS)	SMITH(1984) SHARK BAY(W. AUSTRALIA)	0.01*	(LOW)	*NET IMPORT FROM OCEAN ; NO SIGNIFICANT TERRESTRIAL INPUTS' (FIXATION ESTIMATE = 0.32 - 0.6 gN m^{-2} yr^{-1})
	SMITH(1984) CHRISTMAS ISLAND(CENTRAL PACIFIC ATOLL)	0.5*	(LOW)	*NET IMPORT FROM OCEAN ; NO SIGNIFICANT TERRESTRIAL INPUTS' (FIXATION ESTIMATE = 0.87 - 3.00 gN m^{-2} yr^{-1})
	ARMSTRONG(1982) NUECES ESTUARY(TX, USA)	1.52	(LOW)	INPUTS FROM FRESHWATER FLOW, MARSH EXCHANGE & PRECIP
	ARMSTRONG(1982) MISSION-ARANSAS ESTUARY (TX, USA)	2.0	(LOW)	INPUTS FROM FRESHWATER FLOW, MARSH EXCHANGE & PRECIP
	JAWORSKI(1981) THAMES ESTUARY, 1971 (UK)	462.5	(HIGH)	'TOTAL' N?
	JAWORSKI(1981) RIVER-ON-TYNE ESTUARY 1972 (UK)	750.0	(HIGH)	'TOTAL' N?
(COASTAL BIGHTS)	EPPLEY et al. (1979), EPPLEY& PETERSON (1979) SOUTHERN CALIFORNIA BIGHT(USA)	9.9*	(LOW)	APPROXIMATE REGIONAL AVERAGE (>300 m STATIONS), CALCULATED FROM EXTERNALLY SUPPORTED PRODUCTION
	MALONE(1982) NEW YORK BIGHT APEX(NY, USA)	46.7	(HIGH)	DISSOLVED INORGANIC N FROM RUNOFF (MOSTLY SEWAGE) WHICH PASSESS FROM THE LOWER ESTUARY TO THE BIGHT APEX(1250 km ^{2})
OPEN OCEAN EUPHOTIC ZONE	EPPLEY & PETERSON(1979), DUGDALE(1976) CENTRAL NORTH PACIFIC	0.73*	(LOW)	APPROXIMATE VALUE CALCULATED FROM EXTERNAL SUPPORTED PRODUCTION
	McCARTHY & CARPENTER(1983) SARGASSO SEA	1.31-7.06*	(LOW)	CALCULATION:INCLUDES PRECIPITATION(0.117), FIXATION(0.0005), EDDY DIFFUSION(1.150-6.899), AND ADVECTION(0.042 gN m^{-2} yr^{-1})
	EPPLEY&PETERSON(1979) DUGDALE(1976) COSTA RICA DOME UPWELLING	93.57*	(HIGH)	APPROXIMATE VALUE CALCULATED FROM EXTERNAL SUPPORTED PRODUCTION (EXTRAPOLATED TO ANNUAL BASIS)
	EPPLEY&PETERSON(1979) DUGDALE(1976) PERU UPWELLING	169.2*	(HIGH)	APPROXIMATE VALUE CALCULATED FROM EXTERNAL SUPPORTED PRODUCTION (EXTRAPOLATED TO ANNUAL BASIS; IF UPWELLING FOR ONLY 6 MONTHS, ESTIMATE WOULD BE HALVED)
GLOBAL OCEAN AVERAGE	LAWS (1983)	0.077-0.15*		CALCULATION BASED ON PRECIPITATION, TERRESTRIAL RUNOFF (0.077) PLUS FIXATION (ALSO ABOUT 0.077 gN m^{-2} yr^{-1})

*Denotes calculated value

developed water transport systems to remove much of its wastes, and an extra burden of nutrient inputs thus commonly is received by many aquatic systems; those systems very high on the input scale (Figure 1) receive the majority of their nutrients via this route.

There are examples, both freshwater and estuarine, where input levels approach 10^3 gN m^{-2} yr^{-1}, a value exceeding even intensively fertilized agricultural fields (Table 1) by about an order of magnitude. Human activity can profoundly influence the nutrient runoff to the ocean on the local, or "point", scale at the level of coastal bays, estuaries, and significant portions of large bights. Some aquatic sites may be underestimated where heavy non-point agricultural or urban runoffs (surface or groundwater) are significant and neglected. However, as opposed to diffuse discharges (as well as uniform field fertilizations), when point sources are large, it can be misleading to express inputs on the basis of the entire, arbitrarily defined "ecosystem". The initial receiving zone may be orders of magnitude smaller, and nutrient loads to a receiving basin, harbor, or inlet correspondingly higher. This concern would apply equally when considering point discharges of an effluent to forest, marsh, or any system which has a biological response time faster than the time it takes to achieve uniform dilution.

Although we have not included intentional enriched upland sites (such as those receiving effluent-- cf. D'Itri, 1982), the upper end of the range for non-agricultural terrestrial systems, just over 2 gN m^{-2} yr^{-1} is also associated with human activity. It is well known that higher concentrations of nitrogen in precipitation are found with some of society's industrial and population centers (e.g., Likens and Bormann, 1974); for neighboring sites receiving primarily precipitation inputs, input values are the highest reported for forested areas (Table 1).

If we use the standard conceptualization to derive inputs to the oceans on a global basis, the N input value seems about equally determined by precipitation and land runoff, both of which yield uncertain values (cf. Laws, 1983). Although coastal areas can be dominated by wastewater discharges of nutrients, globally it appears that changes in precipitation, a diffuse source, might have as much potential impact on the budget as changes in runoff. However, using the conceptualization we indicated for open ocean surface waters one sees that the range in nitrogen "externally" supplied varies across a range similar to that seen for agricultural systems. In contrast to all other systems, human wastes do not directly contribute to the upper values in this range--inputs are a function of fluxes from nutrient rich bottom waters. Rather than direct discharges, anthropogenic alterations of climate and oceanic circulation patterns are the more relevant concern with regard to inducing change in surface waters. Here, relatively local situations may even be expressed globally. For example, world fisheries yield for a year can be affected by the intensity of upwelling in one area (cf. Ryther, 1969; Gulland, 1976). A concern for identifying factors capable of directly or indirectly altering local nutrient supplies is relevant to the open ocean.

3.2. Nitrogen and phosphorus relationships in inputs

The relationship between nitrogen and phosphorus in inputs varies, although one general rule is that larger N inputs are accompanied by larger P inputs. Across the individual sites depicted in Figure 2, the range in N/P input ratio is considerable. Many values are very different from a Redfield model, presented as a frame of reference and because of its historical importance in open ocean and coastal marine ecological research (Redfield, 1958; Ryther and Dunstan, 1971; Nixon et al.,1980; Boynton et al., 1982; Smith, 1984). For coastal systems, many lakes, and agricultural systems-- those systems at intermediate to high loading levels-- there are examples surrounding a 10 to 20:1 (atoms) range, with many above, below and within that band; but no particular pattern is evident. Open ocean

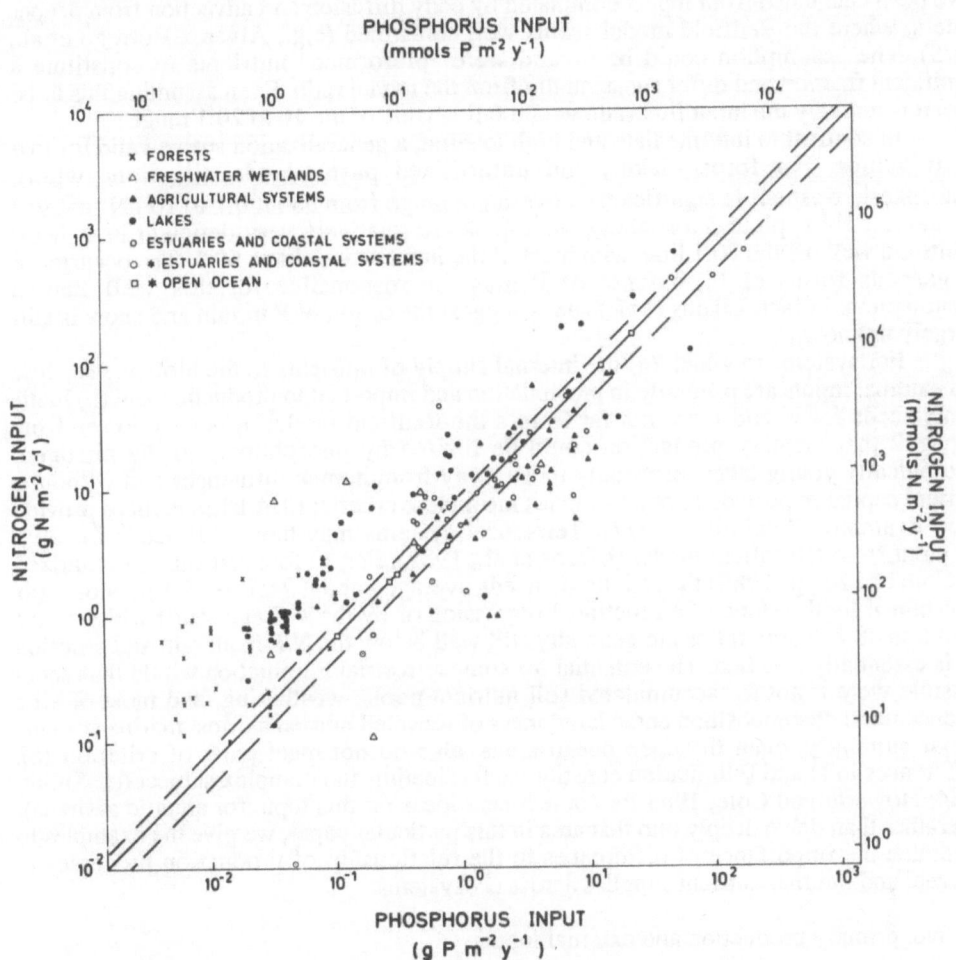

Figure 2: The relationship between nitrogen and phosphorus inputs to ecosystems.

The solid diagonal line indicates a Redfield ratio (16 N : 1 P,atoms). The dotted lines indicate isopleths for N:P ratios of 20:1 (above solid line) and 10:1 (below solid line).

surface systems are contsrained by our methodology to lie on the 16:1 line, because they have been calculated from inputs dominated by eddy diffusion and advection from deeper waters, where the Redfield model seems well confirmed (e.g., Alvarez-Borrego et al., 1975). The assumption could be invalid were "preformed" nutrients to constitute a significant fraction and differ substantially from the model ratio. Even assuming this to be true, it is unlikely the input flux ratio would fall outside of the 10 to 20:1 range.

In contrast to intermediate and high loading, a generalization seems valid for low input values. For forest, lakes, and unfertilized pasturelands--situations where atmospheric washout is significant-- covering a range from about 0.1 to 10 gN m^{-2} yr^{-1} (N and not *N is plotted for lakes), the inputs are relatively very deficient in P. Most points are well off the 20:1 line, with most of the input ratios above 60:1. The occurrence of gaseous forms of N, but not of P must be responsible for this well known phenomenon; in fact, Likens et al. (1981) suggest the origin of P in rain and snow is still "largely unknown".

For systems in which (a) the internal supply of nutrients to the biota is very low and external inputs are primarily in precipitation and important to production and (b) biotic tissue needs for N and P are not far from a the Redfield model, it is easy to see from Figure 2 that primary production could be limited by phosphorus, not by nitrogen. Geologically young lakes, especially those away from human influences and without a nutrient capital to provide internal supplies meet these criteria; ELA lakes perhaps provide good examples (Schindler, 1977). Terrestrial systems may have N/P needs for NPP different from a Redfield model (Likens et al., 1981). For 30 forested sites summarized by Cole and Rapp (1980) the N/P tissue needs averaged about 21:1 to 25:1 by atoms (as determined by the slope of a functional regression or by the average ratio) with a range from 5 to 48:1. These ratios are generally still well below the N/P from rain and criterion (b) is essentially satisfied. The potential for some terrestrial P limitation would thus seem possible were it not for accumulated soil nutrient pools, weathering, and most of all a reliance upon decomposition and a large mass of recycled nutrients. Most marine systems of our summary, even the open ocean areas, also do not meet parts of criterion (a). Differences in N and P limitation constitute a fascinating and complex subject (cf. Smith, 1984; Howarth and Cole, 1985 for some recent ideas on this topic for aquatic systems). But rather than delve deeply into that area in this particular paper, we give the examples to introduce the importance of differences in the relationship of production processes to external and internal nutrient supplies across ecosystems.

3.3. Net primary production and external inputs

Using nitrogen (and *N) as a "generic" nutrient, we plot NPP against external nutrient inputs (Figure 3). For NPP, a range of 300 to 800 gC m^{-2} yr^{-1} produced in terrestrial sites covers the majority of cases. Forested sites range mostly from 500 to 800 (about 300 to 700 without including litterfall), with the highest value being over 1400; according to data in Cole and Rapp (1980), below ground NPP (not included) would add 30 to 300 gC m^{-2} yr^{-1}. With one exception the North American grasslands fell between 350 and 450. The entire terrestrial NPP range reported in the literature is, of course, larger; for example, a meadow-steppe in the USSR exceeded 1700, wet meadows may exceed 800 (Bazilevich & Titlyanova, 1980) and Valiela's (1984) summary shows rain forests up to 2 kgC m^{-2} yr^{-1}. At the other extreme, some deserts may be as low as 1.3 (Hadley and Szarek, 1981), but when irrigated can be similar to other terrestrial sites (upper value for desert range, Figure 3).

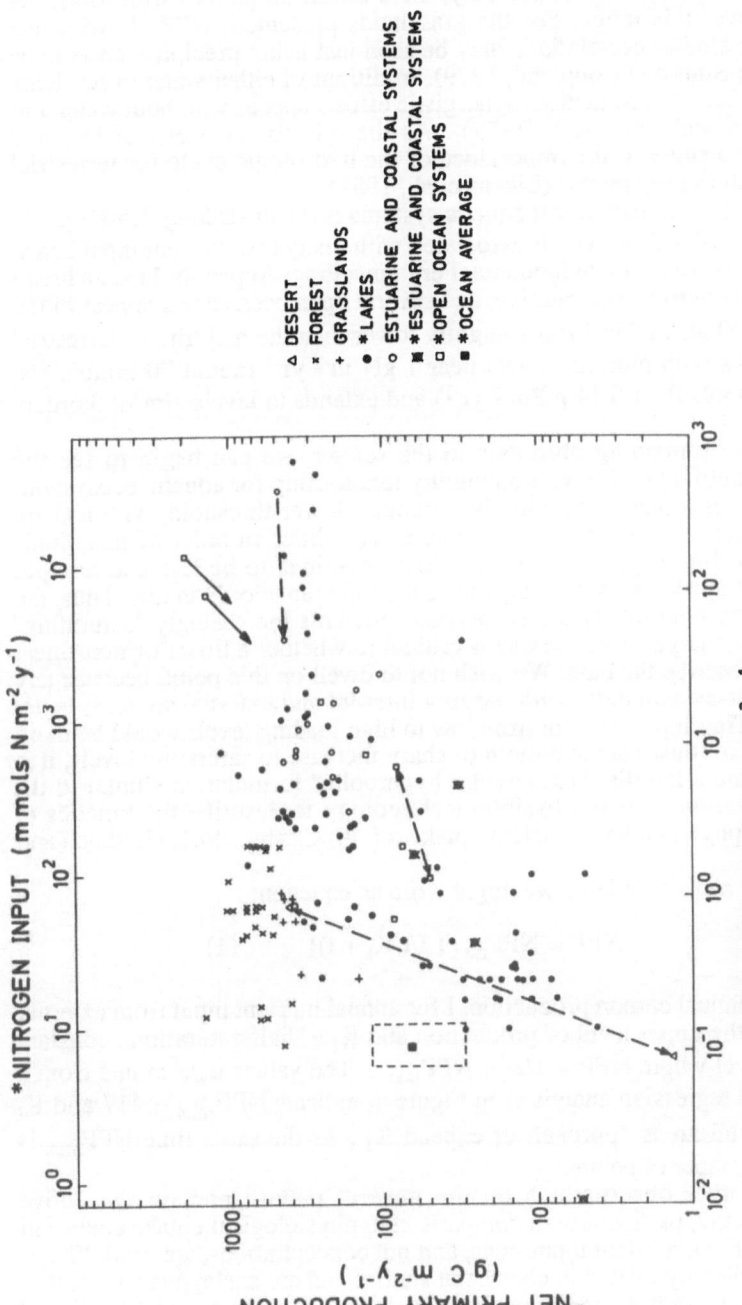

Figure 3: Nutrient inputs and primary productivity in terrestrial and aquatic ecosystems.

The asterisks denote where a value has been calculated from data or is dissimilar to the rest of the data set. Four estuarine systems did not report a standard primary production estimate (see Appendix), other manipulations and definitions of NPP are given in the text. The double-headed arrows show range estimated for deserts and for Sargasso Sea. Open ocean points with single-headed arrows show range assuming upwelling (Peru and Costa Rica Dome) occurs for entire year or only 6 mo. The dotted line with single-headed arrow connects points moving from New York Harbor (higher) to New York Bight Apex. The box around the average ocean value gives a range assuming plus or minus 100% around the estimate.

Accompanying the range in terrestrial NPP is an input range from about 0.1 to slightly over 2.0 gN m^{-2} yr^{-1}. Not unexpectedly, there seems no pattern with forest net biomass production over this range. For the grasslands presented, NPP shows some correlation to N input; similar correlations may be seen just using precipitation volume (e.g., Lauenroth, 1979; Sims and Coupland, 1979). Additions of either water or nutrients to terrestrial sites can increase production; synergistic effects appear with both water and nutrients added (Dodd and Lauenroth, 1979) . As the rain forest, wet meadow and irrigation examples also suggest, the importance of the hydrologic cycle for terrestrial systems seems "difficult to exaggerate" (Likens et al., 1981).

In contrast, the general pattern for aquatic systems is rather striking. Low aquatic NPP on the order of 5 to 50 gC m^{-2} yr^{-1} is associated with many low nutrient input lakes. NPP reaches levels above 1000 (Lake Mendota, Peru upwelling; Appendix 1) yet a broad band of lakes, as well as most coastal marine, and several open ocean areas appear in the narrow range 200 to 600 gC m^{-2} yr^{-1} (only slightly lower than the majority of terrestrial sites). This band begins with nutrient inputs near 1 gN m^{-2} yr^{-1} (about 70 mmols *N, which equals 4.375 mmols P, or 0.14 g Pm^{-2} yr^{-1}) and extends to levels almost 3 orders of magnitude higher.

For the moment remaining oblivious to the scatter, we can begin to see the emergence of a broad nutrient delivery: productivity relationship for aquatic ecosystems (Figures 3,4). Depicted is a pattern seemingly without a lower threshold, with a sharp initial rise in NPP of about an order of magnitude across about an order of magnitude increase in external loading. Apparent NPP increases continue to be less and less per additional increment of nutrients proceeding across the total range of N inputs. Thus, the rapid rise is followed by continually lesser increases towards increasingly "saturating" levels of nutrients. The sharpness of the rise is critical to whether a linear or non-linear formulation can best describe the data. We wish not to dwell on this point, because any formulation using cross-system data, with varying internal and external sources, is not entirely appropriate. "Titrating" a system from low to high loading levels would be more appropriate. However, to illustrate the pattern of sharp increase to saturating levels, it is possible to fit to the data a familiar "rectangular hyperbolic" formulation similar to the Michaelis-Menten equation used in physiological ecology to describe the kinetics of freshwater and marine phytoplankton nutrient uptake (cf. McCarthy, 1981; Goldman and Glibert, 1983).

For the general aquatic pattern, we might write an equation:

$$NPP = NPP_{max} [I/(K_I + I)] \qquad (1)$$

where NPP stands for annual carbon production, I for annual nutrient input from external sources, NPP_{max} for the upper level of production and K_I a "half-saturation" constant indicating the input level where NPP = 1/2 of NPP_{max} . The values determined from a standard linear plot and regression analyis as in Figure 4, indicate NPP_{max} = 417 and K_I =13. Thus, many coastal areas approach or exceed K_I . At the same time NPP_{max} is clearly exceeded by a number of points.

A number of other observations on the general pattern and on the above formulation are noteworthy, particularly in comparison to physiological uptake studies in laboratory chemostats. First, nutrient input rates, and not concentrations, are used. Fluxes and concentrations are directly related in chemostat studies, but are analogous only across ecosystems under certain specific conditions. Important conditions that need be satisfied for supply and concentration to be equivalent across systems are that both water depth and

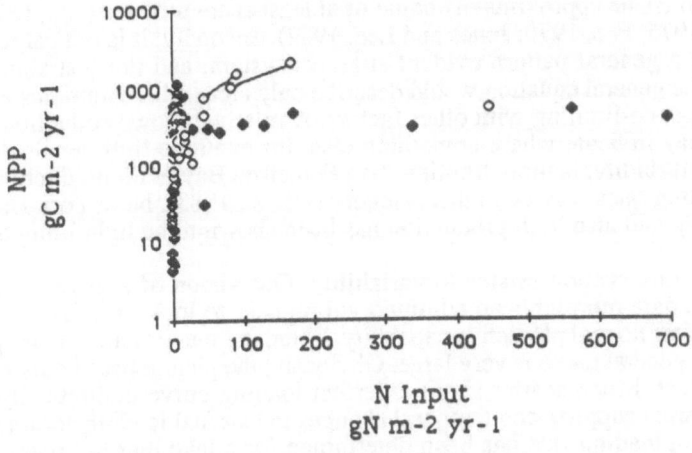

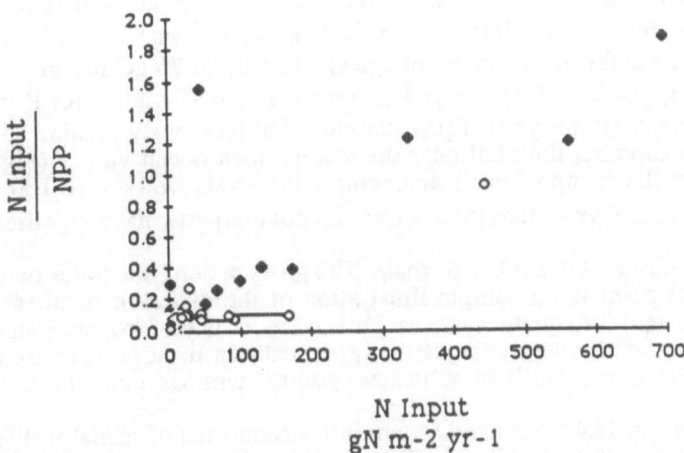

Figure 4: Nutrient inputs and primary productivity in aquatic ecosystems.

Same aquatic data as in Figure 3, with the following exceptions: the four estuarine systems (*) are not included; the average estimate for the Sargasso Sea is used; the average value for global ocean is not included. Closed circles = lakes. Open circles = estuarine and open ocean areas. Lines join the two upwelling sites (1 yr and 6 mo estimates). Note the change, from Figure 3, from a log-log to a semi-log plot in upper panel. The bottom panel shows a linear plot to estimate parameters of a hyperbolic function to describe NPP in terms of N input. Least squares regression ($r^2 = 0.72$, n=110) yields the estimates: $NPP_{max} = 417$, $K_I = 13$ (see text).

water renewal times be approximately equal or at least accounted for (e.g., Dillon, 1975; Vollenweider, 1975; Fee, 1979; Jones and Lee, 1982). Secondly, it is critical to recognize that we suggest a general pattern evident at an ecosystem, and not just single-species, level. Thirdly, the general equation would describe only areas where nutrients are limiting growth, or at least co-limiting with other factors. A relatively low production at a given loading level may indicate where something else, for example light levels due to self-shading or high turbidity, is more limiting. San Francisco Bay, with production less than 200 at high loading rates may be in this category (Nixon, 1983). New York Harbor, with very high loading and also high production has been shown to be light limited (Malone, 1977).

A fourth observation relates to variability. Our vision of a gross pattern in the amassed aquatic data represents an admitted willingness to look far beyond detail. The gross relation offers no real predictive capability. There are many "anomalous" points, the scatter about the general curve is very large. Obviously, the picture based only on external inputs is imperfect. Much scatter about a perfect loading curve undoubtedly is due to variations in internal supplies and temporal changes in external loading across years. For example, where a loading rate has been determined for a lake that has received higher loads in the past, or one that has received a sustained intermediate-to-high level for some time, internal supplies probably become significant. In fact, the suggested initial rise in "response" to loading increases between 0.1 and 1 g*N m^{-2} yr^{-1} probably appears more acute for just this reason (Figure 3). N and P enrichments at ELA (Fee 1979) with background conditions near 0.04 gP (= 0.29 g*N) m^{-2} yr^{-1} and 17.1 gC m^{-2} yr^{-1} (average of n= 20 unfertilized annual budgets) rise only to 76 gC m^{-2} yr^{-1} when inputs increase to 0.4 gP (= 2.9 g*N) m^{-2} yr^{-1} (averages for n=10 values for P input ranges between 0.2 and 0.5 gP m^{-2} yr^{-1}). These enriched, but previously pristine lakes, lacking large internal reservoirs, thus fall near the lowest open ocean values (Figure 3). We hypothesize that the group of lakes achieving NPP levels well above 100 at loadings below 1 to 4 g*N m^{-2} yr^{-1} , like their terrestrial counterparts, have significant internal sources.

The last observations relate to scale. The gross pattern illustrates two significant points. The first point is the simple illustration of the differing degrees of nutrient "limitation" apparent for aquatic systems. Due to the variety of trophic states evident in both freshwater and seawater ecosystems, great caution must be exercised in blindly comparing, for example, "fresh vs. saltwater systems" without regard to the the level of nutrient supply.

Also illustrated by the aquatic pattern is a phenomenon of spatial scaling involving the conceptualization of the ecosystem. That a variety of arbitrary "sizes" of "ecosystems" occur is obvious, and as previously indicated, such aspects influence perception of loading rates. To what extent may we be able to consider the loading: production relationship, as read "backwards" (i.e., from high to low values), to be analogous to a constant point source load delivered to progressively larger areas? As an example, the dotted line with single-headed arrow in Figure 3 gives values moving from the smaller area of New York Harbor to the larger area of the New York Bight Apex. Additionally, using values of O'Reilly et al. (1976), it is suggested that some small areas of the Apex-lower Harbor may achieve even higher production rates, in excess of 600 gC m^{-2} yr^{-1}. Is the pattern thus a tool to extrapolate generally from small to large areas? In many cases, there are problems with this, although perhaps these are not insurmountable. First, in the case of New York as the point source, as we extend the analysis to the entire New York Bight, another input starts to become significant. Malone (1984) calculates that at the level of the entire Bight the lower Hudson-New York Harbor point source of external N becomes dwarfed by an input of "new" nitrogen from the Gulf of Maine and adjacent continental slope which is an order of magnitude larger. Thus, not all point sources can be simply diluted to the larger area even when the nutrients are transported nearly conservatively (cf. Garside et al., 1977; Malone, 1982; 1983) and a general relationship to loading is evident; on regional scales there are sometimes overlapping sources that also must be included. Secondly, the "grand" extrapolation would extend to the global ocean average (Figure 3), which lies well to the left of the sharp initial rise in production. In this case, the simple explanation is that the conceptualization of the system boundaries has been changed such that now "internal" nutrient sources are significant compared to an area like the ELA lakes, and yet these sources are not included in the value on the x-axis.

4. THE IMPORTANCE OF EXTERNAL AND INTERNAL SUPPLIES ACROSS TERRESTRIAL AND AQUATIC ECOSYSTEMS

Obviously, as for globally averaged ocean situation, the internal supply of nutrients in terrestrial situations is key to their productivity. Thus, there has been emphasis on disturbances which cause a loss of the accumulated nutrients available as internal stores (e.g., Vitousek et al., 1979). In general, much effort has been directed towards the identification of nutrient "availability" within terrestrial ecosystems. Since "available" pools and fluxes are imperfectly known, many surrogate measures have been developed to assess the potential fertility of sites. In essence, the surrogate measures are indices which attempt to estimate a major portion of the internal nutrient supply that biota experience. Studies may use a "soil mineralization potential" estimate, litter decomposition rate, or the concentration of N in litterfall as an index of available nutrient (e.g., Keeney, 1980; Birk and Vitousek, in press). Perhaps because of this strong recognition of differences in "availability" and internal sources, land plant physiological concepts relating to the "efficiency" of nutrient use, expressed in terms of net plant biomass (or carbon) production per unit of N used, recently have emerged (e.g., Chapin, 1980; Vitousek, 1982; Shaver and Melillo, 1984).

Analogous attempts to identify the nutrients available for autotrophic production exist for aquatic systems, especially in oligotrophic open ocean surface waters where the question of how production at any level can be supported without detectable nutrient concentrations is of interest. Perceptive work by Dugdale and Goering(1967), Dugdale(1967), and others, including Eppley and Peterson (1979) focused on how to distinguish between internal and external nutrient supplies available to phytoplankton. In spanning the range from oligotrophic to eutrophic systems in the open sea (Figure 3) Dugdale (1976) calculated that external (or "new") N supported about 6 to 66% of production and thus "the amount of new nutrient injected into the photic zone controls the productivity status of an ocean area." Internal supplies are also seen thus to contribute, providing from 94% (low production) to 34% (high production) of the autotrophs' needs.

The common challenge to ecologists in all fields is to determine the total available nutrient supply; or, in ecotoxicological terms, the nutrient "exposure regime" experienced by biota is key. This topic has been broached indirectly by development of empirical tools for lake management. For example, refinements to early predictive models of loading and productivity that correct for features like morphometry and flushing are essentially efforts to account for differences, across lakes, in what nutrient supplies the autotrophs in the epilimnion actually "see".

If we were able to characterize ecosystems in terms of both external and internal fluxes, or "total" supply, terrestrial and aquatic systems might come closer to fitting on one supply: production curve. For example, if one calculates, from the forested site data the N that appears in all measured forms of net biomass production (i.e., incremental growths of aboveground tree tissues and litterfall), the N supply that must occur at minimum can be shown to range from at least 5 to over 20 gN m^{-2} yr^{-1} . Such values are in the intermediate range of values for external nutrient supplies (Figure 3), accompanied by many aquatic production values only a bit less than the terrestrial production values. Calculated values are mimima for the actual supply--nutrient withdrawal from senescing leaves (for example, if as much as 50% of leaf tissue N were retranslocated before abscission a value equal to N in litterfall, or a range of 0.11 to 8.70 gN m^{-2} yr^{-1}) and perhaps stemflow and leaf wash of nutrients (usually less than 1 gN m^{-2} yr^{-1}) might also be included. Even so, it seems that terrestrial carbon production may always lie slightly above an aquatic derived "total" supply:production curve based primarily on phytoplankton production. The C/N ratio in the tissues of net production are much higher than planktonic values because of structural carbon components, especially cellulose. Whereas the Redfield average is 6.625 C:N (atoms) and newly produced tissue of assemblages of plankton probably varies by about a factor of 2 around this value, a functional regression slope (Ricker, 1973) for C and N in forested ecosystem NPP indicated a value close to 60: 1 (atoms).

In spite of the difficulty in determining availability or total supply for any system, some inference on the relative importance of internal supply can be made. This is done (Table 2) by using (1) the ratio of net carbon production in primary production and the external N inputs and (2) the ratio of net N needs for net carbon production and the external N input.

Several points emerge from this exercise. Carbon production per unit externally applied nitrogen (exclusive of N fixation) appears greatest where inputs are low. The range is great in all systems because, as suggested by Figure 3, increases in production range do not keep pace with nutrient inputs. Net production varies by a much smaller factor across increasing nutrient input in each of the systems so ratios of carbon produced per unit of N input decrease dramatically with most enrichments. Values for gC/gN in unfertilized terrestrial systems are greater than most aquatic systems for the reason explained above. However, adding in N fixation in terrestrial areas would probably lower the ratio significantly in many cases. The range for the ratio of needs/inputs (and thus the range in % of needs potentially available from external sources) is overlapping across the systems (except the ocean average). All traditional classes of ecosystems contain examples which exist at each level of trophic status with regard to internal and external nutrient supplies, an observation running counter to some classical aquatic-terrestrial distinctions.

While most systems seem to have less than 50% of their net primary production needs available from outside sources, the most heavily enriched aquatic systems have external supplies that alone seem to exceed plant uptake needs many times over. Note that in some cases a low needs/inputs value could be arbitrarily low if the inputs were not actually within the domain of the producers. For example, nutrient injections into the hypolimnion can represent an input, by and large, not experienced by autotrophs (Fee, 1979); yet, for the production level attained in that situation, the needs/input ratio would appear low. Low values in Table 2 are not of that ilk. In these obvious cases where nutrients are not "limiting" it thus is easily demonstrated that not all "available" nutrient has to be used by an ecosystem. Where the external supplies are not so high, the distinction of nutrient limitation and the ability of the ecosystem to use its available nutrients is not so easily made. Nevertheless, backcalculating the supply of nutrients from the needs of autotrophs can only provide a minumum estimate. Where inputs are less than required, for example, a backcalculation of recycling need not represent the total quantity of a nutrient recycled (cf. Kemp et al., 1982). The efficiency of ecosystem nutrient use of

TABLE 2: INPUTS AND PRODUCTION NEEDS: RANGE OF CONDITIONS IN DIFFERENT ECOSYSTEMS

SYSTEM	gC PRODUCED PER gN (*N) EXTERNALLY SUPPLIED[1]	RATIO OF N NPP NEEDS DIVIDED BY N (*N) EXTERNALLY SUPPLIED[1]	% N NPP PROVIDED BY N EXTERNALLY SUPPLIED[1]
FOREST[2]	8456 TO 220	121 TO 3.6	0.8 TO 28%
GRASSLAND[3]	1097 TO 434	28.9 TO 7.24	3.5 TO 13.8%
LAKE[4]	862 TO 0.5	152 TO 0.09	0.7 TO 1111%
COASTAL[5]	59 TO 1.08	10.3 TO 0.19	9.7 TO 526%
OPEN OCEAN[4,6]	114 TO 11.4	20 TO 2	5 TO 50%
GLOBAL OCEAN[5,7] (AVERAGE)	818 TO 420	144 TO 74	0.7 TO 1.35%

FOOTNOTES:

1 OMITTING N FIXATION, FROM DATA IN APPENDIX 1 (NOTE: HIGHER VALUES WHERE N INPUTS ARE LOWER)

2 NPP FOR C INCLUDES ABOVEGROUND INCREMENTS PLUS OVERSTORY LITTERFALL ; N NEEDS IS N CONTENT IN THOSE SAME FLUXES ; C/N (WEIGHT) IN THESE FLUXES AT TWO EXTREME SITES IS 60.8 AND 69.8 (COLE & RAPP, 1980).

3 NPP INCLUDES ABOVE AND BELOWGROUND PRODUCTION FOR SITES IN APPENDIX 1, ASSUMES CONVERISON FROM C TO N IN NPP OF 38 TO 60 (WEIGHT)TO CREATE EXTREMES OF RANGE (C.F. COUPLAND AND VAN DYNE, 1979; BAZILEVICH AND TITLYANOVA, 1980)

4 NPP AND *N (*N CONVERTED FROM P FOR LAKES, CALCULATED FROM MODEL FOR OPEN OCEAN) C/N = 5.68 (WEIGHT, 6.625 BY ATOMS) ASSUMED.

5 NPP AND N. C/N = 5.68 (WEIGHT, 6.625 BY ATOMS) ASSUMED

6 FROM EPPLEY & PETERSON(1979), ORIGINAL CALCULATIONS OF DUGDALE (1976) ARE SLIGHTLY DIFFERENT (RANGE FROM 6 TO 66%) BECAUSE EPPLEY & PETERSON INCLUDE UREA UPTAKE IN CALCULATING NEW PRODUCTION.

7 RANGE IS WITH AND WITHOUT N_2 FIXATION, SEE APPENDIX 1.

available nutrients, even under "non-limiting" situations where more added nutrients would result in more net production, need not be 100%.

5. ENRICHMENTS OF COASTAL AREAS: EXPERIMENTAL AND FIELD OBSERVATIONS

In coastal areas, where many marine disposal issues, including eutrophication, are focused, what does the picture of production, internal supply, external supply, and nutrient use look like? Boynton et al. (1982), Nixon (1983) and Nixon and Pilson (1983) have looked across systems for relationships between inputs and NPP. Although both within a system (Chesapeake Bay) across years and across different systems, their analyses suggest weak relationships between external supplies and production, the patterns are not particularly striking. In fact, little relationship seemed evident across the U.S. coastal sites of Nixon's summary, which included areas of low-to-high dissolved inorganic nitrogen loading (Figure 5); presented here on an areal basis, corrections for differences in flushing and mean depths improved the relationship slightly (Nixon 1983). External sources, although quite large compared to most systems (Figure 2) seem clearly not the only factor controlling productivity levels (Nixon, 1981; Boynton et al., 1982; Kemp et al., 1982; Nixon and Pilson, 1983).

Recycled (or "regenerated" or "remineralized") nutrients help coastal areas to sustain high levels of production (Furnas et al., 1976; Nixon et al., 1976; Nixon, 1981; Kemp et al., 1982). Like open ocean and deep water pelagic areas in lakes, water column recycling provides an important internal nutrient source. In analogy to terrestrial situations with recycling by litterfall to soil nutrient pools, and in contrast to the open sea or thermally stratified lakes with small areas of epilimnetic sediments (Fee, 1979), sedimentation of plankton and other autochthonous organic detritus to shallow water sediments results in a significant amount of benthic recycling of both N and P back to the water column in partial or well-mixed estuarine and coastal areas (cf. reviews of Zeitschel, 1980; Nixon, 1981; Klump and Martens, 1983; Nixon and Pilson, 1983).

The analogy to the terrestrial system seems apt in a quantitative sense. For example, in Narragansett Bay, the sedimentation of N is on the order of 20 to 22 gN m^{-2} yr^{-1}, representing an amount almost equal to the total N input and much greater than the dissolved nutrient input to the Bay (Kelly and Nixon 1984; Nixon and Pilson 1984). Of this, however, about 12 to 14 gN m^{-2} yr^{-1} returns to the water column in forms available to autotrophs. In comparison, litterfall "sedimentation" ranges from 0.11 to 8.7 gN m^{-2} yr^{-1} for boreal and temperate forests (Cole and Rapp 1980) and 3.4 to 22.4 gN m^{-2} yr^{-1} for tropical forests (Vitousek 1982). Perhaps the terrestrial-coastal analogy could be carried further, if data were available: the retranslocation from leaves could be likened to pelagic remineralization, because both have the effect of retaining nutrients within a sphere of the autotrophs that is most readily available to the primary producers, and outside of a sphere where "immobilization" by soil or sediment microbes can offer competition for the nutrients.

As yet, pelagic remineralization rates seem difficult to quantify fully in time and space. However, a substantial body of information has accumulated for benthic recycling. Benthic metabolism seems directly related to primary production in many areas (Nixon, 1981) and experiments confirm that high rates of organic deposition result in high rates of benthic fluxes of nutrients back to the water column (Nixon et al., 1980; Kelly and Nixon, 1984). Additionally, from the MERL enrichment experiment, the magnitude of benthic fluxes over a summer period generally was related to the magnitude of NPP.

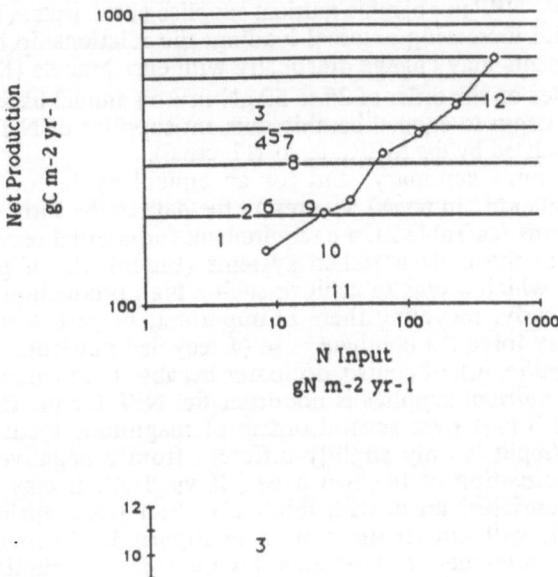

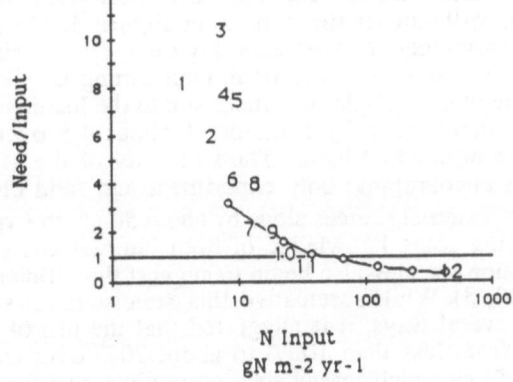

Figure 5: Primary production, nitrogen inputs, and the necessity of recycling: coastal marine ecosystems and the MERL experiment.

The numbers indicate values for U.S. coastal areas of summarized by Nixon (1983) and Nixon and Pilson (1983): 1= Kaneohe Bay, 2 = Long Island Sound, 3 = Chesapeake Bay, 4 = Apalachicola Bay, 5 = Barataria Bay, 6 = Patuxent River Estuary, 7 = Pamlico River Estuary, 8 = Narragansett Bay, 9 = Delaware Bay, 10 = Southern San Francisco Bay, 11 = Northern San Francisco Bay, 12 = New York Harbor. Points with connecting lines are values from second year of the MERL experiment; NPP data are system production calculated from oxygen changes (Oviatt et al., in press) and loadings are slightly modified from Nixon et al., 1984 (see Kelly et al., in press). The bottom panel shows the calculated nitrogen needs for production (assuming Redfield ratio for C/N) divided by inputs, an index of the necessity of recycling to satisfy autotrophic NPP nitrogen needs. The C/N ratio may vary across the input range, but perhaps not by more than a factor of two. The line at 1.0 is to illustrate where N for NPP can be met by external inputs; above 1.0 recycling is required.

Importantly, neither NPP nor benthic nutrient supplies to the water (Figure 6) appeared to keep pace 1:1 with increasing external loading; the relationship between benthic and pelagic compartments may change drastically with enrichments (Kelly et al., in press). Above loading rates on the order of 25 to 50 gN m^{-2} on annual basis, external deliveries of nutrients may begin to exceed benthic nutrient supplies of N for natural situations similar to that simulated by the MERL tanks (Figure 6).

If, for Nixon's summary, and for an annual cycle for NPP in the MERL experiment (Oviatt et al., in press) we present the data in the form of autotrophic needs per input of nutrients (as Table 2), the requirement for internal recycling becomes quite apparent in all but the most enriched systems (Figure 5). Of particular interest is Chesapeake Bay, which seems to achieve such a high production rate for its external loading level. Clearly, recycling there is important; in part, a relatively long water residence time may force the continued use of recycled nutrients. The requirement for recycling (i.e., need/input), of course, decreases because production response to loading at these levels of nutrient supplies is not dramatic; NPP for the field data varies only within a factor of 3 to 4 over several orders of magnitude loading increase and the decrease in need/input is only slightly different from a negative exponential decay indicating autocorrelation of the two axes (X vs. 1/X). It may be that other areas, particularly less enriched areas with relatively short water residence times (unlike Chesapeake Bay), will not fit the pattern in Figure 5. However, results from the controlled MERL enrichment experiments (Figure 5) show a similar pattern, as well as suggest the importance of external loads in maintaining an NPP level. NPP markedly increased, but the reponse is slight in comparison to the loading required to produce the reponse-- NPP changed only by a factor of about 4.5 over nearly two orders of magnitude increase in external loads. The similarity of the MERL results and cross-system analyses is encouraging; both experiment and field observation suggest NPP needs being met by external sources alone by about 50 gN m^{-2} yr^{-1}.

By combining some knowledge of both internal and external supplies in the experimental situation we may also begin to suggest the efficiency of use of nutrient in coastal waters (Table 3). While speculative, this exercise raises some interesting notions. As calculated in several ways, it is suggested that the use of nitrogen "available" to autotrophs drops from less than 100% to about 20% over an approximate range of external nutrient loading which coastal areas currently experience. The values would be a bit low in certain situations where denitrification and the loss of N$_2$ gas were important and inappropriately included in the recycled "available" pool, although the supply calculations (Table 3) take this into account. The observation, especially at levels where internal sources seem required because external sources are insufficient, presents us with a curious situation--nitrogen supplies may be present to excess, but additional nutrients (N, P and Si were added) increase production. Perhaps curious and hidden interrelationships between the internal supplies of different nutrients become involved here. Additionally, the timing of nutrient loads during the year would be important to consider, but summer (the period for these data) is when we often see nitrogen in low concentration and phytoplankton demands most acute (Furnas et al., 1976). This tentative finding may also be surprising to those attempting to add up individual sources of nutrient supply, including bacterial remineralization, meso and microzooplankton excretion, etc. and sometimes having difficulty finding "enough" nitrogen to satisfy the apparent autotrophic needs (cf. Nixon, 1981). We need to examine this topic over annual cycles and for natural ecosystems, especially unenriched ones; but it seems that the relative inefficiency can also be simply inferred from the fact that dissolved inorganic nitrogen is detectable in most estuaries, and that the DIN concentration is related to the external loading across systems (Nixon, 1983; see also the MERL experiment-- Kelly et al., in

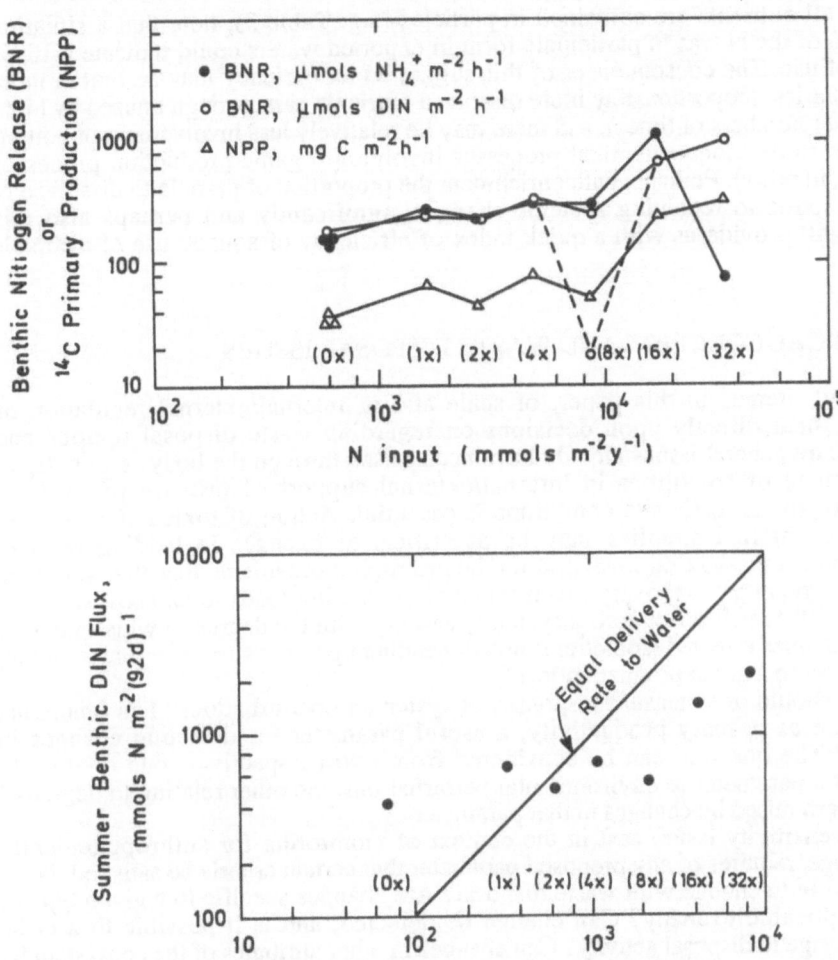

Figure 6: Benthic nitrogen recycling and primary production: average values for a summer period in the MERL enrichment gradient experiment.

Modified from Kelly et al., (in press). Treatment designations are denoted by control (0x), and a geometric series of input levels (1x, 2x,...32x). The top panel shows rates of dissolved inorganic nitrogen (NH_4^+, or DIN = $NO_3^- + NO_2^- + NH_4^+$) released to water column; the pattern across the gradient is similar to, and driven by, the primary production in the water column. The bottom panel shows the relative input to the water from external (feedwater plus nutrient additions) and internal (benthic feedback) sources across the gradient. At the upper treatment levels, benthic fluxes become difficult to measure and estimates are extremely uncertain, but rates are considerably less than the external input rate (see Kelly et al., for detailed discussion).

press). Not all nutrients are contained in particles (e.g. Table 3); note that a situation where 100% of the N was in particulate form in exported waters could indicate a 100% efficiency of use. The consequences of this suggested inefficiency may be that in more enriched estuaries, proportionately more dissolved nutrients pass through unused by biota (or used fewer numbers of times), and there may be relatively less involvement of bottom sediments in those biogeochemical processes involving organic production processes (Kelly et al., in press). Perhaps, with enrichment the proportion of particle to dissolved N forms in exports to receiving systems changes significantly and perhaps also this proportion will provide us with a quick index of efficiency of aquatic use of available nutrients.

6. ECOLOGICAL CONCEPTS AND WASTE DISPOSAL ISSUES

The recurrent themes in this paper, of scale and of internal/external regulation of ecosystems, bear directly upon decisions on regarding waste disposal options and practices. Many general issues already have been raised through the body of this paper. Given the range of conditions in internal/external support of primary production, knowledge of the specific site conditions is essential. Action of toxicants and other chemicals on internal supplies may be as critical as changes in loading rates to maintenance of a production level and the heterotrophic organisms that that supports. Additionally, the rate of recovery given removal of existing loads to an ecosystem (cf. Oviatt et al., 1984) will differ across systems, varying with the degree to which internal sources, sometimes affected (sometimes not, depending upon the stress) by that removal, provide support to a given production level.

How should one measure responses of systems to perturbations? Is a functional measure, such as primary productivity, a useful parameter for detecting changes in ecosystems? The question can be considered from two perspectives, one relating to sensitivity of a parameter to environmental perturbations, the other relating to degree of societal concern raised by changes in that parameter.

The sensitivity issue, cast in the context of monitoring for anthropogenically induced change, requires of any proposed parameter that certain criteria be satisfied. Is the parameter likely to change with waste disposal? Are changes specific to a given type of waste, or applicable to many? Can change be detected, and is it possible to ascribe detectable change to disposal activity? Can changes in other attributes of the ecosystem be used to signal upcoming change in this parameter?

For many marine areas, primary production would be a poor choice if the concern were only to find the most sensitive indicator. To begin with, not all disposals might be expected to change production levels. Indeed, even in cases where coastal external nutrient loads are increased by two orders of magnitude, NPP changes are relatively slight. This insensitivity occurs when production rates exceed values on the order of 150-300 gC m^{-2} yr^{-1}; in other words, at levels except those reported for all but a few very low input systems. Another confounding paradigm is that at any level of production, a similar level of production (either across systems, or across time within an individual system) can result from different mixtures of sources. The peculiar physical circulation and morphometric features, differing across ecosystems, is in part responsible for the occurrence of similar production at different external loading rates. We therefore must expect responses to human interventions will differ somewhat across these systems, where non-biological features play differing roles in shaping biological responses. Nevertheless, for all systems, a change in one source that eventually alters either internal or external supply rates, be it directly from nutrients added in waste discharge or as an

Table 3: AUTOTROPHIC USE OF AVAILABLE NITROGEN SUPPLIES IN THE MERL EUTROPHICATION
GRADIENT EXPERIMENT DURING A SUMMER PERIOD

		TREATMENT		
	0X	1X	16X	32X
SOURCE[1]				
WATER, mmols N m^{-2}				
(DIN)	19	86	1278	1891
(PARTICULATES)	9	28	166	96
EXTERNAL INPUT mmols N m^{-2}	72	341	4314	8588
NET DIN FLUX FROM SEDIMENTS[2] mmols N m^{-2}	415	656	1493	2184
PELAGIC RECYCLING	?	?	?	?
DEMAND[1,3]				
'NET' PRODUCTION NEEDS mmols N m^{-2}	256	514	1886	2586
SUPPLY CALCULATIONS				
(1) EXTERNAL INPUT +NET SEDIMENT FLUX + AMOUNT IN WATER[4] mmols N m^{-2}	515	1111	7251	12759
(2) EXTERNAL INPUTS + 'RECYCLED N' FROM NPP + AMOUNT IN WATER[5]				
a) assume 50% of NPP recycled within system[6] mmols N m^{-2}	228	712	6701	11868
b) assume 90% of NPP recycled within system[6] mmols N m^{-2}	330	918	7455	12902
'NET' PRODUCTION NEEDS AS PERCENTAGE OF CALCULATED SUPPLY				
USING (1) ABOVE[4]	50%	46%	26%	20%
USING (2a) ABOVE[5]	112%	72%	28%	22%
USING (2b) ABOVE[5]	78%	56%	25%	20%

FOOTNOTES:

[1] all expressed per 92 day summer period, data from Kelly et al. (in press)

[2] uncertain values, could be plus or minus at least 50%, especially at higher treatments, Kelly et al. (in press)

[3] Needs converted from hourly to daily ^{14}C production using conversion of Oviatt (1981), Volenweider (1966), see Oviatt et al. (in press), conversion to N by assuming C/N = 6.625 (atoms)

[4] Omits pelagic recycling, but includes only useful DIN (NH_4, NO_3, NO_2), not N_2 gas

[5] Omits benthic regeneration of organic N deposited prior to this period, and could include N_2 gas

[6] Range based on Nixon and Pilson (1984) NEP estimates, and calculations of Malone (1984). For Nixon and Pilson (1984) about 56% of annual production seemed regenerated as useful forms of DIN, 73% of production regenerated in all forms of N. Malone (1984) suggests about 60% of production is regenerated in useful forms on annual basis, with maximum of 80% during summer. The two values above bound these estimates.

indirect consequence of a discharge to either the marine or terrestrial environment may not necessarily significantly change production if the other sources are still sufficient to sustain a given level. Even if a change is expressed, it probably will be relatively small. Finally, given the additional problems of adequately sampling pelagic environments in time and space, it may be difficult to suggest that apparent productivity changes in the field from one year to the next are more real than methodological.

Of course, the situation may be rather different for low production systems where very much of the production occurs from internal recycling. Here, sustaining a level of production seems more critically dependent upon the maintenance of recycling processes; yet, changes in external loads can produce more dramatic changes in production. Schindler et al. (1985) have summarized results of eutrophication and acidification studies in a group of the lowest production aquatic systems studied, the ELA lakes. A conclusion was that compositional change in the biotic community was more sensitive in indicating response to these two anthropogenic disturbances than was a change in productivity.

The higher productivity, the less easily detected will be a productivity change even if it actually occurs, unless one is dealing with a dramatic reduction of external loads to a system with little accumulated internal supply. Is this also true for biotic compositional changes? In other words, is there a "saturation curve" for biotic changes, suggesting incrementally less change per unit of increased loading as one moves towards higher loading levels? Research into compositional changes deserves a great deal more experimental study. But one could begin by constructing loading curves from cross-system studies of biotic structure.

The "baseline" for many coastal areas, the ones most actively studied by coastal ecologists, is one already much affected by human activity, and we need to remain aware of this as we consider changes to ensue from discharge activities. As an example of biotic change across a loading gradient we might begin by a focus on recent studies on the Chesapeake, a system, which , as a whole, is somewhat low-to-intermediate on the scale of reported external nutrient loading to coastal areas (Figures 3-5). Data collected over the last several decades offer evidence on the decline, especially in the more enriched upper reaches of the Bay, of submerged aquatic vegetation, of some species of fish (especially freshwater spawners) and shellfish (oysters), of the numbers of migrating waterfowl, and changes in water quality including a spreading of the area covered by anoxic bottom water (Chesapeake Bay Program, 1982; 1983a,b; Orth and Moore, 1983; Kemp et al., 1983; 1984). Does the sequence of changes we are seeing now in this area suggest a historical phase which many areas have long since passed through and that we did not document nor appreciate? If so, then perhaps we also need to try to read the past history of our enriched estuaries and coastal areas. As Nixon (1983) suggests, this may give evidence of the future of less enriched areas such as lower Chesapeake Bay as population pressures and waste discharges continue to rise. At the other end of the scale, of course, we should be more actively studying estuarine and coastal areas that are not as heavily influenced by human activities.

Although productivity may be a poor measure from the standpoint of sensitivity, it may be valuable to monitor for other reasons. One possible attribute of an indicator measure is that it conveys a sense of importance to humans. Do we care, for example, that a certain sensitive species, perhaps itself neither harvested nor a large component of the community, won't live in a particular area? For many species, importance to the biotic web is slight, and more significance must be assigned to lower trophic levels, which set the table for the rest of the ecosystem. The most sensitive indicators might forewarn of other changes, but do not necessarily provide a basis for management action. Change in such early warning indicators thus is a necessary but not sufficient correlate of fundamental system change (cf. Levin et al., 1984; Kimball and Levin, 1984; Limburg et al., 1984). We have not yet identified sensitive indicators appropriate for all wastes; however, our efforts must be directed beyond the development simply of sensitive indicators. We need a hierarchy of measures ranging from those which respond most rapidly to perturbation to those which are associated with major shifts in ecosystem processes. The determination of when changes in such measures justifies regulatory or management actions is an extrascientific one; but the development of a clear understanding of how various measures respond to perturbations represents an important challenge for scientists.

Finally, our cross-system analyses raise many questions relative to issues of dispersion versus containment of wastes that offer another challenge. Primary among these questions would be a clarification on scale. We need to eventually determine if and when a change, say even clearly unacceptable, to a certain percentage of a given area is equivalent to a lesser change spread over a larger portion of the same area. Such concerns are critical to consideration of disposal options. For ecological problems that must deal with different species and processes operating on different temporal and spatial scales (Steele, 1985 and this volume), in contrast to some chemical problems perhaps addressable by the relatively simple notion of a "assimilative capacity", this represents a most significant problem. We usually can't suggest even what proportion of an area will be affected, nor agree whether that is or is not acceptable, much less comment on the desirability of various scenarios of scale.

As a simple example, consider, as we have for much of this paper, how human waste disposal activities involving primarily "non-toxic" nutrients may influence the coastal areas of the ocean. We now have a rough picture on nutrient enrichments in terms of NPP, but even this will vary greatly based on individual characteristics and past history of a given system. However, it seems safe to suggest, given a nutrient saturation effect, and because it appears that a great percentage of nutrients added to coastal areas may not be bound irretrievably within the sediments, that the spatial scale of this influence is likely to extend to less enriched offshore areas. This is to be expected given a steady or increasing population level in most coastal zones. Nonetheless, it is not simple to predict, by dilution alone, all aspects of a NPP stimulatory effect.

From an aquatic pollution perspective, one of the most important aspects of net autotophic production may be that the autochthonous particles can provide a mechanism for transport of some elements and compounds that are of concern because of their ecotoxicological potential and that are highly partitioned towards particle phases in water or sediments (e.g., Olsen et al., 1982). Changes in NPP, as controlled by flows of nutrients within ecosystems, may therefore affect fate and transport questions. Importantly, particles move differently from fluids, are subject to heterotrophic trophic transfers, and often tend to be concentrated as determined by certain features unique to an individual system (Young et al., 1985). In many cases where dispersion is the goal, an aggregation of particles (either autochthonous or allochthonous) with undesired consequences, in locations not predicted and perhaps of highest biological activity, may be the unavoidable result.

Without biological activity, nutrient dispersion and its attendant effects could be calculated by simple dilution; the reality, unfortunately, is that biology affects many elements and compounds. Even once nutrients become used and hence become "internal" to the ecosystem they are still relatively mobile, available for reuse or passage to another "system". Curiously then, this reuse may act many times over as an agent for the creation of heterogeneous distributions of chemicals particularly attracted to particles, where many of those chemicals have considerably less freedom to escape the grasp of particles than do nutrients. Because of such problems, and due to both ecological and geochemical pecularities, we seem often left without the knowledge to suitably address many issues of scale even though we know them to be critical to waste disposal practices.

7. ACKNOWLEDGEMENTS

We thank R. Amir and T. Butler for compiling an excellent body of information on lakes and grasslands; without their patient efforts the exasperation of the data compilation task would have been insurmountable. D.Weinstein, E. Birk, S. Levine and J. Ford pointed us towards valuable concepts and data contained in papers outside of the marine and estuarine literature, but they are not responsible for the interpretations offered in this synthesis. C. Oviatt, S. Nixon and others involved in the MERL experiment at the University of Rhode Island kindly provided pre-prints of papers. Primary funding for this research was provided by the Office of Research and Development, U.S. Environmental Protection Agency, under Cooperative Agreement No. CR811060. Additional funding was provided by Cornell University. This publication is ERC-091 of the Ecosystems Research Center (ERC), Cornell University. The ERC was established in 1980 as a unit of the Center for Environmental Research at Cornell University. The work and conclusions published herein represent the views of the authors, and do not necessarily represent the opinions, policies, or recommendations of the Environmental Protection Agency or of Cornell University. EPA and Cornell do not endorse any commercial products used in the study.

8. REFERENCES

Alvarez-Borrego, S. , D. Guthrie, C.H. Culberson, and P.K. Park. 1975. 'Test of Redfield's model of oxygen-nutrient relationships using regression analysis.' Limnol.Oceanogr. **20**(5):795-805.

Armstrong, N.E. 1982. ' Responses of Texas estuaries to freshwater inflows.' In: V.S. Kennedy (ed.), Estuarine Comparisons , Academic Press , New York, pp. 103-120.

Bazilevich, N.I. and A.A. Titlyanova. 1980. 'Comparative studies of ecosystem function.' In: A.I. Breymeyer and G.M. Van Dyne (eds.), Grasslands.Systems Analysis. and Man. IBP19, Cambridge University Press, New York,pp. 713-758.

Berounsky, V.M. and S.W. Nixon. In press.'Eutrophication and the rate of net nitrification in a coastal marine ecosystem.' Est. Coast. Shelf Sci.

Birk, E.M. and P.M. Vitousek. 1985. In press. 'Nitrogen availability and nitrogen use efficiency in loblolly pine.'Ecology.

Boynton, W.R., W.M. Kemp, and C.W. Keefe. 1982. 'A comparative analysis of nutrients and other factors influencing estuarine production.' In: V.S. Kennedy (ed.), Estuarine Comparisons, Academic Press, New York, pp. 69-90.

Capuzzo, J.M.,J.M. Teal, and R.K. Bastian. This volume. Ecological and human health criteria for cross ecosystem comparison of impacts of waste management practice. pp. xx-xx.

Carpenter, E.J. and D.G. Capone. 1983. Nitrogen in the Marine Environment , Academic Press, New York, 900pp.

Chesapeake Bay Program. 1982. Technical Studies: A Synthesis. U.S. Environmental Protection Agency, Washington, D.C., 635 pp.

Chapin, F.S. 1980. 'The mineral nutrition of wild plants.' Ann.Rev.Ecol.Syst. 11:233-260.

Chesapeake Bay Program. 1983a. Chesapeake Bay: A Framework for Action, U.S. Environmental Protection Agency, Region 3, Philadelphia, PA, 186 pp + appendices.

Chesapeake Bay Program. 1983b. Chesapeake Bay: A Profile of Environmental Change, U.S. Environmental Protection Agency, Region 3, Philadelphia, PA, 200pp + appendices.

Cole, D.W. and M. Rapp. 1980. 'Elemental cycling in forested ecosystems.' In: D.E. Reichle (ed.) Dynamic properties of forest ecosystems. IBP23 , Cambridge University Press, New York, pp. 341-409.

Correll, D.L. 1981. 'Nutrient mass balances for the watershed, headwaters intertidal zone, and basin of the Rhode River Estuary.' Limnol. Oceanogr. 26(6): 1142-1149.

Coupland, R.T. and G.M. Van Dyne. 1979. 'Systems synthesis.' In: R.T. Coupland (ed.), Grassland Ecosystems of the World. IBP 18, Cambridge University Press, New York, pp. 97-106.

Crawford, C.S. and J.R. Gosz. 1982. 'Desert ecosystems: Their resources in space and time.' Environ.Conserv. 9(3):181-195.

Crisp, D.T. 1966. 'Input and output of minerals for an area of Pennine Moorland: the importance of precipitation, drainage, peat erosion and animals.' J. Appl. Ecol. 3:327-348.

Day, J.W.,Jr. and G.P. Kemp. 1985. 'Long-term impacts of agricultural runoff in a Louisiana swamp forest.' In: P.J. Godfrey, E.R. Kaynor, S. Pelczarski, and J. Benforado (eds.), Ecological Considerations in Wetlands Treatments of Municipal Wastewaters , Van Nostrand Reinhold, New York, pp.317-324.

Dillon, P.J. 1975. 'The phosphorus budget of Cameron Lake, Ontario: the importance of flushing rate to the degree of eutrophy of lakes. ' Limnol. Oceanogr. 20:28-39.

D'Itri, F.M. (ed.) 1982. Land Treatment of Municipal Wastewater, Ann Arbor Sci., Ann Arbor, MI.

Dodd, J.L and W.K. Lauenroth. 1979. 'Analysis of the response of a grassland ecosystem to stress.' In: N. French (ed.), Perspectives in Grassland Ecology, Springer-Verlag, New York, pp.43-58.

Dugdale, R.C. 1967. 'Nutrient limitation in the sea: dynamics, identification, and significance.' Limnol. Oceanogr. 12: 685-695.

Dugdale, R.C. 1976. 'Chapter 7. Nutrient cycles.' In: D.H. Cushing and J.J. Walsh (eds.) The Ecology of the Seas, W.B. Saunders Co.,Philadelphia, pp. 141-172.

Dugdale, R.C. and J.J. Goering. 1967. Uptake of new and regenerated forms of nitrogen in primary productivity. Limnol. Oceanogr. 16:741-751.

Eppley, R.W. & B.J. Peterson. 1979. 'Particulate organic matter flux and planktonic new production in the deep ocean.' Nature 282: 677-680.

Eppley, R.W., E.H. Renger, and W.G. Harrison.1979. 'Nitrate and phytoplankton production in southern California coastal waters.' Limnol. Oceanogr. 24:483-494.

Fee, E.J. 1979. 'A relation between lake morphometry and primary productivity and its use in interpreting whole lake eutrophication experiments.' Limnol. Oceanogr. 24(3):401-416.

Frissel, M.J. (ed.) 1977. 'Cycling of mineral nutrients in agricultural ecosystems.' Agro-Ecosystems 4(1,2): 1-354.

Furnas, M.J., G.L. Hitchcock and T.J. Smayda. 1976. 'Nutrient-phytoplankton relationships in Narragansett Bay during the 1974 summer bloom.' In: Wiley, M (ed.), Estuarine Processes Vol. 1. Uses. stresses and adaptation to the estuary, Academic Press, New York, pp.118-134.

Garside, C., T.C. Malone, O.A. Roels, and B.A. Sharfstein. 1976. 'An evaluation of
    sewage derived nutrients and their influence on the Hudson Estuary and New
    York Bight.' Est. Coast. Mar. Sci. **4**: 281-289.
Goldman, J.C. and P.M. Glibert. 1983. 'Kinetics of inorganic nitrogen uptake by
    phytoplankton.' In: E.J. Carpenter and D.G. Capone (eds.), Nitrogen in the
    Marine Environment , Academic Press, New York, pp.233-274.
Goldman, J.C., J.J. McCarthy, and D.G. Peavey. 1979. 'Growth rate influence on the
    chemical composition of phytoplankton in oceanic waters.' Nature **279**:210-215.
Gulland, J.A. 1976. 'Chapter 12. Production and catches of fish in the sea.' In: D.H.
    Cushing and J.J. Walsh (eds.) The Ecology of the Seas, W.B. Saunders
    Co.,Philadelphia, pp. 283-314.
Hadley, N.F. and S.R. Szarek. 1981. 'Productivity of desert ecosystems.' Bioscience
    31 (10): 747-753.
Howarth, R.W. and J.J. Cole. 1985. 'Molybdenum availability, nitrogen limitation, and
    phytoplankton growth in natural waters.' Science **229**:653-655.
Jaworski, N.A. 1981. ' Sources of nutrients and the scale of eutrophication problems in
    estuaries.' In: B.J. Neilson and L.E. Cronin (eds.), Estuaries and Nutrients,
    Humana Press, Clifton, NJ, pp. 83-110.
Jonasson, P.M. 1981. 'III. Lakes. 2. Europe. Energy flow in a subarctic, eutrophic
    lake.' Verh. Internat. Verein. Limnol. **21** : 389-393.
Jones, R.A. and G. F. Lee. 1982. 'Recent advances in assessing impact of phosphorus
    loads on eutrophication-related water quality.' Water Res. **16**:503-515.
Kaul, L.W. and P.N. Froelich, Jr. 1984. "Modeling estuarine nutrient geochemistry in a
    simple system.' Geochim. et Cosmochim. Acta **48**:1417-1433.
Keeney, D.R. 1980.'Prediction of soil N availability in forest ecosystems: a literature
    review.' Forest Science **26**: 159-171.
Kelly, J.R. and M.A. Harwell. 1982. ' I. Nutrients.' In: Kelly, J.R., M.A. Harwell, and
    A.E. Giblin, 'Comparisons of the processing of elements by ecosystems.' ERC-
    021. Ecosystems Research Center, Cornell University, Ithaca, NY 14850.
Kelly, J.R. and M.A. Harwell. 1985. 'Comparisons of the processing of elements by
    ecosystems. I. Nutrients.' In: P.J. Godfrey, E.R. Kaynor, S. Pelczarski, and J.
    Benforado (eds.), Ecological Considerations in Wetland Treatment of Municipal
    Wastewater, Van Nostrand Reinhold Co., New York, pp.137-157. ERC-021a.
Kelly, J.R. and S.W. Nixon. 1984. 'Experimental studies of the effect of organic
    deposition on the metabolism of a coastal marine bottom community.' Mar. Ecol.
    Prog. Ser. **17**:157-169.
Kelly, J.R., V.M. Berounsky, S.W. Nixon, and C.A. Oviatt. In press. 'Benthic-pelagic
    coupling and nutrient cycling across an experimental eutrophication gradient.'
    Mar. Ecol. Prog. Ser.
Kemp, W.M., R.L. Wetzel, W.R. Boynton, C.F. D'Elia and J.C. Stevenson. 1982.
    'Nitrogen cycling and estuarine interfaces: some currents concepts and research
    directions.' In: In: V.S. Kennedy (ed.), Estuarine Comparisons, Academic
    Press, New York, pp. 209-230.
Kemp, W.M., W.R. Boynton, J.C. Stevenson, R.R. Twilley and J.C. Means. 1983.
    'The decline of submerged vascular plants in Chesapeake Bay: summary of results
    concerning possible causes.' Mar. Tech. Soc. Journal **17**(2): 78-89.
Kemp, W.M. , W.R. Boynton, R.R. Twilley, J.C. Stevenson, and L.G.Ward. 1984.
    'Influences of submerged vascular plants on ecological processes in upper
    Chesapeake Bay.' In: V.S. Kennedy (ed.), The Estuary as a Filter , Academic
    Press, New York, pp. 367-394.

Kimball, K.D. and S.A. Levin. 1985. 'Limitations of laboratory bioassays and the need for ecosystem level testing.' Bioscience 35 (3):165-171.

Klump, J.V and C.S. Martens. 1983. ' Benthic nitrogen regeneration.' In: E.J. Carpenter and D.G. Capone (eds.), Nitrogen in the Marine Environment , Academic Press, New York, pp. 411-457.

Lauenroth, W.K. 1979. 'Grassland primary production: North Americam grasslands in perspective.' In: N. French (ed.), Perspectives in Grassland Ecology , Springer-Verlag, New York, pp. 3-24.

Laws, E.A. 1983. ' Man's impact on the marine nitrogen cycle.' In: E.J. Carpenter and D.G. Capone (eds.), Nitrogen in the Marine Environment , Academic Press, New York, pp. 459-485.

Levin, S.A., K.D.Kimball, W.H. McDowell, and S.F. Kimball (eds). 1984. 'New perspectives in ecotoxicology.' Environ. Manangement 8 (5) : 375-442.

Lieth, H. and R. H. Whittaker (eds.). 1975. Primary Productivity of the Biosphere , Ecol. Studies 14 , Springer-Verlag, New York, 339 pp.

Likens, G.E. and F.H. Bormann. 1974. "Acid rain: a serious regional environmental problem. Science 184: 1176-1179.

Likens, G.E., F.H. Bormann, and N. M. Johnson. 1981. 'Interactions between major biogeochemical cycles in terrestrial ecosystems.' In: G.E. Likens (ed.), Some Perspectives of the Major Biogeochemical Cycles , SCOPE Report XX, pp. 93-112.

Likens, G.E., F.H. Bormann, R.S. Pierce, J.S. Eaton and N.M. Johnson. 1977. Biogeochemistry of a Forested Ecosystem , Springer-Verlag, New York, 146pp.

Limburg, K.L. C.C. Harwell, and S.A. Levin. 1984. 'Principles for estuarine impact assessment: lessons learned from the Hudson River and other estuarine experiences.' ERC-024. Ecosystems Research Center, Cornell University, Ithaca, NY 14853.

McCarthy, J.J. 1981. 'Uptake of major nutrients by estuarine plants.' In: B.J. Neilson and L.E. Cronin (eds.), Estuaries and Nutrients, Humana Press, Clifton, NJ, pp. 139-163.

McCarthy, J.J. and E.J. Carpenter. 1983. 'Nitrogen cycling in near-surface waters of the open ocean.' In: E.J. Carpenter and D.G. Capone (eds.), Nitrogen in the Marine Environment , Academic Press, New York, pp. 487-572.

Malone, T.C. 1977. 'Environmental regulation of phytoplankton productivity in the lower Hudson estuary.' Est. Coast. Mar. Sci. 5:157-171.

Malone, T.C. 1982. 'Factors influencing the fate of sewage-derived nutrients in the lower Hudson estuary and New York bight.' In: G. Mayer (ed.), Ecological Stress and the New York Bight: Science and Management, Estuarine Research Federation, Columbia, South Carolina, pp. 389-400.

Malone, T.C. 1984. 'Anthropogenic nitrogen loading and assimilation capacity of the Hudson River estuarine system, USA.' In: V.S. Kennedy (ed.), The Estuary as a Filter , Academic Press, New York, pp. 291-311.

Medina,E. 1982. 'Nitrogen balance in the Trachypogon grasslands of Central Venezuela.' Plant and Soil 67:305-314.

National Academy of Sciences. 1975. Productivity of World Ecosystems. Proceedings of a Symposium, NAS, Washington, 166 pp.

Nixon, S.W. 1981. 'Remineralization and nutrient cycling in coastal marine ecosystems.' In: B.J. Neilson and L.E. Cronin (eds.), Estuaries and Nutrients, Humana Press, Clifton, NJ, pp. 111-138.

Nixon, S.W. 1983. 'Estuarine ecology-- a comparative and experimental analysis using 14 estuaries and the MERL microcosms.' Final report to the U.S. Environmental Protection Agency, Chespeake Bay Program, Annapolis, MD.

Nixon, S.W. and M.E.Q. Pilson. 1983. 'Nitrogen in Estuarine and Coastal Marine Ecosystems.'In: E.J. Carpenter and D.G. Capone (eds.), Nitrogen in the Marine Environment , Academic Press, New York, pp. 565-648.

Nixon, S.W. and M.E.Q. Pilson. 1984. 'Estuarine total system metabolism and organic exchange calculated from nutrient ratios: an example from Narragansett Bay.' In: V.S. Kennedy (ed.), The Estuary as a Filter , Academic Press, New York, pp. 261-290.

Nixon, S.W., C.A. Oviatt, and S.S. Hale. 1976. 'Nitrogen regeneration and the metabolism of coastal marine bottom communities.' In: Anderson, J.M. and A. Macfadyen (eds.), The Role of Terrestrial and Aquatic Organisms in Decomposition Processes, Blackwell Scientific Publishers, London, pp. 269-283.

Nixon, S.W., C.D. Hunt and B.L. Nowicki. In press. 'The retention of nutrients (C, N, P), heavy metals (Mn, Cd, Pb, Cu), and petroleum hydrocarbons in Narragansett Bay.' In: P. Lasserre and J.M. Martin (eds.) Biogeochemical Processes at the Land Sea Boundary, Elsevier Press, pp. xx-xx.

Nixon, S.W., J.R. Kelly, B.N. Furnas, C.A. Oviatt and S.S. Hale. 1980. 'Phosphorus regeneration and the metabolism of coastal marine bottom communities.' In: K.R. Tenore and B.C. Coull (eds.), Marine Benthic Dynamics, Univ. South Carolina Press, Columbia, S.C. , pp. 219-243.

Nixon, S.W., M.E. Q. Pilson, C.A. Oviatt, P.Donaghay, B. Sullivan, S. Seitzinger, D. Rudnick and J. Frithsen. 1984. 'Eutrophication of a coastal marine ecosystem- an experimental study using the MERL microcosms.' In: M.J.R. Fasham (ed.), Flows of Energy and Materials in Marine Ecosystems, Plenum Press, London, pp.105-135.

Nyholm, N. 1978. 'A simulation model for phytoplankton growth and nutrient cycling in eutrophic, shallow lakes.' Ecol. Modell.4:279-310.

Olsen, C.R., N.H. Nutshall and I.L Larsen. 1982. 'Pollutant-particle associations and dynamics in coastal marine environments: A review.' Mar. Chem. 11: 501-533.

O'Reilly, J.E., J.P. Thomas and C.Evans. 1976. 'Annual primary production (nannoplankton, netplankton, dissolved organic matter) in the Lower New York Bay.' In: W.H. McKeon and G.J. Lauer (eds.), Fourth Symposiun on Hudson River Ecology, Hudson River Environmental Society, Inc. New York, Paper #19.

Oviatt, C., B.Buckley and S. Nixon. 1981. 'Annual phytoplankton metabolism in Narragansett Bay calculated from survey field measurements and microcosm observations.' Estuaries 4(3): 167-175.

Oviatt, C.A., M.E.Q. Pilson, S.W. Nixon, J.B. Frithsen, D.T. Rudnick, J.R. Kelly, J.F. Grassle and J.P. Grassle. 1984. 'Recovery of a polluted estuarine system: a mesocosm experiment.' Mar. Ecol. Prog. Ser. 16:203-217.

Oviatt, C.A., A. Keller, P. Sampou, L. Laffin-Beatty. In press,a. 'Patterns of productivity during eutrophication: a mesocosm experiment.' Mar. Ecol. Prog. Ser.

Oviatt, C.A., D.T. Rudnick, A.A. Keller, P.A. Sampou, G.T. Almquist. In press,b. 'A comparison of system (O_2 and CO_2) and C-14 measurements of metabolism in estuarine mesocosms.' Mar. Ecol. Prog. Ser.

Orth, R.J. and K.A. Moore. 1983. 'Chesapeake Bay: an unprecedented decline in submerged aquatic vegetation.' Science 222:51-53.

Peterson, B.J. 1980. 'Aquatic primary productivity and the ^{14}C-CO_2 method: a history of the productivity problem.' Ann. Rev. Ecol. & Sys. II, pp. 359-386.

Rast, W. and G.F. Lee. 1978. 'Summary analysis of the North American (U.S. portion) OECD eutrophication project: nutrient loading-lake response relationships and trophic state indices.' EPA 600/3-78-008. Corvallis, Oregon.

Redfield, A.C. 1934. 'On the proportions of organic derivatives in sea water and their relation to the composition of plankton.' James Johnstone Memorial Volume , Liverpool University Press, Liverpool, pp.176-192.

Redfield, A.C. 1958. 'The biological control of chemical factors in the environment.' Am. Sci. 46: 205-222.

Redfield, A.C., B.H. Ketchum and F.A. Richards. 1963.' The influence of organisms on the composition of seawater.' In: M.N. Hill (ed), The Sea: Volume 2. The Composition of Seawater , Interscience Publishers, John Wiley and Sons: New York. pp.26-77.

Richardson, C.J., D.L. Tilton, J.A. Kadlec, J.P.M. Chamie and W.A. Wentz. 1978. 'Nutrient dynamics of northern wetland ecosystems.' In: R.E. Good, D.F. Whigham and R.L. Simpson(eds.) Freshwater Wetlands: Ecological Processes and Management Potential , Academic Press, New York, pp.321-340.

Ricker, W.E. 1973. 'Linear regressions in fishery research.'J.Fish. Res. Bd. Can. 30: 409-434.

Riley, G. A. 1972. 'Patterns of production in marine ecosystems.' In: J.A. Wiens (ed.), Ecosystem Structure and Function, Proceedings of the Thirty-First Annual Biology Colloqium. Oregon State University Press, Corvallis, pp. 91-112.

Ryther, J.H. 1969. 'Photosynthesis and fish production in the sea.' Science 166:72-76.

Ryther, J.H. and W.M. Dunstan. 1971. ' Nitrogen, phosphorus, and eutrophication in the coastal marine environment.' Science 171: 1008-1013.

Schindler, D.W. 1974. "Eutrophication and recovery in experimental lakes: Implications for lake management.' Science 184: 897-899.

Schindler, D.W. 1976. 'Biogeochemical evolution of phosphorus limitation in nutrient-enriched lakes of the precambrian shield.' In: J.O. Nriagu (ed.) Environmental Biogeochemistry. Vol.2. Metals Transfer and Ecological Mass Balances, Ann Arbor Science Publishers, Ann Arbor, Mi, pp. 647-664.

Schindler, D.W. 1977. 'Evolution of phosphorus limitation in lakes.' Science 195: 260-262.

Schindler, D.W. 1981. 'Studies of eutrophication in lakes and their relevance to the estuarine environment.' In: B.J. Neilson and L.E. Cronin (eds.), Estuaries and Nutrients, Humana Press, Clifton, NJ, pp. 71-82.

Schindler, D.W. H.E. Welch, J. Kalff, G.J. Brunskill, and N. Kritch. 1974. 'Physical and chemical limnology of Char Lake, Cornwallis Island.' J. Fish. Res. Bd. Can. 31(5):585-607.

Schindler, D.W., K.H. Mills, D.F. Malley, D.L.Findlay, J.A. Shearer, I.J. Davies, M.A. Turner, G.A. Linsey, D.R. Cruikshank. 1985. 'Long-term ecosystem stress: the effects of years of experimental acidification on a small lake.' Science 228 : 1395-1401.

Schlesinger, W.H. 1978. 'Community structure, dynamics and nutrient cycling in the Okefenokee cypress swamp-forest.' Ecol. Monogr. 48:43-65.

Seitzinger, S.P. and S.W. Nixon. In press. 'Eutrophication and the rate of denitrification and N_2O production in coastal marine sediments.' Limnol. Oceanogr.

Shaver, G.R. and J.M. Melillo. 1984. 'Nutrient budgets of marsh plants:efficiency concepts and relation to availability.' Ecology 65(5):1491-1510.

Sims, P.L. and R.T. Coupland. 1979. 'Producers.' In: <u>Grassland Ecosystems of the World, IBP18</u>, R.T. Coupland(ed.), Cambridge University Press, New York, pp.49-72.

Smith, S.V. 1984. ' Phosphorus versus nitrogen limitation in the marine environment.' Limnol. Oceanogr. **29**(6): 1149-1160.

Steele, J.H. 1985. ' A comparison of terrestrial and marine ecological systems.' Nature **313**: 355-358.

Titlyanova, A.A. and N.I. Bazilevich. 1979. 'Ecosystem synthesis of meadows- nutrient cycling.' In: <u>Grassland Ecosystems of the World, IBP18</u>, R.T. Coupland (ed.), Cambridge University Press, New York, pp. 170-180.

Ulehlova, B., E. Klimo, and J. Jakrlova. 1976. 'Mineral cycling in alluvial forest and meadow ecosystems in southern Moravia, Czechoslovackia.' Int. Jour. Ecol. Environ. Sci. **2**:15-25.

Valiela, I. 1984. <u>Marine Ecological Processes</u>. Springer-Verlag, New York. 546 pp.

Valiela, I. and J.M. Teal. 1979. 'The nitrogen budget of a salt marsh ecosystem.' Nature **280**: 652-656.

Van Cleve, K. and V. Alexander. 1981. 'Nitrogen cycling in tundra and boreal ecosystems.' In: F.G. Clark and T. Rosswall (eds.), <u>Terrestrial Nitrogen Cycles.</u> Ecol. Bull. **33**, Stockholm, pp. 375-404.

Van Cleve, K. L. Oliver, and R. Schentner. 1983. 'Productivity and nutrient cycling in taiga forest ecosystems.' Can. J. For. Res. **13**:747-766.

Vitousek, P. 1982. 'Nutrient cycling and nutrient use efficiency.' Amer. Nat. **119**(4):553-572.

Vitousek, P.M., J.R. Gosz, C.C. Grier, J.M. Mellilo, W.A. Reiners, and R.L.Todd. 1979.'Nitrate losses from disturbed ecosystems.' Science **204**:469-474.

Vollenweider, R.A. 1966. 'Calculation models of photosynthesis depth curves and some implications regarding rate estimates in primary production measurements.' In: C.R. Goldman (ed.), <u>Primary Production in Aquatic Environments</u>. Mem. 1st Ital. Idrobiol., 18 Supple., University of California, Berkeley, pp.427-457.

Vollenweider, R.A. 1975. ' Input-output models, with special reference to the phosphorus loading concept in limnology. Scweiz. Hydrol 37: 53-84.

Vollenweider, R.A. 1976. "Advances in defining critical loading levels of phosphorus in lake eutrophication.' Mem. Ist. Ital. Idrobiol. **33**:53-83.

West, N.E. 1978. 'Physical inputs of nitrogen to desert ecosystems.' In: N.E. West and J.J. Skujins (eds.), <u>Nitrogen in Desert Ecosystems.</u> (US/IBP Synthesis Series 9) Dowden, Hutchinson & Ross, Stroudsburg, Pennsylvania, pp.165-170.

Woodmansee, R.G. 1979. 'Factors influencing input and output of nitrogen in grasslands.' In: N.R. French(ed.), <u>Perspectives in Grassland Ecology</u>, Springer-Verlag, New York, pp.117-134.

Young, R.A., D.J.P. Swift, T.L. Clarke, G.R.Harvey and P.R. Betzer. 1985. 'Dispersal pathways for particle-associated pollutants.' Science **229**:431-435.

Zeitschel, B.F. 1980. 'Sediment-water interactions in nutrient dynamics.' In: K.R. Tenore and B.C. Coull (eds.), <u>Marine Benthic Dynamics,</u> Univ. South Carolina Press, Columbia, S.C., pp. 195-212.

APPENDIX 1: DATA BY ECOSYSTEM

TYPE	NAME OR LOCATION	REFERENCES [1]	NPP g C m^{-2} yr^{-1}	N INPUT g N m^{-2} yr^{-1}	P INPUT g P m^{-2} yr^{-1}	*N INPUT g *N m^{-2} yr^{-1}
LAKE [2]	MIRROR (USA)	DEVOL & WISSMAR(1978)	43.0 [3]	1.53	0.15	1.08
	LAWRENCE (USA)	"	171.0 [3]	4.44	0.1	0.72
	FINDLEY (CANADA)	"	5.0 [3]	7.43	0.2	1.44
	ELA240 (CANADA)	FEE (1979)	14.1	2.42	0.08	0.58
	ELA240 (CANADA)	"	11.4	2.23	0.06	0.43
	ELA114 (CANADA)	"	23.7	0.85	0.04	0.29
	ELA114 (CANADA)	"	19.8	0.95	0.03	0.22
	ELA120 (CANADA)	"	8.2	1.11	0.04	0.29
	ELA223 (CANADA)	"	12.9	1.01	0.04	0.29
	ELA223 (CANADA)	"	11.0	1.12	0.04	0.29
	ELA223 (CANADA)	"	16.3	0.8	0.02	0.14
	ELA224 (CANADA)	"	11.1	0.73	0.03	0.22
	ELA224 (CANADA)	"	7.7	0.81	0.03	0.22
	ELA302S (CANADA)	"	30.0	0.65	0.03	0.22
	ELA302S (CANADA)	"	22.1	0.85	0.04	0.29
	ELA302S (CANADA)	"	18.5	0.95	0.03	0.22
	ELA302S (CANADA)	"	30.2	0.73	0.02	0.14
	ELA303 (CANADA)	"	30.6	0.89	0.04	0.29
	ELA305 (CANADA)	"	13.7	0.79	0.04	0.29
	ELA239 (CANADA)	"	14.2	1.05	0.05	0.36
	ELA239 (CANADA)	"	17.1	1.12	0.04	0.29
	ELA239 (CANADA)	"	14.1	1.33	0.05	0.36
	ELA239 (CANADA)	"	14.8	1.24	0.05	0.36
	WASHINGTON (USA) [4]	EDMONSON&LEHMAN(1981), RAST&LEE(1978)	766.0	7.8	2.3	16.6
	WASHINGTON (USA) [4]	"	354.0	4.4	0.43	3.10
	GEORGE (USA)	RAST&LEE(1978)	7.2	1.8	0.07	0.51
	WINGRA (USA)	DEVOL&WISSMAR(1978), RAST&LEE(1978)	870.0 [3]	8.83	0.88	6.36
	WINGRA (USA)		600.0 [3]	23.6	0.96	6.94
	CHAR(CANADA)	SCHINDLER et al.(1974)	4.2	0.31	0.02	0.14
	MERETTA(CANADA)	KALFF&WELCH(1974)	11.3	0.6	0.2	1.44
	CAYUGA(USA)	RAST&LEE(1978)	58.0	14.2	0.81	5.85
	CANADARAGO(USA)	"	195.0	18.0	0.8	5.78
	CANADARAGO(USA)	"	236.0	—	0.49	3.54
	SAMMAMISH(USA)	"	238.0	13.0	0.66	4.77
	MENDOTA(USA)	"	1100.0	13.0	1.2	8.77
	WEIR(USA)	"	36.0	1.92	0.07	0.51
	SHAGAWA(USA)	"	220.0	7.82	0.68	4.91
	TAHOE(USA)	"	5.6	0.52	0.05	0.36
	WESTTWIN(USA)	"	576.0	—	0.3	2.17
	EASTTWIN(USA)	"	474.0	—	0.7	5.06
	GYRSTINGE(DENMARK)	SIMONSEN& DAHL-MADSEN(1978)	416.0	27.0	2.4	17.34
	NORRVIKEN(SWEDEN)	AHLGREN(1978A,B)	320.0	17.1	2.1	15.17
	NORRVIKEN(SWEDEN)	"	410.0	5.6	0.42	3.03
	NORRVIKEN(SWEDEN)	"	430.0	6.7	0.45	3.25
	NORRVIKEN(SWEDEN)	"	480.0	6.4	0.45	3.25
	NORRVIKEN(SWEDEN)	"	380.0	19.5	0.75	5.42
	NORRVIKEN(SWEDEN)	"	390.0	9.3	0.48	3.47
	NORRVIKEN(SWEDEN)	"	290.0	2.4	0.09	0.65
	TIBBS RUN(USA)	SHELLITO&DeCOSTA(1981)	57.6	—	0.5	3.61
	LONG LAKE (USA)	PERKINS et al.(1979)	165.7	—	0.28	2.02
	CLEAR(CANADA)	SCHINDLER& NIGHSWANDER(1970)	250.0	—	0.04	0.29
	MINNETONKA(USA)	RAST&LEE(1978)	440.0	—	0.5	3.61
	MINNETONKA(USA)	"	320	—	0.1	0.72
	MYVATN(ICELAND)	JONASSON(1981)	118.0	1.4	1.5	10.84
	SUPERIOR(US/CAN)	VOLLENWEIDER et al. (1974), N.AMERICAN PROJECT(1977), CHAPRA &SONZOGNI(1979), CHAPRA&DOBSON(1981)	50.0	—	0.05	0.36
	HURON(US/CAN)	"	90.0	—	0.08	0.58
	MICHIGAN(US/CAN)	"	150.0	1.3	0.12	0.87
	ONTARIO(US/CAN)	"	190.0	—	0.35	2.53
	ERIE(US/CAN)	"	250.0	—	0.72	5.20
	ERIE(WEST)(US/CAN)	"	310.0	—	3.02	21.82
	ERIE(EAST)(US/CAN)	"	160.0	—	0.4	2.89
	ERIE(CENTRAL)(US/CAN)	"	210.0	—	0.3	2.17
	CHATAUQUA(N)(USA)	BLOOMFIELD(1978,1980)	275.0	—	0.4	2.89
	CHATAUQUA(S)(USA)	"	375.0	—	0.4	2.89
	OTSEGO(USA)	"	267.7	2.0	0.13	0.94
	SARASOTA(USA)	"	225.0	17.72	1.86	13.44
	PLUSS SEE(FRG)	OHLE(1977)	128.0	—	0.45	3.25

Location	Reference				
SCHLUENSEE(FRG)	"	54.3	—	0.18	1.30
SCHONSEE(FRG)	"	14.7	—	0.15	1.08
GROSSER PLONERSEE((FRG)	"	460.8	—	1.97	14.23
ALPNACHERSEE	IMBODEN& GACHTER(1978)	305.0	—	2.2	15.89
MOSSO1(DENMARK)	RIEMANN(1977)	235.0	—	1.06	7.66
MOSSO2(DENMARK)	"	298.0	—	3.92	28.3
MOSSO3(DENMARK)	"	373.0	—	96.59	697.76
HVIDKILDE(DENMARK)	NYHOLM(1978)	242.0	27.0	0.77	5.56
SOERUP(DENMARK)	"	160.0	13.0	1.58	11.41
NEILSTRUP(DENMARK)	"	30.0	194	6.4	46.23
OLLERUP(DENMARK)	"	263.0	200.0	9.3	67.18
BRAENDEGAARD(DENMARK)	"	45.0	11.0	.021	1.52
NOERRE(DENMARK)	"	139.0	5.8	0.11	0.79
ARRESKOV(DENMARK)	"	153.0	9.3	0.41	2.96
SALTON(DENMARK)	"	159.0	23.0	3.1	22.39
BRAS(DENMARK)	"	279	713.0	47.0	339.53
OERN(DENMARK)	"	322.0	35.0	14.0	101.14
SILKBORG1(DENMARK)	"	320.0	35.0	18.0	130.03
SILKBORG2(DENMARK)	"	300.0	56.0	7.2	52.01
TANGE1(DENMARK)	"	472.0	1209.0	78.0	563.47
TANGE2(DENMARK)	"	437.0	326.0	24.0	173.38

ESTUARY & COASTAL

Location	Reference				
KANEOHE BAY	NIXON(1983), NIXON & PILSON(1983)[5]	165.0	3.6[5]	0.78[5]	—
LONG ISLAND SOUND	"	205.0[6]	6.0	—	—
CHESAPEAKE BAY	"	445.0	7.6	0.5	
APALACHICOLA BAY	"	360.0	8.4	0.59	
BARATARIA BAY	"	360.0	8.6	—	
PATUXENT RIVER	"	210.0	8.8	7.44	
PAMLICO RIVER	"	350.0	12.5	3.63	
NARRAGANSETT BAY	"	310.0	13.7	2.14	
DELAWARE BAY	"	206.0	18.6	2.58	
S. SAN FRANCISCO BAY	"	150.0	22.6	8.15	
N. SAN FRANCISCO BAY	"	108.0	28.4	9.95	
NEW YORK HARBOR	"	483.0	447.0	48.3	
OCHLOCKONEE BAY	KAUL&FROELICH(1984)	32.0[7]	5.28[8]	1.55[8]	—
SHARK BAY	SMITH(1984)	5.3[9]NEP	0.01[9]IMPORT	0.045[9]IMPORT	0.327[9]IMPORT
CHRISTMAS ISLAND	"	26.3[9]NEP	0.5[9]IMPORT	0.192[9]IMPORT	1.39[9]IMPORT
CANTON ATOLL	"	61.3[9]NEP	1.9[9]IMPORT	0.588[9]IMPORT	4.25[9]IMPORT

(INPUT ONLY)[10]

Location	Reference				
BAY OF BREST	MONBET et al.(1981)[11]	—	46.1	0.62	—
PAMLICO1976	JAWORSKI(1981)[12]	—	12.3	3.9	—
PATUXENT1969-71	"	—	8.1	1.8	—
CHESAPEAKE1969-71	"	—	9.5	1.3	—
DELAWARE1976	"	—	100.0	18.9	—
NARRAGANSETT1977	"	—	4.9	1.8	—
POTOMAC1969-71	"	—	20.2	4.3	—
ALBEMARLE SOUND1974	"	—	7.1	0.8	—
ARCHIPELAGO STOCKHOLM1971	"	—	28.5	3.5	—
THAMES1971	"	—	462.5	109.8	—
TYNE1972	"	—	750.0	190.3	—
YORK1969-71	"	—	5.6	0.76	—
RAPPAHANNOCK1969-71	"	—	3.8	0.45	—
JAMES1969-71	"	—	15.6	2.7	—
SABINE-NECHES	ARMSTRONG(1982)[13]	—	54.8	6.82	—
TRINITY-SAN JACINTO	"	—	8.4	2.66	—
LAVACA-TRES PALACIOS	"	—	4.3	1.52	—
GUADALUPE	"	—	10.7	2.00	—
MISSION-ARANSAS	"	—	2.0	0.18	—
NUECES	"	—	1.52	1.20	—

COASTAL BIGHT&OPEN OCEAN[14]

Location	Reference				
NEW YORK BIGHT	MALONE(1977,1982)	430.0	46.7	5.26	—
S.CALIFORNIA BIGHT	EPPLEY et al.(1979) EPPLEY&PETERSON(1979)[15]	125.0	8.1	—	—
SARGASSO SEA	McCARTHY&CARPENTER (1983)	50.0-80.0	1.31-7.06[16]	0.18-0.98	—
NORTH PACIFIC GYRE	"	—	3.23-26.79[16]	0.45-3.71	—
E. MEDITERRANEAN	EPPLEY&PETERSON(1979), DUGDALE(1976)	72.0	2.28	—	—
NORTH PACIFIC	"	83.0	0.73	—	—
EASTERN TROPICAL PACIFIC	"	253.0	19.6	—	—
COSTA RICA DOME	"	1476.0	93.57	—	—
PERU UPWELLING	"	1922.0	169.2	—	—

	AVERAGE FOR GLOBAL OCEAN	BUNT(1975), LAWS (1983)	63.0	0.077-0.15[17]		
DESERT	GENERIC RANGE	CRAWFORD&GOSZ(1982), HADLEY&SZAREK(1981)	1.3-408[18]	0.09-0.89	--	--
FOREST	PAPER BIRCH (ALASKA, USA)	COLE&RAPP(1980)[19]	391	0.21	0.009	--
	DOUGLAS FIR PLANTATION(WA, USA)	"	580	0.17	0.03	--
	DOUGLAS FIR PLANTATION(WA, USA)	"	609	0.17	0.03	--
	DOUGLAS FIR FOREST(WA, USA)	"	611	0.20	0.03	--
	SPRUCE FOREST(USSR)	"	443	0.11	--	--
	SPRUCE PLANTATION (SWEDEN)	"	976	0.82	0.01	--
	SPRUCE PLANTATION (FED.REP.GERMANY)	"	571	2.18	0.05	--
	SPRUCE PLANTATION (FED.REP.GERMANY)	"	613	2.18	0.05	--
	SPRUCE PLANTATION (FED.REP.GERMANY)	"	480	2.18	0.05	--
	SHORT LEAF PINE FOREST (TENN,USA)	"	600	0.87	0.006*(+0.048)	--
	WHITE PINE PLANTATION (NC, USA)	"	836	0.55*(+0.33)	--	--
	WESTERN HEMLOCK FOREST (OREGON, USA)	"	782	0.57	0.09	--
	YELLOW POPULAR FOREST (TENN, USA)	"	516	0.77	0.006*(+0.07)	--
	YELLOW POP-MIXED HARDWOODS(TENN, USA)	"	649	0.87	0.006*(+0.048)	--
	OAK-HICKORY FOREST (TENN, USA)	"	760	0.87	0.006*(+0.048)	--
	CHESTNUT-OAK FOREST (TENN, USA)	"	758	0.87	0.006*(+0.048)	--
	OAK-HICKORY FOREST (NC, USA)	"	617	0.49*(+0.33)	--	--
	NORTHERN HARDWOOD FOREST(NH, USA)	"	687	0.65	0.004	--
	MIXED OAK FOREST (BELGIUM)	"	679	0.87	--	--
	MIXED OAK FOREST (BELGIUM)	"	1003	1.3	--	--
	BEECH FOREST (FED.REP.GERMANY)	"	762	2.18	0.05	--
	BEECH FOREST (FED.REP.GERMANY)	"	685	2.18	0.05	--
	BEECH FOREST (FED.REP.GERMANY)	"	707	2.18	0.05	--
	MIXED DECIDUOUS (UK)	"	516	0.58	0.02	--
	BEECH FOREST (SWEDEN)	"	1040	0.82	0.01	--
	RED ALDER (WA, USA)	"	1438	0.17	0.03	--
	MED. EVERGREEN OAK FOREST(FRANCE)	"	547	1.56	0.10	--
(INPUT ONLY)	BLACK SPRUCE(TAIGA) (AK,USA)	VAN CLEVE et al.(1983)	74.0(not plotted)	0.23	0.007	--
	BLACK SPRUCE(TAIGA) (AK,USA)	"	50.8(not plotted)	0.16	0.005	--
	PAPER BIRCH(TAIGA) (AK,USA)	"	245.1(not plotted)	0.21(w/o snow)	0.006(w/o snow)	--
	QUAKING ASPEN(TAIGA) (AK,USA)	"	380.1(not plotted)	0.21(w/o snow)	0.006(w/o snow)	--
	COSHOCTON(OH,USA)	LIKENS et al.(1977)	--	2.0	0.018	--
	SILVERSTREAM (NEW ZEALAND)	"	--	0.2	0.02	--
	TAUGHANNOCK CREEK (NY, USA)	"	--	0.97	0.007	--
	CARNATION CREEK (CA, USA)	"	--	0.27	0.011	--
	ELA(CANADA)	"	--	0.64	0.032	--
	FINLAND	"	--	0.6	0.01	--
	STORSJON(SWEDEN)	"	--	0.1	0.014	--
	W.CASCADES(OREGON)	"	--	0.25	0.029	--
	RIO NEGRO(BRAZIL)	"	--	0.56	0.02	--
	KUOKKEL(SWEDEN)	"	--	0.115	0.006	--

GRASSLAND						
	PAWNEE(IBP) (CO, USA)	ALL IBP FROM, WOODMANSEE(1979), SIMS&COUPLAND(1979)	370.0 [20]	0.6	–	–
	PANTEX(IBP) (TX,USA)	"	445.5	0.6	–	–
	OSAGE(IBP) (OK,USA)	"	443.5	1.0	–	–
	COTTONWOOD(IBP) (SD,USA)	"	391.0	0.9	–	–
	JORNADA(IBP) (NM,USA)	"	147.5	0.3	–	–
	SAN JOAQUIN(IBP) (CA,USA)	"	452.5	1.0	–	–
	MATADOR(IBP) (SASK,CANADA)	"	351.0	0.32	–	–
WET? GRASSLAND	MEADOW STEPPE (W.SIBERIA,USSR)	TITLYANOVA& BAZILEVICH(1979), BAZILEVICH& TITLYANOVA(1980)	1735.0(not plotted)	0.35+?	–	–
	MESOPHYTIC MEADOW(CZECH)	TITLYANOVA& BAZILEVICH(1979), ULEHOVA et al. (1976)	845.0(not plotted)	?0.8+FLOOD?	?	
	WET MEADOW (CZECHOSLOVAKIA)	"	1205.0(not plotted)	?0.8+FLOOD?	?	–
WETLANDS (INPUT ONLY) FRESHWATER						
	ALLUVIAL FOREST (CZECH)	ULEHLOVA et al.(1976)	–	0.12	0.19	–
	MOIST MEADOW (CZECH)		–	8.8-13.8	0.033-0.123	–
	OKEFENOKEE SWAMP FOREST(GA,USA)	SCHLESINGER(1978)	–	0.332)	0.022	–
	BOG(UK)	CRISP(1966)	–	0.8	0.06	–
	SWAMP FOREST LA(USA)	DAY&KEMP(1985)	–	15.54[21]	4.2[21]	–
	FRESHWATER MARSH FLA(USA)	ZOLTEK et al.(1979), BAYLEY(1985)	–	3.83-9.15[22]	0.43-0.48	–
	PEATLAND MICHIGAN(USA)	RICHARDSON et al.(1978)	–	0.5	0.03	–
SALTWATER	INTERTIDAL MARSH (MD,USA)	CORRELL(1981)	–	10.94	2.76	–
AGRICULTURAL SYSTEMS		ALL FROM FRISSEL(1977)[23]				
	MIXED LIVESTOCK FARM(NETHERLANDS)		–	2.3	0.41	–
	HIGH ELEVATION MOORLAND,SHEEP(UK)		–	0.8	0.02	–
	TRADITIONAL HILL, SHEEP(UK)		–	1.00	0.04	–
	IMPROVED HILL GRASSLAND,SHEEP(UK)		–	1.2	0.07	–
	IMPROVED HILL PADDOCKS & GRASSLANDS, SHEEP(UK)		–	1.2	0.58	–
	IMPROVED HILL PADDOCK,SHEEP(UK)		–	1.0	1.64	–
	MEATHOP WOOD(UK)		–	0.6	0.04	–
	[24]DOUGLES FIR(WASH,USA)		–	1.0	0.03	–
	[24]LOBLOLLY PINE(MISS,USA)		–	1.1	0.03	–
	[24]CONIFEROUS FOREST(FDR)		–	2.9	0.08	–
	[24]DECIDUOUS FOREST(FDR)		–	2.8	0.08	–
	[24]DECIDUOUS&CONIFEROUS [24]FOREST(N. HEMISPHERE)		–	.1.8	0.08	–
	GRASS & CLOVER, SHEEP(UK)		–	15.3	3.32	–
	MIXED LIVESTOCK (NETHERLANDS)		–	1.3	0.21	–
	DAIRY FARM (NETHERLANDS)		–	1.4	0.1	–
	MOUNTAIN FARM, MIXED (CZECHOSLOVAKIA)		–	6.62	1.31	–
	DAIRY FARM (NETHERLANDS)		–	3.3	0.5	–
	SHEEP FARM,GRASS(UK)		–	34.3	5.51	–
	FRENCH INTENSIVE,					

MIXED(FRANCE)	–	10.0	5.0	–
SMALL FARM,TROPICAL & SUBTROP.,ARABLE(S.AMERICA)	–	4.2	0.8	–
WHEAT(KANSAS,USA)	–	4.0	1.31	–
FRENCH INTENSIVE, IRRIGATED BEANS(FRANCE)	–	5.9	7.9	–
GRAZED BLUEGRASS (NORTH CAROLINA,USA)	–	17.8	2.43	–
MIXED ARABLE FARM, TEMPERATE(S.AMERICA)	–	6.1	0.64	–
SHRUB&TREE SAVANNAS, 50% ARABLE (BRAZIL)	–	5.57	1.1	–
MIXED LIVESTOCK FARM (NETHERLANDS)	–	4.3	0.61	–
FRENCH INTENSIVE, MIXED SOME GRAZED(FRANCE)	–	20.0	7.5	–
FRENCH INTENSIVE,MIXED SOME GRAZING(FRANCE)	–	40.0	8.0	–
DAIRY FARM(NETHERLANDS)	–	67.2	3.8	–
WINTER WHEAT(UK)	–	11.5	2.53	–
COTTON(CALIFORNIA,USA)	–	23.2	1.4	–
POTATOES(MAINE,USA)	–	17.2	10.11	–
FRENCH INTENSIVE, MIXED(FRANCE)	–	80.0	16.0	–
FRUIT TREES(JAPAN)	–	20.9	4.82	–
PLANTATION,ARABLE TROP.&SUBTROP.(S.AMERICA)	–	10.6	1.87	–
CORN FOR GRAIN, (INDIANA, USA)	–	12.2	3.03	–
RICE, WHEAT,BARLEY, SWEET POTATO(JAPAN)	–	7.2	3.79	–
SOY BEAN FOR GRAIN, (ARKANSAS,USA)	–	1.0	1.93	–
PADDY RICE(JAPAN)	–	13.8	4.81	–
LOWLAND FARM,ARABLE (CZECHOSLOVAKIA)	–	21.1	11.4	–
DAIRY FARM(NETHERLANDS)	–	78.3	8.7	–
ARABLE FARM,CROPS ROTATED(NETHERLANDS)	–	31.9	2.5	–
TEA PLANTS(JAPAN)	–	29.3	4.99	–
VEGETABLES,MEAN OF 10 LEADING CROPS(JAPAN)	–	26.5	8.11	–
ARABLE FARM, CROPS ROTATED(NETHERLANDS)	–	34.6	3.1	–
GRASS-LEGUME FORAGE CROP(JAPAN)	–	10.5	5.0	–
FRENCH INTENSIVE, GRASS (FRANCE)	–	80.0	15.0	–

Footnotes:

[1] Major reference(s) or a summary paper is given, other papers have been consulted to derive annual values; but not all are given in the interest of keeping an alrea(minimum. Where same lake appears more than once in listing data are for different years.

[2] Inputs are generally total P, sometimes DIN, sometimes total N.

[3] Total NPP includes plankton, macrophytes, epiphytes.

[4] Before and after sewage diversion.

[5] Summary of U.S. coastal systems; amended to include precipitation inputs. N is DIN and P is PO_4.

[6] NPP from oxygen changes not ^{14}C incubation.

[7] NPP calculated from net uptake estimated from model.

[8] N is nitrate +nitrite plus estimate for ammonium. P is phosphate + particulate P.

[9] Net ecosystem production, net import values (less than NPP, and inputs)

[10] Where same system appears twice, the value of Nixon(1983) appears in tables and figures.

[11] Dissolved inorganic N and P.

[12] Values are for Total nutrients??

[13] Form of nutrients?

[14] All can be assumed to be essentially inorganic N and P.

[15] Calculated from model(see text), all values extrapolated to annual basis.

[16] Calculated, range depends on eddy diffusion coefficients primarily. Value of Dugdale for N.Pacific, not this value, is used in table and figures

[17] Land runoff, prec.(see text)

[18] High NPP where irrigated.

[19] NPP calculated from annual increments aboveground plus litterfal fluxes(see text). Most inputs are bulk precipitation, particles indicated in parenthesis .

[20] NPP is above and belowground increments, bulk precip (see text).

[21] Total N and P.

[22] Wet and dry years.

[23] N fixation not included here, but estimates are given foir many sites.

[24] Repetition of some forested sites later summarized by Cole and Rapp(1980)

AN INTEGRATED APPROACH TO ANALYSE THE NORTH SEA ECOSYSTEM BEHAVIOUR IN
RELATION TO WASTE DISPOSAL

R. Klomp, J.A. v. Pagee, P.C.G. Glas
Delft Hydraulics Laboratory
Water Resources and Environment Division
P.O. Box 177
2600 MH Delft
The Netherlands

ABSTRACT. As a part of the drawing up of a water quality management
plan for the Dutch part of the North Sea, studies have been carried
out to assess the present (1980) and future (1990) state of North Sea
pollution. This paper emphasizes the use of modelling techniques to
quantify the anthropogenic increase of heavy metal and nutrient
concentrations and the related impacts on the North Sea ecosystem.

1. INTRODUCTION

Over the past decades, an increase of conflicts between interests in
the North Sea area occurred and still more are anticipated. As a substantial part of these conflicts is related to the deteriorating water
quality of the North Sea, the Dutch Government asked the agency
responsible for water quality management, the Directorate North Sea of
the Rijkswaterstaat (Ministry of Transport and Public Works), to draw
up a water quality plan for the Dutch part of the North Sea [1]. In
this plan, covering all sources and all types of pollution, a coherent
and strategic approach for the water quality management of the North
Sea was to be presented. It was envisaged that such a plan would act
as a basis for identifying and evaluating measures to be taken by
Dutch governmental agencies, as well as for international negocia-
tions, concerning the coordination of approaches with respect to North
Sea water quality management.
    The planning area coincides with the Dutch exclusive economic
zone of the North Sea (DPCS). (Figure 1)
    As all activities affecting the water quality in this area, and
all functions affected by water quality, were to be considered, the
actual area under consideration included the drainage areas of rivers
discharging into the North Sea, covering a substantial part of Europe.

G. Kullenberg (ed.), The Role of the Oceans as a Waste Disposal Option, 205–231.

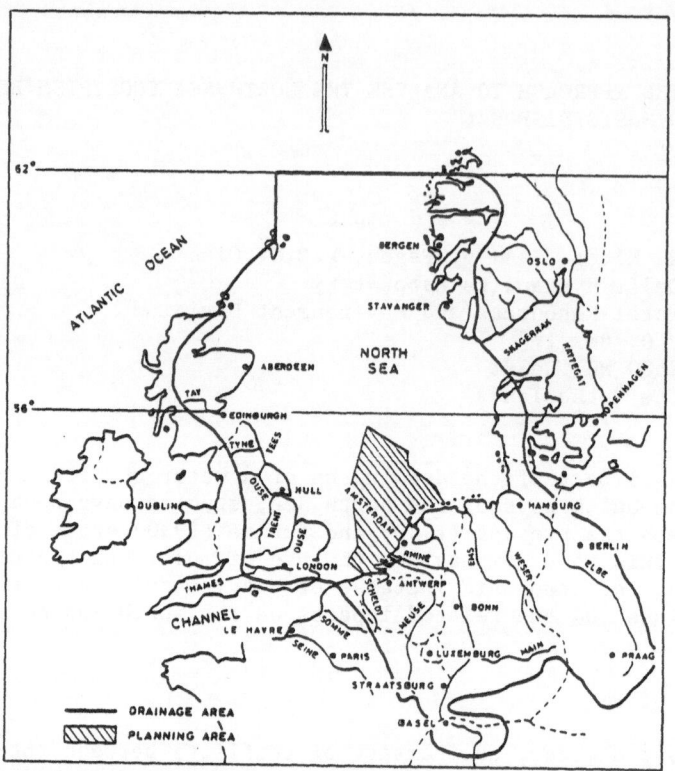

Figure 1. Study and planning area

      The objectives of the study were:
To develop a framework for assessing the actual state of North Sea
  pollution and for formulating and analyzing strategies for water
  quality management.
- To apply such a framework in the formulation of a first water quality plan.
The general framework for analysis is given in Figure 2. The underlying paper primarily deals with the analysis and methods for
assessment of North Sea pollution. [2].
      The North Sea is a rather shallow shelf sea in which wind and
tides induce a counter clock wise circulation, with major inflows from
the Channel and North Atlantic Ocean.
      The North Sea contains a highly productive biotic community,
among other things due to the existence of many estuaries and intertidal wetlands such as the Dutch Wadden Sea.
      The North Sea marine resources are intensively used in many dif-
ferent ways. It acts as a recipient for wastes, generated by activities in the planning area itself, such as shipping and offshore mining

operations, or entering through rivers, atmospheric depositions or sea currents. The geographic location of the North Sea, bordering densely populated and highly industrialized parts of Europe, not only explains the vast waste loads entering the North Sea from the main land, but also results in intensive shipping traffic and harbour activities, with all discharges associated. The immediate coastal zone acts as an important recreation area. The North Sea also contains important natural resources, i.e. oil and gas, sand and gravel and other minerals. As these raw materials become depleted on land, the incentives to explore the North Sea increase. Last but not least the North Sea acts as a major food resource, providing fish and shell-fish and other products.

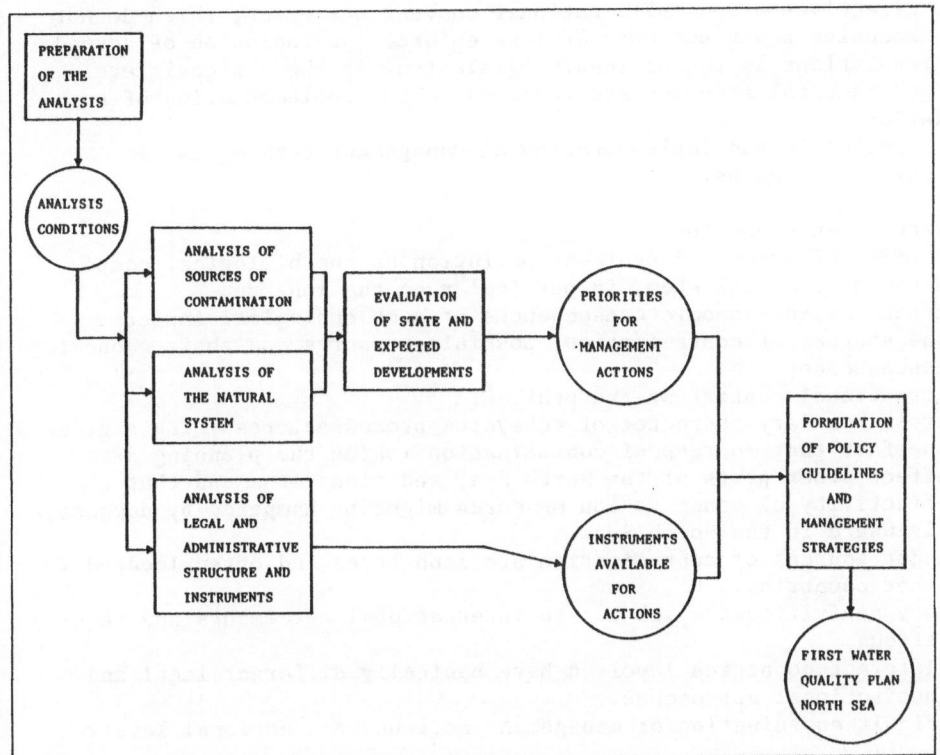

Figure 2. General framework for analysis

Contaminating substances which enter the North Sea, disperse according to the stream pattern and accumulate in estuaries and inter-tidal sedimentation areas, like the Wadden Sea area, or in the deep northern parts of the North Sea. Contaminants are taken up by algae and benthic organisms and are transferred to and accumulated in higher

levels in the food chain. As a result of this combined accumulation,
effects become apparent at the top of the food chain (mammals, birds).
Examples of recent changes in the ecosystem which, a.o., might be
related to effects of pollution include: the significant decline in
numbers of the harbour seal in the Dutch Wadden Sea, probably due to
reduced fecundity; the increase of skin diseases in fish and the
enormous reduction between 1940 and 1965 of the Sandwich tern. In
addition substantial loads of phosphorus and nitrogen enhance algal
growth, which may have promoted oxygen depletion near the bottom under
stratified conditions and changes in abundance and diversity of
species throughout the food web [3].

Combatting pollution requires governmental or management actions.
All human activities involved are subject to some form of regulation,
subject of national legislations. In an international context, specific EC directives a and multi-national conventions exist, which do not
have executive power but more or less enforce the inclusion of agreed
upon regulations in the national legislation. At the national level,
many governmental agencies are involved in the implementation of such
regulations.

Formulation and implementation of management strategies, is
seriously hampered by:

- A lack of knowledge of
  - impacts of water and sediment pollution on the biological compo-
    nents of the ecosystem, in particular on the long run
  - actual socio-economic consequences of such ecological impacts
  - the sources of contamination, possible measures and their economic
    consequences.
- International context of the problem:
  - cross boundary character of ecosystem processes, resulting e.g. in
    the fact that sources of contamination inside the planning area
    affect other parts of the North Sea, and vice versa and that the
    effectivity of conservation measures might be hampered by damages,
    elsewhere in the North Sea.
  - major sources of contamination are land based and often located in
    other countries.
  - many activities are subject to international agreements and regu-
    lations.
  - different countries involved have basically different legal and
    institutional approaches.
- Difficult coordination of management actions on a national level.

In the Netherlands a process of continuous harmonization is
started of which the so-called Water Quality Plan is a major component.

The plan concentrates mainly on the formulation of general policy
guidelines and desired levels of emissions and of water and sediment
quality. Less attention in this first plan was dedicated to concrete
management actions to obtain these levels and to day to day operational measures.

2. ANALYSIS

2.1. Waste loads entering the North Sea

The southern part of the North Sea receives large quantities of pollutants by inflowing rivers such as the Scheldt, Rhine, Meuse, Elbe and Thames, direct discharges of sewage and more or less treated waste water, dumps of dredging materials and atmosphere. Not only the quantity of individual sources, but also the location and local physical characteristics of the receiving water has to be taken into account.

Based on the situation of 1980 an inventory was made of the yearly pollutant loads entering the North Sea [4]. In Figure 3 an overview of the available information on pollutant loads is presented. The overview concerns discharges from the coastal zone between 56° NB and the Strait of Dover. The data are limited to nutrients (N, P) and heavy metals (Cd, Hg, Pb, Cu, Zn, Cr) since data on other pollutants such as organic micro pollutants are too scarce to use them for load estimation.

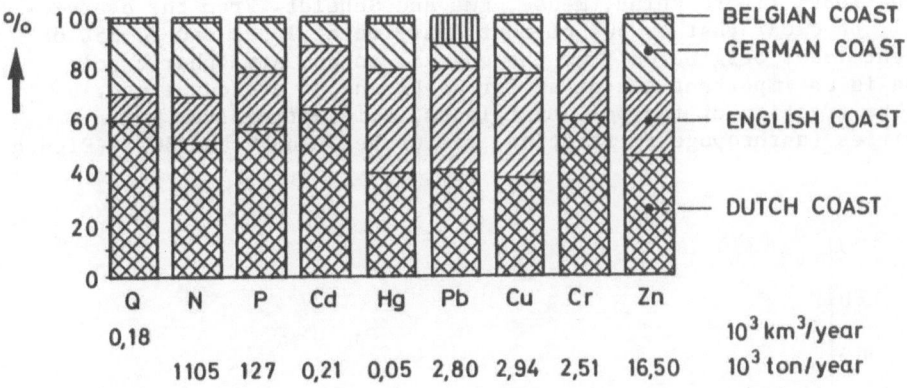

Figure 3.   Input by point sources of fresh water, nutrients and heavy metals

Based on calculated water circulations and measured concentrations in the inflowing waters from the Channel and North Atlantic Ocean, the cross boundary input has been estimated for the nutrients and heavy metals considered. Atmospheric inputs are estimated from monitoring data on the chemical composition of precipitation and related air pollution studies [5]. In order to compare the contribution of point sources, diffuse atmospheric loads and inflowing water from the Atlantic Ocean an overview is given in Figure 4.

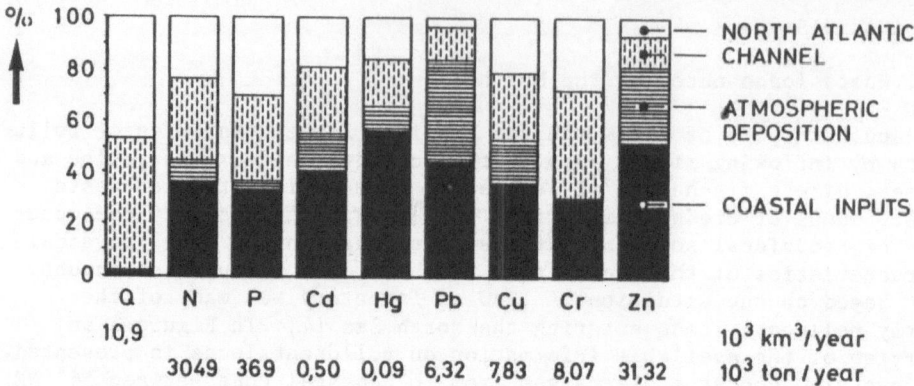

Figure 4. Contribution of different sources (%)

The overview shows that the Dutch coastal zone contributes considerably (40-50% of all sources from coastal zones) to the pollution of the North Sea. This is closely connected to the loads by the international rivers Rhine, Meuse, Ems and Scheldt. From the overview it will be clear that 40-50% of most heavy metal loads and 30-35% of nutrient loads originates from the coastal zones. Atmospheric pollution is an important source as far as Pb and Zn are concerned.

The fraction of the pollutant loads, originating from human activities (anthropogenic fraction) varies between 30 and 50%. (Figure 5)

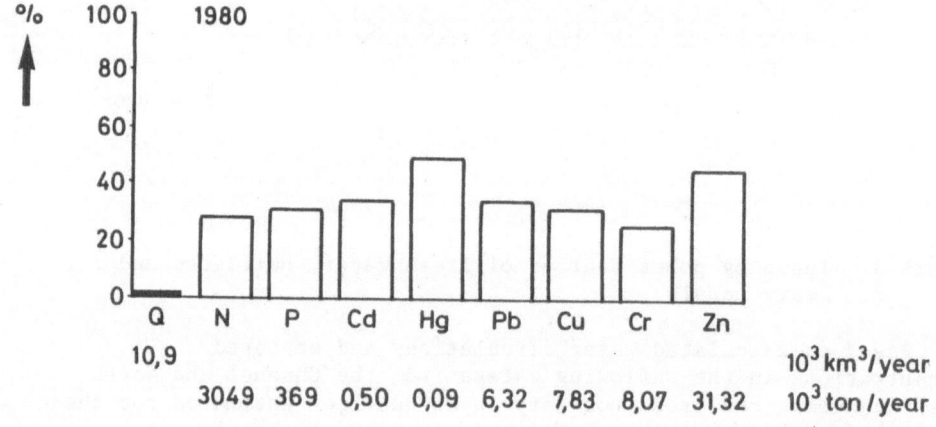

Figure 5. Anthropogenic fraction (%) of total input from all sources

2.2. Water quality

The hydrodynamic behaviour of the North Sea is determined by factors
as tides, winds, morphology, density gradients, inflowing rivers and
the cross boundary flows from the Channel and the North Atlantic
Ocean.

   Although the instantaneous water movement is to a high extent
determined by the tidal mechanisms, the long term water movement and
transport of dissolved solids are mainly determined by the tidal
averaged or so-called residual currents.

   As a basis for the calculations of mass transport a 2-dimensio-
nal, depth averaged hydrodynamic model – ESTFLO – was used to calculate tidal and wind induced flow velocities and water levels, long-
term flows were calculated by averaging the tidal flows over a 25 hour
tidal cycle. Summer and winter situations were considered with corresponding average wind conditions.

   The results of the simulated residual flows (Figure 6) show that
the residual flow rates are between 3-8 cm/s with minimal rates near
the central Dutch coast and with maximum rates near the Wadden
Islands. The circuation pattern near Doggersbank indicates a long
retention time. Similar patterns were already suggested by Boehnecke
in 1922. [6]

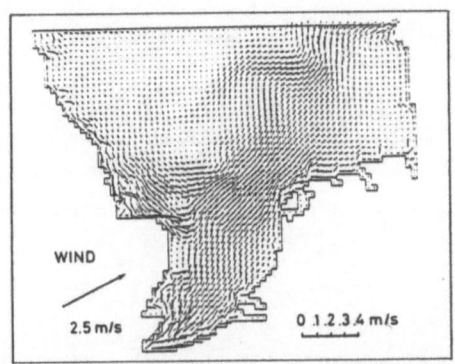 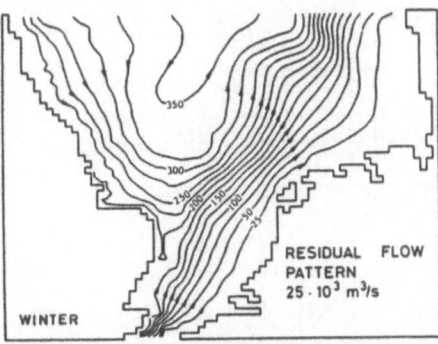

Figure 6. Simulated residual flows in the southern North Sea

   In addition to the calculated water movement by ESTFLO, the
transport of dissolved and suspended solids can be calculated by the
multi dimensional water quality model DELWAQ. Coupled to ESTFLO the
advection/diffusion equation for 2199 segments of the Southern North
Sea is solved. Although the model allows a time dependent simulation,
the inputs and residual flows were assumed to be time independent. In
this way the spatial distribution of pollutants can be calculated on
basis of an implicit numerical scheme. The advective transport is

determined by the residual flow field calculated by ESTFLO whereas the
dispersive transport is derived from long term salinity distributions
calculated by DELWAQ and observed salinity concentrations. A dispersion coefficient D = 150 m^2/s provided good results with respect to
the actual salinity gradient along the Dutch coast.
          Calculations of the transport of conservative tracers entering
the North Sea from the Channel (Strait of Dover), from the North
Atlantic Ocean (southward along the Scottish and English coast) and
from important river inflows lead to more insight in the characteristics of mass transport in the southern North Sea.
From the ratio between calculated concentrations from conservative or
non-conservative (first order decay) tracers, age functions can be
derived representing the time since its entrance in the area
considered. The age distribution for a certain inflow shows the speed
of transport of pollutant through the southern North Sea, as shown by
Figure 7.

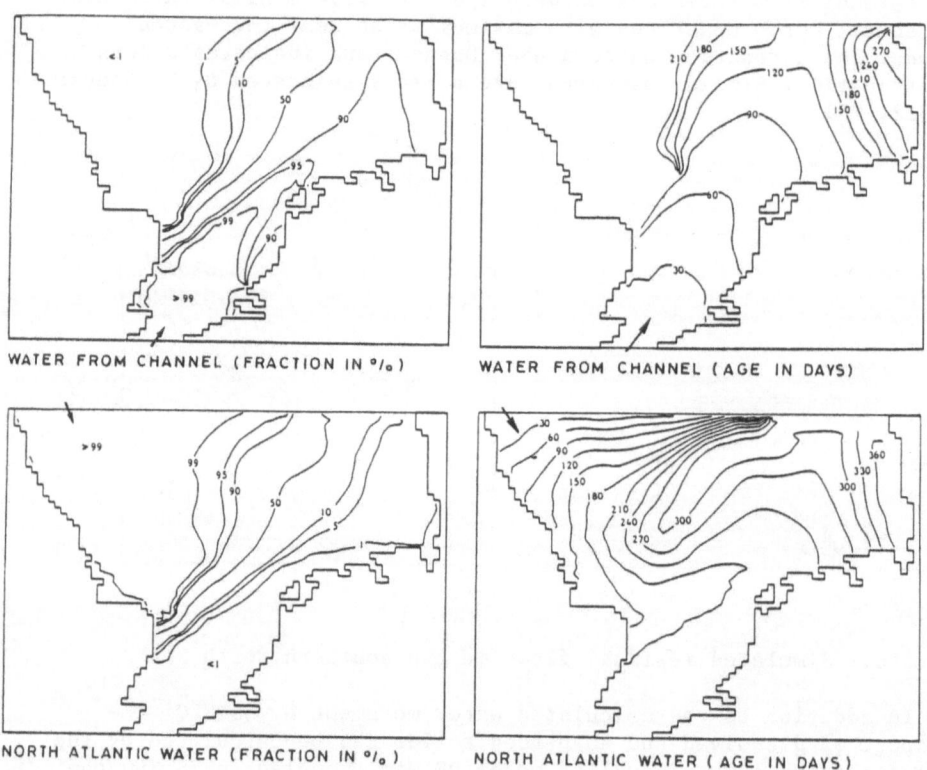

WATER FROM CHANNEL (FRACTION IN °/o)   WATER FROM CHANNEL (AGE IN DAYS)

NORTH ATLANTIC WATER (FRACTION IN °/o)   NORTH ATLANTIC WATER (AGE IN DAYS)

Figure 7.   Overview of the fractions and age functions of water from
                  the Channel and North Atlantic Ocean

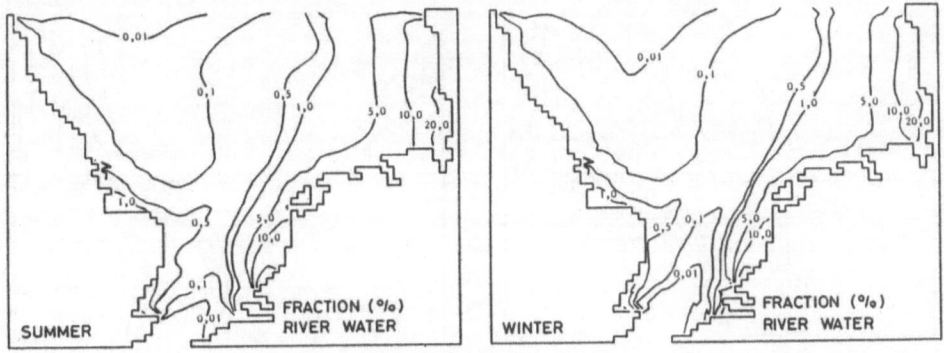

Figure 8. Fraction of water from rivers (%) in the southern North Sea

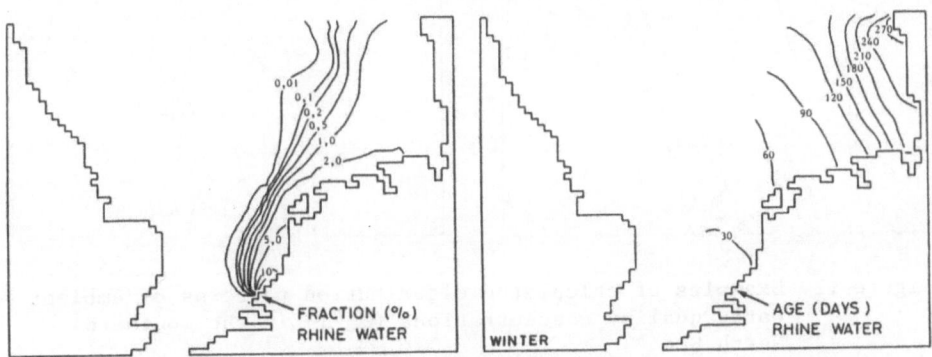

Figure 9. Fraction and age function of water from the river Rhine in the southern North Sea

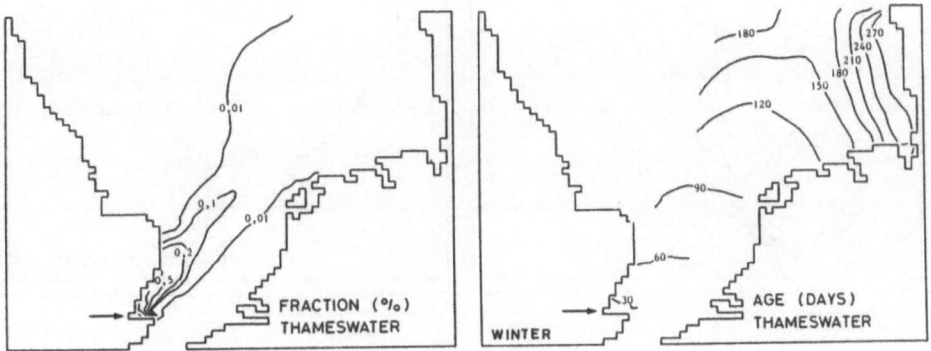

Figure 10. Fraction and age functions of water from the river Thames in the southern North Sea

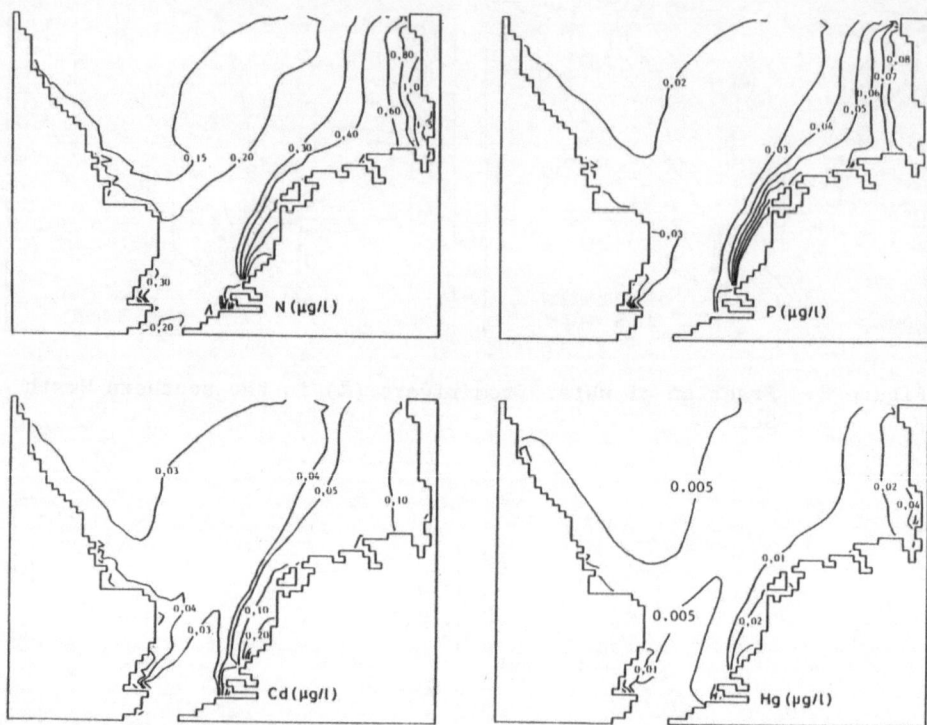

Figure 11. Examples of calculated distribution patterns of ambient water quality concentrations (μg/l) in the southern North Sea

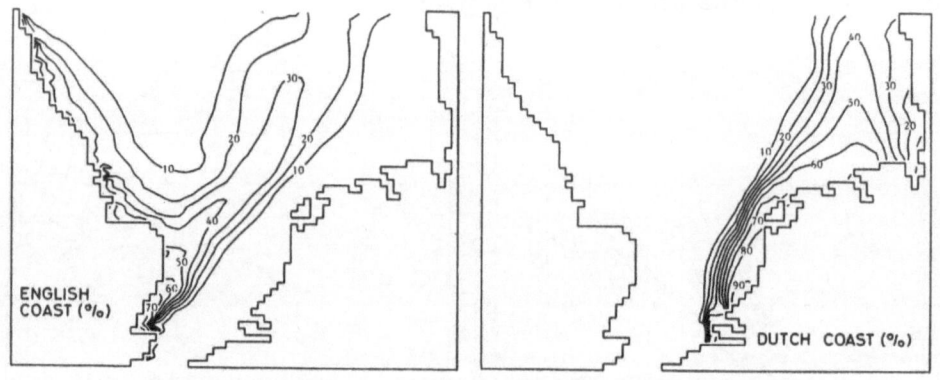

Figure 12. The relative contribution of the Dutch and English coast to the cadmium concentration in the southern North Sea

Similar to the distribution of water from the North Atlantic Ocean and the Channel the distribution of river water entering the North Sea can be considered. In Figure 8 the fresh-water fraction coming from inflowing rivers is shown. From this figure it is clear that the highest percentage of river water can be found along the Dutch coast and in the German Bight (up to 25% depending on the season).

For individual rivers the distribution and age function has been calculated as well (Figures 9 and 10). Figure 9 shows that the main part of fresh water from the river Rhine flows northward along the Dutch coast and the German Bight. The age of the water from the river Rhine, when it leaves the area considered, ranges from 270-360 days.

Due to relatively low fresh water discharge and considerable dilution by the receiving water only a small part of the North Sea contains more than 1% water from the river Thames (Figure 10).

Since the concentrations of various pollutants in rivers are high compared to the natural background concentrations in the North Sea, the transport of fresh water indicates the dispersion of river pollution inputs through the North Sea. This is not the case for pollutants in particulate form, since this transport is strongly related to the sediment or suspended solids transport. In this respect interactions between water and bottom sediment (erosion and sedimentation) play an important role. However, because of a lack of knowledge of sedimentation and erosion on one hand and the lack of data on both dissolved and particulate pollutant loads on the other hand, only total (dissolved + particulate) concentrations were considered in this first analysis.

Keeping in mind that in this way the exchange between water and bottom sediments is completely ignored the agreement between observed and calculated concentrations is surprising (Table I).

Simulation results for various substances are summarized in Figure 11. The calculated concentrations for the southern Dutch and Belgian coast seem too low. This difference may be caused by the use of time independent residual currents preventing a southward transport of water from the river Rhine and/or the accumulation of water from the river Scheldt in the Delta area.

The relative contribution of various pollutant loads to the ambient water quality was calculated from the results of the DELWAQ model for individual inputs. The results for Cd are shown in Figure 12.

From this figure it can be seen that the relative contribution of sources along the Dutch coast to the Cd concentration in the Dutch coastal waters ranges from 50-90%. The remaining contribution can be ascribed to the cross boundary inflow from the Channel. For the German Bight the picture is slightly different: 10-30% from the Channel, 30-50% from the Dutch coast and 10-70% from the German coast. A more detailed overview of the concentration build up is given in Figure 13.

	observed	calculated
N	0.2 -1.0	0.25 - 1.0 mg/1
P	0.03-0.2	0.03 - 0.2
Cd	0.09-0.15	0.04 - 0.20 g/1
Hg	0.01-0.06	0.005- 0.03
Pb	1.0 -1.8	0.5 - 2.0
Cu	1.0 -2.2	0.5 - 2.0
Zn	2.1 -9.0	2 -10
Cr	0.5 -2.3	1 - 3

Table I.   Comparison between observed and calculated pollutant
            concentrations in Dutch coastal waters (10-40 km)

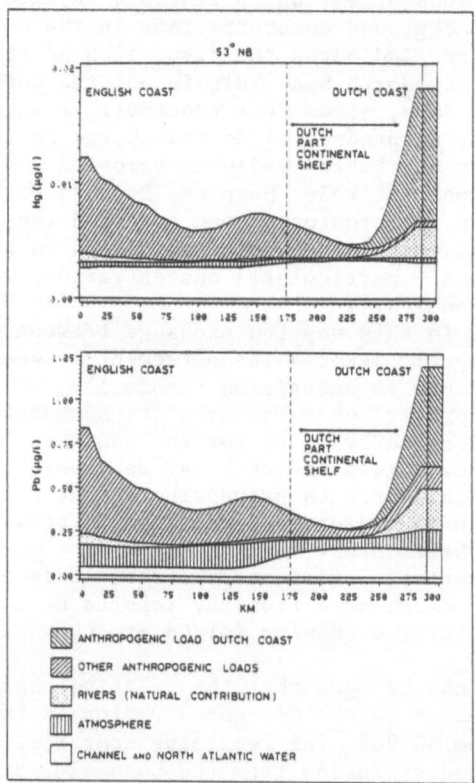

Figure 13. Overview of the Hg and Pb concentration build up in a
            cross section between the Dutch and English coast (53 N)

From the viewpoint of management an important question is related
to the contribution of human activities to the ambient water quality
concentrations. In answering this question a distinction has to be
made between reference concentrations and the anthropogenic concentration. Reference concentrations can be calculated by taking the
nutrient and heavy metal concentration of the English Channel, North
Atlantic Ocean and the "natural background" concentration of the inflowing rivers [7]. Based on the calculated reference concentrations
(C reference) and the calculated concentrations scenario 1980 (C 80)
the anthropogenic fraction (A) can be defined as:

$$A(x,y) = \frac{C\ 80\ (x,y)\ -\ C\ reference\ (x,y)}{C\ 80\ (x,y)}$$

From Figure 14 it can be seen that in large parts of the southern
North Sea the anthropogenic fraction of Cd, Hg, Pb and Zn is more than
50%. In this respect it has to be stressed that the real anthropogenic
fraction may be even higher since no anthropogenic fraction of the
pollutant input by the atmosphere deposition is considered, while the
concentration in Rhine river sediments from 1900 is assumed to be
representative for natural particulate concentrations.

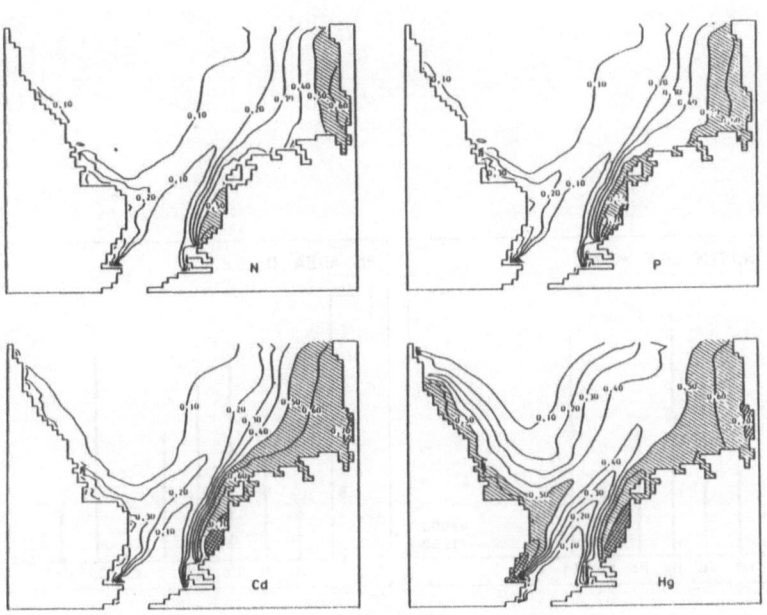

Figure 14.   The anthropogenic fraction of N, P, Cd and Hg in the
             southern North Sea

Based on regional water quality plans, sanitation measures in the drainage area of various rivers it is expected that the pollutant load will be reduced considerably (Table II). Simulations for scenario 1990 result in the ambient water quality concentrations and the reduction in anthropogenic fraction to be expected in 1990 (Figure 15).

	1980	1990	ref.	
N	583	558	165	1000 t/j
P	73	65	11	1000 t/j
Cd	135	72	7.7	t/j
Hg	21	11	4.4	t/j
Pb	1165	629	373	t/j
Cu	1125	731	308	t/j
Zn	7629	4505	1540	t/j
Cr	1523	523	330	t/j

Table II.   Overview of the total pollutant input from the Dutch coast in 1980, in 1990 (estimated) and natural background load (reference load)

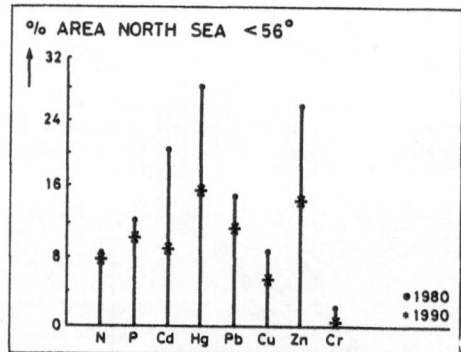 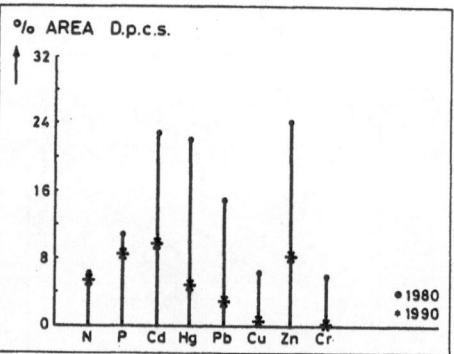

Figure 15.   Surface area (%) where concentrations exceed 50% anthropogenity in 1980 and 1990 (completed pollution control measures in the Netherlands according to planning 1990)

2.3. Sediment contamination

Most contaminants can be absorped to suspended solids, implying that the transport of these substances is not only determined by water transport but also by exchange processes with the bottom. Dependent on the behaviour of currents and turbulence induced by tide, wind and waves suspended solids may settle or come into suspension again. Especially heavy metals, organics radionuclides and phosphorus are influenced significantly by these processes. The sorption to or release from sediments is highly determined by pH, salinity, temperature, etc.

In order to analyse sedimentation and erosion characteristics for particulate forms of pollution, a detailed analysis of measured suspended solids in front of the Dutch coast has been worked out. Based on biweekly monitoring of water quality parameters over a period from 1975 till 1983 at 76 locations, mass balances for suspended solids have been established in which the gains and losses indicate influences of net erosion and sedimentation. The transport of suspended solids through the coastal zone has been calculated from the simulated residual flows by the ESTFLO model and quarterly averaged concentrations.

From this analysis it was concluded that erosion of sediment mainly takes place in the southern part of the Dutch coast, whereas sedimentation occurs in the Wadden Sea and north of the Wadden Islands.

Characteristic sedimentation rates lie between 0.5-5 mm/year. To evaluate the development of sedimental pollution the accumulation of pollutants in the surface part of the bottom has been estimated for a specified 'active' bottom surface layer by the following mass balance equation:

$$\frac{dC_b}{dt} = \frac{S}{H} (C_{ss} - C_b)$$

in which,

C_b	concentration of "active" bottom layer
C_{ss}	concentration of suspended sediment
S	sedimentation rate
H	thickness of "active" bottom layer

The active layer is defined as the benthic layer which has a biological active interaction with the aquatic ecosystem. This includes both physical (diffusion, turbation), chemical (adsorption, desorption) and biological (bioturbation, benthal organisms) activity with a more or less direct relation with the chemistry and biology of the water phase. Based on the depth of benthic organisms the active layer thickness varies between 5-50 mm. With sediment rates of 0.5-5 mm/year the characteristic time scale to reach and equilibrium between pollution of suspended sediments and pollution of the bottom layer equals:

$$\frac{H}{S} = 10 - 1000 \text{ year}$$

Based on developments in cadmium pollution of sediments from the river Rhine since 1900, a tentative calculation of bottom pollution has been carried out. For the Dutch coast it is estimated that the suspended sediment has an origine of almost 90% marine sediment and 10% sediment from the river Rhine.

Figure 16 shows the pollution of suspended marine sediment, suspended sediment from the river Rhine and suspended sediment and suspended sediment in front of the Dutch coast.

The calculated pollution of the active layer for H/S = 50 and 100 year shows that despite a standstill of the pollution in the river Rhine from 1980 on, the pollution of the bottom still increases.

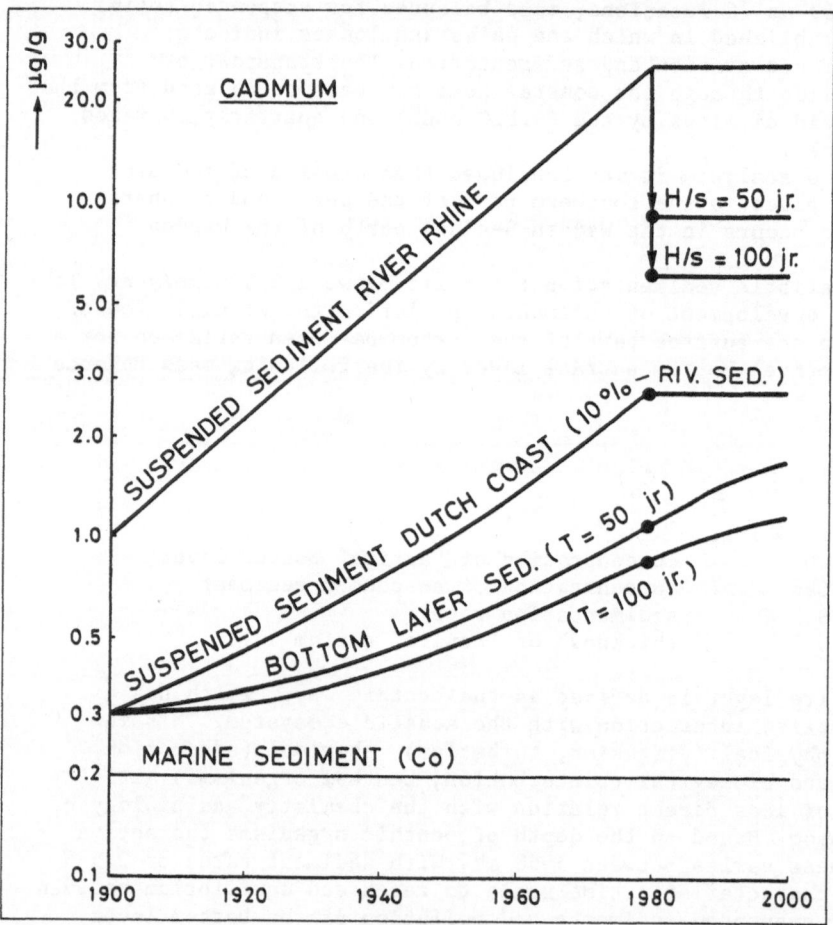

Figure 16. Developments in sediment pollution (1900-2000)

2.4. Nutrient load and primary production

Phytoplankton can be considered to be the most important primary food-source for the foodweb in the North Sea. Seasonal fluctuations, long term changes in biomass and primary production, shifts in species compositions affect consumers directly and indirectly.

Observations during the last decennia indicate increased nutrient concentrations and less pronounced a shift in species composition [8].

To get a more quantitative insight in the relation between nutrient loads, phytoplankton biomass dynamics and species composition, a mathematical model – SEAWAQ – was developed and applied [9].The main components of the model are given in Figure 17.

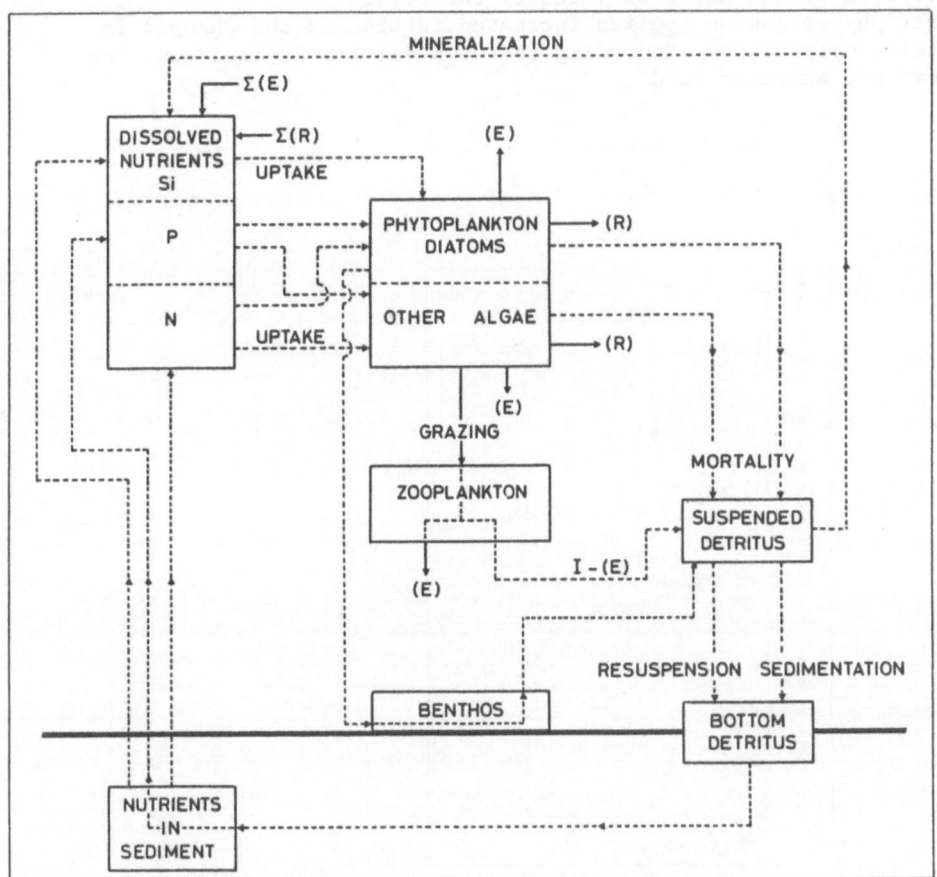

Figure 17.   Schematization of the primary production model "SEAWAQ"

Steering variables in the model are irradiation, temperature, background extinction and grazing functions. The phytoplankton module contains assimilation, respiration, excretion, grazing and mortality. The influence of nutrients on the growth rate is described by a minimum function of 3 Michaelis Menten relations. The mortality rates in the model are age dependent.

Simulations for 1930 and 1980 led to the following conclusions (Figure 18):

• increased nutrient loads (P and N) during 1930-1980 show a minor effect on the biomass of diatoms in the central North Sea. A more careful look at the modelling results indicates that even a slight decrease in biomass of diatoms occurred. This phenomenon may be due to the fact that Si is a limiting factor to diatoms which build this element into an external silica skeleton, the concentration of which decreased due to lower Si loads by the rivers
• other phytoplankton species increased in biomass and changed in species composition. According to the model results this was due to increased nutrient loads.

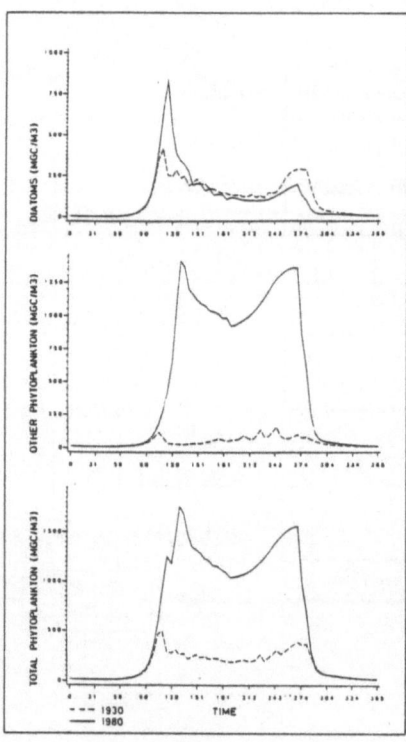

Figure 18.   Simulated phytoplankton biomass during 1930 and 1980 using the model "SEAWAQ"

In the near shore zone the diatom biomass is presently not limited by nutrient concentrations. Due to Si limitation in offshore and more central parts of the North Sea and the effect of dispersive mixing, however, biomass concentrations never reach maximum levels.

Other phytoplankton species in the coastal zone were formerly limited by phosphate concentrations. In present days this is not the case any more. As with diatoms, however, phosphate and nitrogen limitation in offshore waters result in lower biomasses in those parts and through dispersive mixing also puts an upper limit to the attainable biomass in the near shore areas.

In periods of calm weather the dispersion is limited. Under such conditions excessive blooms may occur, especially during the summer half year.

2.5. Accumulation of toxics

In order to establish the influence of the anthropogenic load of micropollutants on the ecosystem of the North Sea it is important to know whether toxics are accumulated via the foodchain, taken up by the surrounding environment or by a combination of both.

This ratio can be estimated by calculating the minimum amount of water flowing along the gills, that is necessary to supply the amount of oxygen needed for assimilation. Based on such a calculation the ratio between the exposure from water ands the exposure from food can be estimated.

Between heavy metals and organic microcompounds substantial differences in the pathway of toxics exist. For heavy metals and consumers as benthic fauna and fish, water seems to be the most important pathway for exposure. For organochlorines as DDT, Endrin and PCB's, food is by far the most important pathway for exposure.

3. INTEGRATED ASSESSMENT OF WATER POLLUTION

In order to formulate strategies that will ultimately meet water quality objectives for the North Sea, it will be essential to quantify the objectives and subsequently to estimate the necessary reduction of pollutant loads.

The assessment of water pollution for the present and future situation in relation to the present and future pollution load has to be based on:
- observed negative effects on ecosystem and other functions to be fulfilled by the North Sea
- comparison between present toxic levels and standards or critical levels
- anthropogenic part of the ambient water pollution
- affected areas and their importance for the ecosystem behaviour of North Sea.

Thereby taking into account:
- the availability of toxics for marine organisms

- the toxicological properties of the pollutant and related risk for the ecosystem
- persistency of the toxic pollutant
- accumulation of toxics in sediments, mobilization and effects on benthos
- accumulation/bioconcentration of toxics in organisms and the effect on the food chain.

The process of analysis, evaluation and subsequently formulation of measures is given in Figure 19.

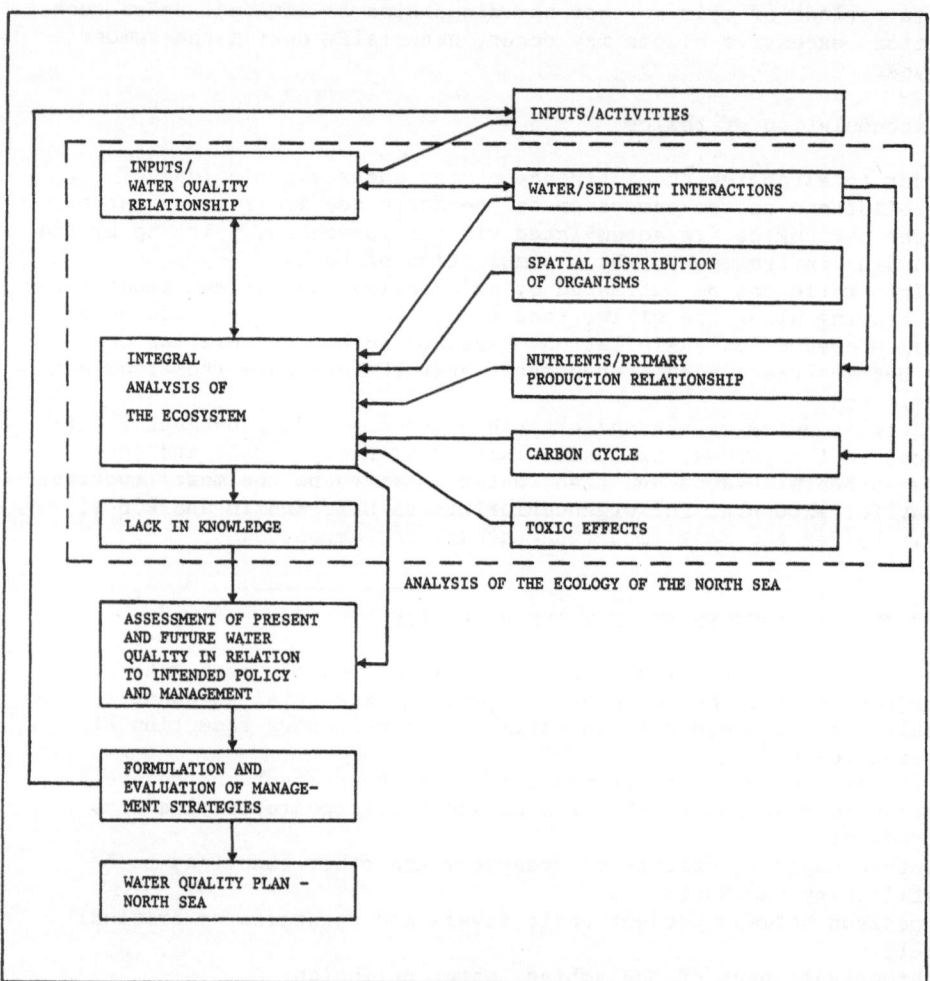

Figure 19.   The process of analysis, assessment and formulation of measures within the frame of the Water Quality Plan North Sea

3.1. Observed negative effects

Various investigations to the North Sea ecosystem have pointed out
that:
- a considerable reduction of the number of harbour seals in the
  Wadden Sea and Delta waters in the south-western part of the
  Netherlands has taken place [10]
- low oxygen concentrations occur in the German Bight during stratification in summer [11]
- skin diseases in fish have been noticed [12]
- long term shifts in phytoplankton species composition have been
  observed [8].

It is not clear whether the observed effects are related directly
to increased pollution of the North Sea or which substances are
responsible for these effects. There are indications that PCB pollution is causing the decline of seals and that increased nutrient loads
are responsible for shift in phytoplankton species composition. So far
these observations cannot be used for the integrated assessment of the
water quality of the North Sea, however, they can be used as signals.

3.2. Comparison between toxic levels and standards or assessment levels

Except bacteriological standards for recreational water, no accepted
quality standards for Dutch marine waters are available.
In this respect it was necessary to estimate the influence of
increased nutrient loads on primary production and species composition
by modelling, whereas an overview of critical levels of toxics was
made based on EPA standards [12] and results of dose effect experiments (Table III).

Pollutant	No-effect level	Sublethal concentration	Lethal concentration	EPA-standard
Cd	< 1	1	6.3	5
Cr	1,000	–	33,000	–
Cu	0.2	8	14	1.4
Hg	0.05	0.1	1	0.1
Pb	< 150	150	340	–
Zn	< 120	120	25,000	250

Table III. Critical concentrations in experiments (μg/l) and
proposed EPA-standards

Based on this overview and the spatial pattern of toxic substances in the North Sea, calculated by DELWAQ, a comparison could
take place. Despite limitations such an approach can be used as a
basis for a spatial evaluation of the risks of these toxic levels for
organisms and ecosystems behaviour.

Before executing this comparison a choice with respect to the critical level of the toxic substance has to be made. A possibility in this respect is using the lowest lethal concentration, the lowest sub-lethal concentration, the lowest no-effect level and the EPA stan-dards. Because of the great uncertainty related to the determination of no-effect levels and sub-lethal concentrations, the lowest lethal concentration including a safety factor 100 was selected as the assessment level.

A comparison between these assessment levels and calculated con-centration patterns for Cd, Hg, Cu, Pb, Zn, Cr shows that calculated concentrations of Cd and Hg exceed the assesment levels in large parts of the North Sea, whereas the concentrations of Pb, Zn and Cr remain lower than the assessment level (Figure 20).

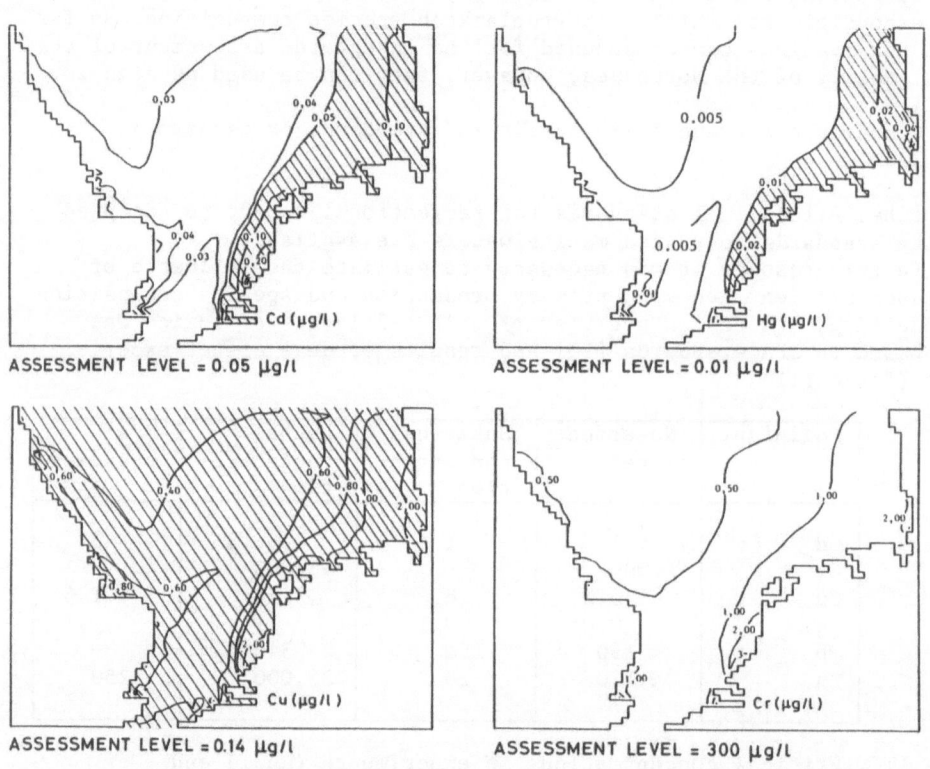

Figure 20.   Overview of areas with higher concentrations than 1%
             of the lethal concentrations in the southern North Sea
             (Cd, Hg, Cu and Cr)

The calculated concentration of Cu is higher than the assessment level for the North Sea as a whole. It has to be mentioned in this

respect that:
1) the applied safety factor is probably too high since the concentration of the inflowing water from the Atlantic Ocean already exceeds the critical level, based on 1/100 of the lethal concentration (Tables IV and V) and
2) different speciations and related differences in bio-availability are not considered.

3.3. Anthropogenic part of the ambient water pollution

In this paper is already indicated in what way the anthropogenic part of the concentration in the North Sea is estimated.

The calculated anthropogenic fractions in the North Sea show that in large parts of the southern North Sea and in particular near the Dutch coast more than 50% of the concentration of N, P, Cd, Hg, Pb, Cu, Zn and Cr originates from anthropogenic discharges (Figure 14, Table IV).

Pollutant	lethal concentration (μg/1)	Safety factors				
		1/10	1/20	1/100	1/200	1/1000
Cd	5	0	1	25	100	100
Hg	1	0	1	21	59	100
Pb	340	0	0	0	3	58
Cu	14	2	7	100	100	100
Zn	25,000	0	0	0	1	7
Cr	30,000	0	0	0	0	0

Table IV. Part of the southern North Sea (%) in which concentrations exceed critical levels given different safety factors

Pollutant	lethal concentration (μg/1)	Safety factors				
		1/10	1/20	1/100	1/200	1/1000
Cd	5	0	1	27	100	100
Hg	1	0	0	20	78	100
Pb	340	0	0	0	4	84
Cu	14	6	24	100	100	100
Zn	25,000	0	0	0	0	0
Cr	30,000	0	0	0	0	0

Table V. Part of the Dutch part of the Continental Shelf (%) in which concentrations exceed critical levels given different safety factors

3.4. Affected areas and their importance for the ecosystem of the North Sea

Based on the computed concentration distributions, the comparison between concentrations, their assessment levels and the anthropogenic fraction, can be made and a spatial distinction can be made between more or less threatened and affected areas.

Although it may be possible to make a spatial distinction, based on the distribution of organisms and the function of the areas in the North Sea ecosystem it is at present impossible to translate this information into differentiated assessment levels.

Consequently the assessment will be limited to the determination of (geographic and functional) subsystems of the North Sea ecosystem, which are threatened based on criteria related to concentrations/standards, assessment levels and anthropogenic fractions.

From the exceedance of the assessment level as well as the anthropogenic fraction it can be concluded that the coastal zones are most severely polluted. This concerns especially the Dutch coast and the German Bight. Also because these zones play an important role in propagation of fish, extra attention has to be paid to the reduction of pollutants in these areas.

In Figure 21 the anthropogenity and ratio between calculated concentrations and lethal concentrations for Cd, Hg, Pb, Cu, Zn and Cr is presented in a diagram for the Dutch coastal zone.

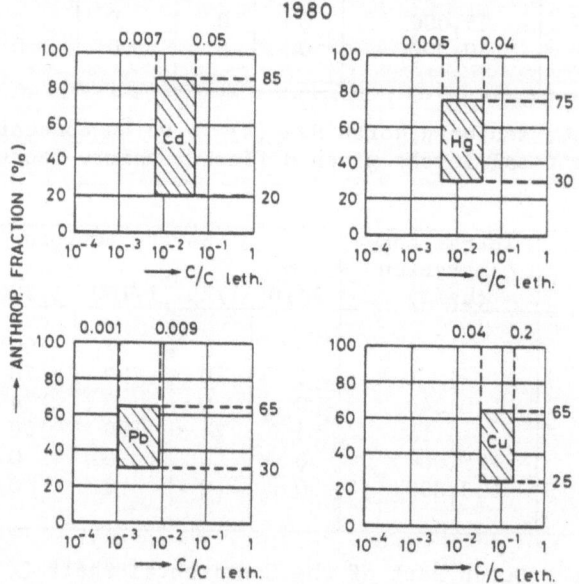

Figure 21.   Anthropogenity versus C/Cleth. in the Dutch coastal zone (0-60 km)

From the diagrams it was concluded that the highest priority in the reduction of pollutant loads should be given to Cd, Hg and possibly Cu. Also scarce observations on Fe, PCB and HCH concentrations indicate a relatively high risk for effects on organisms.

4 EVALUATION OF TRENDS IN THE CONTAMINATION OF WATER, SEDIMENT AND ORGANISMS

Based on the expectations with respect to the pollutant loads from the Dutch coast in 1990 [4] the concentration pattern and anthropogenic fraction have been calculated. Tables VI and VII show the influence of sanitation up to 1990 on the area of the considered part of the North Sea where the calculated concentration exceeds the assessment level in combination with the areas where the anthropogenic part of the concentration exceeds 50%. The tables show that despite sanitation of pollutant sources the concentration levels for certain substances (Cd, Hg) still remain critical to the marine environment especially when the accumulation of pollutants in the bottom sediments are also taken into account.

Pollutant	Surface area (%) of:			
	Dpcs		North Sea < 56°	
	1980	1990	1980	1990
N	6	6	8	8
P	10	8	12	11
Cd	23	9	21	9
Hg	22	5	29	16
Pb	15	3	16	12
Cu	6	1	9	6
Zn	24	8	25	15
Cr	6	0	2	1

Table VI. Areas where concentrations exceed 50% anthropogenity

Pollutant	Critical level (μg/1)	Surface area (%) of:			
		Dpcs		North Sea < 56°	
		1980	1990	1980	1990
Cd	0.05	27	20	25	21
Hg	0.01	20	9	21	14
Pb	3.4	0	0	1	1
Cu	0.14	100	100	100	100
Zn	250	0	0	0	0
Cr	300	0	0	0	0

Table VII. Areas where concentrations exceed critical levels

With respect to the anthropogenic input of nutrients (P, N) in the North Sea ecosystem, it has to be emphasized that not only the primary production and algae species composition is influenced, but also the higher trophic levels will be affected. Increased biomass production will result in increased mineralization, which may cause oxygen depletion in stratified systems of the North Sea. Due to lower oxygen concentrations the exposure of organisms to toxics will be enlarged.

This illustrates, that it will be necessary to dedicate more research and modelling effort to a quantitative or at least semi-quantitative description of the marine ecosystem of North Sea including the higher trophic levels in the near future.

5. ACKNOWLEDGEMENT

The authors wish to thank Rijkswaterstaat - the Directorate North Sea, the Delta Department and the Government Institute for Waste Water Treatment (RIZA) - for their cooperation during the study and permission for publication of this part of the study.

6. REFERENCES

[1] Anonymus; Waterkwaliteitsplan Noordzee (draft), Ministry of Traffic and Public Works, The Netherlands, 1985

[2] Anonymus; Waterkwaliteitsplan Noordzee; verslag en onderzoek deel 2b, Analyse en het natuurlijk systeem. Rijkswaterstaat, Waterloopkundig Laboratorium, maart 1985

[3] Anonymus; Quality status of the North Sea; contributions by experts from the reparian states of the North Sea, International Conference on protection of the North Sea, (FRG), Hamburg, 1984

[4] Anonymus; Waterkwaliteitsplan Noordzee; verslag en onderzoek deel 3: activiteiten en bronnen van verontreiniging; Rijkswaterstaat en Waterloopkundig Laboratorium, 1985

[5] Aalst, R.M. van et al.; Vervuiling van de Noordzee vanuit de atmosfeer, TNO-rapport CL82/152a, augustus 1983

[6] Boehnecke, G.; Salzgehalt und Stroemungen der Nordsee. Veroeff. Inst. Meereskunde, Berlin A 10, 1922

[7] Eck, B. van, Turkstra, E., Sant, H. van 't; Voorstel referentiewaarden Nederlandse zoute wateren; discussiestuk Ministerie van Volkshuisvesting, Ruimtelijke Ordening en Milieu-Beheer, october 1983

[8]   Reid, P.C.; Continuous plankton records; large scale charges: the
      abundance of phytoplankton in the North Sea from 1958-1973; Rapp.
      P. v. Reunie Cons. Int. Explor., mei 172 (1978) p. 384-389

[9]   Anonymus; The phytoplankton-nutrient model SEAWAQ and its appli-
      cation to the Southern Bight of the North Sea, R 1908, Delft
      Hydraulics Laboratory, December 1984

[10] Rat von Sachverständigen fuer Umweltfragen; Sondergutachten Juni
      1980; Umweltprobleme der Nordsee, Kohlhammer Verlag Stuttgart,
      Mainz, 1980

[11] Dethlefsen, V., Westerhagen, H. von; Oxygen deficiency and
      effects on bottom fauna in the eastern German Bight 1982;
      Meeresforschung / Rep.Mar.Res. 30 (1), 1983

[12] Train, R.E.; Quality Criteria for Water, EPA, Castle House Publi-
      cations Ltd., ISBN 0719400236, 1979

EXPERIENCES OF THE FEDERAL REPUBLIC OF GERMANY
WITH DUMPING OF SEWAGE SLUDGE

V. Dethlefsen
Bundesforschungsanstalt für Fischerei
Institut für Küsten- und Binnenfischerei
Toxikologisches Laboratorium Cuxhaven
Niedersachsenstraße
2190 Cuxhaven
Federal Republic of Germany

ABSTRACT. From 1961 to 1980 annually 350 000 m³ of sewage sludge
of the City of Hamburg were dumped into an area located in the
outer stretches of the Weser and Elbe estuary in the inner German
Bight of the Southern North Sea.
Long term investigations on macrobenthos revealed a steady downward
trend of the species diversity mass development of single species
at certain periods of the year and a recurrent collaps of the
macrofauna during summer. Low dissolved oxygen in the water column
and the sediments were considered to be prime causes.
Impairment of trawling was one further nuisance. Increased
prevalences of diseases of cod and dab in the vicinity of the area
were further biological effects ascribed to the pollution in the
estuary. The dumping was shifted to a location further offshore in
the German Bight from there into an area in the Atlantic and it was
finally terminated at April 15, 1983. Investigations were continued
in the area, slight changes of sediments were demonstrated, an
indication for recovering benthic communities was present but a
collaps of macrofaunal elements in the summer of years 1983 and
1984 occurred, which were interpreted as indication for the general
degraded water quality within the estuary.
After cessation of dumping the wastes were stored in land fills and
it is presently attempted to develop alternative uses for the sewage
sludge material, which is mainly objected because of its high
contamination with organochlorines and heavy metals.

1. INTRODUCTION

Opinions on possible effects of marine dumping of sewage
sludges range from beneficial to detrimental. This is depending
on properties of the wastes on the one hand side and the recipient
on the other. While it is generally accepted that uncontaminated
sewage sludges can be safely disposed in marine areas with
dispersive character, problems might arise when contaminated
sludges are dumped into accumulating areas.

G. Kullenberg (ed.), The Role of the Oceans as a Waste Disposal Option, 233–242.
© 1986 by D. Reidel Publishing Company.

Examples are available in the literature characterizing the behaviour
of dumped sewage sludge material at different dump sites. The Lyme Bay
on the southeast coast of the United Kingdom can be taken as one
example for an accumulative dump site, where increased levels of
carbon, nitrogen, trace metals, such as mercury, copper, zinc, lead,
nickel, chromium, were detectable despite the relatively low quantities
dumped.

Accordingly effects on benthos are detectable (Eagle et al., 1978).
At dumping areas with high dispersion as in the Humber estuary ele-
vations of organic matter and trace metals are slight and localized
(Norton, 1980).

In case of sewage sludge dumping in the outer Thames estuary it was
found that current rates of dumping have exceeded the dispersive capacity of the area resulting in the readily identifiable regions where
organic matter and metals have accumulated. These have placed the
natural fauna under stress and promoted the growth of pollution indicators species (Norton et al., 1981).

Frequently contaminated sewage sludges and industrial wastes are
dumped together as this is the case in the Humber estuary, which is
considered to be grossly polluted. Here it is difficult to isolate the
influence of dumping from that of the estuary (Lindsay et al.,1980).

Often contaminated sewage sludges are dumped into grossly polluted
estuaries, where faunal changes are clearly detectable but impossible
to separately ascribe them to the dumping or to the background
pollution of the estuary.

From 1961 to 1980 annually 350 000 m³ sewage sludge were dumped into
the German Bight. Details will be given on investigations carried out
in the dumping area. The results of which finally lead to the decision
to sease dumping.

2. THE AREA

The dumping area is located between the island of Helgoland and the
coast within the outer stretches of the Elbe estuary of the German Bight.
It is situated within the Elbe-Weser-triangle receiving heavily contaminated estuarine water from both river systems. Sediments in the
dumping area were formerly sandy/muddy (Caspers, 1979) and in the 1980s
muddy (Rachor, 1980).

The dumping site covers the eastern edge of a sedimentary area (muddy)
(Figge, 1981) with two different types of predominating sediments.
According to this sediment distribution the bottom fauna east of the
site is a near shore Macoma baltica community whereas to the west a
Nucula nitidosa community is found (Rachor, 1984). Due to this location
in a faunistic transition area interpretations of benthic changes are
difficult. The situation is further complicated by wide fluctuations of
environmental factors such as salinity, temperature, tidal currents,
variability in run off from the rivers. Despite the shallowness
stratification in summer frequently leads to oxygen depletion (Rachor,
1984).

Medium depth of water is 20 - 30 m. Sediments within the immediate

vicinity of the dumping area are characterized by a small oxidative
layer and the development of H_2S (Mühlenhardt-Siegel, 1981).

While the general sedimentary properties of material from the dumping
area proper were not different from those of the nearby sedimentary
area the presence of tomato seeds, matches and plastic hairs indicate
an input of anthropogenious origin.

3. THE WASTE

The sewage sludge of the City of Hamburg resulted from a mixed
industrial and urban runoff and was dumped into an area of 30 km^2. The
material was released into the wakes of a dumping vessel on a west-
northwesterly transect of a length of 15 km.

Nauke (1974) who also investigated the heavy metal contents in
sediments of the dumping area concluded that the typical heavy metal
contents were detectable with no enrichment due to the dumping of sewage
sludge. He covered a relatively small area of only few kilometres within
the vicinity of the dumping area.

According to Caspers (1979) the sewage remains relatively concentrated
for one hour after the dumping and during the sedimentation fine
material is transported away while only the corser material of the
sewage sludge is sedimented in the dumping area.

4. IMPACT ON MACROBENTHOS

Investigations on effects of the dumping of sewage sludge on macrobenthos were started in 1970 (Caspers, 1979). In 1970 and for some
years thereafter Caspers detected a mass development of Abra alba.
At times up to 1 750 individuals per 1/10 m^2 bottom were counted. The
highest population density of this mollusc was always found in areas
which were through presence of tomato seeds characterized to be the
sewage sludge dumping area.

Zones less frequently covered by sewage were characterized by the
presence of Nucula turgida. In areas where sediments were partly
characterized by the presence of sewage transitions between the
occurrence of Abra alba and Nucula turgida were found.

In April 1974 one macrobenthos species were found in the centre of
the dumping area, while in the neighbouring Nucula area only four
macrobenthos species occurred. Caspers took this finding as an expression for the manyfoldness of macrobenthos species in the dumping
area.

In 1977 and 1978 macrobenthos catches revealed a very negative
picture. At this time Abra alba was almost completely absent.
The benthos was reduced to a few remaining elements. Surrounding areas
were characterized by the normal species and individual diversity.
Caspers' interpretation of his results of the long term studies within
the dumping area is that certain species are favoured by the dumping
but this development is not ecologically negative. He does not agree
that the benthic communities found in the dumping area should be
characterized as of inferior value.

Rachor (1977; 1980) started his investigations on abundance and species diversity of macrobenthos in an area 8 - 9 km west of the dumping area in 1969. He found a continuous reduction of faunal elements at his station. One of his striking results was that populations rich in individuals repeatedly collapsed during summer months. The summerly reductions did result in an impoverishment of the fauna. 1977 Rachor considered his results as being preliminary but he concluded that the dumping area was not suited for a continuation of inputs. In 1980 the interpretation of his results was that the continuous reduction of macrobenthos fauna at his stations indicates the ecological sensitivity of the inner German Bight. He stresses that 1976 much of the area investigated was devoid of macrofauna. Although macrofauna organisms inhabiting natural sedimentation areas are adapted to high inputs of organic substances and low DO conditions, the compensation capabilities of the fauna within the area under consideration should not be overestimated. Based on his 10 years study his conclusion is that the area of the inner German Bight should not be used for input of high quantities of wastes.

Mühlenhardt-Siegel (1981) took samples in July, August and November 1978 at 5 stations situated on a transekt including central and peripheral areas in the dumping region. He demonstrated that apart from a low species diversity at the central station the lowest biomass was measured here as compared to neighbouring areas.

5.   LOW DISSOLVED OXYGEN

Rachor in his paper from 1980 suspected low dissolved oxygen in bottom near water to be amongst the causes for reduction of macrofauna in the dumping area. Beginning in 1981 recurrent low dissolved oxygen conditions were detected in the German Bight (Dethlefsen and von Westernhagen, 1983). For the sewage sludge dumping area concentrations around 50 % saturation were detected in the summer months (Rachor and Albrecht, 1983).

6.   HEAVY METALS

Although increased contents of heavy metals in sediment were not detected in the dumping area within the German Bight (Nauke, 1974), Karbe (1980) described increased concentrations of mercury in Echiurus spec. of the dumping area. Since only singular analysis are existing and the investigation has not been persued this is the only indication for increased heavy metals within that area.

Naukes' (1974) finding that no elevated heavy metal concentrations were found in sediments of the dumping area is consistent with results from further sedimentological studies where it was also demonstrated that contamination of sediments of the inner German Bight were not different from those of other parts of the German Bight (Schwedhelm and Irion, 1984; Deutsches Hydrographisches Institut, 1984), the general levels being high.

Mart et al. (1984a) describe heavy metal concentrations in the sea-

water in the dissolved and particulate state of on- and offshore stations
of the German Bight. For lead, copper, nickel and cobalt they were also
able to show that very high concentrations, especially in the particulate
phase, existed in onshore waters close to the sewage sludge dumping area.
Decreasing concentrations in the particulate as well as in the dissolved
phase were analyzed in further offshore areas.

For cadmium it was demonstrated that also in relatively offshore station
high cadmium concentrations in the seawater in both phases existed indicating the transport of cadmium from the estuary into the open North Sea
and from thereon north up the Norwegian coast until Spitzbergen with no
further significant dilution (Mart et al., 1984b).

7. INTERFERENCE WITH FISHERIES

In 1976 first complaints from fishermen were heard when especially in
summer in certain regions within the vicinity of the dumping area fishing
became impossible. Frequently nets were lost because they were clogged
by fine sediments or they could only be hauled with great difficulties.
Systematic investigations following these complaints resulted in the proof
of severe impairments of fishing, especially when small mesh sized fishing
gear as necessary for the catching of eels in summer were used. The
locations were identical to those that have been shown to be the deposition
area for corser material of the sewage sludge. Sludge samples obtained from
the nets were analyzed by the German Hydrographic Institute, but it was not
possible to demonstrate that these sludges were identic to sewage material
dumped. The question had to remain open whether the difficulties for fisheries were due to natural circumstances in the area or due to the dumping.

8. FISH DISEASES

In 1977 investigations started on occurrence and abundance of fish
diseases in the German Bight. During these studies it could be demonstrated that in the area between Helgoland and Cuxhaven certain diseases
were quite frequent, especially those probably caused by bacteria.
Infected fish species were cod, flounder, plaice and dab (Dethlefsen,
1980; 1984). The diseases most frequently occurring on cod were skeletal
deformities, pseudobranchial tumours and ulcerations and those most frequently found on dab were lymphocystis, epidermal papilloma, hyperplasia
and ulcerations.

9. PLEA FOR CESSATION

The combination of results obtained during studies in the dumping
area for wastes of sewage sludges within the German Bight lead the
Federal Research Board of Fisheries to the conclusion that environmental
impact of the dumping cannot be excluded.
The responsible body for dumping in the Federal Republic of Germany,
the German Hydrographic Institution, could not dispel this concern,
which was aggrevated by further institutes involved in the process of
issuing licences for dumping. The City of Hamburg was therefore re-
quested to shift the dumping into an area 30 nm west of Helgoland by

June 1980 and to finally stop dumping by January 1981.

The City of Hamburg protested claiming it would be impossible to install landbased facilities to dispose of the sewage at such as short notice. Altogether the City of Hamburg did not share the concern of the scientific institutions, but did not resist the request to stop dumping in the North Sea mainly in reaction to public pressure.

Beginning at April 15, 1981, sewage sludges of the City of Hamburg were dumped in an area 922 nm off Helgoland in a position 47° 30'N 11° 00'W at a depth of water of 2 400 m. It was the intention of the City of Hamburg to continue dumping until April 15, 1983, and afterwards to use landbased disposal possibilities which should be operational at this time.

10. INVESTIGATIONS AFTER CESSATION OF DUMPING IN THE GERMAN BIGHT

Changes in the colour of sediments from dark grey to brown have occurred in wide stretches of the dumping area. No further complaints of fishermen were heard although during fisheries with research vessels the old clogging problem still seemed to exist, but lots of fishing gear seemed to be less frequent than in former times. According to Irion (pers. comm.) several metres of sewage are presently stored in the former dumping area and it is not very likely that this material will be drifted away rapidly.

Rachor continued his investigations on macrobenthos communities in the former dumping area and in 1983 he found 37 species, an unusually high diversity, the best result obtained in his long term study. While the species diversity remained high until the summer of 1984 still represented by 23 species giving rise to some hope for the indication of a recovery of the area the bentic community finally collapsed in autumn of 1984.

In January 1985 seven species were present. Since already in 1976 and 1980 high species diversity were obtained, Rachor does not interprete his 1983 data too optimistic, but he claims that for example Echinocardium, a species never seen in the dumping area during his long term studies, was present in 1983 (Rachor, pers. comm.).

11. FURTHER FATE OF THE WASTES

At the time of this writing the City of Hamburg is still struggling for an optimum disposal method. Since the original intention to mix the sewage sludge with lime and to use it as fertilizer did not materialize the wastes were stored on land based storage sites in Hamburg. Due to the heavy metal contents and the general contamination of sewage sludge this storage site was not considered to be safe and the sludge is at present after partial desiccation transported into a land disposal site, which is considered safe located in the German Democratic Republic. Since various tests for the further use of the sewage sludge material are presently carried out, also this disposal method is not considered to be final.

12.   WAS THE DECISION TO STOP DUMPING CORRECT?

In view of the present sludge disposal method the question arises
whether the decision to stop dumping in the Elbe estuary was correct.
To answer this question the overall ecological situation of the inner
German Bight should be considered. Here a number of changes in the
ecosystem occurred in recent years which taken together are interpreted as ecological dysfunctions.

a) The most striking phenomenon is eutrophication. Recurrent low
dissolved oxygen situations in the German Bight were first detected
in the summer of 1981. Increased nutrient concentrations have been
demonstrated (Anonymous, 1981) for a station in the vicinity of the
area under consideration. When data from 1936 of nutrient concentrations
in the German Bight were compared with data obtained in 1976
(Weichart, 1984) it also became obvious that nutrient concentrations
have increased in the waters of the open German Bight by factors
between 2 and 3.

Phytoplankton biomass has increased accordingly (Anonymous, 1981)
thus giving the prerequisits for the occurrence of low dissolved oxygen
situations at times of high water temperature and stable water stratification.

Effects on fish and benthos were clearly demonstrated (Dethlefsen
and von Westernhagen, 1983).

b) Fish diseases
First information on increased prevalences of externally occurring
diseases on flatfish in the North Sea is from 1977. Since then in a
number of systematic studies it has been established that increased
disease rates occur throughout the southern North Sea with disease hot
spots being located in the German Bight, Danish, Dutch and British
coastal waters and the Doggerbank area (Möller, 1977; Dethlefsen, 1984).
In one of the preceding chapters it has already been mentioned that
certain diseases on flatfishes and gadoids occurred in increased
frequencies in the inner Elbe estuary.

c) Benthic reductions in the Elbe estuary have already been mentioned.

d) Changes in fish species composition in coastal regions over the
last 28 years have been investigated by Tiews (1983). While traditional
Wadden Sea species have decreased in their abundance a loss of biomass
was compensated by the intrusion of non-traditional Wadden Sea fishes,
so that the total biomass remained more or less constant. Changes in
species composition could not be explained by hydrographical and
climatological variations nor by changes in fishing effort. It is
therefore discussed that pollution might be one of the causative agents.

e) Malformation rates of pelagic fish embryos in the southern North
Sea have been investigated with the result that relatively high malformation rates were detectable. The results are not yet fully evaluated
but might be taken as one further indication for pollution related
abnormalities (Dethlefsen, unpublished data).

It is therefore that the question whether the decision to stop dumping
was correct or not, is answered with yes. It was correct to stop the
dumping in the German Bight since the changes observed indicated that
some undesirable ecological changes are occurring.

13. POLLUTION MANAGEMENT CONSIDERATIONS

Since the process of the termination of the dumping of sewage
sludges of the City of Hamburg in the inner German Bight much doubt
has been expressed as to the validity and the interpretation of the
monitoring data. Some general points on management of pollution in
coastal areas seem to be relevant.

Most of the dumping occurs in grossly polluted areas. This certainly
holds true for the coasts of the United Kingdom, the Netherlands,
Belgium and the Federal Republic of Germany. It is also accepted that
the additional load of pollution introduced via dumping is only minor
compared to the loads already present.

The aim of present monitoring concepts is to show changes beyond
natural variability. It is clearly evident that effects due to dumping
would only peak slightly out of those effects present through background
pollution. They would therefore not be separatable from the background
variability. It is therefore that present monitoring concepts are
conceptionally wrong (Dethlefsen, 1985, in press).

So the inevitable conclusion based on results obtained in traditional
monitoring programme is that the effects are only local and management
necessities derived from this often result in shifting of the dumping
area into more dispersive grounds. The site by site monitoring will
not be able to detect larger scale changes and if these changes occur
it will not be possible to clearly link them to any specific causes.
Examples for this destiny of monitoring data are manyfold.

Following the debate on monitoring results where all changes so far
detected were either local or debated the question arises as to what
effects do we expect. What would be an acceptable endpoint to define
the assimilative capacity of a marine system. The disease debate
clearly demonstrates that even the presence of some 50 % of diseased
flatfishes in an area can be debated on grounds that this might be
due to natural causes.

The same happened to mortality demonstrated in low dissolved oxygen
areas where dead fishes were present, but it was not possible to
clearly state whether they were killed by low dissolved oxygen or by
fisheries (Dethlefsen and von Westernhagen, 1983). And having in mind
the periodicities, as for example demonstrated in the Russel Cycle,
where after 30 to 40 years of absent certain species reoccurred in the
samples, this clearly casts further doubt on the validity of monitoring
data.

So the present pollution management techniques do not seem to be
very soundly based and they contain an element of risk that dysfunctions
from pollution loads would not be detectable due to their slow and
chronic development. This uncertainty led the Federal Republic of
Germany to establish pollution management strategies which as a
precautionary measure reduce the input of wastes which are suspected
to be harmful to marine ecosystem.

On the basis of this precautionary principle dumping of sewage sludge
in the inner German Bight has been terminated and the dumping of wastes
from titaniumdioxide production in the outer German Bight will be terminated in the second half of the 1980s.

14. REFERENCES

Anonymous, 1981. Jahresbericht 1980. Biologische Anstalt Helgoland.
   Biol. Anst. Helgoland, Hamburg, 25, 57-58.
Caspers, H., 1979. Die Entwicklung der Bodenfauna im Klärschlamm-
   Verklappungsgebiet vor der Elbe-Mündung. - Arbeiten des
   Deutschen Fischerei-Verb., 27, 109-134.
Dethlefsen, V., 1980. Observations on fish diseases in the German
   Bight and their possible relation to pollution. - Rapp. P.-v.
   Réun. Cons. int. Explor. Mer 179, 110-117.
Dethlefsen, V., 1984. Diseases in North Sea fishes. - Helgoländer
   Meeresunters. 37, 353-374.
Dethlefsen, V.; Westernhagen, H. von, 1983. Oxygen deficiency and
   effects on bottom fauna in the eastern German Bight 1982.
   Meeresforsch. 30, 42-53.
Deutsches Hydrographisches Institut, 1984. Überwachung des Meeres.
   Bericht für das Jahr 1983. 1-30.
Eagle, R.A.; Hardiman, P.A.; Norton, M.G.; Nunny, R.S., 1978. The field
   assessment of effects of dumping wastes at sea: A survey of the
   sewage sludge disposal area in Lyme Bay. - Fish. Res. Techn. Rep.
   49, 1-22.
Figge, K., 1981. Sedimentverteilung in der Deutschen Bucht. Karte
   Nr. 2900 mit Beiheft. - Deutsches Hydrographisches Institut.
Karbe, L., 1980. Schwermetalle und Spurenelemente. - In: Umwelt-
   probleme der Nordsee. Sondergutachten 1980. Der Rat von Sachver-
   ständigen für Umweltfragen. 156-173.
Lindsey, A.M.; Norton, M.G.; Nunny, R.S.; Rolfe, M.S., 1980.
   The field assessment effects of dumping wastes at sea: The
   disposal of sewage sludge and industrial waste of the River
   Humber. - Fish. Res. Tech. Rep. 55, 1-35.
Mart, L.; Nürnberg, H.W.; Nützel, H., 1984a. Heavy metal levels in
   the German Bight. - Submitted to Estuarine, Coastal and Shelf
   Science. 1-24.
Mart, L.; Nürnberg, H.W.; Nützel, H., 1984. Comparative studies on
   cadmium levels in the North Sea, Norwegian Sea, Barents Sea and
   the Eastern Arctic Ocean. - Fresenius Z. Anal. Chem. 317, 201-209.
Möller, H., 1977. Distribution of some parasites and diseases of
   fishes from the North Sea in February, 1977. - C.M./ICES E:20.
Mühlenhardt-Siegel,U., 1981. Die Biomasse mariner Makrobenthos-
   Gesellschaften im Einflußbereich der Klärschlammverklappung
   vor der Elbemündung. - Helgoländer Meeresunters. 34, 427-437.
Nauke, M., 1974. Die Schwermetallgehalte der Sedimente im Klärschlamm-
   Verklappungsgebiet vor der Elbmündung. - Deutsche Hydrogr. Zeitschr.
   27, 203-213.
Norton, M.G., 1980. Monitoring of areas receiving dumped wastes
   around England and Wales. - ICES C.M./E:28.
Norton, M.G.; Eagle, R.A.; Nunny, R.S.; Rolfe, M.S.; Hardiman, P.A.,
   Hampson, B.L., 1981. The field assessment of effects of dumping
   wastes at sea: Sewage sludge dumping in the outer Thames Estuary.
   Fish. Res. Techn. Rep. 62, 1-62.

Rachor, E., 1977. Faunenverarmung in einem Schlickgebiet in der Nähe
    Helgolands. - Helgoländer wiss. Meeresunters. 30, 633-651.
Rachor, E., 1980. The Inner German Bight - an ecological sensitive
    area as indicated by the bottom fauna. - Helgoländer Meerresunters.
    33, 522-530.
Rachor, E.; Albrecht,H., 1983. Sauerstoff-Mangel im Bodenwasser der
    Deutschen Bucht. - Veröff. Inst. Meeresforsch. Bremerh. 19, 209-227.
Rachor, E., 1984. Monitoring of the macrofauna near the former disposal
    site of sewage sludge in the inner German Bight. - Standing Advisory
    Committee for Scientific Advice of Oslo Commission. 1-5.
Schwedhelm, E.; Irion, G., 1983. Heavy metal distribution in tidal flat
    sediments of the German Part of the North Sea. - In: Proc. Intern.
    Congress Heavy Metals in the Environment. 1038-1040.
Tiews, K., 1983. Über die Veränderungen im Auftreten von Fischen und
    Krebsen im Beifang der deutschen Garnelenfischerei während der Jahre
    1954-1981. - Ein Beitrag zur Ökologie des deutschen Wattenmeeres
    und zum biologischen Monitoring von Ökosystemen im Meer. - Arch.
    FischWiss. 34, 1-156.

THE IZMIT BAY CASE STUDY

S. Tuğrul, M. Sunay, Ö. Baştürk and T.I. Balkaş

TUBITAK, Marmara Research Institute
P.O.Box 21 Gebze-Kocaeli,
TURKEY.

The polluted waters of Izmit Bay were studied to measure the variations
of biochemical and physical characteristics monthly. The spatial and
temporal changes of nutrients (nitrate, nitrite, ammonia and o-phosphate),
chlorophyll-a, BOD_5 , total coliform, total mercury, dissolved petroleum
hydrocarbon (PAH), secchi disc depth, temperature and salinity in the
bay were monitored during the period of May 1984 - April 1985. Also,
some reliable data obtained from previous studies is included. The total
amounts of wastes discharged into the bay and types of polluting sources
were identified, as well as governmental regulations on discharges of
industries being discussed. The evaluated results were compared with
the results obtained from the relatively unpolluted waters of Marmara
Sea. The degree of pollution in the bay was assessed.

INTRODUCTION

In recent decades, the remarkable technological progress of industries in developed countries have accelerated the transfer of technical
knowledge to developing countries. As a consequence of heavy invest-
ments in the industrialization of developing countries, pollution problems of the environment have increased enormously due to inadequate
legislation. In Turkey, as a developing country, the environmental
pollution problems have particularly increased since 1960, due to rapid
growth of industry and widespread increase of population. The large
volumes of wastes from domestic and industrial sources have been mostly
discharged to the environment without any pre-treatment. Along with the
land and the atmosphere, the coastal waters of Turkey have been receiving
the majority of these wastes. Accordingly, the pollution of the Marmara
Sea, particularly of Izmit Bay, has caused great concern to local and

243

G. Kullenberg (ed.), The Role of the Oceans as a Waste Disposal Option, 243–274.

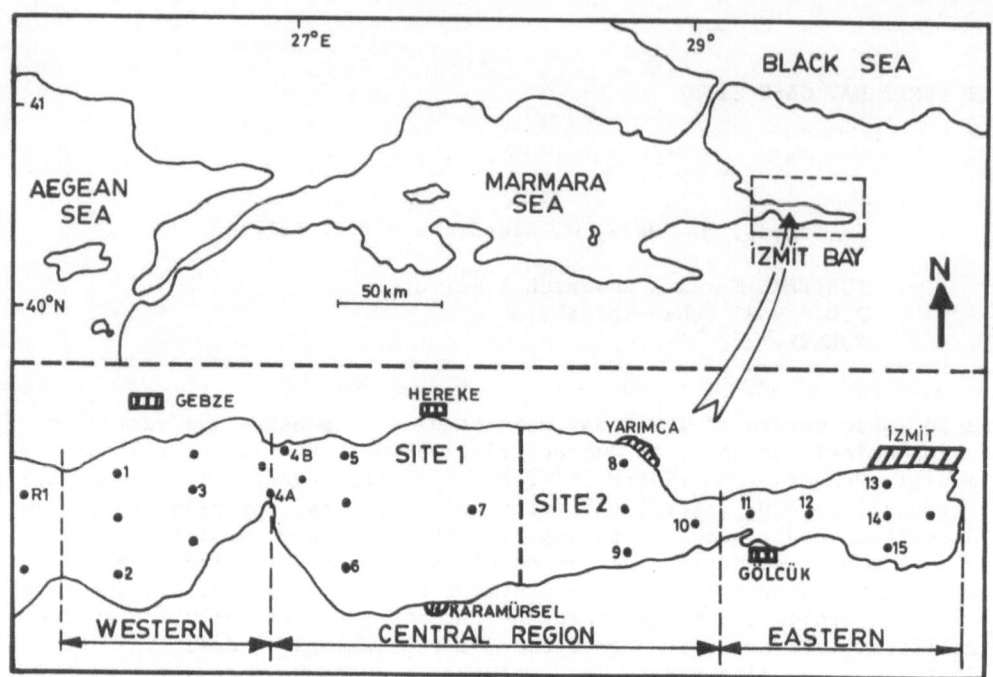

Figure 1. Sampling   locations in İzmit Bay

national authorities due to the environmental deterioration which this area has undergone. Up till the present time, it was assumed that dumping waste water into the bay was a practical solution. In the area surrounding the bay, more than 140 large industrial plants have arisen since 1965. Their solid and liquid wastes mostly find their way directly into the bay without any pretreatment (1-5). The volume of unpurified waste waters, both domestic sewage and industrial, is now so great that the biological self-purification capacity of the bay is no longer sufficient to restore the equilibrium to its normal state. It is well known that the self-purification capacity of bay water is directly related to the processes that control the oxygen balance of that water.

In order to improve water resources and prevent and control pullution of the bay, it is necessary to limit the quantities of organic matter, nutrients, and other toxic substances discharged into the bay. Before deciding what restriction must be placed upon the discharges and what concentrations are acceptable by the receiving water, it is necessary to assess the assimilative capacity of the bay. The selection of treatment methods and the extent of treatment are also firmly related to the assimilative capacity of the receiving water. This can be achieved by obtaining adequate and reliable oceanographic and pollutional data about Izmit Bay.

Although, so far, many studies have been accomplished on the characterization and treatment alternatives of waste effluents (1-5), very limited work has been performed on measuring the oceanographic characteristics of the bay systematically (6-8). Without having knowledge of the physical, chemical and biological properties of the bay, finding a reasonable solution to the pollution is unlikely. Therefore, reliable temporal and spatial measurements of oceanographic parameters in the bay water are of vital importance to assess the assimilative capacity of the bay. It has been suggested that the Bay of Izmit is in an entirely polluted state (4-8). However, all of these previous studies have been focused on the coastal areas, which are influenced directly by waste waters. After obtaining the systematic reliable data for physico-chemical parameters which control the bioassimilation capacity of the bay, the mathematical model being developed for the two-layer water masses of the bay would say the amount of biodegradable organic matter that can be assimilated by the water body. Then, the comparison of the volume of waste waters to be treated with the estimated allowable load of the model study would make a beneficial usage programme for the bay possible to initiate.

In order to assess the degree of pollution and the assimilative capacity of Izmit Bay, a monthly measurement programme was carried out during the period May 1984 - April 1985. The preliminary results obtained will also be submitted in detail (9), to the NATO Scientific Affairs Division, which has provided financial support for this work.

EXPERIMENTAL

Sampling stations. The measurements of physical and biochemical parameters were carried out at the points given in Figure 1. The numbered stations represent biochemical measurement points while dissolved oxygen, temperature and conductivity measurements were also done at the additional points marked on the map.

Temperature and conductivity. Temperature and conductivity were measured, in situ, by a Kahlsico Surveyor Model 6D probe.

Sampling. The water samples were collected by Hydro-Bios model nansen bottles.

Sample preservation. The water samples taken for nutrient analyses are preserved as given in standard methods (10). The samples for Hg-analysis were preserved by adding HSO_4 + $K_2Cr_2O_7$ to the samples (11).

Chlorophyll-a and total coliform. The water samples collected from different depths were filtered on board and frozen following the addition of $MgCO_3$ solution (12). The filter papers were homogenized and the absorptions at different wavelengths were measured using a Varian Model UV/VIS spectrophotometer. The water samples for total coliform were collected by the special sterilized glass bottles. Treatments of the samples were completed on board by using a Millipore Model incubation system.

Nutrient analysis. Phosphate, nitrite, nitrate, ammonia and silicate analyses were accomplished by a Technicon Model autoanalyzer system (13).

Total mercury. The samples digested with NHO_3 + $KMnO_4$ were analyzed by cold vapour-AA technique after concentrating the inorganic mercury in a 8 ml of dilute KMnO$_4$ solution *(11)*.

Dissolved petroleum hydrocarbon (PAH). PAH extracted into the CCl$_4$ was dried and dissolved in hexane again. The measurements were achieved using a fluorescent module of the UV/VIS spectrophotometer at 310 nm of excitation, e.g. characteristics excitation max. of chrysene *(14)*.

RESULTS AND DISCUSSION

    Before discussing the oceanographic results of the bay, a brief summary of the population distribution around the bay is given to indicate the expected population increase in the future and to allow the comparison of amounts of pollutants entering the bay originating from domestic, industrial and surface run-off of drainage areas (rural, urban and agricultural areas) are given to evaluate the annual pollutional loads reaching the bay waters. The basic polluting industries are defined with their contributions to the pollutional load in terms of BOD$_5$, nitrogen, phosphorus and other toxic substances. Finally, the present pollutional level of the bay water is estimated and the oceanographic parameters of the area are evaluated to assess the assimilative capacity of the bay and the limiting nutrient element in these waters.

Population Distribution Around The Bay

    The settlement area along the northern coast of Izmit Bay is severely limited becasue of the sharp topography of the steep coastal hills to the north and south. Most of the residential and commercial settlements are in the east-west direction along the transportation routes. There are four major urban centers around the bay namely; Gebze area on the north, Izmit metropolitan area on the northeastern part, Gölcük and Karamürsel areas on the southern banks of the bay. About 70% of the population live on the northern, 20% on the southern and 10% on the eastern banks of the bay area (Table-1). 56% of the total population live close to the industrial areas *(5)*. Rural areas, which are mostly inhibited by workers and their families, are in the neighbourhood of the metropolitan areas. The estimated population growth rates of major urban centers are given in Table-2 *(5)*. It is evident that the most drastic population growth is observed in the Gebze area, which acts a buffer-zone against the migration from rural areas to the Istanbul metropolitan area. The projections of the population distributions illustrated in Table-1 were obtained by the logistic and geometric extrapolation methods *(5)*. As is seen in the table, the expected population of the northern part of the bay will be more than two-fold of the present population in 20 years. About 10% of the present population are working in 140 major industries.

Sources And Amounts Of Pollutants Entering The Bay

    The amounts and types of wastes discharged into the bay waters from both domestic and industrial sources within the area have been well documented (Tables 3-7), whereas those from agricultural areas and surface

Table 1. Expected populations (x1000) around Izmit Bay

LOCATION	1980	1985	Y E A R S 1990	1996	2000	2005
NORTHERN PART						
Gebze	84.5	116.0	170.0	227.0	280.0	321.0
Izmit	192.5	230.0	261.0	286.0	305.0	319.0
Yarımca	19.0	29.0	39.0	48.0	56.0	61.0
Derince	67.0	117.0	164.0	204.0	232.0	248.0
Other small towns	36.0	52.0	73.0	94.0	120.0	149.0
SUB-TOTAL	399.0	544.0	707.0	859.0	993.0	1098.0
SOUTHERN PART						
Gölcük	46.0	57.0	64.0	69.0	73.0	75.0
Karamürsel	28.0	31.0	36.0	40.0	45.0	49.0
Other small towns	28.0	34.0	40.0	45.0	54.0	70.0
SUB-TOTAL	102.0	122.0	140.0	154.0	172.0	189.0
EASTERN PART (Total)	23.0	31.0	42.0	59.0	84.0	123.0
OVERALL TOTAL	524.0	697.0	889.0	1072.0	1249.0	1410.0

Table 2. Population growth rates of major urban areas around the bay (%)

YEARS	GEBZE	IZMIT	GÖLCÜK	KARAMÜRSEL
1945-50	2.5	3.2	2.9	2.3
1950-55	2.1	3.9	5.1	2.3
1955-60	3.7	3.8	3.9	2.8
1960-65	2.1	3.0	1.9	2.0
1965-70	6.9	4.6	5.5	1.2
1970-75	8.3	5.4	1.5	0.2
1975-80	19.5	3.5	6.0	3.5

run-off from urban and rural areas are not well known (5). However some figures have been estimated for surface run-off (Table 7). The estimated domestic waste water discharged into the bay and their pollutional loads are given in Tables 3 and 4, respectively.

As can be seen from Table 4, a major part of domestic wastes discharged into the bay waters are along the northern coasts. Contributions of industrial wastes to the pollutional profile of the bay are depicted in Tables 5 and 6. The pulp and paper industry complexes (SEKA), which are located within the area between Derince and Izmit, have enormous pollutional pontential in terms of its BOD_5(biodegradable   organic load).

Table   3.   Expected domestic waste water discharges into the Bay of
             Izmit (x 10^3 m^3/day)

LOCATION	Y E A R S				
	1980	1985	1990	1995	2000
Northern part	75	110	150	195	240
Eastern part	4.4	6.5	9	14	20
Southern part	17	22	27	61	37
TOTAL	92.4	138.5	186	270	297

Table   4.   Pollutional loads of domestic waste water sources (tons/day)

LOCATION and PARAMETER	Y E A R S				
	1980	1985	1990	1995	2000
Northern part					
BOD_5	24	33	42	52	60
TSS	36	49	64	77	90
T-N	4.8	6.6	8.5	10.3	11.9
T-P	1.2	1.6	2.1	2.6	3.0
Eastern part					
BOD_5	1.4	1.9	2.6	3.6	5.1
TSS	2.1	2.8	3.8	5.3	7.6
T-N	0.28	0.39	0.52	0.71	1.0
T-P	0.074	0.096	0.13	0.18	0.26
Southern part					
BOD_5	6.2	7.4	8.4	9.3	10.4
TSS	9.2	11.0	12.6	13.9	15.5
T-N	1.2	1.5	1.7	1.9	2.1
T-P	0.31	0.37	0.42	0.46	0.52
GRAND TOTAL					
BOD_5	31.6	42.3	53	64.9	75.5
TSS	47.3	62.8	80.4	96.2	112.6
T-N	6.3	8.5	10.7	12.9	15.0
T-P	1.6	2.2	2.7	3.2	3.8

BOD_5 : Biochemical oxygen demand ;   TSS :   Total suspended solid
T-N :   Total nitrogen   ;   T-P :   Total phosphorus

About 50% of the organic load reaching the bay are discharged from SEKA.
The total BOD_5 loads of 6 plants around the bay are 92% of the grand
total entering the bay waters *(5)*.   The biodegradable organic load of
industrial discharges is equivalent to 105 tons of dissolved oxygen per

day as BOD. The 80% of nitrogen loads from industries originate from
two fertilizer plants (5). The total volume of waste waters of industries directly discharged into the bay is about 1.9×10^5 m³/day. Three
industries,SEKA (pulp and paper industry), PETKIM (petrochemical industrial complexes) and IPRAS (a rafinery) contribute 58% of the total
industrial waste water discharges.

Table 5. Estimated pollutional loads of industrial waste waters
(tons/day)

TYPE OF INDUSTRY	TOTAL FLOW (m³/d)	BOD₅	TSS	T-N*	NH₃	T-P	OIL AND GREASE
Food	9320	20.3	1.4	0.94	–	0.18	0.4
Textile	555	0.30	0.19	0.001	0.007	–	–
Leather	1479	0.30	0.23	–	0.006	0.40	–
Pulp and paper	102628	61.7	42.4	–	–	–	–
Chemicals	51019	17.5	10.9	4.3	9.8	1.66	3.3
Rafineries	13765	3.4	1.1	–	0.21	–	0.84
Cement, glass and ceramics	2870	0.34	18.8	–	0.042	–	0.09
Metal finishing and machinery	5472	1.7	1.2	–	0.026	0.021	1.44
TOTAL	187108	105.5	76.9	4.24	10.1	2.26	6.07

* T-N : Total organic nitrogen

Table 6. Toxic dissolved substance discharges of industrial origin
into the Bay.

PARAMETER	kg/day	PARAMETER	kg/day
Lead	6.6	Nickel	2.2
Copper	11.9	Mercury	5.1
Zinc	43.2	Cynide	0.74
Chromium	209	Phenol	200
Cadmium	1.8	Fluoride	2000

The estimated pollutional loads of different industrial settlements
and urban areas are complied in Table 7. As can be seen from the results,
organic loads of industrial activities, 105 tons of BOD per day, are
much higher than that of domestic wastes. A similar pattern is seen in
the nitrogen content of industrial wastes as the phosphorus load of domestic waters is two-fold that of industrial origin.

Currently, There exists no sewerage facility in the concerned area.
Domestic and industrial waste waters are mostly discharged directly into
the bay or dragged by storm run-off. The projected sewerage system was
prepared by SWECO and BUE in 1975 (8). The proposed system does not involve all of the urban and industrial areas. Although the sewerage pro-

Table 7 . Estimated pollutional loads of domestic and industrial
wastewaters and surface run-off waters of drainage area
reaching the bay waters (tons/day)[a]

LOCATION and SOURCES	BOD$_5$	TSS	T-N	T-P
Northern Part				
Industrial	83	70	11.4	0.61
Domestic	24	36	4.8	1.2
Run-off	0.1	42	0.08	0.005
Eastern Part				
Industrial	21.7	5.6	2.2	0.19
Domestic	1.4	2.1	0.29	0.07
Run-off	0.1	43.6	0.08	0.005
Southern Part				
Industrial	0.78	1.26	0.9	-
Domestic	6.1	9.2	1.23	0.31
Run-off	0.8	85.9	0.2	0.02
Industrial Total	105.5	76.9	14.5	0.80
Domestic Total	31.5	47.3	6.3	1.58
Run-off Total	1.0	171.5	0.36	0.03
OVERALL TOTAL	138.0	295.7	21.16	2.41

(a) : Data from reference 5.
T-N : Total kjeldahl nitrogen (organic-N + NH_3)

ject has been authorized, no attempt at construction has been made so
far. The studies performed up to 1984 defined the chemical compositions
of industrial wastes with their daily flow rates and applicable treatment systems for the most polluting industries (5). Following that project, some discharge standards were developed for major polluting plants
(14). According to these standards, non-hazardous waste waters of the
industries can be discharged into the domestic sewerage system to be
constructed after the pretreatment of industrial waste waters in the
areas of the plants. Then, the collected wastes will be discharged into
defined points in the bay by following the physico-chemical and biological treatment.

Physical And Meteorological Properties Of The Bay

Izmit Bay, located on the northeastern part of the Sea of Marmara
has a surface area of approximately 310 km². It is an elongated semi-
enclosed body of water, about 50 km in length and 2 - 10 km in width,
and has considerable depth except for the eastern part. Topographically
and in terms of its oceanographic characteristics, Izmit Bay may be se-

parated into three distinct regions; briefly western, central and eastern regions which are connected to each other through narrow openings
(5, 8). The bottom of the western section slopes upward in an easterly
direction from about 100 meter-depth contour which bounds the bay to the
northeastern Marmara Basin. As seen from Table 8, the surface area of
this section is about 100 km^2 and it is connected to the northeastern
Marmara Basin through a 5.5 km wide opening. Continuing in an easterly
direction, there exists a sill about 55 meters in depth and 3 km in width,
which separates the western region from the central part of the bay.
The central region, which has about 170 km^2 surface area, constitutes
the largest part of the bay. The eastern part of the bay is the smallest
section of the system (44 km^2) and is connected to the central region by
a 2 km wide opening. It is relatively shallow with a maximum depth of
about 35 m. Although this section of the bay is small in area and shallow in depth, it receives the majority of domestic and industrial wastes.
This portion also receives some inflow of fresh water from several creaks
which are mostly polluted by industrial wastes.

Table 8. Basic physical properties of hydrographic regions of Izmit
Bay *(5-8)*

REGION	LENGTH(km)	WIDTH(km)	MAX.DEPTH(m)	SURFACE AREA(km^2)	VOLUME (km^3)
Eastern	16	2-5	35	44	850
Central	20	3-10	180	166	12420
Western	17	3-5.5	1000	100	?

The total drainage area of Izmit Bay for its central and eastern
sections is about 1205 km^2, excluding the area of the bay itself. The
drainage area of the eastern part is about 230 km^2 whereas that of the
central section is about 975 km^2. The annual precipitation is about
700 mm, as the evaporation from the surface of the bay is about 600 mm
annually. Estimated fresh water inputs into the eastern and central
sections are about 12 and 3 m^3/ sec, respectively. The total input of
the fresh water can be ignored when compared with the volume of the bay.

The regional wind regimes influencing Izmit Bay are dominated by
northerlies from the Black and occasional southeasterly winds from the
Aegean Sea. The prevailing wind from the north known as Etesian winds
are particularly pronounced in July and August. Cold northerly outbreaks
are also effective in the winter. The winter season is further characterized by the occasionally pronounced southerly winds with the local
name of Lodos. They frequently produce storm waves and surges within
the bay, and bring warmer and humid weather to the region. Long-term
mean air temperatures around the area vary from about 3°C in winter to
about 26°C in summer *(5)*.

Temperature And Salinity Variations In The Bay

As a part of the Turkish Straight System, the hydrological regime
of the bay is, to a large extent, governed by the exchange of water

between the Black Sea and the Mediterranean (Figure 1). More specifically, the bay exchanges its water with the Sea of Marmara through its
opening to the northern Marmara Basin. This exchange has, presumably,
a considerable variation throughout the year, even though inflow and
outflow rates and their temporal variations are not known precisely.
In terms of flow and stratification characteristics, the basic nature of
İzmit Bay is the existance of a two-layer current system associated with
two-layer stratification held throughout the year (T.Oğuz, per.comm.).
In general, spring and summer seasons are characterized by the inflow
of the low saline waters of Black Sea origin with 21 - 24 ppt salinity
over the high saline waters of the Mediterranean origin (Figure 2). This
period corresponds to times in which precipitation and fresh water flow
into the Black Sea increase substantially. As seen in Figure 2, the surface salinity of the bay waters rises 26 −•28 ppt in November due to inflow of high saline bottom waters into the bay. The increase in the
surface salinity continues during the winter months and reaches 30 ppt
in March. The subsequent decrease in the surface salinity appeared in
April due to inflow of the low saline waters of Black Sea origin into
the bay again. The surface salinity distribution data indicates that
the replacement of the surface waters in the bay starts in April. The
surface water temperature in the bay stays almost constant during the
summer months. It starts decreasing in October, reaching $11^{o}C$ in January and $7.5^{o}C$ by March. On the other hand, the winter season is characterized by the inflow of bottom waters of the northeastern Marmara Basin
into the bay and the subsequent increase in surface layer salinity. Thus,
the layer of the high saline waters, which have 38 - 38.5 ppt salinity
and $14.4 - 15.0^{o}C$ temperature, rises inside the bay. Also in relation
with bottom water, it was concluded that, while the surface water inflow
is seen in April (Figure 2) the high saline waters of the northeastern
Marmara Basin enter the bay from bottom during the same period. Consequently, as the surface salinity decreases the 38 ppt salinity contour
along the bay rises more than 10 meters with respect to the depth of
that contour in March.

Typical seasonal variations of salinity and temperature within the
bay showing the permanent existence of two-layer stratification are illustrated in Figure 2, at a station located in the central region of the
bay. The longitudinal variations of salinity with depth in summer and
winter months are depicted in Figure 3, which show the temporal variations of water masses of different densities within the different regions
of the bay, the times of exchange of the bay water masses with the northeastern Marmara waters and the extent of the vertical mixing in the bay.
As seen in Figure 3, the thickness of the transition zone between two
water masses of different salinity during the summer months is very limited as it becomes larger in the winter months due to the strong turbulent
mixing across the interface. However, there exists sharp temperature
and salinity stratifications throughout the year with the exception of
the eastern region during the winter months.

Some short-term peaks in the surface salinity can appear due to
strong wind episodes in the winter months. A typical example is shown
in Figure 4, indicating an almost 5 ppt increase in the surface salinity
value due to the 10 m/sec northeasterly wind blown in February (7). It
is proposed that this type of wind episode could be the major factor

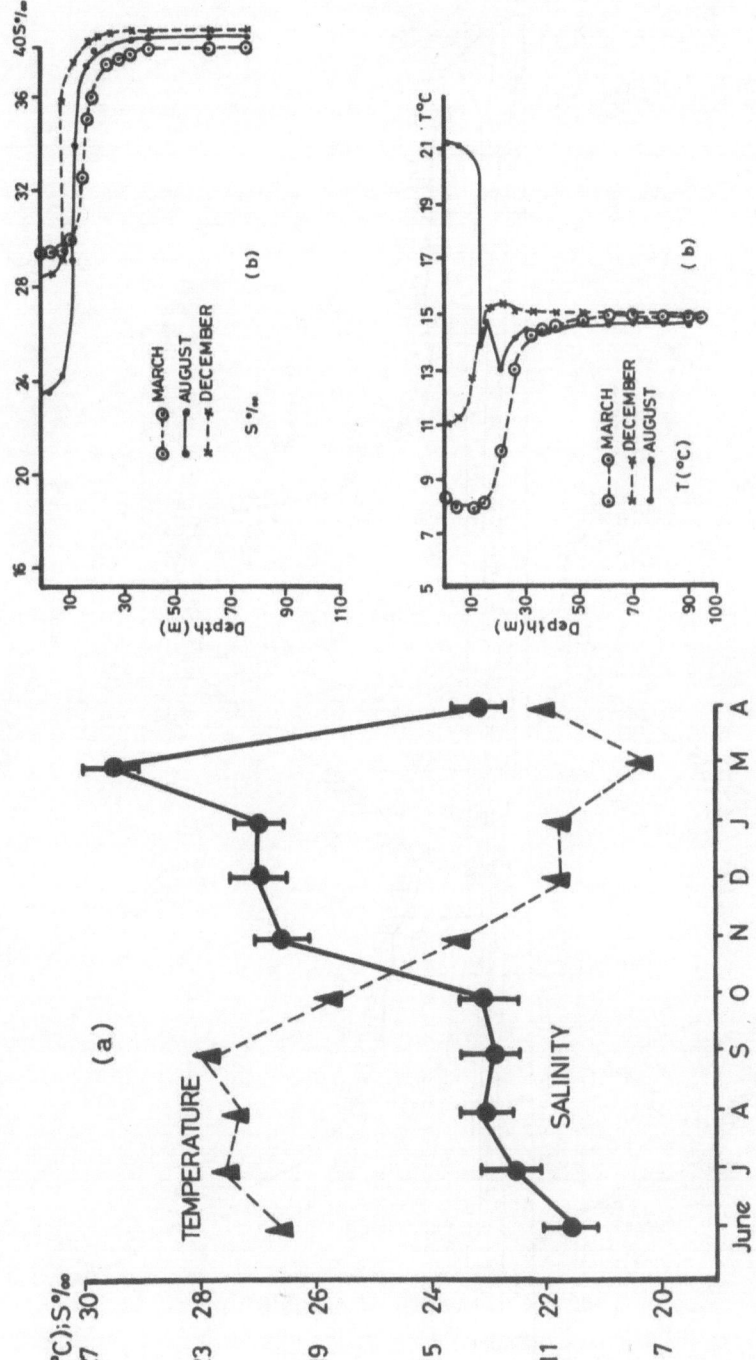

Figure 2. Variations of temperature and salinity ;(a) at the surface waters of the bay and(b) at Station 7.

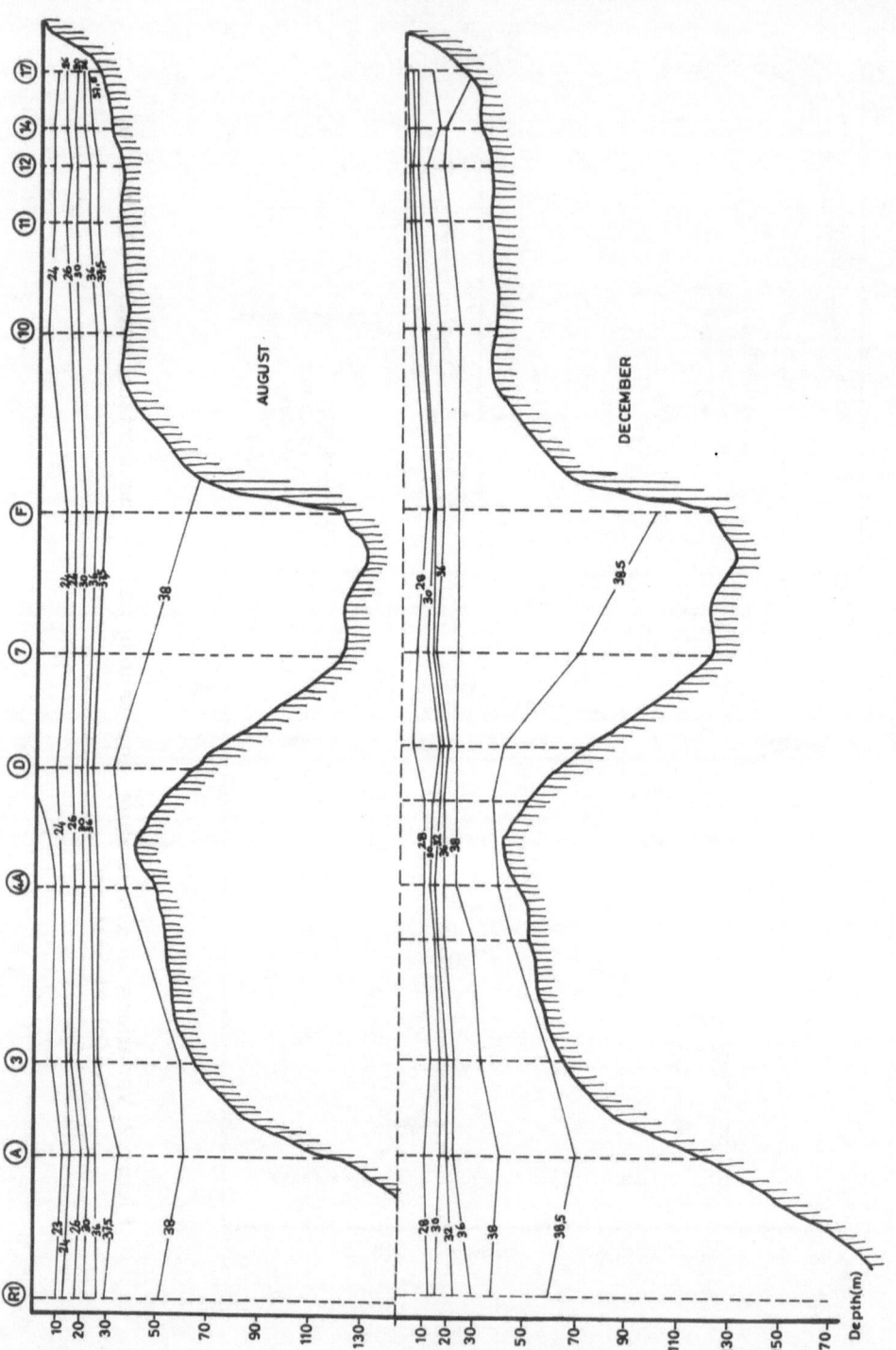

Figure 3 . Longitudinal variations of salinity in İzmit Bay.

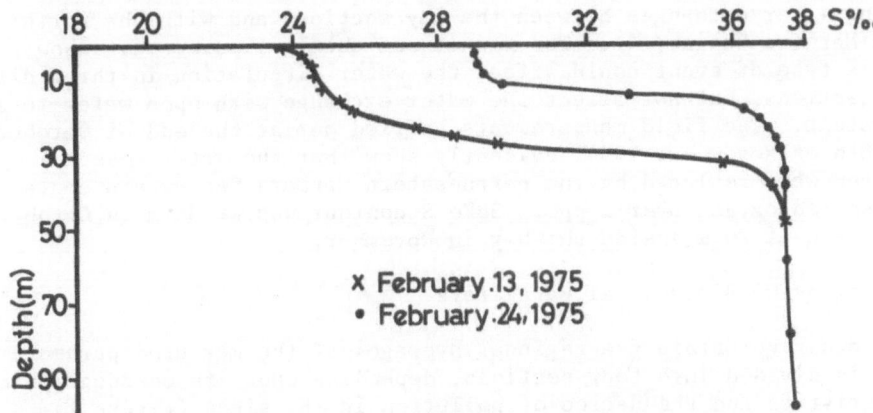

Figure 4.Effect of northeasterly wind episode on the
salinity distribution (7).

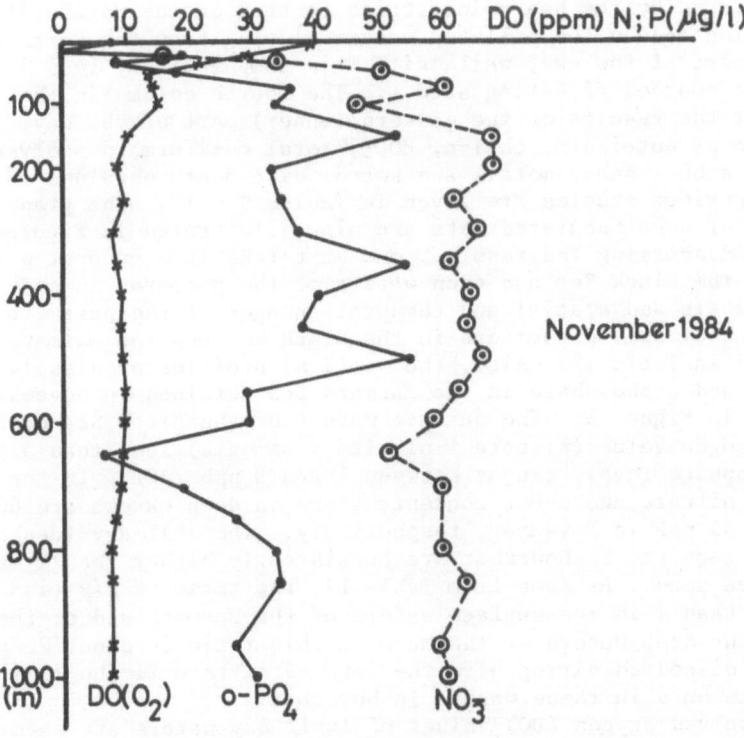

Figure 5 . Nitrate, o-phosphate and dissolved oxygen
distributions in the Marmara Sea .

affecting water exchanges between the bay sections and with the northeastern Marmara Basin (16). The results of chemical parameters show
that this type of event could affect the water circulation in three different sections, but not affect the water exchange with open water to a
great extent. The field measurements carried out at the end of October
and on 8th of November, 1984, evidently show that the water masses in
the bottom were replaced by the northeastern Marmara Sea waters containing dissolved oxygen over 2 ppm. 38‰ S contour was at 35 m in October
and was seen at 20 m inside the bay in November.

Distributions Of Biochemical Parameters

        In order to obtain the regional averages of the measured parameters
the bay is divided into four sections, depending upon its oceanographic
characteristics and the degree of pollution in the sites (Figure 1).
The western region of the bay is the least polluted site in the area,
since it is under the direct influence of the Marmara Sea through a wide
and deep channel, and in addition in the region, there is no industry
with high polluting potential. The central part of the bay is divided
into two sections. Sites 1 and 2 represent the western and the eastern
sections of the central region of the bay. Site 2 of the central region
is highly influenced by the heavy industries located on the northeastern
part of the region and by the polluted waters flowing from the surface
of the eastern part of the bay, while site 1 is comparatively less influenced from those sources affecting site 1. The fourth column in the
tables represent the results of the eastern (inner) part of the bay.
        The results of nutrients, chll-a, BOD_5, total coliform, dissolved
petroleum hydrocarbon, heavy metals and secchi disc depth obtained in
this work and previous studies are given in Tables 9 - 18. The graphical
representations of some tabulated data are also illustrated in Figures
5 - 11. Before discussing the results, one must take into account nutrient levels in the Black Sea and open waters of the Marmara, in order
to assess the levels and spatial and temporal changes of the parameters.
The distributions of some parameters in the Black Sea and the Marmara
Sea are compiled in Table 12. Also, the vertical profiles of dissolved
oxygen, nitrate and o-phosphate in the Marmara Sea obtained in November
are illustrated in Figure 5. The surface waters of the Black Sea have
a dissolved nitrogen value (nitrate + nitrite + ammonia) less than 5 ppb
while ortho-phosphate level, ranges between 1 and 8 ppb (17). In the
Marmara Sea the nitrate and o-PO_4 concentrations in deep waters are 60 -
70 ppb and 25 - 55 ppb in November, respectively. The chll-a values of
both sea waters measured in November are consistently higher than 1 mg/m^3
in the productive zone. As seen from Table 12, the ratio of N/P (in
weight) is less than 1 in the surface waters of the Marmara and of the
Black Sea. In the deep waters of the Marmara this ratio is about 2.
In other words, dissolved nitrogen in the form of nitrate is the limiting nutrient elements in these waters in November.
        As the dissolved oxygen (DO) values of Izmit Bay waters are examined
(Table 9 and Figure 6), it is seen that the upper layer waters of Black
Sea origin (21 - 30 ppt salinity) have oxygen concentrations mostly at
their saturation levels throughout the year. However, there is a net

Table 9.   Seasonal variations of dissolved oxygen and BOD_5 in the
           Izmit Bay waters (mg/1)

MONTH/LOCATION	WESTERN		CENTRAL (Site 1)		CENTRAL (Site 2)		EASTERN	
	U	B	U	B	U	B	U	B
May, 1984	9.1	1.7	8.4	1.5	9.0	0.9	8.7	0.5
June	7.5	1.8	7.0	1.4	6.2	0.9	4.9	0.7
July	5.9	1.8	6.4	1.1	4.6	0.7	3.8	0.4
August	8.7	1.7	6.9	1.0	6.5	0.7	3.3	0.4
September	7.6	1.6	7.7	0.6	8.5	0.6	9.8	0.5
October	9.9	(0-2)*	9.6	(0-0.8)	11.3	(0-0.6)	7.3	0.0
November	9.1	3.0	9.0	1.6	9.4	2.0	8.7	1.0
December	7.8	2.0	7.8	1.2	8.4	1.2	6.2	0.9
January, 1985	8.6	1.6	8.1	1.1	7.6	1.3	7.7	1.0
March	8.7	3.4	8.3	2.7	7.9	2.0	8.6	2.1
April	9.3	1.6	9.3	1.3	9.2	1.3	8.5	1.0
BOD_5								
September, 1984	0.3	–	0.8	–	2.4	–	5.1	–
October	1.2	–	1.2	–	2.7	–	5.0	–
November	2.0	–	2.0	–	1.4	–	4.1	–
December	0.8	–	1.4	–	1.1	–	1.1	–
January, 1985	0.7	–	0.9	–	1.0	–	2.7	–
March	1.2	–	1.1	–	0.7	–	1.3	–

(*)   : Range values of the bottom waters;   (-) : not measured
U   : Upper layer (20-30 ppt salinity) ;   B   : Bottom layer
                                                 (37-38.5 ppt salinity)

decrease in the DO levels of surface waters up to 3.8 ppm in the eastern
region of the bay in July. In August, it increases again, except in the
eastern part.   In October, there is a remarkable increase in the surface
water values of DO as the opposite is seen in the bottom waters in which
all available DO was consumed by the biochemical decomposition of organic
matter falling down from the productive upper layer.   The anoxic bottom
waters appeared in October and were replaced in November by the oxygenated water masses. Consequently, the DO concentrations of the bottom layers reach the values of 1.6 - 2.0 and 1 ppm for the central and eastern
regions of the bay (Table 9). The DO levels of the bottom waters stay
constant during the winter months, and tend to increase in March, which
corresponds to the least productive time of the year (Table 10), and in
addition, the transition zone between the two layers is significantly
larger to allow the diffusion of dissolved oxygen in the overlying
waters to the bottom waters. Then, in April, the DO values of bottom
layers start to decrease due to the strong stratification in the bay
which render difficult oxygen diffusion through the interface. This situation was created by the inflows of low saline waters of Black Sea

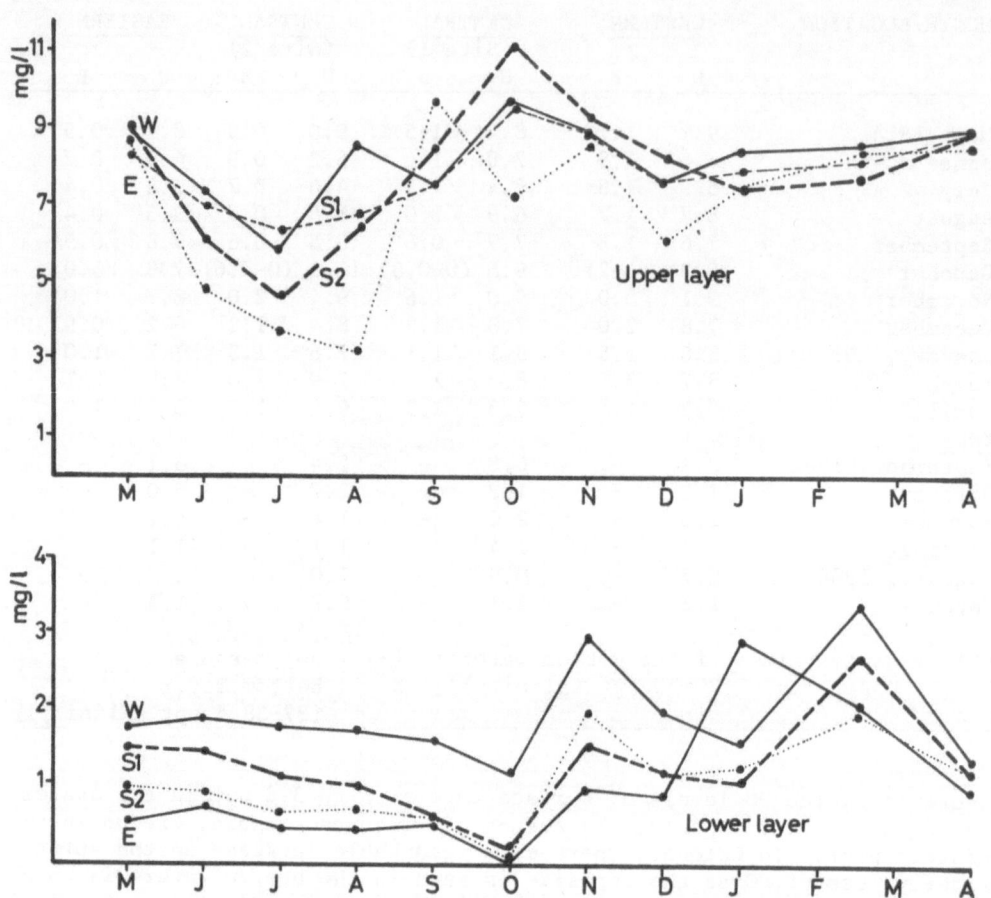

Figure 6· Dissolved oxygen variation in İzmit Bay

Table 10. Longitudinal variations of chlorophyll-a (mg/m^3) and
          secchi disc depth (m) in the Bay.

MONTH/LOCATION	WESTERN	CENTRAL (Site 1)	CENTRAL (Site 2)	EASTERN
Ch11-a				
May, 1984	2.6	5.3	12.7	10.7
June	2.2	5.0	7.4	5.6
July	1.7	13.8	24.2	9.1
August	1.0	2.4	2.3	5.0
September	0.85	2.2	6.3	21.7
October	2.9	3.3	0.9	6.5
November	1.2	5.4	12.4	11.5
December	1.8	2.5	3.5	4.0
January, 1985	2.6	1.9	0.35	0.63
March	0.66	0.55	0.85	1.0
April	2.2	4.5	5.0	5.3
S.Disc Depth				
May, 1984	3.5	2.7	1.9	1.5
June	3.5	3.2	2.7	2.0
July	5.5	2.5	1.5	1.1
August	10.0	5.8	3.6	3.0
September	9.5	7.2	3.7	2.5
October	4.7	3.6	3.2	2.0
November	3.5	2.8	2.7	1.8
December	3.9	4.0	2.4	1.8
January, 1985	5.0	3.6	3.6	3.0
March	4.5	4.0	4.2	1.7
April	3.0	1.8	2.0	2.2
December, 1969[a]	10.0	6.0	−	4.5
January, 1970[a]	11.0	5.5	−	3.5
August, 1970[a]	12.0	9.0	−	6.0

[a] : Data from reference 7.

origin (22 - 24 ppt) and high saline waters (38 - 38.5 ppt) into the bay
in April.

The biodegradable organic matter contents of the productive layer
(0 - 10 meters) of the bay water in terms of BOD_5 shows spatial and temporal variations. The highest values during the summer months were observed in the eastern part (Table 9). This coincides with the DO variation pattern in the same region. In March and November, BOD values of
the western region are almost at the same level with those measured in
the Site 1, and higher than those observed in the Site 2 of the central
region. Before renewal of the bay water, there exists an increasing
pattern in BOD values toward the eastern part. In November , the inflow
of water from the bottom into the bay creates an opposite current system

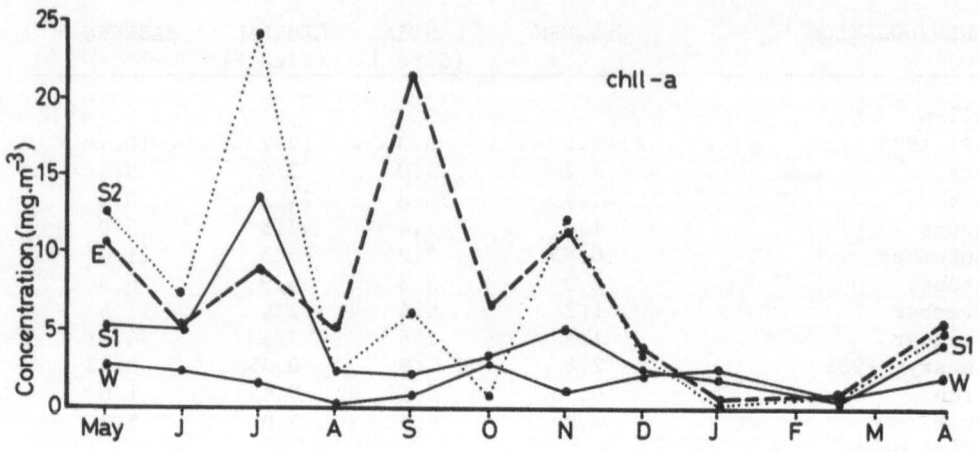

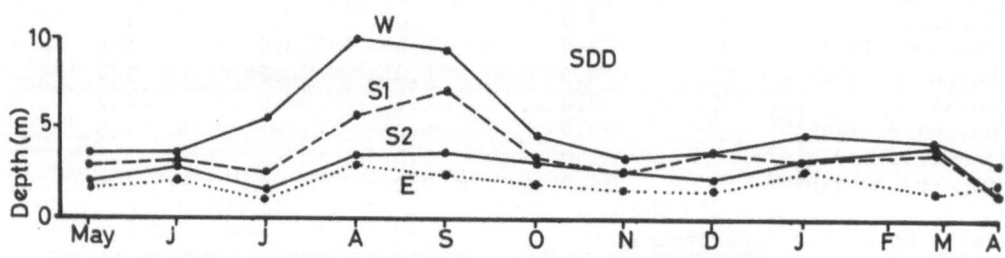

Figure 7. Chlorophyll-a and secchi disc depth variations in İzmit Bay

in the surface layer, which carries the pollutants at the surface waters
towards the outside of the bay. For this reason, no significant difference in the BOD results of the western and central regions was detected.

As expected, the light transparency in the bay (secchi disc depth)
decreases towards the inside of the bay depending on the productivity
and the amount of terrestrial solid materials in the water (Table 10 and
Figure 7). It reaches 10 meters in the western region in August and
start to decrease in October as results of increase in the primary productivity. As can be seen from the chll-a results in Table 10, the August - September period is the less productive time of the year in the
area. There is also a correlation between BOD and Secchi disc depth (SDD)
results of the western and the central regions. SDD increases in the
regions as BOD_5 (biodegradable organic matter) drops in September and the
winter months, December-January. As Black Sea waters enter the bay in
April, some remarkable decreases in the SDD values were seen and significant increases in the chll-a (productivity) values. As can be seen in
Figure 7, the SDD, one of the indicators of primary production and pollutional levels in the aquatic environment, was comparatively high in 1970
and independent of the season in the western region of the bay. The comparison of the results of 1970 and 1984 - 1985 appear in the 10 year -
pollutional level in the bay, particularly in site 2 of the central and
eastern regions.

The primary production in the bay shows seasonal and regional difference (Table 10). In general, during the May - December period the chll-a
values of the western region are always lower than those found inside the
bay waters, although an opposite trend is seen in the winter months.
There exists three major chll-a peaks for the regions, which appear in
different months (Figure 7). In the western part, the maxima values are
seen in May, October and January as the peaks of the central part appear
in May, July and November. When the November data of the bay is compared
with the chll-a results of the Marmara Sea and the Black Sea obtained in
November, the productivity is four-fold in site 1 and ten-fold in the
open waters in site 2 of the central region. In the eastern part of the
bay, no correlation has been found between chll-a and secchi disc depth
due to the large amounts of terrestrial input of suspended solid wastes
to this shallow area. The chll-a and BOD_5 results measured in the same
sites do not show a positive correlation due most probably to the biodegradable organic matter of anthropogenic origin in the surface layer.
The primary production values of May 1984 calculated from the results of
chlorophyll-a are 0.08, 0.24 and 0.38 $g-C/m^2$-day in the western, central
and eastern regions of the bay, respectively. In August, 0.16, 0.35 and
0.43 $g-C/m^2$-day were found in the same regions, respectively (19). These
calculated figures just allow us to compare the productivity level with
other seas, rather than to use the data in the calculation of annual primary production of the system.

The total coliform numbers per 100 ml of water are found to be below
the swimming standards, which is accepted by many countries, in the western and central (site 1) regions during June 1984 - March 1985 (Table 11).
But, the central (site 2) and eastern parts of the bay mostly have such
large numbers of coli that swimming is prohibited. The lowest numbers
of coli in the site 1 and eastern part were observed in November which
corresponds to the renewal time of the bay water as pointed out in the

Table 11. Total coliform distribution in the bay waters(number/100 ml).

Month/Location	Western		Central (Site 1)		Central (Site 2)		Eastern	
	S	5m	S	5m	S	5m	S	5m
June,1984	215	87	535	555	1980	1360	6800	5610
July	150	105	1530	970	>10^4	>10^4	>10^4	>10^4
August	905	560	85	95	3680	1040	375	330
September	90	195	190	115	4370	1230	–	6040
October	500	162	390	325	275	165	7665	5235
November	925	292	155	245	–	140	–	350
December	367	345	435	385	250	175	1435	1025
March, 1985	20	30	35	80	1-500	20-1000	>10^4	>10^4
April	136	110	113	625	10^3-10^4	100-10^4	3500	1950

Table 12. The levels of some parameters in the Black Sea
and the Marmara Sea.

Parameter	Black Sea[a]	Marmara Sea	
Nitrate (µg-N/1)	<1-4 (0-100 m)	<2	(surface, Nov.1984)
		60-75	(75-1000 m,Nov.1984)
O-PO₄(µg-P/1)	2-40 (0-100 m)	1-8	(surface, Nov.1984)
		20-55	(75-1000 m,Nov.1984)
		10-40	(200-1000m,June 1984)
Silicate (µg/1)	10-150(0-800 m)	1300-1750	(200-1000m,June 1984)
Ch11-a (mg/m³)	1.1-1.6(Nov.1984)	1.3-1.7	(Nov.1984)
pH	8.5-7.5	8.45-7.95	

(a): Some of the Black Sea data are from reference 17.

discussion of the dissolved oxygen results. In July, the transport of
dissolved nutrient elements of the bottom waters to the surface layer by
strong physical évents, increased the ch11-a level in the bay. Similar
increases in the coli concentrations of the surface waters were seen
due presumably to the strong currents carrying the waste waters from
coast to off-shore. In July, as seen in Figure 2a, the surface salinity
inside the bay rises 1 ppt and it was also observed that there exists
a water flow inside the bay from the bottom,which is easily followed
from the salinity variations of the bottom waters. The coliform distributions in different parts of the bay can help us to understand the current system affecting the surface waters. If one looks at the March
result, it easily be understood that the surface currents in site 2 and
the eastern part of the bay are not strong enough to carry the pollutants
of the inner waters towards the bay entrance. This event occured in
November. The inflow of the Marmara waters from the bottom created a
current system strong enough to carry the pollutants far from their
sources.
The reactive silicate in the surface waters of the bay is generally

high enough not to limit primary productivity throughout the year(Table
13). However, as seen from the results, the surface water levels in
May, June and September are so significantly low that this could be a
critical constituent of the photosynthetical production in the photic
zone of the bay, particularly of the central region. The lowest values
of silicate in the bottom waters were observed in October which corresponds to the oxygen deficiency time in the bay (Figure 8). In other
words, the exchange of water masses between different sections of the
bay, the bay and the northeastern Marmara are very restricted. So,
although the chll-a is significant inside the bay in October, the organic
matter settles down before it is decomposed within the water column. In
November, silicate rich bottom waters of the northeastern Marmara Basin
enter the bay. Consequently DO, chll-a and silicate contents of the
bottom waters of the bay increased. Some of the nutrients are carried
to the surface waters by vertical mixing in the bay. The silicate distribution with depth always shows a uniformly increasing pattern in the
bay water and the Marmara Sea. The silicate concentrations of the Marmara are much higher than of the Black Sea.

The nutrient elements, phosphorus and nitrogen in the forms of
$o-PO_4$, NO_3, NO_2 and NH_3, show seasonal changes in the bay waters (Tables
14 - 16, Figures 9 - 11). Ortho-phosphate concentration in the surface
waters of the central and western regions reaches its lowest levels in
August and March (Figure 9). The maxima in the regions appear in May and
January. These high concentrations correspond to the times in which the
nutrient rich waters of Black Sea origin flow into the bay from the surface and the phosphate rich bottom waters entered the bay during November
-December period mix vertically inside the bay and as a consequence of
the continuous vertical mixing, some of the dissolved phosphorus is transported to the surface waters. As seen in Table 14, $o-PO_4$ becomes the
limiting parameter in the productive layer of the western region in August. However, it is high enough not to be a limiting factor upon the
photosynthesis in the inner regions of the bay throughout the year. If
the nitrate values are compared with the results of $o-PO_4$ in the surface
waters, it is obvious that the N/P ratio never approaches the natural
value, 7, in weight. In the bottom waters of the bay, nitrate and ortho-
phosphate do not show similar seasonal distribution patterns. Nitrate
in the western region is comparatively lower than those of the inner
parts during the May - October period as the changes of $o-PO_4$ shows fluc-
tuation during this period. In November, which corresponds to the inflow
of the Marmara bottom waters, NO_3 and $o-PO_4$ concentrations increase in
the bay waters. In January, $o-PO_4$ is at its maximum levels in the western and central regions while the NO_3 concentration approaches its minimum value in March. The expected nitrate enrichment in the bay water
appeared in April after the nutrient rich Black Sea waters entered the
bay (Figures 2 and 10). The nitrate measurements recorded in 1976 by
Artüz and Kor(7) are in agreement with the values of this study with the
exception of the winter results. Although both seasonal changes have the
same pattern, the magnitues of the concentrations are different in the
winter data. The nitrate concentrations are within the ranges of 40-50
and 1 - 10 μg at/l for measurements in 1975 and 1985, respectively.
The nitrite distributions in the bay are depicted in Figure 11. No re-

Table 13. The distribution of reactive silicate in the Bay Waters(µg/l).

MONTH/LOCATION	WESTERN		CENTRAL (Site 1)		CENTRAL (Site 2)		EASTERN	
	U	B(a)	U	B	U	B	U	B
May,1984	8.6	729	<1	496	9.8	796	6.9	64
June	11.8	1060	7.0	1097	6.7	1306	78	690
July	42	1008	51	1003	12.2	1071	22	888
August	32	709	28	741	45	757	77	764
September	4.4	942	4.9	1007	1.0	1007	33	141
October	24	562	34	619	60	626	91	560
November	275	968	142	1010	209	878	212	918
December	142	768	193	810	186	783	178	776
January,1985	181	955	295	974	397	944	502	544
March	31	807	43	841	46	923	66	959
April	88	981	72	954	127	974	127	971

(a) : This column represents the average silicate results of the bottom
waters up to 200 meters.

markable change was observed in the nitrite concentrations of the western
regions. The ammonia formation, particularly in the surface layers, was
only seen in the winter season. It was found as 6.7 and 8.3 - 28 ppb
for the western and central regions in December. In January, it ranges
between 10 - 15 ppb in both upper and lower layers of the regions. No
absolute data exists for March month due to contamination problems encountered in the analytical method. In April, the ammonia concentrations
are as high as 83 - 108 ppb in the bottom waters of the bay as the surface values range between 17 and 43 ppb (Table 16).

The nutrients, o-PO_4 and NO_3, are always available in the bay and
keep the photosynthetic reaction of the bay over the natural level of
the marine environment. Thus, organic loads from man-made sources and
high organic matter produced in the bay create anoxic conditions in the
bottom waters and transition zone between two layers during the stagnant
period of the bay, particularly between August - October.

Petroleum Hydrocarbons (PAH), particularly polynuclear aromatic
hydrocarbon in surface waters of different seas, are tabulated in Table
17. The PAH concentration in Izmit Bay ranges between 3 and 32 µg/l
depending on the location and sampling times. The average values of
five months in the bay are comparatively higher than in the open waters
of Marmara and the northeastern Mediterranean. Such relatively higher
values might be attributed to the dense ship and tanker traffic in the
bay. However the average PAH concentrations were generally far from
being critical. For a better evaluation of the distribution and effects
of petroleum hydrocarbons in Izmit Bay, the PAH's in sediment and organisms should also be analysed as a further study.

The toxic metals, Hg, Cd and Pb, were measured in the water, sediment

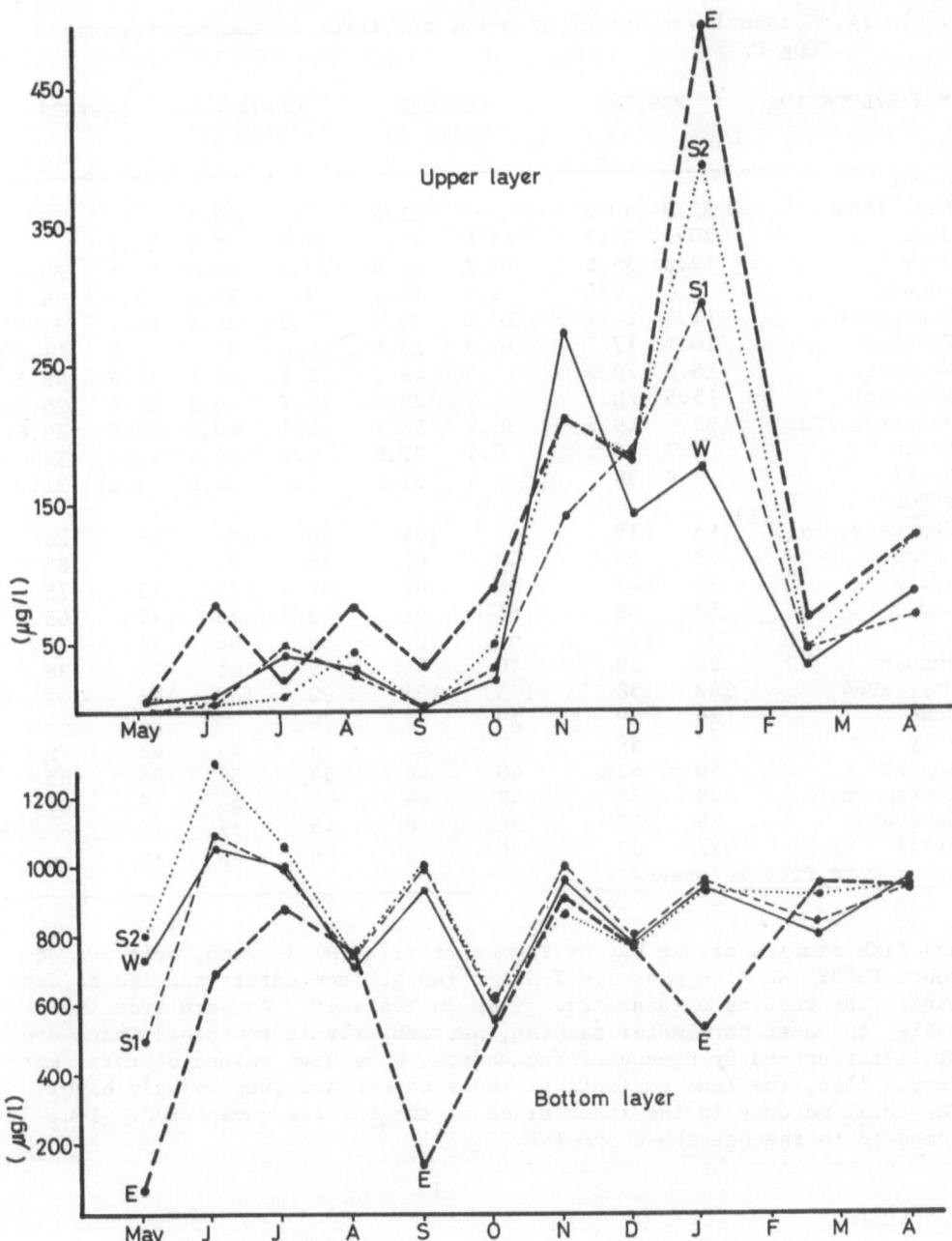

Figure 8. Silicate variation in the bay waters.

Table 14. Seasonal variations of o-PO₄ and t-PO₄ in the bay waters
(μg-P/1).

MONTH/LOCATION	WESTERN		CENTRAL (Site 1)		CENTRAL (Site 2)		EASTERN	
	U	B	U	B	U	B	U	B
o-PO₄								
May, 1984	32.5	30.0	–	31.3	–	33.5	–	–
June	20.7	25.7	23.1	27.8	24.4	30.0	19.7	–
July	12.0	35.6	10.9	38.8	22.5	42.1	15.5	30.1
August	< 0.5	8.5	3.9	11.8	3.5	18.4	13.4	39.1
September	13.4	24.7	10.1	39.4	30.2	38.9	20.1	31.5
October	10.1	17.7	10.0	23.5	28.4	25.9	8.8	11.0
November	15.5	28.3	14.9	46.3	20.8	42.5	26.3	46.7
December	15.5	20.5	17.5	28.2	16.7	25.5	21.6	26.3
January, 1985	35.4	49.5	38.9	52.7	42.1	60.3	42.0	39.6
March	7.7	18.8	6.4	27.6	9.0	34.4	22.8	32.0
April	6.9	30.2	6.4	32.2	7.7	33.6	6.2	31.0
t-PO₄								
February, 1975[a]	95	110	81	104	68	102	104	188
March	58	60	47	67	46	73	82	85
April	20	67	36	83	27	71	33	75
May	52	39	61	91	53	73	77	165
June	33	101	67	107	38	96	71	138
August	26	82	46	91	59	144	77	136
May, 1984	182	258	115	136	92	180	145	342
June	26	29	29	32	28	39	24	–
July	57	59	66	65	56	59	62	79
August	50	65	46	48	54	58	35	55
September	28	38	42	69	43	54	49	45
October	38	39	30	58	45	55	35	–
April	27	62	30	72	68	86	43	109

(a) : Data from reference 7

and fish samples of the bay by *Taymaz et al. (19)* in June, July and October 1980. We also measured T-Hg in the surface waters sampled in June
1984. The results obtained are given in Table 18. As seen from the
table, the nearshore water samples, particularly in the areas which are
still influenced by Hg-containing wastes, have high values of total mercury. Also, the lead contents of these waters are surprisingly high.
The total mercury in the inner sites of the bay are unexpectedly low,
compared to the nearshore samples.

CONCLUSIONS

The Bay of Izmit has a permanent salinity stratification created by
the low saline waters of the Black Sea origin overlying the high saline
Mediterranean waters. The thicknesses of the layers change seasonally
depending upon the current systems in the area, i.e., the inflow times

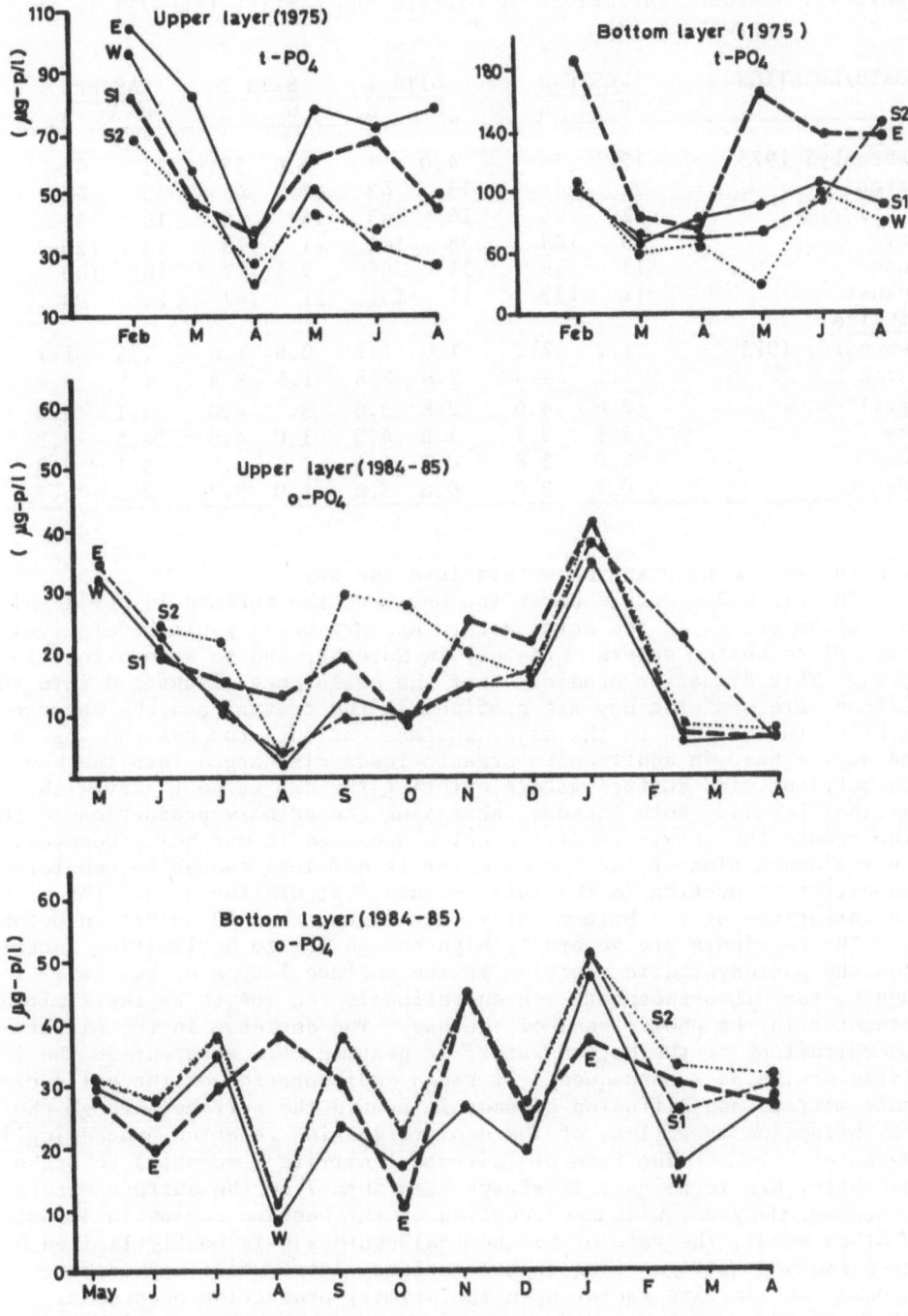

Figure 9. Phosphate distribution in İzmit Bay.

Table 15. Seasonal variations of nitrate and nitrite (μg-N/1) in the
         bay waters (7).

MONTH/LOCATION	WESTERN		SITE 1		SITE 2		EASTERN	
	U	B	U	B	U	B	U	B
Nitrite
Nitrate
February, 1975	12	55	4.0	51	5.6	55	11	41
March	24	47	13	43	20	45	15	41
April	12	84	10	43	14	46	19	54
May	20	108	38	102	11	113	13	120
June	11	89	11	86	9.4	117	18	100
August	11	137	11	151	11	147	24	98
Nitrite
February, 1975	1.7	2.0	1.0	1.7	0.6	1.6	1.5	1.7
March	1.3	4.0	2.8	2.8	1.6	2.3	1.9	3.4
April	2.8	4.0	2.8	3.6	3.1	4.0	4.1	4.8
May	1.6	3.1	1.9	4.3	1.0	4.5	4.5	7.3
June	2.6	3.8	0.7	4.3	1.3	3.5	3.1	5.8
August	0.1	3.0	0.6	3.8	1.0	3.7	2.5	4.7

of both low and high saline waters into the bay.

The Black Sea waters enter the bay from the surface in April while
the bottom waters of the northeastern Marmara Basin replace relatively
less saline bottom waters of the bay in November and to some extent in
April. This situation predicts that the pollutants discharged into the
surface waters of the bay are confined in the central and the eastern
parts of the bay due to the major surface currents towards the inside
the bay. Thus, in addition to organic loads discharged into the bay,
the nutrient rich surface waters entering the bay raise the available
nutrient levels. Both factors increasing the primary production in the
area create the anoxic condition which appeared in October. However,
the residence time of the bottom water is not long enough to complete
the nitrate reduction in the water column. It was found that the nitrate
concentrations of the bottom waters are higher than 40 μg-N/1 in October.

The nutrients are generally high enough not to be limiting factor
upon the photosynthetic reaction in the surface layers of the bay. In
August, the ortho-phosphate concentration is too low to be the limiting
parameter in the photic zone of the bay. The decrease in the nitrate
concentrations of the bottom waters it becomes more apparent in the
winter months as a consequence of rapid sedimentation of the solid organic matters and diffusion of ammonia toward the surface through the
weak halocline as well as of the denitrification reaction proceeding in
the water column. The rate of (nitrate + nitrite + ammonia) to ortho-
phosphate, N/P in weight, is always less than 7 in the surface waters
throughout the year, with the exception of the western region in August.
In other words, the rate of biochemical synthesis is mostly limited by
the dissolved nitrogen ions in the surface waters while o-phosphate
becomes the limiting factor upon the primary production occasionally
between July and October. However, in order to see the consistency of
the temporal changes of the measured parameters and to have more reliable

Table 16. Seasonal variations of nitrate, nitrite and ammonia
           in izmit Bay (μg-N/1).

Month/Location	Western		Central (Site 1)		Central (Site 2)		Eastern	
	U	B	U	B	U	B	U	B
Nitrate								
May, 1984	10.1	135	4.9	123	2.2	114	20.6	5.2
June	3.0	108	2.9	160	5.2	126	6.0	6.8
July	5.9	141	12.4	143	15.5	121	13.5	6.7
August	14.9	124	20.6	119	9.8	130	14.2	11.9
September	5.0	84	4.0	81	5.1	81	11.9	–
October	2.5	49	2.1	40.2	2.3	41.6	2.3	3.0
November	34.3	118	32.8	104	40.8	95	32.4	92
December	4.3	17.3	3.4	18.3	4.0	16.9	5.5	20.6
January, 1985	3.7	7.2	4.7	5.2	5.0	5.3	5.3	5.0
March	1.2	1.4	1.1	2.6	1.3	1.5	1.2	1.3
April	5.6	110	6.8	113	7.3	97	9.7	91
Nitrite								
May, 1984	8.2	9.9	2.1	11.4	1.2	14.1	9.6	–
June	11.3	12.3	12.6	11.7	10.3	16.0	10.3	13.2
July	1.3	2.7	2.3	3.1	1.9	7.3	1.8	4.3
August	2.8	3.2	2.7	3.6	3.2	5.8	3.7	5.1
September	2.0	2.8	2.6	2.7	3.0	14.2	3.5	4.0
October	4.7	6.8	4.3	6.8	6.2	7.4	6.5	6.0
November	3.9	3.7	4.3	2.5	2.0	1.5	4.3	3.7
December	6.4	6.7	7.3	7.0	6.5	6.9	7.4	7.9
January, 1985	2.9	3.2	4.6	5.5	8.0	6.6	8.1	9.8
March	3.7	3.8	4.1	4.4	5.5	4.7	5.2	7.1
April	3.6	2.9	4.5	3.1	4.5	5.8	4.6	7.5
Ammonia								
September	<1-14	<1	<1	<1	<1	<1-14	12	–
October	12	<1	<1	<1	9	15	<1	–
November	<1-10	<1	4	<1	18	<1	42	3.2
December	6.7	<1	8.3	<1	28	<1	15	<1
January	12.5	10.7	12	9.5	15	10	27	32
April	17	87	19	83	23	103	43	108

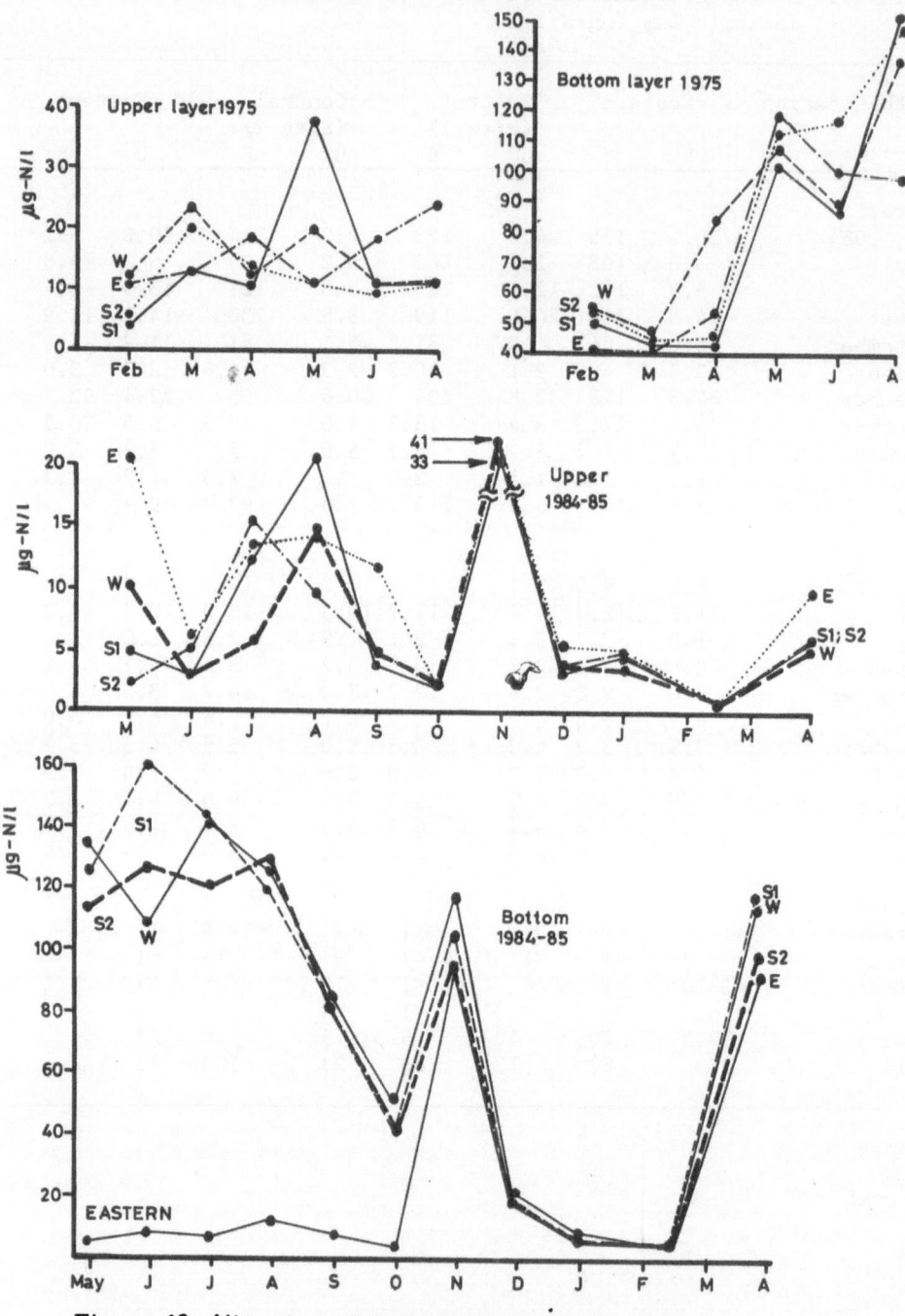

Figure 10 · Nitrate distribution in İzmit Bay

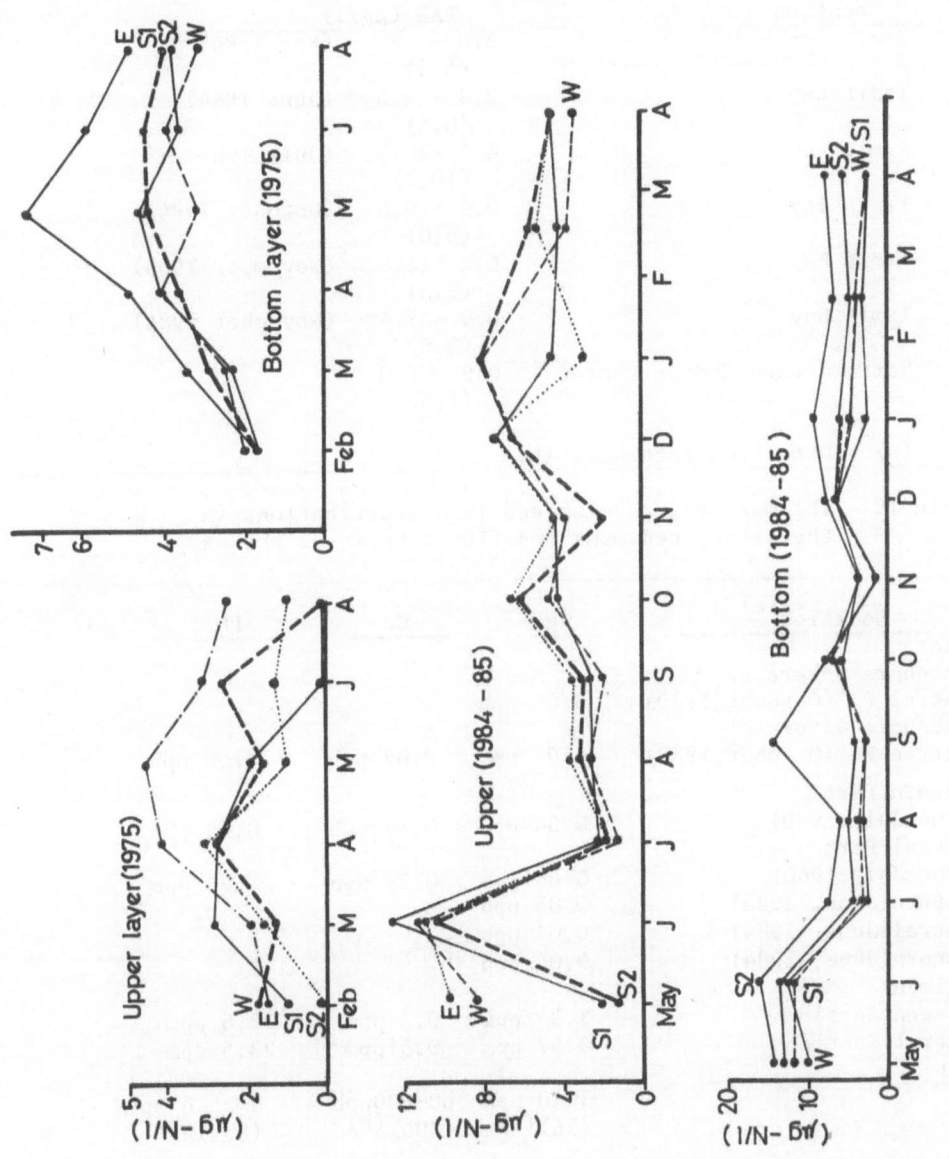

Figure 11. Nitrite distribution in izmit Bay .

Table 17. Dissolved petroleum hydrocarbons in the surface waters
of Izmit Bay and the northeastern Mediterranean.

Location	PAH (µg/1)	
Izmit Bay	3.0 - 32 (5.9)	(May 1984)
Izmit Bay	2.8 - 7.5 (6.6)	(June 1984)
Izmit Bay	4.3 -27.5 (10.7)	(July 1984)
lzmit Bay	0.9 -10.0 (5.0)	(October 1984)
Izmit Bay	6.2 -11.9 (9.3)	(November 1984)
Izmit Bay	0.9 - 7.4 (3.6)	(November 1984)
Northeastern Mediterranean[a]	0.9 - 5.0 (1.5)	

(a) : Data from reference 18.

Table 18. The mercury, cadmium and lead distributions in
the water, sediment and fish samples of the Bay.

Location[a]	Hg	Cd	Pb
Water			
Nearshore waters of central P. (October,1980)	3.0 ppb	–	–
Nearshore waters of eastren P. (October,1980)	6.9 ppb	0.02 ppb	1750 ppb
Eastern Part (June–July,1980)	0.34 ppb	0.18 ppb	0.62 ppb
Central Part (June–July,1980)	0.68 ppb	0.39 ppb	20 ppb
Eastern(June, 1984)	0.05 ppb	–	–
Central(June, 1984)	0.04 ppb	–	–
Western(June, 1984)	0.04 ppb	–	–
Sediment			
Eastern Section	5.3 ppm	0.8 ppm	32.6 ppm
Central Section	0.57 ppm	0.31ppm	23.5 ppm
Fish			
	3-40 ppb (16)[b]	60-600ppb (300)[b]	4.4-15.6 ppm (8.5)[b]

(a) : 1980 data, fish and sediment results from reference 20.
(b) : Average results of different fish species

data this work will be conducted seasonally for one year.

Acknowledgements: This work was fully supported by Scientific Affairs Division of NATO within the framework of "Science for Stability Programme". We also thank O.Sağlamer, E.Morkoç and A.Bozyap, Chemistry Department of MRI, who performed the nutrients, chlorophyll-a and total coliform analyses in the water samples.

REFERENCES

1. Timur,A., Kınayyiğit,G., Dumlu,G., Ilhan,R., Çiler,M., Kavaklı,M., Armağan,Z. and Bozyap,A., 1982, *The prevention and removal of the water pollution in Izmit Bay:Determination of technological aspects,* TUBITAK-MRI, Chem.Dept. Publ., 411 pp. (in Turkish).
2. Timur,A., Dumlu,G., Timur,H., Çiler,M., Ilhan,R. and Balkaş,T. 1982, *The prevention and removal of the water pollution in Izmit Bay: Determination of technological aspects,* TUBITAK-MRI, Chem.Dept.Publ. No.106, 383 pp. (in Turkish).
3. Gönenç,E., Tünay,O., Saybay,S. and Orhon,D. 1983, *The prevention and removal of the water pollution in Izmit Bay: Determination of technological aspects,* ITU-Civil Eng.Publ., 374 pp. (in Turkish).
4. DAMOC. 1971, *Istanbul region drinking water and swerage master plan and feasibility studies,* UNDP/WHO project.
5. Orhon,D., Gönenç,E., Tünay,O. and Akkaya,M. 1984, *The prevention and removal of the water pollution in Izmit Bay: Determination of technological aspects,* ITU-Civil Eng.Publ., 373 pp. (in Turkish).
6. Kor,N., (1974), *The control of pollution in Izmit Bay,* TUBITAK Publ., No. MAG 211/A (in Turkish).
7. Artüz,I. and Kor,N. 1971, *A preliminary work on the control of pollution in Izmit Bay,* Hydrobiology Inst. Publ., (in Turkish).
8. SWECO and BUE. 1976, *İzmit swerage project: Master plan,* (in Turkish).
9. TUBITAK-MRI 1985, *Determination of characteristics and assimilative capacity of Izmit Bay,* TUBITAK-MRI, Chem.Dept. (to be published).
10. APWA-AWWA-WPCF 1980, *Standard methods for the examination of water and wastewater,* 15th ed.
11. EPS 1981, *Mercury: Methods for sampling, preservation and analysis,* Economic and Technical Review Report EPS 3-EL-81-4, 108 pp.
12. Yılmaz,A., 1982, *Fluorescence measurements in Marine Environments,* METU-MSI, Master Thesis.
13. TIS 1978, Industrial Methods No: 253-80 E and 100-70 W/B, Technicon Industrial Systems.
14. IOC 1975, *Guide to operational procedures for the IGOSS pilot project on marine pollution (petroleum) monitoring,* UNESCO, Manuals and Guides No:7, pp. 28-31.
15. TUBITAK-MRI 1984, *Wastewater discharge quality standards and application guidelines for Izmit Bay,* TUBITAK-MRI, Chem.Dept. Publ. No.129, 51 pp.
16. Sümer,M. 1983, *Water movements in Izmit Bay,* ITU-Civil Eng. Publ. (in Turkish).

17. Grasshoff,K. 1971, *The hydrochemistry of land-locked basin and fjords,* in Chemical Oceanography, ed. J.P.Riley and G.Skirrow, Academic Press, vol.2.
18. Sunay,M. 1982, *Distribution and source identification of petroleum hydrocarbons in the marine environment,* METU-MSI, Ph.D. Thesis, pp. 99-100.
19. Morkoç,E. 1984, *Primary productivity in Izmit Bay,* IU-MSGI, Master Thesis, pp.44-51.
20. Taymaz,K., Yiğit,V., Özbal,H., Ceritoğlu,A. and Müftügil,N. 1984, *Heavy metal concentrations in water, sediment and fish from Izmit Bay, Turkey,* Intern.J. Env.Anal. Chem., 16, pp.253-265.

SITES FOR EFFLUENT OUTFALL AND SLUDGE DUMPING IN THE SARONIKOS GULF, GREECE

N. Friligos
National Centre for Marine Research
GR 166 04
Hellinikon
Greece

ABSTRACT

The variation of nutrient concentrations and physical characteristics were studied in the vicinity of the Athens sewage outfall during 1974-1976 and 1982. Suggestions are made as to the extent of the treatment as well as the choise of the length of the outfall and sludge dumping locations.

INTRODUCTION

The construction of the Athens untreated sewage effluent in Keratsini and the great urbanization and industrial development in the area has affected the equilibrium of the marine ecosystem of the Saronikos Gulf. The present discharge at Keratsini (Fig. 1) is most unsuited. However, engineering consideration (Harremoes, 1976) of cost of relocation of trunk sewage tends to favour an outfall extending into the Bay from this point. In view of the fact that the adjacent island of Psitalia is a bare island suited for a possible site for an eventual treatment plant it is reasonable to select the area off this island for discharge.

This paper deals with the nutrient levels and the hydrological parameters, during the period 1974 to 1982 in two possible locations from the outfall. It also aims at helping the engineers make an assesment on the effluent treatment and the sludge dumping locations.

HYDROGRAPHICAL FEATURES

The Saronikos Gulf (Fig. 2) is typical of many semi-enclosed seas in the Mediterranean, with exposure to pollution in the northern end. Pollutants consist mainly of (a) the untreated sewage and runoff from the Athens-Piraeus urban area at an approximate rate of $7 m^3 s^{-1}$ (Table 1) and (b) an obvious but unknown amount of runoff sewage and industrial effluents discharged into Elefsis Bay (Friligos, 1979), the most industrial area in Greece. The present outfall is located in 100m off-shore

G. Kullenberg (ed.), The Role of the Oceans as a Waste Disposal Option, 275–284.

at a depth of about 30 m. This outfall has been in use for some 25 yr.
The area of sea floor covered by black anaerobic sediments has increased
7-fold, to nearly 9 km² in the past 2,5 yr (Griggs & Hopkins, 1976).
    The Gulf is basically a 50 km square, with a topography complicated
by a spine of peninsulas and islands, dividing the gulf into an eastern
and a western part. The Eastern gulf tapers from approximately 40 km
width at the mouth, where it connects with the Myrtoon sea and the Aegean system. The Western gulf presents the greater depths with a great
depression more than 400 m in Epidavros providing the possibility of
trapping of nutrients. As shown in Figure 2, there is a distinct change
in the sea bed level between the Inner and Outer sections of the Eastern
gulf. Depths in the Outer gulf generally range between 150 and 300 meters, whereas the depth of the Inner gulf is effectively about 80 to 90
meters. Psitalia island, the northern limit of the Inner Saronikos gulf,
is one of the busiest shipping areas in Greek waters, near the Piraeus
harbour and Elefsis Bay.
    Previous work, using hydrographic data (Coachman & Hopkins, 1975;
Coachman et al. 1976) or a wind driven model (Hopkins & Coachman, 1975),
revealed the water masses and the circulation pattern in the Saronikos
gulf. The flow field in the Saronikos is wind driven and there are no
appreciable tidal effects. The winds develop either cyclonic or anticy-
clonic circulation. During cyclonic circulation Aegean oligotrophic (Outer Gulf) water enters from the eastern side of Aegina, mixes with water
between Aegina and Salamis, exposes this mixture to the outfall effluents
and moves it out to the southeast. The opposite case was found during an-
ticyclonic circulation.

MATERIALS AND METHODS

Two stations were chosen, station B, near Psitalia, with a depth of 50m
and station A, 6km off Psitalia, with a depth of 90m (Fig. 1). Between
December 1972 and March 1976, 19 sets of data were gathered and 9 in summer 1982.
    Water samples were collected with a Nansen type water bottle. Tem-
perature was measured with a deep-sea reversing thermometer. Samples of
water were drawn from 1,10,20,30,40,50,75 and 90m. Measurements of hydrological parameters and inorganic nutrients were made by the methods described by Friligos (1982).

RESULTS AND DISCUSSION

Hydrological Parameters

Friligos (1984) reporterd that the upper waters in the Inner Saronikos
Gulf underwent a well defined annual cycle of thermal stratification.
From a minimum average termperature of about 14°C in February and March,
temperature rose during the spring and summer to a maximum in July and
August of about 25°C. The same author indicated that the deeper waters

also showed a temperature increase during the spring summer periods, but much more slowly and the maximum temperature (about 16°-18°C) was recorded in October. Peak summer temperatures of the near surface waters of the two stations recorded during the summer 1982 cruises again reached 26°C (Fig. 3)

Friligos (1984) reported that the salinity range was 38-39°/∞ in the Inner Saronikos Gulf and the density lay in the range 26-29σ_t. The density increased with depth and pycnocline was related to the thermocline. The situation was not similar during the 1982 summer cruises: the salinity range was 36.9-38.4°/∞ and low salinities were observed in the upper layers (Fig. 3). These probably indicate changes of a regional nature in the Aegean source waters. Bottom water salinities are less variable and similar to those reported by Friligos (1984). It should be noted that, with the decrease of salinity in the upper layers of the water column, the density difference increased (25-29σ_t) between bottom and surface waters during the summer 1982 cruises (Fig. 3).

Measurements of dissolved oxygen taken during summer 1982 at the two stations indicated a fairly constant level of around 5 ml.1^{-1} at all depths (Fig. 3). During some cruises, at stations A and B, the surface waters contained higher dissolved oxygen concentrations (with 84-235% saturation) than the bottom waters (with 75-104% saturation). The highest surface values of oxygen at both stations might be due to the effect of planktonic algae.

Nutrients

Table 1 shows that the content of silicon, nitrogen and phosphorus of the inorganic nutrients in the Keratsini waste water is high, as expected. The inorganic nitrogen is almost all present as ammonia. The inorganic N_i:P ratio in Keratsini domestic sewage is typically low (N_i:P=9.6).

Because of the length of time between sampling trips and differences in the dates of sampling in different years, it is difficult to compare the results. However, it is possible to distinguish similarities or differences, when comparing the 1974-1976 summers with that of 1982. For the purpose of comparing the two stations, we have simply averaged was parameter value over the depth of sampling on each sampling day and then taken the mean of these integral average over all the sampling days. Mean nutrient and oxygen concentrations for these periods are shown in Tables 2A & 2B. The phosphate content did not change significantly from year to year, this enabled Friligos (1981) to use it as a conservative parameter and define through it as an index of marine pollution. Also, Ryther and Dunstan (1971) have noted that in N-limited systems inorganic nitrogen concentrations are kept uniformly low by phytoplankton uptake, and in such systems P_i provides a much better tracer of eutrofication. Their results are consistent with the present findings. Thus, the inorganic nitrogen falls by about half, owing to the diminution of nitrate and nitrite, since ammonia remains nearly constant. Silicates are reduced even more, to one third. Especially, in the case of station A, the mean values did not differ except in the case of nitrite. On the contrary, at station B nitrate, silicate and dissolved oxygen values differed significantly. Moreover from this statistical examination the mean water column of ammonia and phosphate concentrations were rather constant between summers, and that nitrate plus nitrite were actually reduced, suggests that the phytoplankton stripped virtually all of the added ammonia from the water and then took up additional nitrite and nitrate, to compensate for the low N_i:P ratio in the sewage. Silicate concentrations were lower also during the 1982 summer owing to silicate being required to support the additional biomass produced from the N and P in the sewage.

In areas such as the Saronikos Gulf, where the surrounding waters are naturally unproductive and where the bottom shoals rapidly, primary treatment may be preferable to secondary treatment. In such unproductive areas, the adverse effect of sewage discharge into the sea on the phytoplankton community apparently can be made insignificant if the outfall is properly designed. The separation of domestic wastes from industrial effluents should be planned for the future to enable biological treatment to take place later on. Partial biological treatment is a desirable step-before committing to full biological treatment. A 2000m long outfall diffuser at a depth of 60 m or more between stations A and B is needed for either primary or secondary effluent to take advantages of stratification.

Sludge Dumping Locations

Disposal of the sludge to sea is suggested by the consulting firm Watson (1978) to be economically viable, and therefore the early implementation of studies to choose a sludge dumping ground is recommended. They concluded that the marine area for the disposal of sludge is very large in comparison with the volume of the sludge to be discharged during the 25 years of the economic life of the transporting vessels. On the contrary, the land area for the disposal of sludge will be available only for ten years after 1985, which is much less than the acknowledged economic life of the dehydration installation. Also, they emphasized that the lanol to be used for the disposal of sludge will no longer be available to receive municipal garbage and transport by ship causes less inconvinience to the public than transportation by land, while consuming less energy. Moreover in case of high levels of toxic substances, the danger is more immediate in the instance of the land drainage.

The only area within the Saronikos Gulf which could be considered for sludge dumping is the deep Epidavros Basin. However, in view of the middly eutrophic conditions which are thought to exist within this area (Friligos, 1983), it would appear unwise at this stage to assume that the Western Gulf will be able to absorb the major increases in nutrient input that would result.

The most promising area for the disposal of sludge seems to be the open waters of the Myrtoon Sea with depths 300 to 900 m (Fig. 4). We have been able to base our assesment only on very broad assumptions, because of the lack of any existing data concerning the water movements, surface drift patterns and sediment transport.

It is unlikely that the enrichment of the waters in the Myrtoon Sea will give rise to undesirable side effects and, provided suitable limits are imposed on the dumping grounds, shoreline conditions should not be adversely affected. The prevailling winds even across the open sea in this region are predominantly light and variable. Therefore, sludge solids should not be carried out prior to settling into the lower waters. The nearest coast line would be at some 30 kilometres. Under stronger winds lasting several days, surface solids might travel in unknown concentrations over such distances, but these winds would normally be from the north, creating a drift through the channel between Krete and the Peloponnese some 80 kilometers away. It should be remembered that this drift to the south-west is the dominant direction for water movement in this area.

Provided that the concentration of toxic materials in the sludge is limited to safe levels, we are of the opinion that the addition of sludge to the opean sea should be beneficial to fish production and hence enable the recycling of the nutrients with benefit to the economy, while at the same time maintaining satisfactory environmental conditions.

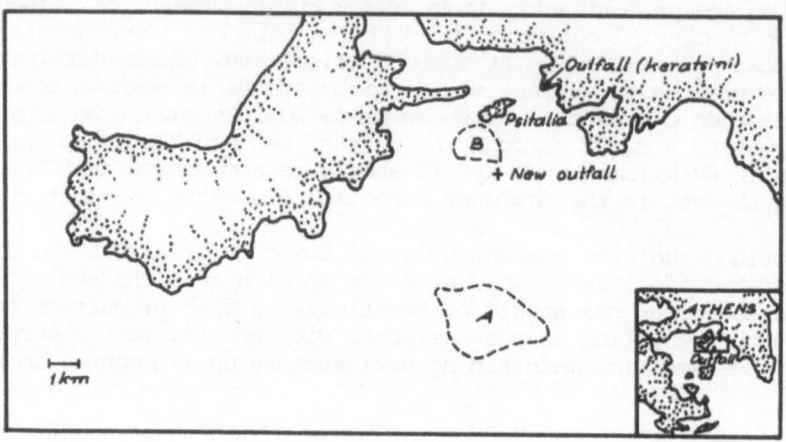

Fig. 1.   Location of stations in the Saronikos Gulf.

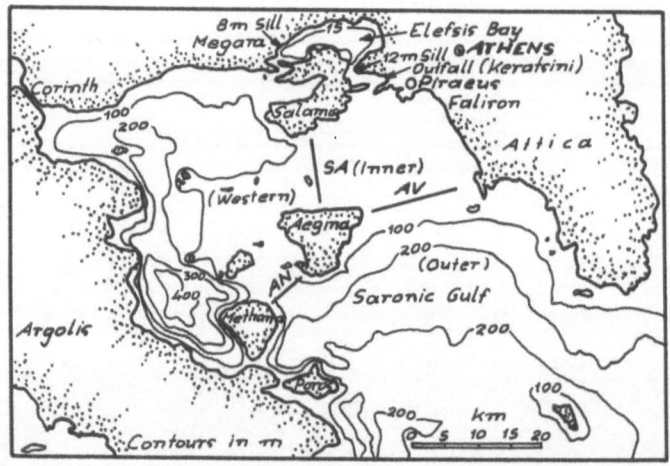

Fig. 2.   Oceanographic subregions and bathymetry of the Saronikos Gulf.
Main passages designated at follows: SA, Salamis-Aegina;
AV, Aegina-Vouliagmeni; AM, Aegina-Methana.

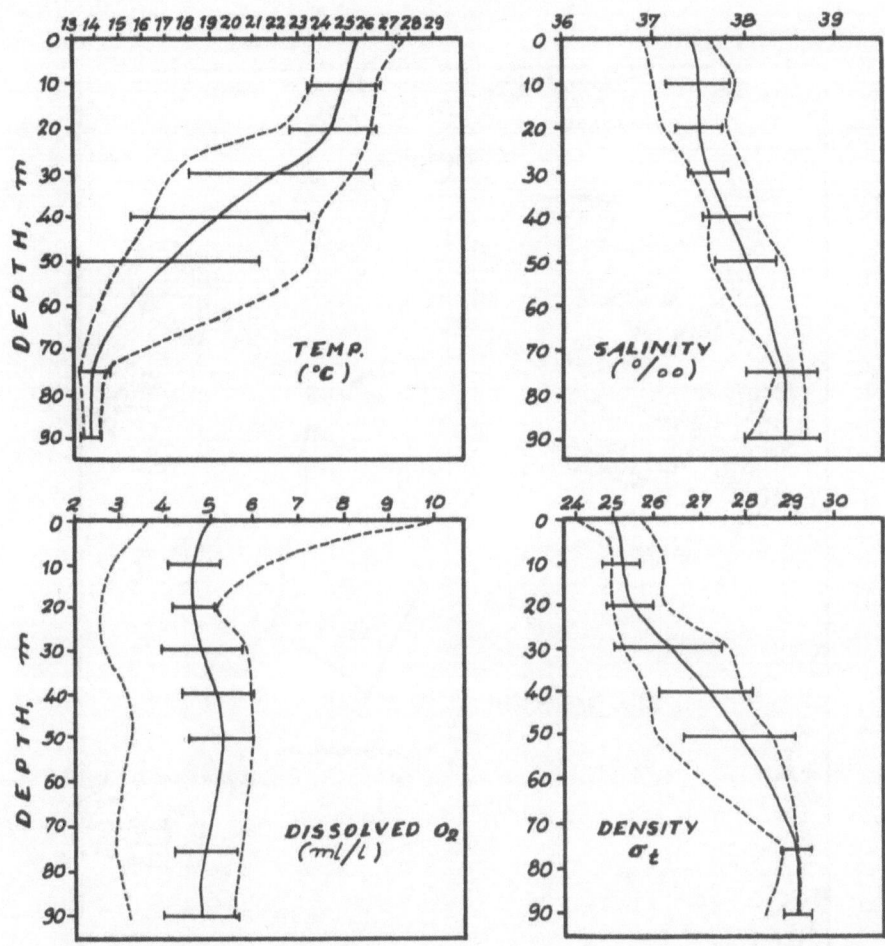

Fig. 3. Vertical distributions of water properties at stations near site of proposed sewage outfall. Note: Dashed lines indicate range of values; ├──•──┤ indicate width of two standard deviations; and • indicates mean value.

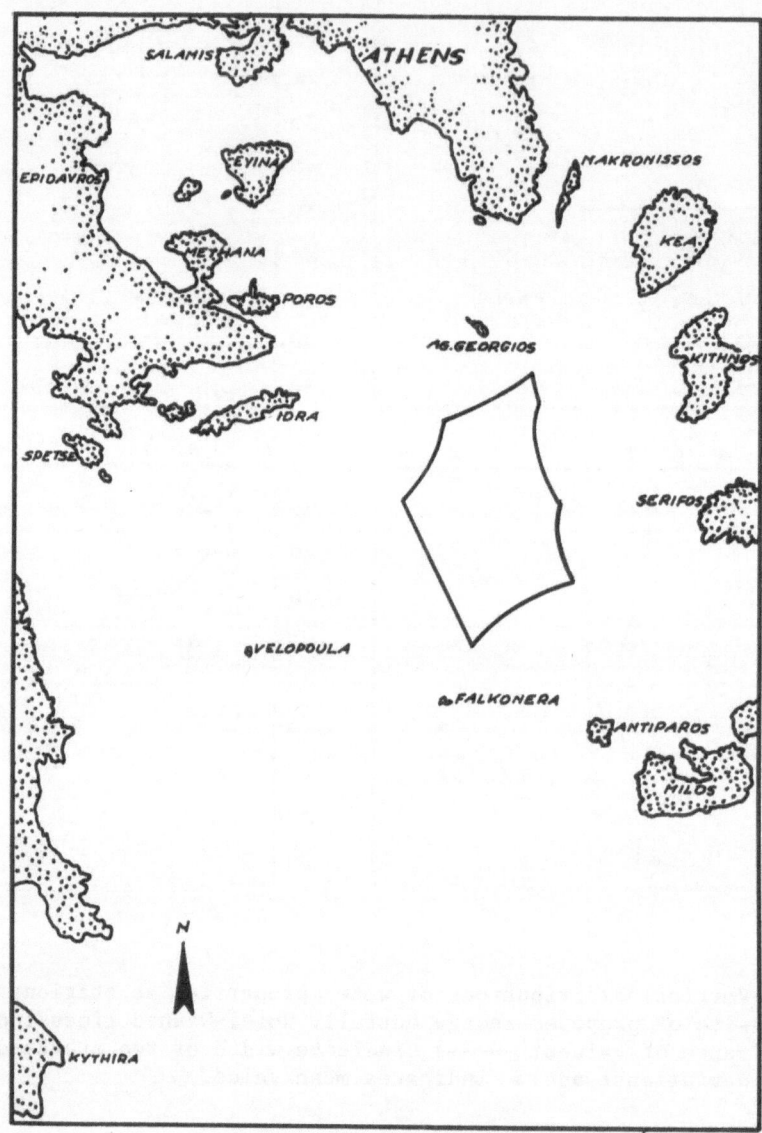

Fig. 4. Proposed sludge dumping area in the Myrtoon Sea.

Table 1. Concentration of chemical parameters in a composite sewage sample. August 1982 (the composite sample was made up of 24 hourly samplings)

Ammonium	$1580 \mu M$
Nitrite	$75 \mu M$
Nitrate	$136 \mu M$
Phosphate	$187 \mu M$
Silicate	$332 \mu M$
Suspended matter	730 mg l^{-1}
BOD,	360 mg l^{-1}
COD	820 mg l^{-1}

Table 2. Mean values at Sta. A during the summers 1974-1976 and 1982

	1974–1976 mean ± 95%	1982 mean ± 95%	% change
NH_4 (μM)	$0.67 \pm 0.43 (5)$	$0.50 \pm 0.25 (8)$	-26
NO_2 (μM)	$0.15 \pm 0.05 (3)$	$0.07 \pm 0.02 (8)$	-53^*
NO_3 (μM)	$1.41 \pm 1.05 (5)$	$1.01 \pm 0.44 (8)$	-28
PO_4 (μM)	$0.15 \pm 0.76 (5)$	$0.14 \pm 0.03 (8)$	-7
SiO_4 (μM)	$2.71 \pm 2.01 (5)$	$1.57 \pm 0.55 (8)$	-42
DO ($ml \, l^{-1}$)	$5.34 \pm 0.47 (5)$	$4.97 \pm 0.18 (7)$	-7

*Difference significant at $P < 0.05$.

Table 3. Mean values at Sta. B during the summers 1974–1976 and 1982

	1974–1976 mean ± 95%	1982 mean ± 95%	% change
NH_4 (μM)	$1.08 \pm 0.76 (5)$	$1.19 \pm 0.47 (9)$	$+10$
NO_2 (μM)	$0.21 \pm 0.17 (3)$	$0.14 \pm 0.05 (9)$	-33
NO_3 (μM)	$1.41 \pm 1.21 (5)$	$0.51 \pm 0.29 (9)$	-64^*
PO_4 (μM)	$0.31 \pm 0.10 (5)$	$0.35 \pm 0.14 (9)$	$+13$
SiO_4 (μM)	$2.27 \pm 1.71 (5)$	$0.99 \pm 0.19 (9)$	-56^*
DO ($ml \, l^{-1}$)	$5.40 \pm 0.39 (5)$	$4.80 \pm 0.26 (8)$	-11^*

*Difference significant at $P < 0.05$.

REFERENCES:

Coachman L.K. and Hopkins T.S. 'Description analysis and conclusions on water masses in the Saronikos Gulf'. Interim Technical Report 2. Environmental Pollution Control Project, Athens. Greece, 1975.

Coachman L.K., Hopkins T.S. and Dugdale R.C. 'Water masses of the Saronikos Gulf in winter'. Acta adriatica 18, 131-161, 1976.

Friligos N. 'Influence of industries and sewage on the pollution of Elefsis Bay. Revue Internationale d'Oceanographie Medicale 55, 3-11, 1979.

Friligos N. 'An index of marine pollution in the Saronikos Gulf'. Marine Pollution Bulletin, 12, 96-100, 1981.

Friligos N. 'Some consequences of the decomposition of organic matter in Elefsis Bay, an anoxic basin'. Marine Pollution Bulletin, 13, 103-106, 1982.

Friligos N. 'Nutrients of the Saronikos Gulf in relation to environmental characteristics (1973-1976)'. Hydrobiologia, 112, 17-25, 1984.

Griggs G.B. and Hopkins T.S. 'The delineation and growth of a sludge field'. Water Research, 10, 1-6, 1976.

Harremoes P. 'Report for the Greater Athens Environmental Pollution Control Project, 10 pp., 1975.

Hopkins T.S. and Coachman L.K. 'Circulation patterns in the Saronikos Gulf in relation to the winds'. Interim Technical Report I. Environmental Pollution Control Project. Athens, Greece, 1975.

Ryther J.H. and Dunstan W.M. 'Nitrogen, phosphorus and eutrophication in the coastal marine environment'. Science, 171, 1008-1013, 1971.

Watson J.D. and D.M. 'Report for the Greater Athens Environmental Pollution Control Project', 152 pp., 1978.

A COMPARATIVE STUDY OF COPPER CYCLES IN TWO FRESHWATER ENVIRONMENTS

Carlos Vale and M.L.S. Simões Gonçalves*
Instituto Nacional de Investigação das Pescas
Av. Brasília, 1400, Lisboa Portugal
* Centro de Quimica Estrutural, C. Interdisciplinar
Instituto Superior Técnico, Lisboa Portugal

ABSTRACT. A comparison of copper distribution in two fresh-water environments is presented. Based on field results, it is discussed the role of the sediments as an internal source, the particulate/dissolved interaction at different chemical and biological conditions, the changes of copper speciation and its ecological significance.

INTRODUCTION

The geochimical cycle of trace elements in natural water systems is complex and dependent on several factors, namely chemical composition of the water, chemical form of the trace element, and the biological dynamics which may interact with the solution composition.

Anthropogenic contributions of the global flux of copper to the world hydrologic cycle are not significant [1] however, local perturbations on freshwater systems due to human activities can have a major impact on the surrounding ecology. Copper and biota are known to be strongly inter-related and too high concentrations of copper ion may cause inhibition of photosynthetic activity of some phytoplankton [2]. Settling of inorganic and biogenic particles, on the other hand, by taking up metal ion [3] tends to reduce their residual concentration.

In this report we compare the copper content in a oligotrophic water receiving a mine effluent rich in copper, with the concentration in a mesotrophic water reservoir. Based on field results, a discussion about the particulate/ /dissolved copper interaction and the influence of copper speciation on the ecological significance of the pollutant input is presented.

G. Kullenberg (ed.), The Role of the Oceans as a Waste Disposal Option, 285–305.

TABLE I - Morphometric characteristic of the reservoirs

	volume 10^6 m^3	area 10^4 m^2	max depth m	mean depth m
Venda Nova	95	400	96	23
Maranhão	205	1257	55	16

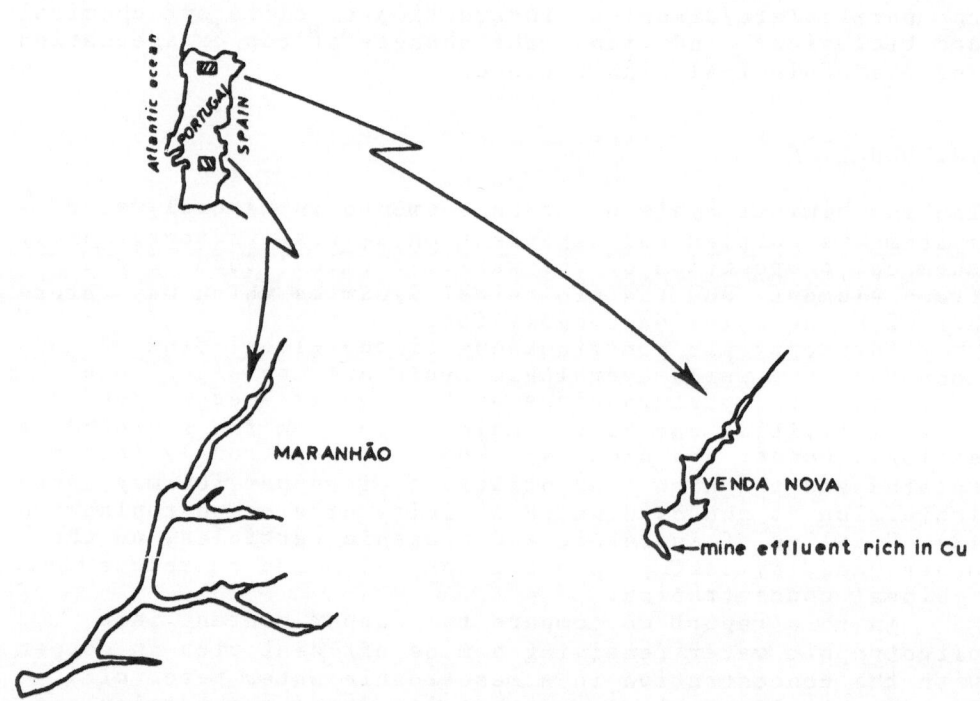

Figure 1 - Location of the reservoir of Maranhão and Venda Nova

STUDY AREA

The present field study was carried out in the water retained in two impoundments - Venda Nova and Maranhão (Fig. 1).
The Venda Nova reservoir is located in the north of Portugal and belongs to the hydrological system of Rabagão and Cávado rivers. This nutrient-poor reservoir is characterized by a well oxygenated status, very low alkalinity and hardness, and by the input of a mine effluent rich in copper. Planktonic communities are under stress and fish is depleted in this reservoir, contrasting with others from the same area (4).
The Maranhão reservoir is located in the south of Portugal and is a part of the hydrological system of Sorraia river. As opposed to Venda Nova, this water reservoir is comparable to mesotrophic lakes. Anoxic conditions are seasonally developed in the water column and planktonic blooms are often recorded (5). In this impoundment fish farming activities are conducted. The main morphometric characteristics of the reservoirs are given in Table I.

METHODOLOGY

Copper concentration has been determined in bottom sediments, pore water, particulate suspended matter and in the water column at both reservoirs by atomic spectrometry. Two box--cores were collected at Maranhão and Venda Nova in September 83 and April 84, respectively. Sediments were sliced at about 5 mm layers in nitrogen atmosphere and pore water was removed by centrifuge and filtration (0.45 um). The solid phase was analysed following sequential extraction (6). The total particulate copper, iron, manganese and aluminium were determined after a strong acid attack (7). Dissolved copper iron and manganese (< 0.45 um) concentrations were obtained by direct measurements. Suspended particulate matter concentrations (SPM) at different depths were determined by passing through Nucleopore filters (0.45 um) between 250 and 500 ml of water samples.
Dissolved iron, manganese, calcium and magnesium were analysed by atomic absorption spectrometry. Potassium and sodium in the waters were determined by flame photometry.
An estimate of the organic matter of these waters was obtained by spectral measurements, assuming the correlation between U.V. absorbance at 280 nm and the dissolved organic carbon (8).

RESULTS

1. Venda Nova - copper results

TABLE II - Copper concentration in pore water (nmol Kg^{-1}) and at different sediment fractions (ug g^{-1}) in a box-core collected at Venda Nova.

depth mm	diss Cu nmol Kg^{-1}	extractable fraction ugCu g^{-1}	carbonate fraction ugCu g^{-1}	oxide fraction ugCu g^{-1}	organic fraction ugCu g^{-1}	total copper ugCu g^{-1}
0- 2	630	135	6.0	219	160	2162
2- 4	710	145	7.0	213	180	2303
4- 6	240	126	7.2	167	270	2304
6- 8	2800	100	3.0	148	410	2841
8-10	11500	44	0.7	111	430	3230
10-12	790	22	0.5	93	450	3053
12-14	-	19	0.5	84	200	2733
14-16	1260	88	1.1	138	200	2230

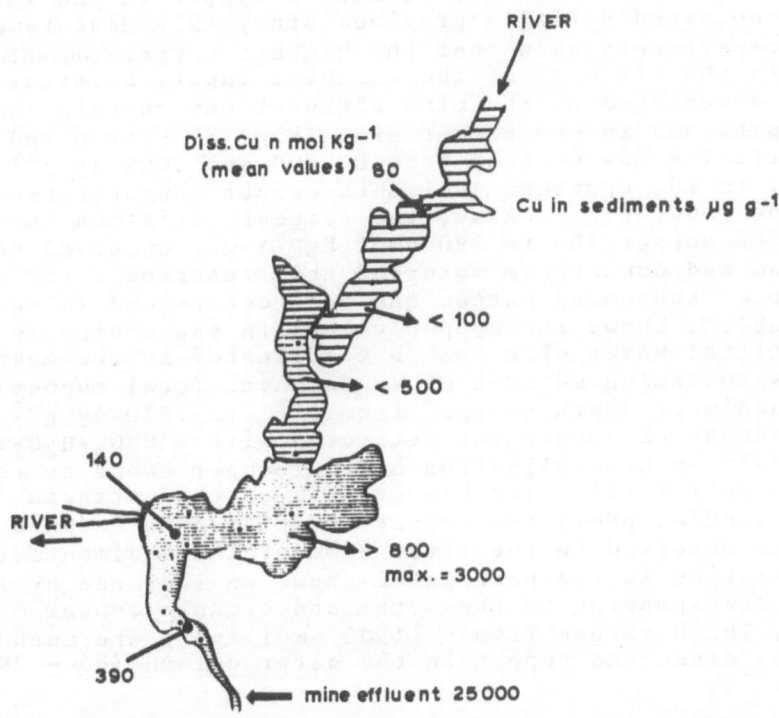

Figure 2 - Distribution of copper in the sediments of Venda Nova reservoir; sampling stations for dissolved copper (large dots) and for sedi ents (small dots).

Figure 2 shows the distribution of total copper in the
superficial sediments of Venda Nova reservoir, as well as
the annual mean values for dissolved copper in the water
column obtained during a previous study (9). Measurements in
both compartments show that the highest copper concentrations
occur in the vicinity of the effluent input. Particulate
matter discharged by the mine effluent may contain about 2.5%
of copper, and in the closer area (Fig. 2) bottom and suspended particles may present between 800 and 3000 ug g^{-1} of
copper. In the upstream sediments copper concentration is
less than 100 ug g^{-1}. Also, the largest variation in the
dissolved copper (80 to 390 nmol Kg^{-1}) was observed between
upstream and downstream water of this reservoir. Low concentration of suspended matter has been determined (4 mg/l).

Table II shows the copper values in the sediments and
interstitial water of a box-core collected in the most
heavily contaminated area of Venda Nova. Total copper in the
16 mm sediment depth changed from 2162 to 3230 ug g^{-1}. A
large amount of copper was extracted with a HNO$_3$/H$_2$O$_2$ mixture,
presumably by mineralization of the copper bound to the
organic matter (6). Also high values of the extractable
copper (MgCl$_2$, pH=7) and copper associated to Fe/Mn oxides
(6) were observed in the first 8 mm of the sediments. Copper
concentration in the pore water shows an increase by 10 mm
depth corresponding to the total and organic copper maximum
layers. These values (240 - 11500 nmol Kg^{-1}) are much higher
than the dissolved copper in the water column (80 - 390
nmol Kg^{-1}).

2. Maranhão - copper, iron and manganese results

In contrast to the other reservoir, chemical and biological
water conditions at Maranhão are quite variable during the
year (10,5).

Fig. 3 illustrates the temperature, dissolved oxygen,
iron, manganese and copper depth profiles in winter (February
1984) and at early autumn (October 1983) occurring at Maranhão. The February survey showed that dissolved oxygen
changed from 11 mg/l near surface to 2 mg/l at the bottom
water. A sharp increase of Mn^{2+} also reflects the low oxygenated water near the sediment surface. Iron and copper
also showed an increase with the depth. Results of the
October survey evidenced that the anoxic conditions may be
expanded to the water wolumn during the summer. Together
with the sharp decrease of dissolved oxygen (< 1 mg/l) an
increase of dissolved manganese and iron in the hypolimnion
was recorded. An increment of dissolved copper at the anoxic
bottom waters was also measured.

Table III the dissolved Fe, Mn and Cu concentrations and
the ratios to Al of Fe, Mn and Cu contents in the suspended
particles in February and October at the Maranhão reservoir.

TABLE III - Dissolved copper (nmol Kg^{-1}), iron (umol Kg^{-1}) and manganese (umol Kg^{-1}), Cu/Al, Fe/Al and Mn/Al ratios at suspended matter of the Maranhão waters (February and October surveys).

Survey of February 84

depth m	Cu nmol Kg^{-1}	Cu/Al x10^4	Fe umol Kg^{-1}	Fe/Al	Mn umol Kg^{-1}	Mn/Al
0	140	6	25.1	0.74	-	0.011
5	80	11	48.4	0.66	0.4	0.010
10	130	10	21.5	0.80	0.4	0.009
15	190	10	60.9	0.75	0.4	0.012
20	90	10	52.0	0.84	0.5	0.012
25	110	10	44.8	0.81	1.2	0.021
30	330	12	84.2	0.73	1.8	0.021
39	160	8	98.6	0.63	10.0	0.019

Survey of October 83

depth m	Cu nmol Kg^{-1}	Cu/Al x10^4	Fe umol Kg^{-1}	Fe/Al	Mn umol Kg^{-1}	Mn/Al
0	3	210	1.8	1.53	0.1	0.676
5	3	489	1.8	4.39	0.1	1.345
10	3	194	7.2	3.56	48.3	0.127
15	3	400	21.5	3.13	31.7	0.070
20	27	733	19.7	4.25	31.7	1.316
22	9	-	19.7	-	33.0	-

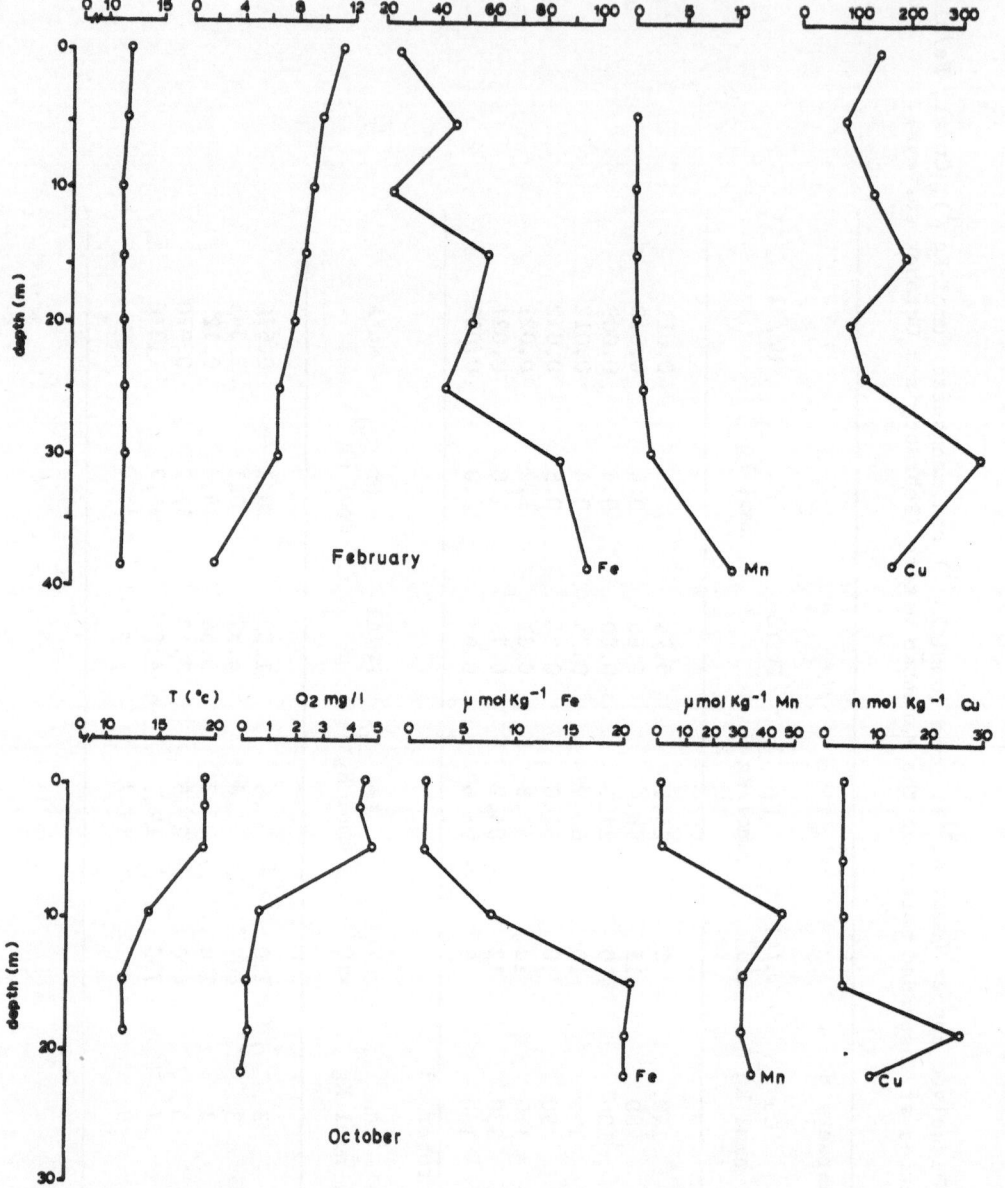

Figure 3 - Profiles of temperature, dissolved oxygen, iron, manganese and copper in February 1984 and October 1983.

These results point out the changes of Mn^{2+} and Fe^{2+} concentrations with the oxygenation status of the water column. An increase of the Mn^{2+} (0.1 to 48.3 umol Kg^{-1}) with the depth in October survey was remarkable. Also higher Fe^{2+} concentration in the anoxic layer (Fig. 3) was recorded. By comparing values from both surveys (Table 3) one may observe, more highly dissolved Fe and the absence of sharp variation of Mn^{2+} in the water column during the winter survey, as well as a general trend to an increase of dissolved metals with the depth of water. In contrast to the dissolved Fe^{2+}, Fe/Al ratio in the suspended particles changed in an opposite way. Higher Fe/Al and Mn/Al ratios were measured in early autumn particularly at mid-water layers. Copper distribution in the water column follows about the same pattern as iron: a decrease of dissolved copper from winter to latesummer, with an increase at bottom waters, and an enrichment of copper in particles in suspension during the warmer periods.

Table IV shows the copper content in the sediments and interstitial water of a box-core collected in September 1983 at the central zone of Maranhão reservoir. The results indite that from the sequential extration (6) the largest amount of copper in the sediments is clearly associated with the organic matter. Copper in pore water showed the highest concentration in the first centimeter sediment and a sharp decrease below 12 mm depth.

3. Organic matter

Absorbance at 280 nm of interstitial waters is closely correlated with the values of dissolved organic carbon (DOC) and provides a quick way of estimating DOC (8). At Venda Nova bottom waters the 280 nm absorbance was about 0.045 increasing by a factor of 2 far away from the mine effluent input. Oxygenated water at Maranhão presented a much higher absorbance, between 0.115 and 0.150. Using the correlation with the DOC proposed by those authors, about 15 mg C 1^{-1} can be estimated for Maranhão. The low absorbance at Venda Nova is probably outside of the linearity (11) and it may be less than 5 mg C 1^{-1}.

4. Calcium, magnesium, sodium and potassium

Table 5 presents these major element concentrations in both water reservoirs. As one can observe, water from Venda Nova contains lower Ca, Mg, Na and K. Values of chloride, sulphate and alkalinity (10) are also presented.

5. Copper speciation calculations

From the results obtained in both water reservoirs the copper speciation has been attempted. On the basis of the

TABLE IV - Copper concentration in pore water (nmol Kg^{-1}) and at different sediment fractions (ug g^{-1}) in a box-core collected at Maranhão.

depth mm	diss Cu nmol Kg^{-1}	extractable fraction ugCu g^{-1}	carbonate fraction ugCu g^{-1}	oxide fraction ugCu g^{-1}	organic fraction ugCu g^{-1}	total copper ugCu g^{-1}
0- 4	145	0.9	0.5	2.4	6.9	53
4- 8	192	0.1	0.2	3.6	6.4	53
8-12	235	0.1	0.2	3.9	5.3	55
12-16	98	0.1	0.2	5.1	7.9	58
16-20	98	0.3	0.2	4.1	5.8	59
20-30	33	0.1	0.2	3.3	6.4	54

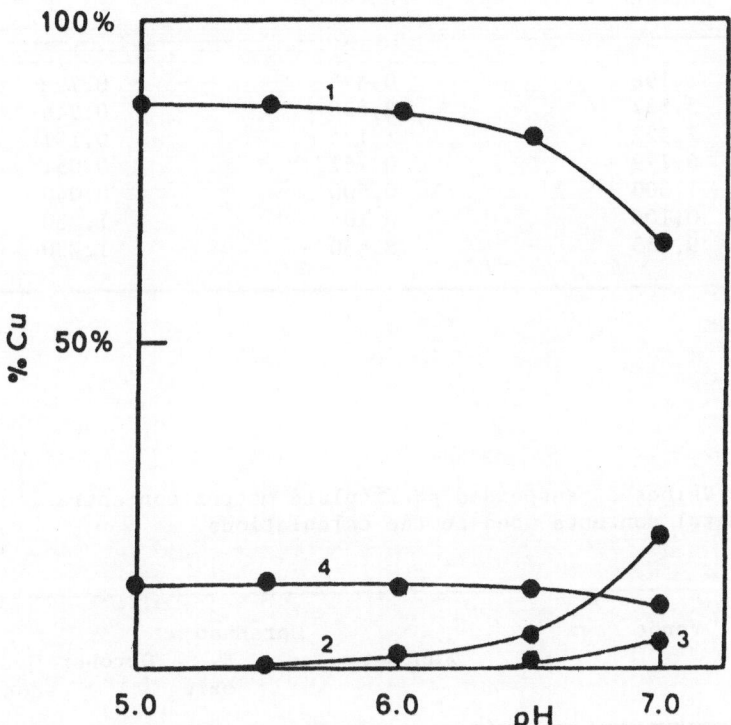

Figure 4 - Percentage of copper (1), hydrolysed copper (2), copper carbonate (3) and copper sulphate (4) versus pH for a total copper (II) concentration of 0.390 uM in the experimental condition of Venda Nova reservoir.

TABLE V - Concentration (mM) of cation and anion macro-constituints of Venda Nova and Maranhão.

	Maranhão		Venda Nova
	October	February	
Ca^{2+}	1.198	0.873	0.798
Mg^{2+}	3.537	2.838	0.946
Na^+	1.522	1.174	0.191
K^+	0.179	0.767	0.051
HCO_3^-	1.500	0.500	0.040
SO_4^{-3}	0.104	0.104	1.230
Cl^-	9.463	8.450	1.230

TABLE VII - Values of suspended particulate matter concentration (SPM) and heavy metal contents used in the calculations.

	Venda Nova	Maranhão		
		February	October	
			oxic	anoxic
SPM (mg/l)	4	50	50	100
Cu (ug g^{-1})	3000	100	400	1000
Fe (%)	10	5	4	15
Si (%)	20	20	20	20

TABLE VI - Estimated dissolved organic carbon (280 nm absorbance) and range of concentration of the following ligands: fulvic acid (FA), oxalic acid (Oxal), glycine (Gly), citric acid (Cit), sulfide (S^{2-}), polysulfide (S_4^{2-}) and cysteine (Cyst) and total copper concentration used in speciation calculation.

Oxic conditions	DOC mgC l^{-1}	FA (M)	Oxal=Gly=Cit (M)	Cu (M)
	~15	10^{-6} to 10^{-5}	5×10^{-7} to 5×10^{-6}	1.5×10^{-7}

Anoxic conditions	DOC mgC l^{-1}	S^{2-} (M)	S_4^{2-} (M)	Cyst (M)	Cu (M)
	15-60	1.8×10^{-4} to 5.0×10^{-3}	2.0×10^{-5} to 3.0×10^{-4}	1.5×10^{-4} to 7.0×10^{-5}	3×10^{-8}

main anion and cation concentrations (Table V), the ionic
strength of the medium has been calculated and stability
constants from the literature (12), have corrected according
to Davies expression. From total concentrations and steche-
ometric stability constants valid to the medium, speciation
calculation has been done by using the Comics computer program (13). Among the several limitations of such a copper
speciation approach, these calculations reflect only equilibrium conditions, which may be quite different from those
existing in natural waters.

5.1 Venda Nova
Due to low organic matter estimated at Venda Nova waters,
only inorganic ligands have been computed. The copper speciation model predictions are show in Figure 4 for total Cu of
390 nmol Kg^{-1} and a pH range of 5.0 to 7.0. Under the conditions given, free copper is the dominant species and the
hydrolysed, carbonate and sulphate species are negligible at
low pH. As pH increases these species are more abundant and
at pH 7.5 their presence is comparable to free copper. By
changing calcium and magnesium concentrations followed by a
decrease of copper in a simulation of upstream conditions of
Venda Nova reservoir (Fig. 2) we found no appreciable effect
in copper speciation.
        From Figure 4 one may observe that copper speciation
changes rapidly within the 6.5 - 7.0 pH interval. Such alterations may have a major importance on toxicity effects.
Although several inorganic and organic complexes can be toxic
(14), several works (15) have pointed out that toxicity is
mainly dependent on free copper.

5.2 Maranhão
In contrast to Venda Nova, seasonal alterations at Maranhão
waters are superimposed on geographical changes. The abundant
organic matter is decomposed and anoxic conditions in the
water column are frequently developed. These fluctuations
can drastically change copper speciation due to the high
affinity of this element to the organic ligands.

Oxic conditions
So, at Maranhão we considered the usual inorganic cations
and anions (calcium, magnesium, sodium, potassium, chloride,
sulphate, hydroxide and carbonate) as well as organic matter
from aquagenic and pedogenic origin (Table VI).
        As other workers (16,17) we assumed the following
dominant species: fulvic and humic acids, citric acid, oxalic
acid and glycine as a model of aminoacids. Concentrations of
these organics are also estimated from the 280 nm absorbance
and based on the literature values (17). As demonstrated by
BUFFLE (18), stability constants for fluvic acid depend not
only on the ionic strength and pH, but also on the ratio

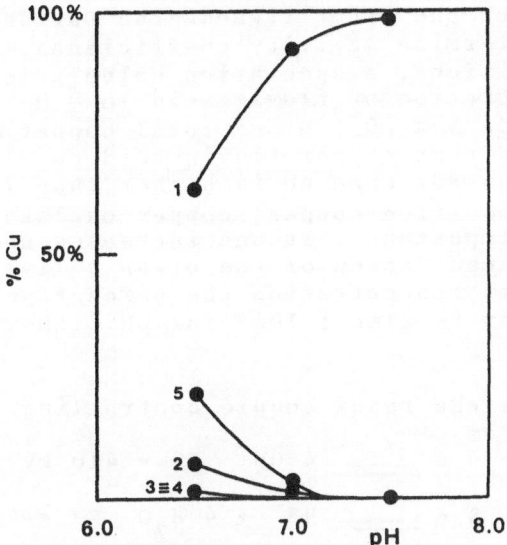

Figure 5 - Percentage of copper complex with fulvic acid (1), oxalic acid (2), citric acid (3), glycine (4) and the copper (5) versus pH for a total copper (II) concentration of 0.150 uM in the experimental conditions of Maranhão (winter situation).

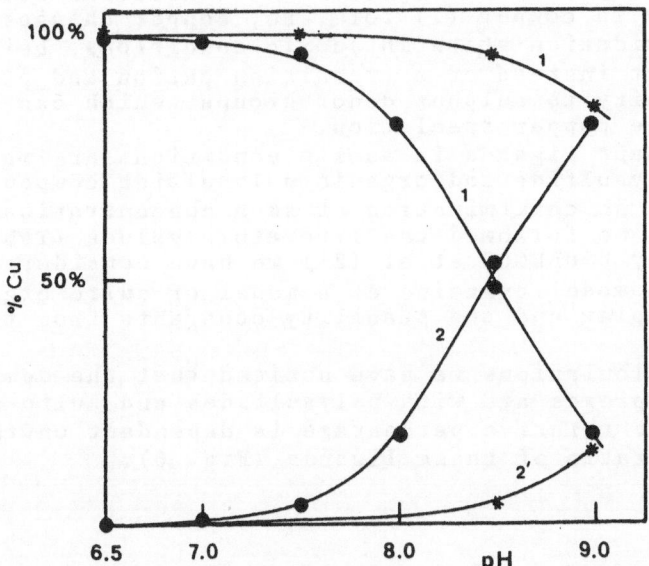

Figure 6 - Percentage of copper (I) complexes with cysteine and polysulphide versus pH for a total copper concentration of 0.003 uM, a total sulphide concentration of 10^{-4} M and a ratio $C_{S_4^{2-}}/ C_{cyst}$ equal to 0.1 (●) or 0.5 (*) being the total concentration of cysteine 4×10^{-5} M ; 1=Cu polysulphide , 2=Cu cysteine; experimental conditions of Maranhão in anoxic conditions.

C. Metal/C. Ligand. For the other ligands the Davies equation has been used to determine activity coefficients. Considering the field conditions, a speciation calculation has been done, with a concentration of fluvic acid 10^{-6} M, for the other organic ligands 5×10^{-7} M and total copper 1.5×10^{-7} M. It has been observed that copper-fluvic acid complex axists at levels higher than 80% when pH is higher than 7.0 (Fig. 5). The other species (free copper, copper-oxalate, etc.) showed a much less importance. If one increases one order of magnitude the concentration of the organic ligands for the same total copper concentration the percentage of copper- fulvic acid complex is almost 100% for pH higher than 6.5.

Anoxic conditions
In anoxic conditions the redox couple controlling the redox potential changes from

$$O_2 + 2 H_2O + 4 e \underline{\qquad} 4 OH^- \quad E^o = 410 \text{ mV}$$

to

$$SO_4^{2-} + 9H^+ + 8 e \underline{\qquad} HS^- + 4 H_2O \quad E^o = 251 \text{ mV}$$

So, if consider as anoxicity limit conditions $SO_4^{2-}/H\bar{S} < 3 \times 10^4$ (19,20), it can be observed for the system

$$Cu^{2+} + e \underline{\qquad} Cu^+$$

that $Cu^{2+}/Cu^+ = 2.2 \times 10^{-6}$ for pH= 7.0 and $Cu^{2+}/Cu^+ = 1.2 \times 10^{-8}$ for pH= 8.0. This means that at anoxic conditions all copper is practically in copper (I) form. So, copper changes completely its oxidation state in anoxic conditions, being a class B element instead of a transition cation and presents a higher affinity to sulphur donor groups, which can drastically influence copper speciation.

The dominant ligands in anoxic conditions are mainly bisulfide, polysulfide and organic polysulfide compounds (19,20). Since no determination of such concentrations has been obtained, we followed the literature values (Table VI) As suggested by BOULEGUE et al (21) we have considered in our speciation model cysteine as a model of sulfo-organic copper (I) complex and the stability constants from the literature.

In our calculations we have noticed that the dominant copper (I) complexes are with polysulfides and sulfo-organic ligands and the relative percentage is dependent on the concentration ratio of these ligands (Fig. 6).

DISCUSSION

Two different situations about copper input in freshwater systems are reported here. In one case particles with high copper content (0.3 to 2.5%) from mine activities are continuously discharged into a reservoir. By reaching this water reservoir, most of the particles settle and a layer

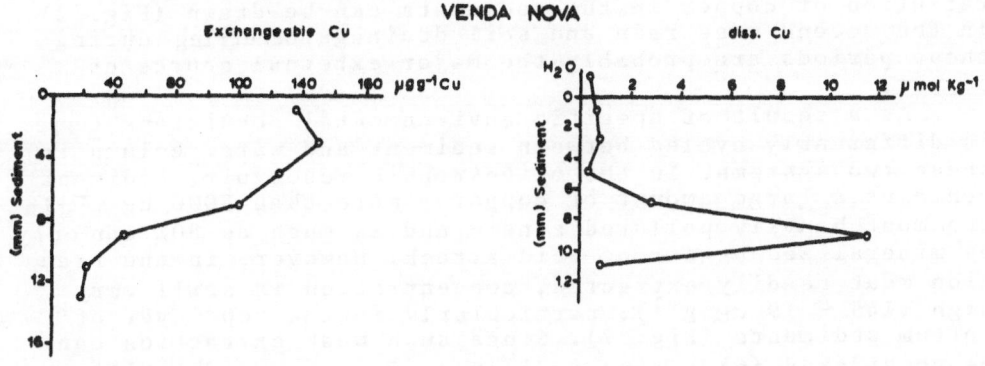

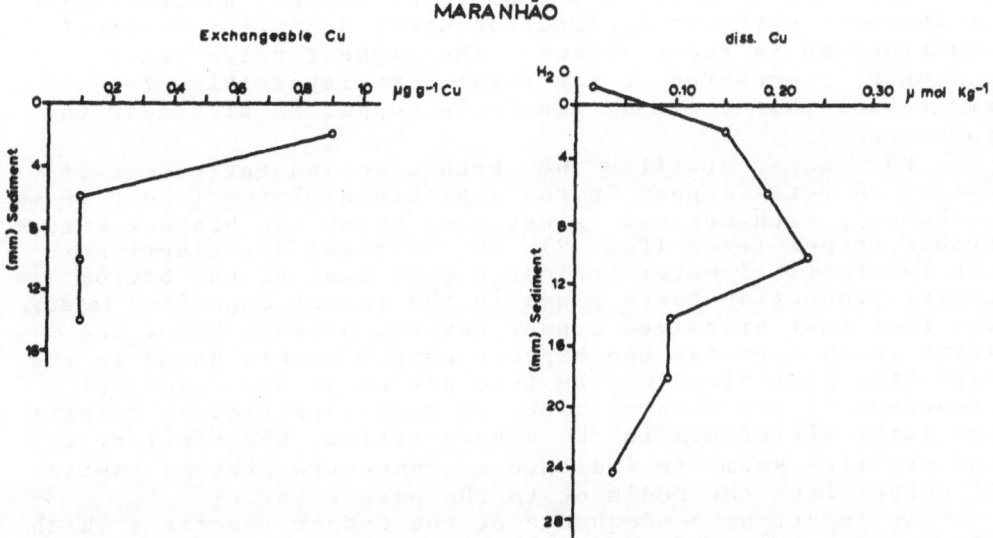

Figure 7 - Profiles of sediment extractable copper ($MgCl_2$) and dissolved
total copper in pore water at Venda Nova and Maranhão.

of variable thickness covers the bottom (9). From both grab
and core sampling, geographical areas of preferential
retention of copper in the sediments can be drawn (Fig. 2).
In the second case rain and soil drainage occurring during
these periods are probably the major external source of
copper.

As a result of specific environmental conditions, copper
is differently cycled between sediment and water column in
these two systems. In the oligotrophic reservoir, sediment
contains a large amount of copper - more than 2000 ug g^{-1} in
the most heavily polluted zone - and as much as 80% can only
be mineralized by strong acid attack. However, in the fraction most readily extracted, concentration is still very
high (145 - 19 ug g^{-1}), particularly in the top layer of
bottom sediments (Fig. 7). Since such weak extraction can
be considered (6) a good indicator of copper availability
in the sediments, one may conclude that a large amount of
copper is weakly preserved in the first centimeter of the
bottom sediments. At the mesotrophic system, sediments do
not show more than 60 ug g^{-1} of total copper, however, copper
in the most extractable fraction shows a similar vertical
distribution in the sediments. The highest value (only
0.9 ug g^{-1}) occurred in the first 4 mm layer (Fig. 7),
suggesting that the most available copper is also near the
interface.

Pore water profiles show both cases an enrichment of
dissolved total copper in the superficial layer (about 10 mm
tickness), with maximum values just below the highest extractable copper layer (Fig. 7). The vertical distribution in
the interstitial water indicates that most of the bottom
copper production takes place in the recent deposited layers.
The fact that dissolved copper maximum appears below the
layer which contains the highest copper weakly bound to the
deposited particles, may be interpreted as the result of
breakdown of the weakest links of copper-particles. Despite
the large difference in the concentration, the similary in
the profiles seems to indicate a common transfer mechanism
of copper from the sediment to the pore water.

An important consequence of the copper reactions which
occur close to the sediment-water interface is the influence
which they may exert upon the copper content in the water
column. According to Fick's first law, the difference of
dissolved copper between pore water and water column indicates a flux of Cu out of the sediments (22). This Cu flux
may be particularly high in the oligotrophic reservoir,
since the gradient across the sediment-water interface is
very strong due to the high copper pore water concentration.
Also, physical perturbation in the bottom layers, as it has
been occasionally observed, may instantaneously increase
the flux out of both dissolved and particulate copper and
consequently increase the stress on the biological communities.

In the oligotrophic reservoir, due to low suspended matter concentration and productivity, a few carriers are comparatively present and the large portion of dissolved copper is probably dispersed inside the reservoir and to the hydrological system. A complete different situation occurs in the mesotrophic water reservoir. A large number of potential carriers are present in the water column, namely algae and inorganic particles. This and the absence of located source of copper explains that seasonal variations of copper (Table III) are superimposed on geographical changes. Total dissolved copper may change two orders of magnitude between summer (3 nmol Kg^{-1}) and winter (330 nmol Kg^{-1}). A similar drastic but opposite variation, occurs in the suspended phase (Cu/Al ratio varies from 6 x 10^{-4} to 733 x 10^{-4}).

Because anoxic conditions are developed in the water column, intense migration of Fe^{2+} and Mn^{2+} occurs in the water as well as precipitation of hydroxides of these elements near the anoxic/oxic layer (Table III). The high content of iron and manganese measured in the suspended phase particularly near the oxic/anoxic boundary reflects the precipitation of iron and manganese (23). However, vertical fluctuations in the particulate metals (iron, manganese and copper) may be due to different kinetics at several water depths (24) and/or due to short-temporal migration of the oxic/anoxic boundary. This indicates the non-equilibrium situation occurred in such natural systems. During the warm period blooms of phytoplankton are also developed(5). So, both biogenic and inorganic suspended particles may incorporate from the water column a large amount of copper (25) presumably by absorption, precipitation or uptake, for which such particles could serve as nuclei. The high content of particulate copper and the decrease on dissolved phase in later summer indicate that such particles are important scavengers for copper. Pore water profiles show that copper after reaching the bottom sediment is transfered from the particles to the interstitial water inside sediments (Fig. 7) and a fraction which as been transported to the bottom may be released to the water column again.

After the turnover and oxygenation of the water, these elements (Fe, Mn and Cu) increase their concentration in the "dissolved phase" and decrease it in the suspensions. As proposed by some authors (3,24) such changes may correspond to the in situ dissolution of the particulates in the case of copper, or to the existence of colloids with iron and manganese. These fluctuations in both dissolved and particulate phase indicate that in mesotrophic systems, particularly with anoxia, some elements are extremely mobile between the water column/bottom sediments and vice versa. Practically no copper in the suspended matter of the oligotrophic reservoir can be explained by absorption on goethite and silica. Indeed, calculations based on surface equilibrium

constants (26) show that only 0.2% of copper is absorbed on
goethite and 0.7% adsorbed on silica considering the mean
field values used (Table VII). In the mesotrophic water the
same calculations show that, in winter, most of the particulate copper can be explained by absorption on these inorganic surfaces (~ 7.5% on goethite and ~63% on silica).
However, the large amount of particulate copper in warmer
periods is probably related to the copper association with
organic surfaces. In especial humic acids that dominate the
chemistry surface there may be increase of trace metal
uptake (3). So, particles may play a more decisive role of
copper scavenger as productivity of the environment increases.

Speciation computation has show that copper forms are
controlled by the presence of a few ligands. At oligotrophic
water due to the low organic dissolved matter, free copper
(II) tends to be the dominant species, although hydrolysed
copper becomes more abundant as pH increases. This indicates
that inorganic ligands have a minor effect and, on the other
hand, a decrease of pH or increase of copper input (due,
for example, to the mine effluent input) may result in a
higher percentage of free copper in the water reservoir.

At the mesotrophic reservoir, speciation in oxic conditions is strongly dominated by the copper fulvic acid complex,
particularly at pH higher than 7.0. Since these are the most
representative pH field conditions, one may conclude that
free copper is pratically absent and speciation is greatly
controlled by the copper/fulvic acid concentrations. In
water levels where degradation of organic matter consumes
all the dissolved oxygen, speciation changes drastically,
because copper is mainly in copper (I) forms. The dominant
copper (I) complexes are principally with polysulfides and
sulfo-organic ligands. Other sulfur ligands like hydrogen
sulfide and sulfide have a negligible effect. The more
abundant presence of polysulfides and organics at certain
water levels may explain the increase of copper (I) solubility. At strong anoxia levels bisulfide and chlorides should
be the dominant ligands and so solubility of copper (I)
should decrease. The observed copper concentrations in the
water column of the mesotrophic water (Fig. 3) compare quite
well with this anticipated prediction. These field results
seem to confirm (19) the importance of polysulfide complexation to the solubility of copper in anoxic environments.

So, sporadic alterations on oligotrophic systems such
as lowering pH or increase of copper flux from the sediment
cause a greater increase of free copper in the water than
in more productive environments. Such changes, in those
systems may thus have a decisive importance on toxicity
effects.

For similar dissolved copper concentrations, the presence of natural organic ligands may be responsible for a

significant decrease of the most toxic species. So, for a proper waste-disposal option, there is a need to evaluate the assimilative capacity of each natural system.

REFERENCES

1. SOYLE E.A. In: Copper in the environment, part 1: ecological cycling.Ed. Jerome O.Niagu (1979)
2. ANDERSON, D.M. and MOREL F.M.M. Limnol. Oceanogr. 23 (1978) 283
3. TIPPING E., GRIFFIT, J.R. and HILTON, J., - Croat chem Acta 56 (1983) 613
4. OLIVEIRA, R., MONTEIRO, T., CABEÇADAS, G., VALE, C. and BROGUEIRA, M. J. - Verh. Internat. Verein Limnol. 22 (1985) 2395
5. OLIVEIRA, R. - Hydrobiol. (in press)
6. TESSIER, A., P.G.C. CAMPBELL and M.BISSON - Anal Chem 51 (1979) 844
7. RANTALA, R.T.T. and LORING, D. - Atomic Absorption Newsletter 14 (1975) 117
8. KROM, M.D. and SHOLKOVITZ, E.R. - Geochim et Cosmochim Acta 41 (1977) 1565
9. VALE, C. - Bol. Inst. Nac. Invest. Pescas 7 (1982) 5
10. BROGUEIRA, M.J. and CABEÇADAS, G. - (in preparation)
11. ELDERFIELD, M. - Amer. J.Science 281 (1981) 1184
12. SMITH, R.M. and MARTELL, A.E.- Critical stability constants, Plenum Press, New York (1976,1982)
13. PERRIN, D.D. and SAYCE, I.G. - Talanta 14 (1967) 833
14. GIESY, D.P., NEWELL, A., and LEVERSEE, G.J. - The Science of the total environment 28 (1983) 23
15. FLORENCE, T.M. - Talanta 29 (1982), 345
16. GUY, R.D., and ROSSEAN, - Wat.Res. 14 (1980), 891
17. MCKNIGHT, D.M., FEDER, G.L.,HURMAN, E.M., WERSHAW,R.L. and WESTALL J. - The Science of the total environment 28 (1983), 65
18. BUFFLE, J. - In:Metal ions in biological systems. Ed. by H. sigel, Marull bekker (1984)
19. EMERSON, S., JACOB, L. and TEBO, B. - In: trace metals in sea water. NATO Conf. Serv. IV:9 (1983) 579
20. KREMLING, K. - Mar. Chem. 13 (1983), 87
21. BOULEGUE, D., LORD, C.L. and CHURCH, T.M. - Geochim and Cosmochim Acta 46 (1982) 453
22. BERNER, R. - Early diagenisis. Princeton Press. Princeton (1980)
23. DAVISON, W. and WOOF. C, - Water Res. 18 6 (1984) 727
24. DAVISON, W. and SEED, G., - Geochim and Cosmochim. Acta 47 (1983) 67
25. WAGEMANN, R. and BARICA, J. - Water Res. 13 (1978) 15
26. SIMÕES GONÇALVES, M.L., SIGG, L, and STUMM, W. - Env. Sci. Tech. 19 (1985), 141

A CASE STUDY OF WASTE INPUTS IN THE TAGUS ESTUARY

Margarida C. de Barros
Instituto Nacional de Investigação das Pescas
Avenida de Brasília
1400 Lisboa
Portugal

ABSTRACT. The Tagus Estuary, the largest in the European occidental coast, has been receiving large quantities of wastes from agricultural and industrial activities as well as from large urban centers. The estuary is a well mixed body of water in which tidal energy plays a dominant role in the currents regime and the salinity distribution, unless unusual very high influxes from the River Tagus are occurring. Sediments are mainly of fluvial origin and deposition rate or export to sea are low. A continuous process of erosion, suspension and redeposition is taking place. The estimated pollution loads from the very large number of sources draining their effluents to the Estuary show that the most important feature of the discharges into the Tagus Estuary is the simultaneous input of suspended solids, organic matter and contaminants, in particular Hg, As, Pb and other heavy metals. Though inputs of persistent organic compounds have never been estimated organochlorine insecticides and PCBs are detectable in sediments and biota at levels indicative of local pollution. A complete survey of pollutants in water, sediments and organisms has never been made but available data show that present levels of Hg, As and Pb are among the highest detected in similar environments. Biological effects at species level are suspected but the ecosystem as a whole seems to be adapted to the alterations in water quality that have been occurring.

1. INTRODUCTION

Estuaries have always been in the history of mankind areas of choice for settlement and have developed into the preferred location for large urban and industrial developments. They have been used as receptacles for solid and liquid wastes from all sources, as harbours and as recreational areas while maintenance of their capacity as fishing grounds was also sought.

The Tagus Estuary is no exception, a complex ecosystem supporting wildlife and commercial fisheries, surrounded by agricultural and urban areas, has multiple uses which have caused some degradation of the water quality. The extent of this degradation was not assessed

307

G. Kullenberg (ed.), The Role of the Oceans as a Waste Disposal Option, 307–324.
© 1986 by D. Reidel Publishing Company.

until recently though there was a public awareness that pollution was
very high in some areas and that some mollusc and fish species were no
longer found in the Estuary.

2. DISCRIPTION OF THE STUDY AREA

The Tagus Estuary, the largest in the european occidental coast,
extends through an area of 320 Km^2 and receives inputs of freshwater
from the Tagus and from other smaller rivers. Several towns under
expansion are located around the Estuary with a population of more than
2.5 million. The area is also heavily industrialized and agriculture is
intense.

The inner part of the Estuary is a delta with channels and islands
surrounded by land of intense agricultural activity. In the middle part,
rather wide (10 Km), are found salt marshes, shoals and mud flats which
constitute in part an intertidal area, important habitat of migratory
birds and a Natural Reserve. Commercial fisheries of molluscs,
crustacea and several fish species are still existing. Harbour
facilities, a large shipyard, and chemical and metalurgical complexes
are located on the adjoining banks. The outer part of the Estuary is
a deep, narrow channel and the immediate coastal zone is essentially
a beach area. (Figure 1).

The River Tagus has a mean annual discharge of 10.10^9 m^3
corresponding to a modular flow of 300 m^3s^{-1} (Loureiro, 1979). The rate
of influx of freshwater varies greatly, the influence of rainy periods
in the large watershed (80 600 Km^2) of the Tagus is not controlled in
spite of the number and size of the dams that have been built along
this River and its main tributaries. Variations of flow between
50 m^3s^{-1}, as occurred in the summer of 1981, and 3000 m^3s^{-1} during the
November 1983 flood have been reported (Martins et al, 1984). The
contribution of the other rivers to the total input of freshwater to
the Estuary is small (Quintela, 1976).

The mean volume of the Estuary is $0.19x10^{10}$ m^3 with a mean tidal
prisma of $0.075x10^{10}$ m^3. (Drena 1979). The tide is semidiurnal, with
a maximum amplitude of 4m and the duration of the flood period is
longer than that of the ebb. The intrusion of seawater reaches Vila
Franca de Xira, 50 Km inland from the mouth of the River. The intertidal area is about 40 per cent of the whole area.

The currents inside the Estuary are essentially tidal currents.
The ebb current, more intense than the flood, has the highest velocity
at the surface, while the maximum intensity of the flood current is
found at the lower layers. Residual circulation is an important feature
in the hydrodinamics of the Estuary (Oliveira, 1967). The residence
time of freshwater in the Estuary is estimated to vary between 140 days
for a freshwater influx of 50 m^3s^{-1} and 6 days when the Tagus flow
increases to 2000 m^3s^{-1}. For the modular flow of 300 m^3s^{-1}, the
residence time is estimated to be 23 days (Martins et al, 1984).

The distribution of salinity is highly influenced by the inflow
of freshwater. In Figure 2 is shown the general character of the mean
salinity distribution at the surface under mean low or high flow

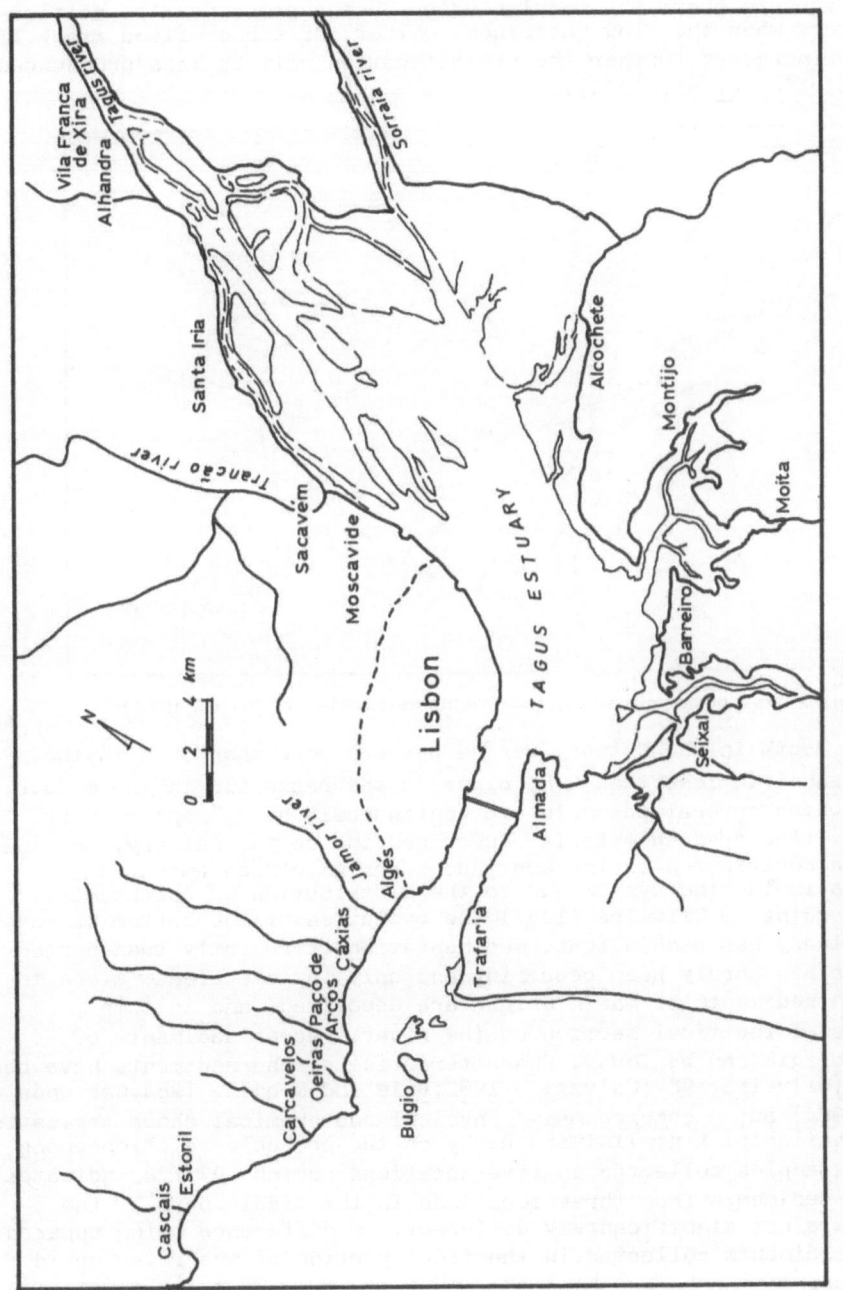

Figure 1 —The Tagus Estuary. Main towns and rivers

conditions (CNA, 1977). The vertical distribution of the mean salinity during different tidal conditions is relatively uniform when the flow of the Tagus is about the modular value, but a pronounced stratification occurs when the flow increases, either for ebb or flood tide. If the flow increases further the stratification will be less pronounced again (Martins et al, 1984).

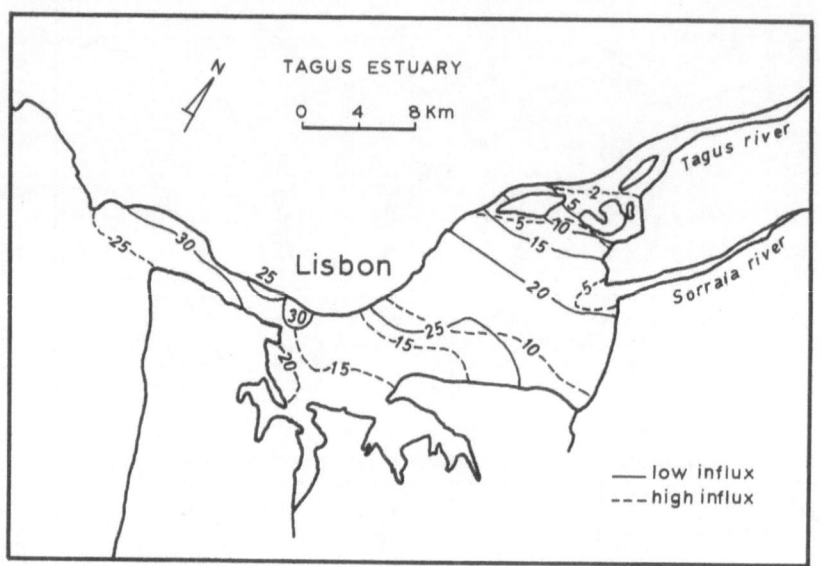

Figure 2-Distribution of mean surface salinities (adapted from CNA, 1977)

The depth in the Estuary varies between more than 30 m in the inlet channel to less than 5 m, close to the banks and in the middle part in which several channels are continuously being kept open by dredging. (The dredged material is dumped inside the Estuary, and this operation contributes to the continuous spread of contaminated sediments inside the system and to the distribution of pollutants).

According to Oliveira (1967) the evolution of the bottom of the Tagus Estuary has been slight. Sedimentation of recently transported materials has hardly been occurring and only in some areas; close to the mouth sediments of marin origin are deposited, and in some locations of the tidal section of the river, recent sediments of fluvial origin can be found. Characteristics of the sediments have been studied (Oliveira,1967;Calvário, 1982;Vale and Mendes, 1984;Carrondo et al, 1984) but a comprehensive physical and chemical characterization is not available. A preliminary study on the probable relation among sediment samples collected in five locations during 1977/78 indicates that the sediments from three locations in the middle part of the Estuary are not significantely different, a difference being apparent between sediments collected in the tidal portion of the river or in areas close to the banks where the input sources are located (Almeida, 1979). More studies are needed for confirmation of this hypothesis.

The solid load carried by the Tagus into the Estuary, at mean river flow, is lower than could be excepted because a large percentage of the suspended matter settles in the dams built along the River and its main tributaries. Important temporal variations are known to occur in the amount and characteristics of the River borne material, essentially related to the influx of freshwater. According to Vale (1981), at low river flow the concentration of suspended matter has been reported as 5 mg l^{-1} while during an extremely high river flux the concentration of suspended sediments at the limit of the river tidal section was 300 mg l^{-1} decreasing to 20 mg l^{-1} as the flow decreased to values closer to the mean winter flow. The studies conducted during the flood of 1979 by this author, indicate that the transport of suspended material during the short period of a week may have represented 85 per cent of the year's load – a total of 0.9×10^6 tons has been estimated. The particles transported during periods of flood or of normal seasonal flow show considerable differences in composition: with the increase of freshwater influx, there are larger inputs of particles containing higher mineralogical fraction and lower metal concentrations. The suspended particulate matter that reaches the Estuary will become part of the estuarine circulation. Its characteristics and composition will vary due to physical and chemical processes and due to man-made inputs of pollutants. The tidal energy, in particular that associated with the neap-spring tidal cycle is an important factor to intensify these processes (Vale, 1984).

The turbidity is mainly influenced by the tidal energy (provided no abnormal freshwater inflow occurs) and the entire Estuary at periods of spring tide is much less transparent than at neap tide situations. The extent and intensity of areas of greater turbidity are difficult to define precisely and so is the lower and upper limit of the maximum turbidity zone due to the simultaneous and variable influence of the tide amplitude and of the river flow (Vale and Sundby, 1982). The occurrence of a sediment layer of about 3 mm thick at slack water periods, which disappears with the increase of tidal currents has been affirmed (Vale, 1984) meaning that about 10^4 to 10^5 tons of sediments are continuously eroded, resuspended and redeposited within a period of 14 days.

The work of Vale (1984) and of Vale and Sundby (1984) do not show a clear export or import of sediments through the mouth of the Estuary (Figure 3) though these authors consider that it is reasonable to assume that, under certain conditions, the stock material leaves the Estuary, reaches the coastal zone and settles there. This export is most likely influenced by the river flow since tidal amplitude variation appears not to be a determinant factor for the escape of sediments from the Estuary.

Several studies have been conducted, in particular from 1979 onwards, identifying benthic organisms, plankton and fish species that are found in the Estuary (Rodrigues and Moita, 1979; Sobral, 1982; Calvario, 1982; Saldanha, 1980; Re, 1983; Ramos, 1983; Vilela, 1975; Marques and Costa, 1983). Investigations on the biology of some of the more important or abundant species of fish and crustacea have been carried out (Ferreira et al., 1982) or are under way. The role of the Tagus Estuary as a nursery for several species of fish found in the

In this paper is presented a review of available information on the Tagus Estuary about waste inputs and water quality. Several studies are still under way and many questions about the basic physical, chemical and biological processes prevailing in the Estuary, still need to be answered. Nevertheless, the overall situation can now be appraised and the data available may be usefully used in the future management of the Estuary and surrounding areas.

3. POLLUTION INPUTS

The first comprehensive study undertaken in 1976/77, to determine the sources of pollution inputs to the Tagus Estuary concluded that, in addition to the discharges of a population of more than 2 million people, many industries, that range from small artisanal factories to a very large chemicals producing complex and a metalurgical plant, drain their liquid effluents with scarcelly any treatment, to the Estuary, directly or through its tributaries (Table 1). Six hundred

TABLE I. Most important types of industries draining
          effluents to the Tagus Estuary.[1]

Activity/product	Activity/product
acids*	glue manufacture
animal feedlots	leather products
artificial fibers	metal plated products
brewed and distilled beverages	oil refining*
canned and preserved foods	paint and ink
ceramics and china	plastics and resins
chemicals*	paper products
chloroalcaline industry*	pesticides*
cooking oils and margarines	phosphate and phosphorus*
cork products	petrochemicals*
dairy products	pyrite roasting*
detergents	shipyards*
distilleries	steel*
electric and electronic equip.	sugar
explosives	tanning
ferrous and non ferrous metal	textyles
products*	tyres
fertilizers*	yeast

(1) adapted from Costa et al, 1982
 * main contributors of pollution

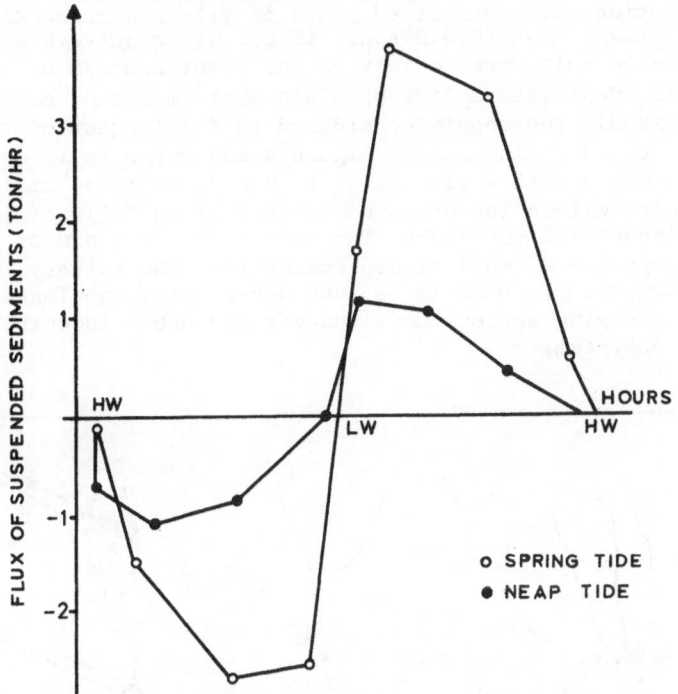

Figure 3 – Horizontal sediment transport across a
section of the Estuary inlet (Vale 1984)

coastal areas of·Portugal has been appraised (Costa, 1982). It is
expected that the data now available on the estuary biota and on other
parameters of biological significance such as measurements of primary
productivity, will be appraised to assess if biological effects are
detectable in the Estuary which may be related to the levels of
pollution that are determined.

Studies to determine pollution levels started in the late 60's
and by 1980 an integrated project, sponsored by UNDP was initiated.
This project, the "Environmental Study of the Tagus Estuary", was
designed as a comprehensive interdisciplinary study aiming at
contributing for a rational management of the estuary's water resources
harmonizing its multiple uses with the socio-economic development of
the area and public health protection (Santo and Martins, 1983).

As part of the "Environment Study of the Tagus Estuary" mathematical modelling for several situations in the Estuary has been
developed. In particular, a two-dimensional numerical model was
developed (Rodrigues et al, 1982). Other models are now being used for
specific studies such as the one dimensional model for the study of the
BOD and D.O. (Costa and Camara, 1982). These models were developed
using data from several synoptic sampling surveys in water quality
carried out during 1980/82 (Santo and Martins, 1983).

sources of pollution were identified south of Vila Franca de Xira and
so within the Estuary area (D.G.R.A.H., 1978). Other surveys were
conducted following this work, mainly on the point sources or at the
largest outfalls identified as the possible most important contributors
(Costa et al., 1983). The inputs considered in the largest of these
surveys are indicated in the map of Figure 4 and refer to 15 outfalls
of urban sewage mixed with effluents of rather diversified small
industries located within the urban center and 11 outfalls of large
industries or industrial complexes. The inputs of all kinds of
discharges through the 9 small rivers coming into the Estuary, which
serve as open sewers, has also to be considered, plus the Tagus and
Sorraia Rivers carrying agricultural run-off and urban industrial
effluents from upstream.

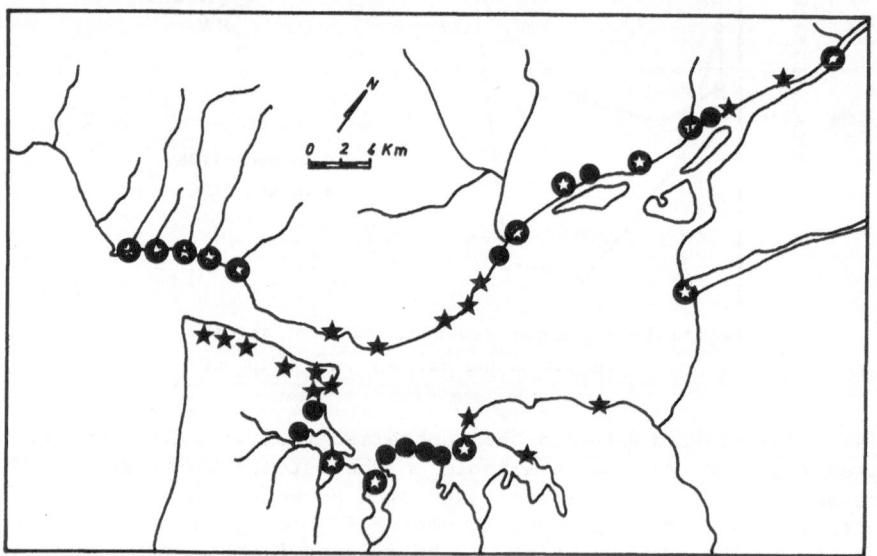

Figure 4 — Main wastes outfalls   in the Tagus Estuary

A first estimate of total inputs of the pollutants of more concern
was presented by Janeiro (1981) and the values reported have latter
been re-evaluated for the most important sources. Their relative
importance has always been confirmed but the quantified estimates of
discharges are significantly different in some cases. In fact, it
must be noted that sampling strategies and analytical techniques have
varied in the different studies and that very often informations
provided by the industries on their technology, organization or
discharges have been incomplete or erroneous usually by lack of
knowledge of the problems or by lack of data. Since the input estimates
were based on both the results of the surveys and information from
industry, the figures presented here should be considered as an
indication of the expected range rather than as the actual discharges.

In Table 2 are included minimum and maximum total inputs estimated by several authors, for the pollutants of higher concern - organic matter expressed as the biochemical oxygen demand (BOD_5), total suspended solids, total nitrogen and phosphorus, and heavy metals. In Figure 5 are indicated the outfalls of areas of major inputs since it has been established that more than 80 per cent of the total discharges are occurring from a very small number of industries or through a few of the urban sewage outlets.

TABLE II.   Range of total input estimates from industrial sources

	Minimum estimated	Maximum estimated
BOD_5	60.6×10^3*	72.6×10^3*
TSS	50×10^3*	66.2×10^3*
N	995*	
P	1700*	–
As	117.6*	
Cd	10.9	31
Cr	9*	21.6
Cu	312	–
Fe	1442	9090*
Hg	0.5	4.2
Mn	11*	39.1
Ni	73.7	81
Pb	198	458
Zn	3435	3592

* main sources only

Persistent organic compounds which are undoubtly discharged have never been analysed for in the effluents. For instance PCBs and organochlorine insecticides have been detected in sediments and biota at levels indicative of local pollution (Barros, 1984) but the main contributors, if any, have not been identified though the distribution of their residues in sediments gives an indication of such sources.

In Table 3 are presented the total inputs of organic matter, suspended solids, nitrogen and phosphorus from urban sewage discharges, calculated by Castanheiro (1983) using the method of specific coefficients and considering a population of over 2 million. An attempt to apply a similar method from industrial sources was not successful since the coeficients available in the literature and derived for more developed countries are not applicable to the situation in the portuguese industry in view of the different socio-economical conditions (Machado et al., 1984).

The most important feature of the liquid discharges into the Tagus

Estuary is the simultaneous input of suspended solids, organic matter
and contaminants. This will influence greatly the dispersion and bioavailability of pollutants such as heavy metals and organochlorines.

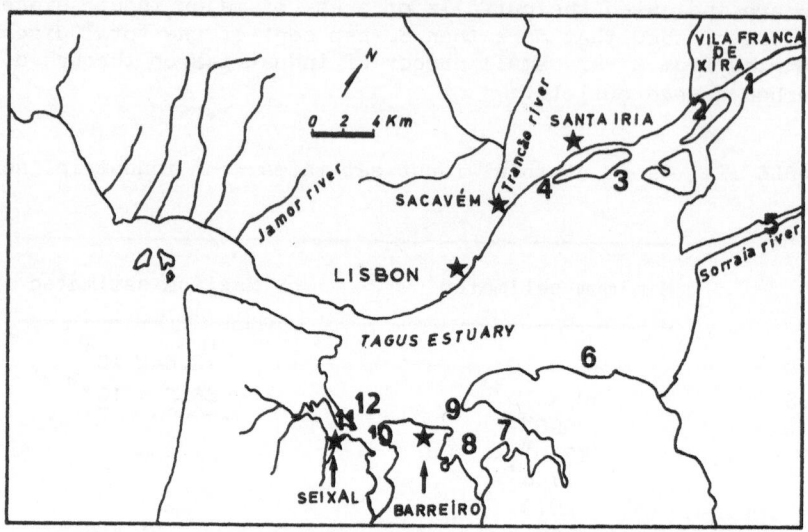

Figure 5 — Location of major input sources and of water sampling,
stations 79/80

TABLE III.   Estimated inputs from urban sewage discharges
             (tons per year).[*]

	South bank	North bank	Total
BOD_S	7 077.1	28 304.1	35 381.2
COD	15 801.6	63 199.0	79 000.6
TSS	7 182.4	28 672.7	35 855.1
TDS	12 796.0	53 233.5	66 029.5
N	1 156.8	4 732.9	5 888.7
P	140.3	573.3	713.6

[*] from Castanheiro (1983)

In addition to direct discharges of aqueous effluents, atmospheric
deposition may be an important source of some contaminants but its
relative importance cannot be assessed since very few studies on air
pollution levels have been carried out.

4. THE POLLUTION STATUS

Several studies have been conducted to determine concentrations of some pollutants in water, sediments and biota but the data are scattered over different years and very seldom the different compartments involved in the dynamics of the contaminants have been studied at the same time. So, only partial information is available and a comprehensive assessment of the pollution in the Estuary can not be made yet.

4.1. Nutrients and other water quality parameters

The extensive data collected through the synoptic sampling surveys shows that quality of the water in the Tagus Estuary though acceptable is influenced by the inputs of pollutants. The level of dissolved oxygen is high in most of the Estuary and though lower in areas close to the major discharges of urban sewage, the depletion is minimal because these are made into the area of maximum tidal flow. Nutrients seasonal cycles are within normal but levels are, at some periods, high without being excessive. The effect of urban sewage is detectable in particular in regard to ammonia concentrations.

An assessment of the nutrients behaviour in the Estuary concludes that silica and the oxidative forms of nitrogen are conservative, while phosphate concentration depends on the freshwater influx from the Tagus, on sewage inputs and on sediment water exchange. Ammonia concentrations, highly variable, are not influenced by the influx of freshwater and, are essentially dependent on effluent discharges (Martins et al.,1984).

4.2. Metals

Very few analysis have been performed on water samples. In Table 4 are presented the results of a survey, from several areas in the Estuary, which was conducted in 1979/80. The analysis were carried out on unfiltered samples taken at the locations indicated in the map of Figure 5.

The studies by Vale and co-workers (1982,1984) have shown that there is an enrichment of the suspended particles in Zn, Cu and Pb from the inner part to the inlet of the Estuary which the authors consider may be attributed to both anthropogenic inputs and natural processes, the first cause playing a decisive role. For Cd the more significant changes were observed between near-shore and mid-channel suspensions. The results of sediment samples (Vale and Mendes, 1984) show that the highest concentrations of Ca, Zn, Cu, Pb, Hg and Cd are found in the lower estuary being particularly high near the industrial outfalls, and decreasing with the distance from the inputs. The contents of Mg, Fe, Mn, Co, Ni and Cr did not show significant flutuations. Concentrations as high as $246x10^4$ Pb/Al, $149x10^4$ Zn/Al, $64x10^4$ Cu/Al were detected in the vicinity of the inputs (Barreiro multi-chemical complex and Rio Trancão) but high values were also found in the central channel of the Estuary. A plume of about 5 Km with very high trace metal concentrations, that originates at the Barreiro outlets is appearant.

TABLE 4. Results of water samples analysis 1979/80[*]

Stations[1]	BOD$_s$ (O_2 mg/l)	Hg (ug/l)	Pb (mg/l)	Cd (mg/l)	Zn (mg/l)
1	4	<0.2-2.6	<0.01-0.33	<0.01	<0.01-0.08
2	55	<0.2-2.0	<0.01-0.22	<0.01	<0.01-0.22
3	13	<0.2-1.0	<0.01-0.15	<0.01-0.01	<0.01-0.15
4	7	<0.2-1.3	<0.01-0.32	<0.01-0.4	<0.01-0.85
5	33	<0.2-1.7	<0.01-0.08	<0.01-0.09	<0.01-0.62
6	120	<0.2-6.0	0.01-0.33	<0.01-0.09	<0.01-0.36
7	24	<0.2-8.5	<0.01-0.33	<0.01-0.08	0.11-0.71
8	6	<0.2-8.5	0.19-0.64	<0.01-0.07	0.21-0.84
9	65	<0.2-15.2	<0.01-0.36	<0.01-0.07	0.19-0.96
10	119	<0.2-1.2	0.08-0.36	<0.01-0.05	0.04-0.21
11	115	<0.2-1.9	<0.01-0.34	<0.01-0.06	0.04-0.16
12	104	<0.2-18.0	<0.01-0.75	<0.01-0.06	-

(1) Location of stations in Figure 5.
* adapted from Janeiro, 1981.

The mean content in areas where lower concentrations are found is reported as 20×10^{-4} Zn/Al, $3-5 \times 10^{-4}$ Cu/Al, $3-12 \times 10^{-4}$ Pb/Al. The high contents in the sediments deposited in zones around the sewage outfalls indicate that sedimentation and/or precipitation of pollutants tends to occur in their vicinities. It is also noticeable that the metal content in sediments in the tidal part of the river agrees with the composition of river material discharged (Vale, 1981) but an upstream sediment movement due to the Estuarine circulation may also occur and particles from polluted areas may move upstream. This mechanism could explain the somewhat higher concentrations found in areas within the upper Estuary where no direct inputs are known (Vale and Mendes, 1984).

The Tagus Estuary is one of the most contaminated with mercury in the European coast (Figueres et al, 1985). These authors conclude that the dissolved mercury increases from 10 ng/l in the river to 80 ng/l in the lower estuary, and is almost totally organic in the river, inorganic in the middle estuary (the discharges are also in the inorganic form), and again organic in the lower part. Suspended matter is highly contaminated (median value 4.5 ug Hg g^{-1}) and in sediments the median concentrations is 40x the natural background level. Also for this metal the distribution in sediments reflects the point sources of emissions, concentrations over 6 mgKg^{-1} having been detected in these areas (Figueras et al, 1985; Vale and Mendes, 1984).

The contamination by As and Sd was studied by Andrea (1982) and evidence of particular sources of input was found. Total dissolved inorganic arsenic was detected in the range of 2.7-6.65 ug L^{-1} and total dissolved inorganic antimony was determined in the range of 0.102-0.669 ug L^{-1}. In the sediments the As concentrations tends to be high, an average of 139 mg Kg^{-1} was detected by Martin (personnal communication to Andrea, 1982) and the higher values have been found close to the input sources.

Very few data are available on metal contamination in biota. Concentrations of mercury in fish and crustacea do not reflect usually the proximity of the sampling area to the pollution sources and the body burdens in several species can be considered low (<0.3 mg Kg^{-1}f.w) with only a few exceptions in fish collected at the upper limit of the Estuary (Vila Franca de Xira) where no important sources of pollution have been identified but concentration in sediments is high (Lima et al, 1982). The concentration of heavy metals in mussel samples taken at the mouth of the Estuary and on the adjacent coastal area has been studied (Mendes and Vale, 1984) and the results show a pronounced seaward decrease of the content of Fe, Mn, Pb, Cd indicating that the Estuary output is probably the most important heavy metals source to the coastal area. Similar concentrations were reported by Stenner and Nickless, 1975. Geographical and seasonal variation was detected but, for all sampling periods, the concentrations reported are indicative of local low pollution levels (or low bioavailability) with the exception of Pb which concentrations are always higher than those reported from other areas. The Pb concentration was found in the range of 3->10 mg Kg^{-1}, with most of the samples from within the Estuary having concentrations in the upper limit of this range.

4.3. Organochlorine insecticides and polychlorinated biphenyls

As previously noted no input estimates were ever made for organo-chlorine compounds though use of insecticides such as DDT has been intense and formulation of pesticides is done in some plants located in the area. PCBs have been used as everywhere else in the world and contamination with these compounds has been studied in Portugal (Barros, 1980).

In the Tagus Estuary studies on the concentration of these pollutants have been carried out solely by the author of this paper and an overview of the results obtained is about to be published (Barros, 1984). A summary of the conclusions of that work is presented here.

In water (unfiltered samples) organochlorine residues (DDT and TDE) were detected in samples from the middle south Estuary, an area that receives the input from the Sorraia River and from the south bank where a pesticide plant is located. PCBs were never detected in water at the limit of 10 ng L^{-1}.

Analysis of sediments for organochlorine contaminants was carried out in samples taken for several years (1972-80) at different locations and the results show a rather low level of contamination with DDT (median <10 ng g^{-1} dry weight) or other insecticides with the exception of the channel draining the effluents from the multi-chemicals complex of Barreiro, in which pesticides have been and are formulated. In this channel aldrin (34 ng g^{-1} dry weight) was detected in addition to the DDT group, in samples taken at different periods. The contamination with PCBs is higher and more widespread, since these compounds were detectable in 90 per cent of the samples with a median of ca. 20 ng g^{-1} (dry weight). High residue values (100-780 ng g^{-1} d.w.) were registered in sediments from near shore areas within the influence of industrial outfalls in particular the shipyards of Seixal and the Trancão River. High residues (>100 ng g^{-1} d.w.) were also detected in the deeper part of the inlet channel.

Residues in mussels and oysters show important seasonal variations as is shown in Figure 6 for oysters and dieldrin, DDT and its metabolites as well as PCBs were detected in all samples taken at different years. Values in the range of 20-69 ng g^{-1} wet weight are reported for DDT. PCB residues detected were in the range of 33-95 ng g^{-1} wet weight. Dieldrin residues reached a maximum of 20 ng g^{-1} wet weight. Chlordane and lindane were detected in some samples.

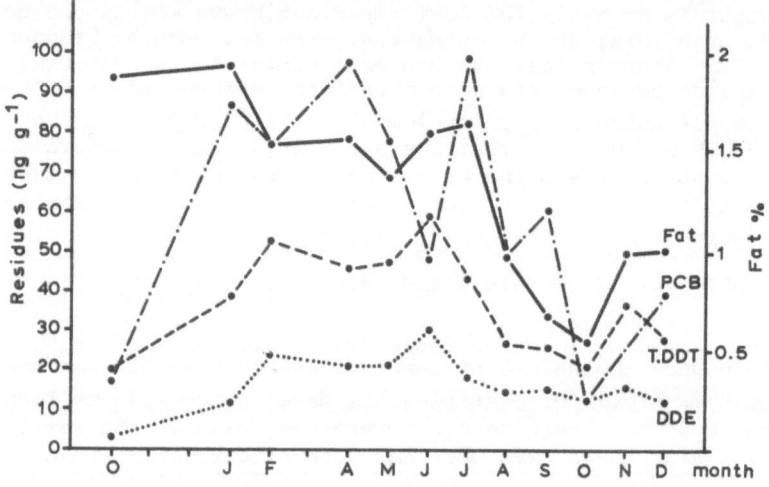

Figure 6 - Seasonal variation of T.DDT and PCB residues, in oysters - 1980/81

From the analysis of several fish samples of different species it can be concluded that the contamination with organochlorines is low since only 25 per cent of the samples showed muscle residues over 25 ng g^{-1} w.w. for T.DDT or over 70 ng g^{-1} w.w. for PCBs. Residues in fish liver are, as expected, higher (median T.DDT 150 ng g^{-1} and median PCBs 500 ng g^{-1}) and levels more consistent with the suspected contamination of the collection area for some species - the highest residues in Solea solea livers was found in the areas where higher sediment concentrations were reported. Residues of dieldrin, chlordane and lindane were detected in some of the samples (Barros, 1984). T.DDT levels detected in fish from the Estuary are similar to those found in fish from upstream in the Sorraia River but the PCB residues are consistently higher in fish caught in the Estuary (Barros, 1980).

The study of the frequency distribution of the concentrations in different compartments of the ecosystem (water, sediments, molluscs and fish) illustrates the bioacumulation trend for these compounds in the Tagus Estuary (Figure 7).

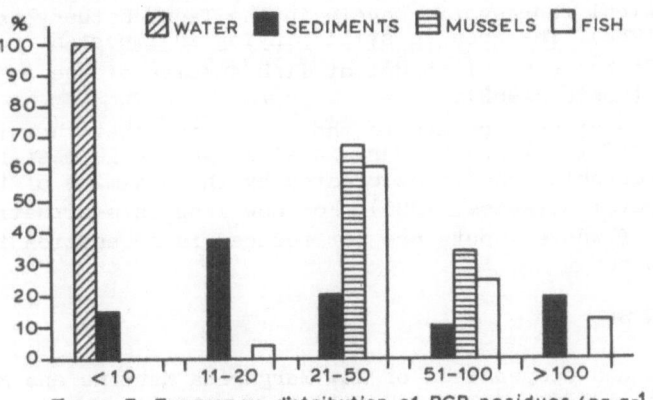

Figure 7—Frequency distribution of PCB residues (ng g⁻¹)

5. CONCLUSION

It seems to be well established that very large quantities of industrial and urban wastes are being drained into the Tagus Estuary. Sewage with high loads of suspended solids and organic matter, industrial effluents carrying in addition to the same pollutants toxic substances of all categories, are continuously entering the estuarine system.

From the studies so far conducted it can be concluded that high concentrations of heavy metals are present in sediments or in suspended particulate matter. More complete studies are available regarding Hg than other metals and the contamination with this metal in the abiotic environment attains levels of extreme concern, though concentrations in biota seem to indicate that most of the mercury distributed is not bioavailable. On the other hand it should be noted that no information is available on the Hg species accumulated in fish or crustacea. Much less data is available on other metals of particular concern and recent studies suggest that high levels may also be present in biota, in particular for Pb. Nutrients and dissolved oxygen or ammonium levels show the influence of the wastes inputs in some areas essentially those near shore, close to the most important sources. For some pollutants, other sinks within the middle Estuary are suspected, as indicated by the analysis of sediments.

Observations carried out during the last five years on several of the components of the complex system of the Tagus Estuary helped to establish a baseline for the study of the biological processes in the Tagus Estuary though these studies still need to be completed. Intertidal communities existing in different biotypes (muds, sands, oyster beds, salt marshes) as well as the subtidal communities are those of an estuary where pollution is not an acute problem but the number of species in some areas is suspected of having been reduced while some particular species are more developed than in a well

balanced environment (Ramos, 1985). Biological cycles of fish or invertebrate species that have been studied show essentially the same
pattern   as in known from undisturbed areas, the example of oysters
transplanted from the Sado Estuary, a less polluted body of water, that
showed a classical reproductive cycle in the Tagus Estuary is encouraging (Ramos, 1982). The area is still playing an important role as
nursery grounds for several important fish species of the adjacent
coastal areas (Costa, 1982).

The data available points to the conclusion that that the biogenic capacity of the Tagus Estuary is still very high most probably
due to the favourable conditions created by the dynamics of the tidal
regime that prevails (Ramos, 1985). For how long this situation will
be maintained if waste inputs are not reduced is a question that
remains unanswered.

ACKNOWLEDGEMENTS

The assistance and cooperation of Ms. Margarida Martins and Ms.
Laudemira Ramos is gratefully acknowledged.

BIBLIOGRAPHY

Almeida, M. (1979). Análise estatística de amostras de sedimentos CNA/
   Tejo nº 2, Lisbon.
Andrea, M.O. (1982). Arsenic and mercury in the estuarine environment
   CNA/UNESCO/IOC Scientific Workshop on Estuarine Processes: an
   application to the Tagus Estuary. Lisbon.
Barros, M.C. (1980). Bifenilos policlorados. CNA. Lisbon.
Barros, M.C. (1980). Residues of organochlorine insecticides and PCBs
   in fish from Lezíria do Ribatejo. DGPPA, PPA(TC). 6/80.Lisbon.
Barros, M.C. (1984). An overview on organochlorine insecticide and
   PCB residues in the Tagus Estuary (to be published).
Calvario, J. (1982). Povoamentos bentónicos intertidais (substratos
   móveis) CNA/Tejo, nº 19. Lisbon.
Carrondo, J.J.T., Reboredo, F., Ganho, R.M.B. & Oliveira, J.F.S. (1984).
   Analysis of sediments for heavy metals by a rapid electrothermal
   atomic absorption procedure. Talanta 31, 561-564.
Castanheiro, J. (1983). Cargas poluentes de origem doméstica afluentes
   ao Estuário do Tejo (estimativa indirecta). CNA/Tejo nº 27. Lisbon.
CNA (1977). Environmental study of the Tagus Estuary. Water study.
   First progress report. Lisbon.
Costa, M.J. (1982). The Tejo Estuary as a nursery. In CNA/UNESCO/IOC
   Scientific Workshop on Estuarine Processes: an application to the
   Tagus Estuary. Lisbon.
Costa, E. & Camara, C. (1984). Application of BOD model to forecast
   D.O. distribution in the Tejo Estuary. In CNA/UNESCO/IOC Scientific
   Workshop on Estuarine Processes: an application to the Tagus Estuary.
   Lisbon.
Costa, A.A.M., Gaspar, M.N. & Pinelas, R.M. (1982). Determinação de
   cargas industriais afluentes ao Estuário do Tejo. In Symposium on
   "Bacia Hidrografica do Rio Tejo: perspectivas para o seu desenvolvi-

mento e para a gestão dos seus recursos hídricos". APRH. Lisbon.

Costa, A.A.M., Pinelas, R.M. & Gaspar, V. (1983) Determinação das
cargas poluidoras industriais as fontes poluidoras mais importantes
do Estuário do Tejo. M.O.P. Lisbon.

D.G.R.A.H. (1978). Poluição do Estuário do Tejo. Elementos para o seu
estudo. Estudos de Poluição de cursos de água nº 17. M.H.O.P. Lisbon.

Drena, Hidroprojecto, Epal (1979). Região de Saneamento Básico de
Lisboa. II - Estudos de Base de Engenharia. Ed. do A. Lisbon.

Elias, N.P. (1982). Application of the physical model of the Tejo
Estuary. In CNA/UNESCO/UNEP/IOC Scientific Workshop on Estuarine
Processes: an application to the Tagus Estuary. Lisbon.

Ferreira, J.P.T., Silva, J.A.G. & Rebelo, M.A.S. (1982). Ciclo bioló-
gico da Scrobicularia plana (lambejinha, molusco bivalve). Relatório
preliminar CNA/Tejo nº 13. Lisbon.

Figuères, G., Martin, J.M., Meybeck, M. & Seyer, P. (1985). A comparative study of mercury contamination in the Tagus Estuary (Portugal)
and in major french estuaries (Gironde, Loire, Rhône). Estuarine,
coastal and shelf science (in press).

Janeiro, A.F.F. (1981). Avaliação da carga poluidora afluente ao
Estuário. D.S.C.P./M.H.O.P. Lisbon.

Lima, C., Martin, J.M., Meybeck, M., Figuères, G. & Seyler, P. (1982).
Mercury in the Tejo Estuary: an acute or absolete problem. In CNA/
UNESCO/IOC Scientific Workshop on Estuarine Processes: an application
to the Tagus Estuary. Lisbon.

Loureiro, J.M. (1979). Curvas de duração dos caudais médios diários no
Rio Tejo. D.G.R.A.H. Lisbon.

Machado, V.H.A.C. & Amaro, A.P.N. (1984). Análise do método dos coefi-
cientes específicos para avaliação da carga poluente de origem industrial para meios hídricos - o Estuário do Tejo. In I Symposium Luso-
Brasileiro de Engenharia Sanitária e Ambiental. Lisbon.

Marques, J.C. & Costa, I. (1983). Crustaceos decapodes do Estuário do
Tejo: distribuição das espécies e estudo da biologia das populações
de Carcinus maenas (Decapodas, Brachyura) Crangon crangon, Palaemon
longirostris e Palaemon serratus (Decapoda, Caridea). CNA/Tejo nº 26.
Lisbon.

Martin, J.M., Figueres, G. & Meybeck, M. (1982). Le mercure et
l'arsenic dans l'Estuaire du Tage. CNA/Tejo nº 5. Lisbon.

Martins, M.C.S., Ferreira, J.G., Calvão, T. & Figueired, H. (1984).
Nutrientes no Estuário do Tejo. Comparação da situação em caudais
médios e em cheia, com destaque para alterações da qualidade da água.
In "I Symposium Luso-Brasileiro de Engenharia Sanitária e Ambiental".

Mendes, R. & Vale, C. (1984). Geographic variation of heavy metal
contents in mussels (Mytilus galloprovincialis) from the coastal
area adjacent to the Tagus Estuary, Portugal. ICES C.M.1984/E:38.

Oliveira, R. (1967). Contribuição para o estudo do Estuário do Tejo
(Sedimentologia) Memo 296. L.N.E.C. Lisbon.

Quintela, A. (1967). Recursos de aguas superficiais em Portugal continental. Ed. Fundação Gulbenkian. Lisbon.

Ramos, L. (1982). Growth, survival and reproduction of oysters in the
Tejo Estuary. In CNA/UNESCO/IOC Scientific Workshop on Estuarine
Processes: an application to the Tagus Estuary. Lisbon.

Ramos, L. (1985). Personnal communication.

Ré, P. (1983). Ictioplâncton no Estuário do Tejo. Resultados de 4 anos de estudo (1978-1982). CNA/Tejo nº 24. Lisbon.

Rodrigues, A. & Moita, T. (1979). Estudo quantitativo do fitoplâncton do Estuário do Tejo (1ª e 2ª Campanhas). CNA/Tejo nº 3. Lisbon.

Saldanha, L. (1980). Povoamentos bentónicos, peixes e ictioplâncton do Estuário do Tejo. CNA/Tejo nº 5. Lisbon.

Santo, T.R. & Martins, M. (1983). Contribution for water resources planning in the Tejo Estuary. In JNICT/NAS/USAID Workshop on Water Resources Planning Process, Techniques and Implementation. Ericeira.

Sobral, P. (1982). Zooplancton do Estuário do Tejo (Resultados relativos ao ano de 1980). CNA/Tejo nº 16. Lisbon.

Stenner, R.D. & Nickless, G. (1975). Heavy metals in organisms of the Atlantic coast of S.W. Spain and Portugal. Mar. Pollut. Bull. 6, 89-92.

Vale, C. (1981). Input of suspended particulate matter in the Tagus Estuary during the flood of February 1979. Recursos Hídricos, 2.

Vale, C. (1982). Geographical and tidal variations in the chemical composition of suspended particulate matter in the Tagus Estuary. In CNA/UNESCO/UNEP/IOC Scientific Workshop on Estuarine Processes: an application to the Tagus Estuary. Lisbon.

Vale, C. (1984). Transport of particulate metals at different fluvial and tidal energies in the Tagus Estuary. In "ICES Symposium on Contaminant Fluxes through the coastal zone", Nantes. 1984.

Vale, C. & Mendes, R. (1983). An approach to the study of the chemical composition of bottom sediments in the Tagus Estuary. ICES C.M.1983/E:32.

Vale, C. & Sundby, B. (1982). The relationship between the turbidity maximum and the tidal amplitude in the Tagus Estuary. In CNA/UNESCO/UNEP/IOC Scientific Workshop on Estuarine Processes: an application to the Tagus Estuary. Lisbon.

Vilela, H. (1975). A respeito de ostras. Biologia, exploração, salubridade. S.E.P. Lisbon.

THE BALTIC SEA: CONDITIONS AND OPTIONS OF MANAGEMENT

G. Kullenberg
University of Copenhagen
Department of Physical Oceanography
Haraldsgade 6
DK-2200 Copenhagen N

INTRODUCTION

The Baltic Sea (Viopio 1981) is a semi-enclosed area which has been subject to considerable human use and influence. There are several examples of waste inputs which have had harmful effects, and which have been restricted or banned on both a national and an international basis. There are other examples of inputs where some restrictions have been made, which, however, have not had the expected or desired effect, and where other options therefore must be sought. There are many other examples where the chosen option has implied replacement of a harmful substance with another substance, which in some cases has also turned out to be harmful. Finally, there are examples where the option of reducing the total inputs is simply not feasible, and where alternative options must be found. One such option could be ecosystem manipulation, e.g. by mariculture.

It is the intention here to discuss these problems in the general framework of the Baltic Sea oceanographic and ecosystem conditions.

The GESAMP definition or marine pollution reads:

"Introduction by man, directly or indirectly, of substances or energy into the marine environment (including estuaries) resulting in such deleterious effects as harm to living resources, hazards to human health, hindrance to marine activities including fishing, impairing of quality for use of sea water, and reduction of amenities."

Implicit in this definition is the understanding that:

(i)   marine pollution is caused by the introduction into the marine environment of substances and energy which have adverse effects;

(ii) marine pollution can be related to its sources, it is created by man and can, in some instances, result

G. Kullenberg (ed.), The Role of the Oceans as a Waste Disposal Option, 325-345.

in the increased flux of substances in existing
natural cycles;

(iii) polluting substances are dispersed through the
marine environment by various processes,
whereupon they affect organisms including man,
particularly as a user of the ocean systems;

(iv) the significance of the pollution depends upon
its effects on different targets, and social
values are often involved;

(v)   the quantification of the effect and hazard of
pollution is a question that must usually be
answered before a judgment can be made as to
whether or not pollution is acceptable.

The definition and its implications give guidance on how to answer
the question: What do we need to know to be able to evaluate the
conditions as regards pollution in the oceans?
Information must clearly be available on:

(i)   sources of substances and energy, their existing
and predicted quantities, and their distribution
in the environment;

(ii) the processes leading to dispersion in the marine
environment, where substances from a particular
source will go, and what targets may be affected;

(iii) effects of pollution on various targets and the
significance of these effects.

A central problem is the distinction between environmental
variations caused by pollution effects, and natural environmental
variability. In this respect the Baltic Sea area has several important
natural characteristics: large positive water balance; brackish water
with large and stable horizontal and vertical salinity gradients; strong
physical constraints on the water exchange and mixing, imposing large
residence times (several decades); and special geochemical systems
caused by oscillations between oxic and anoxic conditions in bottom and
deep waters. Human conflicting uses of the sea area include: waste
disposal, sea transportation, fishing, aquaculture, and recreation.
Large inputs of nutrients, organic material, and metals occur from land
and from the atmosphere.
     In the Scandinavian countries, populations and industrial
activities are concentrated along the coasts of the Baltic Sea, with
many small to medium size harbors along the coast.
     Along the eastern and southern borders of the sea, few but large
population centers and harbors occur, but with generally less population
concentration and industrial developments along the coasts (Bruneau 1980).

However, large rivers enter the Baltic from a highly agricultural
drainage area along the southern coast. During the last 3 decades
marked agricultural developments have occurred in the northern Baltic
countries. Today an effective fertilization is applied so that the
production potential is well utilized.

The Baltic Sea ecosystem is influenced by a large input of nutrients
and other substances from a large drainage area as well as inputs from
the atmosphere. The whole system is coupled through interactions and
response times on various scales.

OCEANOGRAPHIC CONDITIONS

The semi-enclosed Baltic Sea (370000 km^2; 21000 km^3) 54^0N to 66^0N,
an intracontinental mediterranean sea, is one of the largest brackish
water bodies in the world. It is very shallow with a mean depth of
57 m and about 17% of the area is shallower than 10 m (e.g. Kullenberg
1983). The Baltic Sea depression essentially constitutes a long fjord
in the north-south direction ($\sim$ 1500 km) with an average width of 230 km.
The topography divides the sea into a series of relatively deep basins,
with maximum depths in the range of 105 m to 459 m. These are separated
by sills, or shallows, of depths in the range of 25 m to 140 m. The
sea is connected to the North Sea and the open Atlantic via the
Skagerrak and a narrow and shallow transition area, namely the Kattegat,
the Belt Sea of sill depth 17 to 18 m, and the Øresund (Sound) of sill
depth 7 to 8 m. These topographic characteristics are of great
importance in understanding the conditions in the Baltic Sea, i.e. a
lateral exchange of deep and bottom waters can only be accomplished by
water which has come from the North Sea, a distance of about 1000 km,
and has passed over several shallow sills.

The second essential feature of the Baltic Sea is its marked
positive freshwater balance, with an annual river runoff of 440 to
480 km^3 or about 2.2% of the total volume (Ehlin 1981). The runoff
usually has a maximum in May and a minimum in January or February. It
shows large long-term variations, with amplitudes of 10 to 20% of the
long-term mean (100 years), primarily due to climatic variability. The
runoff generates an outgoing surface layer flow of low salinity water
which sets up an inward compensating flow at intermediate depths. A
marked permanent salinity stratification results in a transition layer
at 65 to 75 m.

A third factor of great importance is the influence of the meteorological conditions on the exchange of the deep and bottom waters in the
Baltic deep basins. Favorable inflow situations occur with persisting
westerly winds, high pressure over Jutland, and low pressure over
Scandinavia. Often particularly marked inflows occur at semi-periodic
intervals of 3 to 5 years, which can be related to specific, persisting
meteorological conditions over the Atlantic (Dickson 1971). High saline
water in the Baltic deep basins will remain there until forced away by
an inflow of denser water. The bottom water salinity can only decrease
by vertical mixing which in the permanently stratified Baltic Sea is a
very slow process. A decrease of about 1^0/00 takes about 1 year (e.g.

Kullenberg 1982).

Considerable long-term fluctuations of salinity and temperature
occur in the deep and bottom waters. Trends have also been established
with increases of temperature and salinity since the 1880s from 0.6
to 2.7 C and 0.8 to 1.7 p.p.t., respectively (Melvasalu et al. 1981;
Kullenberg 1981; Matthäus 1980 and 1983a). The fluctuations (periods of
several years) of the salinity are correlated with fluctuations in the
river runoff, at least in the 0 - 100 m part of the water column. The
trend of increasing salinity may be related to a long-term decrease of
the runoff of about 15% seen in observations from 1900 - 1960. However,
over the period 1880 to 1980, no significant decrease of the runoff can
be seen. The increase of the salinity can be caused by other factors,
such as an increase of the salinity of the inflowing water, a gradual
increase of the sill-depths in the Danish Sounds due to the subsidence
(Striggow 1983), and a change in the water exchange through the Sounds
(Nehring 1979). The latter may be due to a change in the meteorological
forcing of the water exchange.

The meteorological forcing shows large long-term fluctuations
(Kullenberg 1977), but no significant trend can be seen over the period
1900 to 1960 from direct wind observations. Pedersen (1982) found a
decreasing trend of the monthly peak-to-peak barometric pressure at
Bornholm and over the period 1900 to 1970, and a 10% decrease of the
average wind strength may have occurred since the end of the last
century. The observations clearly suggest the importance of the river
runoff, the wind and the topographic constraints for the oceanographic
conditions in the Baltic Sea. The wind provides the energy input for
vertical mixing together with the cooling in the fall-winter period,
and the energy for the lateral water exchange.

The depth of the primary halocline fluctuates markedly in the range
of 20 - 30 m. However, the mean depth has changed only slightly, by
5 - 6 m from 77 - 71 m, during the present century (Matthäus 1980).

The density of the water also displays long-term trends of changes
with a general increase during the present century (1900 to 1980) of
about 0.5 - 1 σ_t-unit (Matthäus 1983b). However, over shorter time
periods other trends can occur in different parts of the water column.
Matthäus (1983b) found a decrease of 0.4 - 0.7 σ_t-units at the 150 -
200 m level for the period 1952 to 1980, and an increase of 0.3 -
0.7 σ_t-units in the surface water during the same period. Large
fluctuations of the stability over the primary halocline layer occur
(Fonselius 1969; Kullenberg 1977; Matthäus 1983c). However, no
significant long- term trend can be seen (Kullenberg 1981; Matthäus
1983c). In the deeper waters a decreasing trend is obvious during the
present century of about $0.5 \cdot 10^{-5} m^{-1}$ (Matthäus 1983c).

OXYGEN CONDITIONS

The long residence time, years to decades, of the deep and bottom
waters in the Baltic Sea implies a gradual oxygen depletion due to
oxidation of sinking organic matter and organic matter in the sediments.
Anoxic conditions often occur in large parts of the bottom and deep

waters (Fonselius 1981). During this century there has been an increase
of the volume of anoxic water, although interrupted by aperiodic
decreases (Fonselius 1969, 1981; Jansson 1978). A temporary maximum
occurred in the mid-1970s. The stagnant conditions usually last 3 - 5
years, when the bottom waters are exchanged, a process which itself
takes about 9 months, operating successively in a cascade from the
Bornholm Basin to the inner parts. Occasionally, after particularly
strong or high saline inflows, the stagnant period between inflows can
extend for 6 years, which was the case for the last stagnation period.
The length of these periods depends upon the density (salinity) of the
water. Especially high salinities are found in the European shelf seas
at intervals of 3 - 5 years (Dickson 1971).

A key question in relation to Baltic Sea management is why there
is a trend of increasing oxygen depletion. The trend can be related
to several factors:

(i) Climatic variations, which influence the water balance
     exchange, for instance giving rise to an increased
     residence time of the bottom waters. A decrease of
     wind forcing, which influences both the mixing and
     the water exchange, is suggested from air pressure
     records since 1880; the decreasing trend is especially
     marked in the period 1900 to 1950 (Pedersen 1982).

The depth of the halocline and the stability of the stratification
across the halocline, both factors of importance for the mixing across
the halocline, have not changed significantly. The mean circulation in
the Baltic Sea is weak. It is conceivable that, given a decreasing
energy input from the wind, the fluctuating circulation has decreased,
which may imply increased residence times at intermediate levels (60 -
120 m).

(ii) An increased rate of oxygen consumption, in bottom and
      deep water, which can be due to the increased temperature
      (Kullenberg 1970), to an increased amount of organic
      material in the water column (Shaffer 1979) and an
      increased input of organic matter to the bottom due
      to an increased primary production (Jansson 1978, 1984).
      This may lead to an accumulation of organic material
      in the sediments.

(iii) An increased input of organic matter from land-based
       sources and rivers or from an increase of the primary
       production. Observations of primary production in
       some parts of the Baltic Sea since about 1960 suggest
       that an increase of the primary productions is occurring,
       especially in the southwestern areas, amounting to
       about 50% over the last 20 years (Kullenberg 1983).
       The "average" primary production is about
       100 g C m^{-2} yr^{-1}. An increase in the primary production
       by a factor of 1.5 - 2 has also occurred in parts of

the Kattegat (Anon. 1983a).

(iv)   A change in the ecosystem balance governing the uptake
        nutrients and distribution of organic matter. Observations
        suggest that the occurrence of blue-green algae capable
        of uptake of atmospheric nitrogen has increased over
        the last 1 - 2 decades (Jansson 1978, 1984). In the
        last years, several abnormal plankton blooms have occurred,
        but it is not possible as yet to conclude that the
        structure (length, timing) of the normal plankton blooms
        has changed, although there are indications of this.
        In the Baltic Sea there normally occurs a spring bloom
        in April or May, and a second bloom in July to August,
        whereas in The Gulf of Bothnia, the spring and fall
        blooms have merged (Jansson 1979).

ASPECTS OF THE ECOSYSTEM

    The organisms in the Baltic Sea are adjusted to low salinities.
There are fewer species, they are less specialized and smaller than in
fully marine areas (Jansson 1978; Kullenberg 1983). Temperature governs
the seasonal and vertical variation of the zooplankton fauna, and
salinity affects the species composition. The ecosystem is subject to
large natural and man-made variations, but it is very difficult to
ascertain the impacts of the latter on the system. It appears that
there has occurred a change from oligotrophic to eutrophic conditions in
coastal zones and possibly in parts of the open sea (Dybern and
Fonselius 1981; Jansson 1984). The benthos has shown considerable
changes over the last 50 years. There are indications that oligotrophic
fish such as pike and perch are being replaced by the eutrophic bream
and roach (Jansson 1978). In some coastal areas the fucus has decreased
or disappeared whereas green and brown algae (Cladophora and
Enteromorpha) have increased.
    The increased primary production has led to an increasing sediment-
ation and accumulation, slowly changing hard bottoms to soft bottoms
(Jansson 1984). Large-scale, long-term changes in the benthic soft
bottom macrofauna have been observed in deep parts of the Baltic
(Bornholm - Gdansk - Central Basins and the Gulf of Finland), amounting
to a decline or disappearance of macrofauna or a change of the fauna
composition (Andersin et al. 1978). In the southern parts, the changes
started in the early 1950s, and in the late 1950s in the northern parts
(Andersin et al. 1978). The changes were partly triggered by the very
large salt water inflow in 1951 - 52, described by Wyrtki (1954).
    Cedervall and Elmgren (1980) presented results from benthic
macrofauna biomass investigations in the late 1970s repeated at the same
stations as had been occupied in the 1920s. Above the halocline layer
a large increase in biomass was established whereas in areas below the
halocline depth ($\sim$ 70 m) a very strong decrease in biomass was
established. The average annual increase above the halocline was 2 - 4%
leading to a higher food supply to the benthos than before. The

long-term changes in natural oceanographic conditions (S, T, stability, etc.) have not been large enough to explain the drastic changes in various parts of the ecosystem.

The fish catch has increased steadily since the 1930s; herring, sprat, and cod being the most important species. During the mid-1940s the cod stocks increased. A rapid increase of sprat occurred in the 1950s and continued up to around the 1970s, after which it has decreased. The herring has increased evenly since 1960. The cod stock has shown a strong increase since 1977 and is now stronger than ever (Otterlind 1983; Jansson 1984). The fish catch has doubled during the last 20 years and is now about 0.9 million tons annually. It is possible that an increase in herring and sprat stocks also started before the increase of anthropogenic inputs during the 1950s and 1960s, but it seems likely that the fish stocks have benefitted from the increased input of nutrients (Otterlind 1983).

During recent years frequent observations have been made of various types of fish diseases in many parts of the Baltic Sea. It is not possible to demonstrate to what extent this is due to contamination, except in some localized cases where industrial inputs probably have had an influence.

INPUTS OF SUBSTANCES

The level of phosphate in the surface layer winter water has increased by a factor of three over the last 2 - 3 decades (Nehring 1974, 1981; Fonselius 1980). The increase is also obvious in deep water. Around 1969, the rate of increase phosphate levels showed a marked increase, and since that year an increase of the nitrate concentrations has also been established. It is interesting to note that this increase of nitrate levels in the Baltic surface winter water is more or less mirrored by the increase of nitrate in Danish groundwater (Fig. 1). The ratio N:P has decreased slightly (Nehring 1984), which may imply that nitrogen is becoming increasingly important as the most limiting nutrient for primary production. An important question in respect to environmental management concerns the source of the nutrients. The N and P operate in different chemical cycles. In the case of nitrogen there is the controlling factor of denitrification in oxygen depleted waters (Gundersen 1981), which has been shown to be of great importance in the Baltic (Rönner and Shaffer 1983). During anoxic conditions the phosphate which has been trapped in the sediments during oxic conditions becomes released to the water column. This is obvious from the observations. When the deep and bottom waters are renewed some phosphate is transferred towards the intermediate and surface layers; estimates of 7000 tons during a turnover have been given (Holm 1978). It is clear that considerable amounts of phosphate can reach the surface layer in this way; a few large inflows from the Kattegat may generate the transfer required to explain the increase of phosphate levels in the surface winter water since 1968.

The rivers and the land runoff are very important sources of phosphate and nitrate. Larsson et al. (1984) estimated annual river inputs of $50 \cdot 10^3$ and $640 \cdot 10^3$ tons of phosphorus and nitrogen, respectively.

The total atmospheric input of nitrogen (sum of deposition and nitrogen fixation) was estimated at $456 \cdot 10^3$ tons a^{-1}, while the river input was $610 \cdot 10^3$ tons a^{-1} and the sum of industry and urban inputs of nitrogen was about $80 \cdot 10^3$ tons a^{-1}. Dybern and Fonselius (1981) gave values of $34 \cdot 10^3$ tons a^{-1} of phosphorus and $77 \cdot 10^3$ tons a^{-1} of sewage nitrogen, while Pawlak (1980) estimated a total land-based input of $309 \cdot 10^3$ tons a^{-1} of nitrogen and $26 \cdot 10^3$ tons a^{-1} of phosphorus. A summary of N and P inputs is given in Table 1.

The recent estimates suggest an eight-fold increase of phosphorus and a four-fold increase of nitrogen inputs from land and atmosphere during this century (Larsson et al. 1984). It should be noted that the land-based inputs are very unevenly distributed. About 75% of the nitrogen input occurs at the southeastern shores, and about 11% occurs along the Gulf of Bothnia shores. The input of nitrogen to the Gulf of Finland shows an increasing trend over the period 1975 to 1981, amounting to about $40 \cdot 10^3$ tons, or an increase of about 70%.

The atmospheric input is an important factor, amounting to about half the river input of nitrogen, for which positive trends have also been suggested (Anon. 1983; Larsson et al. 1984). The nitrogen fixation blue-green algae occurring during July-August also yield a considerable input, estimated at $(1 - 1.5) \cdot 10^3$ tons annually (Lindahl 1977; Rinne et al. 1977; Gundersen 1981; Larsson et al. 1984). Pawlak (1980) estimated a total land-based input of BOD_7 of $1.4 \cdot 10^6$ tons a^{-1} whereas Dybern (1974) gave an input of $410 \cdot 10^3$ tons a^{-1} or organic matter expressed as BOD_7 from sewage, agriculture etc., and $780 \cdot 10^3$ tons a^{-1} of BOD_7 from industry. The largest amount of the latter, or $530 \cdot 10^3$ tons, entered the Gulf of Bothnia and came from forestry-based industries. For the Gulf of Bothnia, Sweden and Finland (Anon. 1983b) estimated a total input of $378 \cdot 10^3$ tons and 475 tons of BOD_7 for the years 1980 and 1981, respectively. These numbers suggest a decrease of about 20% over the decade. At the same time they show the annual variability of the inputs and the difficulty of establishing trends, at least over short time periods. The same source gives an annual land-based input of $80 \cdot 10^3$ tons and $123 \cdot 10^3$ tons of nitrogen for the years 1980 and 1981, respectively, compared to the $80 \cdot 10^3$ tons given by Larsson et al. (1984).

Besides the inputs of nutrients and oxidizable organic material, large amounts of anthropogenic inputs of other materials occur. In the sediments an increase of Hg, with more than 200%, of Cd with more than 100% can be seen in many parts of the Baltic since about the time of the start of industrialization, 100 to 200 years ago, and increases have also been established for Pb, Zn, and Cu (Brügmann 1981).

Estimates of the total input of metals vary considerably due to large uncertainties in determinations of atmospheric and river inputs and lack of reliable data. Table 2 gives examples of integrated inputs based on various sources. The variability is clearly brought out.

Regional studies of inputs have been made over a number of years through bilateral cooperations between countries. An example is the Gulf of Bothnia, for which area a summary of land-based inputs for some

metals is given in Table 3. Only in the case of As is there a clear
trend of decrease. Trends are, however, also indicated for Hg and Pb
(Fig. 2).

Other harmful substances well-known from the Baltic Sea area
include most organochlorine compunds. Pawlak (1980) as well as Dybern
and Fonselius (1981) concluded that the information on inputs of these
compounds was so limited that no reliable estimate of the total load
could be made. Estimates of total inputs of PCBs of about 10 - 15
tons a^{-1} have been given (Roots 1981; Andrulewicz and Trzosinska 1985).
The latter authors attempted to estimate the total content in various
compartments (water, suspended matter, sediments, plankton, zoobenthos,
fish), finding about 30 tons of Σ DDT and 190 tons of PCBs. These
estimates are of course rather uncertain.

The total annual input from the atmosphere was estimated by Rodhe
et al. (1980) to 6 (2 - 25) tons of DDT and 8 (2 - 35) tons of PCB, with
the range given in parenthesis.

CONCENTRATION LEVELS AND BIOLOGICAL EFFECTS

Concentration levels of several potentially harmful substances
have been observed in Baltic Sea biota over a nubmer of years. In some
cases, the data have been obtained by intercalibrated common methods and
over large parts of the area, whereas in other cases, data from single
laboratories may be considered in a time-series perspective.

There are indications that the Σ DDT levels are decreasing and that
this may also be the case for PCBs. In the mid-1970s the concentrations
of Σ DDT and PCB in fat from the blubber of ringed seas were about
100 mg/kg and 85 mg/kg, respectively, while in the period 1980 to 1983,
the levels were 54 mg/kg and 76 mg/kg, respectively (Helle and Stenman
1984; Leppäkoski 1980). Dybern and Fonselius (1981) give values in the
range of 100 - 200 mg/kg for DDT and 70 - 100 mg/kg for PCB from the
early to mid-1970s. The average ratio of DDT/PCB in seals has decreased
from about 1.2 in 1973 - 1975 to about 0.8 in 1980 - 1983 (Helle and
Stenman; Leppäkoski 1980). Dybern and Fonselius (1981) present data
from several investigations, giving values of Σ DDT and PCB in Baltic
cod in the range of 0.11 - 0.06 mg/kg and 0.12 - 0.04 mg/kg,
respectively. In herring, the concentrations were about a factor of
10 larger.

Examples of concentration ranges of some metals in fish are given
in Table 4. These were obtained through the ICES Coordinated Baseline
Study in 1976/1977, so the range of values covers most of the Baltic
Sea. A comparable material will become available in 1985/1986 through
the ICES 1985 Baseline Survey.

Effects of contamination from PCB, DDT, and Hg were observed in
the 1960s and 1970s in Baltic seals and birds, primarily on reproduction.
In the 1970s and 1980s, fish diseases have been increasingly observed,
but it has not been possible to relate these to contamination sources
except in localized cases. Fish (herring) appear to be able to avoid
certain types of contaminated water, e.g. by titanium dioxide waste
(Anon. 1983c). Great concern has, however, been expressed about the

large inputs of various contaminants, and efforts have been made to
reduce the inputs. It is clear that a reduction was required, and in
some cases still is.

During recent years great concern has also been expressed about
eutrophication of not only coastal zones, but also the open Baltic Sea.
Indications of eutrophication and increasing production of organic
matter include: the increase in benthic biomass above the halocline
and the sharp decrease of disappearance below the halocline (Cedervall
and Elmgren 1980); the increase of nutrients in the winter surface layer
(Nehring 1981, 1984); an increasing trend in the western parts of the
Baltic Sea, but not in the open Gotland Basin; an increased oxygen
depletion in the intermediate deep water (Shaffer 1979); an increase
of blue-green algae; a tendency towards a change in fish population
(Jansson 1978). These signs have given rise to great concern about the
very large, increasing inputs of nutrients and organic matter.

EVALUATION AND OPTIONS OF ACTION

We can now return to the definition of marine pollution and its
implications. Information, although limited, is available on amounts
and distributions of inputs, on distributions of contaminants in the
environment, and on effects. We know the basic natural conditions of
the Baltic Sea, namely: semi-enclosed with narrow and shallow connec-
tions with the open sea; large, positive water balance; low salinity and
permanent stratification at depths 2 - 4 times below the entrance sill
depths; essentially wind-driven circulation and mixing, not tides; and
an oligotrphic brackish water ecosystem changing to a eutrophic one.

Inputs of substances of concern include nutrients, organic matter,
metals, and organochlorines originating from municipalities, industry,
agriculture, horticulture, rivers, and the atmosphere. Diffuse sources
and long-range transports of some materials are parts of the overall
problem. Certain biological effects have been demonstrated and there
are signs of disturbances in the system over regional scales. At the
same time, the natural fluctuations are considerable. These occur over
a series of time-scales, the most important ones probably being over
several years to many decades. In some cases, trends of changes are
seen, like the increase of salinity and temperature, although over
longer time-scales these changes are presumably fluctuations. Trends
are seen in important fisheries, which are partly due to fishing efforts
and partly due to natural variations.

The circulation of substances within the system and between the
different compartments of the system is known only to a very limited
extent. Brügmann (1984/85) formulated a mass balance for selected
metals, bringing out the role of the sediments and the recirculation
within the water column through biochemical processes. Andrulewicz and
Trzosinska (1985) made a similar attempt on DDT and PCBs, again
emphasizing the role of the sediments and bringing out uncertainties of
the order of 10 or more. The circulation of nutrients and organic
material has been studied more (Fonselius 1969; Jansson 1978; Viopio
1981; Melvasalo et at. 1981; Kullenberg 1982).

The question, then, is how to cope with the human influence on the system so as to avoid unacceptable and irreversible deteriorations. International cooperation is clearly needed so that legal coordination of actions can be achieved. This is catered for through the Helsinki Convention on the Baltic Marine Environment Protection, signed by all Baltic Sea states in March 1974, and which entered into force on May 3, 1980. The governing council of the Convention is the Helsinki Commission with the secretariat located in Helsinki. Scientific and technological cooperation and advice is furnished through expert groups under the Convention as well as through interaction with international organizations such as the International Council for the Exploration of the Sea (ICES) and the Baltic Marine Biologists.

Under the Convention, agreement is made to counteract the introduction of hazardous substances, such as DDT and its derivatives DDE and DDD, PCBs, and to take effective steps to control and minimize inputs from land-based sources, especially substances like Hg, Cd, Cu, Cr, Pb, An, chlorinated hydrocarbons, phenols, cyanides, biocides, radioactive substances, acids, oil, and oily wastes. The Convention also covers pollution from ships, prohibiting any discharge of oil or oily mixtures, and dumping at sea, which is prohibited with the exception of dredge spoils provided they do not contain significant amounts or concentrations of hazardous or noxious substances. The Contracting parties shall harmonize their policies for issuing permits allowing discharge of harmful substances. Both emission and immission standards are applied, it not being practical to settle for only one approach.

It is noted that the Convention deals primarily with the open Baltic Sea environment although information about conditions in the coastal zone is increasingly exchanged. The Convention does not specifically consider inputs of nutrients or organic material. Compilation of information of inputs through land-based sources is generally carried out through bilateral agreements covering sub-areas of the Baltic Sea, including the Kattegat. It is becoming increasingly evident that there is a strong need for more precise and detailed (substance by substance) information and quantification on inputs.

Through the initiative of the Helsinki Commission a first integrated assessment of the effects of pollution on the natural resources of the Baltic Sea was carried out in close cooperation with ICES (Melvasalo et al. 1981). In this the following was concluded: In many coastal areas, changes in the species composition of phytoplankton have occurred along with an increase in primary production and an increase of the supply of organic matter to the benthic community; the accumulation of DDT and PCBs was an acute problem, and because their use has been banned in most countries, a decrease in DDT was observed in sediments and fish; lignin sulfonates are present in Baltic Sea water as a consequence of human activity, the source presumably being the pulp and paper industry; the levels of Zn and Pb in sediments have increased 3-fold and the levels of Hg and Cd do not appear to have increased in open sea sediments; a ban on Hg-compounds used in the pulp and paper industry has resulted in some decrease in Hg levels in fish from the northern areas, but many coastal areas are still seriously

contaminated; an unresolved issue concerns the degree, causes, and
effects of eutrophication and its relation to oxygen depletion in the
deep basins.

The options available to control inputs are related to the
characteristics of the substances, the sources, and the producers of the
substances in question, e.g. industry, municipalities, and agriculture.
In the case of DDT, legal action was chosen in most countries forbidding
its use, following reports on toxic effects already in the 1960s.
Likewise, the use of PCBs has been strictly controlled. In the end of
the 1960s, the use of Hg in the pulp and paper industry was banned in
Sweden, and similar measures have been taken in other countries. These
legal actions have forced industry to find substitues, develop new
technology including recirculation systems, and to establish internal
procedures of purification and control, which has also led to the
reduction of other wastes. Examples of results of such activities are
furnished by the pulp and paper industries of Sweden and Finland, where
emissions of organic material as BOD_7 have been reduced by factors
of 2 - 10 over the last 2 - 3 decades (Brännland 1984/85; Laasonen
1984/85). The release of suspended solids and use of freshwater have
also been considerably reduced. To a large extent these improvements
are due to a change of, or introduction of new technology. The options
used by the industry have also implied that alternative ways of disposal
have not been required. However, substitute substances may still prove
to be environmental problems. At the same time the production has
increased.

The cost for industry has also been considerable. Economical and
technological constraints imply an increasing demand for precise
information on what further reductions are required.

The inputs or wastes considered so far were all related mainly
to well-defined, few, and fairly large sources, as well as to developed
industrial processes. The actions taken have reduced the inputs, but
they have also implied that the remaining input of a given substance
is often presently related to diffuse or widespread small sources. This
is relatively more complicated to cope with. It is yet another matter
in the case of the large inputs of N and P which are related to the
diffuse-type sources, to a large extent. The dipsersion of these
substances is also different. It is relatively more complicated to
predict the response of the system to a certain reduction of inputs of
N and P, than it is to a reduction of inputs of hazardous substances.
Comparisons may be made with the situation in the Great Lakes (see
Nriagu, this volume), but the Baltic Sea is a brackish water system of
a rather special nature. So, such a comparison can only be made with
great care. According to the data presented by Kelly et al. (this
volume), it seems, however, that the input of N has increased, at least
in some parts of the Baltic Sea, to such a level that the sytem has
changed from oligotrphic to eutrophic. This has also been pointed out
by other authors (e.g. Jansson 1984). A considerable reduction of the
input of N may well be required in order to change the trend. The
question is how this can be achieved and what the options are. An
international concerted action is required, and the major sources will
have to be tackled. In Denmark, considerable progress has been made in

identifying the problem and certain options of action have been proposed (e.g. Anon. 1983c).

Table 5 represents an attempt to summarize some of the information discussed here. The options of alternative disposal of wastes on land or otherwise have not been discussed since the problems appear to have been tackled by waste reduction at the source. In the case of N, the problem now concerns both the groundwater and the seawater, at least for some large areas. These will have to be considered together.

Table 1. Summary of nitrogen and phosphorus inputs to the Baltic Sea, in 10^3 t.p.a.

Source	Nitrogen	Phosphorus
Industry	14(5–19)	3.6
Municipal	86(23–86)	18
Rivers	640(258–640)	50
Atmosphere	450	–
Total estimate for years:		
1900	∿250	∿10
1984	1180	72
Without atmosphere, annual input per unit area:		
1900	∿0.6 g N/m^2, a, average	
1984	∿2 g N/m^2, a, average	
1984	∿6 g N/m^2, a, in southern part	

Table 2. Examples of total metal inputs to the Baltic Sea in tons a^{-1}

Metal	Total[A]	Total[B]	Total[C]	Atmospheric B	D
Hg	30*	58	34	13	30[2]
Cd	250*	457	–	260	80
Pb	4300–6200*	3446	$\sim 10^3 - 10^4$	2200	2400
Ni	2100–5100[1]	3390	–	800	700
Cu	4500–8500*	3430	$\sim 10^3 - 10^4$	1000	1400
Zn	15500–18900[1]	26100	$\sim 10^4$	12000	6000

* : 50 – 80% atmospheric;
[1] : 50 – 80% rivers;
[2] : wet deposition only;
A : various sources;
B : Brügmann (1984/85);
C : Dybern and Fonselius (1981);
D : Rodhe et al. (1980), central values

Table 3. Land-based inputs (industrial sources) of some metals to the
          Gulf of Bothnia in tons a^{-1}.

Metal	1976[1]	1979[2]	1980[2]	1981[2]
Hg	0.9	0.4	0.5	0.3
Cd	1.7	2.5	3.0	1.3
Pb	18.5	11.7	9.8	8.1
Zn	299	293	270	205
As	902	189	275	100

[1] : from Pawlak (1980)
[2] : Anon. (1981, 1983)

Table 4.   Concentration ranges of some metals in fish species in mg/kg^{-1}

	Hg	Cd	Pb	Cu
Cod	0.02-0.88	0.002-0.05	0.03-1.3	0.08-2.4
Herring and Sprat	0.004-0.09	0.002-0.07	0.01-1.4	0.3-1.9
Flounder and Plaice	0.1-0.45	0.002-0.04	0.02-0.26	0.10-1.2

Table 5. Summary of information and options

Substance and characteristics	Source characteristic of input	Part of system affected, and effect	Effects proven	Applied option for remedy	Result
PCB, DDT persistent and dispersible	large range: small to large points, and diffusive (atmosphere)	mammals, birds; on reproduction	yes	ban of use and control	decrease in concentration levels of DDT and same trend for PCB introduction of new substances
BOD_7; degradable, not very dispersible	large to medium point sources	water column O_2; water quality	yes	internal industry control; technology change	decrease in input by a factor of 10
Hg; biotransformation	large range of point and diffusive sources	birds and fish disturbances	yes	ban of use and technology change	decrease in levels; now substances
P, N	large range of sources land vs. sea	water column production	yes	not clear treatment, technology change	not clear

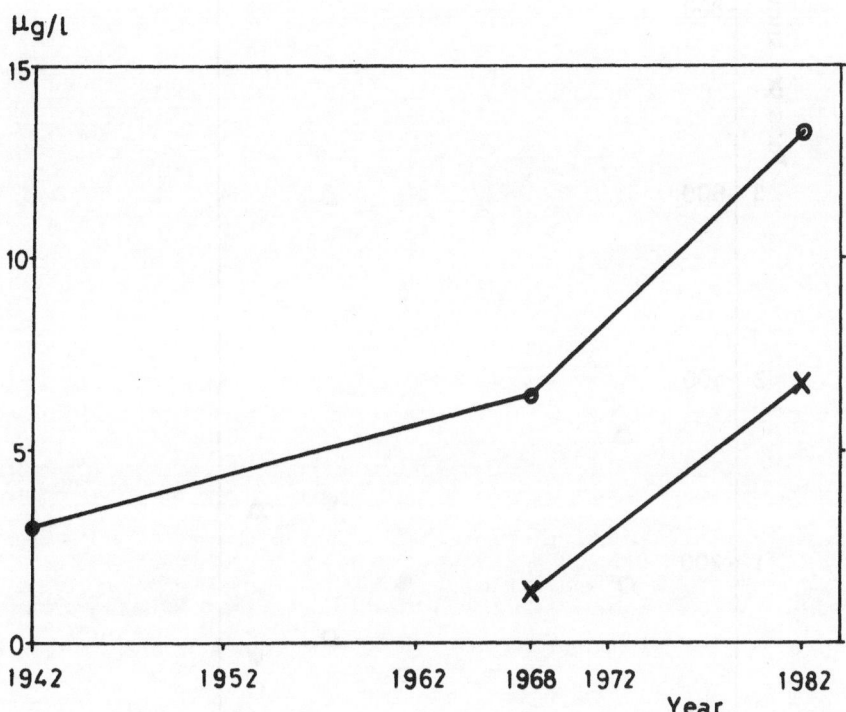

Figure 1. Trends in increase of nitrogen levels in Baltic Sea winter surface water (X) and groundwater in Denmark (0).

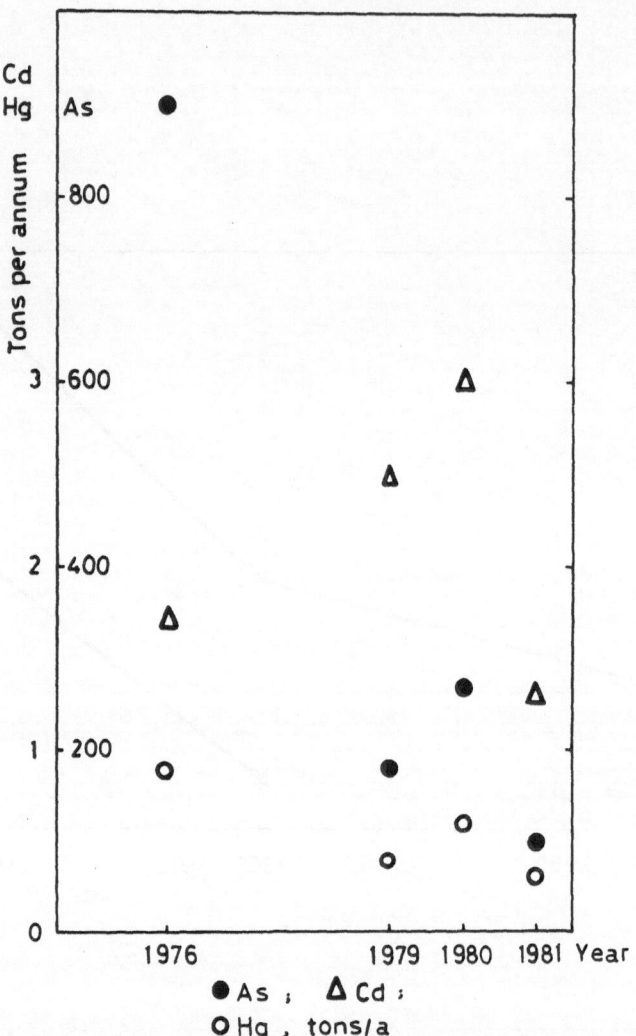

Figure 2.   Input (total) of mecury, cadmium, and arsenic to the Gulf
            of Bothnia in tons per annum for a series of years.

REFERENCES

Andersin, A.-B., J. Lassig, L. Parkkonen and H. Sondler 1978. The
    decline of macrofauna in the deeper parts of the Baltic proper and
    Gulf of Finland. Kieler Meeresforschungen 4, pp. 23-51.
Andrulewicz, E. and A. Trzosinska 1985. Attempts at massbalances of
    selected contaminants in the Baltic Sea. Mimeo, unpubl. manuscript.
Anon. 1981, 1982, 1983a. The Gulf of Bothnia - discharges from land
    and air. National Environment Protection Board, Sweden and
    National Board of Water, Finland.
Anon. 1983b. Causes of diminished trawl catches in the Bothnian Sea
    (Baltic). Institute of Marine Research, MERI, rep. no. 12, 231 pp.
Anon. 1984. Supply of nitrogen, phosphorus and organic material to
    groundwater, fresh and marine waters in Denmark. In Danish.
    Miljøstyrelsen, Strandgade 29, 1401 Cph. K, Denmark.
Brännland, R. 1984/85. The discharge of waste from the pulp and paper
    industry into coastal waters. What must be done nationally and
    internationally. In: The Baltic Sea - an environment worth
    protecting. (In press). Helsinki Commission.
Brügmann, L. 1981. Heavy metals in the Baltic Sea. Marine Pollution
    Bulletin 12, 6, pp. 214-18.
Brügmann, L. 1984/85. Influence of coastal zone porcesses on mass
    balances for trace metals in the Baltic Sea. In: Proceedings of
    Contaminant Fluxes through the Coastal Zone, ICES, Copenhagen
    (In press).
Bruneau, L. 1980. Pollution from industries in the drainage area of
    the Baltic. Ambio, 9, pp. 145-152.
Cederwall H. and R. Elmgren 1980. Biomass increase of benthic macro-
    fauna demonstrates eutrophication of the Baltic Sea. Othelia,
    suppl. 1, pp. 287-304.
Dickson, R.R. 1971. A recurrent and persistent pressure-anomaly pattern
    as the principal cause of intermediate-scale hydrographic variation
    in the European shelf sea. Deutsches Hydorgraphisches Zeitschrift
    24, 3, pp. 97-119.
Dybern, B.I. 1974. Water pollution - a problem with global dimensions.
    Ambio 3, 3-4, pp. 139-145.
Ehlin, U. 1981. Hydrology of the Baltic Sea. In: The Baltic Sea,
    A. Voipio (ed.), pp. 123-134. Elsevier Oceanography Series vol. 30,
    Amsterdam, 1981, 418 pp.
Fonselius, S.H. 1969. Hydrography of the Baltic deep basins III.
    Fishery Board of Sweden, Ser. Hyrdrogr. Rep. 23, pp. 1-97.
Gundersen, K. 1981. The distribution and biological transformations of
    nitrogen in the Baltic Sea. Marine Pollution Bulletin 12(6),
    pp. 199-205.
Helle, E. and O. Stenman 1984. Recent trends in levels of PCBs and DDT
    compounds in seals from the Finnish waters of the Baltic Sea.
    ICES, C.M. 1984/E:3, 8 pp.
Holm, N.G. 1978. Phosphorus exchange through the sediment-water
    interface. Mechanism studies of dynamci processes in the Baltic
    Sea. Ph.D. thesis. Dept. of Geology, University of Stockholm,

Microbial Geochemistry, 3, 149 pp.

Jansson, B.-O. 1978. The Baltic - A systems analysis of a semi-enclosed sea. In: Advances in Oceanography, H. Charnock and G. Deacon (eds) Plenum Press, Oxford, pp. 131-184.

Jansson, B.-O. 1984/85. The Baltic Sea and the nutrients. In: The Baltic Sea - an environment worth protecting. (In press). Helsinki Commission.

Kullenberg, G. 1970. On the oxygen deficit in the Baltic deep water. Tellus 22, p. 357.

Kullenberg, G. 1977. Observation of the mixing in the Baltic thermo- and halocline layers. Tellus 29(6), pp. 572-587.

Kullenberg, G. 1981. Physical Oceanography. In: The Baltic Sea, A. Voipio (ed.), pp. 123-134, Elsevier Oceanography Series vol. 30, Amsterdam 1981, 418 pp.

Kullenberg, G. 1982. Mixing in the Baltic Sea and implications for the environmental conditions. In: Hydrodynamics of semi-enclosed seas, pp. 399-418, J.C.J. Nihoul (ed.), Elsevier, Amsterdam 1982.

Kullenberg, G. 1983. The Baltic Sea. In: Estuaries and Enclosed Seas, pp. 309-335, B.H. Ketchum (ed.), Elsevier, Amsterdam 1983.

Laasonen, E. 1984/85. Types of discharge from the pulp and paper industry to the Blatic Sea. In: The Baltic Sea - an environment worth protecting, (In press). Helsinki Commission.

Larsson, U., R. Elmgren and F. Wulff 1984/85. Eutrophication and the Baltic Sea - causes and consequences, Ambio (In press).

Leppäkoski, E. 1980. Man's impact on the Baltic Ecosystem. Ambio 9, pp. 174-181.

Lindahl, G., K. Wallström and ˃ G.Brattberg 1977. On nitrogen fixation in a coastal area of the northern Baltic. Baltic Marine Biologists 5th Symposium, Kiel 1977, 6 pp.

Matthäus, W. 1980. Zur Variabilität Primären halinen Sprungschict in der Gotlandsee. Beiträge zur Meereskunde 44/45, pp. 27-42.

Matthäus, W. 1983b. Langzeittrends der Dichte im Gotlandbecken 48, pp. 57-71.

Matthäus, W. 1983c. Zur Variation der vertikalen Stabilität der thermohalinen Schichtung im Gotlandtief. Beiträge zur Meereskunde 48, pp. 47-56.

Melvasalo, T., J. Pawlak, K. Grasshoff, L. Thorell and A. Tsiban 1981. Assessment of the Effects of Pollution on the Natural Resources of the Baltic Sea, 1980. Baltic Sea Environment Proceedings no.5B, Baltic Marine Environment Protection Commission, Helsinki Commission, 1981, 426 pp.

Nehring, D. 1974. Untersuchungen zum Problem der Denitrifikation und Stickstaff Entbindung im Tiefenwasser der Ostsee. Beiträge zur Meereskunde 33, pp. 135-139.

Nehring, D. 1979. Relationships between salinity and increasing nutrient concentration in the mixed winter surface layer of the Baltic from 1969 to 1978. ICES C.M. 1979/C:24, pp. 1-8, mimeo.

Nehring, D. 1981. Phosphorus in the Baltic Sea. Marine Pollution Bulletin 12(6), pp. 194-198.

Nehring, D. 1984. Chemical investigations into nitrate reduction in Baltic deep waters. Preprint, Baltic Sea Environment Proceedings.

Otterlind, G. 1983. Long-term trends in the Baltic fish yield. ICES C.M. 1983, mimeo, unpubl. manus.

Pawlak, J. 1980. Land-based inputs of some major pollutants to the Baltic Sea. Ambio 9, pp. 163-167.

Pedersen, F.B. 1982. The sensitivity of the Baltic Sea to natural and man-made impact. In: Hydrodynamics of semi-enclosed seas, J.C.J. Nihoul (ed.), Elsevier, Amsterdam, pp. 385 398.

Rinne, L., T. Melvasalo, A. Niemi, and L. Niemistö 1977. Nitrogen fixation in a coastal area of the northern Baltic. Baltic Marine Biologists 5th Symposium, Kiel 1977, 6 pp.

Rodhe, H., R. Söderlund and J. Ekstedt 1980. Deposition of airborne pollutants on the Baltic. Ambio 9, pp. 168-173.

Shæffer G. 1979. On the phosphorus and oxygen dynamics of the Baltic Sea. Contribution Askö Laboratory 26, University of Stockholm, 90 pp.

Striggow, K. 1983. Die relative Landsenkung im Bereich des Sundes und der Beltsee - eine weitere Ursache der rezenten Salzgehaltsnahene der Ostsee, Gerlands Beiträge·Geophysik, 92, pp. 228-240.

Voipio, A. (ed.) 1981. The Baltic Sea. Elsevier Oceanography Series vol. 30. Amsterdam, 418 pp.

ECOLOGICAL AND HUMAN HEALTH CRITERIA FOR CROSS ECOSYSTEM
COMPARISON OF WASTE DISPOSAL IMPACTS

Judith M. Capuzzo[1], John M. Teal[1] and Robert K. Bastian[2]
[1]Woods Hole Oceanographic Institution
Woods Hole, MA 02543 USA
[2]U.S. Environmental Protection Agency
Washington, DC 20460 USA

ABSTRACT. The risk to human health, degree of ecosystem damage,
recovery time necessary to restore a natural community, and the
resiliency of a system to further damage should be considered in
any comparison of waste disposal options. Engineering
coniderations for waste disposal practices are quite distinct
for land and sea disposal options, yet a common set of criteria,
delineating how different ecosystems respond to perturbations and
identifying by what pathways toxicants are transferred to man,
can be developed. This paper addresses the formulation of such
criteria and their applicability in understanding the relative
sensitivity of different ecosystems to waste disposal impacts and
the relative risks to man of various waste disposal options.

1. INTRODUCTION

The evaluation of human health impacts as a result of waste
disposal requires an understanding of (1) the temporal and
spatial scales of impact; (2) the organizational scale of impact
-i.e., population changes, loss of productivity, loss of habitat,
direct and indirect threats to human health; and (3) the level of
risk imposed by such impacts. In making management decisions
concerning waste disposal options, it is essential that a common
set of criteria be applied to the various media (land, sea and
air) under consideration as waste disposal sites. Development of
such criteria requires an understanding of how different
ecosystems respond to perturbations and identifying by what
pathways toxicants are transferred to man. During 1983 the U.S.
National Research Council conducted a workshop on Land, Sea and
Air Disposal of Industrial and Municipal Wastes, the goal of
which was to develop criteria common to all disposal media by
which the efficacy of each disposal option could be evaluated
(U.S. NRC, 1984a). The topics addressed included the design of
wastewater treatment facilities, features of specific disposal

347

G. Kullenberg (ed.), The Role of the Oceans as a Waste Disposal Option, 347–360.

options, development of ecological and human health criteria,
environmental considerations of the disposal medium, and specific
case studies dealing with management of industrial waste and
sewage sludge disposal.   This paper will review the ecological
and human health criteria developed at that workshop for land
and sea disposal and address the context in which common criteria
can be applied to various ecological systems.

2. OCEAN DISPOSAL OF WASTES

The impact of waste disposal in the marine environment is
dependent on the composition and volume of wastes discharged and
on the dispersal and transport characteristics of the site
selected for disposal.   Coastal environments receive
anthropogenic wastes from ocean dumping activities, pipeline
discharges, atmospheric fallout, riverine inputs, and other non-
point pollution sources (Figure 1). Contaminants of biological
concern such as pathogenic microorganisms, metals, and organic
compounds of both synthetic and natural origin are associated
primarily with particulate matter. Transport of contaminants
within coastal areas coincides with sediment transport processes
and, thus, such material tends to accumulate in depositional
areas.   There are numerous examples around the world where
sediment deposits in coastal areas reflect a waste disposal
history.   The distribution, fate, and effects of wastes
discharged to the sea are governed by the physical, chemical, and
biological processes that generally reduce the concentration,
alter the chemical form and bioavailability of toxicants and
organic materials, and ultimately eliminate them from the water
column.   Transfer of contaminants to marine biota and man and
disturbance of ecological systems are dependent on the
availability and persistence of contaminants in benthic
ecosystems.

In selecting a strategy for ocean disposal of wastes, the
principal options are containment and dispersal.   With the
exception of extremely hazardous wastes, such as high level
radioactive wastes that may be contained before disposal and
dredged material that may be contained to a submarine pit and
capped, the containmant strategy is not feasible for disposal of
most wastes in the ocean.   Resuspension and transport of
materials by bottom currents and degradation and recycling of
materials in biogeochemical cycles are natural dispersal
mechanisms. Degradation of benthic habitats as a result of waste
disposal has been attributed to high levels of organic enrichment
in bottom sediments (Pearson and Rosenberg, 1978; Boesch, 1982).
The delineation between observed benthic effects of waste
disposal at non-dispersive sites and no observed effects at
dispersive sites clearly demonstrates that wide dispersal may be

not only the most feasible disposal option but also the preferred one.

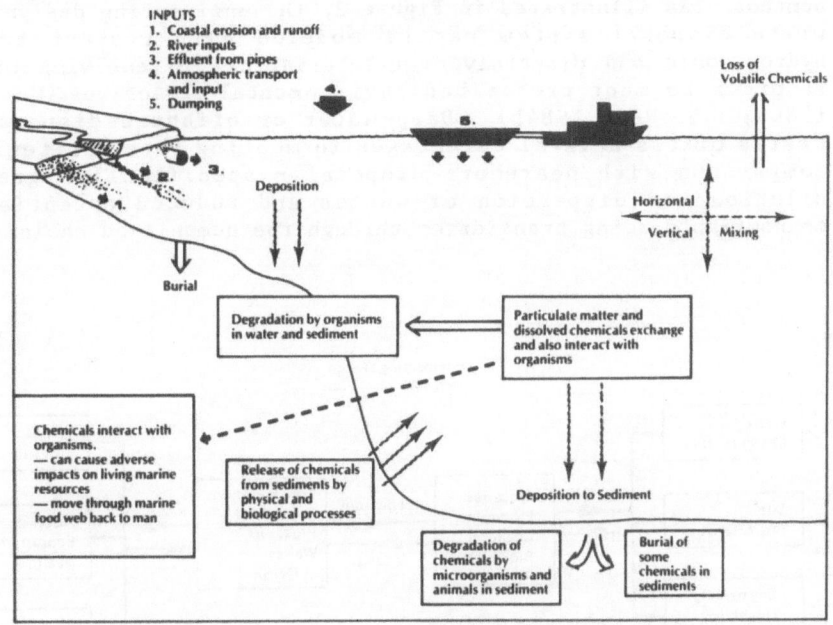

INPUTS
1. Coastal erosion and runoff
2. River inputs
3. Effluent from pipes
4. Atmospheric transport and input
5. Dumping

Figure 1.   Input and transport of wastes in marine
            ecosystems.   From: Farrington et al.
            (1982).   Artist, Kevin King (OCEANUS
            Magazine).

Environmental and human health concerns with waste disposal in the ocean include: (1) uptake and accumulation of pathogenic organisms and chemical contaminants in marine resources destined for human consumption; (2) toxic effects of chemical contaminants on the survival and reproduction of marine organisms and the resulting impact on marine ecosystems; and (3) the release of biodegradable organic matter and nutrients, which under quiescent conditions may result in localized eutrophication, organic

enrichment, and oxygen depletion (Capuzzo et al., 1985). To
minimize organic loading and accumulation of contaminants in
natural resources, disposal of wastes in the sea should occur in
areas where horizontal dispersion distributes the wastes over a
wide geographical area, thus preventing overloading of natural
microbial and biogeochemical processes, severe alterations in
macrobenthic communities, and accumulation of contaminants in the
benthos. As illustrated in Figure 2, the engineering design of a
waste disposal system can be modeled with respect to the
hydrographic and dispersive characteristics of receiving waters
in order to meet prescribed environmental objectives (U.S. NRC,
1984a; U.S. NRC, 1984b). Deep-water or offshore disposal of
wastes offers several advantages in meeting these criteria in
comparison with nearshore disposal - specifically, greater
dilution and dispersion of wastes and reduced potential of
contaminants being transferred through the human food chain.

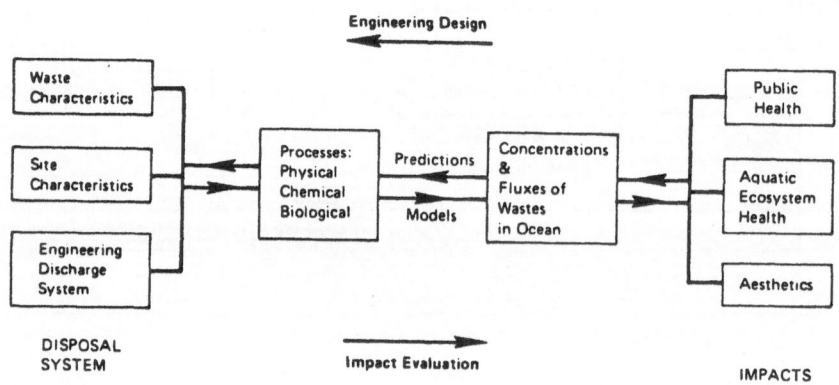

Figure 2.   Engineering considerations for waste disposal
             assessment; this figure may be read from
             "disposal system" to "impacts" for impact
             assessment or in the reverse direction for
             engineering design constrained by limits on
             the impacts. From: U.S. NRC (1984a).

3. LAND DISPOSAL OF WASTES

Impacts on terrestrial and affiliated freshwater systems as a
result of waste disposal are quite distinct from those of ocean
disposal. Because man is in more direct contact with terrestrial
ecosystems, the strategy for land disposal has traditionally been
one of containment. Concern of impacts are less focused on
large-scale contamination of natural resources or destruction of
ecological systems within the disposal site, but rather directed
at export of contaminants to other ecological systems, resulting
in contamination of surface and groundwater resources, secondary
effects on valuable natural and agricultural lands, and direct
threats to human health. Specific concerns for land disposal
include: (1) the transport of toxic organics and heavy metals to
both surface and ground waters, with potential ecological and
human health effects, such as contamination of potable water
supplies through percolation of leachate to groundwater supplies
or through surface run-off; (2) transport of pathogens to man,
through such pathways as crops grown in waste-amended soils and
contamination of groundwater and surface water systems; and (3)
the export of nutrients from landfill use to non-target
ecological systems, such as wetlands and the possible destruction
of wildlife habitats and unique ecosystems. Selection of land
disposal sites, therefore, must consider not only the
hydrological and geological suitability of the site for
containment of wastes, but also the resiliency of that ecosystem
and adjacent ecosystems to damage. Some of the engineering and
environmental considerations of land disposal practices as they
relate to waste migration through groundwater flow and transport
are presented in Figure 3.

There are a large number of terrestrial waste disposal options
currently in use including land spreading and reclamation; sludge
applications to agricultural lands; sewage treatment by wetlands,
cypress domes, and silviculture; and landfill and mine
reclamation (CAST, 1976; Loehr et al., 1979; Bastian, 1981;
Sopper and Kerr, 1981; Ewel et al., 1982). In several instances
waste disposal may be a component of ecosystem management
programs such as enhancing the diversity and productivity of park
lands, or the creation of artificial wetlands. The major concerns
in site selection for land disposal practices are to ensure that:
(1) groundwater and surface water resources in and around the
disposal site do not become contaminated beyond acceptable
levels; (2) land-use patterns are not compromised; (3) unique
ecosystems and habitats are preserved; and (4) soil-amendments do
not result in transfer of contaminants to plants and animals and
to the human food chain (U.S. NRC, 1984a). Concerns also arise
from non-point sources of contamination, such as agricultural

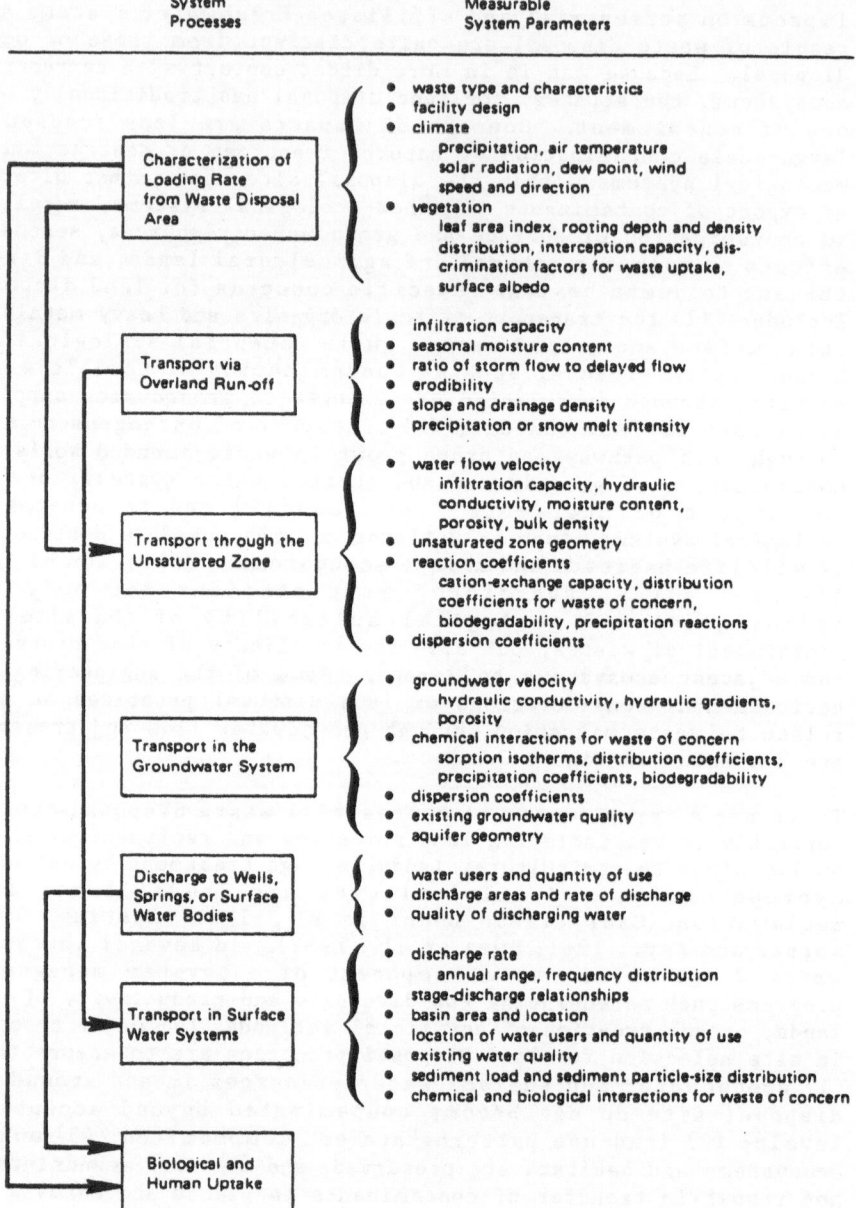

Figure 3.   Engineering and environmental considerations
            related to waste migration from land disposal
            practices.   From: U.S. NRC (1984a).

runoff and septic tank inputs. Predictions of contaminant migration to groundwater and surface water and to non-target ecosystems require an understanding of the processes controlling transport, hydrodynamic dispersion, and the chemical, physical, and biological reactions that affect contaminant distributions at a given site for a given period of time (U.S. NRC, 1984c). Criteria for selection of a disposal site, therefore, are such that the geologic, geochemical, and hydrologic characteristics should isolate the wastes from the biosphere for a long period of time. Deep well injection offers several advantages over shallow burial or surface storage in that the potential for migration to surface waters is reduced and the potential for dilution and dispersion of contaminants is increased.

4. ECOLOGICAL CONSIDERATIONS

Ecological responses to environmental perturbations can be examined at both the population and process levels. The first category includes changes that affect species distributions and abundance, such as habitat alterations; the second category includes alterations in energy flow through the ecosystem and changes in biogeochemical cycles. For most waste disposal practices, concern is focused on population and community effects rather than on processes, due to the limited spatial scale of most population perturbations in comparison to the spatial scale of ecosystem processes. Specific concerns at the population and community levels include loss of species, habitat destruction, interspecific interactions that affect community structure and function, and recoverability. These responses are common features of all ecosystems and can serve as comparative measures of ecosystem impact of ocean-based and land-based waste disposal options. Steele (1985), however, points out that the internal dynamics and structure of marine and terrstrial systems differ considerably, such that the temporal and spatial responses to change in each type of environment may also be quite distinct.

The response of individual ecosystems to perturbations may be highly variable. As an example, nutrient enrichment can have varying effects in different aquatic and terrestrial ecosystems. Nutrient enhancement of nutrient deficient areas will respond favorably to nutrient additions as evidenced by increased diversity and production (Willis, 1963; Specht, 1963). Yet, the Park-Grass experiment and old-field studies revealed changes in biomass, reductions in species numbers, and loss of various life forms as a result of nutrient enrichment (Bakelaar and Odum, 1978; Milton, 1947; Silvertown, 1980). In aquatic systems, nutrient enrichment may result in a variety of impacts depending on the severity of nutrient loading, ranging from small increases in productivity to extensive eutrophication (Nixon et al., 1984).

Disturbance in any ecosystem can be manifested by changes in the physical, chemical and biological properties that govern the structure and function of communities within that ecosystem. Loss of habitat can be viewed as retrogression or reversed succession (Woodwell and Whittaker, 1968). Recovery, therefore, is dependent on the initiation of successional stages and the restoration of ecosystem properties (Holdgate, 1978). For marine benthic communities, community changes along either a spatial or temporal gradient of impact may reflect successional stages and a gradual restoration of stable community caracteristics (Pearson and Rosenberg, 1978; Sanders et al., 1980). From their studies on organic enrichment in shallow benthic communities, Pearson and Rosenberg (1978) suggested that such successional stages include an initial zone of a few, small, short-lived species with high genetic variability, followed by gradual changes in populations with wider ecological and reproductive characteristics, but lower genetic flexibility and contributing to increased community complexity.

The same principles apply in terrestrial environments, where there are numerous examples of successional changes (McIntosh, 1980). Successional rates vary widely in terrestrial ecosystems and in some instances natural succession may not be sufficient to restore a unique habitat (e.g., desert ponds, isolated wetlands, etc.). Restoration of unique habitats in terrstrial and aquatic ecosystems may require artificial reclamation as propagules for recolonization are not available. Creation of artificial wetlands (Race and Christie, 1982), restoration of mangrove communities (Teas, 1977), and the restoration of plant communities in many terrestrial environments, ranging from high alpine to desert ecosystems (Cook, 1976) are a few examples of successful restoration programs.

5. HUMAN HEALTH CONSIDERATIONS

Concerns of human health impacts in both aquatic and terrestrial ecosystems focus on the potential pathways of pathogen and toxicant transfer to man. The transfer of toxic chemicals through various components of marine and terrestrial food chains can induce specific changes at each trophic level or result in bioaccumulation and transfer to man. Of specific concern is the uptake and transfer of metals, halogenated hydrocarbons, and other organic contaminants. Defining the potential risk of food chain contamination, or groundwater contamination in the case of land disposal options, requires an understanding of the flux and bioavailability of contaminants within each environment and elucidation of the potential routes to man. Contaminants that demonstrate carcinogenic, mutagenic, and teratogenic potential must be carefully evaluated before a disposal option is selected.

Land application of wastes may result in the transfer of contaminants to soil surfaces, to plants grown in sludge-amended soil, to groundwater, and to surface water runoff that can transport contaminants to other terrestrial or aquatic sites. Contaminant bioavailabiltiy and mobility may vary among different soil types as a result of differential cation-exchange properties, partititon coefficients, and biodegradation rates for the soil/contaminant system (U.S. NRC, 1984a). For organic contaminants both parent compounds and the by-products or metabolites of biodegradation may be of human health concern. In soil-amendment applications, transfer of contaminants to man may result from consuming contaminated plants (Chaney, 1982), or indirectly from consuming milk or meat of grazing animals that have ingested sludge-amended soils (Bergh and Peoples, 1977; Collett and Harrison, 1968; Hansen et al., 1981; Harrison et al., 1969, 1970).

In the marine environment, toxicant transfer to man is primarily through the consumption of contaminated fish and shellfish resources. Bioaccumulation of contaminants in such resources may occur through aqueous, sedimentary, and dietary pathways. Bioaccumulation of metals is dependent on metal speciation and bioavailability. O'Connor and Rachlin (1982) reviewed metal uptake processes in marine organisms and demonstrated that metal burdens in tissues of marine animals were dependent on multiple routes of uptake and distribution - i.e., food, water, and sediments. Some metals can be regulated to a greater extent than other metals and exposure to heavily contaminated areas does not always result in increased bioaccumulation of contaminants in commercial resources. For organic contaminants that are lipophilic in nature (PAHs, PCBs, etc.), bioaccumulation by marine animals is dependent on both chemical factors, such as solubility, adsorption-desorption kinetics, and the octanol-water partition coefficients of specific compounds, and biological factors, such as the transfer of such compounds through food chains and the amount of body lipid (Neff, 1979). The availability of such compounds from sedimentary reservoirs are dependent on chemical and microbial processes within the sediments. Both parent compounds and metabolites are of concern, as the latter may be linked with carcinogenic and mutagenic potential.

Pathogen transfer to man from waste disposal in both terrestrial and marine environments is dependent on the survival and persistence of pathogens at the disposal site and the mobility of pathogens to other sites through either contaminantion of groundwater and surface water runoff from land applications or through sediment transport and contamination of bathing waters and edible shellfish from ocean disposal. One of the problems associated with estimating pathogen persistence and mobility is our reliance on indicator organisms, such as members of the coliform group, rather than enumeration of specific pathogens (bacteria and viruses) to predict potential risks to human health. Information gaps exist for both disposal options for the development of predictive models relating pathogen contamination to health risks. Specific information needs include: (1) enumerating bacteria and viruses in marine and terrestrial environments and development of more sensitive indicators, so that more accurate predictions of pathogen survival and persistence and human health risks can be made; and (2) understanding processes that control pathogen mobility, such as surface and subsurface drainage patterns that result in contamination of groundwater and surface waters, and the dynamics of particle associated-pathogen transport in marine environments that result in contamination of bathing waters or edible shellfish.

6. CRITERIA

The management objectives for waste disposal options are to ensure that a specific waste disposal practice is functioning with minimum risks to human health and ecological systems. To this aim, specific goals should be focused at understanding (1) the functional characterisitcs of an ecosystem that govern both impact and recovery; and (2) defining pathways for pathogen and toxicant transport to man. Table I summarizes the types of concerns for both ecological damage and human health impacts that effective management of waste disposal systems must consider for a variety of ecosystems (U.S. NRC, 1984a). Such a framework can be useful in evaluating the biological concerns in cross ecosystem comparisons of waste disposal impacts and should become an integral component of the decision-making process.

Table I. Ecological and Human Health Concerns of Waste Disposal Impacts in Aquatic and Terrestrial Ecosystems (U.S. NRC, 1984a).

Type of Environment	Species Extinction	Habitat Loss	Elevated Nutrients	Recoverability	Containment	Remedial Action	Uncertainty	Visibility	Pathogen Routes to Society	Toxicant Routes to Society
Land										
Disturbed lands[a]	1	0	1	0	1	0	1	5	5	5
Remnants[b]	0	5	1	1	1	1	1	5	5	5
Temperate forest	1	1	1	2	1	1	1	3	2	2
Temperature grassland	1	1	1	1	1	1	1	3	3	3
Pasture	0	0	0	1	2	1	1	5	5	5
Agricultural land	0	1	0	2	2	1	1	5	5	5
Arid land	3	2	1	3	1	3	2	2	1	1
Arctic land	0	1	1	5	1	5	4	1	1	1
Freshwater										
Lake	1	5	5	3	5	4	2	5	4	4
Stream	5	5	3	2	5	4	3	3	5	5
Wetland	5	5	5	3	5	5	4	2	3	3
Groundwater	3	1	5	5	5	5	5	0	5	5
Marine										
Wetlands (U.S. East Coast)	1	4	3	3	5	5	3	5	5	5
Wetlands (U.S. West Coast)	5	5	3	3	5	5	2	5	5	5
Estuaries	5	5	3	3	5	5	2	5	5	5
Coastal areas	1	3	1	1	5	5	3	1	3	4
Open ocean	1	1	0	5	5	5	5	0	1	1

NOTES: Species extinction: 5 = greatest concern.
Habitat Loss--loss of a significant portion of a habitat type: 5 = greatest concern.
Elevated Nutrients: 5 = highest probability of change to ecosystem.
Recoverability--ability of system to repair itself after input ceases: 5 = slowest recover, decades to centuries; 1 = rapid recovery, years.
Containment--ability of unmodified system to restrict spread of inputs: 5 = greatest difficulty.
Remedial action--ease with which we can repair damage to ecosystem: 5 = greatest difficulty.
Visibility: 5 = most visible.
Pathogen Routes to Society: 5 = highest probability of reaching society.
Toxicant Routes to Society: 5 = highest probability of reaching society.

[a]Disturbed Lands--land highly modified by human activities.
[b]Remnants--isolated natural spots within developed or otherwise highly modified areas.

7. REFERENCES

Bakelaar, R. and E. Odum. 1978. 'Community and population level responses to fertilization in an old-field ecosystem.' Ecology 59: 660-671.

Bastian, R.K. 1981. 'EPA's role and interest in using wetlands for wastewater treatment.' Paper presented at the Midwest Conference on Wetland Values and Management, St. Paul, MN, June 17-19, 1981.

Bergh, A.K. and R.S. Peoples. 1977. 'Distribution of polychlorinated biphenyls in a municipal wastewater treatment plant and environs.' Sci. Total Environ. 8: 197-204.

Boesch, D.F. 1982. 'Ecosystem consequences of alterations of benthic community structure and function in the New York Bight region.' In: G.F. Mayer (ed.), Ecological Stress and the New York Bight: Science and Management. Estuarine Research Federation, Columbia, SC, pp. 543-568.

Capuzzo, J.M., W.V. Burt, I.W. Duedall, D.R. Kester, and P.K. Park. 1985. 'Future strategies for nearshore waste disposal.' In: B.H. Ketchum, J.M. Capuzzo, I.W. Duedall, W.V. Burt, P.K. Park, and D.R. Kester (eds.), Wastes in the Ocean. Vol. 6, Nearshore Waste Disposal. Wiley-Interscience, New York.

CAST (Council for Agriculture Science and Technology). 1976. 'Application of sewage sludge to cropland.' U.S. Environmental Protection Agency, Office of Water Programs, Washington, DC, EPA-430/9-76-013.

Chaney, R.L. 1982. 'Foodchain pathways for toxic metals and toxic organics from wastes.' Manuscript, 50pp.

Collett, J.N. and D.L. Harrison. 1968. 'Lindane residues on pasture and in the fat of sheep grazing pasture treated with lindane spills.' N. Z. J. Agr. Res. 11: 589-600.

Cook, C.W. 1976. 'Surface-mine rehabilitation in the American west.' Environ. Conserv. 3: 179-183.

Ewel, K.C., M.A. Harwell, J.R. Kelly, H.D. Grover, and B.L. Bedford. 1982. 'Evaluation of the use of natural ecosystems for wastewater treatment.' Ecosystem Research Center Report No. 15, Cornell University, Ithaca, NY, 55 pp.

Farrington, J.W., J.M. Capuzzo, T.M. Leschine, and M.A. Champ. 1982. 'Ocean dumping.' Oceanus 25(4): 39-50.

Hansen, L.G., P.K. Washko, L.G.M.T. Tuinstra, S.B. Dorn, and T.D. Hinesly. 1981. 'Polychlorinated biphenyl, pesticide, and heavy metal residues in swine foraging on sewage sludge amended soils.' J. Agr. Food Chem. 29: 1012-1017.

Harrison, D.L., J.C.M. Mol, and W.B. Healy. 1970. 'DDT residues in sheep from the ingestion of soil.' N. Z. J. Agr. Res. 13: 664-672.

Harrison, D.L., J.C.M. Mol, and J.E. Rudman. 1969. 'DDT and lindane: 'New aspects of stock residues derived from a farm environment.' N. Z. J. Agr. Res. 12: 553-574.

Holdgate, M.W. 1978. 'Final discussion.' In: M.W. Holdgate and M.J. Woodman (eds.), The Breakdown and Restoration of Ecosystems. Plenum Press, New York, pp. 465-473.

Loehr, R.C., W.J. Jewell, J.D. Novak, W.W. Clarkson, and G.S. Friedman. 1979. Land Application of Wastes. Van Nostrand-Reinhold Co., New York, 431pp.

McIntosh, R.P. 1980. 'The relationship between succession and the recovery process in ecosystems.' In: J. Cairns (ed.), The Recovery Process in Damaged Ecosystems. Ann Arbor Science, Ann Arbor, MI, pp. 11-62.

Milton, W. 1947. 'The yield, botanical and chemical composition of natural hill herbage under manuring, controlled grazing and hay conditions. I. Yield and botanical.' J. Ecol. 35: 65-89.

Neff, J.F. 1979. Polycyclic Aromatic Hydrocarbons in the Aquatic Environment. Sources, Fates and Biological Effects. Applied Science Publishers Ltd., London, 262pp.

Nixon, S.W., C.D. Hunt, and B.L. Nowicki. 1984. 'The retention of nutrients (C, N, P), heavy metals (Mn, Cd, Pb, Cu), and petroleum hydrocarbons in Narragansett Bay.' Paper presented at SCOR Seminar on Biogeochemical Processes at the Land-Sea Boundary, Roscoff, France, October 22-24, 1984.

O'Connor, J.M. and J.W. Rachlin. 1982. 'Perspectives on metals in New York Bight organisms: Factors controlling accumulation and body burdens.' In: G.F. Mayer (ed.), Ecological Stress and the New York Bight: Science and Management. Estuarine Research Federation, Columbia, SC, pp. 655-673.

Pearson, T.H. and R. Rosenberg. 1978. 'Macrobenthic succession in relation to organic enrichment and pollution in the marine environment.' Oceanogr. Mar. Biol. Ann. Rev. 16: 229-311.

Race, M.S. and D.R. Christie. 1982. 'Coastal zone development: Mitigation, marsh creation, and decision-making.' Environ. Management 6: 317-328.

Sanders, H.L., J.F. Grassle, G.R. Hampson, L.S. Morse, S.Garner-Price, and C.C. Jones. 1980. 'Anatomy of an oil spill: Long-term effects from the grounding of the barge Florida off West Falmouth, Mass.' J. Mar. Res. 38: 265-380.

Silvertown, J. 1980. 'The dynamics of a grassland ecosystem: Botanical equilibrium in the Park-Grass experiment.' J. Appl. Biol. 17: 491-504.

Sopper, W.E. and S.N. Kerr. 1981. 'Revegetating strip-mined land with municipal sewage sludge.' U.S. Environmental Protection Agency, Municipal Environmental Research Laboratory, Cincinnati, OH, EPA-600/52-81-182.

Specht, R.L. 1963. 'Dark Island heath (Ninety Mile Plain, South Australia). VII. The effect of fertilizers on composition and growth, 1950-1960.' Austral. J. Bot. 11: 62-66.

Steele, J.H. 1985. 'A comparison of terrestrial and marine ecological systems.' Nature 313: 355-358.

Teas, H.J. 1977. 'Ecology and restoration of mangrove shorelines in Florida.' Environ. Conserv. 4: 51-58.

U.S. National Research Council. 1984a. Disposal of Industrial and Domestic Wastes. Land and Sea Alternatives. National Academy Press, Washington, DC, 210pp.

U.S. National Research Council. 1984b. Ocean Disposal Systems for Sewage Sludge and Effluents. National Academy Press, Washington, DC, 126pp.

U.S. National Research Council. 1984c. Groundwater Contamination. National Academy Press, Washington, DC, 179pp.

Willis, A. 1963. 'Braunton burrows: The effects on the vegetation of the addition of mineral nutrients to the dune soils.' J. Ecol. 51: 353-374.

Woodwell, G.M. and R.H. Whittaker. 1968. 'Effect of chronic gamma irradiation on plant communities.' Quart. Rev. Biol. 43: 42-55.

ORGANIC CHEMICAL POLLUTANTS IN THE OCEANS AND GROUNDWATER: A REVIEW
OF FUNDAMENTAL CHEMICAL PROPERTIES AND BIOGEOCHEMISTRY

John W. Farrington
Chemistry Department and
Coastal Research Center
Woods Hole Oceanographic Inst.
Woods Hole, MA, USA 02543

John Westall
Chemistry Department
Oregon State University
Corvallis, Oregon, USA 97331

ABSTRACT. Physical-chemical parameters derived from equilibrium partitioning are described and discussed within the context of biogeochemical cycles of pollutant organic chemicals in ground water and
marine ecosystems. A few examples of actual field data are used to
illustrate current state-of-knowledge regarding application of these
parameters (K_{ow}, K_{oc} BCF-biological concentration factors) and concepts to ground water, estuarine and open ocean ecosystems. Further research and assessment considerations are discussed within the
context of comparison of land and ocean disposal of wastes containing
organic compounds of environmental concern.

1.   INTRODUCTION

There are several examples of organic compounds which are of environmental concern because of evidence from actual cases of release to
the environment and subsequent adverse effects, including, lethality
on man, or valuable living resources. These examples include chlorinated pesticides such as DDT (egg shell thinning in certain populations of birds), polychlorinated biphenyls (contamination of rice
cooking oil by PCBs, or more probably contaminants in PCBs), tetra-
chloroethylene in drinking water, kepone in the James River, Chesapeake Bay, USA ecosystem (NAS, 1979; UN, 1979; CEQ, 1984; Bender
et al., 1977). We offer these as a limited subset of the larger set
of examples. Furthermore, extensive experimental testing has identified a large number of organic chemicals as having known or potential
adverse impacts on human health or living natural resources. This
situation requires the development of a sufficient body of knowledge
to provide, at the very least, rudimentary predictive capability concerning: 1) the routes of entry of compounds to the environment and
routes of movement through ecosystems - especially back to man;
2) reactions of these compounds in the environment - either chemical,
photochemical, or biological transformations/degradations; 3) rates
of movement and reaction; 4) reservoirs of short-term (days-months)
and long term (years to decades) accumulations. These are the 4 R's

361

G. Kullenberg (ed.), The Role of the Oceans as a Waste Disposal Option, 361–425.

of biogeochemistry: Routes, Reactions, Rates, Reservoirs and they are
incorporated into quantitative and qualitative descriptions of biogeochemical cycles of the environment as depicted in a general way in
Figures 1 and 2.

We will discuss some aspects of the state-of-knowledge regarding
physical-chemical and bioaccumulation aspects of the biogeochemical
cycles in a way that facilitates their incorporation into models of
the physical dynamics/hydrodynamics of ground water aquifers, riverine, lacustrine or oceanic ecosystems to be discussed in other papers
at this workshop. This focus is on the properties of organic chemicals and derived parameters that are useful to predicting the behavior
of organic compounds in aquatic systems. We discuss a few examples
of application to ground water and coastal pollution problems. We
also discuss future research needs and the present state-of-the-art
for monitoring to confirm predictions of environmental behavior.

2. COMPOUNDS OF CONCERN

There are a wide range of organic compounds synthesized by man for
various uses in modern society. The total number synthesized exceeds
100,000 with an estimated 60,000 in common use and approximately 1,000
being added per year as of 1978 (Maugh, 1978; CEQ, 1984). These
60,000 compounds are potentially of local, regional, or global
interest from the perspective of environmental concerns. Many of
these compounds enter the environment by various pathways via accidental, unintentional or intentional releases. Fortunately, the number of compounds of real environmental concern are a small percentage
of the total released but these are not a trivial problem. Butler
(1978) estimated that approximately 1,000 substances are manufactured
in quantities that potentially could pollute the globe if released.
His estimate took into account a wide ranging definition of pollution,
but does, nonetheless indicate the magnitude of the real and potential problem. The compounds of environmental concern share all or
some of the following characteristics: persistence, mobility, short
term or long term adverse biological activity (to non-target organisms
in the case of pesticides) (Miller, 1984). There have been national
and international efforts to assemble lists of organic chemicals of
most environmental concern via assessment of the preceding characteristics. For example, the U.S. E.P.A. priority pollutant list
(Table I) illustrates the range of compounds to be considered.

We fully recognize that lists of this type (Table I) are only
initial attempts in need of periodic reassessment, expansion (and
perhaps deletion on occasion) as more information becomes available.
It is extremely difficult within the context of multi-national com-
panies, varying worldwide socio-economic-political arrangements to
gain, or mandate, access to production data and practically impossible
at this time to do more than estimate release to the environment for
the myriad chemicals of concern. Other authors or organizations have
attempted this herculean task (e.g. OAD, NOAA, 1984). These efforts
should continue since a very important input to biogeochemical assess-

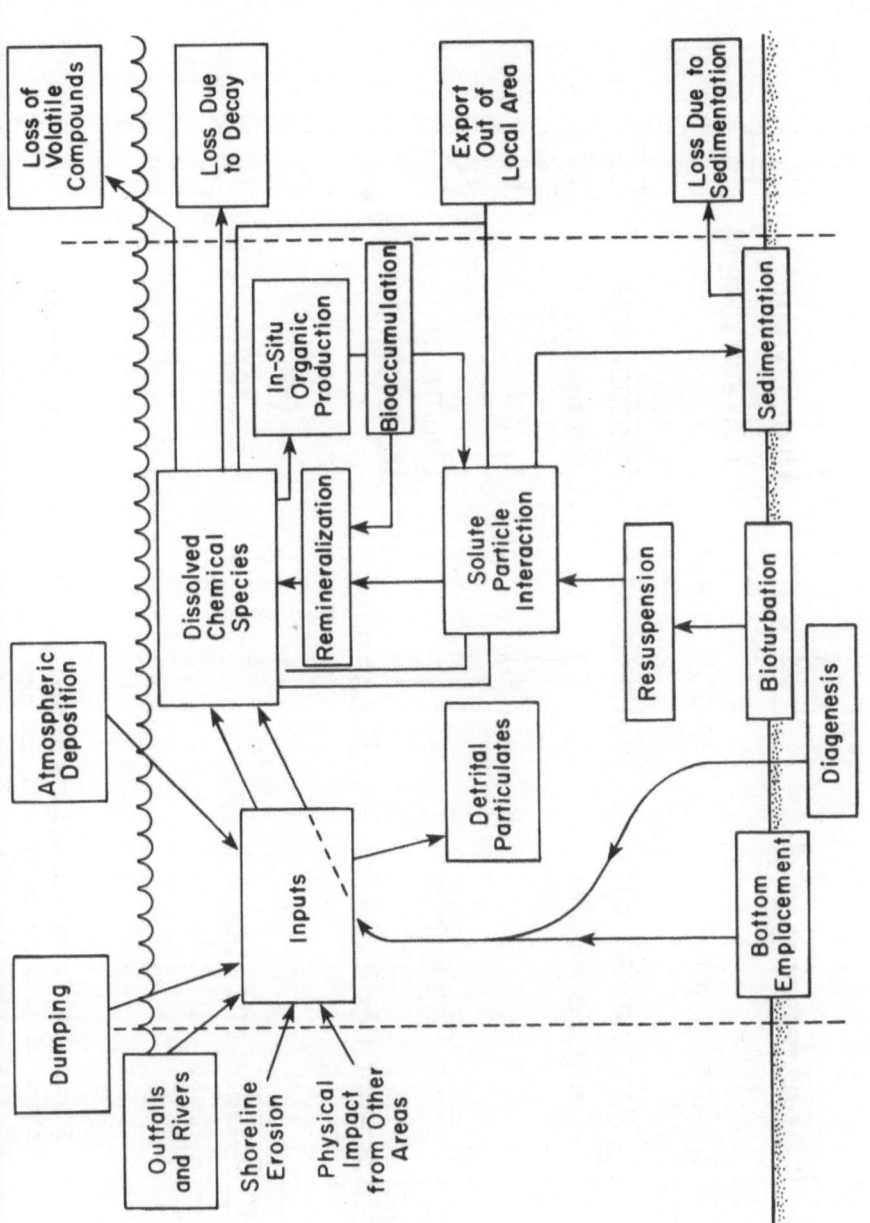

Figure 1. Important biogeochemical processes affecting organic chemical pollutants in marine ecosystems.

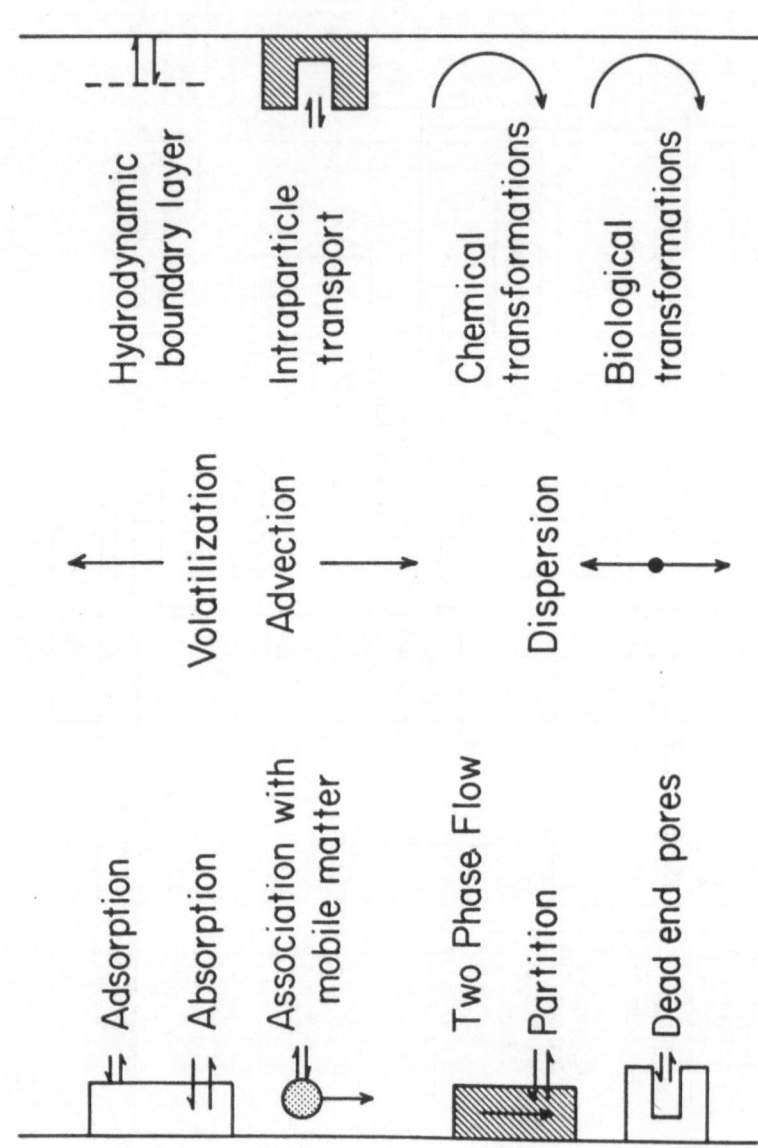

Figure 2. Important biogeochemical processes affecting organic chemical pollutants in soil and groundwater.

ment of organic compounds in wastes is the obvious requirement of knowing with what and with how much we are challenging the receptor system. This becomes a significant problem if biogeochemical (and biological/ecological) assessment of complex samples such as urban sewage sludge or dredged urban harbor sediment are considered. Examples of types and concentrations of synthetic organic chemicals and petroleum and pyrogenic byproducts reported in industrial wastes, sewage sludge, and urban harbor sediments are given in Tables II and III. It is important to note that the lists of compounds are not complete and contain only parent compounds and not metabolites or chemical/photochemical reaction products which in some cases can be considered equally hazardous or more hazardous than their precursors (Sheehan et al., 1984; NAS, 1981; NAS, 1983; Malins et al., 1984; Gosset et al., 1984).

3. PHYSICAL-CHEMISTRY CONSIDERATIONS

There has been significant progress in research concerned with identifying and quantifying physical-chemical parameters of utility in predicting the environmental behavior of organic chemicals. Until recently, fundamental properties such as solubilities in water were available for relatively few organic compounds of environmental concern, in particular for higher molecular weight non-polar compounds such as PAH, PCBs, and chlorinated pesticides. This situation is improving rapidly (e.g. Mackay and Shiu, 1977; Mackay et al., 1980; May and Wasik, 1978; May et al., 1978; May, 1980, among others). The influence of salinity on solubility and solubility in open ocean seawater have been less extensively investigated (e.g. Sutton and Calder, 1975; Eaganhouse and Calder, 1976; May, 1980; Whitehouse, 1984) and requires much further research. Calculation or estimation of water solubility of organic compounds from theoretical and empirical considerations using molecular surface area and volume or activities of the organic solute in the organic phase and activity coefficients in aqueous phase continues to be an active area of research (e.g. Banerjee, 1985). The solubilities for some compounds of interest are given in Table IV. Two important points are illustrated by this data; (i) the wide range of solubilities to be considered for these essentially hydrophobic compounds and (ii) the very low aqueous solubilities for several of the compounds which indicate that these compounds have an affinity for surfaces in aquatic ecosystems.

The interaction of an organic compound with the surface of particles is one of the key factors influencing the transport and fate of the compound in the aquatic environment. Binding of the compound to a surface has a direct influence on the transport of a compound, since the water and the solid usually have different velocities. In addition, many chemical and biological transformations take place on solid surfaces.

Binding of a compound to solid materials as occurs in the aqueous environment is generally referred to as sorption. Distinction can be made between adsorption, or binding at a two dimensional interface or

Table I.   Organic compounds in the U.S. Environmental Protection
           Agency's list of 129 unambiguous priority pollutants.

Volatiles-33	Pesticides and PCBs-28
Acrolein	Endosulfan, α
Acrylonitrile	Endrin
Benzene	Endrin aldehyde
Bis(chloromethyl)ether	Heptachlor, epoxide
Bis(2-chloroethyl)ether	Heptachlor
Bis(2-chloroethoxy)methane	Hexachlorocyclopentadiene
Bromoform	Hexachlorobutadiene
Carbon tetrachloride	Hexachlorobenzene
Chlorobenzene	PCB 1221
Chlorodibromomethane	PCB 1232
Chloroethane	PCB 1242
Chloroform	PCB 1248
Dichlorobromomethane	PCB 1254
Dichlorofluoromethane	PCB 1260
Ethyl benzene	PCB 1016
Methyl bromide	Toxaphene
Methyl chloride	4,4'-DDD
Methylene chloride	4,4'-DDE
Tetrachloroethylene	4,4'-DDT
Toluene	
Trichloroethylene	
Trichlorofluoromethane	Other Neutrals-10
Vinyl chloride	
1,1-Dichloroethane	Bis-2-chloroisopropyl ether
1,1-Dichloroethylene	Hexachloroethane
1,1,1-Trichloroethane	1,2,-Dichlorobenzene
1,1,2-Trichloroethane	1,2,4-Trichlorobenzene
1,1,2,2-Tetrachloroethane	1,3-Dichlorobenzene
1,2-Dichloroethane	1,4-Dichlorobenzene
1,2-Dichloropropane	2-Chloronaphthalene
1,2-Trans-dichloroethylene	4-Bromophenyl phenyl ether
1,3-Dichloropropene	4-Chlorophenyl phenyl ether
2-Chloroethyl vinyl ether	Isophorone

Pesticides and PCBs-28	Phenols-11
	p-Chloro-m-cresol
Aldrin	Pentachlorophenol
α-BHC	Phenol
β-BHC Hexachlorobenzenes	2-Chlorophenol
Δ-BHC	2-Nitrophenol
γ-BHC Lindane	2,4-Dichlorophenol
Chlordane	2,4-Dimethylphenol
Dieldrin	2,4-Dinitrophenol
Endosulfan sulfate	2,4,6-Trichlorophenol
Endosulfan, β	4-Nitrophenol
	4,6-Dinitro-o-cresol

Table I. (continued)

N-Containers-4

1,2-Diphenyl hydrazine
Nitrobenzene
2,4-Dinitrotoluene
2,6-Dinitrotoluene

Hetero-Carcinogens-6

N-nitroso-di-n-propyl amine
N-nitroso-diphenyl amine
N-nitroso-dimethyl amine
2,3,7,8-Tetrachlorodibenzo(p)dioxin
Benzidine
3,3'-Dichlorobenzidine

Phthalates-6

Bis(2-ethyl hexyl)phthalate
Butyl benzyl phthalate
Di-n-butyl phthalate
Diethyl phthalate
Dimethyl phthalate
Di-n-octyl phthalate

Hydrocarbons-16

Acenaphthylene
Acenaphthene
Anthracene
Benz(e)acephenanthrylene
Benzo(k)fluoranthene
Benzo(a)pyrene
Chrysene
Fluoranthene
Fluorene
Indeno(1,2,3-cd)pyrene
Naphthalene
Phenanthrene
Pyrene
Benzo(ghi)perylene (1,12 benzo-
    perylene)
Benzo(ghi)anthracene (1,2 benz-
    anthracene)
1,2,5,6-Dibenzanthracene

SOURCE: Reprinted with permission from Environ. Sci. Technol., vol. 13, L. H. Keith and W. A. Telliard, ES&T special report priority pollutants: I. A perspective view, copyright 1979, American Chemical Society.

Table II.   Concentrations of organic compounds in a composite indus-
            trial waste dumped off Puerto Rico (from Brooks et al.,
            1983).

Compound	Concentration mg/L of waste
Acetone	6,000
Dichloromethane	110
Chloroform	5
Benzene	2
Methylisobutylketone	20
Toluene	50
Ethylbenzene	0.4
m,p-Xylenes	2
o-Xylene	1.5
Propylbenzene	0.4
Cumene	2
Mesitylene	3
C_3-Benzene	1
C_4-Benzene	0.9
C_4-Benzene	1
C_4-Benzene	0.8
Cymene	1
Butylbenzene	1
Naphthalene	0.3

Table III. Concentrations of some organic compounds in urban sewage sludge and urban harbor sediments of the New York area (from MacLeod et al., 1981).

	Dredged Material (3 samples) 10^{-9} g/g			Sewage Sludge Samples (2 samples) 10^{-6} g/L	
C_3-Benzenes	—	400	100	100	200
C_4- "	—	100	100	70	600
C_5- "	—	50	60	300	200
C_6- "	—	100	—	100	40
C_7- "	—	—	—	—	30
C_8- "	—	—	60	30	—
C_9- "	—	—	—	—	—
C_{10}- "	—	—	—	—	—
C_{11}- "	—	—	—	—	—
C_{12}- "	—	—	—	—	—
C_2-Naphthalenes	600	1000	400	400	600
C_3- "	400	3000	600	600	800
C_4- "	400	4000	600	300	500
C_5- "	100	800	500	50	80
Methylbiphenyls	30	1000	100	50	200
C_2-Biphenyls	—	4000	80	30	80
C_3- "	60	700	80	90	100
C_4- "	90	200	400	—	30
C_5- "	—	100	30	—	—
C_6- "	—	—	50	—	—
Acenaphthene	500	500	80	—	—
Methylacenaphthenes	—	—	—	—	70
C_2-Acenaphthenes	—	200	—	—	60
Acenaphthylene	30	—	—	—	—
Fluorene	400	300	40	30	30
Methyl fluorenes	200	700	40	30	200
C_2-fluorenes	80	2000	100	70	200
C_3- "	—	700	500	30	30
C_4- "	—	—	50	—	—
Methylenephenanthrene	—	2000	200	—	—
Me-Phen./anthracenes	800	200	500	200	300
C_2- " "	500	3000	800	200	300
C_3- " "	200	1000	400	50	90
C_4- " "	—	300	100	20	20
C_5- " "	—	—	—	5	—
Dibenzoheptafulvene	100	400	60	—	10
Phenylnaphthalenes	80	—	—	20	—
Benzylnaphthalenes	—	—	—	—	—

Table III. (continued)

	Dredged Material (3 samples) 10^{-9} g/g			Sewage Sludge Samples (2 samples) 10^{-6} g/L	
Me-Fluoranth./pyrenes	200	1000	400	10	5
C$_2$- " "	--	400	40	3	--
C$_3$- " "	—	40	—	3	--
C$_4$- " "	—	10	30	--	--
Benzofluorenes	100	—	50	--	--
Benzo(g,h,i)fluoranthene	—	--	40	—	20
Benzo(c)phenanthrene	—	100	—	--	--
Triphenylene	40	40	60	--	--
Me-Benzanthr./chrysenes	30	300	100	8	20
C$_2$- " "	--	50	—	8	—
Terphenyl	—	--	—	--	--
Benzo(k)fluoranthene	—	60	70	5	20
Benzo(b) "	40	60	80	5	20
Benzo(j) "	--	10	20	--	--
Binaphthyl	—	30	—	--	--
Benzo(c)chrysene	--	6	5	--	--
Dibenz(a,b)anthracene	--	—	5	--	--
Indeno(1,2,3-cd)pyrene	--	—	20	--	--
Benzo(g,h,i)perylene	--	—	10	--	--

Table IV. Water solubilities of selected organic compounds at 25°C
          (from Banerjee, 1985, who compiled the data from several
          sources).

Compound	log S
Acenaphthene	-3.86
Acrylonitrile	0.15
Aniline	-0.15
Benzene	-1.68
Benz(a)pyrene	-6.26
Biphenyl	-3.88
1,3-Butadiene	-1.86
Carbon tetrachloride	-2.31
Chlorobenzene	-2.35
Dibenz(ah)anthracene	-6.28
m-Dichlorobenzidine	-3.01
3,3'-Dichlorobenzidine	-3.84
2,4'-Dichlorobiphenyl	-5.32
1,2-Dichloroethane	-1.09
7,12-Dimethylbenzanthracene	-5.71
Ethylbenzene	-2.80
Hexachlorobenzene	-5.48
2,2',4,4',5,5'-Hexachlorobiphenyl	-7.66
3-Methylcholanthrene	-6.54
Naphthalene	-3.04
Pentachlorobenzene	-4.85
Phenanthrene	-4.42
Styrene	-2.57
1,2,3,5-Tetrachlorobenzene	-4.33
2,2',5,5'-Tetrachloroethane	-1.75
Toluene	-2.22
1,3,5-Trichlorobenzene	-4.09
1,1,-Trichloroethane	-2.00

S = molar solubility

surface, and absorption, or partitioning into the three dimensional
bulk of the sorbing phase. Attempts have been made to develop methods
of estimating degree of sorption (distribution ratio) for any organic
compound on any sorbing substrate. While these global estimation
methods work surprisingly well in many cases, the physical basis for
the methods are understood better and the estimates are generally
more precise if compounds with similar chemical characteristics are
considered together. Here we present examples for hydrophobic neutral
and ionic organic compounds.

3.1 Sorption of Neutral Hydrophobic Compounds

In recent years, many studies have been conducted to investigate the
sorption behavior of various classes of neutral hydrophobic organic
compounds. For several groups of compounds, including polycyclic
aromatic hydrocarbons (Karickhoff et al., 1979; Means et al., 1980),
halogenated hydrocarbons (Chiou et al., 1979; Schwarzenbach and
Westall, 1981), and certain pesticides (Briggs, 1981), it has been
found that sorption can be described in terms of a simple partitioning
model, i.e., the phase transfer reaction is viewed as transfer of the
neutral compound from the aqueous phase to a bulk non-aqueous phase:

$$A \xrightleftharpoons{\hspace{1cm}} \bar{A} \tag{1}$$

where A represents a neutral compound, and the overbar indicates species in the non-aqueous phase. In these cases, sorption equilibrium
can be described by an equilibrium partition coefficient for a particular species

$$K_p = [\bar{A}] / [A] \tag{2}$$

The usefulness of these partition relationships to problems of
environmental geochemistry is greatly increased if the value of K_p
in a particular system can be related to fundamental properties of
the sorbent and the sorbate. Fig. 3 shows, exemplified by 1,4-
dichlorobenzene, that the organic constituents of natural sorbents
are primarily responsible for sorption of hydrophobic organic compounds, if the organic carbon content exceeds about 0.1% (f_{oc} =
fraction of organic carbon < 0.001). For such nonpolar compounds,
the partition coefficient K_p can be expressed in terms of the
organic carbon content of the sorbent, and in terms of a partition
coefficient K_{oc} between water and a hypothetical sorbent of 100%
organic carbon representing an average natural organic material:

$$K_p = f_{oc} \cdot K_{oc} \tag{3}$$

Linear free energy relationships as described by Leo et al. (1971)
and recently discussed by Westall (1984) can then be used to calculate
partition coefficients K_p from partition coefficients derived for a
reference solvent/water system, e.g., the n-octanol water system (K_{ow},

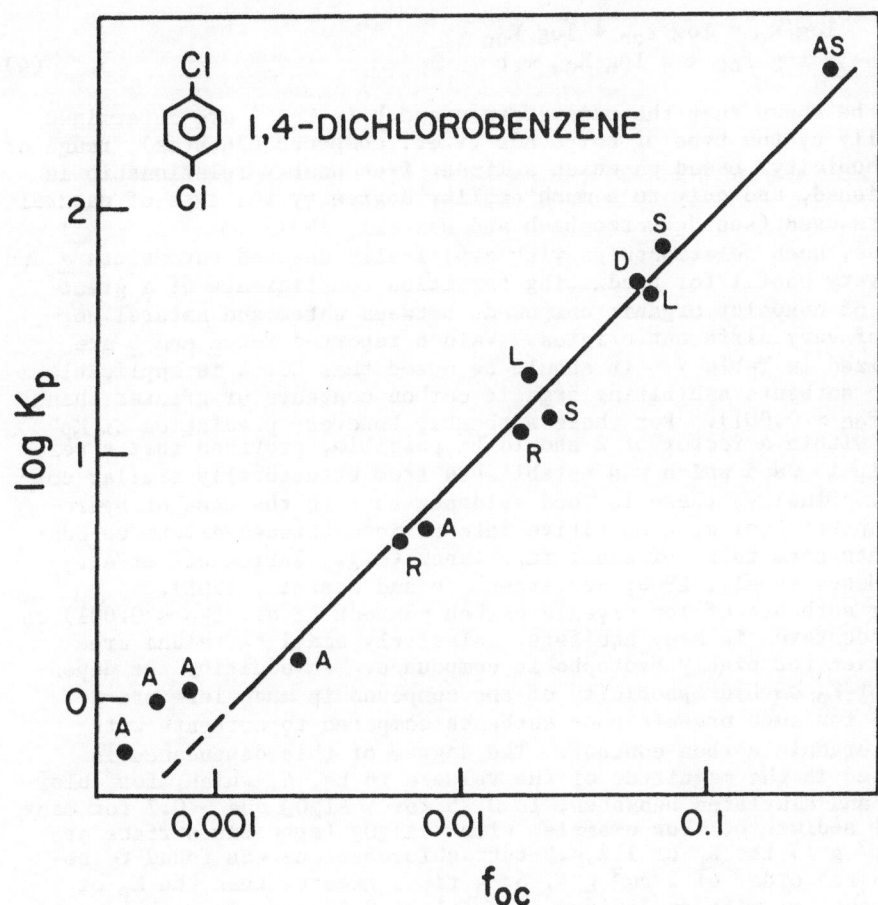

Figure 3. Solid/water partition coefficients of 1,4-dichlorobenzene as a function of the organic carbon content of the sorbent (A = aquifer material, AS = activated sludge, D = detritus, L = lake sediment, R = river sediment, S = sea sediment) (Schwarzenbach and Westall, 1985).

see example given in Fig. 4):

$$\log K_p = \log f_{oc} + \log K_{oc}$$
$$= \log f_{oc} + a \log K_{ow} + b \tag{4}$$

It can be shown that the parameters $\underline{a}$ and $\underline{b}$ in Eq. 4 are determined primarily by the type of compounds (i.e., compound class(es), range of hydrophobicity) based on which a linear free energy relationship is established, and only to a much smaller degree by the type of natural sorbents used (see Schwarzenbach and Westall, 1981).

Thus, such relationships with empirically derived parameters $\underline{a}$ and $\underline{b}$ are very useful for predicting partition coefficients of a great number of nonpolar organic compounds between water and natural sorbents of very different origins. Values reported for $\underline{a}$ and $\underline{b}$ are summarized in Table V. It should be noted that Eq. 4 is applicable only to sorbents exhibiting organic carbon contents of greater than 0.1% ($f_{oc} > 0.001$). For these sorbents, however, prediction of K_p values within a factor of 2 should be possible, provided that a relationship is used which was established from structurally similar compounds. Finally, there is good evidence that in the case of hydrophobic partitioning, competitive interactions between dilute co-contaminants seem to be of minor importance (e.g., Karickhoff et al., 1979; Means et al., 1980; Schwarzenbach and Westall, 1981).

For sorbents of low organic carbon content (i.e., $f_{oc} < 0.001$) as are encountered in many aquifers, relatively small K_p values are found even for highly hydrophobic compounds. In addition the dependence of K_p on hydrophobicity of the compound is much less pronounced for such organic poor sorbents compared to sorbents with higher organic carbon content. The degree of this dependence is reflected in the magnitude of the value $\underline{a}$ in Eq. 4, which, for chlorinated and alkylated benzenes, is 0.25 for γ-Al$_2$O$_3$ and $\sim$ 0.7 for many natural sediments. For example, with γ-Al$_2$O$_3$ (specific surface area = 120 m^2 g^{-1}) the K_p of 1,2,4,5-tetrachlorobenzene was found to be only on the order of 2 cm^3 g^{-1}, or 4 times greater than the K_p of chlorobenzene; with an aquifer material of 0.7% organic carbon and a specific surface area much smaller than that of γ-Al$_2$O$_3$ (4 m^2 g^{-1}), the K_p of the tetrachlorobenzene was 40 cm^3 g^{-1} or 40 times greater than that of chlorobenzene. For more details of this study, see Schwarzenbach and Westall (1981). Further work is necessary to establish quantitative relationships for estimating partition coefficients of non-polar organic compounds with organic-poor sorbents. Another area which needs more research, is the influence of colloidal and/or dissolved organic material (e.g., fulvic and humic materials, solvents, detergents, etc.) on the transport of non-polar organic contaminants, in particular, of highly hydrophobic compounds. The results of the few studies on this topic (Gschwend and Wu, 1984; Brownawell and Farrington, 1985a,b, and references cited therein) clearly indicate that, in some cases (e.g., when studying leachates), interactions between the contaminant and the bulk of organics present in the groundwater cannot be neglected. We will present some results from Brownawell and Farrington, 1985a,b in a later section.

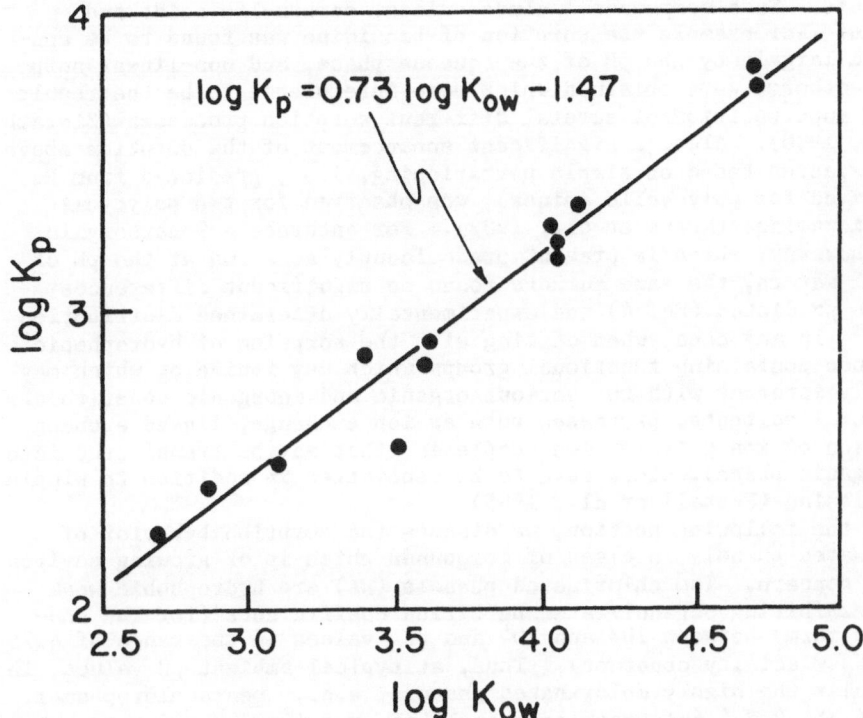

Figure 4. Linear free-energy relationships: Correlation between K_p and K_{ow}. The compounds used were methyl- and chlorobenzenes ranging from toluene to tetramethylbenzene and chlorobenzene to tetrachlorobenzene (Schwarzenbach and Westall, 1985).

3.2 Sorption of Ionizable Hydrophobic Compounds: Chlorinated Phenols

The simple partitioning model used to describe the sorption of neutral
hydrophobic organic chemicals (Eq. 2) is applicable only to a limited
degree to compounds which are fully or partially ionized at natural
pH values. Such compounds include amines, carboxylic acids and
phenols. For example the sorption of benzidine was found to be controlled largely by the pH of the aqueous phase, and non-linear sorption isotherms were obtained which were interpreted to be the result
of the superposition of several different sorption processes (Zierath
et al., 1980). Also, a significant enhancement of the sorption above
that expected based on simple partitioning, i.e., predicted from Eq.
4 (derived for polycyclic amines), was observed for two polycyclic
aromatic amines (Means et al., 1982). For anthracene-9-carboxylic
acid, however, which is present predominantly as anion at the pH of
natural waters, the same authors found no significant differences
between predicted (Eq. 4) and experimentally determined distribution
ratios. In any case, when dealing with the sorption of hydrophobic
compounds containing functional groups which may ionize or which may
strongly interact with the various organic and inorganic constituents
of natural sorbents, processes such as ion exchange, ligand exchange,
formation of ion pairs or ion complexes (that may be transferred into
the organic phase), etc., have to be considered in addition to simple
partitioning (Westall et al., 1985).

In the following section, we discuss the sorption behavior of
chlorinated phenols, a class of compounds which is of growing environmental concern. The chlorinated phenols (HA) are hydrophobic weak
acids exhibiting octanol/water partition coefficients (for the non-
ionized form) between 10^2 and 10^5 and pK_a values in the range of 4.75
to 9 (K_a = acidity constant). Thus, at typical ambient pH values, in
particular the highly chlorinated phenols, e.g., pentachlorophenol,
pK_a = 4.75; 2,3,4,6-tetrachlorophenol, pK_a = 5.40; 2,3,4,5-tetrachloro-
phenol, pK_a = 6.35 are present in the water predominantly as phenolate anions (A^-). From the results of a recent study by Schellenberg et al. (1984), the following conclusions concerning the sorption
behavior of chlorinated phenols can be drawn:
(i)   For concentrations up to at least 1 μmol L^{-1} of chlorina-
        ted phenols and at fixed pH and ionic strength, a linear
        isotherm can be used to describe the overall distribution
        between water and sorbents with organic carbon contents
        greater than about 0.1% (see example given in Fig. 5). The
        distribution ratio, D, is defined for the total analytical
        concentrations:

$$D = \frac{[\ \overline{HA}\ ] + [\ \overline{A}\ ]}{[\ HA\ ] + [\ A^-\ ]} \tag{5}$$

where the overbar indicates species in the non-aqueous

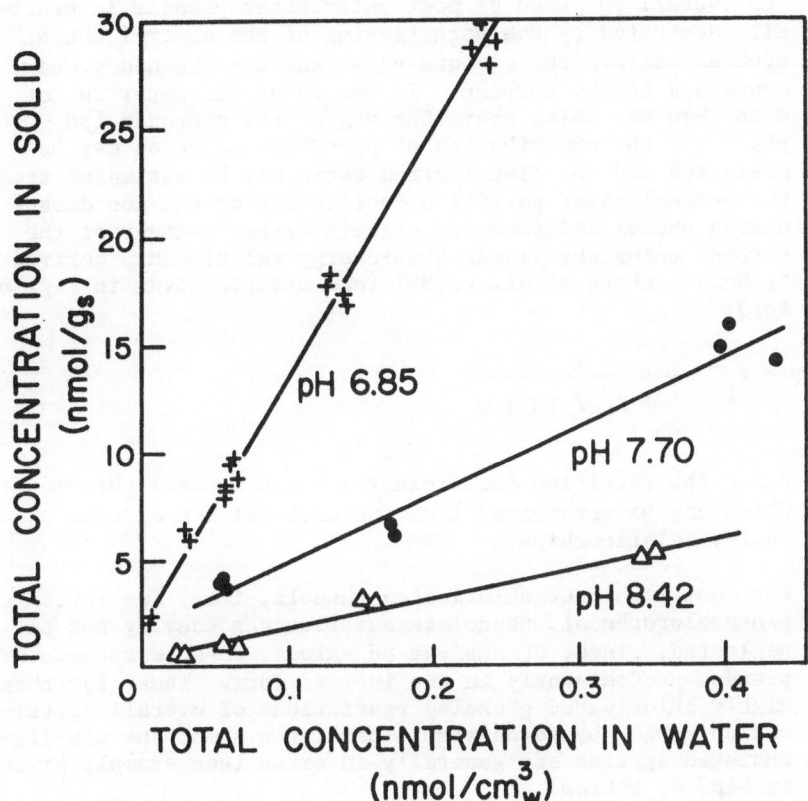

Figure 5. Sorption isotherms for 2,3,4,5-tetrachlorophenol (pK_a= 6.35) at different pH values. Sorbent: river sediment (f_{oc}=0.026). Adapted with permission from Schellenberg et al. (1984). Copyright 1984, American Chemical Society.

phase. As is seen from Fig. 5, the value of D is strongly pH dependent.

(ii) In waters of low ionic strength (i.e., I ≈ 5 x 10⁻³ M), the overall sorption of most chlorinated phenols is generally dominated by the partitioning of the neutral phenol species between the aqueous phase and the organic phase contained in the sorbent. If the pH of the water is not more than two units above the pK$_a$ of the compound (pH − pK$_a$ > 2), the contribution of phenolate sorption may be neglected and the distribution ratio may be estimated from the octanol/water partition coefficient of the non-dissociated phenol and from the organic carbon content of the sorbent using the linear free-energy relationship derived by Schellenberg et al. (1984) (see example given in Fig. 6, top):

$$D = K_p \cdot \frac{1}{1 + K_a / [H^+]_w} \tag{6}$$

K$_p$ is the partition coefficient for the neutral phenol which can be determined from conventional linear free energy relationships.

(iii) For some important chlorinated phenols, i.e., for tetra- and pentachlorophenol, phenolate sorption can usually not be neglected, since, at ambient pH values, these compounds are present predominantly in the ionized form. Thus, for these highly chlorinated phenols, predictions of overall distribution ratios based on simple partitioning of the non-dissociated species are generally in error (see example given in Fig. 6, bottom).

There is evidence that the sorption of phenolate species is predominantly a partitioning process between the aqueous phase and the organic phase present in a natural sorbent, and that the sorptive processes are strongly influenced by the ionic strength in the aqueous phase. Thus, for the description of the overall sorption of highly chlorinated phenols, more complete mechanisms have to be considered. Some of the mechanisms involved in the distribution of hydrophobic ionizable organic compounds between the aqueous phase and the non-aqueous phase are discussed in a recent study by Westall et al. (1985).

In the experiments discussed here, the distribution of chlorinated phenols between water and the reference solvent n-octanol was studied as a function of pH and ionic strength. Experiments carried out at fixed ionic strength (0.2 M NaCl) in the aqueous phase and varying pH revealed three domains for the distribution ratio as a function of pH (see Figure 7). At low pH, the neutral form of the chlorinated phenol predominates in both the aqueous and nonaqueous phases, and the observed distribution ratio is approximately independent of pH. At pH values up to a few units above the pK$_a$ of the chlorinated phenol,

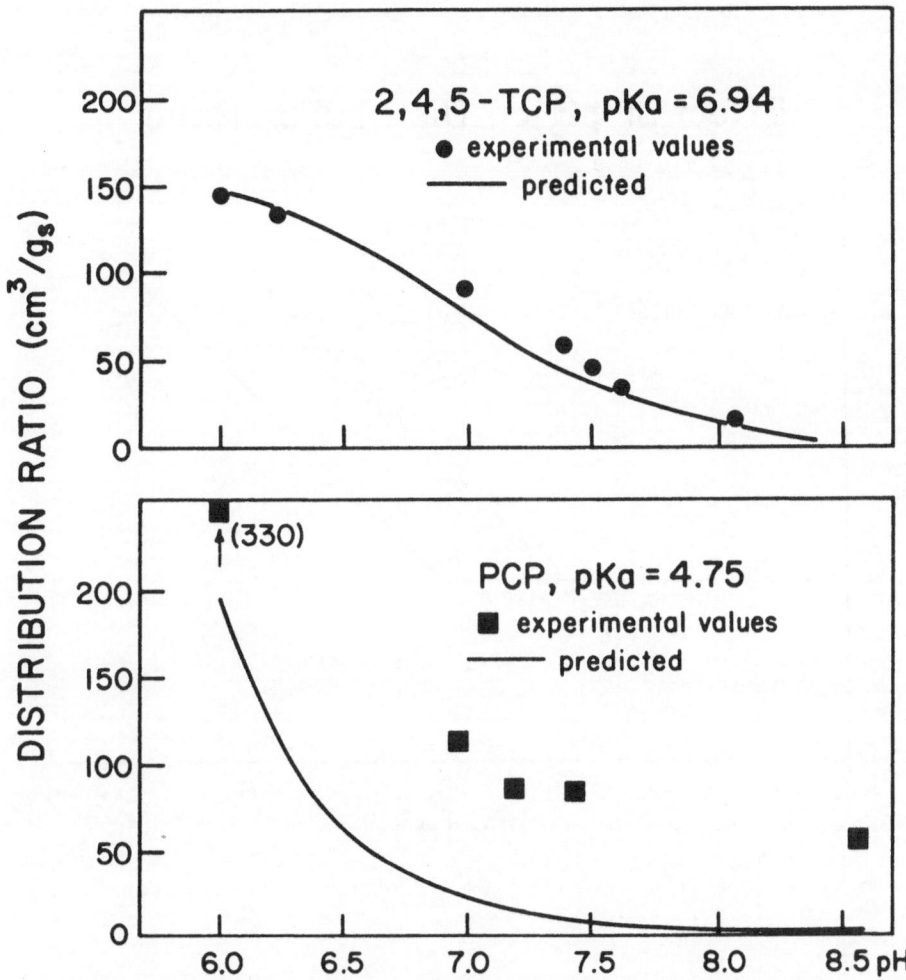

Figure 6. Predicted (Eq. 6) vs. experimentally determined overall distribution ratios for 2,4,5-trichlorophenol (top, pK_a = 6.94) and pentachlorophenol (bottom, pK_a = 4.75) as a function of pH. Sorbent: lake sediment (f_{oc} = 0.094). The pH is controlled by a $CaCO_3/CO_2$ buffer. Note that the Ca^{2+} concentration decreases with increasing pH. Adapted with permission from Schellenberg et al., 1984. Copyright 1984, American Chemical Society.

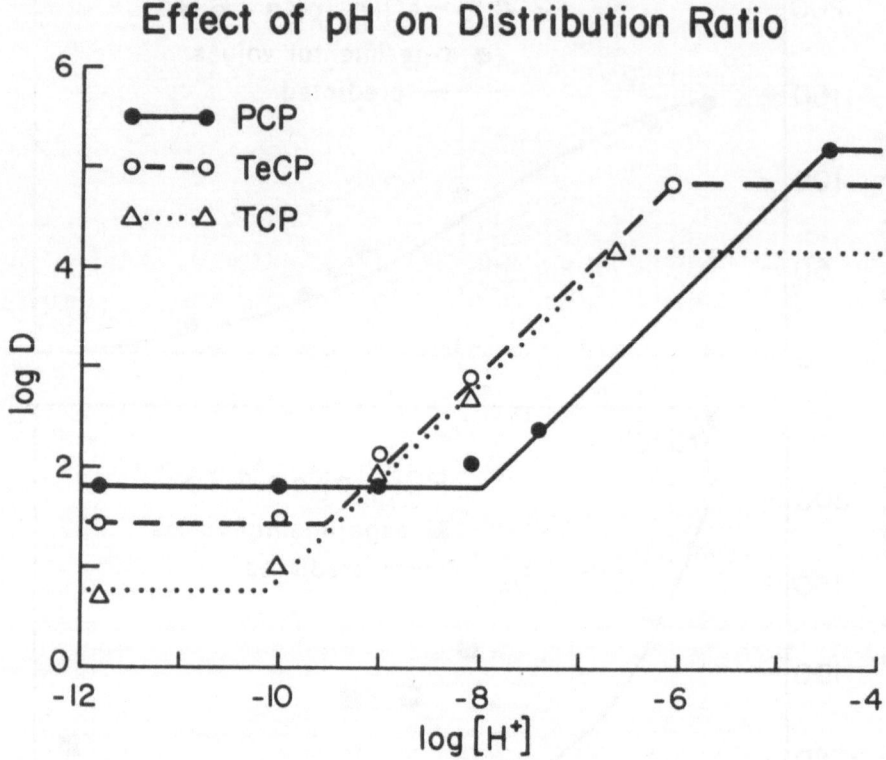

Figure 7. Octanol/water distribution ratio of pentachlorophenol
(PCP), 2,3,4,5-tetrachlorophenol (TeCP), and 2,4,5-trichlorophenol
(TCP) as a function of pH. Ionic strength of the aqueous phase is
0.2 M (NaCl).

the phenolate ion A$^-$ is the predominate species in the aqueous phase
and the neutral species is predominant in the nonaqueous phase. In
this region, the observed distribution ratio is proportional to the
hydrogen ion concentration. At high pH values, the phenolate ion is
the predominate ion in both the aqueous and nonaqueous phases, and
the distribution ratio is again independent of pH, but much less
favorable for the nonaqueous phase than at low pH, reflecting the
relative compatibility of the neutral and ionized species with the
aqueous and nonaqueous phases. In the high pH domain the distribution
ratio is also dependent on ionic strength, since the cation of the
salt must be transferred to the bulk of the non-aqueous phase to compensate the charge of the phenolate anion.

3.3 Field Studies

A rough estimate of the retention behavior of a compound in a groundwater aquifer may be obtained if the partition constant K_p is known.
For porous media it has been shown that only the fine fraction of the
aquifer material (fraction passing through a 125 μm sieve) predominates for sorption (Karickhoff et al., 1979; Schwarzenbach and
Westall, 1981). Thus, on the assumption of a homogeneous distribution of the fine material in a given aquifer segment, an average
retardation factor R_f = ratio of residence time of the compound τ
to the residence time of the water τ_w can be calculated:

$$R_f = \frac{\tau}{\tau_w} = 1 + fK_p \, \rho \, (1 - \varepsilon)/\varepsilon \qquad (7)$$

where f is the fraction of the aquifer material responsible for sorption (e.g., $\phi < 125$ μm), K_p is the equilibrium partition coefficient
between that fraction and water, ρ is the density of the aquifer
material and ε is the total porosity.

3.4 Application of Estimated Retardation Factors to Field Studies

To date, there are not enough field data available to allow a general
assessment of the validity of model calculations of retardation factors from relationships such as given in Eqs. 4 and 7. Furthermore,
many of the few studies reported (Roberts et al., 1982; Schwarzenbach
et al., 1983; Mackay et al., 1983; Curtis et al., 1984) were confined
mostly to contaminants of low to medium hydrophobicity, i.e., compounds exhibiting octanol/water partition coefficients between 10^2
and about 10^4. For these compounds, however, which included predominantly halogenated C_1- and C_2-compounds as well as chlorinated
benzenes, satisfactory agreement was found between model predictions
and actual retardation of the compounds in the aquifer, if the organic
carbon content of the aquifer exceeded 0.1%.

Fig. 8 gives an example from a year-round study at a river water/
groundwater infiltration system, where it was possible to determine
experimentally an average field retardation factor for tetrachloroethylene (log K_{ow} = 2.88) for the aquifer segment between the river

and an observation well located at a distance of 4 m from the river
bed (for details of this study see Schwarzenbach et al., 1983). Based
on temperature measurements (Fig. 8a), the average residence time of
the water between the river and the observation well was approximately
half a month (assuming a retardation factor of $\approx$ 2 for the temperature). Fig. 8b shows that the response to the seasonal variation of
the tetrachloroethylene concentration in the river (peak in the winter
months) occurred in the groundwater about four to five months later,
which suggests an average retardation factor on the order of 8 to 10
for this compound. The f_{oc} values determined for the fraction < 125
µm of various samples from the quite heterogeneous aquifer between
the river and the observation well ranged from 0.002 to 0.008, and
the f values were found to be between 0.2 and 0.3. Thus, when assuming a density of 2.5 g cm^{-3} and a porosity of 0.2, an average retardation factor of between 4 and 10 would be predicted from Eqs. 7 and
4, which is in reasonable agreement with the field observation. This
example demonstrates the applicability of the retardation factors to
a reasonably well controlled field system. Another example involving
migration of chemicals from waste disposal sites is described in the
next section.

3.5 Migration of Chlorophenols from a Chemical Waste Disposal Site

Johnson et al. (1985) describe the migration of chlorophenols from a
chemical waste disposal site at Alkali Lake in South Central Oregon,
U.S.A. The wastes included a mixture of chlorophenol and chloro-
phenoxyphenol by-products from the manufacture of the herbicide 2,4-
dichlorophenoxyacetic acid (2,4-D), and 4-methyl-2-chlorophenoxyacetic
acid (MCPA). The groundwaters in the Alkali Lake region are at pH
10. In solutions of this pH value, the chlorophenols are present
primarily as the anionic species, and thus highly mobile.
    In November, 1976, over twenty-five thousand 210 liter drums which
had been stored at the site were crushed and buried in shallow unlined
trenches. Since that time an array of 49 wells has been used for
monitoring the movement of chemicals in the ground water. Among the
chemicals monitored were 2,4-DCP, 2,6-DCP, 2,4,6-TCP, 2,3,4,6-TeCP,
and PCP. The hydrology is described by Pankow et al. (1984). The
concentration contours for the two dichlorophenols and the trichloro-
phenol are similar, while the tetra- and pentachlorophenols are significantly retarded (Figure 8'). The mobility of the compounds were
interpreted quantitatively in terms of the distance from the site at
which the concentration in the groundwater was 2% or 25% of the maximum value (Table VI). Since the mobility of the water in this system
is not known, it is impossible to calculate absolute retardation factors. However, one can calculate the retention of the phenols relative to each other. These relative retention factors are presented
in Table VI.
    These values of relative retention factors observed in the field
can be compared to those predicted from Equation 7. It is seen that
the predicted relative retention factors are about a factor of two
greater than the field values for TeCP and PCP. The authors discuss

Table V. Estimation of K_{oc} based on K_{ow}; log K_{oc} = a log K_{ow} + b

Correlation Coefficients a	b	Correlation Coefficient r²	Number of Compounds	Range of log K_{ow}	Type of Chemicals	Reference
0.544	1.377	0.74	45	-3.0-6.6	primarily agricultural chemicals	Kenaga and Goring (1980)
1.00	-0.21	1.00	10	2.1-6.3	polycyclic aromatic hydrocarbons	Karickhoff et al. (1979)
0.937	-0.006	0.95	19	2.1-6.3	compounds of Karickhoff et al. (1979), triazines, nitroanilines	Brown and Flagg (1981)
1.029	-0.19	0.91	13	0.4-6.3	herbicides, insecticides	Rao and Davidson (1980)
1	-0.317	0.98	13	1.6-6.5	compounds of Karickhoff et al. (1979), heterocyclic aromatic compounds	Means et al. (1980)
0.72	0.49	0.95	13	2.6-4.7	chlorinated hydrocarbons, alkylbenzenes	Schwarzenbach and Westall (1981)
0.52	0.64	0.84	30	0.5-3.3	substituted phenyl ureas and alkyl-N-phenyl carbamates	Briggs (1981)

Table VI. Distance traveled (X) and relative retardation (R_r)
values for Alkali Lake soil/water system (pH = 10)
(from Johnson et al., 1985).

Compound	X (2%)	X (25%)	R_r (2%)[a]	R_r (25%)[b]	R_r (pred)[c]
2,4–DCP	400.	250.	1.0	1.1	1.0
2,6–DCP	420.	270.	1.0	1.0	1.0
2,4,6–TCP	400.	270.	1.0	1.0	1.0
2,3,4,6–TeCP	330.	210.	1.2	1.3	3.5
PCP	260.	40.	1.6	6.8	13.5

[a]R_r (2%) = retardation factor relative to 2,6–DCP = $X_{2,6-DCP}$
(2%)/X(2%).

[b]R_r (25%) = retardation factor relative to 2,6–DCP = $X_{2,6-DCP}$
(25%)/X(25%).

[c]R_r (pred) = R (pred)/$R_{2,6-DCP}$ (pred). Retardation factors
predicted from equation 7. Laboratory batch experiments were used
to determine fK_p; ε and ρ were determined from observation of
material from the field.

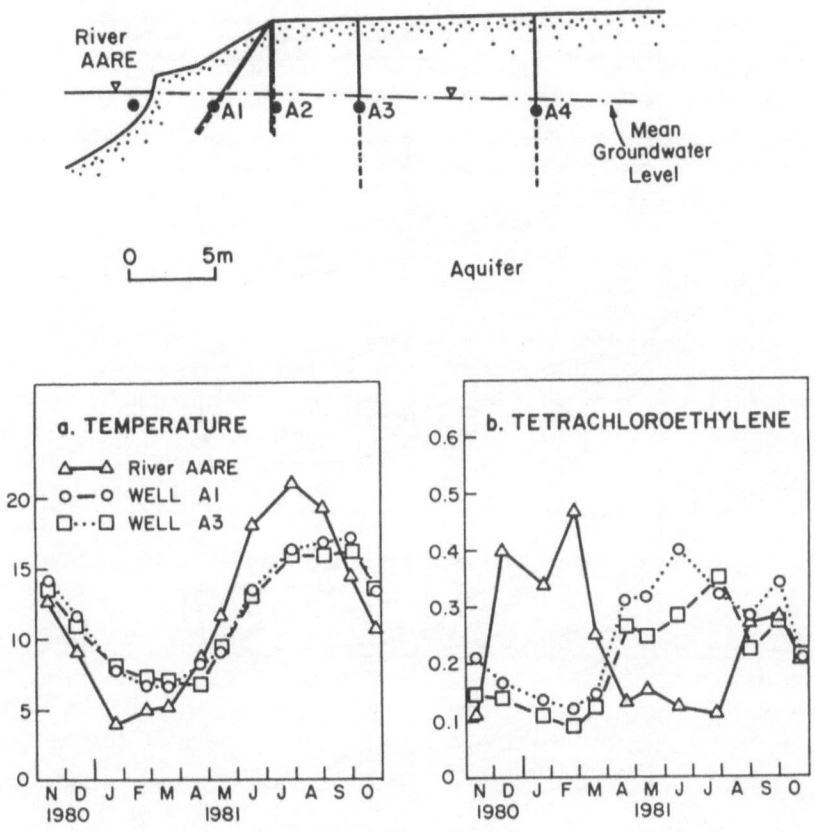

Figure 8. Layout of a field study site in the Lower Aare Valley, Switzerland, and monthly determined values for temperature, °C (a) and tetrachloroethylene concentration, g/L (b) in the River Aare and in observation wells Al and A3. Adapted with permission from Schwarzenbach et al. (1983). Copyright 1983, American Chemical Society.

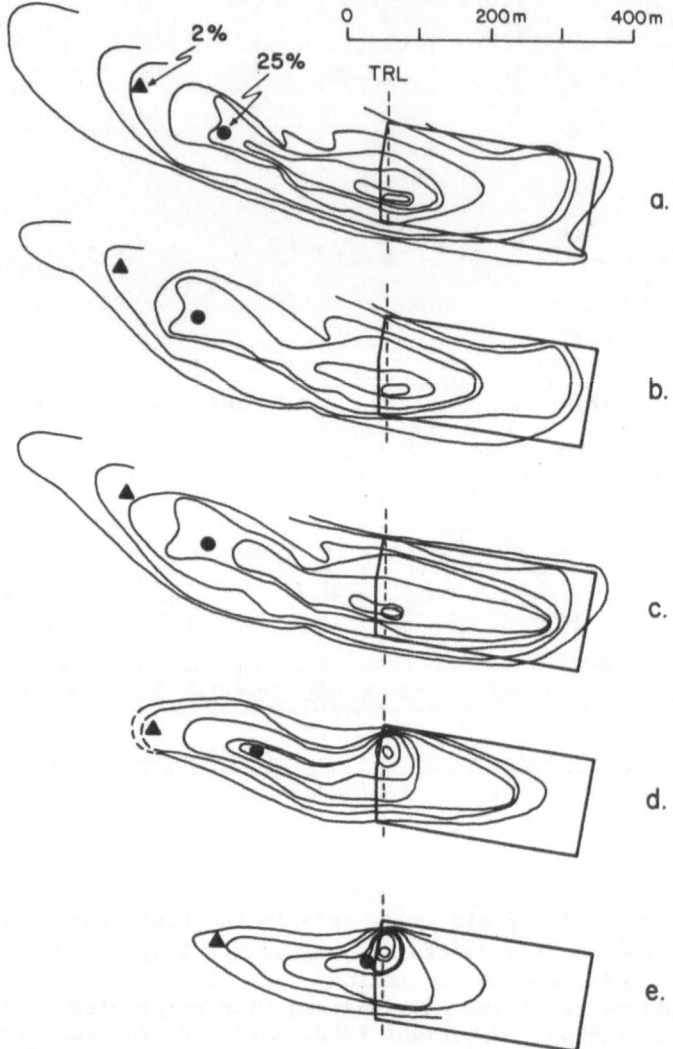

Figure 8'. Isopleths of concentration of chlorinated phenols in groundwater near Alkali Lake disposal site: a. 2,4-dichlorophenol; b. 2,6-dichlorophenol; c. 2,4,6-trichlorophenol; d. tetrachlorophenol; e. pentachlorophenol. Adapted with permission from Johnson et al., 1985. Copyright 1985, Waterwell Journal Publishing Co.

several possible causes for this apparent enhanced mobility of TeCP
and PCP in the field: (i) co-solvent effects within the plume itself;
(ii) spatially or temporally non-uniform release of the chemicals;
(iii) fractures in the aquifer; (iv) non-uniformity of groundwater
flow. Evidence is presented to support the last reason.

3.6 Leaching of Chemicals from a Landfill Site

The leaching of organic compounds from a landfill site near North Bay,
Ontario has been described by Reinhard et al. (1984). The site lies
on a complex deposit of glaciofluvial sand and has been receiving
primarily domestic and commercial wastes with small quantities of
industrial waste since 1962. The aquifer is approximately 20 meters
deep with a 0.5 m unsaturated zone. The groundwater aquifer discharges a number of springs within 800 m of the dump site.
    The plume of the aquifer has been monitored in both the vertical
and horizontal direction. The extent of the leachable plume is indicated in Figure 8'' top by the concentration of a conservative tracer,
chloride ion. However, as is shown by the concentration profiles in
Figure 8'' (bottom) there appears to be a secondary source of some
organic compounds near the bottom of the aquifer. The authors consider possible origins of this source: (i) transport of the compounds
to the bottom of the aquifer in an organic medium that is more dense
than water, e.g. tri- or tetra-chloroethylene; (ii) differences in
distributions of sources of the compounds within the landfill; and
(iii) differences in degradation rates vertically over the aquifer
due to different conditions (redox potential biological activity,
etc.). Several classes of compounds appear to have migrated from the
landfill site to the springs 800 m away. Biological or chemical
transformations of some compounds in the aquifer were detected. This
study by Reinhard et al. (1984) illustrates the complexity of transport in the subsurface terrestrial environment and the necessity for
three-dimensional monitoring.

3.7 PCBs in Coastal Estuarine Areas

The EPA Superfund Site in the Acushnet River Estuary - Buzzards Bay,
Massachusetts, USA provides a field example of a PCB contaminated
coastal estuarine area to illustrate applications of the relationships
discussed in previous sections. Polychlorinated biphenyls are examples of the types of hydrophobic compounds of environmental concern
in the marine environment. Brownawell and Farrington (1985a,b) and
Farrington et al. (1985) have measured concentrations of total PCBs
and several individual chlorobiphenyls in water, sediment, interstitial water and several species sampled in the area. We discuss a few
examples from these studies.
    Brownawell and Farrington (1985a,b) have presented results from
two box cores showing the distributions of chlorobiphenyls in interstitial waters and sediment. Data for one of the cores are given in
Table VII. D in Table VII are calculated from measured concentrations
according to equation 8.

Table VII. Profiles of TOC, DOC, total PCBs, and K'$_d$s of selected isomers at station 67 (from Brownawell and Farrington, 1985b).

| Depth | TOC | DOC | Total PCBs | | D X 10^{-3} (L/kg) | | |
| | | | Sediments | Pore Waters | Chlorobiphenyls | | |
cm	%Org C	mg/L	(µg/g)	(µg/L)	52	101	153
0-3	6.06	14.4	16.6	1.31	12.3	11.6	---
3-5	5.23	17.4	16.0	3.61	4.11	3.63	3.42
5-7	3.74	27.7	14.6	4.72	2.84	2.17	2.00
7-9	4.98	40.3	16.0	9.97	1.68	1.34	1.25
9-11	4.40	47.9	21.6	20.1	1.32	0.950	0.746
11-13	4.14	32.6	24.6	12.7	2.36	1.68	1.48
15-17	4.02	39.2	33.0	13.8	3.08	2.31	1.90
17-19	4.74	55.1	27.8	14.8	2.45	1.59	1.48
19-21	4.98	42.0	30.7	14.9	2.04	1.72	1.62
21-23	5.96	87.0	30.3	13.6	1.73	1.78	1.92
23-25	5.68	50.9	26.1	12.3	2.68	2.03	1.74
25-27	4.78	81.4	27.4	9.68	3.31	2.63	2.51
27-29	4.86	43.7	27.3	8.37	4.88	3.33	2.73
29-31	5.33	54.2	25.2	11.0	3.51	2.66	2.29
35-41	4.75	41.4	13.4	8.18	1.67	1.93	1.53

$$D = \frac{\text{concentration in sediment}}{\text{total concentration in pore water}} \tag{8}$$

$$(D \text{ in units of } L/kg)$$

The data were modeled assuming an approach to equilibrium. Three principal phases - dissolved, colloidal, and solid are assumed and are described by equation 10.

$$D = \frac{f_{oc}s \, K_{oc}s \, C}{C + f_{oc}c \, K_{oc}c \, C} \tag{9}$$

where C is concentration of dissolved chemical, $f_{oc}s$ and $f_{oc}c$ are fraction of organic carbon of sediment and colloidal phases respectively. $f_{oc}c$ is some fraction α of the total measured DOC. If we simplify for purposes of modeling and assume $K_{oc}s = K_{oc}c$ equation 9 becomes

$$D = \frac{f_{oc}s \, K_{oc}s}{1 + \alpha DOC \, K_{oc}s} \tag{10}$$

Figure 9 presents log D vs. log K_{ow} for several individual chloro-biphenyls for two intervals from a core of sediment at Station 67 of Brownawell and Farrington (1985b). The data strongly indicate the importance of a colloidal organic matter phase because the three-phase model predicts the observed distribution much better. However, there are still apparent departures from prediction for the measured K'd which could be due to departures from equilibrium due to sorption kinetics, steric effects on sorption, and microbial degradation. The exact reasons are unknown at this time and need further investigations.

Similar treatment of water column data from particulate matter and filtrate analyses for filtered water column samples from the same area yielded the plots shown in Figure 10 (Brownawell and Farrington, 1985b). It is apparent from this and data from other samples in this study that colloidal organic matter is less important to water column distributions of chlorobiphenyls in comparison to the interstitial water-sediment components of this ecosystem.

The association with colloids has obvious importance for understanding key aspects of biogeochemical cycles related to calculating or predicting the flux from polluted sediments to the overlying water column and bioavailability.

3.8 Chlorinated Pesticides and PCBs in the Open Ocean Water

Most measurements of chlorinated pesticides and PCBs in open ocean waters were made prior to 1976 (Farrington and Vandermeulen, 1983). Problems associated with suspected contamination during sampling have made measurements below the upper 10 meters of surface waters a prob-

lem. De Lappe et al. (1983) have discussed these problems and reported on a polyurethane foam adsorbent used to isolate chlorinated
hydrocarbons from 100 to 1,000 liters of seawater. However, the
deployment of such a pumping system in waters deeper than 60 meters
has yet to be undertaken, although it appears to be feasible technically.

Despite the problems with conventional sampling with hydrocast
bottles, Tanabe and Tatsukawa (1983) have undertaken the much needed,
difficult task of measuring chlorinated pesticides and PCB depth profiles in the open ocean. Their data and interpretations are instructive; but we must keep in mind, as stated by those authors, that data
from deep waters probably suffered from some degree of artefact or
contamination during sampling.

Despite this problem, the comparisons of the upper ocean depth
profiles, particulate matter/filtrate partitioning and calculation of
residence time in the euphotic zone are of interest to the theme of
our paper. They report measurements for samples from depths ranging
to 1500 to 5000 meters at six stations in the western Pacific Ocean
and near Antarctica. Their major findings were:

i)   for the deepest sample at each station: total PCBs = 0.035 –
0.49 x 10^{-9} g/L; total DDT = 0.007 – 0.67 x 10^{-9} g/L; total HCH
(hexachlorocyclohexane) = 0.079 – 1.8 x 10^{-9} g/L;

ii)   a plot of log solubility versus log of the ratio of suspended
matter concentrations divided by filtered water concentrations yields
a negative correlation in accord with equilibrium partitioning
(Figure 11);

iii)   residence times for PCBs, HCHs and DDT compounds were estimated using extrapolations from large particle flux of organic matter
from other studies, suspended particulate matter chlorinated hydrocarbon data from their study, and adjustments for primary productivity
differences between stations. Results are presented in Table VIII.
The more soluble HCH compounds have much longer residence times in
the upper ocean euphotic zone than the less soluble DDT and PCBs as
expected from physical–chemical considerations based on solubilities.

4. BIOACCUMULATION; UPTAKE AND RELEASE OF ORGANIC COMPOUNDS

There has been a school of thought that adhered to the theory
that interaction between water and aquatic organisms was the major
factor controlling body burdens of non-polar organic pollutants
(Hamelink et al., 1971; NAS, 1979). This led to consideration of an
equilibrium partitioning theory and application of parameters such as
Kow to predict bioconcentration factors.

Mackay (1982) has recently presented an assessment of data from
the literature illustrating the relationship of Kow to biological
concentration in aquatic organisms as shown in Figure 12. This relationship applies in several cases and in a general manner as a useful
guide for further considerations. Factors such as differential uptake
and release rates, food input, different degrees of bioavailability
and metabolism and excretion by the organism are not taken into

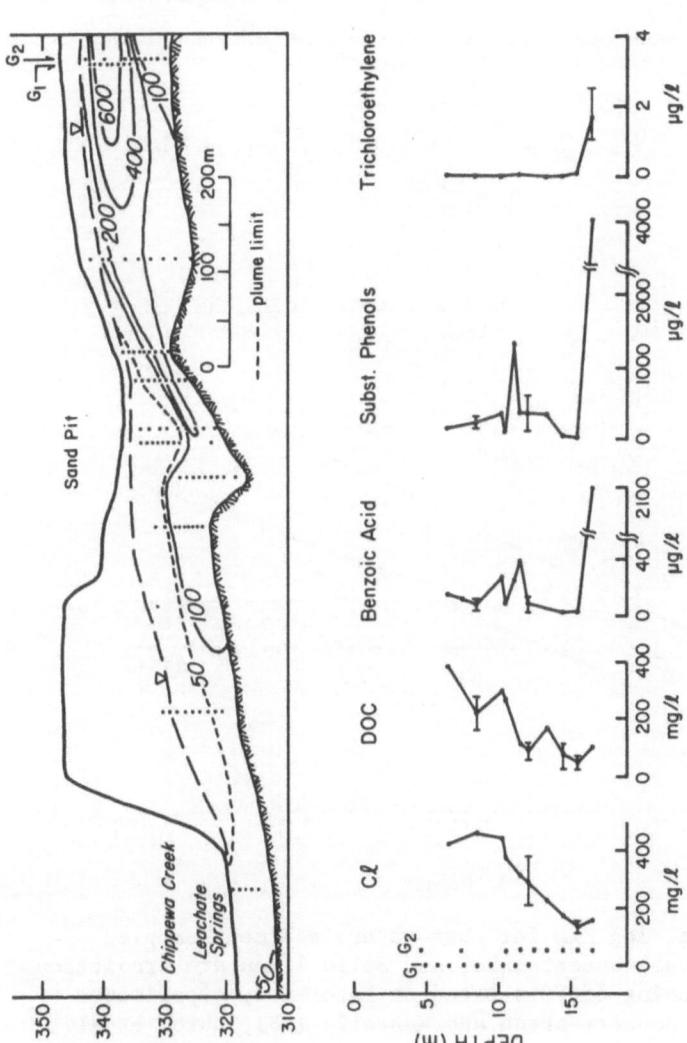

Figure 8''. a. Cross-section of aquifer showing vertical distribution of Cl⁻ plume from landfill leachate. Concentration in mg/L. Points represent sampling sites. b. Vertical concentration profiles for Cl⁻ and selected organic compounds in grounwater near edge of landfill at points G-1 and G-2 shown in a. Adapted with permission from Reinhard et al., 1984. Copyright 1984, American Chemical Society.

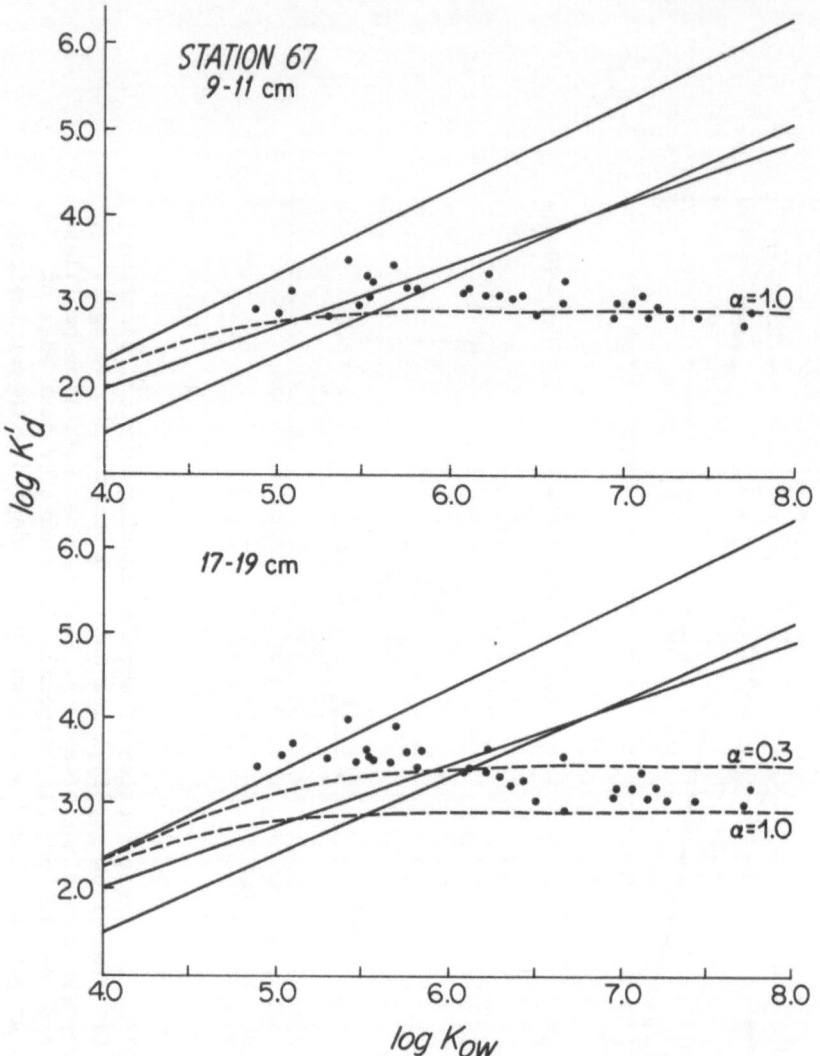

Figure 9. Log K'd vs. log K_{OW} for pore water, sediment samples
from Buzzards Bay, Massachusetts, U.S.A. Solid lines are predictions
of two-phase partitioning of PCBs based on laboratory experiments
(Means et al., 1980; Schwarzenbach and Westall, 1981; Chiou et al.,
1979). Dotted line represents calculations using three phase, solid,
colloid, solution model.   is the fraction of dissolved organic
carbon assumed to be colloidal. Solid dots are data points (from
Brownawell and Farrington, 1985b). Note: IUPAC convention is still
not firmly established and Brownawell and Farrington have used log
K'd in the manner of an apparent K_D (distribution constant) since
equilibrium is <u>assumed</u>, only.

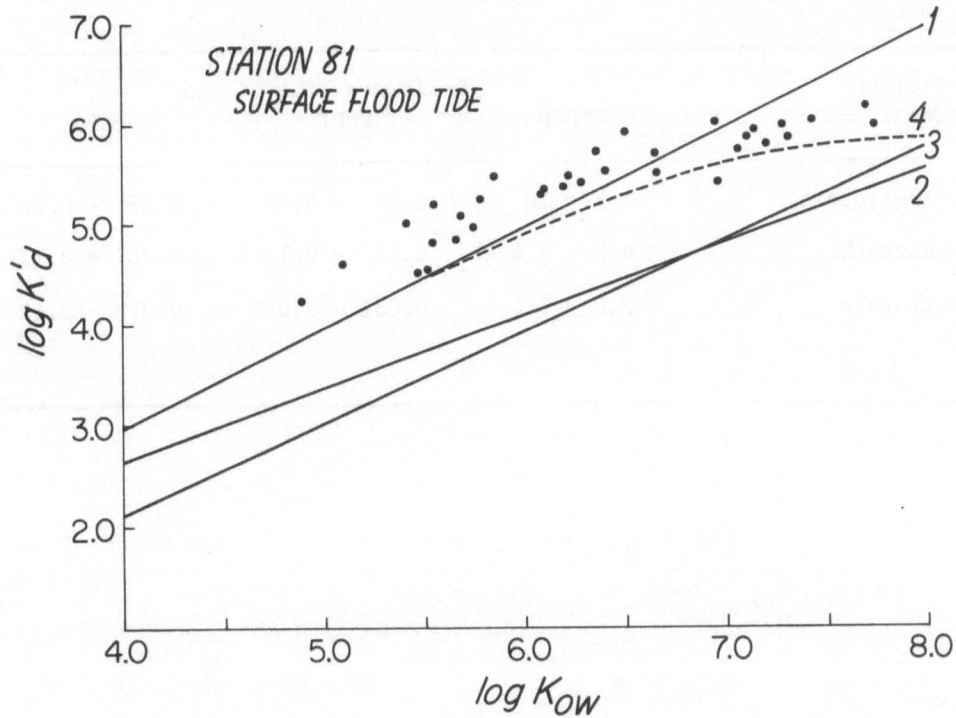

Figure 10. Log K'd vs. log K_{OW} for filtered particulate matter and filtrate for water sample from New Bedford Harbor, Buzzards Bay, Massachusetts U.S.A. (Brownawell and Farrington, 1985b). Numbers refer to predicted values from other studies (1, 2, 3 - see legend, Figure 9 and references, Brownawell and Farrington, 1985b) and (4) three phase partition model - solid, colloid, solution assuming all of the dissolved organic carbon is colloidal.

Table VIII. Estimated residence times for HCH, DDT, PCBs in open
ocean euphotic zone (from Table 3 of Tanabe and
Tatsukawa, 1983, in modified version).

Station Area Type	Residence Time (years)		
	Σ HCH	Σ DDT	Σ PCBs
Oligotrophic	5.1 – 10	0.19 – 0.37	0.38 – 0.76
Mesotrophic	4.3 – 6.4	0.08 – 0.12	0.082 – 0.12
Eutrophic	2.0 – 3.4	0.031 – 0.052	0.070 – 0.12

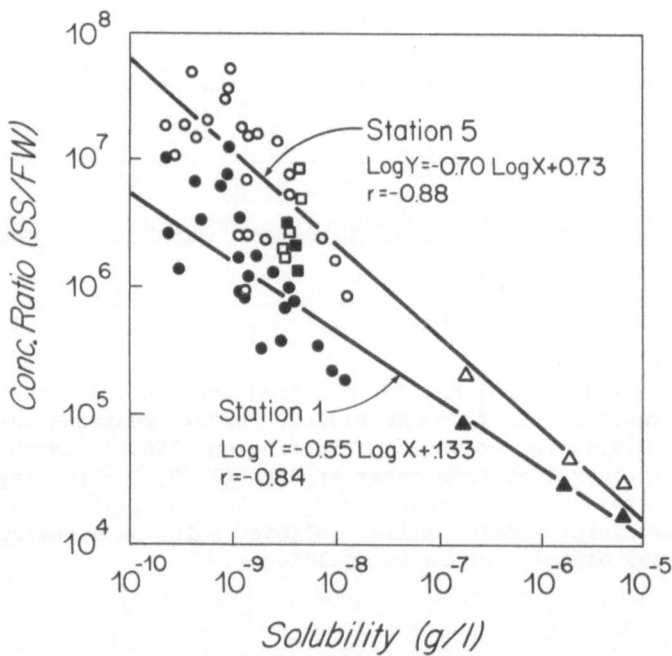

Figure 11. Relationship between the water solubilities of PCBs (● ○),
DDT compounds (■ □) and HCH isomers (▲ △) and the ratios of their
concentrations in suspended solids to those in filtered water. The
closed and open symbols indicate the Stns. 1 and 5, respectively
(from Tanabe et al., 1982). Adapted with permission.

account by this approach as discussed by O'Connor and Pizza (1985).
We use data from studies in the Acushnet River Estuary (Farrington
et al., 1985a) as an example.

Table IX presents representative data for individual chlorobiphenyls which have been measured for various species in the study
area: mussel (Mytilus edulis), lobster (Homarus americanus) muscle
tissue and viscera, muscle tissue for flounder (Pseudopleuronectes
americanus). Bioconcentration factors have been calculated using
water column data for individual chlorobiphenyls. Chlorine positions
on the biphenyl ring are given in Table X for chlorobiphenyls relevant to this paper. Plots of the log BCF (bioconcentration factors)
vs. log K_{OW} are given in Figures 13, 14, 15. Several chlorobiphenyls exhibit significant departures for measured values of BCF
compared to values expected from the empirical log K_{OW}, BCF relationships reported by Mackay (1982). Furthermore, log K_{OW}, BCF
relationships vary between species and vary between tissues within a
given species. Various factors such as lipid concentration and even
quality of the lipid composition are probably controlling factors
(Chiou, 1985) although we do not have data from this specific study
to support this suggestion. Figure 16 illustrates that there is
extensive alteration of the mix of chlorobiphenyls presumably by
metabolism of organic pollutants in crustacea (and for fish, not
shown) most likely by microsomal enzymes referred to as mixed function oxidases (NAS, 1979; Stegeman, 1981). Some chlorobiphenyls
appear to be more recalcitrant to metabolism or microbial degradation
than others. PCBs with chlorines in any or, a combination of
3,4,5,3',4',5' positions appear to be more recalcitrant than other
chlorobiphenyls as a result of blockage of adjacent C atoms at the
end of the biphenyl molecule preventing enzymatic formation of an
epoxide as the first step in metabolism (Schulte and Acker, 1974).

A variety of laboratory field studies provide evidence for the
need to extend log K_{OW}/BCF equilibrium theory to include kinetics.
A starting point for such a model is given in Table XI. The model
could be extended for other compartments of the environment or other
compartments (specific tissues) of organisms by including additional
coupled differential equations. Equations of the form of C and D
have been used to model uptake and release of hydrocarbons and PCBs
by bivalves (Fossato and Cazonier, 1976; Ernst, 1977; Farrington
et al., 1982, among others).

An illustration of field data is presented in Figure 17 for
Mytilus edulis transplanted from a relatively pristine site off Nantucket Island, Massachusetts, USA, to sites in Buzzards Bay: (i) New
Bedford Harbor-Acushnet River Estuary previously referred to as the
site with extensive PCB pollution; (ii) Penikese Island; and (iii)
Cleveland Ledge – both less polluted sites in the bay. The uptake
kinetics are described in a general manner by equations similar to
those in Table X as shown by the parameters in Table XI calculated
using those equations. Data from a second set of transplants are
also given in Table XII as are data for the reverse type of experiment for release of PCBs after several months exposure to high concentrations in the habitat and transplant to the relatively clean

water of Vineyard Sound, Massachusetts, USA. It is apparent from
closer examination of the data in these and other experiments that
the relatively simple equations of Table XI are not providing a good
fit of prediction with observed results after the first several weeks.
For example the release of petroleum compounds from bivalves has been
shown to be a function of the exposure concentration, and duration of
exposure. The higher the exposure concentrations and the longer the
exposure, the slower the overall release kinetics and the longer the
time to reach pre-exposure concentrations. The release kinetics have
suggested that the process should be modelled using the concept of
multiple compartments within the bivalves which, of course, is easy
to visualize given the multitude of tissue type and function in bivalves (and fish) (Stegeman and Teal, 1973; Farrington et al., 1982,
among others). In addition to variable exposure concentrations and
duration of exposure, other factors such as spawning exert influences
on body burdens (Phillips, 1980). The four year record of body burdens for _Mytilus edulis_ at a station location in Narragansett Bay,
Rhode Island, USA, is given in Figure 18. Although there are year to
year variations of average body burdens, perhaps reflecting different
exposure concentrations in the habitat, there are consistent reductions in body burdens from June through July which correspond to the
periods of spawning for the populations in this general area. We
have used total PCBs in these preceding examples to illustrate the
points considered. We recognize the importance of having individual
chlorobiphenyl data but have yet to complete that aspect of the work.

     O'Connor and Pizza (1985), Pizza and O'Connor (1983) have applied
similar mathematical models to treatment of data for PCB burden in
striped bass in the Hudson River. O'Connor et al. (1985) have also
applied similar models to kepone body burdens in fish in the Chesapeake Bay region. They have had success in these efforts, but extension of these models is limited by inability to satisfactorily predict water and sediment contaminant burdens and food web transfer. A
detailed discussion of these problems is beyond the scope of our
paper. However, in the next section we address some specific aspects
of these problems.

4.1 Polluted Sediment-Organism Interactions

     Sediments are an accumulation site for several medium to higher
molecular weight organic compounds of environmental concern (Baker,
1980). A major concern for ocean dumping is the accumulation of such
compounds in surface sediments and then subsequent release to the
overlying water and uptake by benthic organisms (O'Connor et al.,
1983). For these reasons, we briefly discuss the conclusions of two
recently published studies which provide significant new information
about these issues. Karickhoff and Morris (1985) have shown the
Tubificid oligochaetes transported more than 90% of hexachlorobenzene, pentachlorobenzene and trifluralin to the sediment surface in
laboratory microcosm experiments initiated with a uniform depth distribution of the test chemicals in the sediment. Diffusive transport
of chemicals to the surface from depth was much less than sediment

Table IX. Octanol-water partition coefficients and bioaccumulation concentration factors on a wet wt. basis of selected chlorobiphenyls.
Log BCF

IUPAC NO.	Log K_{ow}	Mussel	Clam	Lob A		Lob B		P. Amer.
				Musc.	Visc.	Musc.	Visc.	
28	5.69	3.74	3.66	4.14	5.37	3.87	4.95	3.35
52	6.09	4.42	4.06	3.95	5.44	3.47	4.86	3.67
49	6.22	4.47	4.08	3.88	5.35	3.13	4.50	3.95
44	5.81	4.48	4.11	3.80	5.09	3.11	4.44	3.72
70	6.23	4.84	4.17	4.20	5.63	3.69	5.07	4.66
95	6.55	4.78	4.07	4.75	6.28	4.44	5.81	4.56
101	7.07	5.10	4.41	4.70	6.28	4.10	5.72	5.01
87	6.37	5.12	4.54	4.57	6.13	4.13	5.56	5.29
60	5.84	4.74	4.16	4.72	6.15	4.28	5.71	4.49
153	7.75	5.48	4.83	5.75	7.46	5.37	7.08	5.90
138	7.44	5.45	4.93	5.72	7.38	5.26	7.01	5.85
128	6.96	5.42	4.81	5.78	7.39	5.42	7.12	5.33

Table X.   Individual Chlorobiphenyls

IUPAC No.	Chlorine Substitution
8	2,4'
28	2,4,4'
29	2,4,5
44	2,2',3,5'
49	2,2',4,5'
52	2,2',5,5'
60	2,3,4,4'
70	2,3',4',5
86	2,2',3,4,5
87	2,2',3,4,5'
95	2,2',3,5',6
101	2,2',4,5,5'
105	2,3,3',4,4'
110	2,3,3',4',6
118	2,3',4,4',5
128	2,2',3,3',4,4'
129	2,2',3,3',4,5
137	2,2',3,3',6,6'
138	2,2',3,4,4',5
143	2,2',3,4,5,6'
153	2,2',4,4',5,5'
156	2,3,3',4,4',5
180	2,2',3,4,4',5,5'

Table XI. Simple model for consideration of uptake and elimination kinetics.

$$1 \underset{k_2}{\overset{k_1}{\rightleftharpoons}} 2$$

C_1 concentration in compartment 1 (e.g. food, water, sediment, etc.)

C_2 concentration in compartment 2 (body, tissue, etc.)

. First order rate equations:

$$\frac{dC_1}{dt} = - k_1 C_1 + k_2 C_2 \qquad\qquad\qquad A$$

$$\frac{dC_2}{dt} = + k_1 C_1 - k_2 C_2 \qquad\qquad\qquad B$$

. Time course of concentrations:

$$C_1 (t) = [C_1 (o) - C_1 (\infty)] \exp [-(k_1 + k_2)t] + C_1 (\infty) \qquad C$$

$$C_2 (t) = [C_2 (o) - C_2 (\infty)] \exp [-(k_1 + k_2)t] + C_2 (\infty) \qquad D$$

$C (o)$: initial concentration

$C (\infty)$: equilibrium concentration

$$\qquad\qquad\qquad\qquad\qquad\qquad\qquad\qquad\qquad\qquad E$$

. At Equilibrium

$$\frac{C_1 (\infty)}{C_2 (\infty)} = \frac{k_2}{k_1}$$

Table XII.   Uptake and elimination rates for PCBs in <u>Mytilus</u> edulis
             transplanted in Buzzards Bay (Farrington et al., 1985b).

Experiment I – Uptake June–September, 1981 (Day 17 to Day 108)	$-k_a$*	r^2	$t_{1/2}$*
New Bedford	.01	.98	69
Cleveland Ledge	.01	.96	69
Penikese Island	.01	.75**	69
Experiment II – Uptake September–October, 1982 (Day 6 to Day 27)			
New Bedford	.03	.99	23
Cleveland Ledge	.02	.99	25
Penikese Island	.02	.93	35
Experiment III – Elimination February – June, 1982 (120 days)	$-k_e$*	r^2	$t_{1/2}$*
New Bedford transplanted to ESL-WHOI (Vineyard Sound)	-0.1	.66**	69

*Calculated from Equations in Table X under the assumption that the
system is far from equilibrium and the back reaction is negligible:

Absorbtion (assume $k_1 C_1 \gg k_2 C_2$

$$\frac{dC_2}{dt} = k_1 C_1$$

$$C_2(t) \tilde{=} C_2(o) \exp(k_1 t)$$

$$\ln C_2(t) = \ln(C_2(o)) + k_1 t$$

Elimination (assume $k_2 C_2 \gg k_1 C_1$)

$$\frac{dC_2}{dt} \tilde{=} k_2 C_2$$

$$C_2(t) = C_2(o) \exp(-k_2 t)$$

$$\ln C_2(t) = \ln C_2(o) - k_2 t$$

**Poor fit of data to equations.

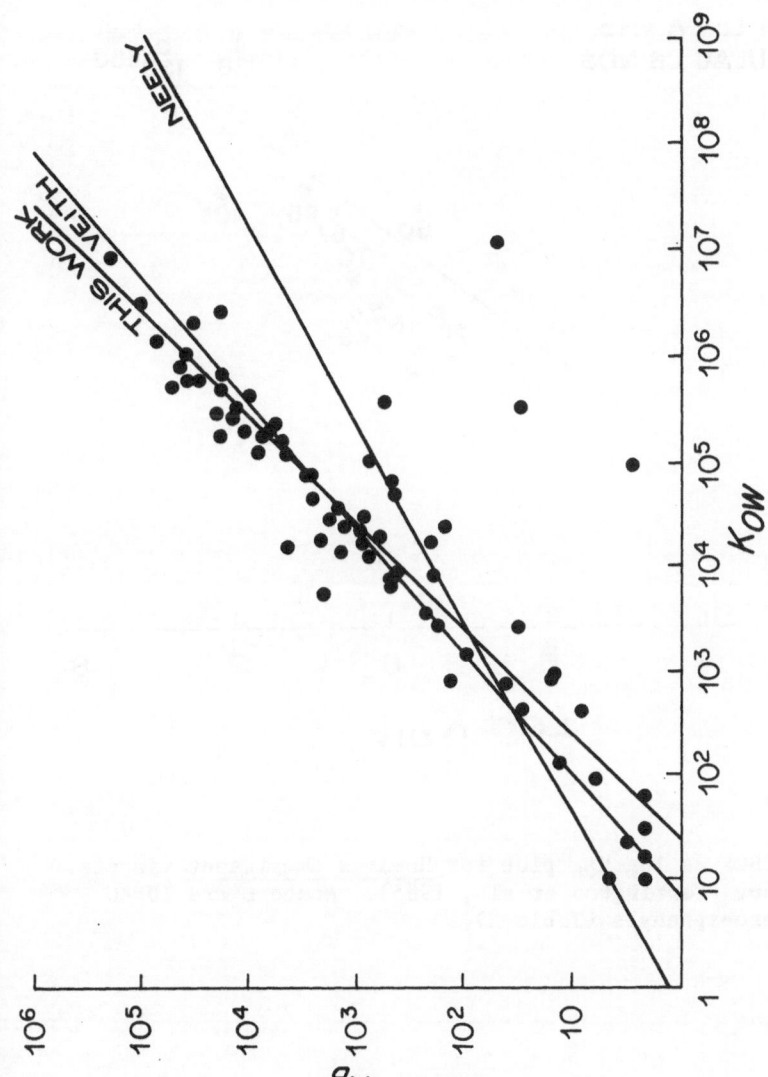

Figure 12. Plot of K_B versus K_{OW} showing the relationship between K_{OW} and biological uptake (from Mackay, 1982). Adapted with permission from Mackay, 1982. Copyright 1985, American Chemical Society.

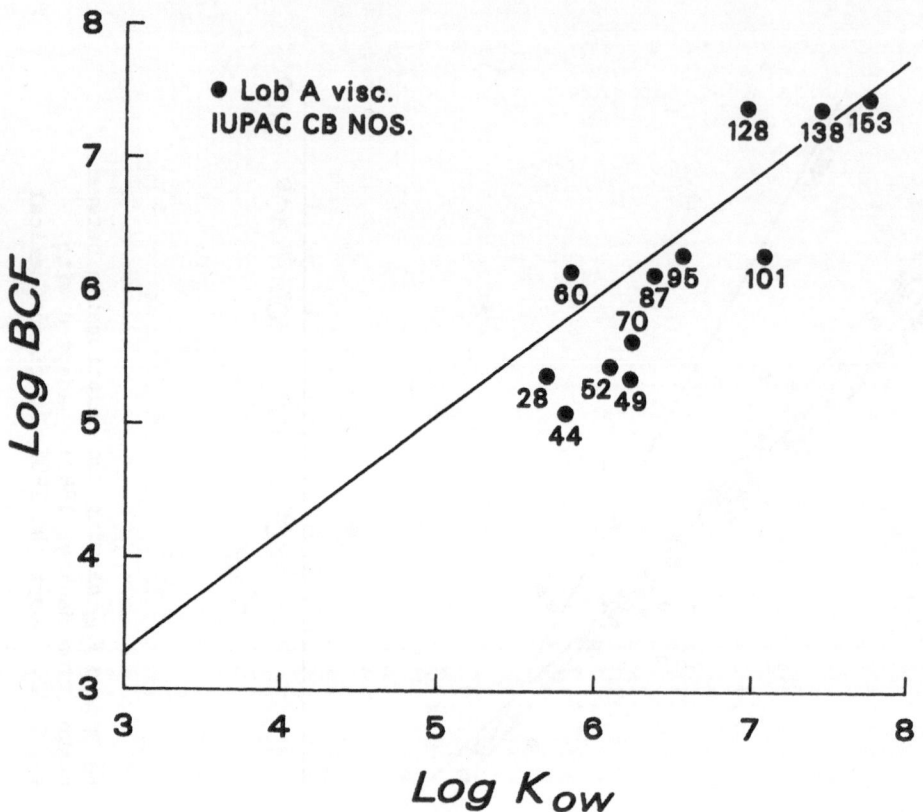

Figure 13.   Log BCF as log K_{OW} plot for <u>Homarus</u> <u>americanus</u> viscera.
New Bedford Harbor (Farrington et al., 1985).   Numbers are IUPAC
numbers for chlorobiphenyls (Table X).

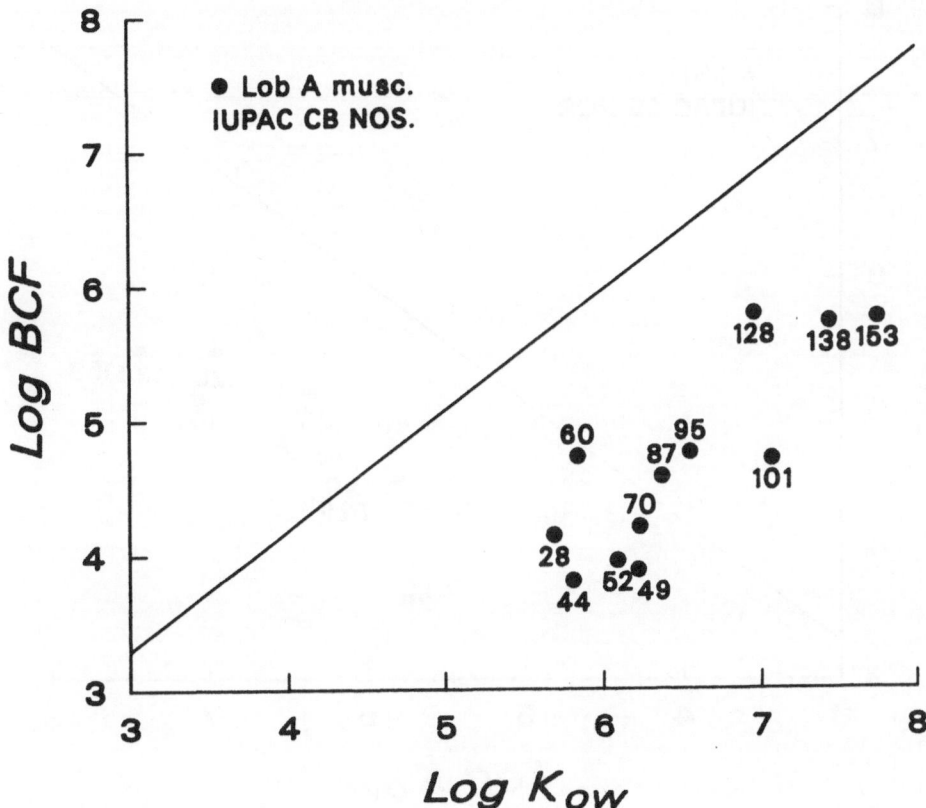

Figure 14. Log BCF vs. log K_{OW} plot for <u>Homarus</u> <u>americanus</u> muscle tissue. New Bedford Harbor (Farrington et al., 1985). Numbers are IUPAC numbers for chlorobiphenyls.

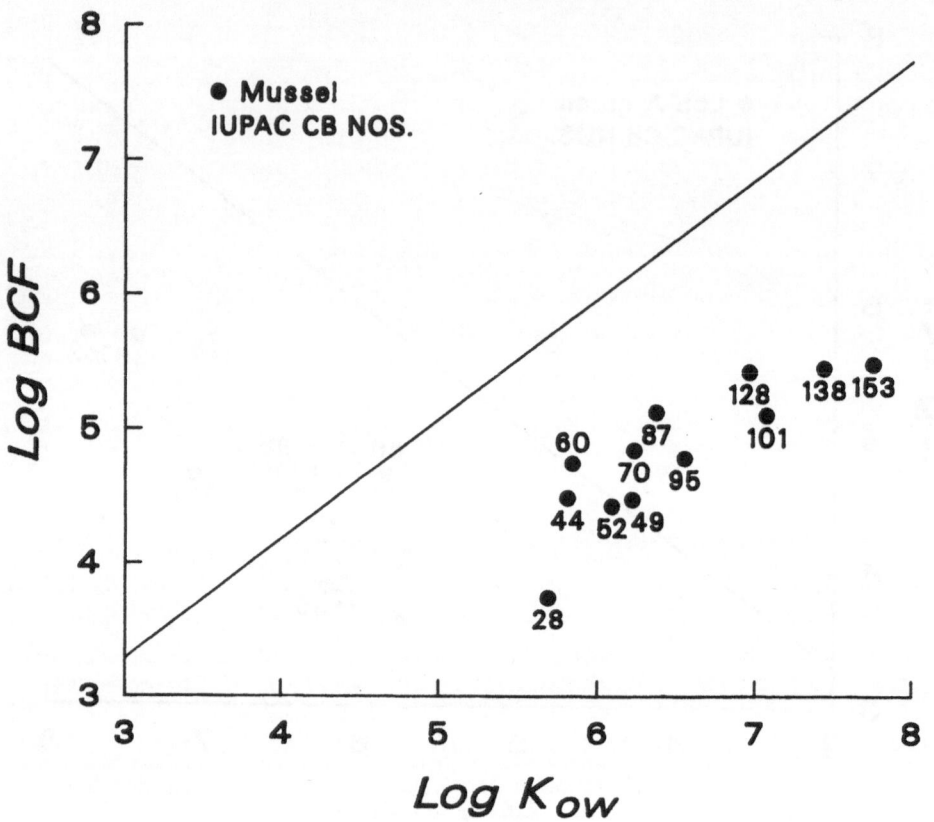

Figure 15. Log BCF vs. log K_{ow} plot for <u>Mytilus edulis</u>. New Bedford Harbor (Farrington et al., 1985). Numbers are IUPAC numbers for
chlorobiphenyls (Table X).

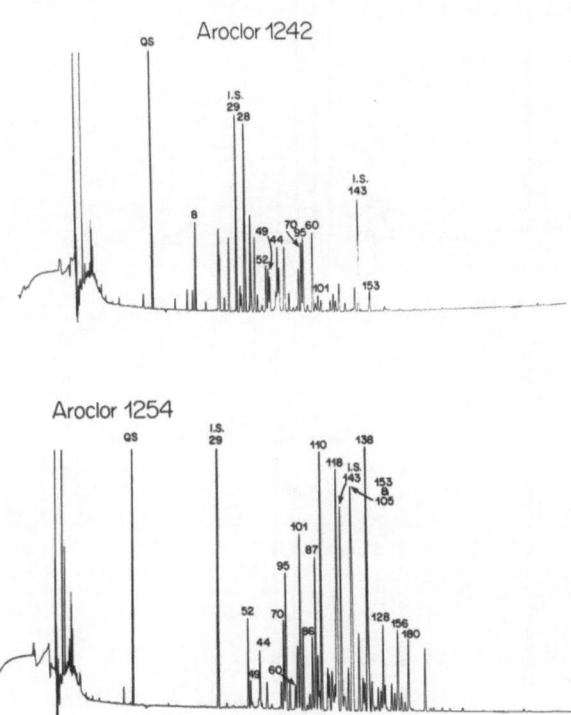

Figure 16a. High resolution gas chromatograms of PCBs – New Bedford Harbor. Numbers refer to IUPAC numbers for chlorobiphenyls (Table X). I.S. = internal standard.

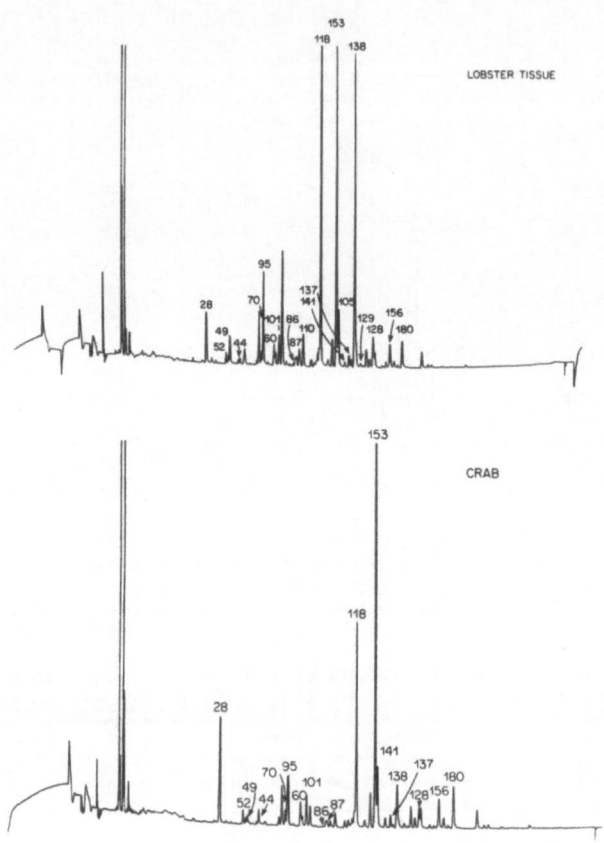

Figure 16b. High resolution gas chromatograms of PCBs – New Bedford Harbor. Numbers refer to IUPAC numbers for chlorobiphenyls (Table X). I.S. = internal standard.

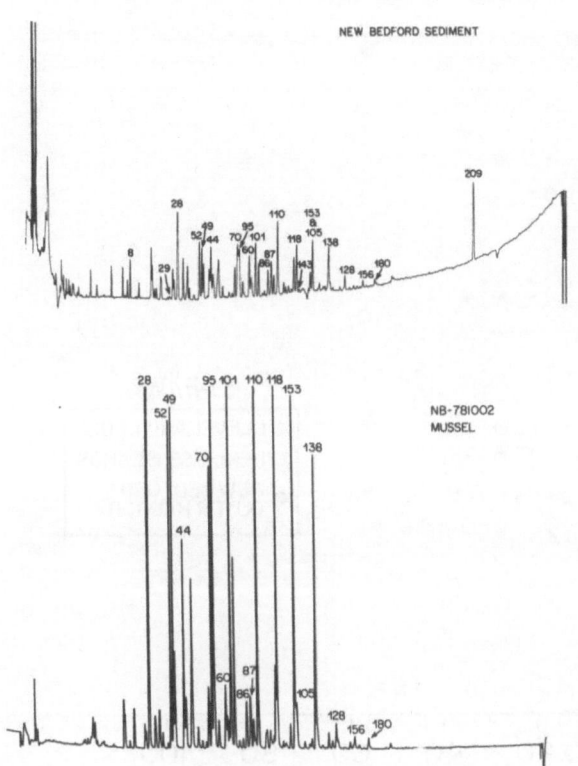

Figure 16c. High resolution gas chromatograms of PCBs – New Bedford Harbor. Numbers refer to IUPAC numbers for chlorobiphenyls (Table X). I.S. = internal standard.

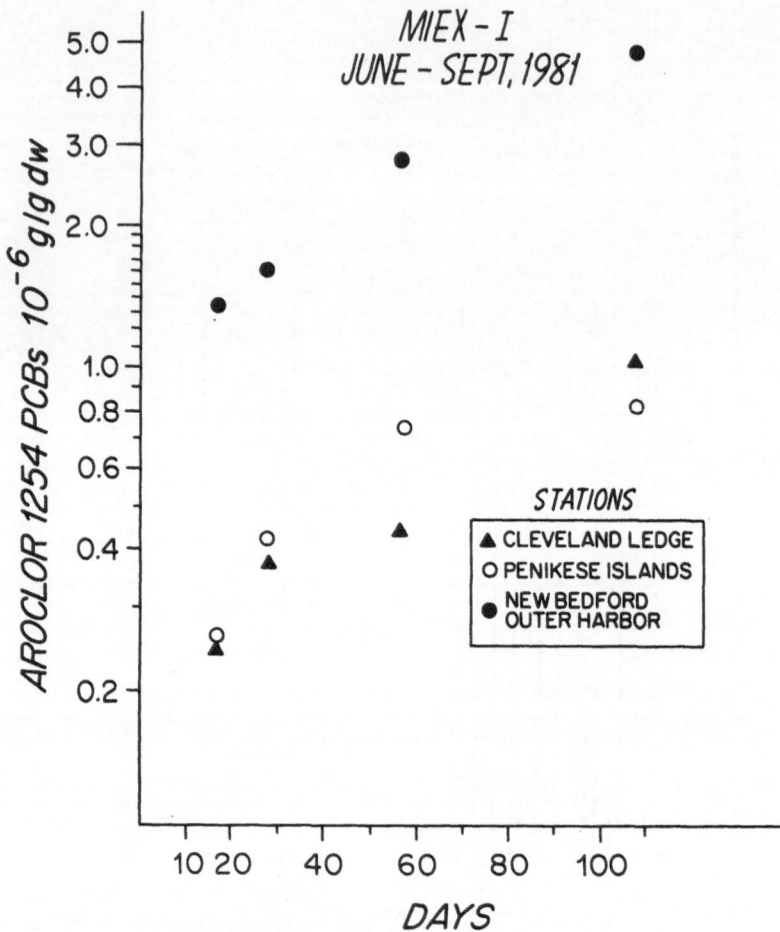

Figure 17. Plot of uptake of PCBs by <u>Mytilus edulis</u> transplanted from Nantucket Island to three stations in Buzzards Bay.

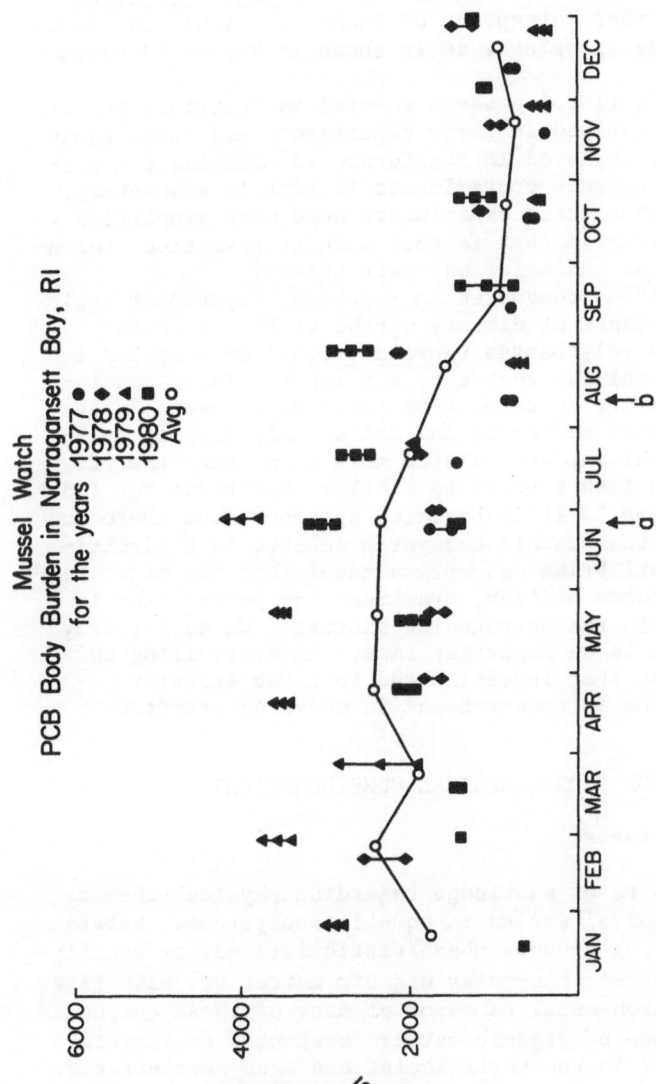

Figure 18. PCBs in *Mytilus edulis* from one location in Narragansett Bay, Rhode Island, U.S.A.; a, b indicate time of slight station shift in 1979 and 1980.

mediated transport via fecal material. The presence of worms increased the release of the chemical from the sediment to the overlying water by 4 to 6-fold compared to the control sediment without
worms. This was less than the orders of magnitude expected based on
the rate of test chemical transport to the surface sediments via
fecal material. These researchers demonstrated by experimentation
with the fecal material that entrapment of chemicals within the fecal
material reduced the rate of release as is shown in Figure 19 taken
from their paper.

Karickhoff and Morris (1985) present several mathematical treatments of the processes involved in their experiments and these equations should be usefully employed in the future in modeling the biogeochemical behavior of organic contaminants in benthic ecosystems.
These authors did note that their experiments were much simplified
versions of natural systems in that factors such as advection, turbulence, and other megafauna and meiofauna were absent.

Rubenstein et al. (1984) conducted an important experiment evaluating the relative importance of dietary uptake of PCBs in fish
(Leiostomas xanthus) fed polychaetes (Nereis virens) as compared to
exposure to only water-sediment routes of PCB input. Their conclusions, illustrated by Figure 20 taken from their paper, were "fish
exposed to PCB contaminated sediments and fed a daily diet of polychaetes from the same sediment accumulated more than twice the PCB
whole body residues than fish exposed to similar conditions but fed
uncontaminated polychaetes." It is becoming apparent from these and
other observations that the benthic ecosystem departs in a significant manner from the equilibrium assumption model that the major factor controlling body burdens of fish, crustacea, or polychaetes is
partitioning with water in the surrounding habitat. We do not deny
that exchange with water is an important factor in controlling body
burdens. Rather we think that ingestion and food web transfer
deserve more consideration in research and in modeling efforts.

5. GENERAL DISCUSSIONS OF BIOGEOCHEMICAL CONSIDERATIONS

5.1 Physical-Chemical Issues

We have discussed the state of knowledge regarding physical-chemical
parameters and biogeochemical cycles in aquatic ecosystems. Substantial progress in describing aqueous phase/particulate matter equilibrium partitioning of neutral, non-polar organic matter has made first
order prediction of environmental behavior of many of these compounds
a reality. The importance of organic matter "coatings" or "layers"
of the particulate matter in the partitioning has been demonstrated.
It appears that organic matter source, "quality", in terms of the
various soils and sediments examined to date have no discernible
influence of the partitioning - perhaps due to the averaging effect
of the various diverse organic compounds and functional groups incorporated into soil, humic and fulvic material and analogous macro-
molecular material in aquatic sediments. We caution that experiments

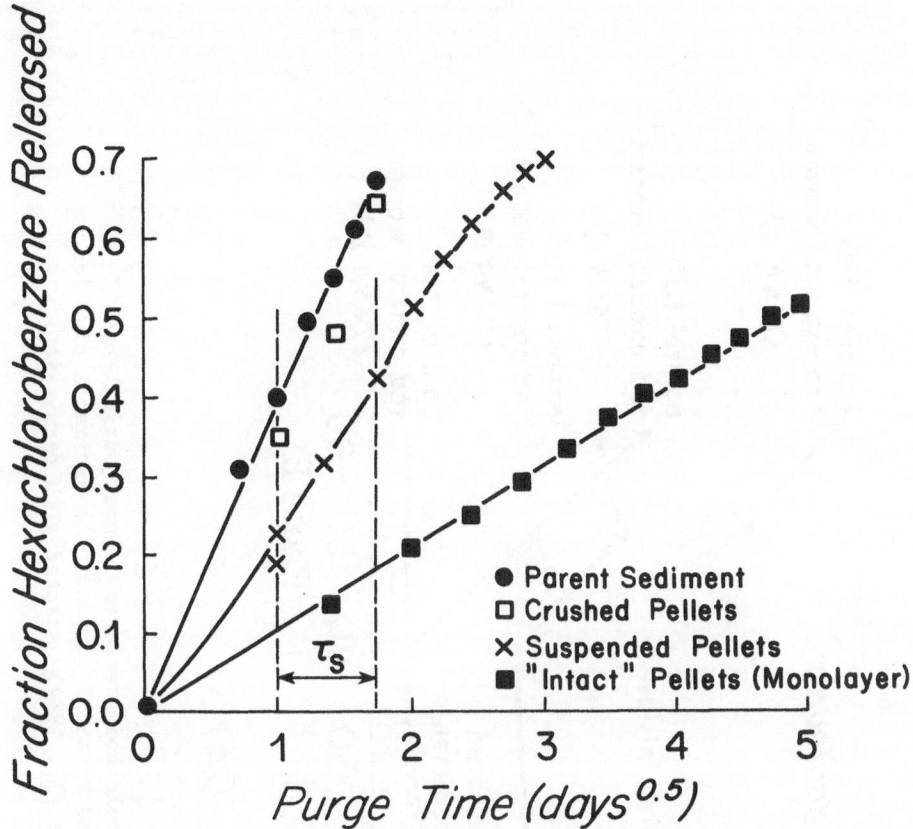

Figure 19. Release of chemicals from fecal pellets. Adapted with permission from Karickhoff and Morris, 1985. Copyright 1985, American Chemical Society.

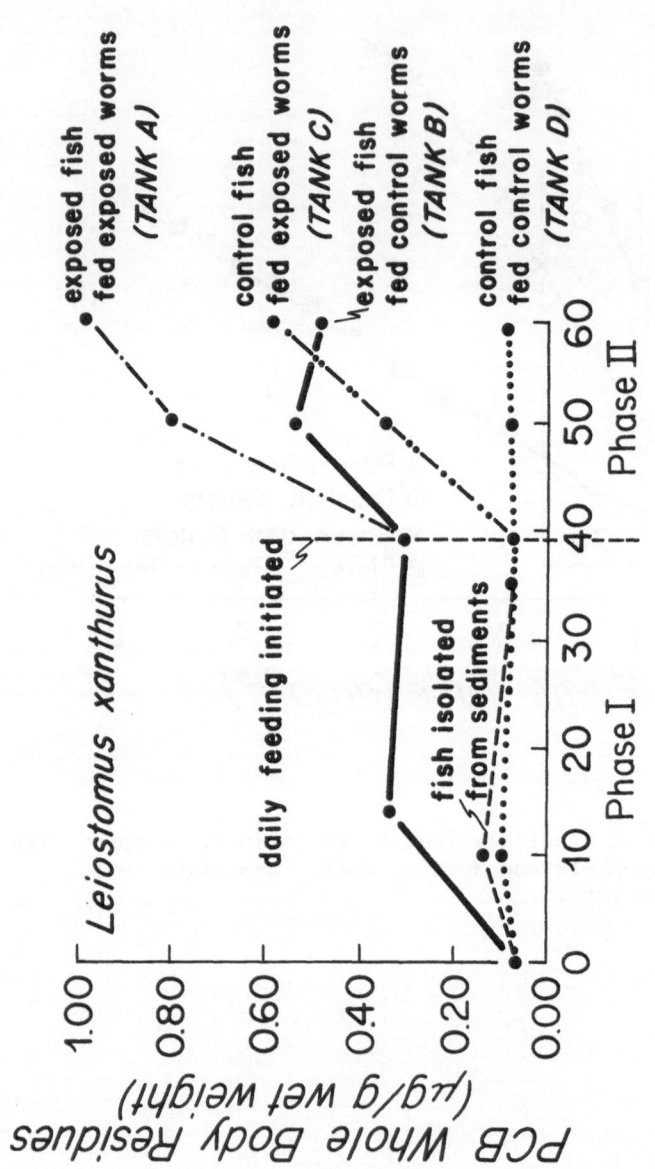

Figure 20. Average PCB whole body residues in spot during Phase I (exposure to sediments) and Phase II (exposure to sediments plus fed worms) (from Rubenstein et al., 1984) Adapted with permission.

have yet to be undertaken with the wide range of sediment-organic matter types in oceanic environments.

A few investigations of the environmental behavior of ionizable organic compounds have been reported and equilibrium based equations have satisfactorily described partitioning when pH and ionic strength of solutions are incorporated. These investigations need to be expanded to other chemicals and environmental conditions.

A major gap in many of the physical-chemistry considerations to be applied to the marine environment involves understanding the influence of salinity or more correctly the ionic strength and composition of seawater. There are predictive tools available, but solid experimental evidence is needed before we can predict, with certainty, such a fundamental parameter as solubility in seawater for the vast majority of organic chemicals of environmental concern. This appears to be a simple matter of experimentation as the technological tools to conduct such research are available and protocols are established from fresh water research.

Another major gap is knowledge of the role of colloids. Colloids appear to have a substantial role in biogeochemical behavior of compounds in pore waters and may have an equally important role in some ground water considerations. The nature and chemistry of colloids in aquatic ecosystems is an important area of future research and especially modelling efforts.

If we consider the issue of sewage sludge disposal we can quickly come to grips with the potential importance of both the role of colloids and the influence of salinity. Most sewage sludge accumulates in treatment plans under conditions of different ionic strength, or at least, different ionic composition than seawater. Some of the organic matter accumulated from the sewage and generated by microbes in the sludge process is of sufficient molecular weight and low aqueous solubility to form colloids. A better than first order prediction of physical-chemical behavior of organic chemicals of environmental concern when sludge is dumped in the marine environment requires an understanding of the influence of salinity on solubility, on colloids, and on the colloid/compound interactions. This is also the case when considering disposal of wastes to estuaries where salinity gradients are a potential major influence.

5.2 Biological Aspects of Biogeochemical Cycles

The utilization of an equilibrium approach to predicting body burdens of organic chemicals of environmental concern gives a useful first order starting point but does not offer anywhere near the required predictability for most environmental compounds due to differential metabolism or degradation rates in the environment. It has become clear that the concept that exchange with water is the overwhelming control of body burdens needs to be re-examined. This is especially the case with respect to benthic ecosystems where the major long term "source of input" is probably release of material delivered to the sediments.

The re-evaluation of the importance of ingestion as a significant source of control on body burdens raises the important issue of food web transfer. If, as suggested by recent experiments, ingestion is a major control, then we will need a better description of food webs to model biogeochemical cycles. This is important not only from the perspective of biological transfer but also from the perspective of the influence of biology on the speciation and physical form of organic chemicals. A simple example of this is the transfer of small particles to packages of larger particles - fecal matter, which first principle arguments would indicate to have an influence on sorption as has been shown to be the case experimentally by Rubenstein et al. (1984) as previously discussed.

We have considered results of research concerned with open ocean particulate matter from a variety of areas as assembled by Wakeham et al. (1984), with due acknowledgement to numerous investigators cited therein, and we present the diagram in Figure 21. This figure illustrates what we think is a major unknown in our ability to model, in a predictive manner, the biogeochemical cycle of organic chemical pollutants disposed of at sea in deep water and that is the lack of knowledge of mesopelagic, and bathypelagic biology; particularly feeding strategies and food web transfer.

Assessment of bioavailability is another issue of importance. Most chemical measurements, especially of sediment or sludge, do not provide a useful index of the proportion of the analyte that is available for biological uptake. Recent experience with polynuclear aromatic hydrocarbons (Farrington et al., 1983; McElroy, 1985, among others) illustrate the importance of these considerations.

5.3 Boundary Scavenging

Spencer et al. (1981) and Bacon et al. (1980) have discussed the importance of horizontal advection of ^{210}Pb to slope and shelf regions where it is scavenged by sorption processes. We note that K_d of partitioning for ^{210}Pb and several other metals are in the range of K_d and K_{oc} for organic chemicals such as DDT, PCBs, HCHs, PAHs. Thus, models and research concerned with biogeochemical cycles of these compnounds in ocean waters below the euphotic zone will need to evaluate the influence of boundary scavenging in a manner similar to the ^{210}Pb distribution models and measurements discussed by Spencer et al. (1981).

We emphasize this because the legitimate excitement associated with significant advances in measuring and modelling large particle flux (Tanabe and Tatsukawa, 1983; Wakeham et al., 1984, and references therein) focus attention on the vertical transport mechanisms and the potential influence of horizontal processes coupled with boundary scavenging has not been mentioned, to our knowledge, when assessing organic chemical pollutant biogeochemical cycles in the oceans.

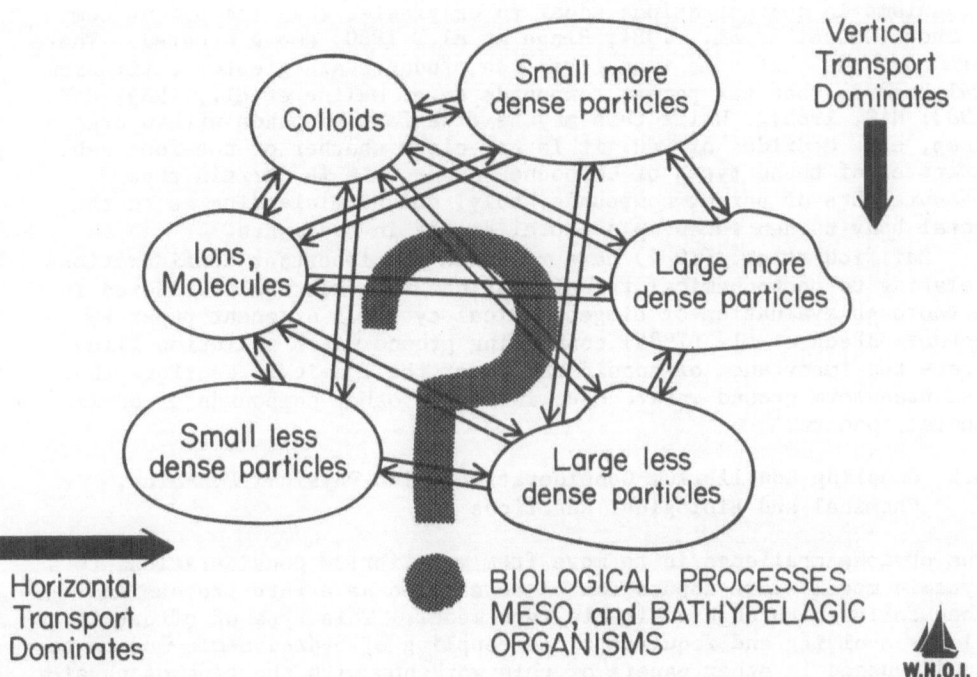

Figure 21. Conceptual Model.

5.4 Microbial, Chemical and Non-microbial Biological Transformations

We have not discussed the extensive literature of microbial transformations and non-microbial transformations of organic chemicals because
such a detailed discussion is beyond the scope of this paper. However, several investigations have unequivocally demonstrated that
metabolites or reaction products often are present in sediments and
organisms in concentrations equal to or greater than the parent compounds (Gosset et al., 1984; Hinga et al., 1980, among others). There
are concerns that some transformation products are greater environmental hazards than the parent compounds (e.g. Malins et al., 1984; NAS,
1983; NAS, 1985). While this may be true for compounds within organisms, e.g. epoxides of PAH, it is not clear whether or not food web
transfer of these types of compounds occur. It is certain that
measurements of parent compounds, only, can be misleading as to the
total body burden taken up or total amount in sediments.

Zafiriou et al. (1984) have reviewed the important considerations
relating to photochemical transformations that must be considered in
a thorough evaluation of biogeochemical cycles. A recent paper by
Schwarzenbach et al. (1985) concerning ground water pollution illustrate the importance of consideration of the chemical reactions that
can transform ground water contaminants to other compounds of environmental concern.

5.5 Coupling Equilibrium Considerations with Physical Dynamics, Chemical and Biological Reactions

The obvious challenge is to move from equilibrium considerations to a
dynamic model where inputs can be considered as a rate process and
then followed through a kinetic type model. This type of effort is
slowly evolving and requires close coupling of hydrodynamic models to
be discussed in other papers of this workshop with the type of physical-chemical, biochemical, and food web information we have discussed
in preceding sections.

A key issue is to discern what are the next critical steps in
these efforts at coupling between the various predictive modelling
efforts. A body of generic knowledge and modelling capability can
develop, but it is apparent that the application of comparative risk
for various options of waste disposal, e.g. land or ocean, will require site specific information and site specific tuning of models.

The application of models presents a paradox; simplicity, complexity. This is an important consideration in the use of biogeochemical models in assessment of disposal options from a policy and
management perspective. A complex model allows us to consider a wide
array of competing processes simultaneously, to perform sensitivity
analysis, and to estimate probabilities; but a complex model by virtue
of its own complexity often lends unwarranted credibility to its results, particularly when delivered with elegant tables and multi-
colored graphs. We believe that complex models are appropriate for
researchers. At the same time it is a major responsibility of the
researcher to interpret results of a model in terms of a few dominant

processes which can be understood in a simple intuitive basis. This subjective intuitive assessment of "is it reasonable" is often one of the most valuable tests for establishing the validity of a model. In this way the uncertainty in the presumed fate of the compound does not become masked in a numerical solution to a complex set of equations. Our ability to describe systems mathematically and to solve systems of equations is far ahead of our understanding of fundamental physical, chemical and biological processes.

The intuitive approach is so important in biogeochemical modeling because there are still very few biogeochemical systems that can be modeled based on first principles. Therefore most of our models are not fully deterministic, although this fact is overlooked often. Sophisticated statistical tests while helpful, are not entirely adequate in situations where we have great uncertainty not only in the data that goes into the model but also in the description of the processes within the framework of the model. Therefore the step of interpreting models in terms of dominant processes should be employed, since this allows a researcher to apply years of experience to interpretation of the results.

The biogeochemical processes that are commonly considered in assessing the transport and fate of toxic organic compounds in the environment are listed in Table XIII. Many of the processes are common to both terrestrial and ocean environments. A more detailed description of the processes can be found in the background material for mathematical models of transport and fate which are based on these processes. The processes in Table XIII are listed roughly in order of decreasing knowledge about them. The only processes that we consider to be reasonably well understood in general for many different compounds under many types of environmental conditions are physical transport and transfer; advection, dispersion, volatilization and sorption.

6. PREDICTION AND VERIFICATION

An issue which seems to have been set aside over the past decade while working towards the difficult challenge of predictive capabilities is the vexing problem of verification. Should a comparative assessment of the relative risks of land or ocean disposal of wastes containing organic chemicals of concern include a comparison of ability to verify predictions? If the answer is yes – and we suspect most scientists and the general public would answer in some way in the affirmative – then there is a serious shortfall in present capabilities in regard to measurements in the deep ocean.

Tanabe and Tatsukawa (1983) and co-workers, as previously discussed, have one of the very few sets of data for chlorinated pesticides and PCBs in the deep ocean below 100 meters. Most other data predates 1976 (NAS, 1979; Farrington, 1985) and most of this is suspect because of contamination problems (DeLappe et al., 1983). Even Tanabe and Tatsukawa (1983) in their impressive efforts were restricted in their data interpretation due to this problem. They stated

Table XIII.   Biogeochemical processes commonly considered in
              transport and fate of toxic organic compounds.

Groundwater	Ocean
Advection	Advection
Dispersion	Dispersion
Sorption	Sorption
Volatilization	Volatilization
	Photolysis
Chemical Transformations	Chemical Transformations
Biological Transformations	Biological Transformations

"There are some difficulties in sample collection and the chemical
analysis because of the very low concentrations. . . . However, contamination from the large volume water sampler employed could not be
checked because the sampler which is made of plastic was not rinsed
with organic solvents."   There are serious, but not insurmountable,
difficulties to making open ocean seawater measurements for organic
contaminants.   De Lappe et al. (1983) have made several suggestions
of how to make these measurements and built a pump-sorbent system
which has been used in surface waters as discussed earlier.   Technically, it is feasible to transfer this to deep water measurements
but it has yet to be proven that it will work.   This restriction on
verification has to be acknowledged when assessing deep ocean disposal
options.
    We think that a careful assessment of present and projected capa-
bilities and needs for verification of predictions of biogeochemical
behavior of organic chemicals in groundwater and the oceans is needed
as part of a comparative risk assessment.

7.   ACKNOWLEDGEMENTS

We are indebted to Dr. Rene Schwarzenbach for suggesting that we write
this paper together and for many enlivened discussions of environmental chemistry.   Bruce Brownawell kindly allowed us to use portions of
his Ph.D. thesis research to illustrate points of discussion.   We
gratefully acknowledge the support of the Andrew W. Mellon Foundation,
the Richard King Mellon Foundation for providing financial support
via the Coastal Research Center, W.H.O.I., and the U.S. Environmental
Protection Agency for partial support through Cooperative Agreement
Number CR 811894-01-1 for our collaboration on this paper.   This is
Woods Hole Oceanographic Institution Contribution No. 6006.

8.   REFERENCES

Bacon, M. P., P. G. Brewer, D. W. Spencer, J. W. Murray and J. Goddard (1980). 'The behavior of lead-210, polonium-210, manganese and iron in the Cariaco Trench'. Deep-Sea Research, 27A, 119-135.

Baker, R. A. (ed.) (1980). Contaminants and Sediments. Ann Arbor Science Publishers, Inc., Ann Arbor, Michigan.

Banerjee, S. (1985). 'Calculation of water solubility of organic compounds with UNIFAC-derived parameters'. Environ. Sci. Technol., 19, 369-370.

Bender, M. E., R. J. Huggett and W. J. Hargis, Jr. (1977). 'Kepone residues in Chesapeake Bay biota'. In: Proceedings of the Kepone Seminar II, September 19-21, 1977. U.S. Environmental Protection Agency, Region III, U.S. E.P.A., Washington, D. C., USA, pp. 14-65.

Briggs, A. A. (1981). 'Theoretical and experimental relationships between soil adsorption, octanol/water partition coefficients, water solubilities, bioconcentration factors, and the parachor'. J. Agric. Food Chem., 29, 1050.

Brooks, J. M., D. A. Wiesenberg, G. Bodennec and T. C. Sauer, Jr. (1983). 'Volatile organic wastes at the Puerto Rico dumpsite'. In: Wastes in the Ocean, Vol. 1, Industrial and Sewage Wastes in the Ocean, I. W. Duedall et al. (Eds.). John Wiley and Sons, New York, N.Y.

Brownawell, B. J. and J. W. Farrington (1985a). 'Biogeochemistry of PCBs in interstitial waters of a coastal marine sediment'. Submitted to Geochim. Cosmochim. Acta.

Brownawell, B. J. and J. W. Farrington (1985b). Chapter 7. 'Partitioning of PCBs in marine sediments'. To be published in A.C.S. Symposium volume.

Butler, G. C. (ed.) (1978). 'Principles of Ecotoxicology, SCOPE 12'. John Wiley and Sons, New York, N.Y.

Butler, G. (1984). Chapter 1 in Effects of Pollutants at the Ecosystem Level, SCOPE 22, P. J. Sheehan et al. (Eds.). John Wiley and Sons, New York, N.Y.

CEQ (1984). 'Environmental Quality 1983'. 14th Annual Report of The Council on Environmental Quality Executive Office of the President of the United States, Washington, D. C., USA.

Chiou, C. T., L. J. Peters and V. J. Freed (1979). 'A physical
    concept of soil-water equilibria for non-ionic organic compounds'.
    Science, 206, 831.

Chiou, C. T. (1985). 'Partition coefficients of organic compounds in
    lipid-water systems and correlations with fish bioconcentration
    factors'. Environ. Sci. Technol., 19, 57-62.

Curtis, G., P. V. Roberts and M. Reinhard (1984). 'Sorption of
    organic solutes: Comparison between laboratory estimated retarda-
    tion factors and field observations'. Paper presented at 2nd
    International Conference on Ground Water Quality Research, Tulsa,
    Oklahoma, March 26-29, 1984.

De Lappe, B. W., R. W. Risebrough and W. Walker, II (1983). 'A large
    volume sampling assembly for the determination of synthetic
    organic and petroleum compounds in the dissolved and particulate
    phases of seawater'. Canadian Journal of Fisheries and Aquatic
    Sciences, 40 (Suppl. 2), 322-336.

Eaganhouse, R. P. and J. A. Calder (1976). 'The solubility of medium
    molecular weight aromatic hydrocarbons and the effects of hydro-
    carbon co-solutes and salinity'. Geochim. Cosmochim. Acta, 40,
    555-561.

Ernst, W. (1977). 'Determination of the bioconcentration potential
    of marine organisms. A steady state approach'. Chemosphere,
    11, 731-740.

Farrington, J. W., A. C. Davis, N. M. Frew and K. S. Rabin (1982).
    'No. 2 fuel oil compounds in Mytilus edulis. Retention and
    release after an oil spill'. Marine Biology, 66, 15-26.

Farrington, J. W. and J. H. Vandermeulen (1983). Session IV -
    'Summary and overview: organic synthetics'. Canadian Journal of
    Fisheries and Aquatic Sciences, 40 (Suppl. 2). Proceedings of
    the Conference on Pollution in the North Atlantic Ocean, pp.
    346-348.

Farrington, J. W., E. D. Goldberg, R. W. Risebrough, J. H. Martin and
    V. T. Bowen (1983). 'U.S. "Mussel Watch" 1976-1978: An overview
    of the trace metal, DDE, PCB, hydrocarbon, and artificial radio-
    nuclide data'. Environ. Sci. Technol., 17, 490-496.

Farrington, J. W., A. C. Davis, B. J. Brownawell, B. W. Tripp and
    J. B. Livramento (1985a). 'PCBs in biota of the Acushnet River
    Estuary'. To be published in A.C.S. Symposium volume.

Farrington, J. W., A. C. Davis, C. H. Clifford and L. Stathopoulos (1985b). 'Kinetics of PCB uptake and release and seasonal variability in Mytilus edulis, Buzzards Bay and Narragansett Bay'. To be submitted.

Fossato, V. U. and W. J. Canzonier (1976). 'Hydrocarbon uptake and loss by the mussel Mytilus edulis'. Marine Biology, 36, 243-250.

Gossett, R. W., D. A. Brown, S. R. McHugh, A. M. Westcott (1983-84). 'Measuring the oxygenated metabolites of chlorinated hydrocarbons'. Southern California Coastal Water Research Project. Biennial Report, 1983-84. S.C.C.W.R., 646 West Pacific Coast Highway, Long Beach, CA, USA 90806, pp. 155-169.

Gschwend, P. M. and S.-C. Wu (1984). 'On the constancy of sediment-water partition coefficients of hydrophobic organic pollutants'. Environ. Sci. Technol., submitted.

Hamelink, J. L., R. C. Waybrant and R. C. Ball (1971). 'A proposal: exchange equilibria control the degree chlorinated hydrocarbons are biologically magnified in lentic environments'. Trans. Am. Fish. Soc., 100, 207.

Hinga, K. R., M.E.Q. Pilson, R. F. Lee, J. W. Farrington, K. Tjessem and A. C. Davis (1980). 'Biogeochemistry of benzanthracene in an enclosed marine ecosystem'. Environ. Sci. Technol., 14, 1136-1143.

Johnson, R. L., S. M. Brillante, L. M. Isabelle, J. E. Houck and J. F. Pankow (1985). 'Migration of chlorophenolic compounds stored at the chemical waste disposal site at Alkali Lake, Oregon. 2. Contaminant distribution, transport, and retardation'. Groundwater, in press.

Karickhoff, S., D. S. Brown and T. Scott (1979). 'Sorption of hydrophobic pollutants on natural sediments'. Water Res., 13, 241.

Karickhoff, S. W. and K. R. Morris (1985). 'Impact of tubificid oligochaetes on pollutant transport in bottom sediments'. Environ. Sci. Technol., 19, 51-56.

Keith, L. H. and W. A. Telliard (1979). 'Environmental Science and Technology Special Report: Priority Pollutants. I. A perspective view'. Environ. Sci. Technol., 13, 416-423.

Leo, A., C. Hansch and D. Elkins (1971). 'Partition coefficients and their uses'. Chem. Rev., 71, 525.

Mackay, D. and W. Y. Shiu (1977). 'Aqueous solubility of polynuclear aromatic hydrocarbons'. J. Chem. Eng. Data, 22(4), 399-402.

Mackay, D., R. Mascarenhas and W. Y. Shiu (1980). 'Aqueous solubility of polychlorinated biphenyls'. Chemosphere, 9, 257-264.

Mackay, D. (1982). 'Correlation of bioconcentration factors'. Environ. Sci. Technol., 16(5), 274-278.

Mackay, D. M., D. L. Freyberg, M. N. Golz, G. D. Hopkins and P. V. Roberts (1983). 'A field experiment on groundwater transport of halogenated organic solutes'. Proceedings of the 186th National Meeting of the Division. Vol. 23, No. 2, p. 368.

MacLeod, W. D., L. S. Ramos, A. J. Friedman, D. G. Burrows, P. G. Prohaska, D. L. Fisher and D. W. Brown (1981). 'Analysis of residual chlorinated hydrocarbons, aromatic hydrocarbons and related compounds in selected sources, sinks, and biota of the New York Bight'. NOAA Technical Memorandum, OMPA-6, Boulder, CO. NOAA, U.S. Dept. of Commerce, Washington, D. C., U.S.A.

Malins, D. C., B. B. McCain, D. W. Brown, S. L. Chan, M. S. Meyers, J. T. Landahl, P. G. Prohaska, A. J. Friedman, L. D. Rhodes, D. G. Burrows, W. D. Gronlund, H. O. Hodgins (1984). 'Chemical pollutants in sediments and diseases of bottom-dwelling fish in Puget Sound, Washington'. Environmental Science and Technology, 18, 705-713.

Maugh, T. H., II (1978). 'Chemicals: how many are there'? Science, 199, 162.

May, W. E. and S. P. Wasik (1978). 'Determination of the solubility behavior of some PAH in water'. Anal. Chem., 50, 997-1000.

May, W. E., S. P. Wasik and D. H. Freeman (1978). 'Determination of the aqueous solubility of PAH by a coupled column LC technique'. Anal. Chem., 50, 175-179.

May, W. (1980). 'Aqueous solubilities of PAH'. In: Petroleum in the Marine Environment, L. Petrakis and F. T. Weiss (Eds.), Advances in Chemistry Series 185, Chapter 7, American Chemical Society, Washington, D. C., U.S.A.

McElroy, A. (1985). 'Biogeochemistry and physiological effects of benz(a)anthracene exposure in benthic chambers containing the polychaete Nereis virens'. Ph.D. Thesis, Woods Hole Oceanographic Institution-Massachusetts Institute of Technology Joint Program in Oceanography, Woods Hole, MA 02543, U.S.A.

Means, J. C., S. G. Wood, J. J. Hassett and W. L. Banwart (1980). 'Sorption of polynuclear aromatic hydrocarbons by sediments and soils'. Environ. Sci. Technol., 14, 1524.

Means, J. C., S. G. Wood, J. J. Hassett and W. L. Banwart (1982).
'Sorption of amino - and carboxy - substituted polynuclear aromatic hydrocarbons by sediments and soils'. Environ. Sci. Technol.,
16, 93.

Miller, D. R. (1984). Chapter 2 in Effects of Pollutants at the
Ecosystem Level, SCOPE 22, P. J. Sheehan et al. (Eds.), John
Wiley and Sons, New York, N.Y.

NAS (1979). 'Polychlorinated Biphenyls'. Publications Office,
National Academy of Sciences, Washington, D. C.

NAS (1981). 'Testing for Effects of ·Chemicals on Ecosystems'. U.S.
National Academy Press, Washington, D. C., U.S.A.

NAS (1983). 'Polycyclic Aromatic Hydrocarbons: Evaluation of Sources
and Effects'. U.S. National Academy Press, Washington, D. C.,
U.S.A.

NAS (1985). 'Oil in the Sea, Inputs, Fates and Effects'. U.S.
National Academy Press, Washington, D. C., U.S.A.

OAD (1984). Ocean Assessment Division, N.O.A.A. U.S. Dept. of
Commerce, Washington, D. C., U.S.A.

O'Connor, T. P., A. Okubo, M. A. Champ and P. K. Park (1983). 'Projected consequences of dumping sewage sludge at a deep ocean site
near New York Bight'. Can. J. Fish. Aquat. Sci., 40 (suppl.
2), 228-241.

O'Connor, D. J. and J. C. Pizza (1985). 'Eco-kinetic model for the
accumulation of PCB in marine fish'. Proceedings 4th International Symposium on Ocean Disposal (in press).

O'Connor, D. J., J. P. Connolly and E. J. Garland (1985). 'Mathematical models - Fate, transport and food chain'. Submitted for publication.

Pankow, J. F., R. L. Johnson, J. E. Houk, S. Brillante, W. G. Bryan
(1984). 'Migration of chlorophenolic compounds at the chemical
waste disposal site at Alkali Lake, Oregon. 1. Site description
and groundwater flow'. Groundwater, 22, 593-601.

Pizza, J. C. and J. M. O'Connor (1983). 'PCB dynamics in Hudson River
striped bass. II. Accumulation from dietary sources'. Aquatic
Toxicology, 3, 313-327.

Reinhard, M., N. L. Goodman and J. F. Barker (1984). 'Occurrence
and distribution of organic chemicals' in two landfill leachate
plumes'. Environ. Sci. Technol., 18, 953-961.

Roberts, P. V., J. Schreiner and G. D. Hopkins (1982). 'Field study of organic water quality changes during groundwater recharge in the Palo Alto Baylands'. Water Res., 16, 1025.

Rubenstein, N., W. T. Gilliam and N. R. Gregory (1984). 'Dietary accumulations of PCBs from a contaminated sediment source by a demersal fish (Leiostomus xanthurus)'. Aquatic Toxicology, 5, 331-342.

Schellenberg, K., C. Leuenberger and R. P. Schwarzenbach (1984). 'Sorption of chlorinated phenols by sediments and aquifer materials'. Environ. Sci. Technol., 18, 652.

Schulte, E. and L. Acker (1974). 'Identifizienung und metabolisier-bankeit von polychlorierten biphenylen'. Naturwiss., 61, 79-81.

Schwarzenbach, R. P. and J. Westall (1981). 'Transport of nonpolar organic compounds from surface water to groundwater'. Laboratory sorption studies. Environ. Sci. Technol., 15, 1360.

Schwarzenbach, R. P., W. Giger, E. Hoehn and J. K. Schneider (1983). 'Behavior of organic compounds during infiltration of river water to groundwater: Field studies'. Environ. Sci. Technol., 17, 472.

Schwarzenbach, R. P. and J. Westall (1985). 'Sorption of hydrophobic trace organic compounds in groundwater systems'. Water Science and Technol. In press.

Schwarzenbach, R. P., W. Giger, C. Schaffner and O. Wanner (1985). 'Groundwater contamination by volatile halogenated alkanes: Abiotic formation of volatile sulfur compounds under anaerobic conditions'. Environ. Sci. Technol., 19, 322-327.

Sheehan, P. J., D. R. Miller, G. C. Butler, P. Bourdeau (Eds.) (1984). Effects of Pollutants at the Ecosystem Level, SCOPE 22. John Wiley and Sons, New York, N.Y.

Spencer, D. W., M. P. Bacon, P. G. Brewer (1981). 'Models of the distribution of ^{210}Pb in a section across the North Equatorial Atlantic Ocean'. J. of Marine Research, 39, 119-138.

Stegeman, J. S. and J. M. Teal (1973). 'Accumulation, release, and retention of petroleum hydrocarbons by the oyster Crassostrea virginica'. Marine Biology, 22, 37-44.

Stegeman, J. S. (1981). 'Polynuclear aromatic hydrocarbons and their metabolites'. Chapter 1 in Polycyclic Hydrocarbons and Cancer, Vol. 3. Academic Press, New York, N.Y., pp. 1-60.

Stumm, W., R. Kummert and L. Sigg (1980). 'A ligand exchange model for the adsorption of inorganic and organic ligands at hydrous oxide interfaces'. Croat. Chem. Acta, 53, 291.

Sutton, C. and J. A. Calder (1975). 'Solubility of alkylbenzenes in distilled water and seawater at 25°C'. J. Chem. Eng. Data, 20, 320-322.

Tanabe, S. and R. Tatsukawa (1983). 'Vertical transport and residence time of chlorinated hydrocarbons in the open ocean water column'. J. of the Oceanographical Society of Japan, 39, 53-62.

United Nations (1979). 'DDT and its derivatives'. Environmental Health Criteria 9. World Health Organization, Geneva, Switzerland.

Wakeham, S. G., C. Lee, J. W. Farrington and R. B. Gagosian (1984). 'Biogeochemistry of particulate organic matter in the oceans: results from sediment trap experiments'. Deep Sea Res., 31, 509-528.

Westall, J. (1984). 'Properties of organic compounds in relation to binding by natural materials'. Proceedings of the Conference on Transport of Contaminants in Groundwater, Stockholm, Nov. 1983.

Westall, J., C. Leuenberger and R. P. Schwarzenbach (1985). 'Influence of pH and ionic strength on the aqueous-nonaqueous distribution of chlorinated phenols'. Environ. Sci. Technol., 19, 193-198.

Whitehouse, B. G. (1984). 'The effects of temperature and salinity on the aqueous solubility of polynuclear aromatic hydrocarbons'. Mar. Chem., 14, 319-332.

Zafiriou, O. C., J. Joussot-Dubien, R. G. Zepp and R. G. Zika (1984). 'Photochemistry of natural waters'. Environ. Sci. Technol., 18, 358A-371A.

Zierath, D., J. J. Hassett, W. L. Banwart, S. G. Wood and J. C. Means (1980). 'Sorption of benzidine by sediments and soils'. Soil Sci., 129, 277.

ORGANO-CHLORINES - A REVIEW OF USES, CONTROL AND DISPOSAL OPTIONS

J. E. Portmann,
Ministry of Agriculture, Fisheries and Food,
Fisheries Laboratory,
Burnham-on-Crouch, Essex, CMO 8HA, UK

ABSTRACT. Organo-chlorine compounds tend to be viewed as equally harmful substances. However, individually they have a wide range of properties and uses and certainly do not present the same degree of hazard to
the environment. The background to the common fear which organochlorines engender and the main control options are briefly examined.
A number of individual compounds are then reviewed in terms of the benefits they offer to man and the attendant side effects they may have on
the environment. The relative impacts of alternative control options
are discussed in terms of the examples given and some comments are
offered on the need for, and relative merits of, alternative disposal
options.

1. INTRODUCTION

There can be no doubt that the wide-range of organic compounds loosely
described as organo-chlorines are chemicals which modern man has found
extremely useful. Indeed it is almost impossible to imagine life in the
developed world without the benefits they either directly or indirectly
provide. Nevertheless, as a group of compounds which might escape to
the environment, let alone be actively discharged to it, organo-chlorine
compounds engender considerable public concern. In fact as far as
deliberate release to the marine environment is concerned, the dumping
of chlorinated organic compounds at sea is not allowed under the terms
of the Oslo and London Dumping Conventions and a number of regional
conventions, e.g. Paris, Helsinki and Barcelona, seek to prevent their
discharge.

There are, of course, exclusion clauses which indicate that the
restrictions apply only to those substances which are persistent, toxic
and bio-accumulated and where the concentrations exceed a certain
value. In practice difficulties occur in quantifying these properties
and so the restrictions tend to be applied to all organo-chlorine compounds whatever the concentration and regardless of the real need. How
then does the concern associated with organo-chlorine compounds arise

427

G. Kullenberg (ed.), The Role of the Oceans as a Waste Disposal Option, 427–440.

and does it extend to all chlorinated organic compounds and, if it does, is such concern justified? This paper attempts to examine these questions and in particular the need to control the release of organo-chlorines to the environment and how best to achieve this. Attention is paid to whether, relative to other disposal options, release to the marine environment is justifiable and under what circumstances.

2.   ORIGINAL CAUSE FOR CONCERN

In order to establish the basis for the concern it is necessary to go back some 15 to 20 years. By the early 1960s the organo-chlorine pesticides, such as DDT and Dieldrin, had been in use for some 20 years and had proved their immense value in a wide variety of insecticidal applications. Extensive use of DDT during the Second World war in delousing programmes, widespread use in anti-malaria campaigns against the mosquito vector and as a general insecticide in agriculture had demonstrated its effectiveness and proved that DDT posed a very low risk to man. Dieldrin was proving its worth as a general insecticide in agriculture, as a moth-proofing agent on woollen goods especially carpets, as a seed-dressing agent, e.g. to protect winter sown cereals, and as a once a year control agent for sheep-scab mites.

Unfortunately these wide-scale uses of compounds which had been selected partly because they were persistent and gave long-lasting protection, were not without side-effects and there were extensive mortalities among birds including some sea birds. These mortalities were soon linked with high residue levels of DDT, Dieldrin and other organo-chlorine pesticides in the carcasses of the victims. Intensive investigations established the cause and effect link for birds and indicated possible risks to their prey (Carson, 1963). Other studies showed that crustacea were particularly sensitive to these insecticides and, in the case of Dieldrin, that concentrations in certain species of fish from particular areas exceeded those regarded as safe for human consumption.

Alarm was spread further when in the mid-60s with the chance discovery that a group of compounds, known collectively as PCBs, which had been used in industry and elsewhere for over 30 years, were also very widespread in the environment. They were found to be particularly prevalent in the fatty tissues of sea birds, marine mammals and certain fish, e.g. herring muscle and cod liver. In such tissues, concentrations often exceeded 10 μg g^{-1} fresh weight; levels which it was suggested could affect reproductive success in sea birds and which were regarded as undesirably high in foodstuffs intended for human consumption on grounds of high bioaccumulation potential. The coup de grace, for organo-chlorine compounds, came in 1971 when it was found that certain producers of vinyl chloride were disposing of the production residue known as EDC tar by discharging it in the wake of a vessel. This waste was found to be toxic to a number of marine organisms and was found to contain a wide range of mainly short chain chlorinated aliphatic compounds (Berge et al., 1972).

It thus became obvious that a range of chlorinated organic compounds exhibiting a wide range of physical and chemical properties

(Table I) were accumulating in the marine environment and thus had the potential to cause problems to aquatic life. As a direct consequence the control measures mentioned earlier were introduced.

TABLE I. Properties of typical organo-chlorines for which disposal to sea is restricted

	Solubility	Density	mm Hg @ 25°C	MPt (°C)	BPt (°C)	Log ectagonal water partition coefficient
DDT	1 μg l^{-1}	-	1.9 x 10^{-7}	108.5	185	4-6
Dieldrin	200 μg l^{-1}	1.75	1.8 x 10^{-7}	175	-	?
PCB Aroclor 1254	12 μg l^{-1}	1.54	7.7 x 10^{-5}	-	256-390	4-8
EDC Tars e.g. 1,2-Dichloropropane	2.7 g l^{-1}	-	-	-100	96.8	2.3

3.   JUSTIFICATION OF AND OPTIONS FOR CONTROL OF RELEASES TO THE SEA

In the developed world, use of DDT has been almost entirely superceded by other less persistent compounds. It is, however, still used extensively in the developing world because it is inexpensive, highly effective and safe to both operator and other acutely exposed humans. The adverse effects on wildlife as a side effect of its use are well-known and accepted and there is general agreement that surplus quantities of material or production residues should not be disposed of in the marine environment. A similar situation arises with Dieldrin, except that it is less widely used in developing countries and is only produced at very few locations. PCBs have very largely also been replaced by less toxic substances, at least in new equipment, although appreciable amounts are still released accidentally to the environment by a variety of routes, e.g. in the form of disused condensers (capacitors), worn-out or leaking transformers and as leachate from land-fill sites. The scale of the problems posed by PCBs and the probable changes in concentrations in the marine environment have been reviewed by ICES (1982) and there are unlikely to be proposals for disposal of additional quantities to the marine environment.

There thus seems little point in examining such substances further. The need for controls have been widely accepted in the developed world where more expensive substitutes can be used and/or regulatory measures have already been applied, and there is unlikely to be any serious attempt to change the position. This paper therefore examines a number of other chlorinated compounds chosen to represent particular forms of use and likely routes of entry to the marine environment. HCH - used as an insecticide; HCB and HCBD - chemical intermediates and by-products from other industrial processes; PCP - used mainly as a fungicide; chloroform - used as industrial solvents and chemical intermediates; and para-dichlorobenzene - used extensively as a disinfectant and odour

masking agent. Each is considered briefly in terms of the benefits it
provides to society and its potentially damaging side effects in relation to man and the marine environment. On this basis the need for
limitations on inputs to the marine environment and the application of
control strategies will be assessed.

4.   CONTROL STRATEGIES

Two basic approaches to the problem of control can be taken. On the one
hand, one can consider what it is possible to achieve by using good
housekeeping to minimise effluent concentrations and by employing
effluent treatment technology to remove as much as possible of the substance concerned. The residual discharge is then accepted on the basis
that it is the best that can be achieved. Alternatively one can
initially examine the data base and determine which concentrations do or
do not cause harm to marine organisms and seek a means of ensuring that
the derived quality criteria or standards are met and thus that the
desired level of environmental protection is achieved. Both systems
have advantages and disadvantages. In practice, neither is likely to be
applied in such a simplistic way. In the former, a balance will always
have to be struck between what it is theoretically and technically
possible to achieve (best technical means available - BTMA) and what it
is practicable (best practicable means available - BPMA) to achieve taking account of costs and operational difficulties. Some attention normally will also be paid to the impact any one discharge would be likely
to have on the environment, if treated by the best practicable means
available. Similarly, it is unlikely that any responsible industrialist
or control authority would set out to allow discharges to be made to the
extent that environmental concentrations are maintained only just below
the maximum concentration regarded as necessary to preserve a particular
desired water use or quality. In practice, discharges would be kept as
low as reasonably achievable in order to maintain the environmental concentrations as far below the defined limit as practicable. In some
situations, as illustrated in the examples which follow, it is even conceivable that discharges of wastes containing more than the smallest
traces simply would not be permissible.

5.   "PROBLEM" CHLORINATED ORGANICS

In the illustrations which follow, references will be made to various
criteria and standards which have been proposed or adopted by the US
Environmental Protection Agency and the European Commission. These are
used merely to illustrate the relative impacts of the two basic control
strategies and no attempt will be made to defend either the standards or
the approaches which might be used.

5.1 HCH

Hexachlorocyclohexane (occasionally but inaccurately still referred to as benzene hexachloride) exists as a number of isomers, all of which are present in the technical product but only one of which the γ isomer, has any commercially useful properties. A colourless solid with the physical properties shown in Table II, γ HCH is rapidly lost from aqueous solutions by a combination of volatility and co-distillation ($T_{1/2}$ ca week), (Portmann, 1979). There is some evidence that in the environment the γ isomer can be microbially converted to the α isomer, (Benezet and Matsumura, 1973). Certainly, environmental samples usually contain both α and γ isomers and the relative proportions increase in favour of the α isomer with time and distance from the source. Neither isomer is as persistent in the environment as DDT, Dieldrin, PCBs, etc.

Although used extensively in the past in the technical form, most HCH is now sold as a 99% or greater purity product and marketed as Lindane. This is widely used in agriculture, as an insecticide on crops

TABLE II. Physico-chemical properties of selected organo-chlorine compounds

	HCH	HCB	HCBD	PCP	$CHCl_3$	p DCB
Solubility	2–12 mg l^{-1}	1–5 µg l^{-1}	5 µg l^{-1}	14 mg l^{-1}	8.2 g l^{-1}	80 mg l^{-1}
Density	1.87	2.04	1.68	1.98	1.49	1.46
Vapour pressure (mm Hg)	1.5×10^{-5} @ 25°C	1.09 @ 20°C	22 @ 100°C	0.12 @ 100°C	–	0.4 @ 25°C
Melting point	112.9°C	229°C	21°C	190°C	−63.5°C	53°C
Boiling point	–	326°C	210–222°C	310°C	61.7°C	174°C
Log octanol water partition coefficient	3.72	6.2	4.78	5.01	1.97	3.37
Bioconcentration factor	100–1000	10 000–30 000	1000	1000 max	3–7	60

and in a variety of seed treatment applications. It is also used in a variety of veterinary products against skin parasites and in particular against the sheep-scab mite. A further use in some countries is in the treatment of timber for either preventative or remedial preservative purposes against the common furniture beetle and other timber damaging pests.

These uses and the ready volatility of γ HCH have led to widespread release to the environment and γ HCH is now evident in most marine organisms, although the concentrations found are usually low, typically < 0.01 µg g^{-1}, unless taken from a location close to a point source. It is also found in most other foodstuffs. FAO/WHO have suggested the acceptable daily intake (ADI) should be 0.0125 mg/kg body weight, i.e. 0.875 mg/day would be unlikely to affect a typical adult, an amount which is unlikely to be even approached given present levels of food contamination. Assuming an intake of fish of 500 g per day as the only source of HCH in the diet the fish could contain up to 1.7 µg g^{-1} without adverse impact on the consumer. Such concentrations are far above the maximum normally found in marine species. There have been suggestions that HCH might be a carcinogen and the US Authorities have proposed maximum intake values which are so low as to be unattainable (USEPA, 1980). The cancer risk is however no more than tentative and there is little evidence that past usage of HCH has presented any hazard to man at the rates normally used.

The log-octanol water partition coefficient for HCH (3.72) suggests that it is likely to be accumulated by marine organisms but not to an excessive degree. The highest recorded concentration factor is only about 1000, which is in reasonable agreement with the partition coefficient. A review of the information on the toxicity of HCH to marine species suggests that, provided concentrations in the water do not exceed 10 ng l^{-1}, marine life is unlikely to be affected (Portmann, 1979). This could lead to concentrations of 0.01 $\mu g \, g^{-1}$ or less in fish tissues which is about the maximum normally found in fish. This suggests that present levels in the marine environment are not likely to pose a threat to aquatic life but that the margins for safety are not large. This being so it is not desirable that inputs to the sea be increased and discharges should be restricted accordingly.

So long as it is found necessary to use HCH there will be little that can be done to prevent diffuse inputs and unless use expands dramatically it is unlikely that environmental concentrations will increase. However, it is possible to restrict inputs from point sources in order to ensure that harmful concentrations do not occur in the environment immediately affected by such discharges. The extent to which this is necessary can be defined by the water quality criteria or environmental standard approach, and this has been done in both the United States and the European Community. The feasibility of restricting concentrations has also been considered by the EC and limits have been laid down for the maximum concentrations allowable in discharges from industrial premises handling HCH. These various limits are shown in Table III.

It should be noted that the US water quality criterion (WQC) is a maximum value whereas the EC environmental quality standard (EQS) is a mean concentration. The US figure does not take account of chronic

TABLE III.   Control options and their implications

	HCH	HCB	HCBD	PCP	CHCl$_3$	p DCB
UK proposed EQS	10 ng l^{-1}	500 ng l^{-1}	10 $\mu g \, l^{-1}$	3 $\mu g \, l^{-1}$	1000 $\mu g \, l^{-1}$	-
EC proposed EQS (P)[1] adopted EQS (A)	(A) 10 ng l^{-1} (EC, 1984)	50 ng l^{-1} (tentative)	-	(P) 1 $\mu g \, l^{-1}$	(P) 10 $\mu g \, l^{-1}$	-
US WQC	160 ng l^{-1} (USEPA, 1980)	-	0.77 $\mu g \, l^{-1}$ (USEPA, 1980)	3.7 $\mu g \, l^{-1}$ (USEPA, 1978b)	620 $\mu g \, l^{-1}$ (USEPA, 1978c)	15 $\mu g \, l^{-1}$ (USEPA, 1978d)
EC effluent standard proposed (P)'/adopted (A)	(A) 7-20 mg l^{-1} (EC, 1984)	200 $\mu g \, l^{-1}$ achievable	-	(P) 1000 $\mu g \, l^{-1}$	(P) 800 $\mu g \, l^{-1}$ reducing to 100 $\mu g \, l^{-1}$	-
Dilution of effluent to achieve EQS	2 to 6 x 10^6	4 x 10^3	13	1000	80 decreasing to 10	-

(1) Anon, 1985

toxicity but the EC value does. The effluent standard values differ according to the nature of the plant but in order to achieve a safe concentration in the environment to which the effluent is discharged a dilution of between 2 and 6 x 10^6 would be necessary for an effluent meeting the effluent standard defined by the EC. Even to reach the US WQC level, a dilution of at least 10^5 would be necessary. Typical dilution on discharge from a pipeline is unlikely to exceed a few hundred fold and the initial dilution of discharges into the wake of a moving vessel do not usually exceed 10^4. It is apparent from this that discharges from factories operating to this emission standard are likely to adversely affect marine life. It is also apparent that with presently available effluent treatment technology it will be very difficult for a factory handling HCH to achieve the EQS limit of 10 ng 1^{-1} in the immediate vicinity of the discharge. In practice this means that under an environmental protection based control strategy, direct discharge of a treated effluent would not be permissible.

5.2   HCB

Hexachlorobenzene is a solid at room temperature, is only slightly soluble in water and has a low vapour pressure (Table II). However, like HCH it is readily volatile in the presence of water vapour and is rapidly lost from clear aqueous solutions (Portmann, 1979). In the presence of suspended solids this process is markedly reduced, due to rapid and strong adsorption on particulates. In the environment, concentrations in water (usually less than 0.01 μg 1^{-1}) are well below the solubility maximum, even close to point sources, and the highest residue levels are usually found in sediments, e.g. 0.1-10 μg kg^{-1} in the UK and > 1000 μg kg^{-1} in one heavily contaminated area in the USA. HCB is slowly degraded and its persistence is similar to or less than that of HCH.

Although widely used in the past as a soil fumigant and in seed dressings for grain or as a fumigant in grain storage, use for such purposes is now declining and largely has been discontinued in Europe. Other uses are believed to be minor and there now seem to be alternatives available which avoid the need to discharge HCB to the environment as a result of actual use of HCB substance. However, HCB is produced as a by-product, e.g. in the manufacture of other chlorinated organic chemicals such as carbon tetrachloride and perchloroethylene, and although the final products contain only very low concentrations of HCB, contamination of process waters is unavoidable. Total HCB production incidental to the manufacture of these substances may be as much as 2% of the quantity of the intended product and although most of the by-products are retained as "heavy ends" in the distillation process, aqueous effluents may contain between one and several hundred microgrammes HCB per litre. It is therefore necessary to consider whether such effluents can be discharged to marine water without risk to the environment.

An examination of the available literature on the acute and chronic toxicity of HCB to marine organisms suggests that provided concentrations in the environment do not exceed 500 ng 1^{-1} even the most sensitive marine species would not be affected. To provide an additional margin of safety, the EC has proposed an EQS of 50 ng l^{-1}. It has also indicated that present methods of effluent treatment are readily able to yield effluents not containing more than 200 μg l^{-1} HCB and it seems likely that a concentration of this order will be set as the effluent standard (Table III).

The discharge to marine waters of effluents containing such concentrations of HCB is likely to lead to some mortalities among marine organisms as it is virtually impossible to achieve the 4000-fold dilution necessary to achieve the EQS close to the point of discharge. However, the scale of area so affected would not be large and, based on concentrations in effluents discharged in the UK, it seems likely that most plants can achieve an effluent quality better than the maximum likely to be suggested by the EC. Thus, it seems that only minimal pollution is likely to arise by strict adherence to an effluent quality standard approach. Similarly, it ought to be relatively easy for most plants to achieve an effluent quality which will allow them to meet the EQS value. In short, discharges of most existing aqueous effluents containing HCB should be easily feasible without presenting a risk of pollution of the marine environment.

The concentration factors for HCB as estimated from environmental situations appear to be around 10 000. Thus, if the maximum concentration of HCB in water is set at 50 ng l^{-1}, the highest concentration likely to occur in fish or shellfish is around 500 μg g^{-1}. This is about ten times the concentration normally found in European waters but concentrations of this order are found in areas which are known to be subject to point source discharges. There is no ADI for HCB but it has been suggested that concentrations of HCB in meat should not exceed 5 μg g^{-1} (EC, 1980). If this same standard were applied to fish this would mean that the concentration of HCB in water should not exceed 0.5 ng l^{-1}, i.e. much lower than the level needed to protect aquatic life. However, it seems likely that such a low level would provide a very large safety factor.

Although the discharge of aqueous process effluents seems unlikely to present an unacceptable hazard to marine organisms, it is clear that the margin of safety involved is small. It follows that the discharge of wastes containing higher concentrations of HCB, e.g. the "heavy ends", is likely to cause considerable risk to pollution. Alternative means of disposal of such wastes are therefore essential. At present the most commonly used are storage on land pending discovery of an acceptable means of disposal, deposition underground in old mines, or incineration. All such procedures present a risk to the environment but the risks associated with each option are different in type and scale.

Both of the storage options are likely to present a risk at some future date, the extent of which it is impossible to predict. Such options are also likely to leave the waste in a condition in which it may be extremely difficult to treat in the future. For this reason it is suggested that incineration is much to be preferred, since if conducted under properly controlled conditions and with flue gas scrubbing, the products need pose only a very small and assessible risk to the environment.

5.3 HCBD

Hexachlorobutadiene is a liquid at room temperature (Table II). It has
an extremely low vapour pressure, is only slightly soluble in water and
is stable under normal environmental conditions, although probably less
so than HCB (ICES, 1983). Like HCB it is rapidly and firmly adsorbed to
suspended solids and sediments and concentrations in water rarely exceed
0.01 μg l^{-1}, although sediment concentrations have been reported to be
as high as several thousand μg kg^{-1}.

HCBD has a number of very minor uses in Europe and in the USA and
primary production is now small, probably no more than a few hundred
tonnes per year. It was used as a soil fumigant and is believed to be
still used in the USSR as a viticulture fumigant against Phylloxera
(ICES, 1983). The main source of HCBD to the environment is as a consequence of production of other chlorinated organic compounds, because
like HCB, HCBD is an unavoidable by-product amounting to as much as 3%
of the total intended product yield. Much of this is retained as the
"heavy ends" of the distillation residues, but small amounts escape to
the process waters, leading to concentrations of 1-10 μg l^{-1} in
discharges of effluent from the production of such compounds as
perchloroethylene, triazine herbicides and chlorine by electrolysis.

The available data on the acute and chronic toxicity of HCBD to
marine organisms suggests that 10 μg l^{-1} would not be toxic to most
marine species even after prolonged exposure (Table III). The US
authorities proposed a water quality criteria value of just below
1 μg l^{-1} and although the EC has not yet formally suggested an EQS it
seems likely that its proposal would not be more than a factor of 10
lower, that is a likely minimum of 0.1 μg l^{-1}. On the assumption that a
typical effluent contains 10 μg l^{-1} HCBD, only a 13-fold dilution would
be necessary to achieve the US WQC value of 0.77 μg l^{-1}, or about
100-fold to achieve 0.1 μg l^{-1}. Such dilutions are readily achievable
and it therefore seems that at present levels, discharges of effluent
containing HCBD are unlikely to pose a hazard to the marine environment.
A similar conclusion was reached by the Advisory Committee on Marine
Pollution (ACMP) of the International Council for the Exploration of the
Sea (ICES), in a review which concluded that HCBD posed less of a problem than HCB and that unless the latter was present in unusually high
concentrations it was unlikely that HCBD would give cause for concern
(ICES, 1983).

HCBD is bioaccumulated by marine organisms but the maximum concentration factors measured under experimental exposure conditions are only
around 1000-fold. Thus, observance of an EQS of 1 μg l^{-1} could lead to
a maximum concentration of HCBD, in fish used as food for man of
1 mg kg^{-1}. It has been suggested (USEPA, 1978a) that the safe level of
intake for man is 2 mg d^{-1}. If it is assumed this is solely contributed
from the fish component of the diet, such a limit would allow up to
4 mg kg^{-1} HCBD in the fish (assuming 500 g d^{-1} consumed). This is well
above the maximum likely to be encountered if the EQS is observed.

The margins of safety however are not such that it is possible to
suggest that HCBD can be discharged to the environment without controls.
It certainly seems undesirable that the "heavy ends" residues should be

discharged to the sea untreated. For such wastes, i.e. those containing
concentrations of HCBD in excess of 100 µg l^{-1}., incineration seems to
present the best option.

5.4 PCP

Pentachlorophenol is a crystalline solid and despite its low vapour
pressure is quite volatile (Table II). It is quite soluble in water, is
adsorbed onto particulates and, based on the octanol water partition
coefficient, is likely to be strongly bioaccumulated by aquatic organisms. However, measured concentration factors do not exceed 1000,
probably because PCP is at least partially metabolised and readily
excreted. PCP is also readily degraded in aqueous media especially
under the influence of UV light. Photodegradation is particularly rapid
in the atmosphere (T$_{1/2}$ a few hours). It is therefore unlikely to
persist for long periods once released to the environment. The method
used for the production of PCP can lead to the production of chlorinated
dibenzofurans and dioxins which may be carried over in varying amounts
as trace contaminants in the final product. These impurities are much
more persistent than PCP and are also much more toxic. Some are
recognised as carcinogens and others are suspected to have such
properties. In the discussion which follows, PCP is regarded as a pure
compound and the possible presence of trace impurities is not taken into
account.

PCP is widely used as a fungicide and slimicide especially on
timber but also on products such as paint, paper and adhesives. Some
loss of PCP may occur as a result of these uses but in properly run
plants such losses are likely to be small, and the main inputs seem
likely to be at very low concentrations from diffuse sources or from the
few sites of manufacture.

An examination of the available data on the acute and chronic
toxicity of PCP to marine organisms led the EC to suggest that an EQS of
1 µg l^{-1} would protect all life stages of marine organisms (Table III).
The EC has also suggested that, if emission standards are adopted as the
means of limiting inputs to the aquatic environment, the maximum allowable concentration should be 1000 µg l^{-1}. In effect this will mean that
effluents containing such concentrations might cause pollution close to
the point of discharge although the size of the area affected is likely
to be small in tidal waters. As the effluent standard is based on what
current effluent treatment technology can achieve it is apparent that
industry should have no great difficulty in ensuring that the EQS value
is met in the environment soon after discharge.

When considering the acceptable daily intake of PCP for man, the US
authorities suggested that up to 0.7 mg d^{-1} might come from fish (USEPA,
1978b). If it is assumed that a typical maximum intake of fish is
500 g d^{-1}, it may be inferred that up to 1.4 mg kg^{-1} PCP in fish is
tolerable. As the maximum concentration factor observed is about 1000,
this would imply a maximum concentration in water of 1.4 µg l^{-1}. This
is just above the EQS and well in excess of concentrations typically
found in European coastal waters (< 0.1 µg l^{-1}).

From the foregoing it is apparent that effluents containing PCP can be discharged from reasonably well run treatment plants without causing pollution of the environment. There is evidence that PCP is degraded in the environment and significant accumulation is unlikely under present conditions of use and discharge. It would however be unwise to regard the marine environment as a suitable receptacle for bulk quantities of waste PCP. Alternative means such as incineration or chemical treatment should be adopted.

5.5 $CHCl_3$ Trichloromethane, Chloroform

Although chloroform is a liquid at normal temperatures it is highly volatile (Table II) and emissions to the atmosphere in the course of production and use are substantial. Up to 3% loss is typical for Europe but losses of between 5 and 10% have been quoted for some sites in the USA (Singh et al., 1980). Chloroform is readily soluble in water - approaching 1% - but in view of its volatility it seems likely that vapourisation will take place from the aqueous environment.

Perhaps best known to the general public as one of the early anaesthetics, it still finds very minor use in certain medicines, although in most developed countries such use is now forbidden due to concern over its possible carcinogenic properties. The main use of chloroform is as an industrial solvent and as a chemical intermediate in the production of other substances especially the fluorocarbons. In view of the wide range of benefits presented to modern society by the latter it must be assumed that production and use of chloroform for this purpose will continue.

Even if industrial production and use were to cease or be totally self-contained, chloroform is still worthy of consideration because it is frequently found in freshwater rivers and drinking water as a consequence of chlorination. The sources of chloroform in freshwater are probably mainly via sewage (especially if disinfected by chlorination) but there is also evidence that chloroform is formed in the atmosphere either naturally or through the photodegradation of other chlorinated organic compounds (Lovelock et al., 1973). It has also been identified as a combustion product in automobile exhaust gases (ICI, 1981). Thus, whilst the major loss during production is to the atmosphere, there are abundant diffuse sources of chloroform leading to input to the marine environment. Accordingly, the dangers inherent in chloroform entering the marine environment merit examination.

Rivers in Europe have been found to contain between 0.1 and 30 μg $CHCl_3$ 1^{-1} on average (Rook, 1977), but near to urban areas levels of up to 2000 μg 1^{-1} have been reported (Rook, 1975). In the North Atlantic concentrations are typically 4-13 ng 1^{-1}, but up to 1 μg 1^{-1} has been reported close to the UK coast in Liverpool Bay (Pearson and McConnell, 1975). The solubility of chloroform in water is such that adsorption onto solids does not seem likely and indeed measured concentrations in Liverpool Bay sediments (4 μg kg^{-1}) (Pearson and McConnell, 1975) tend to confirm this. Equally, the octanol water partition coefficient indicates bioaccumulation is likely to be insignificant and in fact the highest measured bioaccumulation factors for aquatic (not marine) organisms are between 3- and 7-fold.

An examination of the available toxicity data led the UK authorities to recommend 1000 μg l^{-1} as the maximum concentration compatible with protection of marine organisms (Table III). The EC has recently proposed that the effluent emission standard should be set at 800 μg l^{-1} decreasing at a later date to 100 μg l^{-1}. Thus if the EQS recommended by the UK is accepted, effluents meeting the EC emission standard will not require any dilution in order to avoid risk to marine organisms. The EC has in fact proposed an EQS value of 10 μg l^{-1}, a factor of 100 below the UK value, but this appears to be due to a desire to protect man through the consumption of drinking water rather than the need to protect aquatic life. Even if this standard were accepted, discharges of effluents meeting the emission standard value would cause minimal pollution as the 80-fold dilution required to achieve the EQS should occur, if not on discharge, certainly within a very short distance of any outfall in marine waters.

On technological grounds, it is unlikely that treatment of process waters containing high concentrations of chloroform could produce effluents containing less than 100 μg l^{-1} chloroform. The need to do so, at least for discharges to the marine environment, seems questionable. However, disposal of waste chloroform solvent is obviously undesirable and in view of the low boiling point of chloroform, recovery and recycling should be readily feasible for most arisings. If disposal is unavoidable, incineration should be considered.

5.6 Para Dichlorobenzene

(1, 4 DCB) is a white crystalline solid at normal temperatures. It is appreciably soluble in water and is slowly volatalised at room temperatures (Table II). Its major use is as an air deodorant for which purpose it is widely found in public and domestic toilet facilities. The main production route is as a by-product in the manufacture of monochlorobenzene and its sale as a deodorant represents an enterprising means of avoiding the need to treat, or otherwise dispose of, the by-products of such manufacture.

The octanol water partition coefficient suggests that para dichlorobenzene might be accumulated by marine organisms and it has been reported to have been detected in fish tissues with a concentration factor of around 60-fold. The US authorities derived water quality criteria which suggest that the maximum concentration of para dichlorobenzene in marine waters should not exceed 15 μg l^{-1} at any time (Table III). The same authorities have also proposed water quality criteria to protect man through the consumption of fish which might have accumulated para dichlorobenzene. These allow a maximum concentration of 0.35 mg l^{-1}, on the assumption of fish consumption at the rate of 18.7 g fish daily. Even assuming a maximum daily intake of 500 g of fish the allowable concentration in water would be 13 μg l^{-1}, i.e. approximately the same as that needed to protect marine organisms.

The highest reported concentrations of para dichlorobenzene reported in industrial and domestic sewage effluents are only a few tens of

μg 1^{-1} (USEPA, 1978d). Accordingly, from the above assessments, there
seems to be little danger that even the lowest WQC values would be
exceeded under present conditions of production and use.

6. CONCLUSIONS

When organochlorines are considered as a class of compounds there is an
implicit assumption that all have properties similar to DDT, Dieldrin
and PCBs. An examination of some examples of other compounds in the
generic grouping shows that they exhibit a wide range of properties. It
is also apparent that entry into the marine environment is unavoidable
in some cases and/or for the time being inevitable because of likely
continued production and pattern of use. Thus, the no-input under any
circumstances approach as an option for control is both impractical and
totally unrealistic.

Other options for environmental protection centre on two basic
strategies. These are either consideration of what it is possible to
achieve by minimising inputs as far as practicable or consideration of
quality standards essential to protect marine resources and man. The
examples chosen illustrate the spectrum of potential conflicts between
these approaches. In one case (HCH) it is apparent that discharge of
effluents treated to feasible emission standard limits could cause considerable pollution, and also that if environmental protection standards
were applied as control measures, no direct discharge of industrial
effluents would be practicable under that strategy. At the other
extreme (chloroform), effluents treated to meet feasible emission standard requirements would contain less than the limit defined to protect
aquatic organisms. Between these two extremes lies a range of options,
but it is clear that simple compliance with emission standards does not
necessarily guarantee full protection of the marine environment. Our
state of knowledge is such, however, that we can say that the sea has a
capacity to accommodate present rates of input of all these substances,
except perhaps HCH from industrial sources, without any adverse effects
occurring. However, the present margin of safety appears least for HCH
and greatest for chloroform and thus the need for restrictions consequent upon increased inputs of any organo-chlorine compound will vary
from substance to substance, and therefore should be considered on a
case by case basis.

7. REFERENCES

ANON., 1985. 'A new round opens in EEC Water Pollution Control Policy.'
    ENDS Report, No. 121 pp. 15-16.
BERGE, G., LJOEN, R. and PALMORK, K. H., 1972. 'The disposal of con-
    tainers with industrial wastes into the North Sea: A fisheries
    problem.' pp. 474-475. In: Ruivo, M. ed. Marine Pollution and Sea
    Life. West Byfleet, England, Fishing News (Books).
BENEZET, H. J. and MATSUMURA, F., 1973. 'Isomerisation of γ HCH to
    α HCH in the environment.' Nature, Lond., 243: 480-481.

CARSON, R., 1963. Silent Spring. London, Hamish Hamilton.

EUROPEAN COMMUNITY, 1980. 'Proposal for Council Directive (2) on the fixing of maximum levels of pesticide residues in foodstuffs of animal origin.' Off. J. Eur. Commun., No. C56, pp. 14-24.

EUROPEAN COMMUNITY, 1984. 'The Council Directive of 9 October 1984 on limit values and quality objectives for discharges of Hexachlorocyclohexane No 84/491/EEC. Off. J. Eur. Commun., No. L274, pp. 11-17.

INTERNATIONAL COUNCIL FOR THE EXPLORATION OF THE SEA, 1982. 'PCBs in the Marine Environment:- An Overview.' Report of the ICES Advisory Committee on Marine Pollution 1981, Annex 4. Coop. Res. Rep., Cons. int. Explor. Mer, No. 112 pp. 44-50.

INTERNATIONAL COUNCIL FOR THE EXPLORATION OF THE SEA, 1983. 'Hexachloro - 1, 3 - Butadiene.' Report of the ICES Advisory Committee on Marine Pollution, 1983, Annex 5. Coop. Res. Rep., Cons. int. Explor. Mer, No. 124 pp. 53-56.

IMPERIAL CHEMICAL INDUSTRIES, 1981. 'Sources of chloroform in urban areas.' ICI, Mond Division, Northwick, Cheshire, Internal Report.

LOVELOCK, J. E., MAGGS, R. J. and WADE, R. J., 1973. 'Halogenated hydrocarbons in and over the Atlantic.' Nature, Lond., 241: 194-196.

PEARSON, C. R. and MCCONNELL, G., 1975. 'Chlorinated C_1 and C_2 hydrocarbons in the Marine Environment.' Proc. R. Soc. Lond., 189: 305-332.

PORTMANN, J. E., 1979. 'Evaluation of the impact on the aquatic environment of Hexachlorocyclohexane (HCH isomers), Hexachlorobenzene (HEB), DDT (+ DDE and DDD), Heptachlor (+ heptachlor epoxide) and Chlordane.' Report prepared for CEC, Brussels, under Contract No. U/78/180.

ROOK, J. J., 1975. 'Headspace analysis of volatile trace compounds in the Rhine.' Vom Wasser, 44: 23-30.

ROOK, J. J., 1977. 'Chlorination reactions of fulvic acids in natural waters.' Environ. Sci. Technol., 11: 478-482.

SINGH, H. B., SALAS, L. J., SMITH, A. J. and SHIGEISHI, H., 1980. 'Measurements of some potentially hazardous organic chemicals in urban environments.' Atmos. Environ. 15: 601-661.

US ENVIRONMENTAL PROTECTION AGENCY, 1978a. 'Ambient water quality criteria - hexachlorobutadiene.' USEPA, Washington, Rep. No. PB-292-435, 49 pp. (mimeo).

US ENVIRONMENTAL PROTECTION AGENCY, 1978b. 'Draft ambient water quality criteria for pentachlorophenol.' USEPA, Washington, 90 pp. (mimeo).

US ENVIRONMENTAL PROTECTION AGENCY, 1978c. 'Draft ambient water quality criteria for chloroform.' USEPA, Washington, 68 pp. (mimeo).

US ENVIRONMENTAL PROTECTION AGENCY, 1978d. 'Draft ambient water quality criteria for dichlorobenzenes.' USEPA, Washington, 119 pp. (mimeo).

US ENVIRONMENTAL PROTECTION AGENCY, 1980. 'Notice of water quality criteria documents.' Fed. Reg., 45 (231), 79318-79346.

METAL POLLUTION IN THE GREAT LAKES IN RELATION TO THEIR CARRYING CAPACITY

Jerome O. Nriagu
National Water Research Institute
Department of the Environment
P.O. Box 5050
Burlington, Ontario, Canada
L7R 4A6

ABSTRACT. Although the Great Lakes may be resilient to moderate inputs of nutrients and some biodegradable organics, they are seriously vulnerable to the persistent toxic metals. This report makes the point that the assimilative capacity of the Great Lakes for toxic metals has already been exceeded in Lakes Michigan, Erie and Ontario, and in the problem areas of Lakes Huron and Superior. Any added use of these lakes as waste space for the disposal of metal contaminated refuse will only further depreciate the health of the Great Lakes ecosystems.

1. INTRODUCTION

The five Great Lakes represent the largest reservoir of freshwater in the world. The Great Lakes basin itself covers about 785,450 km^2 (about 300,000 square miles) and extends for more than 1000 km inland from the Atlantic Ocean to the heart of the North American continent (Figure 1). Because of their enormous size, it is not surprising that these "sweetwater seas" have often evoked comparisons with the oceans as the following analogy by Herman Melville aptly illustrates: "Those grand fresh-water seas of ours -- Erie, and Ontario, and Huron, and Superior, and Michigan -- possess an ocean-like expansiveness with many of the ocean's noblest traits; with many of its rimmed varieties of races and of climes. They contain round archipelagoes of romantic isles, even as the Polynesian waters do; in large part, are shored by two great contrasting nations, as the Atlantic is" (from Twine Line, vol. 8, no. 6, p. 3, Dec. 1984). This unique water resource with extensive mineral wealth in its basin and the agricultural opportunities afforded by the fertile land and an agreeable climate supports one of the largest and most rapidly growing industrial and urban complexes in the world. Today, over 37 million people (which include 40% and 15% respectively of the populations of Canada and the United States) dump their wastes into and drink the water from these lakes. And the population is expected to rise to over 50 million by the year 2020, with attendant sharp increases in industrial and urban development in the lake basin.

G. Kullenberg (ed.), The Role of the Oceans as a Waste Disposal Option, 441–468.

There is every indication that the wastes from intense cultural activities have adversely affected the quality of the Great Lakes and their fish resources. The first comprehensive survey of pollution in the lakes, conducted between 1913 and 1916, only found localized bacterial contamination in the nearshore waters and concluded that industrial pollutants were not discharged in sufficient quantities to seriously affect the water use. By 1960, subsequent studies showed that the pressure on the Great Lakes from human activities had reached the crisis stage. "Lake Erie is Dead" became a rallying cry for the effort to "save" the lakes. As to be expected, the nutrient cycles dominated most of the early work on the pollution of the Great Lakes as evidenced by the 1972 Great Lakes Water Quality Agreement (between the United States and Canada) which dealt primarily with water quality problems in relation to the eutrophication processes.

Since about 1970, there has been a growing public concern about the occurrence of toxic chemicals in air, water, soil and food chains in the Great Lakes basin. The public interest was aroused by the excessively high levels of mercury in fish which forced the closing of commercial fishery in Lake St. Clair and the western arm of Lake Erie in 1970 Fimreite, 1979), and other subsequent warnings issued periodically on the hazards of consuming mercury-contaminated fish from the Great Lakes. The concern was further heightened by the mirex residues in Lake Ontario fish and the PCB residues in Lake Michigan fish which attained concentrations high enough to warrant human health advisories against the consumption of fish from these lakes. At the present time, about 30,000 compounds of commercial and industrial significance are being used in the Great Lakes basin with some 2000 to 3000 new compounds being added to the list each year (IJC, 1978). About 500 toxic compounds have already been identified in the Great Lakes and many new contaminants continue to be discovered as the detection limits for the hazardous substances are lowered by improvements in the analytical capability. Indeed, work done during the past year or so has found that the Great Lakes contain some of the world's highest concentrations of polynuclear aromatic hydrocarbons (PAH) which are of concern as carcinogens and mutagens (Eadie, 1984; Hallett and Brecher, 1984). The growing number of new toxic substances and the lack of understanding about their behaviour and the effects of long-term exposures to them have generated public outrage about the ability of chemical producers, users and governments to ensure the safety and well-being of the lakes and the people living within the basin area.

This report presents a brief and selected overview of the problems of toxic metals, their entry into, movement through and possible effects on the Great Lakes. Additional details as well as reviews of organic pollutants in the Great Lakes can be found in the recent publication Toxic Contaminants in the Great Lakes (Nriagu and Simmons, 1984).

2.   ANTHROPOGENIC SOURCES OF METALS

Contributions from human activities now totally dominate the cycling of
toxic metals in the Great Lakes ecosystems. The pollutant metals enter
the lakes through emissions to the atmosphere, from industrial and
municipal wastewater discharges, urban and rural land runoffs, and from
such adventitious sources as spills, vessel coatings, dredged material,
and leaching of solid and liquid waste disposal sites. Evidence for the
increased flux of pollutant metals into the Great Lakes comes from sharp
enrichments of the trace metals in the most recent sediments (Kemp and
Thomas, 1976; Kemp et al., 1978; Edgington and Robbins, 1976; Nriagu et
al., 1979; Robbins, 1980), and surface microlayers at the air–water
interface (Elzerman and Armstrong, 1979; Owen and Meyers, 1984).
Changes in the isotopic signature of lead in the sedimentary column
further evince the dominating role of anthropogenic inputs on the trace
metal economy of the lakes.

2.1 Atmospheric Inputs

The lake surface area constitutes from 25% (Lake Ontario) to 64% (Lake
Superior) of the drainage area and averages about 46% for the five lakes
(Table I). The low watershed–to–lake area ratios mean that a large
fraction of the water entering the lakes falls directly on their
surface. The atmospheric precipitation in the Great Lakes basin
generally contains elevated levels of metals, implying that the
atmospheric route should be a major contributor to the trace metal
budgets of these lakes.

One of the earliest documentations of the atmosphere as an
important source of metals in the Great Lakes was by Winchester and
Nifong (1971). They estimated that the emissions and subsequent
deposition of metals from the urban–industrial complexes at Gary–
Chicago–Milwaukee was responsible for large fractions of the toxic
metals getting into southern Lake Michigan. Other authors have
subsequently estimated the atmospheric fluxes of metals into the lakes
(for example, see Gatz, 1975; Klein, 1975; IJC, 1977a; Eisenreich, 1980;
Sievering et al., 1979; Dolske and Sievering, 1979), and these data,
which are at times at considerable variance, have recently been reviewed
(IJC, 1980; Schmidt and Andren, 1984). Basically, the quantification of
the atmospheric deposition rates for the trace metals is hampered by (a)
an inadequate data base on their atmospheric concentrations including
the spatial and temporal variations; (b) inadequate knowledge of the
removal processes and rates; (c) absence of pertinent information on the
micro– and macro–meteorological variables over the lake during dry and
wet deposition; (d) lack of data on the episodic nature of the fallout
processes; and (e) improper understanding of the transfer of metals
across the air–water interface.

A selected set of the reported atmospheric loading rates for trace
metals in the Great Lakes is shown in Table II. In spite of the

Table I. Physical characteristics of the Grat Lakes

Feature	Superior	Michigan	Huron	Erie	Ontario
Length (km)	560	490		385	309
Breadth (km)	256	188		91	85
Area (km^2)					
Water surface, USA	53,618	58,016	23,700	13,400	9,324
Water surface, Canada	28,749	–	36,100	13,500	10,360
Drainage basin land, USA	43,900	118,600	42,100	54,200	47,100
Drainage basin land, Canada	84,200	–	92,600	23,000	32,800
Drainage basin land, total	128,100	118,600	134,700	77,200	79,900
Total drainage basin	210,600	176,600	194,500	104,100	99,600
Maximum depth (m)	406	281	229	60	244
Average depth (m)	145	85	59	19	86
Volume of water (km^3)	11,920	4,910	3,515	468	1,630
Mean anual precipitation (mm)	736	787	792	863	863
Mean outflow (km^3/yr)	67	36	161	182	212
Average water retention time (yr)	178	136	22	2.6	7.7

TABLE II. Anthropogenic inputs (t/yr) of metals into the Great Lakes*

	Atmospheric	Industrial Effluents	Municipal Effluents	Others	Total
Lake Superior					
Cd	82	0.62	0.60	0.85	84
Cu	821	15	7.6	8.4	852
Pb	1230	3.8	7.0	12	1250
Ni	328	10	7.6	3.5	349
Zn	1850	20	22	19	1910
Lake Michigan					
Cd	58	19	8.4	8.1	94
Cu	575	420	102	110	1210
Pb	1730	104	98	193	2120
Ni	575	280	108	96	1060
Zn	3510	560	329	439	4830
Lake Huron					
Cd	60	0.26	1.5	0.62	62
Cu	298	6.4	18	3.2	326
Pb	596	1.6	17	6.1	621
Ni	89	4.2	19	1.1	113
Zn	1190	8.4	56	12	1270
Lake Erie					
Cd	75	3.8	19	9.8	108
Cu	151	90	232	47	520
Pb	754	22	216	99	1090
Ni	75	60	238	37	410
Zn	1010	120	706	184	2020
Lake Ontario					
Cd	28	3.0	12	4.3	47
Cu	95	70	144	31	340
Pb	379	17	142	54	592
Ni	76	48	178	30	332
Zn	948	94	470	151	1660

*Compiled from the information available in the literature.

Table III. Total deposition of airborne trace organics to the Great
           Lakes (from IJC, 1980)

Compound	Superior	Michigan	Lake Huron 10^3 Kg yr^{-1}	Erie	Ontario
Total PCB	9.8	6.9	7.2	3.1	2.3
Total DDT	.58	.40	.43	.19	.14
α-BHC	3.3	2.3	2.4	1.1	.77
γ-BHC	15.9	11.2	11.6	5.0	3.7
Dieldrin	.54	.38	.55	.17	.13
HCB	1.7	1.2	1.2	.53	.39
p,p'-Methoxychlor	8.3	5.9	6.1	2.6	1.9
α-Endosulfan	7.9	5.6	5.8	2.5	1.8
β-Endosulfan	8.0	5.6	5.8	2.5	1.9
Total PAH	163	114	118	51	38
Anthracene	4.8	3.4	3.5	1.5	1.1
Phenanthrene	4.8	3.4	3.5	1.5	1.1
Pyrene	8.3	5.9	6.1	2.6	1.9
Benz(a) Anthracene	4.1	2.9	3.0	1.3	.94
Perylene	4.8	3.3	3.4	1.5	1.1
Benzo(a) Pyrene	7.9	5.6	5.8	2.5	1.8
DBP	16	11	12	5.0	3.7
DEHP	16	11	12	5.0	3.7
Total Organic Carbon	2×10^5	1.4×10^5	1.5×10^5	$.66 \times 10^5$	$.46 \times 10^5$

uncertainty about the error ranges, the data nevertheless suggest that
huge quantities of the toxic metals are transported annually via the
atmosphere to the Great Lakes. According to the estimates, the
atmospheric route contributes annually about 8,500 metric tonnes of
zinc, 5,000 tones of lead, 2,000 tonnes of copper and 300 tonnes of
cadmium to the Great Lakes. This huge pollution load with a large
soluble component (Lindberg and Harris, 1983) is delivered directly to
the surface of the lakes so that the potential impacts of the metals are
readily maximized.

It must be emphasized that large quantities of toxic organic
chemicals are also delivered to the lakes via the atmosphere (see
Table III). Notice the continuing influx of large amounts of such
compounds as PCB and DDT which are no longer being used in the Great
Lakes basin. Once can suspect that some of these organic compounds
would potentiate the low-level toxicity of metals in these waters.

2.2 Industrial Discharges

The Great Lakes basin includes a wide diversity of economic conditions
and occupational pursuits. The northern portion of the basin is
characterized by industries dependent on forest and mineral resources.
Agriculture and diversified manufacturing are concentrated in the
southern section of the basin while on the lakeshores, particularly of
the Lower Lakes, are a number of centres for heavy industry with
emphasis on iron, petroleum and chemical production. Wastewater
effluents from many of these industrial processes contain elevated
levels of the toxic metals.

A rough estimate of the loadings of trace metals to the Great Lakes
from industrial wastewaters is shown in Table II. The data suggest that
the contributions from industrial wastewaters amount to about 5-20% of
the annual atmospheric inputs. The estimates are based on annual
wastewater discharges of 25, 5.3, 350, 75 and 60 billion litres into
Lakes Superior, Huron, Michigan, Erie and Ontario, respectively, and
average concentrations of 1.2, 1.6, 0.3, 0.8, 0.1 and 0.05 mg/l,
respectively, for Cu, Zn, Pb, Ni, As and Cd in the wastewaters. It goes
without saying that the wide variation in the industrial processes will
result in highly variable metal concentrations in the effluents and that
few of the industrial operations have been monitored for their metal
emissions. This point is amply illustrated by Table IV which shows the
Cd concentrations in treated wastewaters from several industrial
processes discharging their wastes into the Great Lakes (see IJC,
1980). The simplistic calculations (Table II) merely suggest the
possible magnitude of the loadings from industrial wastewater
discharges, and serves to emphasize the need for measurements to assess
the role of this particular source in the trace metal budgets of the
lakes.

2.3 Municipal Wastewater Discharges

Invariably, municipal wastewaters show elevated levels of trace metals.
The annual flows of municipal effluents into Lakes Superior, Huron,
Michigan, Erie and Ontario have been estimated to be 82, 266, 1140, 2520
and 1680 billion litres, respectively (IJC, 1980). Analyses of primary
and secondary effluents from wastewater treatment plants in several
cities of southern Ontario found the average values (μg/1) to be:
Cd=7.3, Bi=410, Cr=72, Cu=91, Fe=1270, Pb=86, Hg=1.5, Ni=95 and Zn=274
(Oliver and Cosgrove, 1975). The inputs of trace metals have been
estimated (Table II) assuming these data to be representative of the
effluents from other municipalities in the Great Lakes. These estimates
must be regarded as being conservative considering that 20% of the Great
Lakes population live in high-density, non-sewered residential areas
(IJC, 1977b).

Table IV. Cadmium concentrations in industrial wastewaters flowing into the Great Lakes [a,b]

Industry	Treated Wastewater					
	Concentration, µg/l			Loading,[c] kg/d		
	Minimum	Maximum	Mean	Minimum	Maximum	Mean
Coal mining	2	4	2[d]	0	0.053	0.0076
Textile mills	NA	13	6[d]	2.5×10^{-5}	0.17	0.011
Timber produs processing	BDL	7	1	0	0.046	8.6×10^{-5}
Petroleum refining	<1	20	<2	0	0.19	0.012
Paint and ink formulation	BDL	200	24	0	0.0011	3.8×10^{-5}
Gum and wood chemicals	NA	NA	NA	NA	NA	NA
Rubber processing	NA	1,500	760	0.00015	0.65	0.025
Auto and other laundries	<1.0	31	11	1.0×10^{-5}	0.015	0.002
Porcelain enameling	ND	2,000	650	0.0004	0.27	0.052
Pharmaceutical manufacturing	ND	ND	ND	ND	ND	ND
Ore mining and dressing	<0.002	16	<0.3	0	NA	0.001
Foundries	10	840	120	0	0.80	0.32
Iron and steel manufacturing	NA	770	270	0	NA	49
Nonferrous metals manufacturing	ND	3,000	780	0	NA	41

[a]From the IJC (1980) Report
[b]NA = not available; ND = not detected; BDL = below detection limit
[c]The loadings are into the Great Lakes. These data are derived by multiplying the mean Cd concentration by
industrial wastewater discharges (total) into the Great Lakes
[d]Median, not the average value

2.4 Other Sources

Urban runoffs in North America are often contaminated with toxic metals, especially Cd, Pb, Cr and Zn (Table V). An estimate of the metal inputs by this route cannot be attempted since there is little quantitative information on the flow of urban and rural runoffs into the Great Lakes. The metal concentrations in representative runoffs certainly point to this route as an important contributor of metals to the lakes.

There are well over 4000 liquid and solid waste disposal sites in the Great Lakes basin, the most notorious of which is the Love Canal in Niagara Falls. Little is currently known about the significance of such sites in the supply of trace metals to the Great Lakes. Furthermore, they represent a potential source of toxic metals to the groundwater which can then be cycled to the lakes.

A large fraction of the sediments in the harbours and shipping channels are heavily contaminated with trace metals (Thomas and Mudroch, 1979). The average annual volume of dredged material in the Great Lakes basin has been estimated to be about 6.4 million cubic meters (IJC, 1977b). Although about 75% of the polluted material is confined on land, the deepwater disposal of the remainder represents an important pathway of trace metal transfer to the offshore waters.

Spills which occur during the transportation and handling of products contribute unknown quantities of trace metals to the Great Lakes ecosystems. Another potentially important adventitious source of some metals which deserves to be mentioned are the vessel coatings (including antifouling paints, primers and spent fuel residues).

It is believed that these "other sources" contribute the equivalent of about 1% of the anthropogenic inputs of metals getting into Lakes Superior and Huron, and about 10% of the loadings into Lakes Michigan, Erie and Ontario (Table II).

2.5 Relative Importance of the Various Sources

The loadings of trace metals from the various anthropogenic sources are compared in Table II. The atmosphere, by far, is the principal source of trace metals in the Upper (Superior and Huron) Lakes, accounting for over 90% of all the pollutant metals getting into Lake Superior. Even in the Lower Lakes, the atmosphere still accounts for over 50% of the Pb, Cd and Zn inputs. Industrial waste discharges account for 1-20% of the metal flux into the Lower Lakes. By contrast, municipal wastewaters typically supply 20-50% of the pollutant metals getting into the Lower Lakes. It should be noted that industrial and municipal wastewaters supply over 50% of the anthropogenic inputs of Ni, and Cu into Lakes Erie and Ontario (Table II).

Table V. Grand ranges and averages of heavy metals in combined sewer overflow, urban runoff and municipal wastewater treatment Plant Effluent Concentrations (from IJC, 1980)

	Lead (mg/l)	Cadmium (g/l)	Chromium (mg/l)	Copper (mg/l)	Mercury (g/l)	Nickel (mg/l)	Zinc (mg/l)	Arsenic (mg/l)	Iron (mg/l)
Grand range of CSO values from Table 1.	<0.02-33.0	<10.-1600	<0.03-18	0.02-4.2	0.1-28	<0.01-0.50	0.04-5.3	0.049-14.0	0.11-27.7
Grand range of urban runoff values from Table 1.	0.002-3.0	0.4-7000	0.0025-11	0.004-0.5	<0.1-47	0.024-0.29	0.005-4.6	<0.01-0.1	0.01-59.0
Average of CSO flow weighted storm average values from Table 1.	3.9c (2)	28.5 (2)	1.4 (2)	0.26 (2)	-	-	0.67 (2)	0.05 (1)	1.50 (1)
Average of urban runoff flow weighted storm average values from Table 1.	0.16c (9)	5.8 (7)	0.03 (6)	0.08 (6)	-	-	0.11 (9)	0.05 (5)	1.26 (7)
Threshold concentrations that are inhibitory to activated sludge processes b	0.1	10,000-100,000	1-10	1.0	100-5,000	1.0-2.5	0.08-10	0.1	1000
Primary plants effluent concentrations	0.01-1.7	3-40	0.006-2.6	0.01-1.7	0.1-5.0	0.006-1.7	0.03-3.6	-	0.4-5.0
Trickling filter plants effluent concentrations	x=0.27 (37)	x=14 (35)	x=0.40 (40)	x=0.19 (48)	x=1.0 (23)	x=.17 (33)	x=.55 (49)	-	x=1.5 (30)
Activated sludge plants effluent concentrations	0.005-1.8 x=0.12 (45)	1-66 x=11 (41)	0.003-3.2 x=0.24 (52)	0.003-1.8 x=0.13 (54)	0.1-10 x=1.0 (22)	0.007-1.5 x=0.20 (38)	0.04-2.8 x=0.32 (57)	-	0.1-65.6 x=2.9 (34)
Average of primary, trickling filter and activated sludge plants effluent concentrations	0.14 (133)	30 (124)	0.23 (152)	0.46 (170)	3.3 (82)	0.18 (127)	0.35 (172)		1.7 (101)

a Municipal Wastewater Treatment Plants may include wet-weather flows.

b Source: "Federal Guidelines: State and Local Pretreatment Program Volume I", Construction Grants Program Information, EPA-430/9-76-017a, January 1977

c Values in parentheses refer to the number of flow weighted storm averages.

d Values in parentheses refer to the number of plants.

3.   COMPARTMENTALIZATION OF TRACE METALS IN THE GREAT LAKES

The trace metals which enter the Great Lakes from many sources are
distributed throughout the ecosystem at different concentrations in the
suspended particulates, water, sediments and biota.   Such compartmentalization determines the ability of the Great Lakes ecosystems to
assimilate the metal pollution.

3.1 Water Column

The high enrichment of metals and organic pollutants in the surface
microlayers (Table VI) suggests that this interface plays an important
role in the exchange of material between the air and the water.   It
would seem that the accumulation of metals at this zone is driven by the
fallout from the atmosphere.   The metals in the microlayer are
presumably concentrated into the solid phase by bursting of bubbles, or
into sinking particles by the compression of the surface film.   The
uptake of the metals in this zone by the neustonic communities can lead
to the introduction of the metals into the aquatic food webs (Owen and
Meyers, 1984).
      The pH values for the waters of the Great Lakes typically fall in
the range of 7.5 to 8.5.   Thus, the soluble trace metals falling into
the lakes will quickly be hydrolyzed and sedimented out as
particulates.   Any biological packaging of the metals should also assist
in their removal to the sediments.   Indeed, a number of studies (Lum and
Leslie, 1983; Nriagu et al., 1981; Kauss, 1983) show that the suspended
particulates are highly enriched with trace metals (Table VII).   In
general, suspended particulates account for 50-80% of the Cu, 10-40% of
the Ni, 20-60% of the Cd and over 60% of the Pb concentrations in Lakes
Ontario and Erie (Nriagu et al., 1981; Lum and Leslie, 1983).
      In comparing the trace metal concentrations in the Great Lakes with
the levels in the oceans, it should be recognized that (a) the
particulate production rate is much higher in the Great Lakes and (b)
the half-lives for the metals are much shorter in the shallower Great
Lakes.   The high pH and rapid turnover of particulates may explain the
relatively low concentrations of trace metals in the waters of the Great
Lakes (Table VIII).
      As to be expected from the source data, the metal concentrations in
Lakes Erie, Ontario and Michigan are considerably higher than those of
Lake Huron (Table VIII).   The urban and industrial centres in the basin
are concentrated around ports, along shore or near the junctions of
major land and water transportation routes.   The water quality tends to
reflect such cultural development and there is some evidence to suggest
that the trace metal concentrations in the nearshore zones of the Lower
Lakes are higher than those of the offshore waters (Neilson, 1979; 1983;
IJC, 1983.

Table VI.   Average concentrations (µg/l) of heavy metals in
              Lake Michigan surface waters (film pressure > 1
              dyne.cm^{-1})[a]

Environment[b]	Element			
	Zn	Cd	Pb	Cu
Rivers and Harbours				
M	28	1.0	15.2	4.4
S	14	0.40	4.9	2.0
FR	2.0	2.5	3.1	2.2
Mixing Zones[c]				
M	15.2	0.31	7.4	4.1
S	3.8	0.09	1.2	1.0
FR	4.0	3.4	6.2	4.1
Nearshore[d]				
M	10.9	0.24	8.2	2.0
S	3.2	0.08	1.2	0.9
FR	3.4	3.0	6.8	2.2
Midlake[e]				
M	5.6	0.12	3.8	2.4
S	2.0	0.07	1.3	1.1
FR	2.8	1.7	2.9	2.2

[a] After Elzerman and Armstrong (1979); Owen and Myers (1984).
[b] M = Microlayer; S - Subsurface (~ 30 cm); FR = Fractionation ratio.
[c] River plumes, mixing zones and near obvious sources of atmospheric particulates.
[d] < 8 km from shore and > 1 km from river plumes.
[e] > 8 km from shore.

Table VII. Toxic metal levels ($\mu g/g$) in suspended particulates and surficial sediments of Lakes Erie, Ontario and the Niagara River (from Kauss, 1983; Lum and Leslie, 1983)

	Lake Erie	Niagara River	Lake Ontario
Suspended Sediments			
As	–	16	–
Cd	6.6	3.0	10
Cu	170	100	180
Pb	79	170	190
Hg	0.14	0.69	–
Ni	91	53	48
Zn	160	330	410
Surficial Sediments			
As	3.2	2.5	5.1
Cd	2.5	0.72	2.5
Cr	66	15	64
Cu	52	8.8	75
Pb	112	13	155
Hg	0.48	0.19	0.65
Ni	56	7.2	68
Zn	217	63	388

Table VIII. Selected data on trace metal concentrations (ng/l) in waters of the Great Lakes (after Rossman, 1982, 1984; Nriagu et al., 1981; Lum and Leslie, 1983)

	Michigan	Huron	Erie	Ontario
Ag	46	4.4	26	61
As	480	170	430	500
Cd	19	8.7	39	40
Co	38	–	50	67
Cu	410	220	610	–
Hg	7.5	7.4	12	15
Ni	450	310	800	980
Pb	130	38	150	140
Sb	56	–	85	110
Sn	33	–	160	220
V	260	120	230	440
Zn	370	210	550	590

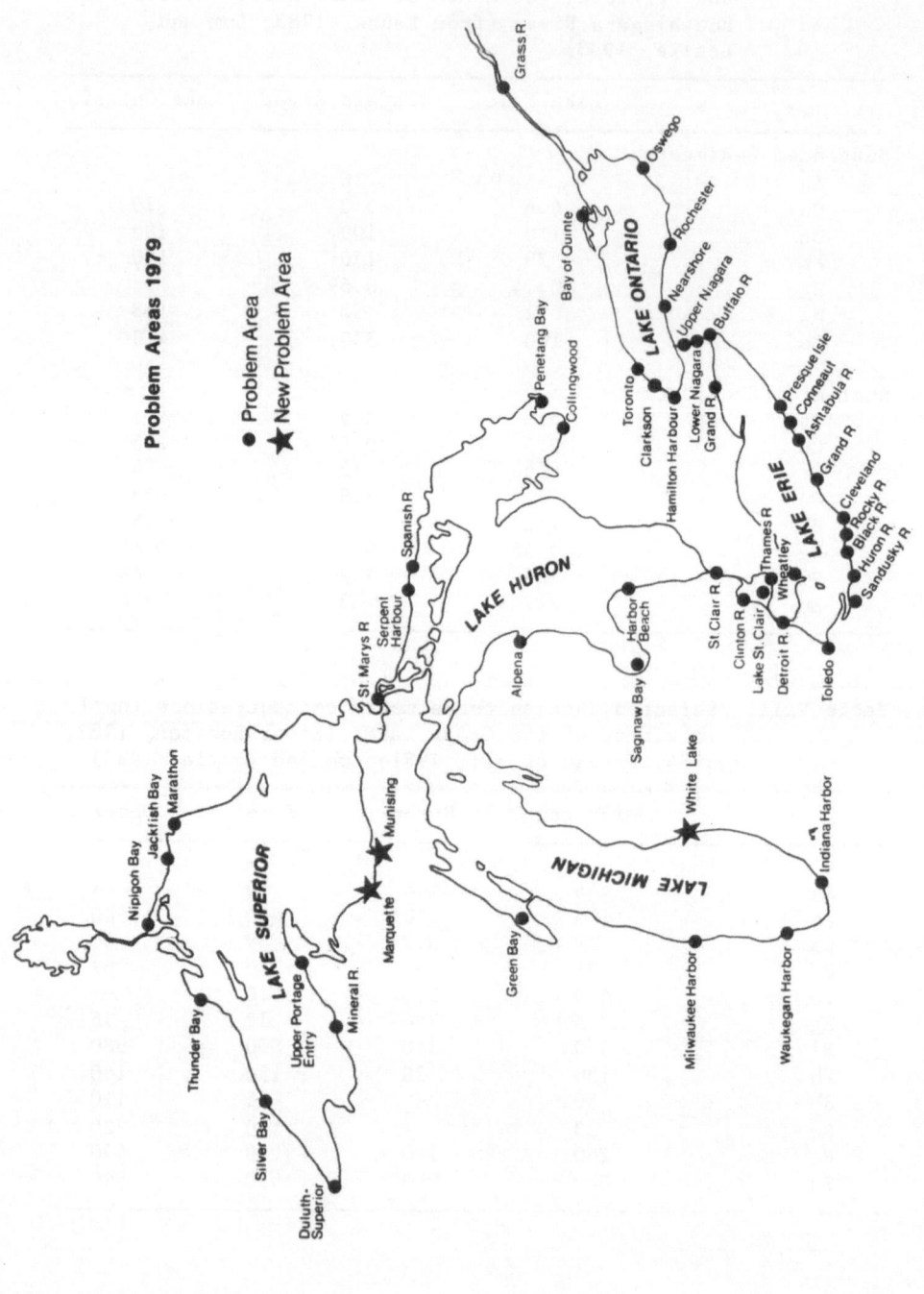

Figure 1   Map of the Great Lakes showing some of the problem areas.

3.2   Sediments

The strong linkage of the trace metal cycle to the particulate dynamics in lakes results in a continuous transfer of the metal pollution to the sediments. Indeed, the distribution of trace metals in the Great Lakes sediments shows a strong positive correlation with the accumulation of the fine-grained sediments. Statistical analyses including correlation, factor and cluster analysis point to the clay-sized material and organic matter as the principal matrices controlling the distribution of pollutant metals in the Great Lakes sediments (see Thomas and Mudroch, 1979; Cahill, 1981; Cahill and Shimp, 1984).

There are, however, marked peculiarities in the distribution of individual metals in the Great Lakes sediments. For example, the Western Basin of Lake Erie shows high Hg loadings from Lake St. Clair and Detroit River (Figure 2a). It should be noted that the decline between 1970 and 1974 in the excessive Hg contents of Lake St. Clair sediments is the result of the migration of the contaminated sediments via the Detroit River to Lake Erie (Thomas et al., 1975). Sediments of the western basins of Lake Ontario also show elevated loadings of Hg which is presumably derived from the Niagara River and its tributaries (Thomas and Mudroch, 1979). The lead distribution shows no obvious point source enhancements (Figure 2b), a feature reflecting the fact that Pb is derived primarily from the atmosphere. However, the massive inputs from the Detroit River and Cleveland are reflected in the higher than normal concentrations of this pollutant in the Western and Central Basins of Lake Erie. The high Cu contents of Lake Superior sediments (Figure 2c) have been attributed to copper mining activities particularly in the Keeweenaw Peninsula in the late 19th and early 20th centuries (Kemp et al., 1978). Apparently, there are no overriding inputs of Cu from major point sources in the Lower Lakes (Figure 2c).

Numerous studies have demonstrated the fact that the sediments often maintain an historical record of recent changes in the flux of metals into the Great Lakes (Shimp et al., 1971; Kemp and Thomas, 1976; Edgington and Robbins, 1976; Kemp et al., 1978; Nriagu et al., 1979; Robbins, 1980). The sediments of the Lower Lakes in particular show a marked enrichment of the trace metals in the most recent sediments in relation to the material deposited in precolonial times (Table IX). The onset of excess metal deposition in these sediments generally coincide with the settlement of the lakes' basins which began in the early to mid 19th century (Figure 3).

The lakewide sediment accumulation rates have been estimated to be 24, 6.9, 10, 13 and 4.4 million tonnes per year, respectively, in Lakes Superior, Huron, Michigan, Erie and Ontario. From the metal contents of the surficial sediments, the tonnage of each trace metal being loaded annually into the Great Lakes sediments has been estimated (Table X). The data suggest that the sediments serve as the sink for about 12% (Cd in Lake Ontario) to 100% of the total anthropogenic trace metal inputs (compare Tables II and X). The variability in the fraction of the

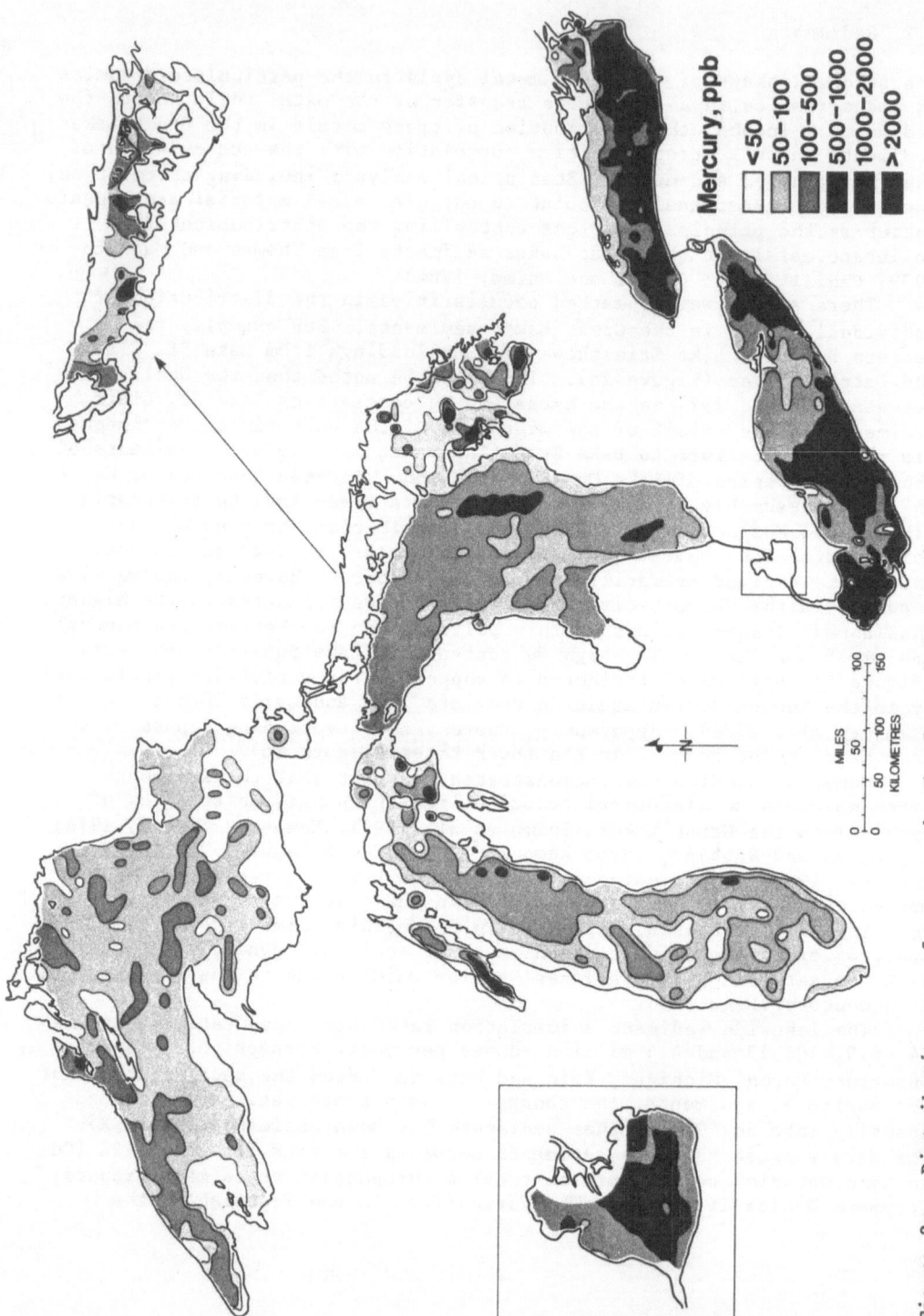

Figure 2a   Distribution map of mercury in the Great Lakes sediments

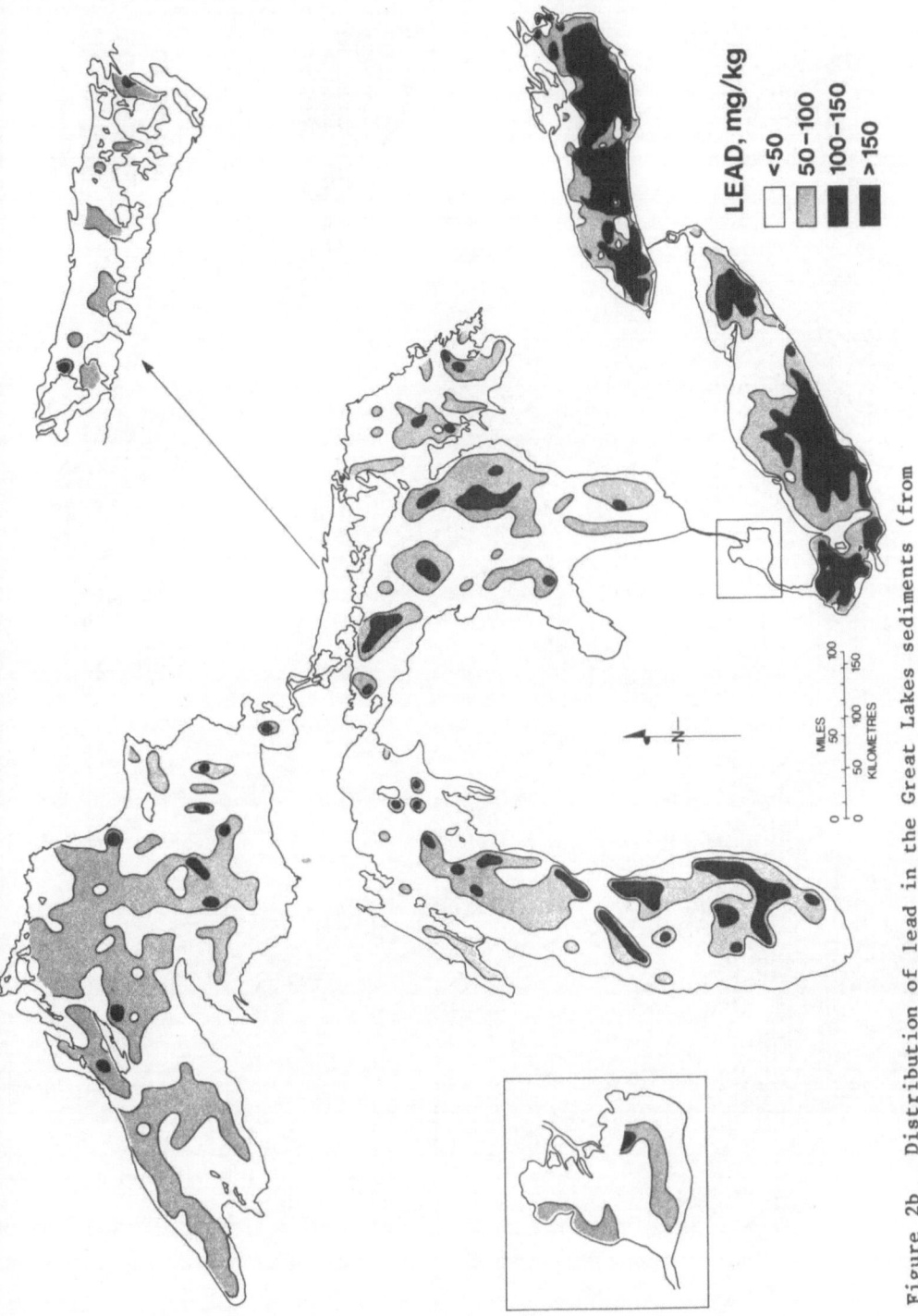

Figure 2b   Distribution of lead in the Great Lakes sediments (from
Thomas and Mudroch, 1979).

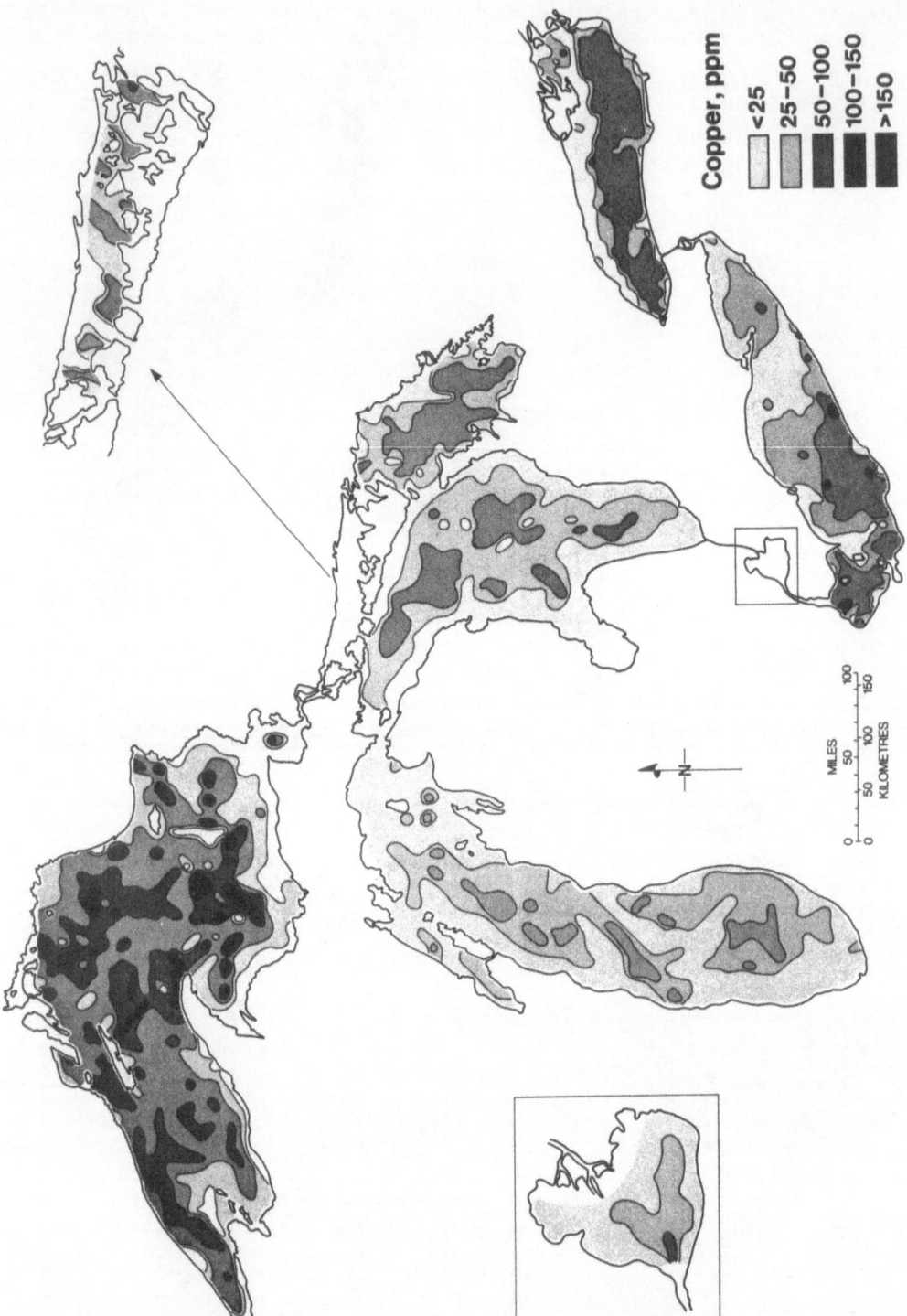

Figure 2c   Distribution of copper in the Great Lakes sediments (from
Thomas and Mudroch, 1979)

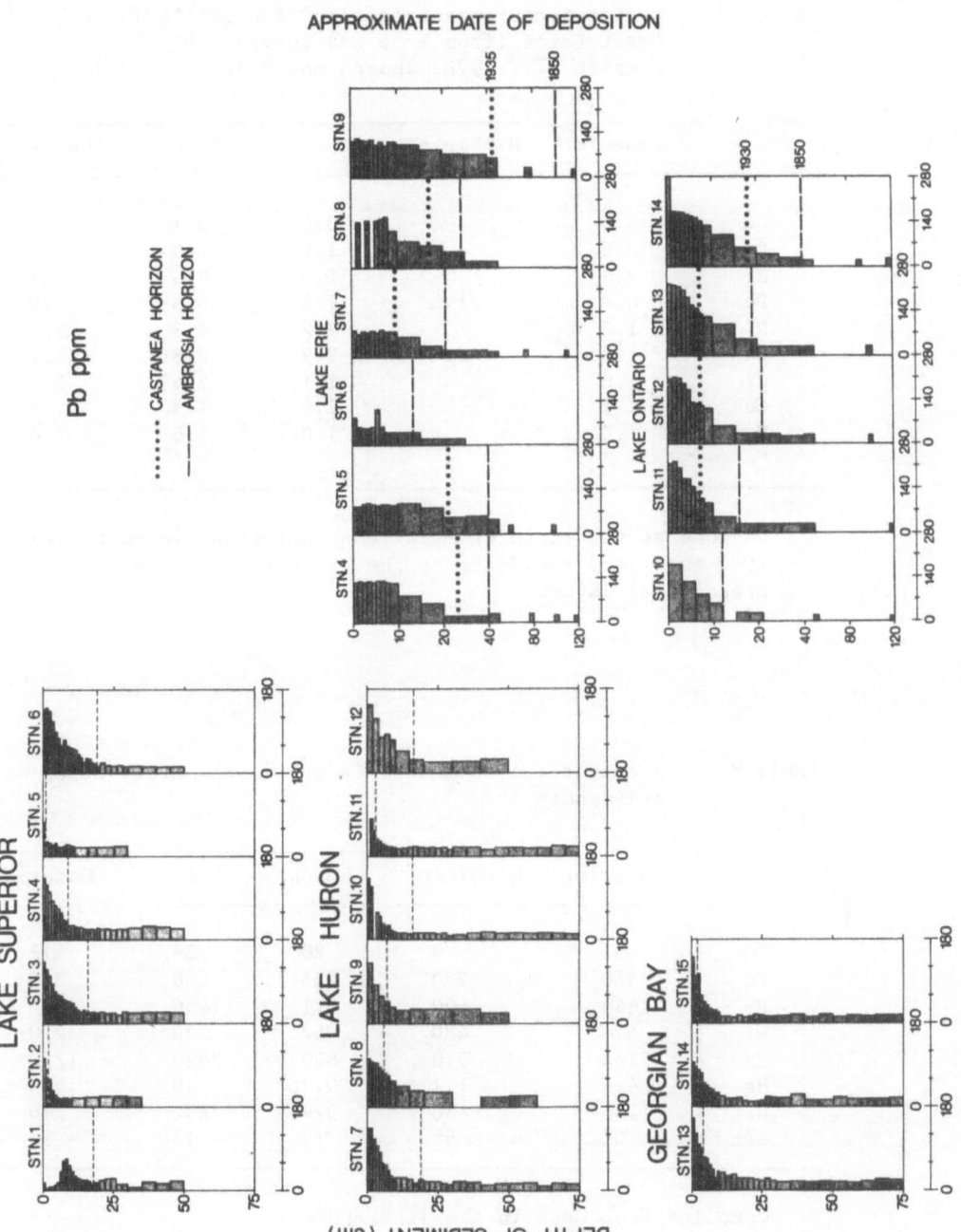

Figure 3   Lead profiles in the Great Lakes sediments (frm IJC 1977b).

Table IX.   Sediment enrichment factors* for metals in the
             Great Lakes (from Kemp and Thomas, 1976;
             Kemp et al., 1978; Thomas and Mudroch, 1979)

	Superior	Michigan	Huron	Erie	Ontario
Cd	3.3		3.0	3.1	7.7
Cu	2.8		1.9	1.9	3.0
Hg	2.7		5.0	9.6	31
Pb	6.8	>2.3	7.2	3.6	10
Ni	1.0		2.2	1.7	2.1
Zn	1.8	3.2	2.9	2.8	5.9
Cr	1.0		1.1	2.4	2.7
Co	1.1		1.4	1.3	1.4
V	1.0		1.0	1.4	1.4

*   Defined as the ratio of metal concentrations in surficial
     (0-2 cm layer) sediments to the concentrations in
     precolonial layers.

Table X.   Accumulation of metals (t/yr) in the Great Lakes
            sediments*

	Superior	Michigan	Huron	Erie	Ontario
Cd	53	9	20	39	12
Cu	960	220	255	676	330
Pb	1440	400	511	1456	682
Ni	624	480	352	728	299
Zn	2160	970	800	2820	1710
Hg	4.6	1.1	0.76	10	4.3
Cr	1130	460	324	832	290
As	36	55	19	13	22

*   Compiled from data in the literature.

metals retained in the sediments points out the inherent problems of using lake sediments to estimate the flux of trace metals into lacustrine environments.

The long-held opinion was that metals were permanently stored in the sediments and were essentially unavailable to the biota in the lakes. Results obtained recently using sediment traps show that the settling flux of metals near the sediment-water interface is many fold higher than the rates in the epilimnion (Rosa et al., 1983; Charlton, 1983). The enhanced fluxes near the mud interface depend on the water depth and can only be attributed to the resuspension of the sediments. The extensive remobilization of metals argues strongly against the disposal of toxic wastes in such an ecosystem.

3.3 Trace Metals and the Health of the Great Lakes Biota

The history of fisheries provides a vivid testimony to the general depreciation of the Great Lakes resources during the past several decades. As a consequence of excessive pollution of the lakes, accelerated degradation of fisheries at an alarming rate began around 1950 (Johnson, 1984). Fish species including Atlantic salmon (Salmo salar), lake trout (Salvelinus namaycush), lake whitefish (Coregonus clupeaformis), sauger (Stizostedion canadense), walleye (Stizostedion vitreum vitreum), and blue pike (Stizostedion vitreum glaucum), which accounted for 70-80% of the early fisheries have now declined to less than 35% of the harvest, and in Lake Ontario only 5% of the catch. Furthermore, the quantity of fish catch in each of the Great Lakes, with exception of Lake Erie, has declined steadily from about 60 in 1979 (Johnson, 1984). Reduced fishery was just one of the many biological evidences of environmental crisis in the Great Lakes. For example, the disappearance of mayflies heralded the excessive eutrophication of Lake Erie while eggshells so soft that they could be dented by simple touch led to the discovery of elevated levels of dioxin in these lakes. Recent assessments of the Great Lakes fish health have reported widepsread incidence of tumour (papilloma, gonadal tumour, thyroid and epidermal hyperoplasia, lymphosarcoma, etc.) in many fish species (Shear, 1984; Leatherland and Sonstegard, 1984). These tumours represent distress signals stemming from disfunctional changes in the Great Lakes ecosystems. The role of metal pollution in the ontogeny of the tumours and other distress signals manifested by the biota has yet to be properly assessed however.

The historical reports point to the Great Lakes as a finely-tuned ecosystem with respect to the biota-toxic metal interactions. The available information leaves no doubt that the flow of metals in the Great Lakes food webs is now dominated by the anthropogenic inputs. High level accumulations of metals have been encountered in the planktonic food webs of these lakes (Table XI), and such plankton may actually be under stress from such metal exposure. Marshall and Mellinger (1980) reported shifts in the species composition of the planktonic communities in the Great Lakes which was attributed to

Table XI.   Lead content of plankton samples from Lake Ontario
(from Hodson et al., 1984)

Sample Location	Pb concentration in µg/g (range)
Net Plankton	
Eastern Basin	4.7 (2.8-7.9)
Point Traverse	7.5 (3.5-12)
Cobourg	6.8 (6.0-8.7)
Port Credit	5.3 (4.8-6.0)
Niagara	29 (17-43)
Zooplankton (Mysis relicta)	
Eastern Basin	2.0 (1.3-2.6)
Point Traverse	1.6 (1.2-2.4)
Cobourg	1.2 (0.9-1.6)
Port Credit	5.1 (4.4-5.8)
Niagara	4.5 (3.9-5.6)

increasing stress due to Cd pollution. Borgmann (1980) showed that
copepod growth was affected by free Cu concentrations as low as 10^{-9} to
10^{-10} mol/1, concentrations that are exceeded in harbours and coastal
areas of the Great Lakes. Borgmann et al. (1981) further found a
seasonal cycle in the toxicity of metals which could be related to the
seasonality in metal concentrations in the biota (Shear, 1984).
Seasonal pulses in metal loading rates and metal toxicity are areas of
concern that have yet to be properly investigated.

The available data suggest that there may be a biodepuration of
metals up the Great Lakes food chain. The levels of metals in whole
fish samples (Table XII) are generally very low compared to those of the
plankton they feed on (see Table XI). One should particularly note the
low levels of Hg in fish samples from Lake St. Clair and the western
Lake Erie. With the discontinuance of mercury discharges into Lake St.
Clair by the offending industry, the Hg in fish has declined from the
excessive levels of 1970 to the current low values. Another interesting
feature is the high Hg levels in Lake Superior fish compared to samples
from the other lakes (Table XII). It would seem that the Lake Superior
fish is still experiencing the effects of Hg inputs from the pulp and
paper industry even though the emissions from this industrial sector
were curtailed in the 1970s (Shear, 1984). This points out the great
susceptibility of this particular lake to toxic metal pollution. It is
a deep, cold oligotrophic lake with long half-lives for pollutant
metals. The natural self-cleansing processes are thus slowed down
including, apparently, the reduction in the body burden of mercury in
the fish.

4.   ASSIMILATIVE CAPACITY OF THE GREAT LAKES FOR THE METAL POLLUTION

All the pollution abatement programs currently employed in the Great
Lakes basin recognize and permit less than 100% removal of pollutants
from waste streams. Besides the obvious economic dictates, this
implicit use of assimilative capacity in waste management is rooted in
the belief that large bodies of water, like the Great Lakes, are capable
of rapid redistribution and dilution of the discharged pollutants in
such a way as to lower their concentrations to acceptable values.
Furthermore, it is believed that with time the pollutant will be
transferred into biologically unavailable forms or to the less
accessible parts of the ecosystem.
        Since about the late 1960s, the use of assimilative capacity as
waste management strategy in the Great Lakes basin has been increasingly
questioned. The negative reaction came about because the "bad actor"
substances were dumped indiscriminately into the lakes with no attempts
made to establish the carrying capacity of the lakes for such
substances. The misuse has since been demonstrated to be tremendously
expensive in terms of the increased amount of water treatment required,
the loss of recreational use of the lakes and the reduction in the fish
harvest. The poor track record of chemical producers, users and
governments in ensuring the safety and well-being of the lakes, the
growing number of hazardous chemicals identified and the lack of
understanding of the behaviour and effects of these toxins tend to
reinforce the public sentiments against any disposal policy that relies
on the assimilative capacity of an ecosystem.
        A number of factors and lake characteristics actually make the
Great Lakes a bad choice for the disposal of metal contaminated wastes.
(i) The lakes, especially the Upper Lakes, still contain extensive
oligotrophic areas with biota particularly sensitive to metal pollution
(see above). (ii) Industrial development in the basin continues to
expand rapidly resulting in increasing waste inputs and enhancing the
prospects for further deterioration in the quality of the lake waters.
(iii) Large fractions of the anthropogenic metals are delivered in
bioavailable forms via the atmosphere. _ (iv) The relatively low
productivity rate (except Lake Erie) and the low sediment load per unit
volume (with the exception of Lake Erie) decrease the rate of removal of
the metals from the water column thereby increasing the potential for
uptake by the biota. (v) Extensive sediment bioturbation and the active
circulation and mixing of the bottom waters result in a significant
regeneration of metals from the lake sediments. (vi) Because of
material transfer via the interconnecting channels, the concentration
and toxicity of metals in the Lower Lakes can be attenuated by
discharges into the Upper Lakes. (vii) The low flow-to-volume ratios
for some of the lakes imply long recovery times by natural flows (over
100 years for Lake Michigan and 500 years for Lake Superior) for
persistent contaminants (Rainey, 1967). (viii) The dense microbial
populations play a particularly active role in trace metal cycling in

the Great Lakes sediments. The alkylation of several metals (Hg, As, Sn and Pb) have been demonstrated in sediments of the Great Lakes (Wong et al., 1975; Chau et al., 1977; 1980; Baker et al., 1981), and thus represents a potentially important pathway for mobilizing the sedimented pollutant metals into the bioavailable and more toxic forms. (ix) The massive burden of toxic organic compounds may attenuate the low-level metal toxicity in these lakes. (x) Model calculations assuming constant Cd input at the present-day rates suggest that within a few decades, the Cd concentration will be raised to the range where deleterious effects on zooplankton communities are readily observed (Marshall et al., 1981; Tissue and Fingleton, 1984).

Unlike the oceans, the chemical processes in the Great Lakes are far from being at a steady state. The ocean-type relationships between the trace metal and nutrient cycles have not been observed in these waters. The Great Lakes are generally more biologically productive, and entertain more extensive circulation and mixing of the waters that the oceans. On a per volume basis, the input rates for toxic metals are already several-fold higher in the Great Lakes compared to the oceans, and the loading rates into the lakes are expected to continue to grow. The added influx of a wide spectrum of organic compounds in relatively large quantitites can further potentiate the low-level toxicity of the pollutant metals in the Great Lakes. The conclusion seems inescapable that any added large-scale disposal of metal contaminated wastes would constitute an unacceptable risk to the health of the Great Lakes ecosystems. One only has to look back to the case history and the ecological problems of the disposal of the supposedly innocuous taconite tailings in Lake Superior for a dramatic illustration of the extreme sensitivity of the Great Lakes ecosystems to contaminant inputs.

References

Baker, M.D., P.T.S. Wong, Y.K. Chau, C.I. Mayfield and W.E. Innis: 'Methylation of Pb, Hg, As and Se in the acidic aquatic environment.' Proc. Internat. Conf. Heavy Metals in the Environ., Amsterdam, Sept. (1981).

Borgmann, U.: 'Determination of the free metal ion concentrations using bioassays.' Can. J. Fish. Aquatic Sci., **38**, 999-1002 (1980).

Borgmann, U., Cove R. and C. Loveridge: 'Effects of metal ions on the biomass production kinetics in freshwater copepods.' Can. J. Fish. Aquatic Sci., **37**, 1295-1302 (1980).

Cahill, R.A.: Illinois State Geological Survey Circular 517. 94 pp. (1971).

Cahill, R.A. and N.F. Shimp: 'Inorganic contaminants in Lake Michigan sediments.' Advanc. Environ. Sci. Technol., **14**, 393-423 (1984).

Charlton, M.N.: 'Downflux of sediment, organic matter and phosphorus in the Niagara River are of Lake Ontario.' J. Great Lakes Res., 9 , 201-211 (1983).

Chau, Y.K., W.J. Snodgrass and P.T.S. Wong: 'A sampler for collecting evolved gases from sediments.' Water Research, 11 , 807-809 (1977).

Chau, Y.K. et al.: 'Occurrence of tetraalkyllead compounds in the aquatic environment.' Bull. Environ. Contam. Toxicol., 24 , 265-269 (1980).

Dolske, D.A. and H. Sievering: 'Trace element loading of southern Lake Michigan by dry deposition of atmospheric aerosols.' Water, Air & Soil Pollut., 12 , 485-502 (1979).

Eadie, B.J.: 'Distribution of polycyclic aromatic hydrocarbons in the Great Lakes. Advanc. Environ. Sci. Technol., 14 , 195-211 (1984).

Edgington, D.N. and J.A. Robbins: 'Records of lead deposition in Lake Michigan sediments since 1800.' Environ. Sci. Technol., 10 , 266-274.

Eisenreich, S.J.: 'Atmospheric input of trace metals to Lake Michigan.' Water, Air & Soil Pollut., 13 , 287-301 (1980).

Elzerman, A.W. and D.E. Armstrong: 'Enrichment of Zn, Cd, Pb and Cu in the surface microlayer of Lakes Michigan, Ontario ad Mendota.' Limnol. Oceanogr., 24 , 133-144 (1979).

Fimreite, N.: 'Accumulation and effects of mercury on birds.' In: Biogeochemistry of Mercury in the Environment (J.O. Nriagu, Editor), Elsevier, Amsterdam, pp. 601-627 (1979).

Gatz, D.F.: 'Pollutant aerosol deposition into southern Lake Michigan.' Water, Air & Soil Pollut., 5 , 239-251 (1975).

Hallett, D.J. and R.W. Brecher: 'Cycling of polynuclear aromatic hydrocarbons in the Great Lakes.' Advanc. Environ. Sci. Technol., 14 , 213-237.

IJC: 'Atmospheric loadings to the Great Lakes.' International Joint Commission on the Great Lakes, Windsor, Ontario (1977a).

IJC: 'Annual Progress Report.' International Reference Group on PLUARG. International Joint Commission on the Great Lakes, Windsor, Ontario (1977b).

IJC: 'A Perspective on the Problem of Hazardous Substances in the Great
      Lakes Basin Ecosystem.' Annual Progress Report, Great Lakes
      Science Advisory Board, International Joint Commission on the Great
      Lakes, Windsor, Ontario (1980).

IJC: 'Report on Great Lakes Water Quality. Great Lakes Water Quality
      Board. International Joint Commission on the Great Lakes, Windsor,
      Ontario (1983).

Johnson, M.G.: 'Great Lakes Water Quality.' Great Lakes Water Quality
      Board. International Joint Commission on the Great Lakes, Windsor,
      Ontario (1984).

Kauss, P.B.: 'Studies of trace contaminants, nutrients and bacteria
      levels in the Niagara River.' J. Great Lakes Res., **9** , 249-273
      (1983).

Kemp, A.L.W. and R.L. Thomas: 'Impact of man's activities on the
      sediments of Lakes Ontario, Erie and Huron.' Water, Air & Soil
      Pollut., **5** , 469-490 (1976).

Kemp, A.L.W., J.D.H. Williams, R.L. Thomas and M.L. Gregory: 'Impact of
      man's activities on the chemical composition of the sediments of
      Lakes Superior and Huron.' Water, Air & Soil Pollut., **10** ,
      381-402 (1978).

Klein, D.H.: 'Fluxes, residence times and sources of some elements to
      Lake Michigan.' Water, Air & Soil Pollut., **4** , 3-8 (1975).

Leatherland, J.F. and R.A. Sonstegard: 'Pathobiological responses of
      feral teleosts to environmental stressors: interlake studies of
      the physiology of Great Lakes salmon.' Advanc. Environ. Sci.
      Technol., **16** , 115-149 (1984).

Lindberg, S.E. and R.C. Harriss: 'Water and acid soluble trace metals
      in atmospheric particles.' J. Geophys. Res., **88** , 5091-5100
      (1983).

Lum, K.R. and J.K. Leslie: 'Dissolved and particulate metal chemistry
      of the Central and Eastern Basins of Lake Erie.' Sci. Total
      Environ., **37** , 403-414 (1983).

Marshall, J.S. and D.L. Mellinger: 'Dynamics of cadmium-stressed
      plankton communities.' Can. J. Fish. Aquat. Sci., **37** , 403-414
      (1980).

Marshall, J.S., Mellinger, D.L. and J.I. Parker: 'Combined effects of
      cadmium and zinc on a Lake Michigan zooplankton community.' J.
      Great Lakes Res., **7** , 215-223 (1981).

Neilson, M.A.:   'Trace metals in Lake Ontario.'   Scientific Report
     Series No. 133, Inland Waters Directorate, Water Quality Branch,
     Burlington, Ontario (1979).

Neilson, M.R.:   'Report on status of the open waters of Lake Ontario.'
     Unpublished Report, Inland Waters Directorate, Water Quality
     Branch, Burlingotn, Ontario (1979).

Nriagu, J.O. and M.S. Simmons, Editors: 'Toxic Contaminants in the
     Great Lakes.' Advanc. Environ. Sci. Technol., 14 , 527 pp (1984).

Nriagu, J.O., A.L.W. Kemp, H.K.T. Wong and N. Harper: 'Sedimentary
     record of heavy metal pollution in Lake Erie.' Geochim.
     Cosmochim. Acta, 43 , 247-258 (1979).

Nriagu, J.O., H.K.T. Wong and R.D. Coker: 'Particulate and dissolved
     trace metals in Lake Ontario.' Water Research, 15 , 91-96 (1981).

Oliver, B.G. and E.G. Cosgrove: 'Metal concentrations in the sewage,
     effluents and sludges of some southern Ontario wastewater treatment
     plants.' Environ. Letters, 9 , 75-90 (1975).

Owen, R.M. and P.A. Meyers: 'The surface microlayer and its role in
     contaminant distribution in Lake Michigan.' Advanc. Environ. Sci.
     Technol., 14 , 127-145 (1984).

Rainey, R.H.:   'Natural displacement of pollution from the Great Lake.'
     Science, 155 , 1242-1243 (1967).

Robbins, J.A.:   'Sediments of Southern Lake Huron.' Report No.
     EPA-600/3-80-080, U.S. Environmental Protection Agency,
     Environmental Research Lab., Duluth, Minnesota, 309 pp (1980).

Rosa, F. J.O. Nriagu and H.K.T. Wong: 'Particulate flux at the bottom
     of Lake Ontario.' Chemosphere, 12 , 1345-1354 (1983).

Rossmann, R.:   'Trace metal chemistry of the waters of Lake Huron.'
     Publication No. 21, Great Lakes Research Division, Univ. of
     Michigan, Ann Arbor, Michigan, 41 pp (1982).

Rossmann, R.:   'Trace metal concentrations in the offshore waters of
     Lakes Erie and Michigan.' Special Report No. 108, Great Lakes
     Research Division, Univ. of Michigan, Ann Arbor, Michigan, 170 pp
     (1984).

Schmidt, J.A. and A.W. Andren: 'Deposition of airborne metals into the
     Great Lakes: An evaluation of past and present estimates.'
     Advanc. Environ. Sci. Technol., 14 , 31-51 (1984).

Shear, H.:   'Contaminant research and surveillance -- a biological
      approach.' Advanc. Environ. Sci. Technol., 14 , 31-51 (1984).

Shimp, N.F. J.A. Schleicher, R.R. Ruch, D.B. Heck and H.V. Leland:
      'Environmental Geology Notes No. 41.' Illinois State Geological
      Survey, Champaigne, Illinois, 25 pp (1971).

Sievering, H., M. Dave, D.A. Dolske and P. McCoy: 'Trace element
      concentrations over midlake Michigan as a function of meteorology
      and source region.' Atmosph. Environ., 14 , 39-53 (1979).

Thomas, R.L. and A. Mudroch: 'Small Craft Harbours -- Sediment survey,
      Lakes Ontario, Erie and Lake St. Clair.' Report to Small Craft
      Harbours Ontario Region, Great Lakes Biolimnology Lab., Burlington,
      Ontario (1979).

Thomas, R.L., J.M. Jaquet and A. Mudroch: 'Sedimentation processes and
      associated changes in surface sediment trace metal concentrations
      in Lake St. Clair, 1970-1974.' Proc. Internat. Conf. on Heavy
      Metals in the Environment, Toronto, Ontario, pp. 691-708 (1975).

Tisue, T. and D. Fingleton: 'Atmospheric inputs and the dynamics of
      trace elements in Lake Michigan.' Advanc. Environ. Sci. Technol.,
      14 , 105-125 (1984).

Winchester, J.N and G.D. Nigong: 'Water pollution in Lake Michigan by
      trace elements from pollution aerosol fallout.' Water, Air & Soil
      Pollut., 1 , 50-64 (1971).

Wong, P.T.S., Y.K. Chau and P.L. Luxon: 'Methylation of lead in the
      environment.' Nature, 253 , 263 pp (1975).

MARINE DISPOSAL OF RADIOACTIVE WASTE: AN OVERVIEW
WITH EXAMPLES FROM THE COASTAL WATER SITUATION

A Preston
Ministry of Agriculture, Fisheries and Food
Fisheries Laboratory
Lowestoft, Suffolk NR33 0HT, UK

ABSTRACT. Basic requirements for controlling the introduction of toxic
materials to the environment are given. The criteria for satisfactory
control are examined in relation to radioactivity. Current radiological
protection requirements are noted and commented upon in an environmental
context. In particular the requirement to reduce radiation exposure to
levels as low as are reasonably achievable is examined and especially
the significance of very low doses delivered far into the future.
Against this background the control of radioactive discharges to coastal
waters, as exemplified by the Sellafield sea discharges, are reviewed.
It is concluded that satisfactory radiological control of coastal discharges is attainable both in terms of short and long term consequences
and that the increasing emphasis on socio-political aspects of such
situations presents a greater potential problem.

1. INTRODUCTION

The controlled introduction of any toxic material to the environment
requires a knowledge of the targets sensitive to the material and the
way in which they are likely to be affected, the elaboration of suitable
standards of protection for the targets, a methodology for relating
those standards to the receiving environment and a basis for setting
acceptable rates of introduction for the materials in question.
   Regulation of the introduction of radioactivity to the environment
enjoys major advantages in these respects when compared to most other
toxic materials in that radiation sensitive targets are known and there
are well defined standards in relation to their protection from radiation exposure[1]. Thus in principle relatively rigorous quantitative
assessment of the impact of environmental introductions of radioactivity
can be made. The protection standards apply to human radiation exposure
and experience to date confirms that limitation of introduction against
these human radiation exposure criteria poses the most limiting
constraint[2,3].

469

G. Kullenberg (ed.), The Role of the Oceans as a Waste Disposal Option, 469–492.
© 1986 British Crown

Given this background the essential problems for the controlled
introduction of radioactive materials to the environment are:

(a) to interpret the basic standards in terms relevant to individual environmental situations;

(b) to regulate introductions such that exposure is limited to
some acceptable fraction of the exposure standard.

Relating basic standards to the environmental situation in a marine
context has become a relatively well understood and straightforward
exercise[4,5]. It needs to be based upon an understanding of the dispersive processes acting to dilute and distribute the introduced material and to take account of those processes tending to localise and
concentrate radionuclides. Against this background the possibilities
for human exposure need to be identified and evaluated. The pathways
likely to result in the most significant degree of exposure are the subject of the most detailed evaluation and it is against their parameters
and the nature of the developing concentration field that recommendations for regulation of the rates of introduction of specified radionuclides are made (Figure 1).

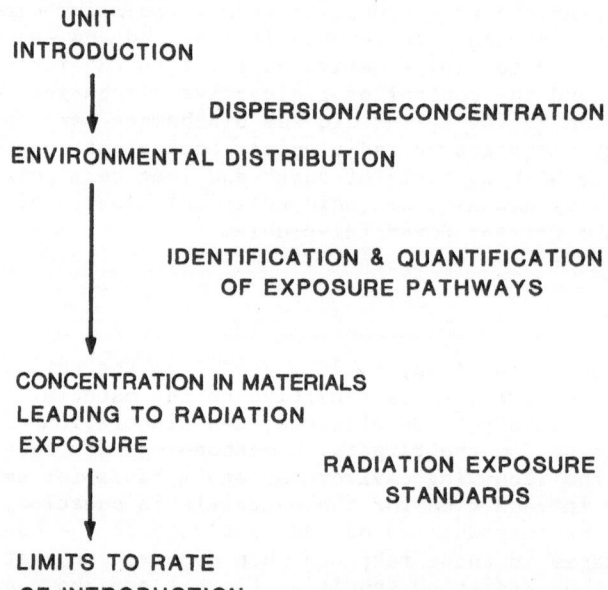

Figure 1. Principal features of a scheme for determining maximum
permissible rates of introduction to the environment for specified
radionuclides.

In some coastal water situations the consequences of actual discharges are susceptible to measurement and thus the overall validity of
the predicted situation can be determined through appropriate monitoring
programmes.

There is now general confidence in our ability to conduct this type of evaluation, to relate rates of introduction to concentrations in selected environmental materials and in turn to relate these to the spectrum of potential human radiation exposure and thus in the short to medium term to predict radiation exposure levels for given rates of introduction. The detailed case histories of Sellafield[6,7,8,9,10] testify to the effectiveness of this type of approach. The evaluation of deep sea disposal carried out under the auspices of NEA and IAEA[11,12, 13,14,15,16] of course remain predictions since no sensible direct measurement of the radiation field resulting from these disposals has yet been possible.

The outstanding unresolved problems in relation to radiation exposure resulting from artificial radioactive materials in the environment relate to prediction of the long term consequences of the introduction of long lived materials and to the definition of what are acceptable levels of exposure within the dose limits defined by the ICRP[1]. To understand the nature of these problems requires a closer look at the way in which the basic radiation standards have been derived and subsequently to look at the significance of these considerations in relation to our understanding of the environmental processes determining the distribution of the longer lived radionuclides.

This paper derives its material primarily from the coastal water situation but the marine and indeed the whole environment is a continuum and many of the points raised are germane to the whole environment and to the regulation of other persistent toxic materials. Case details of assessment procedures, or of the year by year changes in the distribution of radioactivity and of its radiological significance will not be dwelt on since they are well documented[17,18,19,20,21,22]; only such features as are required to illustrate the points being made will be developed.

2. RADIATION EXPOSURE STANDARDS

The basic dose response relationship, from which ICRP develops its radiation protection rationale and recommends its standards, is taken to be one of linear response without threshold and may be represented as a straight line passing through the origin of a graph relating exposure to effect (Figure 2).

However, no effects have ever been observed at the low doses and dose rates akin to those from natural background radiation and likely to be experienced in any regulated regime. Indeed some effects are known to occur only when the radiation dose exceeds a certain threshold. All such data as are available on radiation effects have come from exposures at relatively high dose rate and relate to relatively high accumulated doses, e.g. nuclear explosions, accidental exposure, radiotherapy, or from animal experiments. These data are regularly reviewed and evaluated by UNSCEAR[23] and are utilised by ICRP in its recommendations for the limitation of human radiation exposure.

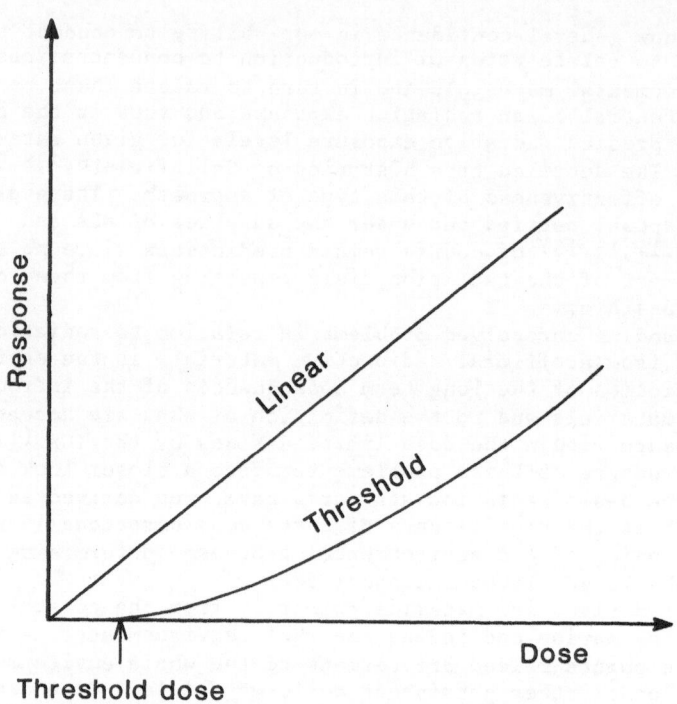

Figure 2.   Possible radiation dose response relationships: linear relationship and relationship with threshold.

        The current ICRP dose limitation framework sets out to limit indi-
vidual radiation exposure in such a way as to prevent non-stochastic
radiation effects in the exposed population, i.e. those effects directly
related to dose above a threshold. It then sets out a basis for limiting stochastic effects, those where the probability of occurrence is
linearly related to dose without threshold, to an acceptable level. To
this end the Commission not only requires that all radiation exposure
should be justified and that dose limits should not be exceeded but that
all doses should be as low as reasonably achievable, economic and social
factors being taken into account: the ALARA principle. The current
dose equivalent limit for exposure of individual members of the public
is 5 mSv (500 mrem) per year but it should be noted that, in terms of
whole body exposure, that where an individual could receive an exposure
at a significant fraction of the dose limit over prolonged periods then
ICRP advises that measures be taken to restrict the lifetime dose to an
average of 1 mSv per year of life long whole body exposure.

3.   CHANGES IN AND IMPLICATIONS OF RADIATION PROTECTION PHILOSOPHY

Management of radiation exposure has thus to be conducted within a framework of dose limits but with a requirement to keep doses as far below these limits as is reasonably achievable.

In this context a current concern in relation to radiological protection and the environment is with long lived radionuclides and has arisen largely since the publication of ICRP 26 when the previous emphasis on the radiation exposure of individuals, exemplified by the dose limits, changed to one of interest in the radiation exposure of populations and in the estimation of collective dose.

In the new regime Dose Limits can be taken to represent the boundary beyond which the estimated risks to the individual are unjustifiable and within which optimisation of the risks and benefits will be used as a guide to the setting of acceptable levels of risk. In order to estimate risk, an estimate of dose is required and whereas under the previous practice an estimate to show compliance with dose limits was adequate this is no longer necessarily sufficient. In order to judge whether or not doses are as low as reasonably achievable a more precise estimate of dose is in theory now demanded even at relatively low doses.

In the previous regime, where the primary emphasis was on the observance of dose limits, the estimation of doses to critical groups could where necessary resort to the use of maximising assumptions and steady state models. In the current regime the need to minimise dose requires realistic estimates of dose incurred by alternative course of action and therefore much more precise environmental data are required in order to derive such improved estimates with the requisite degree of confidence. Clearly where long lived radionuclides are concerned estimates of collective dose equivalent commitment are required to some arbitrary future time, and the definition of this period is of some importance in terms of the requirement to understand the potential behaviour of long lived radionuclides. A clear view of the radiological protection requirement for an understanding of the behaviour of long lived radionuclides is therefore in order.

The changes in emphasis in the dose limitation system create difficulties, for example, in the public perception of the risk associated with stochastic radiation effects, where the probability of occurrence of an effect is linearly related to radiation dose without threshold. Adoption of such a dose response relationship as a basis for protection implies that any exposure, no matter how small, carries with it some finite risk, however remote, that an effect could be induced. It furthermore implies that all doses, large and small, are equally effective in such inductions per unit of radiation absorbed. These implications, which stem logically from the deliberately conservative assumptions of the linear dose-response model, are widely held to be likely to overestimate actual risk at low dose rate and low accumulated doses but they nevertheless constitute the stage on which the current debate on radiation risks is conducted.

This major requirement of current ICRP recommendations to not only keep all radiation exposure within dose limits, but as far below those limits as is reasonably achievable, economic and social factors being taken into account, has introduced a new dimension to radiation protection. In turn this leads to a need to optimise any particular radiation protection practice in such a way as to balance the costs to society, of the detriment incurred from the radiation exposure, with the costs of avoiding the exposure in the first place. In theory when the costs of avoidance (protection) exceed the value to society of the detriment avoided the costs of the avoidance are unjustifiably high. The proper utilisation of these concepts requires assignation of a value to human life and to various forms of health detriment, problems which are being grappled with but which are possibly not subject to a general solution[24].

Some national decisions in relation to regulation of radioactivity will have an international dimension especially in the case of long lived radionuclides released to the environment since by their very persistence they have the potential to become widely distributed and to lead to radiation exposure of individuals and populations in other countries. This international dimension is not new to radiological protection or novel to nuclear energy, witness the case of weapon test fallout, but it does need to be clearly recognised that now, in the context of public concern with the peaceful uses of nuclear energy, there is a steadily increasing international dimension. The provision of scientific evidence of a rigorous character to this debate is of some importance.

To return to the linear dose response relationship and to the implication that even very low doses carry with them an equal probability per unit of dose for the induction of stochastic effects in the exposed population. This interpretation and its equal emphasis on extremely low doses incurred at long times into the future, carried to its logical conclusion, requires us to develop a predictive ability, with respect to the long term behaviour of long lived radionuclides, which is probably beyond reach. If not beyond reach it would require expenditure on R & D out of all proportion to the benefits to be derived from the research, at least in the context of radiological protection.

The linearity assumption can be used to calculate the expected incidence of stochastic effects but there are other considerations beyond these statistically derived effects on populations which merit greater attention than they are currently receiving. Should we not for instance be paying greater attention to doses close to the dose limit than to those far below the dose limit and perhaps to doses delivered over the next few decades rather than to doses theoretically delivered in future millenia? It has been suggested, in the context of costing radiation health detriment as a component in the ICRP optimisation exercise, that a lower unit cost should be assigned to each unit of collective dose equivalent according to a scale such as that indicated in the table (Table I).

TABLE I   Relative monetary values per
unit collective dose for different ranges
of individual dose (Sv)

Individual dose (range Sv)	% ICRP dose limit	Arbitrary unit cost
$< 5.10^{-5}$	< 1	1
$5.10^{-5}-5.10^{-4}$	1- 10	5
$5.10^{-4}-5.10^{-3}$	10-100	25

This suggests that greater importance should be attached to unit
dose at higher dose rate within the range whose upper limit is defined
by the ICRP Dose Limit of 5 mSv (500 mrem) per year, and implies therefore that resources should be concentrated on the reduction of radiation
exposure at higher levels of individual dose and to this extent is analogous to the previously established practice of giving radiological
protection emphasis to 'critical groups'.

Is there a level of dose that could be considered a trivial level
of exposure for the individual? A level of risk of about 1 chance in a
million of harm would probably not be considered as significant when
individuals consider alternative courses of action. This level of risk
corresponds to about 1% of the current ICRP Dose Limit, i.e. 50 µSv per
year or a few per cent of the commonly experienced variations in natural
background radiation exposure[25,26,27].

Even if this were a trivial level of dose for the individual it is
not a simple matter to consider how a trivial level might apply when
large numbers of people are involved in the exposure, i.e. in terms of
collective dose. A reduction by a further factor of ten would certainly
take account of the possibility of multiple sources of exposure which
might be of increasing concern where large numbers of people are involved. Suggestions have been made for source 'upper-bounds', which would
seek to apportion the dose limit between differing sources of radiation
exposure. This is considered necessary·in order to circumvent the
situation where an individual, though within the dose limit in respect
of exposure from a particular source, will remain below the dose limit
even though exposed to several other sources.

In the context of sea disposal one might consider in principle that
the combined effect of say the Sellafield, and Cap de la Hague discharges to the North Sea, is a candidate for such consideration. It has
also been raised in the context of evaluating deep sea disposal and
setting limits to the introduction of material in an ocean basin containing more than one disposal site. However such concepts require
international agreement and a basis for allocating the dose limit
between sources and no machinery yet exists to cope with this
requirement.

    Interestingly enough adoption of a trivial dose level of the order
of 5 μSv would lead to the conclusion that all solid waste disposals to
the North Atlantic to date have not resulted in this level of exposure
having been reached for any individual, despite the conservative assumptions used in the models predicting the dose[12]. Furthermore the
radiological consequences of Sellafield discharges would, on this basis,
largely be confined to critical groups and the average per capita exposure of the population of continental Europe would not enter into the
collective dose calculation.

    However the inescapable logic of the linear dose response relation-
ship would still lead to the derivation of statistical deaths when
applied to large populations, for example a per capita dose of 5 μSv to
the UK population would imply five deaths a year.

    We also have to consider future radiation detriment, that is that
arising as a result of present day practices but from doses delivered
far into the future, perhaps up to 10^9 years from now. Normal accounting approaches would in a general commercial framework assign less
current value to future costs than to current costs. Is it worth
considering therefore whether, in the case of long lived radionuclides
which may deliver doses millions of years into the future, we should
attach less weight to doses delivered in the far future than to doses
delivered in the near future. If it were considered sensible to apply
discounting procedures to the costs for future detriment then for
example a discount rate as low as 0.1% per year would correspond to zero
value at 1000 years and thus concern would centre on radiation exposure
incurred over the next few centuries. Such an approach would suggest
developing a predictive capability in respect of the likely behaviour of
long lived radionuclides that extended with some confidence to a few 100
years. If this were extended to the calculation of collective dose commitment only where the collective dose incurred is made up of annual
individual doses which are greater than 0.1% of the ICRP dose limit of
5 mSv (500 mrem) then this would imply that doses less than about 1% of
the dose incurred from natural background would be ignored in terms of
trying to secure a firm predictive basis. This is also broadly consistent with de minimis dose requirements being considered within IAEA[28].
Looked at from another point of view geochemical considerations suggest
that the life of the more persistent natural trace constituents in sea
water is of the order of 10^3-10^4 years and that global ocean mixing
times are on a timescale of 10^2-10^3 years.

    Furthermore the uncertainties surrounding the prediction of far
future doses resulting from long lived radionuclides makes it virtually
impossible to distinguish the radiological consequences between options
for protection much beyond a few hundred to a thousand or so years. So
in practice only the commitment over some finite time can be used[27].

    Thus it might be argued, on grounds of currently accepted account-
ing criteria in relation to the discounting of costs, in relation to
natural background radiation levels, in relation to the timescale of
chemical and physical events in the ocean which will determine the distribution and persistence of long lived radionuclides in the biosphere,
i.e. where they will lead to significant radiation exposure, a thousand
years might be a useful upper limit to bear in mind when deciding the

scale and nature of research conducted to improve our predictive
ability, within which a hundred years and per capita doses greater than
5 μSv might be the focus of our major concern.

In general therefore though it would remain a requirement to calculate collective dose commitment over all people and all time, wherever
possible identification of that portion of the integral that relates to
doses < 5 μSv per capita and/or which occurs beyond a 1000 years would
serve to indicate where effort might be best deployed in improving the
accuracy of dose estimates or in placing weight when making decisions
about waste disposal.

4. ENVIRONMENTAL CONSIDERATIONS

The key relationship we are seeking in all environmental work with
radioactivity is that between input rate and dose to the critical target, man. For short lived radionuclides, months to a few years, direct
establishment of this relationship from pre-operational estimates or
from monitoring data is quite adequate, and elucidation of the processes
determining the relationship is not called for. For immediate control
purposes in relation to the exposure of critical groups, this is also
probably all that is needed for long lived radionuclides, but in terms
of future collective dose commitment this is not necessarily an adequate
procedure since the material may in years to come be redistributed
within the environment in such a way as to produce new exposure pathways. It becomes necessary therefore, especially in relation to novel
radionuclides such as Pu, Am, Np, etc. where no suitable natural analogues exist on which to base a judgement of likely future behaviour, to
understand in some detail the relationship between the observed distributions and the chemical and physical state of the introduced radionuclides. Without such knowledge it will not be possible to establish the
sets of conditions under which redistribution of environmental contamination may occur and thus not be possible to consider quantitatively the
probability with which such events might arise.

A further consideration must also be to ensure that we have correctly identified the relevant long term transport mechanisms leading to
the redistributions of long lived material already released to the
marine environment. The basic mechanisms are probably well enough
understood, primarily physical in nature and associated with water movement or sediment transport. Biological mechanisms probably play a relatively minor part in the overall transport and distribution process
though they may be of high relative importance in localised areas.
However minor in mass transport terms biological transport mechanisms
may be, they can still be the key processes in providing the pathway
delivering the significant radiation exposure in a particular situation.
For example, in mass balance terms the majority (> 80%) of plutonium
discharged over the last thirty years to the NE Irish Sea remains locked
in bed sediments of that shallow sea area, and most of the rest will
have been flushed from the area entirely[29]. The significant individual radiation exposure from Pu and other actinides occurs via seaweed
grazing gastropod molluscs, winkles and their consumption by the local

population, though almost certainly this relates to the biological
cycling of much less than the equivalent of 0.01% of the annual input.
The doses incurred are significant fractions of the ICRP dose limit[17]
(Table II).

TABLE II   Critical group radiation exposure: NE
Irish Sea fish and shellfish consumption by the
local population (1982)

Consumption rate	Radionuclide	% ICRP dose limit (5 mSv y^{-1})
100 g d^{-1} fish	^{238}Pu	3.3
18 g d^{-1} crustaceans	$^{239/240}Pu$	13.9
45 g d^{-1} molluscs	^{241}Pu	8.7
	^{241}Am	13.9
		39.8

In a separate context it has been estimated that the normalised
collective effective dose equivalent commitment to the world population
from the Sellafield ^{137}Cs discharges is 0.055 man Sv.TBq^{-1}. However
later data (G. J. Hunt and D. F. Jefferies, personal communication) suggest that the value to the European population, where per capita doses
are higher, is probably more nearly 0.078 man Sv.TBq^{-1}. We might therefore conservatively assume that the collective effective dose equivalent
commitment to the world population per TBq^{-1} is 1.10^{-1} man Sv. However
ICRP gives for Sv Bq^{-1} ingested a value of 1.36 10^{-8} or 1.36 10^4 Sv per
TBq^{-1}. Thus only some $7.3.10^{-6}$ of the ^{137}Cs discharged is involved in
the biological cycling causing human radiation exposure via the fish and
shellfish consumption pathway, or in rounded terms one thousandth of 1%.
Thus although their contribution is very small in total activity terms
and though they are only the result of second-order processes superimposed on a concentration field primarily determined by physical
events, they are nevertheless usually the important processes in radiological terms and their evaluation is an essential step in order to
determine the frequency and magnitude of future human radiation exposure
via food consumption, the pathway likely to be of most significance.

5.   THE SELLAFIELD COASTAL WATER CASE

The discharge of low-level liquid radioactive effluent from BNFL via a
2.5 km pipeline has been a continuous practice since 1952. The early
history of this discharge is to be found in papers by Dunster and
others[18,19,20,21] dating from the Geneva Peaceful Uses of Nuclear

Energy Conferences of the 1950s. The early discharges, which were on an experimental basis, were made after extensive studies of local dispersion, local fisheries and habits of the public which might bring them into contact with discharged radioactivity: they were well within limits prescribed by current dose limits and the critical group habits of the period.

5.1. Discharges

The overall discharge picture for the principal radionuclides is given in Table III and the situation in relation to individual years can be gleaned from recent reports[30,31]. Current discharge levels of all the more radiologically significant components are less than half the previous maximum discharge rate and have at all times been within the limits laid down in the legal document, the authorisation for discharge.

TABLE III Sellafield sea pipeline discharges of principal radio-nuclides, 1952-84 (TBq)

	^{90}Sr	$^{95}Zr/$ ^{95}Nb	^{106}Ru	^{137}Cs	^{144}Ce	^{239}Pu	^{241}Pu	^{241}Am
Total 1952-84 TBq	6100	24000	27000	41000	6200	680	21000	530
Mean TBq y^{-1}	180	710	820	1200	190	21	640	16
1984 TBq y^{-1}	72	470	350	430	8.8	8.3	350	2.3
Range TBq y^{-1} upper limit	600	2400	1600	5200	640	66	2800	120
lower limit	16	140	91	14	8.8	0.52	1.0	0

Of particular interest are the reductions in discharge of short lived fission product activity such as $^{95}Zr/^{95}Nb$, ^{106}Ru and ^{144}Ce which though dominant in their impact on human radiation exposure via the Porphyra seaweed/laverbread critical pathway from the late 1950s to the early 1970s have declined in discharge terms (Figure 3) and, more importantly, in radiological terms with the virtual disappearance of the laverbread exposure pathway since the early 1970s. Discharges of long lived more highly mobile fission product radionuclide ^{137}Cs and its neutron activation analogue ^{134}Cs reached peak values between 1974-78 (Figure 4) and have since declined following measures to reduce their concentration in spent fuel holding pond effluents pending provision of new ponds and permanent effluent treatment plant now coming on stream[32]. Discharges of the major transuranic radionuclides have fluctuated but Pu α and ^{241}Pu discharges reached their peak in 1973 and ^{241}Am the following year and all have been declining since (Figure 5).

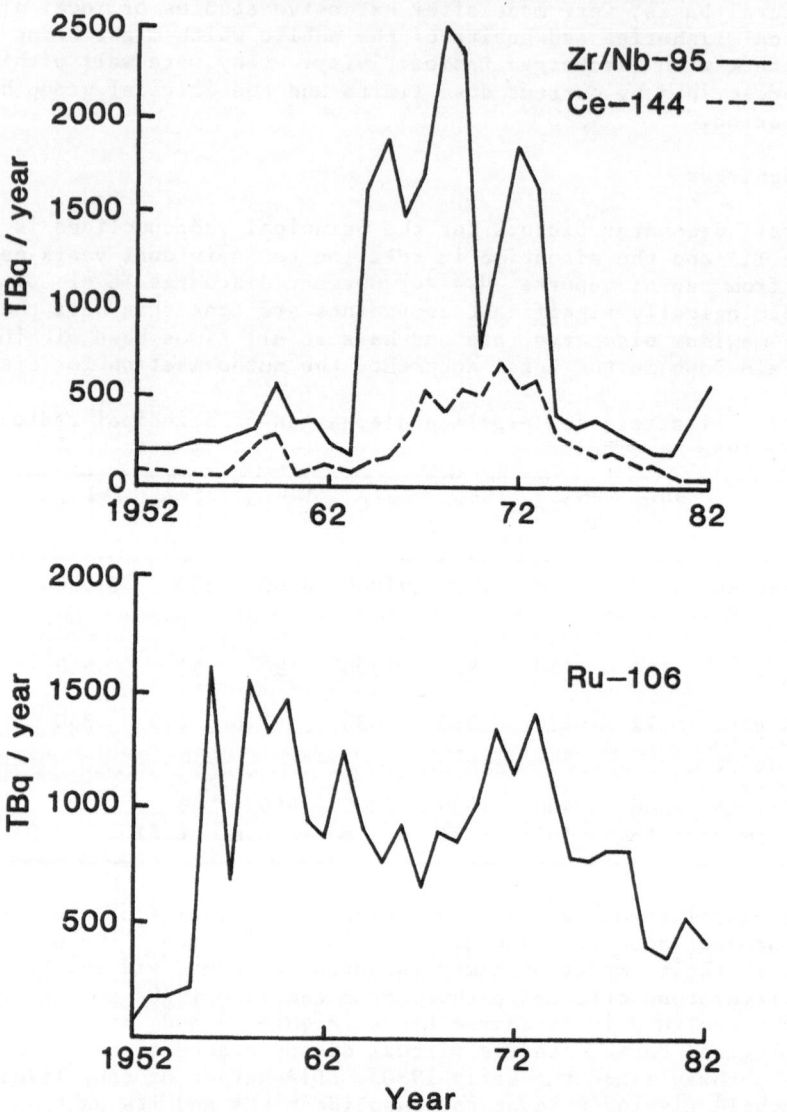

Figure 3. Sellafield sea pipeline discharges, 1952-82: $^{95}Zr/^{95}Nb$, ^{106}Ru and ^{144}Ce (TBq y^{-1}).

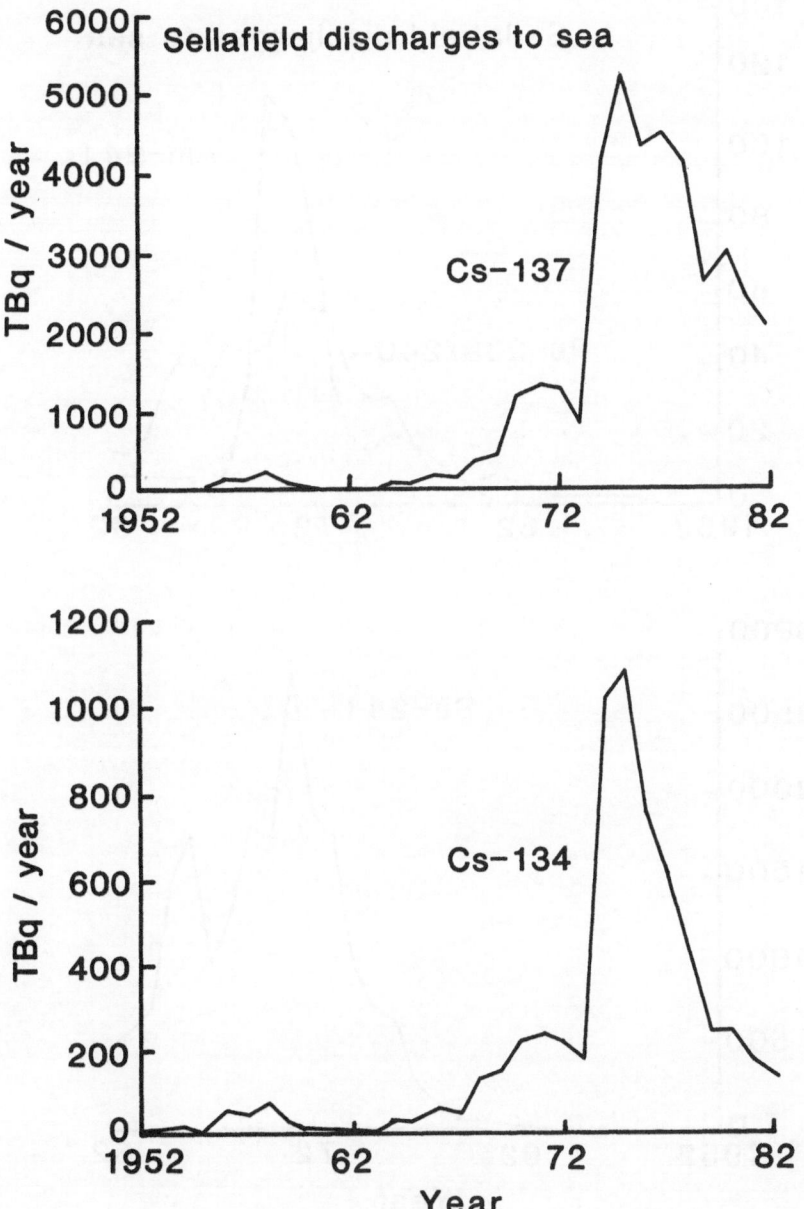

Figure 4.   Sellafield sea pipeline discharges, 1952–82:   [134]Cs and [137]Cs (TBq y[-1]).

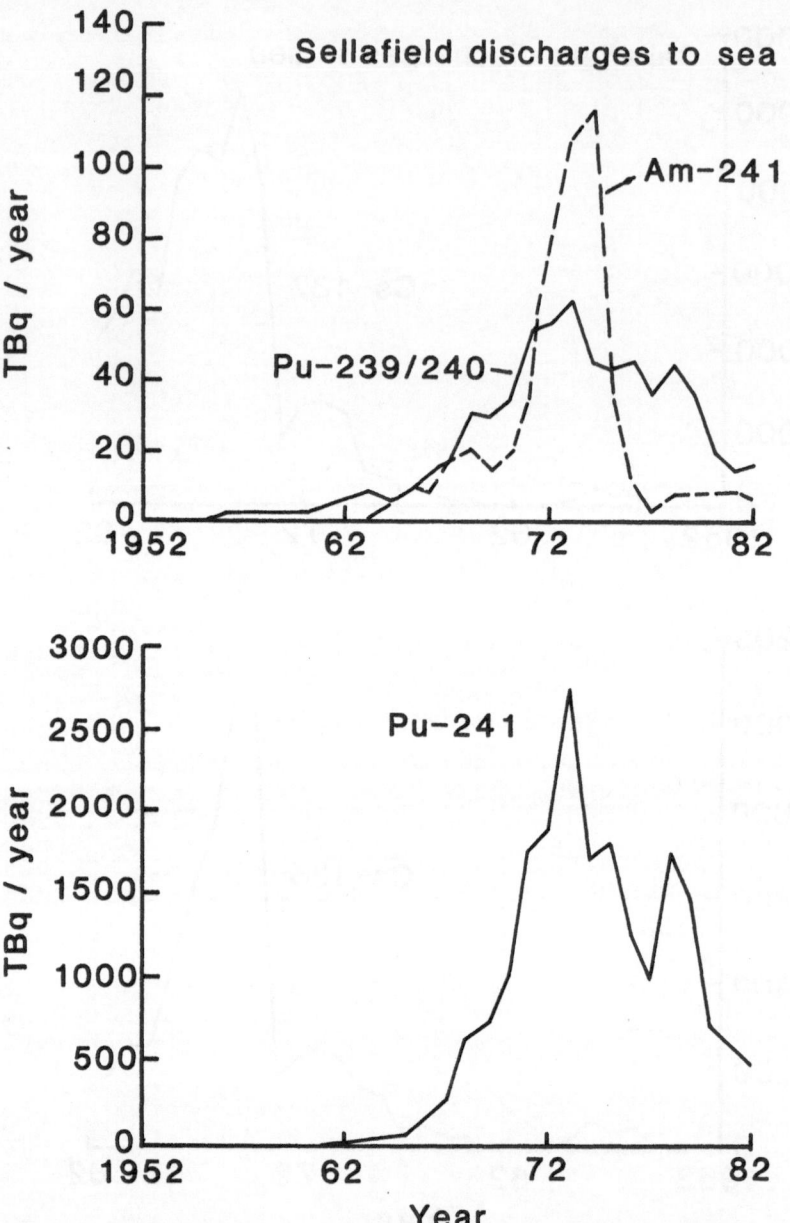

Figure 5. Sellafield sea pipeline discharges, 1952-82: $^{239/240}$Pu, ^{241}Am and ^{241}Pu.

5.2. Individual Exposure

The levels of individual radiation exposure in recent years, utilising relevant critical group exposure criteria, are given in Figures 6 and 7 and Tables IV and V. Further information can be obtained from the annual reports of 'Radioactivity in Surface and Coastal Waters of the British Isles'(17). The Porphyra/laverbread pathway (Figure 6) has not been a significant route of exposure since the early 1970s though it had previously dominated the radiation exposure scenario from the outset.

More recently individual exposure has been via the critical pathway of fish and shellfish consumption with the caesium radionuclides contributing most to the committed dose equivalent (Table IV and Figure 7). In more recent years as the contribution from shellfish consumption has increased the relative contribution of the transuranics and especially $^{239/240}$Pu and ^{241}Am have increased relative to that of the caesium radionuclides.

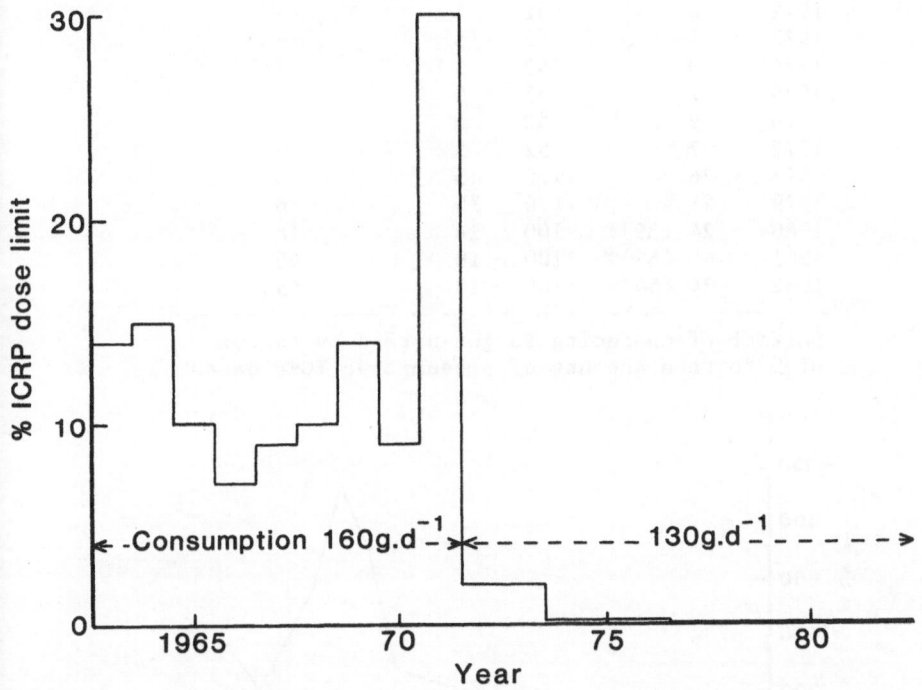

Figure 6. Estimates of public radiation exposure from BNFL Sellafield sea discharges: exposure of critical group of laverbread consumers (% ICRP dose limit).

TABLE IV   Estimates of public radiation exposure from BNFL Sellafield sea discharges:
exposure of critical group of fish and shellfish consumers (% ICRP dose limit)

	%	Consumption rate, g d⁻¹		
		Fish	Crustaceans	Molluscs
1963	0.1	52	5	-
1964	0.2	52	5	-
1965	0.2	52	5	-
1966	0.2	52	5	-
1967	0.2	52	5	-
1968	0.2	52	5	-
1969	0.9	52	5	-
1970	1	52	5	-
1971	2	52	5	-
1972	2	52	5	-
1973	2	52	5	-
1974	3	52	5	-
1975	7	52	5	-
1976	9	52	5	-
1977	7	52	5	-
1978	26	170	15	6
1979	21	170	15	6
1980	24 (39)*	100	18	18
1981	46 (69)*	100	18	45
1982	34 (54)*	100	18	45

*Effect of enhancing Pu gut uptake by factor
of 5 to take account of presumptive ICRP data.

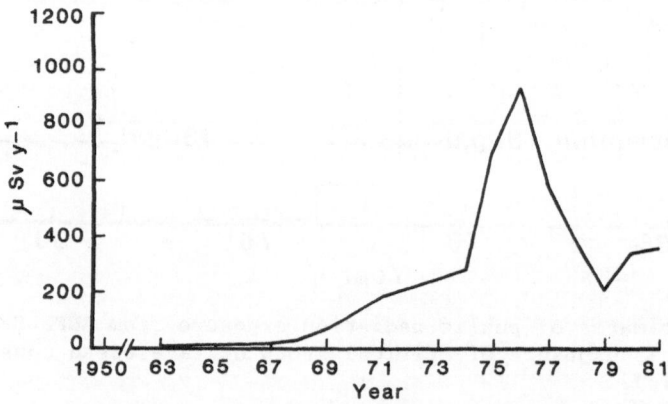

Figure 7.   Estimates of public radiation exposure from BNFL Sellafield
sea discharges: exposure of critical group of fish consumers (plaice
only) annual committed effective dose equivalent.

TABLE V Estimates of public radiation
exposure from BNFL Sellafield sea discharges: exposure of critical group;
external radiation (% ICRP dose limit)

	μGy h^{-1} (less background)	% dose limit	Hours occupancy
1967	1.1	6	300
1968	1.6	8	300
1969	2.5	13	300
1970	1.4	7	300
1971	1.6	8	300
1972	1.7	9	300
1973	1.9	10	300
1974	1.3	7	300
1975	1.6	8	300
1976	1.2	6	300
1977	0.98	5	300
1978	0.87	5	300
1979	1.0	5	300
1980	0.58	11	1 050
1981	0.72	9	710
1982	0.69	8	650
1983	0.61	7	650

External radiation exposure, though never the critical pathway, has
been a persistent feature of the radiological scene at Sellafield
(Table V). Due primarily to the gamma emitting fission product radio-
nuclides ^{95}Zr/^{95}Nb, ^{106}Ru and ^{137}Cs and to a lesser extent ^{134}Cs adsorbed to the surface of particulate material settling on beaches and
creating a gamma radiation field resulting in a degree of whole body
exposure of those using the beaches.

More recently following an inadvertent discharge of solvent and
entrained 'crud' from the reprocessing plant localised prominence has
been given to particulate material on beaches associated with high
energy beta-emitting radionuclides principally ^{106}Ru/^{144}Ce(33,34,35).
The contact doses encountered could in a few instances have been in the
range up to some hundreds of mSv, but would only have posed a risk of
effect (skin erythema) following prolonged contact with the material.
The probability of encountering such material and sustaining the necessary contact were so low as not to constitute a significant radiological
problem. After a few months the beaches concerned returned to near
normal levels and although monitoring of this particular pathway will be
maintained it is not expected to assume any significant radiological
context.

5.3. Collective Dose

Considerations of collective dose resulting from the sea pipeline discharges only came to prominence at Sellafield following the sharp
increase in the discharge of the relatively long-lived, highly mobile
and biologically available radionuclides of caesium in the early 1970s.
The widespread contamination of fish and shellfish that resulted led to
significant collective dose commitment to the UK population and to some
extent that of continental Europe: relevant figures are given in
Table VI and more detail can be obtained by reference to annual
reports[17] and other published material[7,10,36]. These collective
doses are now beginning to decrease, and will do so more rapidly initially for the UK which derives its fish and shellfish from areas
closest to the source and thus sees the initial benefits of reduction in
discharge but this reduction will soon be seen in the collective dose to
populations further afield.

TABLE VI   Estimates of public radiation
exposure from BNFL Sellafield sea discharges: collective effective dose
equivalents (man Sv) from fish and shellfish consumption, United Kingdom and
Europe

	United Kingdom	Europe (including UK)
1974	56	110
1975	99	190
1976	160	330
1977	100	220
1978	130	280
1979	130	300
1980	100	240
1981	130	280
1982	90	190

5.4. The Decade Ahead

Measures already taken or in train will lead to further reductions
in Sellafield sea discharges including, in the early nineties, further substantial reduction in the discharge of the transuranium
radionuclides[37].
     Reductions in the discharge of short lived material or material
that is conservative to sea water such as the radionuclides of strontium
and caesium will soon be reflected in reductions of concentration in
environmental materials either by virtue of radioactive decay or by a
combination of this and dispersion from the vicinity of the discharge.
The behaviour of long lived material that is not conservative in its

behaviour with respect to sea water and which is strongly adsorbed to
particulate material is more difficult to predict in terms of the concentrations likely to be encountered in biological materials following
reductions in discharge. This is so because we do not know to what
extent, for example, the very large reservoir of entrapped plutonium
radionuclides in the bed sediments of the NE Irish Sea will serve as a
source continuing to some extent to top up the otherwise reducing concentration in sea water and biological materials following reductions in
discharge.

Recent estimates[38] of the likely rate of reduction in local concentrations have however been made and show through a time series of
environmental monitoring and discharge data that the transuranics do
have a reducing availability following discharge; estimates of this
mean availability time differ between materials but nevertheless at
least in the short term only appear to be of the order of a few years
(Figure 8). This situation will be followed with interest as discharges
decrease further. However the availability times estimated from this
short term empirical approach may not necessarily describe the situation
beyond a few tens of years as other processes may become predominant in
giving a continued slow release from the sedimentary reservoir. This
may lead to slower rates of decrease in the critical materials, but from
a markedly lower concentration than has obtained in the recent past.

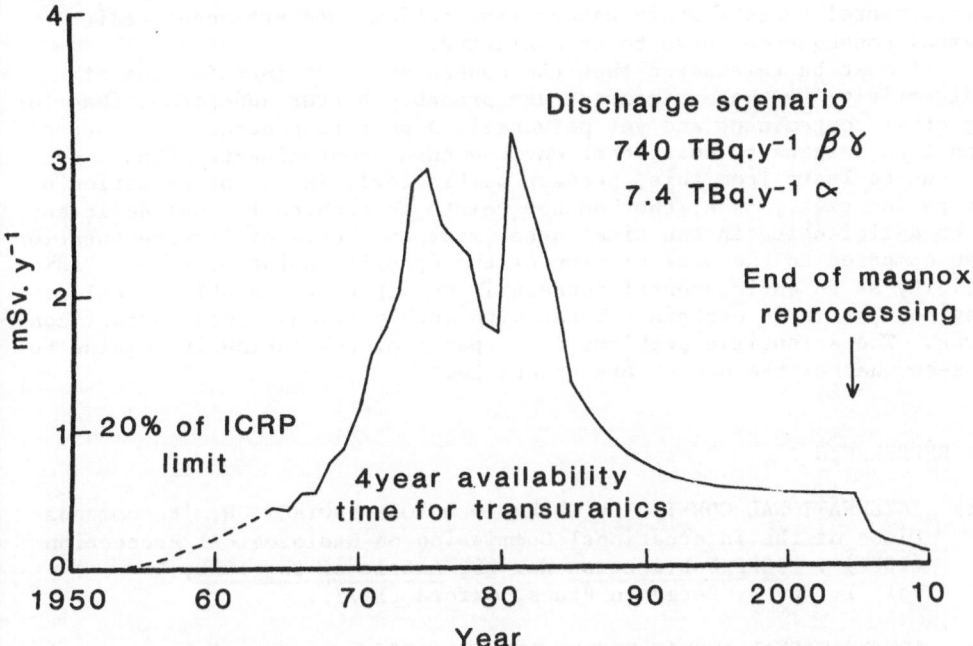

Figure 8. Estimates of public radiation exposure from BNFL Sellafield
sea discharges: critical group of fish and shellfish consumers predicted committed effective dose equivalent (mSv y^{-1}).

The past history of Sellafield sea discharges shows that their radiological impact via either individual dose or collective dose routes are well within the standard recommended by ICRP. Current measures will ensure that both these contributions will fall further. By the strict application of the assumed dose response criteria of ICRP, the calculated collective dose may appear substantial but when viewed against the perspective of risk of significant harm the average per capita UK dose would only equate to 5 μSv per year, equivalent to an individual risk of 1.10^{-7} of severe harm: a probability of effect unlikely to significantly affect individual decision making.

However scientific rationality is not necessarily the uppermost consideration when considering measures to reduce discharges and environmental impact, as any consideration of the BNFL expenditures in pursuit of reduced discharges will attest when the cost per man Sv, $£10^6$, is compared with the range of Sv values deduced on actuarial and related bases viz £2 000–£10 000[24]. The international aspects of the marine, or indeed the whole environmental scenario, with respect to radioactivity have been lifted out of a strictly scientific or radiological protection context and now have to be viewed within a wider socio-political perspective in which the scientific facts and their reasonable interpretation are only the starting point. We need however to take care that the science is well conducted and properly presented not only in relation to the marine environment but in relation to other environmental sectors where alternative options and attendant radiological consequences have to be evaluated.

It must be remembered that the consequences of introduction of radioactivity to the environment are probably better understood than for any other contaminant and yet perversely appear to generate more opposition than relates to most other environmental contaminants. There are lessons to learn from this, perhaps particularly in the presentation of the radioactivity case, that do not relate so much to factual deficiency as to deficiencies in the timeliness, mode and scale of its presentation when compared to the presentation of the opposite point of view. Radioactivity as an environmental concern is not free from problems real or imagined but it is certainly beset with more emotional than factual concerns. The scientific problems are capable of resolution it remains to be seen whether the others are tractable.

6. REFERENCES

(1) INTERNATIONAL COMMISSION ON RADIOLOGICAL PROTECTION, 'Recommendations of the International Commission on Radiological Protection (ICRP)'. ICRP Publication No. 26, Annals of the ICRP, Vol. 1, No. 3, Pergamon Press, Oxford (1977).

(2) INTERNATIONAL ATOMIC ENERGY AGENCY, 'Effects of ionizing radiation on aquatic organisms and ecosystems'. Technical Reports Series No. 172, IAEA, Vienna (1976).

(3)   WOODHEAD, D. S., 'Contamination due to radioactive materials'.
      pp. 1111-1287 In: Kinne, O. (Ed.) Marine Ecology Vol. 5. Ocean
      Management. Part 3. Pollution and Protection of the Seas.
      John Wiley and Sons, Chichester (1984).

(4)   PRESTON, A., 'The United Kingdom approach to the application of
      ICRP standards to the controlled disposal of radioactive waste
      resulting from nuclear power programmes'. Environmental Aspects
      of Nuclear Power Stations (Proc. Symp. New York, 1970), IAEA,
      Vienna (1971) 147.

(5)   MORLEY, F. and BRYANT, P. M., 'Basic and derived radiological pro-
      tection standards for the evaluation of environmental contamina-
      tion'. Environmental Contamination by Radioactive Materials
      (Proc. Symp. Vienna, 1969), IAEA, Vienna (1969) 254.

(6)   HETHERINGTON, J. A., JEFFERIES, D. F. and LOVETT, M. B., 'Some
      investigations into the behaviour of plutonium in the marine
      environment'. Impacts of Nuclear Releases into the Aquatic
      Environment (Proc. Symp. Otaniemi, 1975), IAEA, Vienna (1975)
      193.

(7)   PRESTON, A., JEFFERIES, D. F. and MITCHELL, N. T., 'The impact of
      caesium-134 and -137 on the marine environment from Windscale'.
      Radioactive Effluents from Nuclear Reprocessing Plants (Proc.
      Seminar Karlesruhe, 1977), Commission of the European Communities,
      Luxembourg (1978) 401.

(8)   PENTREATH, R. J., 'Nuclear Power, Man and the Environment'.
      Taylor and Francis, London (1980) 255 pp.

(9)   PRESTON, A. and PORTMANN, J. E., 'Critical path analysis applied
      to the control of mercury inputs to the United Kingdom coastal
      waters'. Environ. Pollut. Ser. B, 2 (1981) 451.

(10) HUNT, G. J. and JEFFERIES, D. F., 'Collective and individual
      radiation exposure from discharges of radioactive waste to the
      Irish Sea'. Impacts of Radionuclide Releases into the Marine
      Environment (Proc. Symp. Vienna, 1980), IAEA, Vienna (1981) 535.

(11) NUCLEAR ENERGY AGENCY OF THE OECD, 'Review of the continued suita-
      bility of the dumping site for radioactive waste in the North-East
      Atlantic', NEA/OECD, Paris (1980) 100 pp.

(12) NUCLEAR ENERGY AGENCY OF THE OECD, 'Review of the continued suita-
      bility of the dumping site for radioactive waste in the North-East
      Atlantic', NEA/OECD, Paris (1985) 485 pp.

(13) INTERNATIONAL ATOMIC ENERGY AGENCY, 'Convention on the Prevention
      of Marine Pollution by Dumping of Waste and Other Matter. The
      Definition Required by Annex I, Paragraph 6, to the Convention,
      and the Recommendation Required by Annex II, Section D, Inf.
      Circ./205/Add. 1, IAEA, Vienna (1975).

(14) INTERNATIONAL ATOMIC ENERGY AGENCY, 'Convention on the Prevention
      of Marine Pollution by Dumping of Waste and Other Matter. The
      Definition Required by Annex I, Paragraph 6, to the Convention and
      the Recommendations Required by Annex II, Section D', Inf.
      Circ./205/Add. 1/Rev. 1, IAEA, Vienna (1978).

(15) INTERNATIONAL ATOMIC ENERGY AGENCY, 'Convention on the Prevention
      of Marine Pollution by Dumping of Waste and Other Matter. The
      Definition Required by Annex I, Paragraph 6, to the Convention and
      the Recommendations Required by Annex II, Section D', Inf.
      Circ./205/Add. 1/Rev. 2, IAEA, Vienna, (Working Group
      Report to IAEA Board of Governors for consideration at meeting
      September 1985).

(16) PRESTON, A., 'Deep-sea disposal of radioactive wastes'.
      pp.107-122 In: Park, P. K. et al. (Eds.), Wastes in the Ocean,
      Vol. 3, Radioactive Wastes and the Ocean. Wiley-Interscience,
      New York and Chichester (1983).

(17) MINISTRY OF AGRICULTURE, FISHERIES AND FOOD, 'Radioactivity in
      surface and coastal waters of the British Isles', Tech. Rep. Fish.
      Radiobiol. Lab., MAFF Direct. Fish. Res., Lowestoft, Nos. 1, 2,
      5, 7, 8, 9, 11, 12, 13, 14 (1967-1978); Aquat. Environ. Monit.
      Rep., MAFF Direct. Fish. Res., Lowestoft, Nos. 3, 4, 6, 8, 9,
      11 (1979-1984).

(18) DUNSTER, H. J., 'The discharge of radioactive waste products into
      the Irish Sea, Part 2, The preliminary estimate of the safe daily
      discharge of radioactive effluent', Proc. 1st Int. Conf.
      Peaceful Uses of Atomic Energy, Vol. 9 (1956) 712.

(19) DUNSTER, H. J., 'The disposal of radioactive liquid wastes into
      coastal waters', Proc. 2nd Int. Conf. Peaceful Uses of Atomic
      Energy, Vol. 18 (1958) 390.

(20) DUNSTER, H. J., GARNER, R. J., HOWELLS, H. and WIX, L. S. U.,
      'Environmental monitoring associated with the discharge of low
      activity radioactive waste from Windscale works to the Irish Sea',
      Hlth Phys., Vol. 10 (1964) 353.

(21) FAIR, D. R. R. and MACLEAN, A. S., 'The disposal of waste products
      in the sea, Part 3, The experimental discharge of radioactive
      effluents', Proc. 1st Int. Conf. Peaceful Uses of Atomic
      Energy, Vol. 2 (1956) 716.

(22) PENTREATH, R. J., 'Behaviour of radionuclides released into
      coastal waters', IAEA-TecDoc-329, IAEA, Vienna (1985) 183 pp.

(23) UNSCEAR, 'Ionizing radiation: sources and biological effects. 1982
      Report to the General Assembly', United Nations, New York (1982).

(24) INTERNATIONAL ATOMIC ENERGY AGENCY, 'Costing transboundary radia-
      tion exposure. (The minimum value applied to unit collective dose
      in differential cost-benefit analysis)', IAEA-Int. alpha (1984) in
      press.

(25) CLARKE, R. H. and FLEISHMAN, A., 'The establishment of de minimis
      radioactive wastes', Proc. International Radiation Protection
      Association 6th Int. Cong., Radiation-Risk-Protection,
      Fachverband für Strahlenschutz e.V., Jülich (1984) 1346.

(26) MEINHOLD, C. B., 'Criteria for a de minimis level', US Health
      Physics Society, New Orleans, 3-8 June (1984).

(27) LINDELL, B., 'Concepts of collective dose in radiological protec-
      tion, a review for the Committee on Radiation Protection and
      Public Health of the OECD', OECD/NEA, Paris (1985) 32 pp.

(28) INTERNATIONAL ATOMIC ENERGY AGENCY, 'Considerations concerning "de
      minimis" quantities of radioactive waste suitable for dumping at
      sea under a general permit', IAEA-TecDoc-244, IAEA, Vienna
      (1981) 28 pp.

(29) HETHERINGTON, J. A., JEFFERIES, D. F., MITCHELL, N. T.,
      PENTREATH, R. J. and WOODHEAD, D. S., 'Environmental and public
      health consequences of the controlled disposal of transuranic ele-
      ments to the marine environment', Proc. Symp. Transuranium
      Nuclides in the Environment, IAEA, Vienna (1976) 139.

(30) LINSLEY, G. S., DIONION, J., SIMMONDS, J. R. and BURGESS, J., 'An
      assessment of the radiation exposure of members of the public in
      West Cumbria as a result of the discharges from BNFL, Sellafield',
      NRPB-R170, NRPB, Chilton, Didcot (1984) 58 pp.

(31) NATIONAL RADIOLOGICAL PROTECTION BOARD, 'The risks of leukaemia
      and other cancers in Seascale from radiation exposure',
      NRPB-R171, NRPB, Chilton, Didcot (1984) 297 pp.

(32) HUNT, G. J., 'Experience of ALARA in controlling radiocaesium dis-
      charges to the Irish Sea', Proc. 2 European Scient. Seminar on
      Radiation Protection Optimisation, EUR 9173 En, Commission of
      the European Communities, Luxembourg (1984) 437.

(33) MINISTRY OF AGRICULTURE, FISHERIES AND FOOD, 'Incident leading to
      contamination of beaches near British Nuclear Fuels Ltd,
      Sellafield, November 1983. Monitoring and assessment of environ-
      mental consequences undertaken by MAFF', MAFF, London
      (December 1983).

(34) MINISTRY OF AGRICULTURE, FISHERIES AND FOOD, 'Incident leading to
      contamination of beaches near British Nuclear Fuels plc,
      Sellafield, November 1983. Monitoring and assessment of environ-
      mental consequences undertaken by MAFF. 2. Supplement for the
      period 1 January-28 February 1984', MAFF, London (March 1984).

(35) WOODHEAD, D. S., JEFFERIES, D. F. and BARKER, C. J., 'Contamina-
      tion of beach debris following an incident at British Nuclear
      Fuels plc, Sellafield, November 1983', J. Soc. Radiol. Prot.,
      Vol. 5, No. 1 (1985) 21.

(36) JEFFERIES, D. F., STEELE, A. K. and PRESTON, A., 'Further studies
      on the distribution of ^{137}Cs in British coastal waters - I. Irish
      Sea', Deep-Sea Res., Vol. 29, No. 6A (1982) 713.

(37) GREAT BRITAIN - PARLIAMENT, 'Sellafield (Discharges),
      Mr Patrick Jenkin's reply to Dr Cunningham', Hansard, 18 December
      (1984), Cols. 93-94.

(38) HUNT, D. J., 'Timescales for dilution and dispersion of transura-
      nics in the Irish Sea near Sellafield', Sci. Total Environ., to be
      published.

THE PRACTICE AND ASSESSMENT OF SEA DUMPING OF RADIOACTIVE WASTE

W.L. Templeton
Pacific Northwest Laboratory
Battelle Memorial Institute
Richland,
Washington
U.S.A. 99352

J.M. Bewers
Dept. of Fisheries and Oceans
Bedford Institute
              of Oceanography
P.O. Box 1006, Dartmouth,
Nova Scotia, Canada B2Y 4A2

ABSTRACT. This paper discusses the practice and assessment of the ocean dumping of low-level radioactive wastes. It describes the international and multilateral regulatory framework, the sources, composition, packaging and rate of dumping and, in particular, the recent radiological assessment of the only operational disposal site in the northeast Atlantic. The paper concludes with a discussion of future ocean disposal practices for radioactive wastes, and the application of the approach to the dumping of non-radioactive contaminants in the ocean.

1. INTRODUCTION

Dumping of low-level radioactive waste in the ocean has been carried out since 1946. The only continuing radioactive waste dumping activity is that in the northeast Atlantic carried out by some European nations under the auspices of the Nuclear Energy Agency (NEA) of the Organization for Economic Cooperation and Development (OECD). Most of the discussion in this paper will centre around this particular practice although reference will also be made to other previous dumping by the United States in both the Atlantic and Pacific Oceans and the planning by the Japanese Government for low-level waste dumping in the western Pacific. The term 'dumping' is used here as implying the irretrievable deposition of packaged radioactive waste by dropping packages overboard from a ship in deep ocean waters. With some minor exceptions, this is the only form of disposal of radioactive waste permitted by the London Dumping Convention (IMO, 1982) and, although not all countries are signatories to this Convention, the activities of all states have been compatible with the Convention since it was promulgated in 1972. This Convention does not apply to land discharges for which there is no current international agreement but does cover incineration at sea of chemical wastes. One option for radioactive waste disposal whose feasibility is being investigated is emplacement of wastes in (or on) deep-sea sediments. However, this method of disposal only holds economic benefits for the disposal of high-level wastes and its eventual legality is in some doubt since it is not clear whether it would fall under the

493

G. Kullenberg (ed.), The Role of the Oceans as a Waste Disposal Option, 493–515.

London Convention or under some other mechanism such as the International Seabed Authority created by the United Nations Law of the Sea Convention (Simmonds, 1983).

In this paper we will describe first the international and multilateral arrangements which circumscribe the practice of low-level waste dumping. Then we will discuss the scale of recent dumping practices and outline the ways in which the impact and safety of this practice has been assessed. Finally, we will discuss the types of materials that might be considered for ocean dumping and future developments that might alter the options available to national authorities for the ocean disposal of radioactive wastes.

2. REGULATORY FRAMEWORK FOR RADIOACTIVE WASTE DUMPING

The first international meeting to deal specifically with issues of radioactive waste disposal into the ocean was held in 1958 during the United Nations Conference on the Law of the Sea. As a result of this meeting, the International Atomic Energy Agency convened a scientific panel to recommend measures to ensure that disposal of radioactive waste into the ocean would not result in undue hazards to man. This panel produced the so-called Bryneilsson Report entitled 'Radioactive Waste Disposal into the Sea' which was published by the IAEA (1961). With the establishment of the Convention on the Prevention of Marine Pollution by Dumping of Wastes and Other Matter in 1972 commonly known as the London Dumping Convention (LDC) (IMO, 1982) the role of the IAEA as the 'competent international body' to formulate criteria for the dumping of radioactive materials under this Convention was formally recognized. This resulted in the IAEA making a series of recommendations to the LDC concerning the 'definition of high-level radioactive matter unsuitable for dumping at sea' (IAEA, 1975, 1978). Such materials are deemed to fall within Annex I of the Convention which contains descriptions of materials proscribed from ocean dumping. Furthermore, the IAEA has formulated related recommendations concerning the manner in which radioactive materials falling outside the definition (i.e., materials of lower activity concentration than those defined as Annex I materials and thereby deemed to fall under Annex II to the LDC) might be dumped at sea without giving rise to unacceptable hazards to man or marine life.

It is useful to briefly digress at this point and outline the basic principles upon which the relevant advice from the IAEA has been formulated. Indeed these same basic principles underlie all of the regulatory and consultative mechanisms put in place for the regulation, surveillance and assessment of consequences of radioactive waste disposal in the sea and elsewhere. These principles are those of the International Commission on Radiological Protection (ICRP, 1977) which justify brief explanation here. Of the three main principles, the overriding one is that the dose-limits to individuals as set out by the ICRP must not be exceeded. Dose limits applicable to both radiation workers and members of the public have been defined by the ICRP in order to ensure that doses from all practices involving radiation, excepting doses from natural background, do not pose unacceptable somatic or genetic effects. The other principles are not stated numerically but

can, nevertheless, involve quantitative calculations in their application. The first of these is called 'justification'. It refers to the
prior assessment of the consequences of a practice, such as an intended
investment in the nuclear power industry by a state, in order to determine that the practice offers some 'net benefit' to the society concerned. The balance that gives rise to this net benefit involves economic, social, political, radiological and other scientific factors. In
many cases, such as that before us here, justification does not apply
because the practice that gives rise to exposures to radiation – the
nuclear power industry as a whole – already exists. The third principle
is that of 'optimization', which applies to both entire practices and
components of a practice, such as waste disposal. It is a relatively
new principle, having first been introduced by the ICRP in 1973, and has
as its objective to keep doses down to levels which are as low as
'reasonably' achievable, social and economic factors being taken into
consideration. This is the reason for optimization often being referred
to as the ALARA (As Low As Reasonably Achievable) principle. One way of
looking at the effects of optimization in the context of radioactive
waste disposal, is that it is a method by which the dose consequences,
or health detriment, is reduced to an extent that is both scientifically
and technically feasible and economically and socially justified. It
should also be noted that 'doses' in this sense refers predominantly to
the collective dose, that is the integral of dose over the entire
exposed population rather than the summation of doses to critically
exposed groups.

The current Definition and Recommendations of the IAEA that determine the Annex assignment of radioactive substances and the methods of
dumping under the LDC were submitted to the Convention in 1978 (IAEA,
1978). The Definition was based upon relatively simple modeling of the
equilibrium concentrations of individual nuclides that would give rise
to doses to the most exposed members of critical populations of 5 milli-
Sieverts per year ($mSv.a^{-1}$). Because of the difficulty of dealing with
sediment/water partitioning for radionuclides released at the ocean
floor, it was assumed that for exposure pathways involving seawater all
the radionuclides become associated with bottom water and in the case of
exposure pathways involving pelagic sediments the radionuclides become
entirely associated with the sediments. Equilibrium release rates were
then calculated which, on the basis of all human exposure pathways
considered, resulted in doses of 5 $mSv.a^{-1}$. These release rates were
calculated for an ocean basin of volume 10^{17} m^3. Finally, these limiting release rates were arbitrarily divided by an assumed upper limit to
the mass dumping rate (10^8 kg) to determine activity concentrations that
establish the lower bound of high-level wastes unsuitable for dumping at
sea and of Annex I.

The Recommendations of the IAEA comprise specifications of the
conditions under which other, Annex II, radioactive wastes may be dumped
in the ocean. These recommendations embody the ICRP principles described above and define how and where Annex II radioactive wastes may be
dumped. Among these conditions are the requirements that these
materials only be dumped in waters of average depth greater than 4000 m
in areas clear of continental margins and away from areas of tectonic

activity.   This effectively restricts dumping to the deep abyssal
plains.   Dumped materials must also be packaged in a manner that ensures
that no release of active matter occurs until the packages have descen-
ded to the ocean floor.

The Recommendations also specify the procedures to be used by
national authorities for the evaluation of the consequences and safety
of the proposed dumping operation, including the comparison of sea dumping with other options such as land disposal.   These various recommendations form the basis upon which national authorities can authorize
dumping operations.   However, national authorities must also consider
the other relevant requirements of the LDC itself, particularly Annex
III which outlines the considerations that must be taken into account
prior to the issuance of a permit for dumping of any substances under
the Convention.

The LDC forms the international legislative background to sea
dumping activities.   It also specifically encourages international and
regional cooperation in the development of procedures for the implementation and promotion of the Convention.   In this latter respect, the
major regional or multilateral agreement established with the specific
purpose of providing a consultative mechanism for issues of sea dumping
of radioactive waste is the OECD/NEA Multilateral Consultation and
Surveillance Mechanism for Sea Dumping of Radioactive Waste (OECD/NEA,
1977).   This mechanism provides a vehicle for consultation within the
OECD with respect to sea dumping activities by certain of its member
countries.   Under the Multilateral Consultative and Surveillance Mechanism, countries directly involved or interested in the regulation of sea
disposal of radioactive waste have involved themselves in a periodic
review of the suitability of the northeast Atlantic Ocean dump site that
has been used by Belgium, the Netherlands, Switzerland and the United
Kingdom for low-level waste dumping during the last decade.   This review
takes place at 5-year intervals with the previous review having taken
place in 1980 (OECD/NEA, 1980).   During the scheduled review of the
northeast Atlantic dump site in 1979 only interim approval was given for
the continuance of dumping operations at this site.   This was largely
due to the insistence of certain countries that a research and surveillance programme be instituted to ensure that assumed conditions in the
vicinity of the dump site were verified and to obtain further insight
into prevailing conditions and processes in order to improve the
reliability of the assessment of consequences of the use of the site for
continued dumping.   The site suitability review at that time constituted
little more than a summary of conditions at the site to demonstrate that
it met the basic criteria specified in the IAEA Recommendations and a
statement as to the proportion that contemporary dumping in the
northeast Atlantic represented of the ocean basin dumping limits specified in the IAEA Definition.   In the following year (OECD/NEA, 1980),
not only was the site suitability assessment substantially improved but
the NEA Coordinated Research and Surveillance Programme (CRESP) was
established (OECD/NEA, 1981).   This programme, which has physical
oceanographic, geochemical, marine biological, modeling and radiological
surveillance components, has continued since that date (Ruegger et al.,
1984).   The results of CRESP activities during the period 1980-1984 have

been incorporated into documents entitled 'An Interim Oceanographic Description of the North-East Atlantic Site for the Disposal of Radioactive Waste', Volumes I and II (OECD/NEA, 1983; 1985a). The scientific quality of this document attests to the commitment made by the nations involved both to CRESP and to the spirit of cooperation engendered within the Consultative Mechanism as a whole. The programme is scheduled for review during the current year to determine whether it requires to be continued.

It is important to the following discussion to specify the difference between the generic ocean basin modeling, undertaken within the IAEA to define wastes unsuitable for dumping at sea under the LDC, and the site-specific nature of the NEA site suitability review and the assessment of consequences of dumping practice at the northeast Atlantic dump site. The IAEA activities were directed to providing generic release-rate limits for radioactive wastes at the ocean floor based upon the oceanographic conditions in a hypothetical ocean basin of volume 10^{17} m^3. These release rate limits were then arbitrarily divided by an assumed upper limit of gross mass of material (including packaging) dumped per year to determine the activity concentration that should be assigned as the definition of high-level radioactive wastes under Annex I of the LDC. It is, however, implicit that the assessment of the consequences of the use of a dumping site in a real ocean basin, such as the northeast Atlantic, be made on the basis of existing oceanographic and demographic conditions in and around that basin. Thus the assessment procedure must be site-specific and a great deal of effort, within the NEA forum, has been devoted to developing site-specific and basin-specific models for assessing the transport of radioactive wastes from the dump site and the radiation exposure of bordering populations during the last five years.

The IAEA Definition and Recommendations under the LDC are currently being revised. The revision process is virtually complete. Both the oceanographic and radiological bases have been revised (IAEA, 1984a) since the formulation of the previous Definition and Recommendations (IAEA, 1978). The oceanographic models have, in particular, been substantially refined and a particularly important change is the introduction of representations of sediment-water partitioning in the oceanographic models to account for the association of radionuclides with aqueous and sedimentary phases following release from the packages. Additional nuclides to those considered in the 1978 revision have been introduced and a more realistic time scale (1,000 years vs 40,000 years) for the practice has been considered. There have also been a number of revisions to annual limits of intake (ALI) of nuclides and the recommended methods of calculating exposures to populations, and this has resulted in some changes in the details of the calculations. It is anticipated that the revised Definition and Recommendations will be submitted to the IAEA Board of Governors during 1985 preparatory to their submission to the LDC. Also, during 1985, the NEA Site Suitability Review process is being repeated in order to determine the suitability of this site for continued dumping activities during the next five years. It is the results of this assessment/review process (OECD/NEA, 1985b) that provide the basis for the assessment of consequences discussed in the next section of this paper.

3. SOURCES, COMPOSITION, PACKAGING AND RATE OF DUMPING OF RADIOACTIVE WASTE

Between 1946 and 1967, the United States disposed of packaged radioactive waste into the oceans. A summary of these disposals has recently been prepared (Johnson et al., 1985). The estimated total amount of radioactivity at the time of packaging was 3.7 10^3 TBq in a total of 49,800 containers of various types. About 95% of this radioactivity was dumped in the North Atlantic at the "2800 m site" located at $38^\circ30'N$, $72^\circ06'W$. This includes about 1200 TBq of activation products in the reactor pressure vessel of the Seawolf submarine propulsion unit. Packaged radioactive waste has also been dumped at 10 sites in the northeast Atlantic by seven western European countries. Since 1949 detailed information on the quantities dumped by specific area, by year and by country is given in OECD/NEA (1985b). A summary of the record amounts dumped in the northeast Atlantic between 1949 and 1982 is given in Table I.

These low-level wastes are derived from nuclear power plants, other nuclear fuel cycle operations, medicine, research and industry and the decontamination and decommissioning of redundant plant and equipment. The waste is similar in form to that arising from non-nuclear industrial, medical and research facilities. However, these materials include items containing radionuclides in various forms, such as surficial contamination, chemically-incorporated nuclides and induced radioactivity and therefore require a range of special handling, treatment and disposal arrangements. The composition of the wastes dumped has varied year by year. Plutonium isotopes and americium-241 account for over 96% of the alpha activity dumped, whereas tritium and plutonium-241 account for over 87% of the beta-gamma activity dumped. The remainder of the long-lived beta-gamma activity mainly comprises the fission products strontium-90 and caesium-137 and the activation product cobalt-60. Packaging guidelines have been published by the OECD/NEA (1979) and the IAEA (1981a). The waste packages are designed to provide shielding and containment of the waste during handling and transportation, and to ensure that the package reaches the seabed (at depths equal to or greater than 4000 m) without losing its integrity. These requirements have been validated by experiments in pressure chambers (Pearce and Vincent, 1963; Seki et al., 1980a, 1980b, 1981), by photographing package descents at sea (King and Hill, 1975; Seki et al., 1980b), by studies on recovered packages (Colombo et al., 1983 and by theoretical calculations (Hunt and Smith, 1982). There are two basic package designs, both based on an outer steel drum or concrete container. One design is the so-called void-containing package, which comprises a concrete lining around an inner waste container for strength and shielding. These packages incorporate a pressure-equalization system to prevent implosion and destruction of the package as it descends to the deep ocean bottom (about 4000 m). The majority of wastes disposed of in the northeast Atlantic have been void-containing packages. In the monolithic packages the waste is incorporated into a matrix which may be cement, bitumen or polymer. The encapsulated waste is usually enclosed within a sealed container surrounded by a concrete lining for shielding.

4. DESCRIPTION OF THE CURRENT NORTHEAST ATLANTIC DUMPSITE

The current dump site, in use since 1971, is located in the foothills of the Mid-Atlantic Ridge close to its junction with the Porcupine Abyssal Plain. Dumping occurs within a rectangular area of about 4250 km^2 which is bounded by latitudes 45°50'N and 46°10'N and by longitudes 16°00'W and 17°30'W. The approximate average depth within the dump site rectangle is 4400 m. The average residual bottom water flow from the area is initially northwest at 1-2 cm.sec^{-1}. The site lies more than 550 km from the edge of the nearest continental shelf. The NEA Coordinated Research and Environmental Surveillance Programme documents, edited by Gurbutt and Dixon (OECD/NEA, 1983) and Dixon, Gurbutt and Kershaw (OECD/NEA, 1985a), provide descriptions of the relevant physical, chemical and biological measurements, made largely at or near the existing NEA dump site up to 1984. A summary of these data is given in the NEA 1985 Site Suitability Review (OECD/NEA, 1985b). These documents indicate that the basic physical, chemical and biological knowledge of the site has improved substantially during the last five years, particularly in respect to water circulation and mixing, bottom and suspended sediment geochemistry, benthic faunal ecology and the rates of sediment bioturbation.

4. RADIOLOGICAL ASSESSMENT

The radiological impacts on man and the marine ecosystem are of primary concern when evaluating the consequences of dumping radioactive waste in the oceans. Radiological surveillance and monitoring can, in principle, provide essential information to assess these impacts. However, where it is necessary to predict these impacts into the future, covering all the relevant radionuclides, and on a global scale, it is necessary to develop and make use of appropriate mathematical models (OECD/NEA, 1985b). The modeling framework used in the recent assessment is shown in Figure 1. The approach taken was to develop a number of major components, each of which was modeled separately, and then to use the output from each model as the input to the next appropriate component. For example, the waste package model provides the rate of release of radionuclides to the ocean as input to the ocean dispersion and sediment interaction model. This composite model provides the rates and patterns of dispersion of radionuclides in the Atlantic and world's oceans, the rates of removal of radionuclides from the oceans by interaction with suspended and bottom sediments, and the rate of return of radionuclides from bottom sediments into the ocean. These predictions then form the basis for the calculation of the dose consequences to man and marine organisms through consideration of exposure pathways. The ocean dispersion model is one of the types of model recommended for development by the United Nations Joint Group of Experts on the Scientific Aspects of Marine Pollution (GESAMP, 1983). In and around the dump site, nested compartments were defined to allow prediction of

Parts of System Included in Model	Model	Major Processes Included in Model
Canister and lining Waste form	Waste Package ↓ rates of release of radionuclides into the ocean, as a function of time	Canister corrosion Degradation of package lining and caps Release of radionuclides from waste forms
Bottom sediments Benthic boundary layer (water and particulates) Open ocean (water and suspended particulates) Coastal waters	→ Ocean Dispersion and Sedimentation radionuclide concentrations in water and sediments, as a function of time	Diffusion and advection Interactions between radionuclides and suspended particulates and bottom sediments

Exposure pathways
- seafoods,
  beaches,
  atmosphere,
  salt, water
Dose to marine organisms
	→
	Dose to Man and Organisms
		Reconcentration of radionuclides in marine organisms, beach sediments, and atmospheric aerosols/vapour
		Radionuclide intake and metabolism by man and organisms

Figure 1. Modelling Framework Used in Radiological Assessment (after OECD/NEA 1985b)

TABLE I   Summary of Recorded Radioactive Waste
          Dumpings in the Northeast Atlantic,
          1948-1982 (OECD/NEA 1985b)

Gross weight	142275 tonnes
Alpha activity	$6.8 \ 10^2$ TBq
Beta/gamma activity	$3.8 \ 10^4$ TBq
Tritium*	$1.5 \ 10^4$ TBq

*1975-1982 only. Tritium in prior years
included in beta/gamma activity

TABLE II   Pathways and Modes of Exposure (after IAEA 1984a)

Pathway	Mode of Exposure
Actual Pathways	
Surface fish consumption	)
Mid-depth fish consumption	)
Crustacea consumption	)
Mollusc consumption	) Ingestion
Seaweed consumption	)
Salt consumption	)
Desalinated water consumption	)
Suspended airborne sediments	) Inhalation
Marine aerosols	)
Boating	)
Swimming	)
Beach sediments	) External irradiation
Deep-sea mining	)
Hypothetical Pathways	
Deep fish consumption	)
Plankton consumption	) Ingestion

radionuclide concentrations in the near field. In the far field of the
ocean, the compartments are defined in terms of isopycnal coordinates
because water is thought to move predominantly along, rather than across
surfaces of equal density. The model is ocean basin specific in the
sense that there are far more compartments in the Atlantic than in the
other oceans. The sediment interaction part of the model is also based
upon the recommendations of GESAMP (1983), and appears to be consistent
with the Appendix VI and IX models. A preliminary comparison was made
between the results of the composite ocean model and those of two other
models: one a compartment model developed by Sandia in the United
States, and the other a model described in GESAMP (1983) as Appendix VI.
The comparison indicated that the two compartment models gave largely
consistent results which are in agreement with the Appendix VI model for
several of the radionuclides of most radiological concern.

It should be noted that the modeling framework, the models, and the
data bases used in the most recent site suitability review, differ
considerably from those employed in the previous assessment (OECD/NEA,
1980). For the current assessment, models for the release of radionuclides from the packages, and for sediment interaction, were introduced. Additionally, radionuclide chains are dealt with by modeling the
behaviour of the long-lived members explicitly and assuming that the
short-lived members are in secular equilibrium with their parents.
Detailed descriptions of the waste package, ocean dispersion and sediment interaction modeling techniques, the data base used and the results
are provided in the site suitability review document (OECD/NEA, 1985b).

One of the less-easily accepted aspects of the practice of low-
level radioactive waste dumping in the oceans is that validation, by
measurement, of environmental concentrations in the critical pathways
and/or radiation dose cannot presently be made. Nor is it likely that
such confirmation could be made in the foreseeable future. We presently
have no consistent evidence from the surveillance programme that radionuclides have been yet released from the packages and are present in
local sediments or biota. The majority of the measurements appear to be
indistinguishable from the concentrations that one would expect from
worldwide nuclear weapons test fallout. This, coupled with the long
time-frame that has to be considered in any assessment of ocean dumping,
makes it necessary to rely upon mathematical models to predict the
radiological consequences. Additionally, it is necessary to consider
not only those exposure pathways that might exist today but also
pathways which might become important in the future. Although the NEA
site assessment is site-specific, it is only so in terms of the basin in
which the dump site is located, since the dispersion model includes all
the world's oceans. It therefore seemed appropriate to use the IAEA
generic pathways (IAEA, 1984a) for modeling human exposures in distal
areas. The pathways included in the calculation of doses to members of
the critical groups are given in Table II. The pathways considered
cover three primary modes of exposure - external irradiation, internal
irradiation through ingestion, and internal irradiation through inhalation. The pathways include those which could exist now (i.e., 'actual
pathways') and those which could apply sometime in the future (i.e.,
'hypothetical pathways'). The values applied to each of the actual

pathways have been chosen on the basis of extrema in current consumption and behavioural habits of potential critical groups. Comparison with the previous assessment (OECD/NEA, 1980) shows that two additional pathways have been added to the group of 'actual pathways' - mid-water fish and deep sea mining - and that the plankton pathway has been transferred to the group of hypothetical pathways with deep-water fish consumption. The consumption rates and occupancy factors applied to each of the pathways are shown in Table III taken from IAEA (1984a).

The radiological assessment for the recent review (OECD/NEA, 1985b) was divided into four sections. The first section was the calculation of the radiation doses to individuals in the critical groups for three scenarios of dumping. These were: A) for past dumping; B) for past dumping plus a further five years dumping at rates typical of those in recent years; and C) for past dumping plus five years at a rate ten times that for recent years. The second section was a calculation of the doses to marine organisms, with comparison to natural background levels and with dose rates at which detrimental effects have been observed. In the third section, the sensitivity of calculated doses to critical groups to variations in key assumptions and parameters was analyzed to indicate uncertainties in the estimated doses. Finally, some preliminary estimates of the collective doses from the dumping at the northeast Atlantic, as a function of time, were made.

4.1 Radiation Doses To Individuals

The peak annual doses via 'actual' pathways for the three dumping scenarios are given in Table IV. First, it will be noted that all the doses are extremely low. The estimation of doses resulting from each of the scenarios are, in all cases, less than 2×10^{-7} Sv.a^{-1}, i.e., less than 0.02 percent of the 1 mSv annual dose objective suggested by ICRP as a maximum for prolonged exposures. Preliminary calculations indicate that even if the practice were continued for 500 years at present rates, doses to individuals would still be more than three orders of magnitude below the relevant limits. Second, all of the predicted peak doses from all of the 'actual' pathways occur via the consumption of Antarctic Ocean species. The concentrations of radionuclides in the Antarctic environment comprise two contributions: that from radionuclides diffusing through the bottom waters of the eastern Atlantic, and that from radionuclides moving north and west from the dump site and then south along the western boundary undercurrent of the Atlantic. The peak annual dose, summed over all radionuclides, occurs 200 years after the start of dumping for all three scenarios, and it is received via the mollusc consumption pathway. The doses from americium-241 are due primarily to ingrowth of this nuclide from its shorter-lived parent plutonium-241. They decrease rapidly after the peak is reached due both to americium-241 decay and its rapid removal from the ocean by particulate scavenging. On the other hand, the doses from plutonium-239 and plutonium-240 remain close to their peak levels for about 1000 years because they have longer half-lives than americium-241 and their sediment/water partition coefficients (k_ds) are lower than that of americium. It should be noted that the radionuclides which give rise to

TABLE III   Consumption and Breathing Rates and Occupancy
            Factors (after IAEA 1984a)

Pathway	Value	Other Parameters
Ingestion Rates		
Fish (Surface, M, Depth)	600 g d^{-1}	
Crustacea	100 g d^{-1}	
Mollusc	100 g d^{-1}	
Seaweed	100 g d^{-1}	
Desalination	2 1 d^{-1}	
Salt	3 g d^{-1}	
Plankton	3 g d^{-1}	
Fish-Deep	60 g d^{-1}	

Inhalation Rates		**Concentrations**	
Rate of Human Air Respiration	23 m^3 d^{-1}		
Suspended Airborne Sediments		10 µg m^{-3}*	
Atmospheric Vapour		10 g m^{-3}	

Occupancy Factors		**Modifying Factors**	
		photons	electrons
Boating	5000 h y^{-1}	0.2	0
Swimming	300 h y^{-1}	1	0.5
Beach Sediments	2000 h y^{-1}	0.5	0.5
Deep-Sea Mining	500 h y^{-1}	0.1	0

*Made up of 0.25 µg m^{-3} fine coastal sediment particles,
3.3 µg m^{-3} dried sea salt particles and 6.6 µg m^{-3} particle-
associated water.

TABLE IV Calculated Peak Annual Doses Via "Actual" Pathways for the Three Dumping Scenarios (after OECD/NEA 1985a)

	Scenario A Past Dumping	Scenario B Past Plus 5 Years Dumping	Scenario C Past Plus 5 Years at 10x Rate
Radionuclide/ Peak annual dose (Sv)	Pu^{239} 9.0×10^{-9} Am^{241} 7.0×10^{-9} Pu^{240} 5.0×10^{-9}	Am^{241} 1.0×10^{-8} Pu^{239} 9.5×10^{-9} Pu^{240} 6.0×10^{-9}	Pu^{239} 6.0×10^{-8} Am^{241} 4.0×10^{-8} Pu^{240} 3.0×10^{-8}
Pathway	Mollusc consumption	Mollusc consumption	Mollusc consumption
Geographical area	Antarctic Ocean	Antarctic	Antarctic
Peak dose	200 years	200 years	Am^{241} – 200 y Pu^{239} – 500 y Pu^{240} – 500 y
Total dose from all radionuclides	2.0×10^{-8}	3.0×10^{-8} Sv/y	1.0×10^{-7} Sv/y
Other significant radionuclides and pathways	Po^{210} crustacean consumption Antarctic Pu^{238} mollusc consumption Antarctic C^{14} fish consumption Antarctic	C^{14} fish consumption Antarctic Po^{210} crustacean consumption Antarctic	Po^{210} crustacean consumption Antarctic C^{14} fish consumption Antarctic Pu^{239} mollusc consumption Antarctic

the highest individual doses are predominantly actinides or actinide daughter products. The calculations for the 'hypothetical' pathways, i.e., consumption of plankton and deep-sea fish, indicate that the deep-sea fish pathway is more restrictive than the plankton pathway. This might be expected since the former is determined by the radionuclide concentrations in deep water in the vicinity of the dump site, whereas the latter is determined by surface water concentrations in the Antarctic Ocean. The deep-fish pathway was introduced into the recent assessment to represent a direct 'short-circuit' from northeast Atlantic bottom waters to man. The results indicate that, even if such a pathway did exist, and even if consumption rates of deep-sea fish were comparable with those of surface and mid-water fish, the doses received would be very much lower than the dose limits and dose objectives.

4.2. Radiation Doses To Marine Organisms

The dose rates to marine organisms within the dump site, and in the surrounding areas, were calculated for each of the three scenarios. Within the dump site rectangle, the major contributors to the dose rates received after 40-50 years are americium-241, plutonium-240, plutonium-239 and plutonium-238. Carbon-14, radium-226 and polonium-210 are additional contributors. Cobalt-60 is the major source of exposure to benthic fish. In the adjacent and regional boxes, the influence of cobalt-60 is reduced because of its half-life and high k_d value. The transuranic nuclides are the major sources of irradiation in these areas. Within the dump site, the dose rates to molluscs under scenarios B and C do exceed the highest doses from natural background. For all other organisms, the dose rates are calculated to be at, or below, the highest doses from natural background. Within the surrounding areas, the calculated dose rates were all in the range of natural background or lower. However, in all cases, the calculated dose rates are well below the dose rates at which deleterious somatic effects on individuals or populations of marine organisms have been observed.

4.3. Collective Doses

The pathways included in the collective dose calculations were the consumption of fish, molluscs, seaweed and plankton and, for tritium and carbon-14, circulation and human exposure on a global scale (UNSCEAR, 1982). The collective dose commitments for scenarios A, B and C were 3.8×10^4 man.Sv, 5.9×10^4 man.Sv and 2.8×10^5 man.Sv, respectively. It should be noted that these collective dose commitments are made up of very small radiation doses to a very large number of individuals. The commitment occurs during the 10,000-year period after dumping commences. The radionuclides which contribute most to annual collective doses and collective dose commitments are carbon-14 and plutonium-239. It may be noted that the collective dose to the world population from naturally-occurring radionuclides in the ocean is estimated to be 3×10^5 man.Sv. Similarly, the collective dose commitment to the public from one year of nuclear fuel cycle operations is about 8×10^4 man.Sv which is about twice that from all previous dumping at the dump site.

5. SENSITIVITY ANALYSES

A limited number of analyses were carried out to investigate the behaviour of separate components of the radiological assessment and to examine the influence of some of the key parameters and assumptions on the assessment results. The areas covered were: 1) the Waste Package Model, 2) the Composite Ocean and Sediment Model, 3) Biological Concentration Factors and Sediment/Water Partition Coefficients, and 4) a Hypothetical Marine Food Chain. In 1) some of the parameters and scenarios were varied to determine the effect on peak doses to individuals from 'actual' pathways. In fact, the doses arising predominantly from long-lived radionuclides remained relatively insensitive to the waste package model assumptions because of their long half-lives and transport times. However, because the maximum deep water concentration occurs at an earlier time than the maximum 'actual' dose, the deep-fish pathway is more sensitive to the release rate. In 2) ten parameters, including ocean mixing coefficients, settling particle fluxes, particle resuspension fluxes and sediment/water partitioning were varied to examine, for a small number of radionuclides, the robustness of the composite ocean model. The predicted peak doses were found to be relatively insensitive to variations in the physical oceanographic parameters, but were affected by changes in the parameters in the sediment interaction part of the model. In 3) the values for the biological concentration factors and sediment/water partition coefficients (k_ds) were varied between maximum and minimum values for individual nuclides (IAEA, 1985). As was to be expected, variations in concentration factors are relatively important in determining radiation doses to critical groups. A reduction in the sediment k_d for plutonium-239 by a factor of ten results in an increase in the peak individual dose by a factor of two. In the case of americium-241 a similar reduction in sediment k_d results in an increase in the peak dose by a factor of twenty. This is because the reference k_d value for americium is sufficiently high that sedimentation acts as an important removal factor. In 4) a number of the controlling biological factors in a hypothetical marine food-chain were varied, as proposed by Pentreath (1983). Calculations were run with the bioturbated layer within the dump site rectangle, at concentrations calculated to result from past dumping, and with the radionuclides caesium-137, americium-241, plutonium-239, plutonium-241 and cobalt-60. The resultant data indicated that while peak individual doses could be higher than those calculated for the 'actual' and 'hypothetical' pathways, they would still be well within the dose limits and objectives.

6. FUTURE OCEAN DUMPING PRACTICES

Predicting the future application of ocean dumping to the disposal of radioactive waste is, at best, difficult. However, such prediction is made even more tenuous when several incomplete international negotiations are considered. Not only are the IAEA Definition and Recommendations under the LDC (IAEA, 1978) being revised but an independent review of the safety of the overall practice is being conducted as an

"Evaluation of Previous and Future Programmes on the Disposal of Radioactive Wastes at Sea" by an expert panel convened under the auspices of
the International Maritime Organization (IMO) and the IAEA on behalf of
the LDC. The report of this review should be available for consideration at the Ninth Consultative Meeting of the LDC in September 1985.
This panel was established following a debate during the Seventh Consultative Meeting of the LDC regarding the safety of ocean disposal of
radioactive waste. This debate resulted in the imposition of a mora-
torium on ocean disposal for a period of two years during which time the
safety of the practice would be reviewed within the LDC. No dumping of
radioactive waste took place during 1984, largely because of this mora-
torium. The likely use of ocean dumping for the future disposal of
radioactive wastes rests heavily on the conclusions of the expert panel
considering the safety of the practice that will be debated during the
Ninth Consultative meeting. The results of this meeting will therefore
have an important impact on the future of ocean dumping as an option for
waste disposal.

The questions regarding the safety of the practice stemmed partly
from the concerns expressed by one European nation over the consequences
of continued dumping in the northeast Atlantic under NEA auspices but
arose principally from the concerns of Pacific island states in respect
to the plans of the Government of Japan for radioactive waste dumping in
the northwest Pacific (RWMC, 1980). The Japanese Government had
prepared an evaluation of possible dump sites in the western Pacific and
presented its draft site suitability and environmental impact assessment
to the NEA in 1982. The NEA reviewed this assessment and suggested
revisions to certain of its contents. However, since that date, there
has been little evidence of additional detailed planning by the Japanese
and this may reflect a decision not to consider this option of disposal
further. During the Conference on the Human Environment in the South
Pacific, which took place in Rarotonga in 1982, concern was expressed
about the dissemination of radionuclides in the region from the testing
of nuclear explosives and the storage and disposal of radioactive
wastes. Following this meeting the United Nations Environment Programme
(UNEP) established a Technical Group on Radioactivity in the South
Pacific Region and assigned it the task of reviewing radioactivity and
its impact in that region (UNEP, 1983).

Assuming that the option of radioactive waste dumping remains open
following the completion of debate within the LDC, there are a number of
materials whose disposal by this route would seem likely to be seriously
considered within the foreseeable future. It should be stressed that
materials currently being dumped in the ocean are largely small quantities of contaminated wastes such as laboratory wastes and pipework for
which ocean disposal is both attractive and economic compared with land
disposal options. Although the vast majority of dumped radioactive
materials are packaged, in order to preclude release of radionuclides
before the packages reach the seabed and to provide containment to
reduce releases further, the IAEA recommendations permit the dumping of
unpackaged materials in which the radionuclides are inherently contained
in an insoluble matrix. This would permit activated metal pipework or
other irradiated materials in which the activation products are

intrinsically associated with the structure of the host material to be
dumped without resort to packaging. A different type of waste that has
recently been considered for ocean dumping is decommissioned nuclear
submarines (U.S. Navy, 1984). These submarines, following removal of
fuel from their reactor systems, were considered for ocean dumping by
the United States but a decision was subsequently made not to proceed
further with the examination of this option. Another class of
materials for which ocean dumping might be both attractive and economic
are some of the materials arising from the decommissioning of electri-
cal-generating reactors. Much of the bulky materials associated with
power reactor installations are active only by virtue of irradiation
during the operating life of the plant. Such materials include pipework
and shielding materials, both of metal and concrete, that contain relatively low levels of activity and can be transported without difficulty.
Sea dumping would appear to offer significant advantages for the disposal of such materials and this may give rise to increased sea-dumping
activity in the future as the number of decommissioned reactor installations increase. However, in talking here loosely of sea-dumping as
'an attractive and economic option' it must be stressed that the viability and attractiveness of this option needs to be carefully evaluated
and demonstrated through the application of the optimization process as
specified by the ICRP [e.g., see IAEA (1984b)].

One additional consideration that bears on the available options
for radioactive waste disposal in the oceans through dumping is the **de
minimis** issue. As already stated, any materials containing radionuclides in concentrations less than those stipulated as unsuitable for
dumping at sea (Annex I materials) may be considered for ocean dumping
in accordance with the LDC and the IAEA Recommendations as Annex II
materials under the Convention. However, all materials are, to some
extent, inherently radioactive and it is clearly not the intention of
either the LDC or the IAEA to consider all candidate materials as Annex
II substances on the basis of radioactivity just because they intrinsically contain some radionuclides albeit of natural origin. The decision
as to whether to treat a prospective dumping material as radioactive or
not is left to the discretion of national authorities and it is made on
the basis of the nature and origin of the material considering the
spirit and intent of the LDC. Such decisions are usually of a de facto
nature in the sense that they are not made deliberately but made intui-
tively on the basis of precedent. Thus, materials such as dredge
spoils, sewage sludge and certain types of industrial and agricultural
wastes are arbitrarily considered as non-radioactive whereas materials
derived from the nuclear industry, or otherwise clearly recognizable as
being radioactive, are considered as radioactive under the Convention.
The IAEA has been concerned about the arbitrary nature of this decision-
making process and has, since 1979, been attempting to develop guidance
to national authorities on how more considered decisions might be made
with respect to material that may be treated as non-radioactive under
the Convention. Unfortunately, while it is possible to devise levels of
individual dose below which it might be reasonable to regard the radiation dose itself, and consequently its effects, as trivial, it is far
more difficult to deal with the problem of defining a trivial collective

dose. Until suitable guidance on this latter issue is received from organizations such as ICRP, it is unlikely that numerical specifications of de minimis will be devised. In the interim, however, qualitative specification of the types of materials that can be regarded as de minimis under the terms of the LDC have been provided by an advisory group to the IAEA (IAEA 1981b). These specifications deal primarily with the origin of the candidate material and are based upon an assumption that only materials that might be suspected of having artificially-augmented concentrations of radionuclides, through contact with some licensed nuclear industrial activity, should initially be considered as potentially 'radioactive'. Such materials should then be subjected to further examination and analysis to determine whether they justify being dealt with as radioactive under the LDC.

One consequence of the eventual establishment of a definition of de minimis levels of radioactivity will be that additional, very weakly active, materials can be considered as non-radioactive under the LDC and therefore considered for dumping under conditions other than those laid down by the IAEA in the Definition and Recommendations (IAEA, 1978). This might permit materials such as contaminated sludges and soils to be disposed of through dumping in shallow waters rather than at greater than 4000 m depth in the pelagic ocean. There have been cases in which ocean dumping has been ruled out for the disposal of relatively small quantities of radionuclide-containing soils partly because of adverse public reaction and partly because the present recommendations would require that these materials be packaged and transported to deep-ocean dump sites which results in costs being exorbitant. While it is difficult to say when resolution of the de minimis issue might be achieved, five years might well be sufficient for a suitable definition relevant to sea dumping to be proposed. It might be added that, should the LDC eventually decide that all radioactive substances be considered as Annex I materials, the relative importance of the de minimis definition for radioactive matter will be heightened since only materials having characteristics falling within such a definition will then be suitable for ocean dumping under the Convention.

7. APPLICATION OF A SIMILAR APPROACH TO OTHER CONTAMINANTS

The application of the approach that has been used for the comparison among disposal options for radioactive waste disposal, and for the assessment of the consequences of the practice of low-level radioactive dumping in the ocean, could, with some modifications, be applied to other contaminants (Templeton, 1983). The modifications will have to take account of the differences between radionuclides and other contaminants and their effects. For example, dose-response relationships assumed for inactive contaminants may well be different from that assumed for radionuclides (i.e. linearity without threshold). In the cases of many of the metallic contaminants, for example, the dose-response relationship is one with threshold which differs from that assumed for radiation effects, namely linearity without threshold. Furthermore, the end point that determines the allowable rate of release of non-radioactive contaminants may not be human exposure but either some effect upon marine organisms (e.g. the effects on shellfish being the most critical with respect to organotin compounds) or the maintenance of environmental quality standards in water or sediments. However, these are generally not reasons to rule out the approach for obtaining greater insight into the consequences of contaminant disposal or for making comparisons among alternative disposal options. The fact that it has seldom been utilized for environmental assessments in respect to non-radioactive contaminants is little more than amazing. The odd instances (Barry, 1979; Preston and Portmann, 1981) in which it has been used have certainly demonstrated its promise. A far more serious attempt to apply this approach to other contaminants should be made to determine its usefulness and strengths and weaknesses. It is disturbing that several naturally-occurring substances still appear on the black-lists of certain international and regional conventions without any concerted effort having been made to justify such assignments by modeling/calculation rather than making them on the basis of previous isolated instances of severe environmental impact and, more recently, precedent.

These questions are particularly timely in the context of current international discussions on an agreement to cover land discharges of contaminants. The major regional agreement respecting the protection of the ocean from land-discharged contaminants is the Paris Convention (PARIS, 1974) to which most of the countries of western Europe are signatories. There have been a number of pleas (e.g. IAEA, 1983) for an international agreement on land-based discharges to complement the LDC in order to ensure that greater holisticism in the use of the ocean for waste disposal be engendered. While many countries have legislation relevant to land discharges of contaminants to the ocean, there is little international coordination on the topic, except multilaterally within the Paris Convention. This, and the existence of the London Convention, results in a preoccupation with sea dumping issues in international discussions without balanced attention to the relative magnitude of other routes of entry of contaminants to the ocean as a whole. For this reason there are now negotiations, taking place under UNEP auspices, towards the formulation of an international agreement on the protection of the ocean from pollution caused by contaminants discharged directly to the ocean from land (UNEP, 1985). This agreement may facilitate the application of a revised approach to the technical aspects of assessment and regulation than that previously used.

8.   REFERENCES

Barry, P.J., 1979. 'An Introduction to the Exposure Commitment Concept with Reference to Environmental Mercury', Report No. 12, Monitoring and Assessment Research Centre, Chelsea College, University of London.

Colombo, P., R.M. Neilson Jr, and M.W. Kendig, 1983. 'Analysis and Evaluation of a Radioactive Waste Package Retrieved from the Atlantic Ocean', In: Wastes in the Ocean, (P.K. Park, D.R. Kester, I.W. Duedall and B.H. Ketchum, Eds.), Vol. 3, Wiley and Sons, New York.

GESAMP, 1983. 'An Oceanographic Model for the Dispersion of Wastes Disposed of in the Deep Sea', Reports and Studies No. 19, IMO/FAO/ UNESCO/WMO/WHO/IAEA/UN/UNEP Joint Group of Experts on the Scientific Aspects of Marine Pollution, International Atomic Energy Agency, Vienna.

Hunt, J.G. and B.D. Smith, 1982. 'Some Physical Impacts of the Marine Environment on Packages of Low-Level Radioactive Waste for Sea Disposal', In Radiological Protection - Advances in Theory and Practice, Third International Symposium, Inverness, SRP, London.

IAEA, 1961. 'Radioactive Waste Disposal into the Sea'. Safety Series No. 5, International Atomic Energy Agency, Vienna.

IAEA, 1975. 'Convention on the Prevention of Marine Pollution by Dumping of Wastes and Other Matter; The Definition Required by Annex I, Paragraph 6 to the Convention and the Recommendations Required by Annex II, Section D', INFCIRC 205/Add.1, International Atomic Energy Agency, Vienna.

IAEA, 1978. 'Convention on the Prevention of Marine Pollution by Dumping of Wastes and Other Matter; The Definition Required by Annex I, Paragraph 6 to the Convention and the Recommendations Required by Annex II, Section D', INFCIRC 205/Add.1/Rev.1, International Atomic Energy Agency, Vienna.

IAEA, 1981a. 'Packaging of Radioactive Wastes for Sea Disposal', TECDOC-240, International Atomic Energy Agency, Vienna.

IAEA, 1981b. 'Considerations Concerning de minimis Quantities of Radioactive Waste Suitable for Dumping at Sea Under a General Permit', TECDOC-244, International Atomic Energy Agency, Vienna.

IAEA, 1983. 'Control of Radioactive Waste Disposal into the Marine Environment', Safety Series No. 61, International Atomic Energy Agency, Vienna.

IAEA, 1984a. 'The Oceanographic and Radiological Basis for the Definition of High-Level Wastes Unsuitable for Dumping at Sea', Safety Series No. 66, International Atomic Energy Agency, Vienna.

IAEA, 1984b. 'Environmental Assessment Methodologies for Sea Dumping of Radioactive Wastes', Safety Series No. 65, International Atomic Energy Agency, Vienna.

IAEA, 1985. 'Sediment k_ds and Concentration Factors for Radionuclides in the Marine Environment', Technical Report Series, International Atomic Energy Agency, Vienna.

ICRP, 1977. 'Recommendations of the International Commission on Radiological Protection (ICRP)', ICRP Publication No. 26, Annals of the

ICRP, 1(3), 1-53.

IMO, 1982. 'Convention on the Prevention of Marine Pollution by Dumping of Waste and Other Matter, London, 1972', reprinted in the Inter-Governmental Conference on the Convention on the Dumping of Wastes at Sea, 1982 edition, International Maritime Organization, London.

Johnson, R.H., M. Kahn and C. Robbins, 1985. 'United States Practices and Policies for Ocean Disposal of Radioactive Wastes 1946-1970', Report EPA 530/1-84-017, U.S. Environmental Protection Agency, Washington, D.C.

King, W.H. and S.S. Hill, 1975. 'Investigations into the Effects of Deep Sea Pressures on Disposal Packages', Report AERE-R7977, United Kingdom Atomic Energy Authority, Harwell.

OECD/NEA, 1977. 'Decision of the Council Establishing a Multilateral Consultation and Surveillance Mechanism for Sea Dumping of Radio-active Waste', Document C(77) 115 (Final), Nuclear Energy Agency of the Organization for Economic Cooperation and Development, Paris.

OECD/NEA, 1980. 'Review of the Continued Suitability of the Dumping Site for Radioactive Waste in the North-East Atlantic', Nuclear Energy Agency of the Organization for Economic Cooperation and Develop-ment, Paris.

OECD/NEA, 1981. 'Research and Environmental Surveillance Programme Related to Sea Disposal of Radioactive Waste', Nuclear Energy Agency of the Organization for Economic Cooperation and Develop-ment, Paris.

OECD/NEA, 1983. 'Interim Oceanographic Description of the North-East Atlantic Site for the Disposal of Radioactive Waste', Vol. 1 (P.A. Gurbutt and R.R. Dickson, Eds.), Nuclear Energy Agency of the Organization for Economic Cooperation and Development, Paris.

OECD/NEA, 1985a. 'Interim Oceanographic Description of the North-East Atlantic Site for the Disposal of Radioactive Waste', Vol. 2, (R.R. Dickson, P.A. Gurbutt and P.J. Kershaw, Eds.), Nuclear Energy Agency of the Organization for Economic Cooperation and Develop-ment, Paris.

OECD/NEA, 1985b. 'Review of the Continued Suitability of the Dumping Site for Radioactive Waste in the North-East Atlantic', Nuclear Energy Agency of the Organization for Economic Cooperation and Development, Paris.

OECD/NEA, 1979. 'Guidelines for Sea Dumping Packages of Radioactive Waste', Nuclear Energy Agency of the Organization for Economic Cooperation and Development, Paris.

PARIS, 1974. 'Convention for the Prevention of Marine Pollution from Land-Based Sources', opened for signature 4 June 1974, International Legal Materials, **13**, 352.

Pearce, K.W. and J.D. Vincent, 1963. 'Investigation into the Effects of Deep Sea Pressures on Waste Materials and Disposal Containers', Report AERE-M1254, United Kingdom Atomic Energy Authority, Harwell.

Pentreath, R.J., 1983. 'Biological Studies'. In: Interim Oceanographic Description of the North-East Atlantic Site for the Disposal of Radioactive Waste, Vol. 1, (P.A. Gurbutt and R.R. Dickson, Eds.), Nuclear Energy Agency of the Organization for Economic Cooperation and Development, Paris.

Preston, A. and J.E. Portmann, 1981. 'Critical Pathway Analysis Applied to the Control of Mercury Inputs to U.K. Coastal Waters', Environmental Pollution, Series B, **2**, 451.

Ruegger, B., W.L. Templeton and P. Gurbutt, 1984. 'The Nuclear Energy Agency Research and Environmental Surveillance Programme Related to Sea Disposal of Low-Level Radioactive Waste', In: Proceedings of the International Conference on Radioactive Waste Management, Vol. 5, 301-313, International Atomic Energy Agency, Vienna.

RWMC, 1980. 'Low-Level Radioactive Wastes Dumping at the Pacific', Radioactive Waste Management Center, Nuclear Safety Bureau, Science and Technology Agency, Japan.

Seki, S., A. Ito and H. Amano, 1980a. 'Integrity Test of Full Size Packages of Cement-Solidified Radioactive Wastes Under Deep Sea Conditions', Nuclear and Chemical Waste Man., **1**, 129-138.

Seki, S., A. Ito, Y. Wadachi and H. Amano, 1981. 'Safety Evaluation of Multistage Type Packages Containing Radioactive Wastes for Sea Disposal', In: Proceedings, Symposium on Impacts of Radionuclide Releases into the Marine Environment, International Atomic Energy Agency, Vienna.

Seki, S., I. Hisa, K. Ouchi and A. Ito, 1980b. 'Integrity Test of Multi-Stage Design Packages of Radioactive Wastes Under Deep Sea Conditions', Journal of Nuclear Science and Technology, **11**, 857-864.

Simmonds, 1983. United Nations Convention on the Law of the Sea. Oceana Publications, Inc., New York.

Templeton, W.L., 1983. 'Lessons from Radioactive waste Disposal Applied to Other Pollutants', In: Wastes in the Ocean, (P.K. Park, D.R. Kester, I.W. Duedall and B.H. Ketchum, Eds.), Vol. 3, Wiley and Sons, New York.

UNEP, 1983. 'Draft Report of the Technical Group on Radioactivity in the South Pacific', United Nations Environment Programme, Geneva, (mimeo).

UNEP, 1985. 'Report of the Third Session of the Ad Hoc Working Group of Experts on the Protection of the Marine Environment Against Pollution from Land-Based Sources', Montreal, 11-19 April 1985, United Nations Environment Programme, Geneva.

UNSCEAR, 1982. Ionizing Radiation: Sources and Biological Effects, 1982 Report of the United Nations Scientific Committee on the Effects of Atomic Radiation, United Nations, New York.

U.S. Navy, 1984. 'Final Environmental Impact Statement on the Disposal of Decommissioned, Defueled Naval Submarine Reactor Plants', Office of the Chief of Naval Operations, United States Department of the Navy, Washington, D.C.

COMPARISON OF LAND AND SEA DISPOSAL OPTIONS FOR LOW AND INTERMEDIATE
LEVEL RADIOACTIVE WASTES

Marion D Hill
National Radiological Protection Board
Chilton
Didcot
Oxon OX11 ORQ
United Kingdom

ABSTRACT. This paper discusses, in general terms, comparisons between
land disposal and sea dumping of radioactive wastes. It begins with an
outline of the scope of comparisons and goes on to discuss techniques
which can be used to aid decisions on appropriate disposal methods,
taking into account the wide range of factors involved. Two examples of
comparisons are given, both focusing on the long term radiological
impacts of disposal, and conclusions are drawn on whether these impacts
are likely to be a major factor in comparisons.

1.   INTRODUCTION

Annex III of the London Dumping Convention (LDC)[1] requires national
authorities who issue special permits for the dumping of matter at sea
to take into account "the practical availability of alternative
land-based methods of treatment, disposal or elimination, or of
treatment to render the matter less harmful for dumping at sea". This
principle is reiterated in the present Definition of radioactive wastes
unsuitable for dumping and Recommendations established by the
International Atomic Energy Agency (IAEA)[2], which state that it is
necessary to consider "the justification for the proposed dumping
operation when weighed against land-based alternatives". Until recently
there was little guidance available as to how to carry out a comparison
of sea dumping of radioactive wastes with land-based alternatives.
However, in 1984 the IAEA issued a Safety Series document[3] which
provides general guidance. The purpose of this paper is to summarise
and discuss those parts of IAEA Safety Series 65[3] which deal with
comparisons between land and sea disposal, and to provide examples of
comparisons for low and intermediate level wastes.

517

G. Kullenberg (ed.), The Role of the Oceans as a Waste Disposal Option, 517–544.
© 1986 NRPB

2.   SCOPE OF COMPARISONS

2.1   Waste Management Systems and Strategies

It is impossible to compare one disposal method for a particular type of
radioactive wastes with other disposal methods without taking into
account the procedures used to treat, immobilise, package, store and
transport the waste prior to disposal. The costs and environmental
impact of these procedures will differ from one disposal option to
another. One obvious example is that when packaging wastes for sea[4]
dumping it is necessary to follow the guidelines issued by the IAEA[4]
and those established within multi-lateral mechanisms[5], and in
particular to ensure that each package has a specific gravity greater
than 1.2. These requirements do not apply to land disposal, where the
package specification may well differ from one waste type to another
(see, for example, reference 6) and from one disposal facility to
another. For this reason, the appropriate comparison to make is between
complete systems for managing a particular type of waste, from the point
at which the waste arises, through its treatment, immobilisation,
packaging, storage and transport, to the disposal operation itself and
the post-disposal period.

While comparisons of the systems for managing each type of waste
are necessary inputs to decisions, there is a further stage to be gone
through before the management option for each waste is chosen. This
stage consists of considering the national waste management strategy as
a whole, including the interactions between the systems for each waste.
Points to be taken into account at this stage include the possible use
of one treatment plant for several wastes, bearing in mind the capacity
of the plant and the rates of waste arisings, and overall limits on the
capacity of each disposal facility in terms of volume and total activity
content. These strategic considerations could show that while one
management system for a waste would be the preferred one in isolation
from all others, another system is preferable because, for example, it
leads to savings in the financial cost of radioactive waste management
as a whole.

2.2   Factors to be Taken into Account

Table 1 (taken from IAEA Safety Series 65[3]) shows a list of the
factors to be taken into account when comparing one waste management
system with another; it is equally applicable to comparisons at the
national waste management strategy level. The list is not exhaustive,
nor are the factors placed in any order of importance; its purpose is to
show that the range of factors is very wide and includes technical,
quantifiable aspects and broader, less quantifiable considerations.
Given the number of factors which may enter into comparisons, an
important stage in the decision-making process is to structure them in a
way which clarifies the areas where judgements are required, and
identifies the points where it will be necessary to assign weightings to
each factor and trade-off one against another in comparisons.

The grouping of factors into a structure can be carried out by carefully considering how to assess each factor, identifying those aspects which are specific to it and those which are common to a number of factors. An example of the kind of scheme of factors which can be produced by this means is given in IAEA Safety Series 65 and is reproduced here as Figure 1. It is evident from Figure 1 that, in addition to grouping factors under specific headings, it is also possible to group factors according to the extent to which they can be quantified and the stage at which they should be taken into account in making decisions. This process was not considered in detail in Safety Series 65 and is therefore discussed further here.

3   DECISION-MAKING

As noted in Section 1, the present IAEA Definition and Recommendations (INFCIRC 205/Add 1/Rev 1)[2] state that national authorities should consider, in addition to other factors, the 'justification' for the proposed sea dumping operation as compared to land-based alternatives. In this context the term 'justification' means the rationale for choosing sea dumping rather than land disposal, not the need for disposing of radioactive wastes. Safety Series 65 states that this recommendation is to be interpreted as meaning that the optimum (in a radiological protection sense) disposal option should be selected. This interpretation is, however, open to question because neither the International Commission on Radiological Protection (ICRP), nor IAEA, have ever defined the scope of optimisation studies, or the way in which the results of such studies should be used as an input to decisions.

In the UK, the issue of what factors are to be included in optimisation (or, synonymously ALARA (as low as reasonably achievable, economic and social factors being taken into account)) studies has been considered in some detail at the Public Inquiry on the pressurised water reactor (PWR) which the Central Electricity Generating Board propose to build at Sizewell. In proofs of evidence to the Inquiry the National Radiological Protection Board (NRPB) clarified its views as to what should be included in ALARA studies. The overall procedure envisaged by NRPB for decisions on radioactive waste management, or any other radiological protection matter of sufficient concern to warrant detailed analysis, is shown in Figure 2. Briefly, this consists of identification of the options which are technically feasible for each stage of a process or system to protect people against radiation. This is followed by identification of the combinations of options which could make up a feasible system, and at this stage obvious constraints are taken into account (for example, regulations for the transport of radioactive materials, plant safety regulations, availability of disposal sites). Preliminary assessments of the potentially suitable systems are then carried out, with the objective of eliminating those which do not meet limits on radiation dose or risk, and which are not cost-effective, in the sense that their financial cost greatly exceeds the savings in radiological health detriment which they would produce. At this point it is also necessary to take into account any "source

upper bounds"[7] which have been established on radiation dose or risk.
The purpose of these upperbounds is to allow adequate margin for the
exposure of the same group of people to several sources of radiation, by
placing constraints on the subsequent optimization process.

The outcome of the actions described above will be the selection of
a small number of systems for detailed analysis and assessment. In the
scheme envisaged by NRPB, this begins with detailed analysis of the
radiological impact of each system, together with refinement of the
estimates of their financial costs. The results of these analyses and
estimates can be used within a framework of cost-benefit analysis to
indicate the "optimum" system from a radiological protection point of
view. For some decisions, however, there are radiological protection
factors which cannot be easily accommodated within the cost-benefit
framework. One obvious example is the existence of low probability,
high consequence events, for which proposed methods of costing
radiological detriment are not appropriate[8,9]. Another class of
factors consists of those which are technical but are not amenable to
straightforward quantification. These include preferences for
operational procedures and plants which are simple and flexible, against
those which are complex and rigid. For these types of situations it is
more appropriate to use decision-aiding techniques which allow
non-quantifiable factors to be more easily considered, for example the
various methods for multi-attribute analysis.

As part of this detailed assessment of systems, it is also
necessary to undertake sensitivity and uncertainty analyses. The
difference between the two types of analyses is that sensitivity studies
are intended to indicate the parameters which will be most important in
influencing decisions, while uncertainty analyses show how realistic
variations in parameters affect assessment results. Thus in sensitivity
analyses it is acceptable to include variations which are well outside
the range which is considered realistic, just to ascertain whether the
conclusions of the study would change if one were to adopt extreme
assumptions. In uncertainty analysis, however, the emphasis is on
quantification of the effects which variations of parameters within a
realistic range are likely to have on the outcome of decisions.

Thus far in the decision procedure only those factors which are
directly related to radiation health detriment have been considered. In
the final stage of decision-making it is necessary to include other,
frequently less quantifiable, but no less important factors. These
include those which are essentially social and/or political in nature
and which lie outside the competence of technical experts. As one who
is not involved in this final stage of making decisions, I can
appreciate the concerns of those who feel that their views have not been
adequately taken into account in the past. Part of the reason for this
is that decision-makers have not made use of the techniques available to
assist them in structuring their judgements and have not provided
adequate explanations of the bases for their decisions. It would, in my
view, be helpful to all concerned if decision-makers worked towards the
transparency and consistency which is the watchword of those who advise
on radiological protection issues. The stage which is designated
"decision on preferred system" in Figure 2 could then be expanded to

show explicitly the procedure followed at the point where the technical preferences are obvious and the consideration of other factors begins.

4   EXAMPLES OF COMPARISONS

In the examples described below two types of waste are considered; the first is the low and intermediate level solid wastes arising during the operation of light water reactors (LWRs) and the second is the very low level solid waste arising from the dismantling of nuclear power plants. Both comparisons focus on the radiological impact of disposal of the wastes, with the objective of determining whether this is likely to be a major factor in decisions on their management.

4.1 LWR Operating Wastes

These wastes consist of the ion exchange resins, sludges, concentrates and filters which arise during the treatment of LWR effluents prior to their discharge into the environment. The quantity of wastes considered is that produced by the operation of 6 PWRs and 2 BWRs (1000 MW(e) capacity) for approximately 6 years. Details of the radionuclide inventory of the wastes are given in Table 2[10]. The wastes are assumed to be incorporated in cement and placed in steel drums prior to disposal.

The two disposal options compared here are burial at shallow depth on land in an engineered facility, and dumping at the designated site in the N.E. Atlantic Ocean. For simplicity, and because incorporation in cement produces a monolithic waste form which is, in principle, suitable for both shallow land burial and sea dumping, it is assumed that the radiological impact of waste treatment, immobilisation and packaging is the same in the two cases. It is also assumed that wastes are disposed of shortly after they are produced, so that the radiological impact of storage prior to disposal can be neglected.

4.1.1 Transport and Disposal Operations. With the assumptions given above, the first point at which differences occur in the two procedures for managing these wastes is in transport from the point of arising to the disposal site. In the case of land disposal transport could be by road or rail and would be direct from the power station to the disposal site. For sea dumping it is necessary to transport wastes by road or rail to the port of loading, and thence by ship to the dumping site. A study of radiation exposure due to transport of radioactive materials in the UK[11] showed that the collective dose to rail staff and dock workers during the 1982 sea dumping operation was about 0.01 man Sv. The total β/γ activity in the wastes was about 4.10^{15} Bq, which is of the same order as that in the LWR wastes considered here. No estimate was made of the collective dose to the public during transport of wastes for the 1982 UK sea dump but studies of transport operations in general indicate that it would have been negligible compared to the occupational collective dose. From this information it seems reasonable to conclude that the radiological impact of transport of wastes prior to sea dumping

is very low. The same conclusion can be drawn for transport to a land
disposal site because, although there may be a difference in the
distances and times of transport, most of the collective dose is
received by workers loading and handling waste packages and, for a given
volume of wastes, would be approximately the same for both land and sea
disposal.

In a full comparison the next stage would be to estimate the
collective dose to the ship's crew during transport to the disposal site
and in the dumping operation itself, and the collective dose to workers
placing wastes in a shallow burial facility. At present it is not
possible to make these estimates because the ship which has been
designated for future sea dumping operations has not yet been used, and
details of the operational procedures at a burial facility of the type
considered here are not yet available. Thus for the present purpose it
will be assumed that the occupational collective doses are approximately
equal for the two types of disposal operation.

4.1.2 Post-Disposal Impact of Sea Dumping.

The radiological impact of
sea dumping has been estimated using the methodology, models and data
base employed in the radiological assessment for the 1984/5 NEA review
of the continued suitability of the N.E. Atlantic dump site[12]. The
modelling framework is shown in Figure 3 and the exposure pathways
considered in calculating individual and collective doses are listed in
Table 3. For consistency with the assumptions made for shallow land
burial (see below), the time before any radionuclides are released from
waste packages after dumping was taken to be 10 y, and the release rate
was taken to be constant at 10^{-2} y^{-1} for all radionuclides. The
sensitivity analysis carried out for the NEA site review shows that
neither of these assumptions has a great effect on predicted doses.

The results of the calculations of annual individual doses are
summarised in Table 4. From these it can be seen that the peak annual
individual dose predicted via an 'actual' exposure pathway (ie one
which, in principle, exists now) is 4.10^{-12} Sv from ^{137}Cs, and the
pathway concerned is external irradiation from spending time on beaches.
This dose is calculated to occur 100 y after the start of dumping.
Radionuclides of secondary importance in terms of peak annual individual
doses via 'actual' pathways are ^{241}Am, ^{239}Pu and ^{240}Pu, which give doses
in the range 5.10^{-13} - 1.10^{-12} Sv y^{-1} through consumption of molluscs.
All these peak doses are predicted to occur in the Antarctic. The
highest peak annual individual dose via a 'hypothetical' pathway is
2.10^{-10} Sv from ^{137}Cs, (deep fish pathway at 50 y after the start of
dumping). This dose is predicted to be received through consumption of
fish caught in the bottom waters of the European Basin of the Atlantic,
outside the dump site itself.

The results of calculations of collective doses are summarised in
Table 5. The collective dose commitment from sea dumping of LWR wastes
(ie the collective dose integrated over all time to the world
population) is estimated to be 12 man Sv, of which 11 man Sv arises from
global circulation of ^{14}C. Most of this collective dose is received
during the first 10,000 years, and all of it consists of very low doses
to a large number of individuals.

4.1.3 <u>Post-Disposal Impact of Shallow Land Burial</u>. The methodology,
models and data base used to estimate the radiological impact of shallow
land burial are described in references 10 and 13. The burial facility
is assumed to be located in a clay formation and to consist of trenches
which are 18 m deep, lined with concrete, infilled with cement, and
capped with alternate layers of cement and rolled clay (see Figure 4).
One trench 25 m wide and 110 m long would accommodate the total volume
of LWR wastes considered here. The clay outcrop is assumed to be under
artesian conditions, with ground-water tending to rise vertically on a
regional scale, because this is typical of many UK clay formations.
Close to the burial facility, the groundwater flow pattern is affected
by minor relief features and by the hydrogeological properties of the
burial trench itself. The local relief assumed for the site is such
that water infiltrating the trench from above will tend to move
downward, then upwards and outwards, eventually entering the streams
assumed to be present on either side of the site (see Figure 4).

Once the burial facility is closed, it is assumed that
radionuclides may be released by two mechanisms : transport in
groundwater or infiltrating rainwater, and disturbance of the waste by
human actions. Water contact is assumed to be certain to occur after an
initial period of total containment, while for human actions both the
probability that disturbance of the site will occur, and the doses
received if it does, are considered. In calculating doses resulting
from water contact it is assumed that release of radionuclides into
water flowing through the trench starts at the same time from all waste
drums (10 y) and occurs at a constant fractional rate (10^{-2} y^{-1}). The
modelling of groundwater flow and radionuclide release and migration
through the surrounding clay strata is discussed in detail in reference
14. The output·of the release and migration calculations consists of
the rates of discharge of radionuclides to nearby streams. It is
assumed that these streams are used as a source of drinking water for
humans and animals, and that fish from the streams are caught and
consumed by a local population.

Once burial operations have ceased and trenches have been capped,
it is likely that restrictions will be placed on the use of the burial
site. The main purpose of these restrictions is to prevent inadvertent
intrusion into the site while the risks of such intrusion are above
acceptable levels. In the studies on which this example is based one of
the objectives was to determine how long this use restriction period
might have to be. The approach adopted was therefore to assume that
there are no restrictions, calculate the risks, then examine the results
to determine whether restricting use of the closed site for any period
would substantially affect its radiological impact. To avoid confusion
with previous results, this approach is retained here.

The two types of human action which could disturb a burial site and
which are included in the assessment are excavation for the purposes of
building, and farming of the land. The probability of the site being
used for building or farming is calculated from current rates of
reclamation of derelict land and by assuming that future land use
patterns are similar to those in the UK today. For building, the doses
calculated are those to an individual worker excavating the trenches,

and arise from external irradiation and inhalation of dust. In order to
calculate the consequences of the site being farmed, the output from the
model for radionuclide migration in the soil is used to derive
radionuclide concentrations in the rooting zone, and hence
concentrations in grass and crops grown on the site. The types of
farming considered are : dairy, beef, sheep, root crops, grain and green
vetetables, and it is assumed that only one of these would be practised
on the site at any given time. The maximum individual dose for each
type of farming is calculated by summing the doses from all the
foodstuffs produced by that type of farming.

Table 6 summarises the results of the calculations of individual
doses, and Table 7 gives the calculated probabilities that the site is
built on or farmed. In Table 8 these sets of results are combined to
derive estimates of overall risk to an individual, that is the
probability that the individual receives a dose, multiplied by the
probability that the dose will lead to fatal cancer. The general
pattern of results is that, in the short term, potential doses and risks
arise mainly from the possibility of excavation for building, and there
is a low probability of relatively high individual doses (for example,
in the 50th year after closure there is a probability of 8.10^{-4} of a
dose of 0.2 Sv). Doses and risks from building decrease with time due
to radioactive decay and migration of radionuclides away from the site
with groundwater. Doses and risks from water pathways and farming are
initially zero, increase to a maximum of about 500 years, then decrease
slowly. The peak annual individual doses from both water pathways and
farming are in the range 0.1 - 5 mSv; the probability that doses are
received via water pathways is unity (by assumption) and the overall
probability that the site will be farmed in a year also approaches unity
at about 250 y after site closure.

Table 9 shows the results of the collective dose calculations for
the drinking water pathway, farming, and global circulation of the
long-lived, mobile radionuclides ^{14}C and ^{129}I. The collective dose
commitment from drinking water is calculated to be about 3 man Sv with
the major contributions arising from long-lived actinides (^{237}Np, ^{234}U).
This dose is to a population using water from the streams and the rivers
into which the streams flow. The collective dose commitment from
farming is calculated to be 18 man Sv (maximum), and is dominated by the
contribution from ^{14}C. The population affected consists of those
consuming produce grown on the burial site. At 11 man Sv, the
collective dose commitment to the world population from global
circulation of ^{14}C released from the burial site is of the same order as
that calculated for farming.

4.1.4 <u>Comparison of Radiological Impacts</u>. Table 10 shows a comparison
of the collective doses from sea dumping and shallow land burial,
integrated over various times, including infinity. For shallow burial,
the collective doses arising from water contact and farming of the
closed site are shown separately because they have different
probabilities of occurrence, and because the farming dose has been
calculated using pessimistic assumptions. At each of the integration
times, the collective doses from sea dumping and from shallow burial

(water contact only) are virtually identical and arise from global circulation of ^{14}C. The key parameters and assumptions in these calculations are the same for both disposal options, and the results are thus subject to the same uncertainties. The collective dose which could be received as a result of farming of a burial site is somewhat higher than that calculated for water contact, and has different uncertainties associated with it, (in particular it depends on the root concentration factors assumed). It should also be noted that the calculated probability that the site will be farmed for root vegetables is about 10^{-3} in a year. The overall conclusion which can be drawn from these results is that it is not possible to discriminate between shallow land burial and sea dumping of LWR operating wastes on the basis of the collective dose received in the post-disposal period.

Turning to doses and risks to individual members of the public, it is clear that there are distinct differences between sea dumping and shallow burial. For sea dumping the highest calculated annual individual dose is 2.10^{-10} Sv, corresponding to risk of fatal cancer of about 3.10^{-12} per year. The hypothetical individual in this case is someone who consumes deep sea fish taken from the region of the Atlantic in which the dump site is located, and who could live in any country which carries out fishing at the assumed location. For shallow land burial, the highest calculated individual risk is 8.10^{-6} per year, and the hypothetical individual is someone who either disturbs the closed burial site for building purposes or obtains a large proportion of their food from a farm established on the site. In the building case the risk is made up of a low probability of a relatively high dose; and for farming the dose is very much larger than the peak one predicted for sea dumping, but much lower than the maximum dose from building.

4.2 Very Low Level Decommissioning Wastes

The waste considered in this second example consists of the reinforced concrete used in the construction of PWRs and which arises as a waste during reactor decommissioning. Only those volumes of concrete which are activated or contaminated at very low levels ($3.7 - 37$ Bq g^{-1}) are included. The choice of upper limit means that about the inner 1 m of the biological shield of a PWR is excluded; the mass of waste falling within the activity limits would be 120 t per PWR, assuming a 5 year delay between reactor shutdown and decommissioning. It is assumed that the concrete is disposed of without any special treatment or packaging.

The disposal options considered are:
- (a) shallow land burial at an ordinary municipal refuse tip;
- (b) shallow land burial at reactors sites;
- (c) sea dumping at the designated site in the N.E. Atlantic;
- (d) sea dumping in coastal seas around the UK.

Options (a) and (d) are included as examples of methods of disposal which would be appropriate if the waste were deemed to be exempt from the usual requirements for radioactive waste.

    Details of the method, models and data used to calculate the
radiological impact of disposal by these four methods are given in
reference 15. The maximum annual individual doses for each disposal
option are given in Table 11, and the collective doses, integrated over
various time periods are given in Table 12. The results show that both
the maximum annual individual dose and the collective dose commitment
are lower for sea dumping in the deep ocean than for any other option.
It must be emphasised, however, that all the calculated doses are very
low indeed, so that differences between them are relatively unimportant
in terms of overall decisions on the management of the wastes. The
maximum annual individual doses from three of the disposal options are
less than 5.10^{-6} Sv. This is the level below which it has been
suggested that doses are of no concern to individuals[16]. The
predicted dose for the other option is 3.10^{-5} Sv y^{-1} and arises
following early intrusion into a burial facility at a reactor site,
which is a very unlikely event. In all four cases the collective dose
commitment from disposal is less than 1 man Sv, and thus could be
regarded by regulatory authorities as of no concern[16].

5   CONCLUSIONS

The examples described above indicate the sort of results which are
likely to be obtained when we attempt to compare sea dumping with land
disposal on the basis of their radiological impacts alone. For low and
intermediate level wastes which contain mostly short-lived radionuclides
the results of the comparison of collective doses are inconclusive.
These doses are dominated by trace amounts of long-lived radionuclides,
and tend to be the same for both land and sea disposal. Maximum
individual risks from sea dumping tend to be much lower, but they could
be incurred by people outside the country producing the wastes, whereas
the maximum individual risks from land burial occur close to the
disposal site. For very low level wastes, the doses and risks from
disposal will be very small for both sea and land options, so that the
wastes could be considered for exemption from some of the regulations
applied to radioactive materials. In the case of both waste types,
decisions on their disposal will need to take into account social,
political and economic factors and these will need to be weighed against
differences in the radiological impact of disposal.
    For other wastes, particularly those containing larger quantities
of long-lived, alpha emitting radionuclides, the comparison would be
between sea dumping and deeper geologic disposal on land. The deeper
the land disposal is, the lower the probability of human intrusion
(although it can never be completely ruled out) and the longer the time
required for a radionuclide release via groundwater. Hence in this
case, the maximum risks to individuals from land disposal would be
expected to be closer to those from sea disposal, and the uncertainties
involved in estimating collective doses would be large.

Table 1

List of Factors to be Considered in Comparing
Sea Dumping with Land-Based Alternatives
(not in order of importance)

Geological suitability	Special materials
Verification of assumptions	Radiological health effects on humans (workers and public)
Societal interests	Individuals
National	Populations
International	Distribution of effects in time and space
	Routine operations and accident conditions
Ability to be monitored	
	Packaging
Transport	
	Cost
Environmental impact	
	Site availability
Ecosystems	
Resources	
	Uncertainty
Funding limitations	
	Conventional safety
Retrievability	

Table 2

Inventory of radionuclides present at site closure in cumulative arisings of
LWR operating wastes from the assumed power programme[1]

Isotope	Half-life (y)	Activity (Bq) assumed to be present in wastes[2]
^{3}H	$1.23 \ 10^1$	$6.2 \ 10^{12}$
^{14}C	$5.69 \ 10^3$	$1.0 \ 10^{11}$
^{54}Mn	$8.55 \ 10^{-1}$	$1.2 \ 10^{13}$
^{55}Fe	2.70	$2.4 \ 10^{14}$
^{59}Fe	$1.23 \ 10^{-1}$	$2.8 \ 10^{12}$
^{58}Co	$1.94 \ 10^{-1}$	$7.1 \ 10^{13}$
^{60}Co	5.27	$3.9 \ 10^{14}$
^{59}Ni	$7.50 \ 10^4$	$1.8 \ 10^{11}$
^{63}Ni	$1.00 \ 10^2$	$1.5 \ 10^{13}$
^{65}Zn	$6.68 \ 10^{-1}$	$1.9 \ 10^{13}$
^{89}Sr	$1.38 \ 10^{-1}$	$1.8 \ 10^{11}$
^{90}Sr	$2.91 \ 10^1$	$1.8 \ 10^{12}$
^{91}Y	$1.60 \ 10^{-1}$	$4.1 \ 10^{10}$
^{95}Zr	$1.79 \ 10^{-1}$	$1.9 \ 10^{10}$
^{99}Tc	$2.13 \ 10^5$	$2.4 \ 10^9$
^{103}Ru	$1.08 \ 10^{-1}$	$8.6 \ 10^9$
^{106}Ru	1.01	$3.8 \ 10^{10}$
^{110m}Ag	$6.84 \ 10^{-1}$	$2.8 \ 10^{12}$
^{124}Sb	$1.65 \ 10^{-1}$	$8.0 \ 10^{11}$
^{125m}Te	$1.59 \ 10^{-1}$	$8.1 \ 10^9$
^{127m}Te	$2.98 \ 10^{-1}$	$1.7 \ 10^{11}$
^{129}I	$1.57 \ 10^7$	$1.0 \ 10^8$
^{134}Cs	2.06	$2.2 \ 10^{14}$
^{135}Cs	$3.00 \ 10^6$	$1.4 \ 10^8$
^{137}Cs	$3.00 \ 10^1$	$4.9 \ 10^{14}$
^{144}Ce	$7.80 \ 10^{-1}$	$1.8 \ 10^{11}$
^{234}U	$2.45 \ 10^5$	$2.2 \ 10^8$
^{235}U	$7.04 \ 10^8$	$5.2 \ 10^8$
^{238}U	$4.47 \ 10^9$	$1.3 \ 10^9$
^{238}Pu	$8.77 \ 10^1$	$5.4 \ 10^{10}$
^{239}Pu	$2.41 \ 10^4$	$1.7 \ 10^{10}$
^{240}Pu	$6.55 \ 10^3$	$1.5 \ 10^{10}$
^{241}Pu	$1.44 \ 10^1$	$7.1 \ 10^{12}$
^{241}Am	$4.32 \ 10^2$	$5.9 \ 10^{10}$
^{242}Cm	$4.46 \ 10^{-1}$	$1.0 \ 10^{10}$
^{244}Cm	$1.81 \ 10^1$	$1.1 \ 10^{10}$

Total volume of processed wastes = $4.0 \ 10^3$ m^3

Notes:
1. Calculated for 6 PWRs and 2 BWRs (1000 MW(e) capacity) operating for
   6.25 y.
2. Calculated from the data given in Table 4. Where a value for
   arisings of a particular isotope is only available for one type of LWR
   it has been assumed that this value is appropriate to both reactor
   types.

Table 3

Exposure Pathways included in Calculations
of the Radiological Impact of Sea Dumping

Exposure Pathway	Symbol	Mode of Exposure
Critical Groups (Doses to Individuals)		
'Actual' Pathways		
Surface fish consumption	FISH-S	
Mid-depth fish consumption	FISH-M	
Crustacea consumption	CRUST	
Mollusc consumption	MOLL	Ingestion
Seaweed consumption	WEED	
Salt consumption	SALT	
Drinking desalinated water	DESAL	
Suspended airborne sediments	SED	Inhalation
Marine aerosols	EVAP	
Boating	BOAT	
Swimming	SWIM	External irradiation
Beach sediments	BEACH	
Deep sea mining	MINE	
'Hypothetical' Pathways		
Deep fish consumption	FISH-D	Ingestion
Plankton consumption	PLANK	
Populations (collective doses)		
Seafood consumption	FISH-S FISH-M CRUST MOLL WEED PLANK	Ingestion
Global circulation of long-lived, mobile radionuclides		Ingestion, inhalation

Table 4

Peak Annual Individual Doses from
Sea Dumping of LWR Operating Wastes[a]

Radionuclide	Peak Annual Individual Dose (Sv)	Time of Peak Dose (y)[b]	Dominant Exposure Pathway[c]
'Actual' Pathways			
^{137}Cs	4.10^{-12}	1.10^2	BEACH
^{241}Am	1.10^{-12}	2.10^2	MOLL
^{239}Pu	6.10^{-13}	5.10^2	MOLL
^{240}Pu	5.10^{-13}	5.10^2	MOLL
^{14}C	3.10^{-13}	5.10^2	FISH-M,S
^{238}Pu	3.10^{-13}	2.10^2	MOLL
^{63}Ni	1.10^{-13}	2.10^2	BEACH
All[d]	4.10^{-12}	1.10^2	BEACH
'Hypothetical' Pathways			
^{137}Cs	2.10^{-10}	50	
^{14}C	$2.10-^{12}$	1.10^2	FISH-D
^{241}Am	1.10^{-12}	1.10^2	
All[d]	2.10^{-10}	50	
^{239}Pu	6.10^{-15}	5.10^2	
^{240}Pu	5.10^{-15}	5.10^2	PLANK
^{241}Am	4.10^{-15}	2.10^2	
All[d]	2.10^{-14}	2.10^2	

Notes

(a) See Table 2 for radionuclide inventory

(b) The "time of peak" is the number of years after dumping started.

(c) All doses via seafood pathways are committed effective dose
     equivalents from one year's intake of radionuclides.

(d) The "All radionuclides" dose is the dose from all radionuclides
     via the pathway shown.

Table 5

Collective Doses from Sea Dumping
of LWR Operating Wastes

a)   Collective Doses summed over radionuclides and pathways

Integration time (y)	Collective Dose (man Sv)
10^2	0.1
10^4	8
10^6	11
Infinity	12

b)   Contributions to collective dose commitment (total 12 man Sv)

Radionuclide	Collective Dose Commitment (man Sv)	
^{14}C	11	(global circulation)
	0.2	(seafood consumption)
^{239}Pu	0.1	(seafood consumption)
^{129}I	0.07	(global circulation)
^{240}Pu	0.04	
^{226}Ra	0.02	
^{237}Np	0.02	seafood consumption
^{210}Po	0.01	

Table 6

Annual Individual Doses from Shallow Land
Burial of LWR Operating Wastes (summed over radionuclides)

Time after Site Closure (y)	Annual Individual Dose (Sv)	Exposure Pathway
10	2	
50	0.2	Excavation for
100	0.01	building purposes
250	1.10^{-4}	(inhalation and
500	2.10^{-6}	external radiation)
500	2.10^{-4}	Consumption of
10^4	2.10^{-9}	freshwater fish
10^5	1.10^{-7}	Drinking water
10^6	2.10^{-8}	Consumption of freshwater fish
500	2.10^{-3}	Dairy farming
10^4	2.10^{-8}	Dairy farming
10^5	2.10^{-7}	Grain farming
10^6	3.10^{-8}	Grain farming

Table 7

Probabilities of a Closed Shallow Land Burial Site being Excavated for Building Purposes of Farmed

a) Probabilities of Excavation for Building

Year after Site Closure (n)	Probability of Excavation in the nth year (total over types of building)
10	3.10^{-3}
50	8.10^{-4}
100	6.10^{-4}
250	5.10^{-4}
500	5.10^{-4}

b) Probabilities of Farming

Year after Site Closure (n)	Probability of Farming (total over farming types)
250	0.8
500	0.7
10^3	0.5

Note

Probabilities in a year decrease with time because the cumulative probability of site disturbance by farming or building increases with time.

Table 8

Individual Risks from Shallow Land
Burial of LWR Operating Wastes

Time after Site Closure (y)	10	50	10^2	250	500	10^3
Risk from water pathways	0	0	0	2.10^{-9}	2.10^{-6}	2.10^{-7}
Risk from building pathways	8.10^{-6}	2.10^{-7}	1.10^{-8}	8.10^{-11}	1.10^{-12}	1.10^{-12}
Risk from farming pathways	0	0	0	2.16^{-9}	8.10^{-6}	4.10^{-6}

Note

Risk is defined as the probability (in a year) that the dose will be received, multiplied by the probability that the dose leads to fatal cancer.

Table 9

**Collective Doses from Shallow Land
Burial of LWR Operating Wastes**

Integration Period	Collective Dose (man Sv)	Radionuclides and Pathways
10^2	0	–
10^4	17 9.	^{14}C, farming ^{14}C, global circulation
10^6	17 10	^{14}C, farming ^{14}C, global circulation
Infinity	3 18 10	^{226}Ra, ^{237}Np drinking water ^{14}C, farming ^{14}C, global circulation

Notes

1) Drinking water dose is from human consumption of water only

2) Global circulation and drinking water doses are assumed to
    be certain to be received. Farming dose is from consumption
    of root vegetables and is a maximum, with a probability of
    occurrence less than unity (see text).

3) No collective dose calculations were carried out for
    building pathways.

Table 10

Comparison of Time Integrated Collective Doses from
Sea Dumping and Shallow Land Burial of
LWR Operating Wastes

	Time Integrated Collective Dose (man Sv)				Dominant radionuclide and pathways
Integration time (y)	10^2	10^4	10^6	Infinity	
Sea Dumping	0.1	8	11	12	^{14}C, global circulation
Shallow Land Burial (a) water contact	0	9	10	10	^{14}C, global circulation
(b) farming	0	17	17	17	^{14}C, consumption of root vegetables

Table 11

Maximum Annual Individual Doses Arising from Disposal of PWR Decommissioning Wastes

Disposal Option	Maximum Annual Individual Dose (Sv)	Dominant radionuclide and exposure pathway
Burial at Municipal refuse tip	4.10^{-6}	^{60}Co, ^{137}Cs, building on closed burial site.
Buried at reactor site*	3.10^{-5}	^{60}Co, ^{137}Cs, building on closed burial site
Sea dumping in coastal waters	8.10^{-9}	^{60}Co, external irradiation on beaches
Sea dumping in deep ocean	2.10^{-17}	actinides, consumption of molluscs and seaweed

*Assumed to be on a river

Table 12

Time Integrated Collective Doses from
Disposal of PWR Decommissioning Wastes

Disposal Option	Time Integrated Collective Dose (man Sv)			
	10^2 y	10^4 y	10^6 y	Infinity
Burial at municipal tip	2.10^{-6}	2.10^{-4}	3.10^{-4}	3.10^{-4}
Burial at reactor site*	9.10^{-6}	2.10^{-4}	2.10^{-4}	2.10^{-4}
Sea dumping in coastal waters	1.10^{-4}	1.10^{-4}	1.10^{-4}	1.10^{-4}
Sea dumping in deep ocean	3.10^{-9}	4.10^{-7}	6.10^{-7}	6.10^{-7}

*Assumed to be on a river

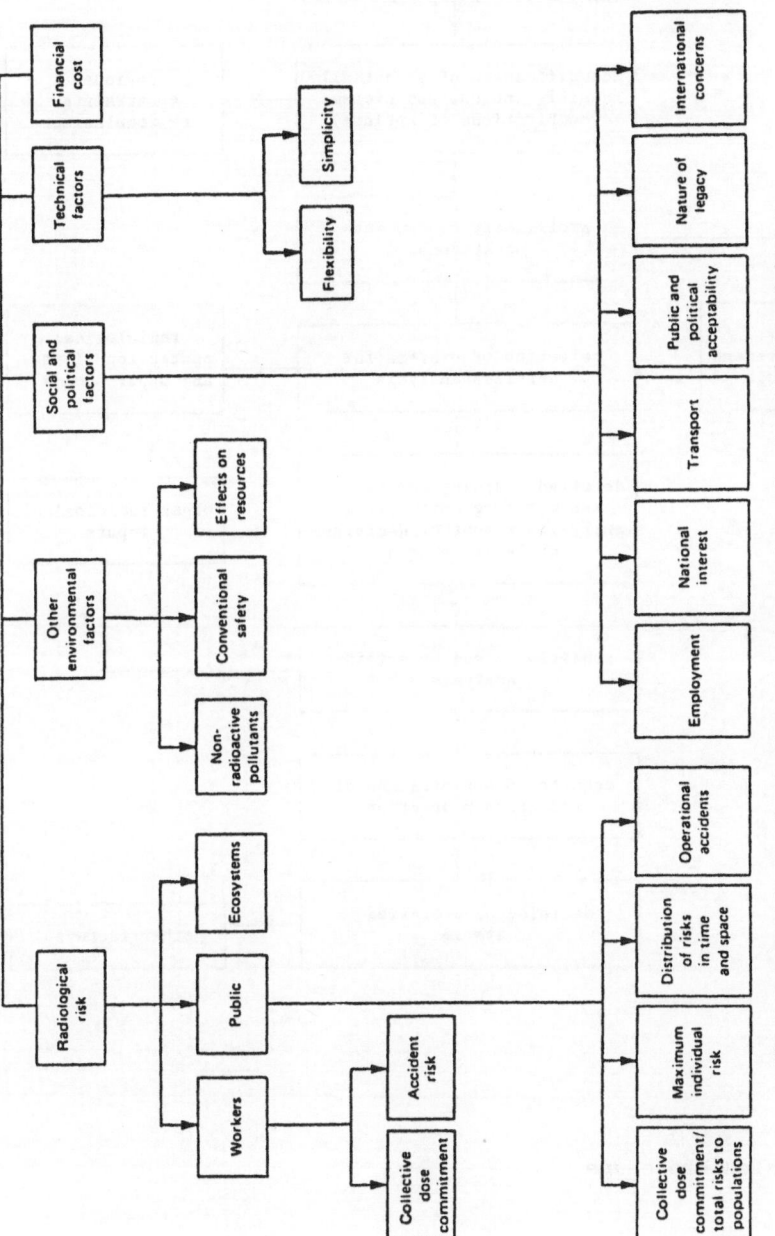

FIG. 1 Scheme of factors to be taken into account in comparing sea dumping with land-based alternatives.

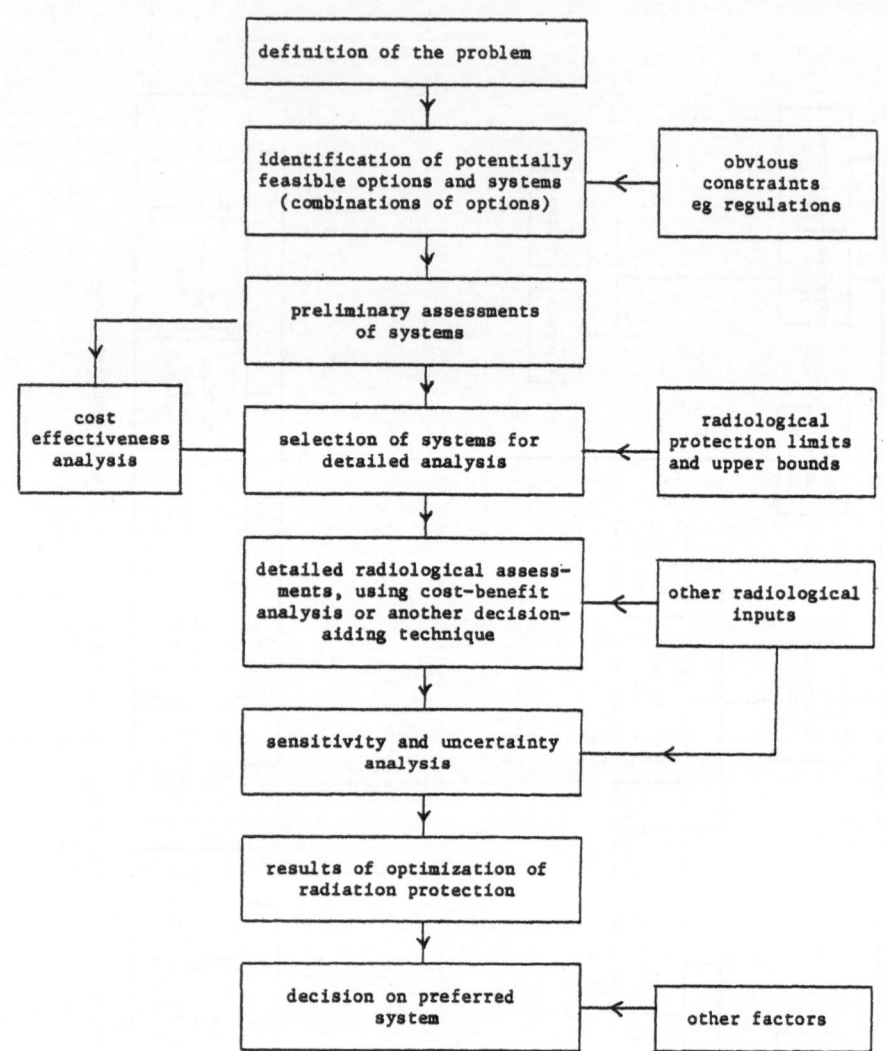

Figure 2.   Decision Procedure

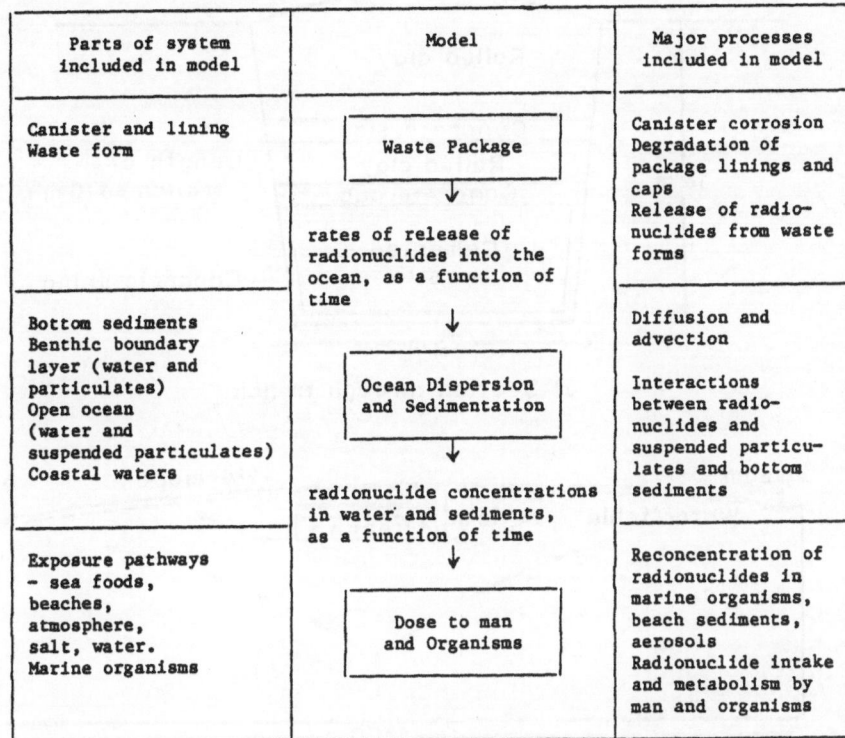

Figure 3.   Modelling Framework used in Radiological Assessment

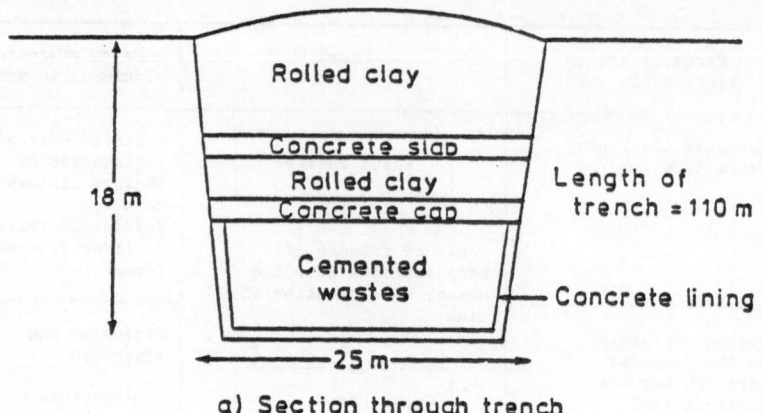

a) Section through trench

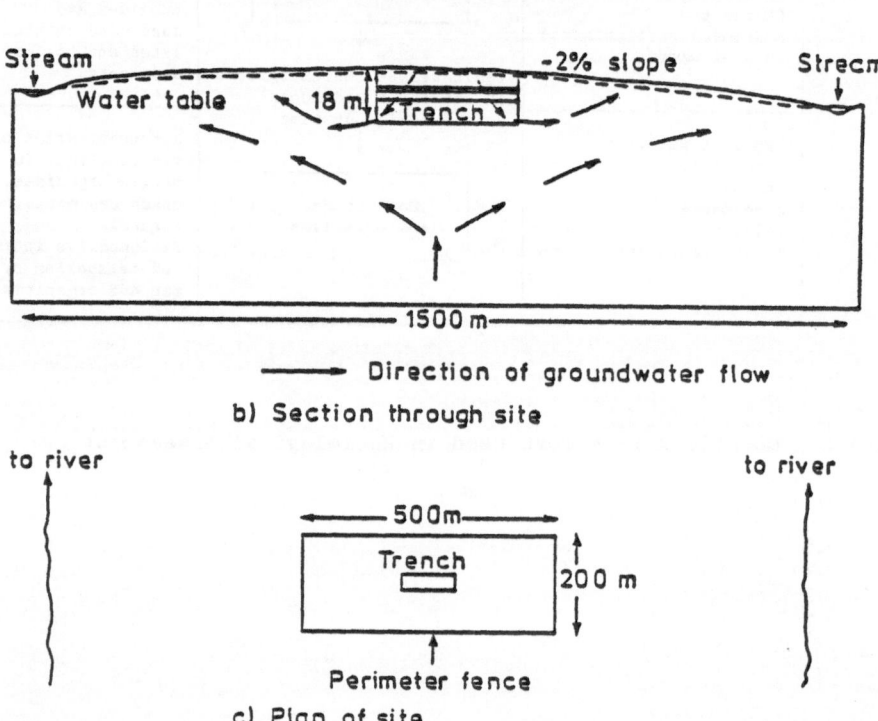

b) Section through site

c) Plan of site

Figure 4.   Design and site characteristics of fully engineered facility

REFERENCES

(1) IAEA, Convention of the prevention of marine pollution by dumping of wastes and other matter. INFCIRC/205, IAEA, Vienna, (1974).

(2) IAEA, Convention of the prevention of marine pollution by dumping of wastes and other matter. The Definition required by Annex I, Paragraph 6 to the Convention, and the Recommendations required by Annex II, Section D. INFCIRC 205/Add 1/Rev 1, IAEA, Vienna, (1978).

(3) IAEA, Environmental assessment methodologies for sea dumping of radioactive wastes, Vienna, IAEA, Safety Series 65 (1984).

(4) IAEA, Packaging of radioactive wastes for sea disposal, IAEA TECDOC-240. IAEA, Vienna, (1981).

(5) NEA, Guidelines for sea dumping packages of radioactive waste. ENEA, Paris (1979).

(6) NUREG 0945 Vol. 2, Final environmental impact statement on 10 CFR Part 61 "Licensing Requirements for Land Disposal of Radioactive Waste". US Nuclear Regulatory Commission (1982).

(7) ICRP, A compilation of the major concepts and quantities in use by ICRP. Oxford, Pergamon Press, ICRP Publication 42. Ann. ICRP, 14, No. 14 (1984).

(8) NRPB, Cost-benefit analysis in optimising the radiological protection of the public: a provisional framework. Chilton, NRPB, ASP 4 (1977). (London, HMSO).

(9) Radiological protection objectives for the disposal of solid radioactive wastes. Chilton, NRPB-GS 1 (1983). (London, HMSO).

(10) Pinner, A V, Hemming, C R and Hill, M D, An assessment of the radiological protection aspects of shallow land burial of radioactive wastes. Chilton, NRPB-R161 (1984). (London, HMSO).

(11) Gelder, R et al, Radiation exposure resulting from the normal transport of radioactive materials within the United Kingdom. Chilton, NRPB-R155 (1984). (London, HMSO).

(12) NEA, Review of the continued suitability of the dumping site for radioactive waste in the North-east Atlantic. (To be published).

(13) NEA, Long term radiation protection objectives in radioactive waste disposal (Expert Group report). Paris, NEA (OECD) (1984).

(14) Pinner, A V and Hill, M D, Radiological protection aspects of
     shallow land burial of PWR operating wastes. Chilton, NRPB-R138
     (1982). (London, HMSO).

(15) Smith, G M, Hemming, C R, Clark, M J and Chapuis, A-M and Garbay, H
     (CEA), Methodology for evaluating radiological consequences of the
     management of very low-level solid waste arising from
     decommissioning of nuclear power plants. CEC (to be published
     1985).

(16) NRPB, Small radiation doses to members of the public. Chilton,
     NRPB, ASP 7 (1985). (London, HMSO).

NUCLEAR WASTE DISPOSAL: OCEAN OR CONTINENTAL CRUST?

W.S. Fyfe
Department of Geology
University of Western Ontario
London, Canada N6A 5B7

ABSTRACT. It is suggested that nuclear waste should be stored at or near surface for periods of 100+ years by which time heat production is small. Critical criteria for choosing a host rock include: isolation from the biosphere, low chance of dramatic changes of the environment, low resource potential, low predictable permeability and freedom from large aperture fractures (finely laminated structures) diverse fine grained mineral phases, good self-sealing properties. On all such grounds, the seabed environment appears attractive. It is further suggested that possible disposal in subduction zone grabens requires careful examination. Given that soil and continental waters will become increasingly precious resources there is good reason to consider the seabed alternative.

Introduction

This meeting is a workshop and as such should provide an opportunity to discuss our concerns openly. What I wish to discuss is some of my present thinking on the problem of nuclear waste disposal, my remarks reflect a long period of study of metamorphic processes in the continental crust and a decade or so of work in collaboration with nuclear waste groups in Sweden and Canada.

The problems under discussion are of vital importance. There is a great and growing concern with the future of the global environment. The International Council of Scientific Unions is now considering a major new international initiative to study global change.[1] There is little doubt that over the next century (or less) the human population presently increasing by 80 million annually will move to ten billion - our continental surface will become crowded and continental water and soil will become precious resources. But it is also certain that if there is to be an approach to equality, tranquility and creative opportunity for man, global energy production must increase by three to five times the present.

At this time, nuclear fission, coal combustion and biomass conversion appear to have the potential for such needs for hundreds of years. Major trials of such options are in progress (nuclear in

545

G. Kullenberg (ed.), The Role of the Oceans as a Waste Disposal Option, 545–550.
© *1986 by D. Reidel Publishing Company.*

France, coal in China, biomass in Brazil). But there are concerns.
The use of biomass is hardly an option given the present world
situation for food and fibre and soil. And fossil carbon combustion
is clearly dubious until we understand and can reliably predict the
climatic impact. Most informed groups who have considered such
problems conclude that use of nuclear processes will expand to meet a
large part of the necessary global need.

The pollution associated with expanded use of nuclear energy must
be assessed and monitored. Already there have been great improvements
in the pollution associated with uranium mining and new concepts for
further improvement are being developed and tested. For power reactor
wastes there are already systems which appear reasonable and have
survived open critical scrutiny (e.g. the Swedish copper-bentonite
concept. In a review of this concept[2], a panel of the U.S.
National Research Council conclude "The KBS research has not only
achieved its purpose of showing that radioactive waste can be disposed
of in Sweden with reasonable assurance of safety for at least one
million years, but it has provided the world with a wealth of basic
data on the corrosion of copper, on the movement of groundwater in
fractured rock, on the properties of bentonite, and on the many
factors that influence radionuclide migration."

Compared to other wastes associated with energy production,
nuclear reactor wastes have several features which make pollution
control relatively simple. First, the waste volumes are very small.
Second, any form of leakage can be detected and monitored with great
precision. Third, the pollutants have finite life and over long time
periods will contribute little to the natural radiation background.

In developing any disposal system the leading consideration is
long term containment. Initial cost must be a secondary consideration
and over the past decades we have seen many examples of the cost
associated with clean up of inadequate waste disposal systems. There
have been strong political pressure to find final solutions to the
nuclear waste disposal problem. The Swedish experience shows that
viable disposal systems exist now. But I think that most would agree
that better solutions may evolve. This places us in the sound
position where alternatives, or even quite new approaches can be
rigorously assessed.

All waste disposal systems involve an array of highly coupled
parts (what rock, where, how to package, how to engineer, etc.), any
of which may have priority in the decision flow. Thus, the research
planning must be as flexible as possible. In these notes I wish to
mainly consider the major question, "where," ocean crust or continental crust?

When?

Spent nuclear fuel produces considerable energy for decades but the
energy production falls rapidly with time (typically an order of
magnitude per decade). Studies of heat production for various storage
designs consider significant thermal perturbations of the environment.[3] Any such perturbations introduce an array of complex

problems (thermal cracking, fluid flow, cavitation, enhanced diffusion) depending on the nature of the host medium. All such problems are reduced by orders of magnitude given long term (100 year +) near-surface storage.

Technology for long term surface or near-surface storage involves well established technologies. Work in Canada and Sweden shows that massive concrete can provide a reasonable host. And there is no question that special concretes can be developed with special matrix phases for retention of problem nuclides and to provide the most desirable physical properties (density, thermal isotropy, etc.). The secure, long term storage of delicate materials was a technology well known to ancient societies, and can certainly be improved today. Long term storage also has the obvious advantage that should new uses be found for such materials, they are available. Given that secure, monitored, long term storage is possible and desirable, it follows that there is ample time to rigorously explore alternatives for ultimate disposal. Any form of "best-guess" or quick alternative is, without doubt, the most irresponsible decision influencing future generations and perhaps ultimately, the most costly.[4] In what follows I will consider that wastes are stored at surface for 100+ years, and that thermal perturbations can be ignored. I will further assume that the packaging will be such that waste form containers and matrix will survive in any chosen environment for at least thousands of years.

Where?

When we consider where to dispose of nuclear waste a number of points are obvious:
(a) The sites should be remote from centres of high human population density and should occur at sites of minimal biomass production.
(b) The sites should be where there is minimal possibility of major natural change or catastrophic change (e.g. volcanism, earthquakes, meteoric impact, ice age fluctuations, heavy erosion, sea level change, high regional stress patterns and induced seismicity, etc., high regional heat flow, deep groundwater flow, new fracture development, etc.).
(c) The sites should be in rocks where there is minimal potential for resource development. Here it should be noted that regions of present desert or cold climates may be subject to change and biomass invasion in times of the order of a few 100 years.[5]

When such very general considerations are listed, sites in the ocean crust appear to have great potential.

Three possible major types of site must include: stable continental regions, stable marine environments and possibly subduction environments. At this time research has concentrated on continental situations but given a longer view, such a priority may require careful reassessment. I think it must be remembered that the problem involves the energy structure of the future, probably the greatest area of technological investment of the next 100 years.

For silicate based waste forms, glass or crystalline, much
remains to be learned before ideal materials can be designed. Most of
our geological experience of mineral alteration is based on situations
of rain water penetration into rocks. Because of the initial purity
of this solvent, all minerals will dissolve to some extent. But for
marine waters, the residue of a mineral solution and precipitation
process, the picture is more complicated. I have been most impressed
by differences in the weathering rates of basalts in the subaereal
versus submarine environments in Hawaii. I am most impressed by the
stability of basalt glass which is extremely common in the Cyprus
ophiolite.[9] As sea water is closer to mineral saturation than
near surface waters, solution rates of appropriate mineral phases
should be lower in sea water.

The Subduction Option

Over the past few years there have been spectacular advances in our
understanding of the subduction process[10] and there is no doubt
that our knowledge of this process will increase by orders of magnitude over the next decade. It now seems clear that when the oceanic
lithosphere bends or descends into the mantle, major horst and graben
topography is created by brittle failure of the upper portions of the
plate.[4] In certain situations there is now little doubt that
light sediments are carried to depth, trapped in the rough surface of
the slab. Recent studies of metamorphic rocks in subduction environments clearly shows that light materials can be carried to depths of
100 km before return to the surface (e.g. the discovery of the high
pressure SiO_2 polymorph, coesite, in alpine rocks).[11]
        In a sense, subduction is the great natural disposal system of
surface debris and the recycling times are of the order of 10^{7-8}
years. But there are problems. During subduction fluids flow up
these thrust surfaces and in some domains extreme deformation and
chemical change occurs.[12] But there are also rocks (eclogite-
pillow lavas) that show little deformation or chemical change.[13]
The fluid mass is quantifiable and careful selection of sites (e.g.
the bottom of grabens) may provide sites where water/rock ratios are
low and deformation minimal. It should also be stressed that as depth
in the mantle increases, the host mantle environment becomes a fluid
absorber. Obviously, there is need for more detailed observations but
the possibility of using such zones should not be dismissed.

Targets for Study

At this time the major effort in nuclear waste disposal technology
involves disposal in continental materials. But there is obvious need
to develop the seabed option. Given that interim storage for periods
of 100 years + is desirable, the time is clearly available to carefully investigate the variety of sites available in ocean environments.
        But particular attention should be given to topics such as:
(1) Water circulation dynamics in seabed situations. Does slow
     thermal convection occur in old ocean crust?[14]

Ideal Rock

Assuming that the geologic barrier is of importance, the factors which
contribute to decisions on an ideal rock will include:
(a) The rock must have a low permeability and a permeability that is
    predictable within narrow limits. The rock must not be subject
    to the development of large aperture fractures (i.e. thinly
    laminated units). In general "hard rocks" are the least predict-
    able.[6]
(b) The medium should have good properties for dispersion - not
    focussed flow.
(c) The constituent minerals should have large surface areas and a
    wide array of cation-anion exchange sites. A diverse chemistry
    and mineralogy is desirable. A layered rock with diverse
    oxidation-reduction environments is desirable to cope with
    species with solubility sensitive to reducing or oxidizing
    conditions.
(d) The rock should have good self-sealing mechanisms to follow
    engineering or other disturbances (a low strength material).
    In a general way when such factors are considered,[7] relative-
ly soft marine sediments appear to have many desirable features and
with their array of clay minerals, zeolites, metal oxide phases,
sulphides and phosphates and flow regimes, are probably unsurpassed by
common igneous and metamorphic rocks. It is perhaps worth noting that
there are few minerals with an anion-cation complexity which can match
marine phosphates (Cl-F-I, U-Th-Sr-lanthanides, etc.). As stressed by
many workers, deep marine sediments are among the most homogeneous,
and hence predictable rocks on earth.[7]

Waste Preparation

There are two extreme views on the importance of waste preparation for
disposal; the main barrier to release can be considered to be the host
rock or the waste package should itself constitute a major long term
barrier. I consider the Swedish concern for the latter of great
importance for given a robust engineered barrier, influences from
thermal and emplacement effects, become less severe. Two examples of
non-corrosive containers given attention in the Swedish program
include copper and Al_2O_3-based ceramics.[8] In certain geologic
environments, both can be shown to have life times in the million year
range. While certain ceramics are extremely unreactive, their
mechanical properties probably make them less predictable than more
ductile metallic materials.
    What I would stress here, is that because of the spectacular
advances in surface corrosion technology, realistic testing of such
rate processes is possible. Via such modern techniques (ESCA, SIMS,
etc.), surface interface processes down to rates of 100A/a (1 cm per
million years) can be accurately studied so that material testing on
appropriate time scales is now possible. For copper in continental
environments (e.g. the billion year old native copper deposits of
Michigan) excellent natural analogues are available for study.

(2) The stability of waste forms, corrosion of possible waste
     containers in various sea floor environments using modern surface
     analysis techniques.
(3) Glass stability in sea floor environments.
(4) The nature of the subduction process and the nature of materials
     that are present in subduction environments.
(5) The types of sedimentary sequences which may be most desirable
     for nuclide retention.
(6) The types of emplacement systems that may be utilized for deep
     seabed disposal (e.g. by drop penetration).[15]

Conclusion

Given that there will be greatly expanded use of nuclear fission
energy over the next 100 years and given that spent nuclear fuel will
be increasingly used in breeder reactors, long term interim storage is
desirable. Such storage virtually eliminates the thermal problems
associated with waste disposal. When many (if not most) of the factors
in waste isolation are considered, disposal in marine sediments appears to
be a potentially more logical place than in continental rocks.

REFERENCES

1. Malone, T.F. and Roederer, J.G. 1985. Global Change, Cambridge
    University Press. 357 pp.
2. National Research Council U.S.A. 1984. Review of the Swedish
    KB5-3 Plan. Natl. Acad. Press, Washington. 69 pp.
3. Wang, T.S.Y., Tsang, C.F., Cook, N.G.W. and Withespoon, P.A.
    1981. J. Geophys. Res., 86, pp. 3759-3770.
4. Fyfe, W.S., Babuska, V., Price, N.J., Schmid, E., Tsang, C.F.,
    Uyeda, S. and Velde, B. 1984. Nature, 310, pp. 537-540.
5. Campbell, P. 1984. Nature, 307, pp. 688-689.
6. Freeze, R.A. and Cherry, J.A. 1979. Groundwater. Prentice-Hall
    Inc., New Jersey. 604 pp.
7. Hollister, C.D., Anderson, D.R. and Heath, G.R. 1981. Science,
    213, pp. 1321-1326.
8. Svensk Karn bransleforsorjning A.B. 1984. Technical Report
    83-77. 155 pp.
9. McCulloch, M.T. and Cameron, W.E. 1983. Geology, 11, pp.
    727-731.
10. Hilde, T.W.C. and Uyeda, S. 1983. Tectonophysics, 99, pp.
    85-400.
11. Chopin, C. 1984. Contrib. Mineral. Petrol, 86, pp. 107-118.
12. Fyfe, W.S. and Zardini, R. 1967. Am. J. Sci., 265, pp.
    819-830.
13. Bearth, P. 1959. Schweiz Mineral. Petrog. Mitt., 39, pp.
    267-286.
14. Anderson, R.H., Hobart, M.A. and Langseth, M.G. 1979. Science,
    204, pp. 828-832.
15. Freeman, T.J., Murray, C.N., Francis, T.J.G., McPhail, S.D. and
    Schultheiss, P.J. 1984. Nature, 310, pp. 130-133.

DISPERSAL OF PARTICULATE WASTE ON AN OPEN CONTINENTAL SHELF

G. T. Csanady
Woods Hole Oceanographic Institution
Woods Hole, Massachusetts   02543 USA

ABSTRACT.   Recent work on hazards associated with ocean dumping points
to the overriding importance of particulate constituents of waste, the
prototype example being the solid content of sewage sludge. Near a
release site on an open continental shelf the background concentration
of sludge particles (that <u>outside</u> freshly released batches or plumes of
waste) may well build up to a level comparable to the concentration of
those naturally existing fine particles which are products of marine
life.
    Bulk parameters characterizing the oceanic dispersal of fine
particles have been estimated for the relatively well explored continental shelf of the Mid-Atlantic Bight. A model of sludge release
nearshore, at a rate equal to the present rate practiced off New York,
shows that the sludge particle background concentration is comparable
to the observed fine particle concentration in the coastal waters of
this region. A sizable fraction of the fine particles present near the
release site are thus likely to have originated from the sludge release.

INTRODUCTION

    Much recent work on ocean dumping, or on the "assimilative capacity"
of the ocean, has pointed to the importance of <u>particulate</u> material
discharged into the ocean, the prototype example being sewage sludge.
There are several reasons:
    1) Toxic substances, such as certain heavy metals, either enter the
ocean already contained in particles, or rapidly become bound to them (i).
    2) Where flow conditions permit, particles deposit on the seafloor
and form permanent accumulations. Demonstrated adverse effects of ocean
dumping of municipal sludge have all been associated with such waste
particle deposits, e.g., in the New York Bight (ii) or in the Southern
California Bight (iii).
    3) Typical sludge particles are in their gross physico-chemical
properties indistinguishable from certain naturally occurring fine
particles and are likely to be ingested by a variety of marine organisms.
    4) A corollary of the above is that the exposure of certain benthic

551

G. Kullenberg (ed.), The Role of the Oceans as a Waste Disposal Option, 551–561.

and pelagic species to the toxic constituents of particulate waste is proportional to the fraction of waste particles in the total particulate load present in the water column or on the seafloor (as recently formed sediment) and not to their concentration in the water column.

In the light of the above, a more realistic measure of environmental impact of, say, sludge discharge, would be the fraction of sludge particles in the total particulate load — rather than the concentration of sludge particles in seawater. At the very least, the quantity of natural organic particles present in seawater at the location of sludge discharge provides a yardstick by which to judge whether the quantity of waste particles present is a little or a lot.

Although much of what is said below applies to all particulate waste discharged into the ocean, illustrative examples will all relate to sewage sludge.

DISPERSAL BY VARIABLE CURRENTS

Whether sludge particles are released in repeated dumps from a barge, or continuously from a pipeline, once in the ocean they are advected by variable ocean currents and mixed by turbulence. An analysis of the process reveals two classes of potential adverse impact (iv, v): one, impact associated with the "initial" dilution of sludge, and two, impact of the large-scale "background" concentration field.

Initial dilution is achieved by distributing the sludge over some large discharge region and by a method of release calculated to enhance mixing. The concentration of sludge achieved at the end of this initial mixing phase is important not only in the immediate vicinity of the dump-site, but also in a larger region reached by relatively swift coherent motions, such as tidal or wind-driven currents. Points within this region are visited by relatively fresh batches of waste-seawater mixture, which have not had time to mix with much more seawater than at the end of the initial mixing phase. While thus extending the range of exposure to moderately diluted waste, the same swift coherent motions also distribute the total waste load over an "extended" discharge region and are thereby instrumental in reducing the maximum background concentration. The problem with the latter arises because the speed of long-term currents is usually much less than that of short-term tidal or wind-driven motions, so that flushing by the mean flow is much less effective than by the fast short-term motions.

Exposure to moderately diluted waste close to the release site can be quantified in various ways, notably by means of the "visitation frequency" distribution. Methods of calculating such distributions from current statistics have been discussed elsewhere (iv), (v), and (vi). In the present article, further attention will be confined to the modeling of the background concentration field of a source of particulate waste, which for simplicity and concreteness will be taken to be sewage sludge. The particles introduced will be regarded as conservative, certainly an oversimplification. The quantity released will be taken to be the sludge release rate off New York, as given by Gross (vii), and the oceanic environment into which the release takes place will be supposed to have characteristics similar to the continental shelf in the Mid-Atlantic Bight. The background concentration field of particles produced by the sludge release will be compared with the concentration of suspended particles observed in the region. The principal objective is to demonstrate how the various bulk quantitative parameters characterizing the large-scale mixing process enter into the determination of the concentration of sludge particles, expressed as a fraction of total particulates present, at various locations on the continental shelf.

BACKGROUND CONCENTRATION OF FINE PARTICLES

The concept of a background concentration field arises from splitting the total field (of concentration or flux) produced by a continuous source into two components, due respectively to "young" (recently released) or "old" batches of contaminant (iv). The background concentration field is that of the old batches. Close to the release site the background concentration is what one observes outside plumes of fresh contaminant; far away it is the total long-term mean concentration. The simplification achieved by considering the field of old batches of contaminant separately is due to the fact that the motion of such batches, viewed in suitably long time steps, can be regarded as random walk, and described by the diffusion equation.

Over some continental shelves fine particles — those that form mud when deposited, i.e., particles less than about 60 μm in diameter — are held in suspension most of the time by turbulence in the relatively fast tidal and wind-driven currents. With some exceptions ("mudholes") this is the case for the entire Mid-Atlantic Bight. The population of these fine particles is primarily generated by biological activity (viii),(ix) over the shelf, kept in the water column through resuspension events associated with storms, to be eventually lost at the shelf edge and to be deposited over the upper and middle continental slope (x),(xi).

A large fraction of sludge particles also falls into the size range of mud. This fraction may be expected to behave as natural fine particles do. Sludge particles substantially heavier are likely to deposit close to the source. Modeling their fate requires a consideration of the behavior of young contaminant batches which will not be pursued here.

Non-depositing fine particles over the continental shelf reside predominantly in the bottom portion of the water column. A realistic assumption is that most of them are contained in a bottom mixed layer of about 30 m depth (x). Within this layer they execute a random hop-scotch motion punctuated by temporary deposition and resuspension events (xii). The resulting concentration field can be calculated from the two-dimensional diffusion equation, given a long-term mean alongshore advection velocity U , an effective horizontal diffusivity K_H , a mass transfer coefficient quantifying particle loss at the shelf edge, and the aggregate intensity and spatial distribution of source strength. It should be noted that an extended source strength distribution must be used in calculating the background concentration field, parameterizing the coherent motions of young batches of contaminant which initially spread out the waste. The coast may be supposed to be a reflecting boundary.

BULK PARAMETERS

   Although only a few bulk parameters are necessary for modeling the
background concentration field, they quantify some complex and not well
understood physical processes. Their choice in most shelf seas has to
be made on the basis of fragmentary information. The east coast continental shelf of the U.S. has been well explored in the course of the
past decade or two so that in this region more reliable estimates can be
made than in most other places.
   The effective horizontal diffusivity K_H has been estimated for
the mixing of freshwater in the Mid-Atlantic Bight by Ketchum and Keen
(xiii) at between 100 and 300 $m^2 s^{-1}$. The diffusivity applying to the
long-term random motion of fine particles may well be somewhat different,
however, because these particles are concentrated in the lower part of
the water column and may not be mixed, e.g., by shear diffusion associated with current shear in the vertical. An independent estimate, however, can be arrived at by separately assessing particle flux and cross-
isobath particle concentration gradient.
   The flux is the more difficult quantity. At the shelf-edge, con-
vergent estimates of freshwater flux, slopewater transfer shoreward, and
nitrate transfer shoreward yield a volume exchange rate per unit length
of coastline of 0.2 $m^2 s^{-1}$(xiv). As Manheim et al. (viii) point out,
fine particle concentration variations on the continental shelf are in a
first approximation attributable to simple binary mixing with slopewater,
which enters at the shelf edge. Slopewater enters near the bottom (at
the rate of 0.2 $m^2 s^{-1}$, as already pointed out) and most of it is likely
to be effective in diluting the particle-laden bottom layer. Thus the
particle flux may be estimated to be 0.2 $m^2 s^{-1}$, times the concentration
difference across the shelf-edge front between shelfwater and slopewater.
   A map of particle concentration in the New York Bight has been given
by Biscaye and Olsen (x). A typical cross-isobath concentration distribution extracted from this map is shown here in Fig. 1. Across the
shelf edge front (near the 100 m isobath) the concentration drops
rapidly: the estimated cross-front difference is about 0.1 mg ℓ^{-1}
(10^{-4} kg m^{-3}). The estimated flux is therefore $2 \cdot 10^{-5}$ kg m^{-1} s^{-1} ,
or about 2 kg per day per meter of shelf-edge front length.
   A rough check on the order of magnitude of this figure is provided
by the deposition rate it implies over the continental slope and rise.
If all these fine particles deposit in a band 100 km wide (as suggested
by the maps of Biscaye et al. (xi)), they should accumulate at the rate
of 0.7 g cm^{-2} per thousand years. According to Turekian (xv), quoted by
McCave (xii), the rate of post-glacial deposition over the continental
rise off the eastern U.S. is "greater than 3 g cm^{-2} per 1000 years".
The result is consistent with the fact that the total seaward flux
across the 1000 km long edge of the shelf of the entire Mid-Atlantic
Bight is about 20 kg s^{-1} (by the above flux estimate) while, over the
continental shelf, the alongshore fine particle transport is about
0.45 $m^3 s^{-1}$ volume transport (xvi) times cross-shelf average concentration or at least four times greater than the flux across the shelf edge.
Near Cape Hatteras all of this shelf water is swept out seaward, and with
it presumably its fine particle load. The shelf water flows along the

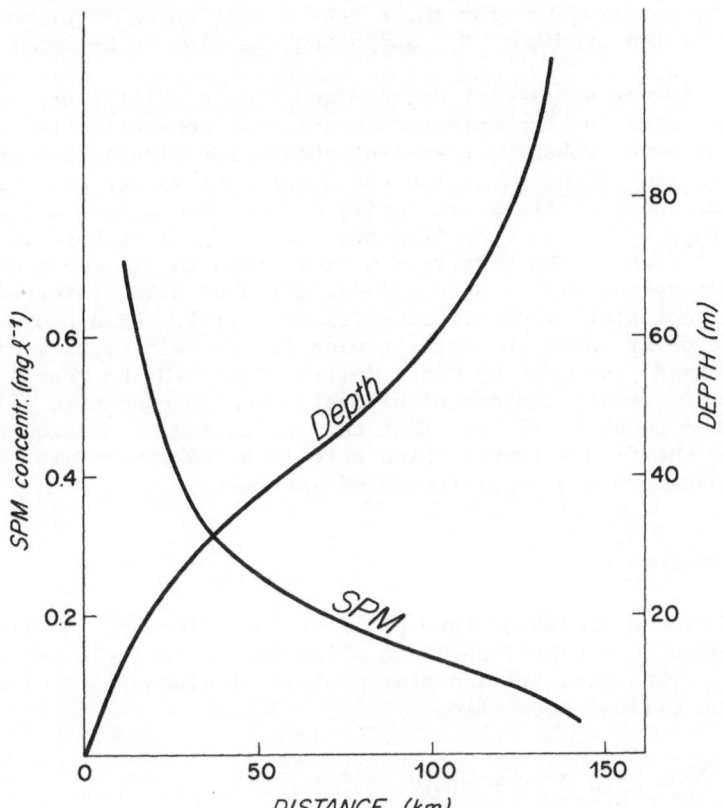

Figure 1. Concentration of suspended particulate matter (SPM) over the continental shelf just sough of New York, in function of distance from shore, derived from data of Biscaye and Olsen (x). The distribution of depth is also shown.

inner edge of the Gulf Stream (xvii) and the fine particles contained in it presumably fall to the seafloor over the slope and rise, to produce most of the recent deposit.

Returning to the estimation of diffusivity over the shelf, Fig. 1 shows a concentration gradient G of about $0.2 \cdot 10^{-5}$ g m^{-4} on the outer shelf. The diffusivity is now calculated from

$$K_H = \frac{F}{hG} \tag{1}$$

where F is particle flux, $2 \cdot 10^{-5}$ kg m^{-1} s^{-1}, as estimated above, h is depth of the particle bearing layer, taken to be 30 m, and G is the gradient just gleaned from Fig. 1. One calculates $K_H = 300$ m^{-1} s^{-1}

over the outer shelf, reducing to about $100 \text{ m}^2 \text{ s}^{-1}$ where the total depth is 30 m, about 35 km from shore (the reduction is inferred from the increase of the gradient G , supposing the flux to be constant for simplicity).

These estimates agree with Ketchum and Keen's (xiii), derived from the freshwater flux and concentration gradient. Apparently the horizontal eddies (or more probably, transient shelf-wide circulation cells) causing the mixing affect particles and freshwater in the same way.

The remaining parameters are easily chosen from direct evidence: a realistic long-term mean mass transport velocity alongshore is $U = 0.05 \text{ m s}^{-1}$, while the effective source strength distribution may be taken to be Gaussian, with a standard deviation of 3 km (inferred from nearshore current statistics off Long Island (iv)). As a prototype aggregate source strength the total sewage sludge dumping rate off New York will be used, as given by Gross (vii): $0.2 \cdot 10^9 \text{ kg year}^{-1}$ or about 6 kg s^{-1} (solid content of sludge). The release site will be chosen so close to shore (12 km) that the limitation of horizontal dispersal by the shelf-edge front (which acts as a leaky boundary) does not affect the concentration at distances of interest.

CALCULATED EXAMPLE

Let the sludge discharge take place at some distance d from the coast of an open continental shelf to which the above estimated parameters apply. The concentration distribution of sludge particles may be found from the diffusion equation

$$\text{U h} \frac{\partial \chi}{\partial x} = K_H \text{ h} \nabla_1^2 \chi + s(x,y) \tag{2}$$

where χ is concentration and $s(x,y)$ is source intensity distribution. With a mirror image term added (to satisfy the reflecting boundary condition at the coast) this will be taken to be:

$$s(x,y) = \frac{m}{2\pi \sigma^2} \exp\left(-\frac{x^2}{2\sigma^2}\right)\left\{\exp\left[-\frac{(y'-d)^2}{2\sigma^2}\right] + \exp\left[-\frac{(y'+d)^2}{2\sigma^2}\right]\right\} \tag{3}$$

where m is the total discharge rate (6 kg s^{-1}), and $\sigma = 3$ km the standard deviation, $d = 12$ km the source distance from shore, chosen to make the visitation frequency at the shore negligible. For the diffusivity this close to shore $K_H = 100 \text{ m s}^{-1}$ will be assumed.

It is convenient to use nondimensional variables defined as follows:

$$\chi^* = \frac{K_H \text{ h} \chi}{m} \qquad\qquad x^*, \ y^*, \ d^* = \frac{U}{2 K_H} \cdot (x,y,d)$$

$$\tag{4}$$

$$s^* = \frac{4\ s(x,y)K_H^2}{m\ U^2}$$

Dropping the stars on the nondimensional variables, the solution can be written down at once:

$$\chi = \frac{1}{2\pi} \int\limits_{-\infty}^{\infty}\int\limits_{-\infty}^{\infty} s(x',y')\ e^{x-x'}\ K_o\left[(x-x')^2 + (y-y')^2\right]^{\frac{1}{2}} dx'dy' \qquad (5)$$

The distance scale with the typical parameters is 4 km, so that nondimensional d is 3, $\sigma = 0.75$. These are the only two numerical parameters left, so that the maximum concentration (nondimensional) is a function of these two. The concentration scale m/K_H h is $2 \cdot 10^{-3}$ kg m^{-3}(2 mg/ℓ) which at once suggests that the concentration of sludge particles will be comparable to naturally occurring fine particle concentrations.

Figure 2 shows the source or maximum (nondimensional) concentration χ_0 versus extended source size σ (also nondimensional). For $\sigma = 0.75$ the background concentration at the source is 0.116, or multiplied by the concentration scale, 0.232 mg ℓ^{-1}, which is rather less than the natural concentration of fine particles in this region (about 0.68 mg ℓ^{-1} according to Fig. 1), but of the same order of magnitude.

Figure 3 shows the nondimensional concentration field over a 120 km long piece of coastline, and out to 40 km from shore. At the outer edge of this region concentrations of about 0.02 mg ℓ^{-1} are reached. The natural fine particle concentration according to Fig. 1 is 10 times greater. Thus the sludge particle release accounts for a sizable fraction of the total suspended particle population, over a large region of the continental shelf. The concentration of sludge particles, expressed as the fraction of the fine particles present, ranges from 10 to 30%. They are "diluted" by naturally occurring particles only by a factor of 3 to 10, whereas, as a fraction of seawater, their concentration is only 3 - 10 times 10^{-8}.

CONCLUDING COMMENTS

The above calculations ignore biological or physico-chemical transformations of the particles and the results can only be used for guidance. They show that on a shelf subject to vigorous coherent motions (σ large) the background concentrations reached are considerably smaller than in more quiescent waters, Fig.2. An additional factor is that the concentration scale varies as K_H^{-1} h^{-1}, and is smaller where current fluctuations are larger. The flushing velocity U mainly determines how wide a region is affected by a given concentration. However, for small U,

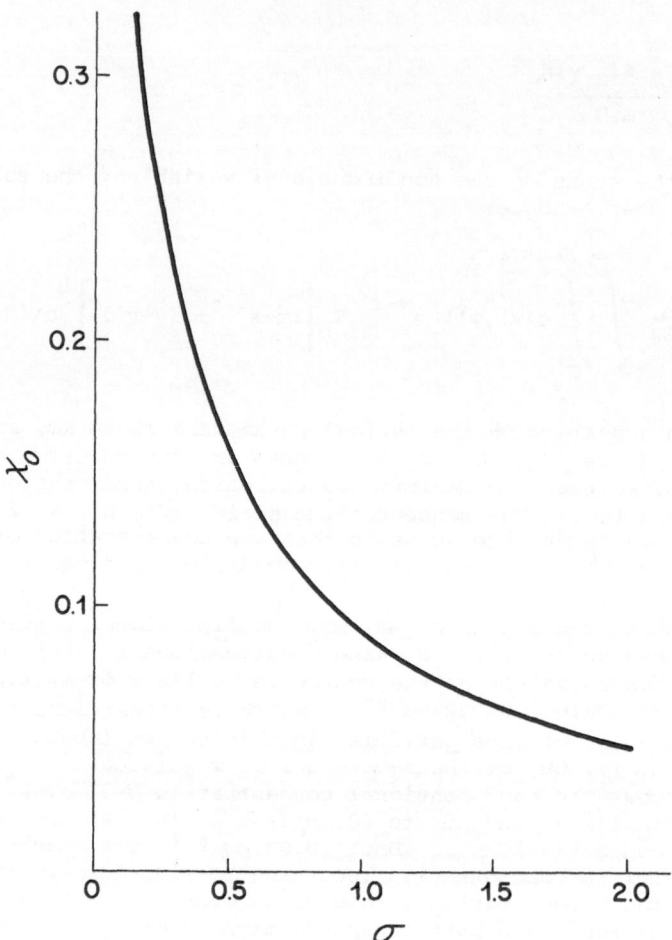

Figure 2.   Maximum (source-center) concentration versus source size,
both in nondimensional units.

the shelf-edge mass exchange becomes important so that a different
solution of the differential equation has to be calculated, subject to
a "radiation" boundary condition at shelf edge.   The significance of
the calculated background concentrations depends on the concentration of
naturally occurring fine particles: sludge particles are presumably more
readily assimilated when they only constitute a small fraction of the
total particle population.   Thus a biologically productive shelf is
presumably better able to assimilate a large sewage sludge load than an
unproductive one, perhaps a counter-intuitive idea.

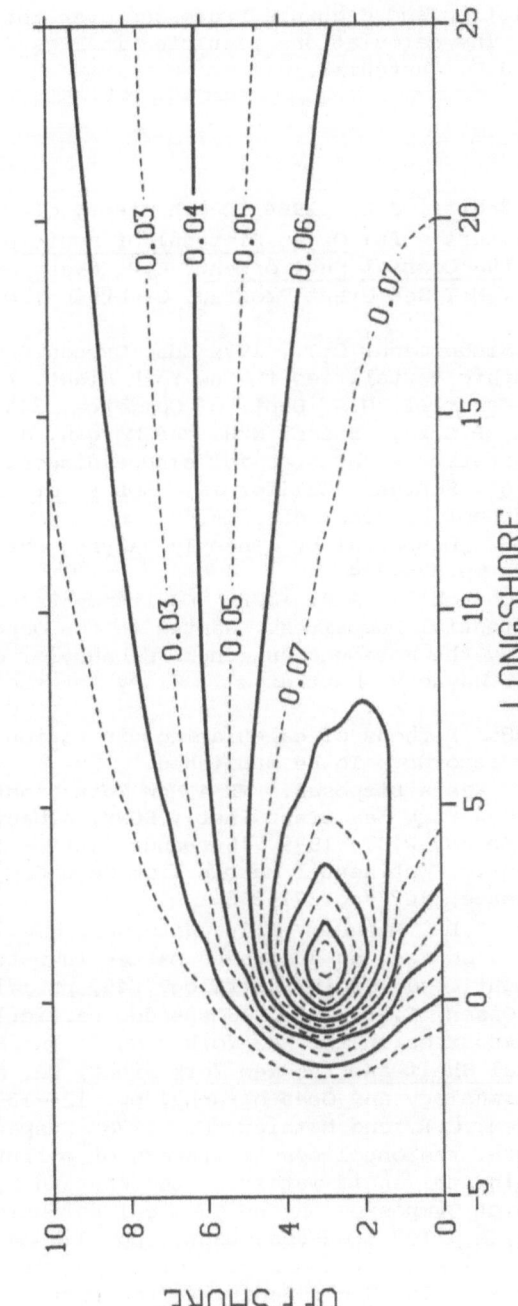

Figure 3. Contours of constant background concentration of sludge particles released 3 distance units offshore over an Atlantic type continental shelf. Under typical conditions, and for a sludge release rate of 6 kg s^{-1}, one concentration unit corresponds to 2 mg ℓ^{-1}, one distance unit to 4 km. The peak value of the concentration, occurring at the source is 0.116 units, corresponding to 0.232 mg ℓ^{-1} under typical conditions. A broad maximum occurs at the coast between 15 and 80 km down-stream of the source, with a peak value of about 0.15 mg ℓ^{-1}.

ACKNOWLEDGEMENTS

     This work was supported by the Department of Energy under a
contract entitled 'Circulation and Exchange Processes over the Continental Shelf and Slope'. The calculations resulting in Figs.2 and 3
have been carried out by J.H. Churchill.

REFERENCES

i.   Morel, F.M.M. and Schiff, S.L. 1984 'Geochemistry of Municipal
        Waste in Coastal Waters' In: <u>Ocean Disposal of Municipal Waste-
        water: Impacts on the Coastal Environment</u>. E.P. Myers and
        E.T. Harding, Eds., MIT Sea Grant Program, Cambridge, Mass.,
        pp. 251-421.
ii.   Swanson, R.L. and Sindermann, C.J. 1979 Eds. Oxygen Depletion
        and Associated Benthic Mortalities in New York Bight, 1976.
        <u>NOAA Professional Paper 11</u>, U.S. Dept. of Commerce, 345 pp.
iii.   Jackson, G.A., Koh, R.C.Y., Brooks, N.H. and Morgan, J.J. 1979
        Assessment of alternative strategies for sludge disposal into
        deep ocean basins off Southern California. <u>EQL Report No. 14</u>,
        Calif. Inst. of Technology, Pasadena, CA.
iv.   Csanady, G.T. 1983 'Dispersal by randomly varying currents'.
        J. Fluid Mech. 132, pp.375-394.
v.   Csanady, G.T. and Churchill, J.H. 1985 Environmental Engineer-
        ing Analysis of Potential Dumpsites. Wastes in the Ocean, Vol.8,
        'Processes affecting the movement and chemical behavior of wastes
        in the ocean.' Eds. Wayne V. Burt et al., Wiley-Interscience.
        In the press.
vi.   Churchill, J.H. 1985 'Methods of calculating visitation frequency
        from current meter records. To be published.
vii.   Gross, M.G. 1976   Waste Disposal. MESA New York Bight Atlas
        Monograph No. 26, New York Sea Grant Inst., SUNY, Albany, 32 pp.
viii.   Manheim, F.T. and Meade, R.H. 1979 'Suspended matter in surface
        waters of the Atlantic Continental Margin from Cape Cod to the
        Florida Keys', Science, 167, pp.371-376.
ix.   Meade, R.H., Sachs, P.L., Manheim, F.T., Hathaway, J.C. and
        D.W. Spencer 1975 'Sources of suspended matter in waters of the
        Middle Atlantic Bight'. J. Sedim. Petrology, 45, pp.171-188.
x.   Biscaye, P.E. and Olsen, C.R. 1976 'Suspended particulate con-
        centrations and composites in the New York Bight' In: <u>Middle
        Atlantic Continental Shelf and the New York Bight</u>, Ed. M.G. Gross,
        American Soc. of Limnology and Oceanography, pp. 124-137.
xi.   Biscaye, P.E., Olsen, C.R. and Mathieu, G. 1978 Suspended
        particles and natural radionuclides as tracers of pollutant
        transports in continental shelf waters of the eastern USA.
        First American Soviet Symposium on the Chemical Pollution of the
        Marine Environment, May 1977, Odessa, USSR. pp. 125-147.
xii.   McCave, I.N. 1972 'Transport and escape of fine-grained sedi-
        ment from shelf areas. In: <u>Shelf Sediment Transport: Process and
        Pattern</u>. Eds. D.J.P. Swift, D.B. Duane and O.H. Pilkey, Dowden,

Hutchinson & Ross, Inc., Stroudsberg, PA. pp. 225-248.

xiii.   Ktechum, B.H., and Keen, D.J. 1955 'The accumulation of river
        water over the continental shelf between Cape Cod and Chesapeake
        Bay', Deep-Sea Res., Supplement to 3, pp. 346-357.

xiv.   Csanady, G.T. 1985 'Circulation of slopewater'. To be published.

xv.   Turekian, K.K. 1968 <u>Oceans</u> Prentice-Hall, New Jersey, 120 pp.

xvi.   Chapman, D.C., Barth, J.A., Beardsley, R.C. and Fairbanks, R.G.
        1985 'On the continuity of mean flow between the Scotian Shelf
        and the Mid-Atlantic Bight'. To be published

xvii.   Kupferman, S.L. and Garfield N. 1977 'Transport of low-salinity
        water at the slope water-Gulf Stream boundary'. J. Geophys. Res.
        82, pp.3481-3486.

HYDRODYNAMICAL MODELLING AS A TOOL IN WASTE DISPOSAL SELECTION.
A CASE STUDY ON SADO ESTUARY.

R.J.J. Neves
Dept. of Mechanical Engineering
IST, Av. Rovisco Pais, 1096 Lisbon, Portugal

ABSTRACT. The proposal of this work is to emphasize and illustrate the importance of the knowledge of the lagrangian residual circulation as a tool in the waste disposal selection in an environment with complex geometry where the transport of pollutants is convection-driven. A model for the computation of the lagrangian residual circulation is also presented. The application to the Sado estuary (Portugal) is made and the results are compared with these of an eulerian residual model. The waste disposal analysis, concerning the location (in space and time) of the discharge of the waste water is made using the results of the model.

1. INTRODUCTION

One of the most important applications of the results of an hydrodynamical model is in the simulation of the transport of pollutants. This is usually achieved by means of the solution of a convective-diffusive equation.

In convection-driven transport of pollutants, the simulation of the concentration field involves difficulties associated with the numerical diffusion generated by the convection term solution, which can be of the same order of magnitude as the real diffusion. In these cases, a simpler and even more realistic approach, mainly in the half closed domains (like estuaries) is to do the waste disposal analysis based on the trajectories of the fluid particules, in order to minimize the residence time inside the estuary.

The lagrangian residual velocity, defined by means of the net displacement of a fluid particule during a time interval T, usually taken as the tidal period, is a good tool in the study of this kind of problems. In fact, in order to neglet diffusion, the time periods concerned must be of the order of the tidal period; therefore, the consideration of the eulerian residual velocity, defined as being the time average velocity at a point, during a time interval T, is not useful in this case since it concerns phenomena with large time scales (Cheng and Casulli, 1983; Neves, 1985a).

563

G. Kullenberg (ed.), The Role of the Oceans as a Waste Disposal Option, 563–576.
© *1986 by D. Reidel Publishing Company.*

In this work a calculation method of the lagrangian residual velocity is presented. Results show that the lagrangian velocity field is quite dependent on the origin of the integration time interval. Comparison of these results with the eulerian velocity field given by Neves (1985a) shows that the Stokes drift cannot be a relation between the two formulations, as proposed by Longuet-Higgins (1969). A similar conclusion was obtained experimentally by Dooley (1974) and Mulder (1983) and numerically by Cheng and Casulli (1982).

The results of the lagrangian model are used to identify the places where to locate the waste disposal effluents, in order to minimize the consequences on the environment and, for each point of the estuary, where convection is the main transport mechanism, to choose the more advantageous tidal moment for the waste discharge.

2. THE CALCULATION METHOD

The lagrangian residual velocity of a fluid particule y_0, that, at the initial instant of time t_0, was at the point (x_0, y_0), is defined as

$$\underset{\sim}{v}_0(x_0, y_0, t_0) = \frac{1}{T} \int_{t_0}^{t_0+T} \underset{\sim}{v}\, dt = \frac{1}{T}\, [(x_F - x_0)\, \underset{\sim}{\ell}_2 + (y_F - y_0)\, \underset{\sim}{\ell}y] \tag{1}$$

where T is the interval of time during which the fluid particule is followed (usually taken as the tidal period), $\underset{\sim}{v}$ is the instantaneous velocity and (x_0, y_0) are the coordinates of the final point of the trajectory.

Experimentally y_0 is calulated using the second equality. The numerical calculation implies the solution of the integral. Since the transient velocity is given by the tidal model at discrete points in time and space, an approximated method must be used. The development of $\underset{\sim}{v}$ using the Taylor series can be applied. In order to reduce the order of the development, the period T is divided in N intervals of length Δt. At $t = (t_0 + (n+1) * \Delta t)$, the position vector of the particle that, at t_0, occupied the position (x_0, y_0) is given by:

$$\underset{\sim}{X}(x_0, y_0, t_0 + (n+1)\Delta t) = \underset{\sim}{X}(x_0, y_0, t_0 + n\Delta t) + \int_{t_0+n\Delta t}^{t_0+(n+1)\Delta t} \underset{\sim}{v}\, dt \tag{2}$$

Developing $\underset{\sim}{v}$ around the point $\underset{\sim}{X}(x_0, y_0)$ at time $t = (t_0 + n(\Delta t))$ using the Taylor series with second order time and space accuracy, replacing the results in (2) one obtains:

$$\underset{\sim}{X}(x_0, y_0, t_0 + (n+1)\Delta t) = \underset{\sim}{X}(x_0, y_0, t_0 + n\Delta t) + \left[\underset{\sim}{v} + \frac{\Delta t}{2}\frac{D\underset{\sim}{v}}{Dt}\right]_{X(x_0, y_0, t_0+n\Delta t)} \Delta t \tag{3}$$

Since the values of the velocity are given by the transient model only at the grid points, an interpolation must be done to obtain the solution at every space point. Since the space step is small (compared with the length scale of the transient velocity variations) a linear interpolation has been used.

The instantaneous velocity is obtained using the shallow water equations (e.g. Nihoul, 1982):

$$\frac{\partial \zeta}{\partial t} + \nabla \cdot (H\underset{\sim}{v}) = 0 \tag{4}$$

$$\frac{\partial \underset{\sim}{v}}{\partial t} + \nabla \cdot (H\underset{\sim\sim}{vv}) + 2\underset{\sim}{\Omega} \wedge \underset{\sim}{v} = -g\nabla\zeta + \nu\nabla^2\underset{\sim}{v} - K|\underset{\sim}{v}|\underset{\sim}{v} \tag{5}$$

where:

    ζ is the free surface elevation over a reference level;
    H is the water column height;
    Ω is the Earth angular velocity;
    g is the gravity acceleration;
    ν is the turbulent viscosity;
    K is the bottom friction coefficient.

These equations are solved by means of an unconditionally stable semi-implicit model using a time splitting method (Neves, 1985b). The time step is divided in two; in each of them, computation is performed implicitly in one direction and explicitly in the other, both in the momentum and continuity equations. When the computation is performed implicitly, the new values of the elevations in the momentum finite difference equations are replaced by the respective expressions obtained from the continuity equation. In this way, in order to advance a time step, no iterative technique is necessary and only two tridiagonal matrices are used. This scheme has second order time accuracy. Exception done for the convective terms, all the space derivatives were calculated with second order accuracy using a staggered type grid. For the convective terms, a first order upstream formulation was preferred since this formulation is conservative and obeys the transportive property of the convection (Roache, 1972).

3. APPLICATION TO THE SADO ESTUARY

The Sado estuary is represented in the figure 1. It has a complex geometry, with non negligible curvature and irregular bathymetry. Physically the estuary can be subdivided in the two regions named region I and region II in the figure 1. Region I is the most important; it is about 5 Km wide and 20 Km long, communicating with the sea by means of a 2 Km wide mouth. The average depth is about 10 m, the maximal depth being greater than 35 m. Sand banks divide this region, downstream, in two channels. Region II is like a sub estuary the river crosses before reaching region I; it is about 15 Km long and 700 m wide, the average depth being about 1.5 m. Mainly due to its

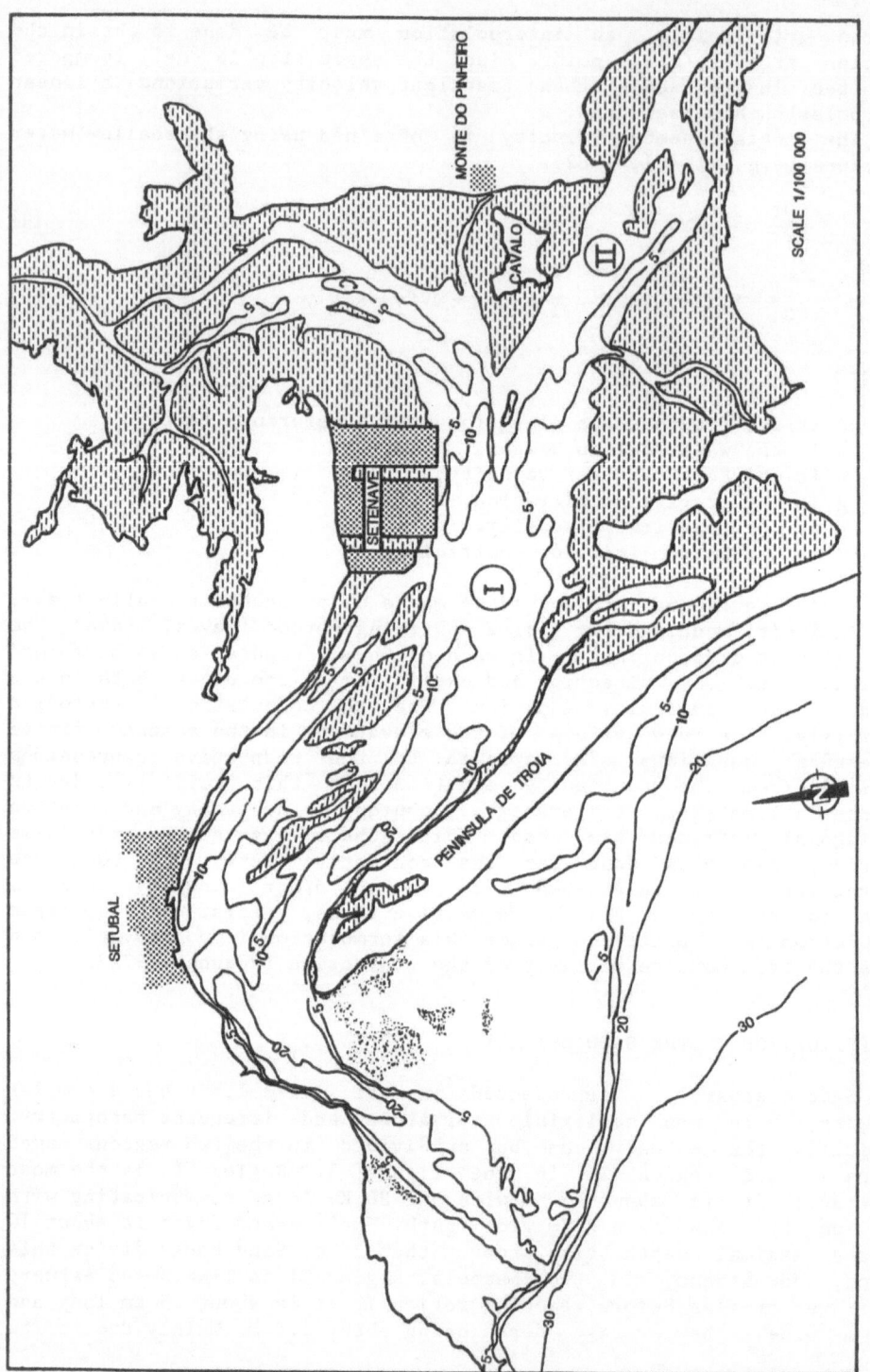

Figure 1 : Sado's estuary. Region II is the region of river's influence.

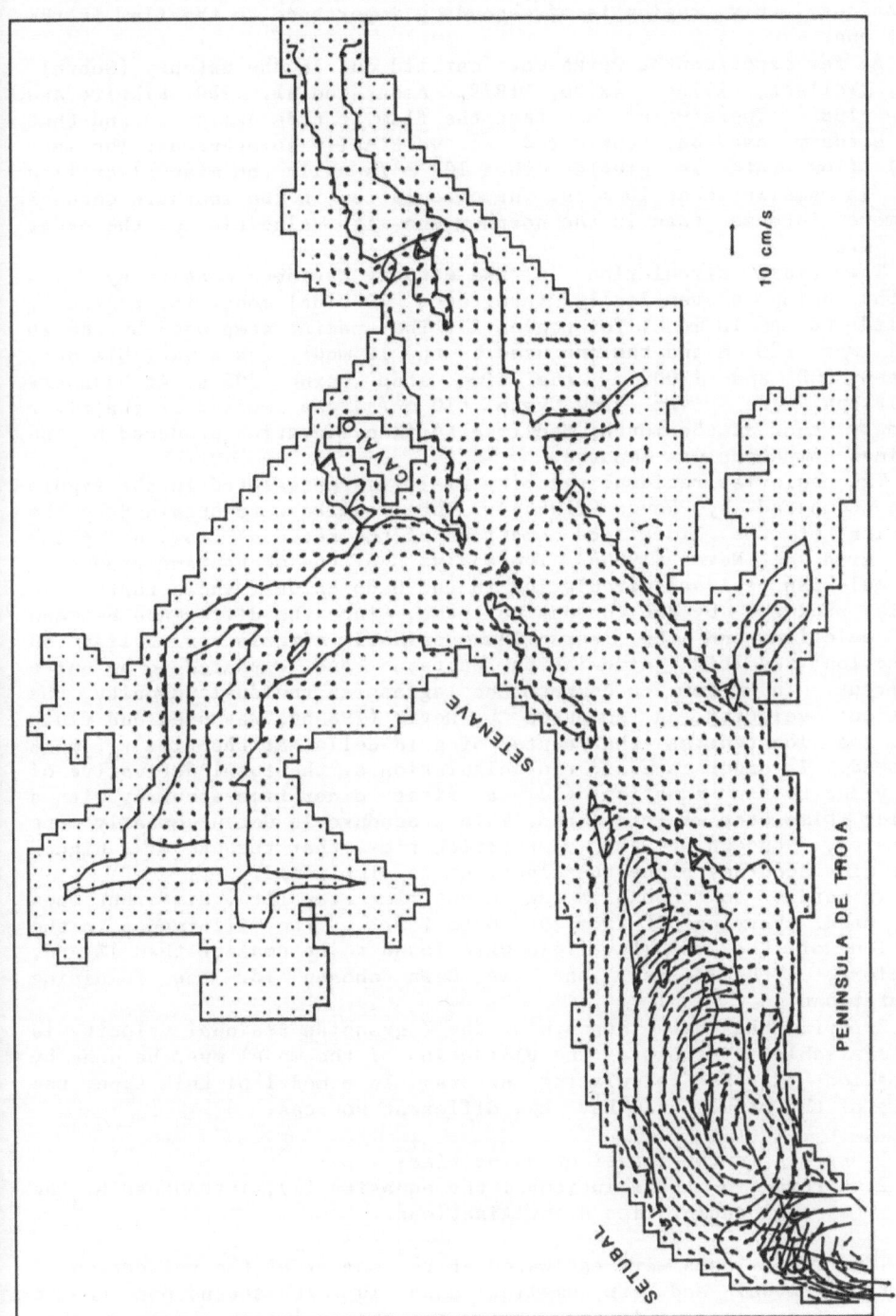

Figure 2 : eulerian residual circulation in the Sado estuary induced by the M2 tidal component as computed by Neves (1985a) using a residual circulation model.

dimensions, this region is of secondary importance to the flow in the first one.

A few experimental works were carried out in the estuary (Sobral, 1977; Wollast, 1978a, 1978b, 1979; Ambar and al. 1980; Ribeiro and Neves 1982). These works show that the flow is tide dominated and that the estuary can be considered as vertically homogeneous. The mean tidal flow rate is greater than 10^3 m^3/s while the mean river flow rate is smaller than 10 m^3/s. The circulation in the southern channel is more intense than in the northen one with velocities of the order of 1 m/s.

The tidal circulation in the estuary has been modeled by Neves (1985b) using a vertically integrated 2D tidal model for region I, coupled to a 1D model for region II. The spatial step used by the 2D model was 250 m and the one used by the 1D model was a variable one, between 500 and 1200 m, the time step being 298 s. As boundary conditions, the river flow rate (10 m^3/s) was imposed at the river boundary and, at the mouth, the free surface elevation produced by the M2 tidal component was imposed.

The eulerian residual velocity field is represented in the figure 2 as computed by Neves (1985a). The results were obtained by the solution of the linearised equations of the residual flow, using the same grid as Neves (1985b). The analysis of the mechanisms producing the eulerian residual circulation in the Sado estuary shows that it is mainly produced by the convective terms. Since the difference between the eulerian and the lagrangian residual circulation relies on convection resulting circulation patterns are expected to be quite different. In order to compute the lagrangian residual velocity, the transient velocity as computed by Neves (1985b), was used and fluid particles located at the center of grid cells, at the time t_o, were followed. In order to avoid the calculation of the total derivative of the velocity in equation (3), a first order time accuracy with a smaller time step was preferred. This procedure is not necessarly more expensive, concerning the computation time, that the use of a higher order time accuracy since the algorithm is simplified.

In order to choose a convenient time step (Δt), different ones were used. Varying Δt from 298/10 to 298/3 s, the differences in the solution of the velocity field were found to be smaller than 1% and, therefore, the last one has been chosen for the remaining computations.

Experimental data concerning the lagrangian residual velocity is not available. Therefore the validation of the model musc be done by estimation of the calculating errors. In a model of this type, the errors of the results can have two different sources:

1. Errors in the transient velocities;
2. Errors in the solution of the equation (2), introduced by the spatial and/or time discretisations.

The first ones were estimated at the moment of the validation of the tidal model and are smaller than 10%. The second ones can be inferred from the differences between the solution obtained using

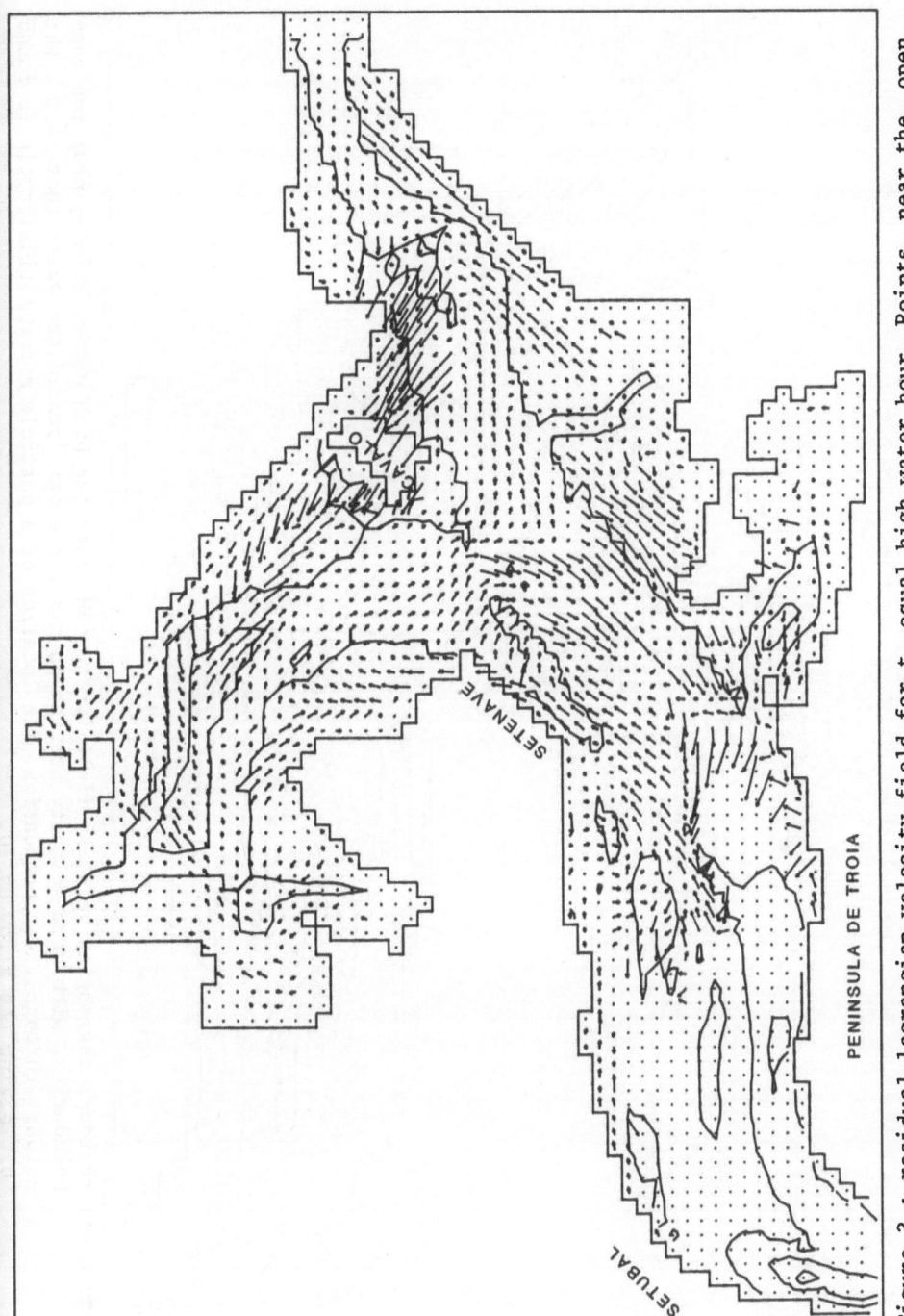

Figure 3 : residual lagrangian velocity field for t_0 equal high water hour. Points, near the open boundaries, with zero velocity are the points where was located the water that left the estuary through those boundaries. The location of a particle after a tidal cycle is found multiplying the arrow by 10.

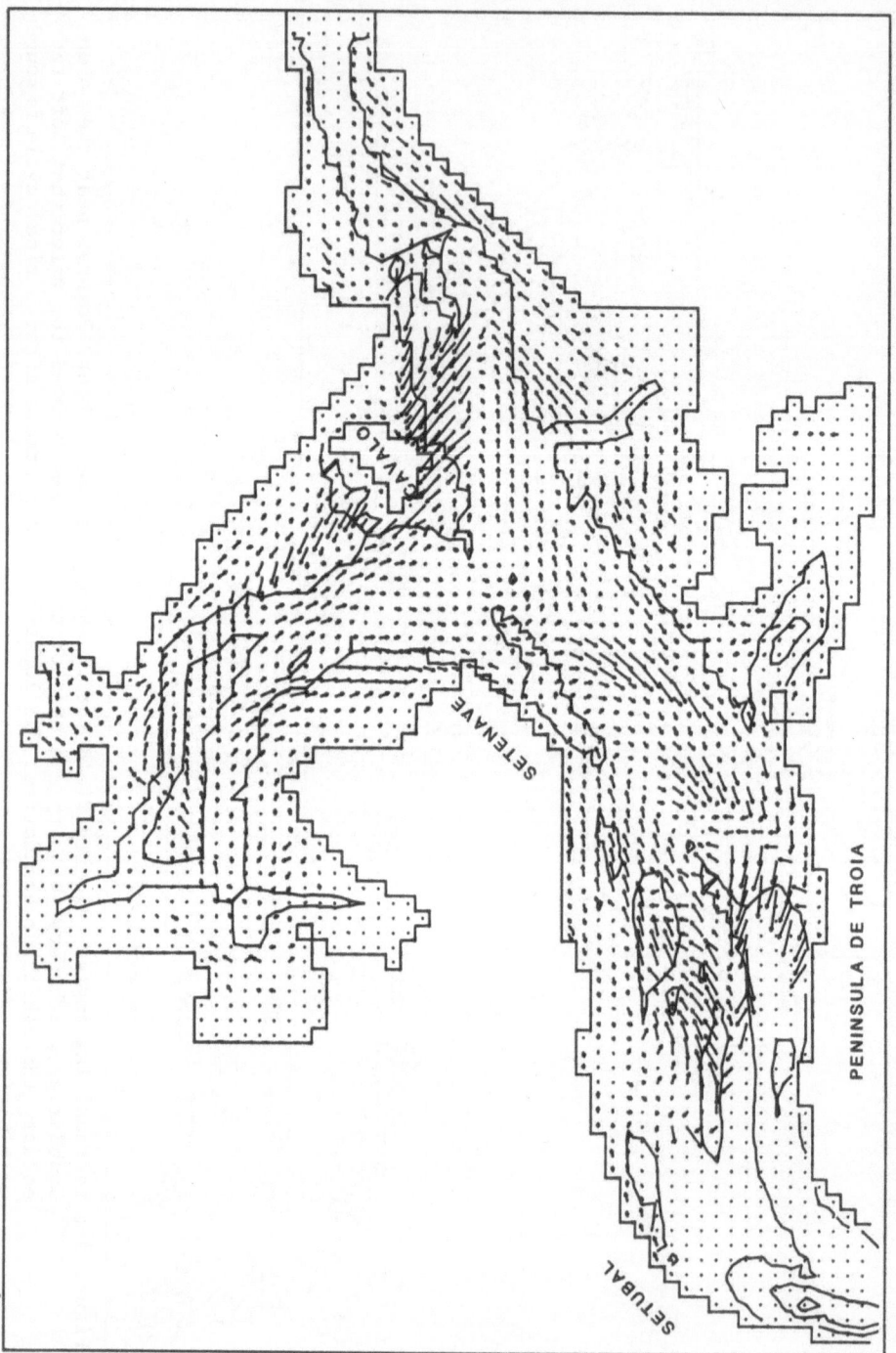

Figure 4 : residual lagrangian velocity fielf for $t_o = 3H$ after the high water. Points, near the open boundaries, with zero velocity are the points where was located the water that left the estuary through those boundaries. The location of a particle after a tidal cycle is found multiplying the arrow by 10.

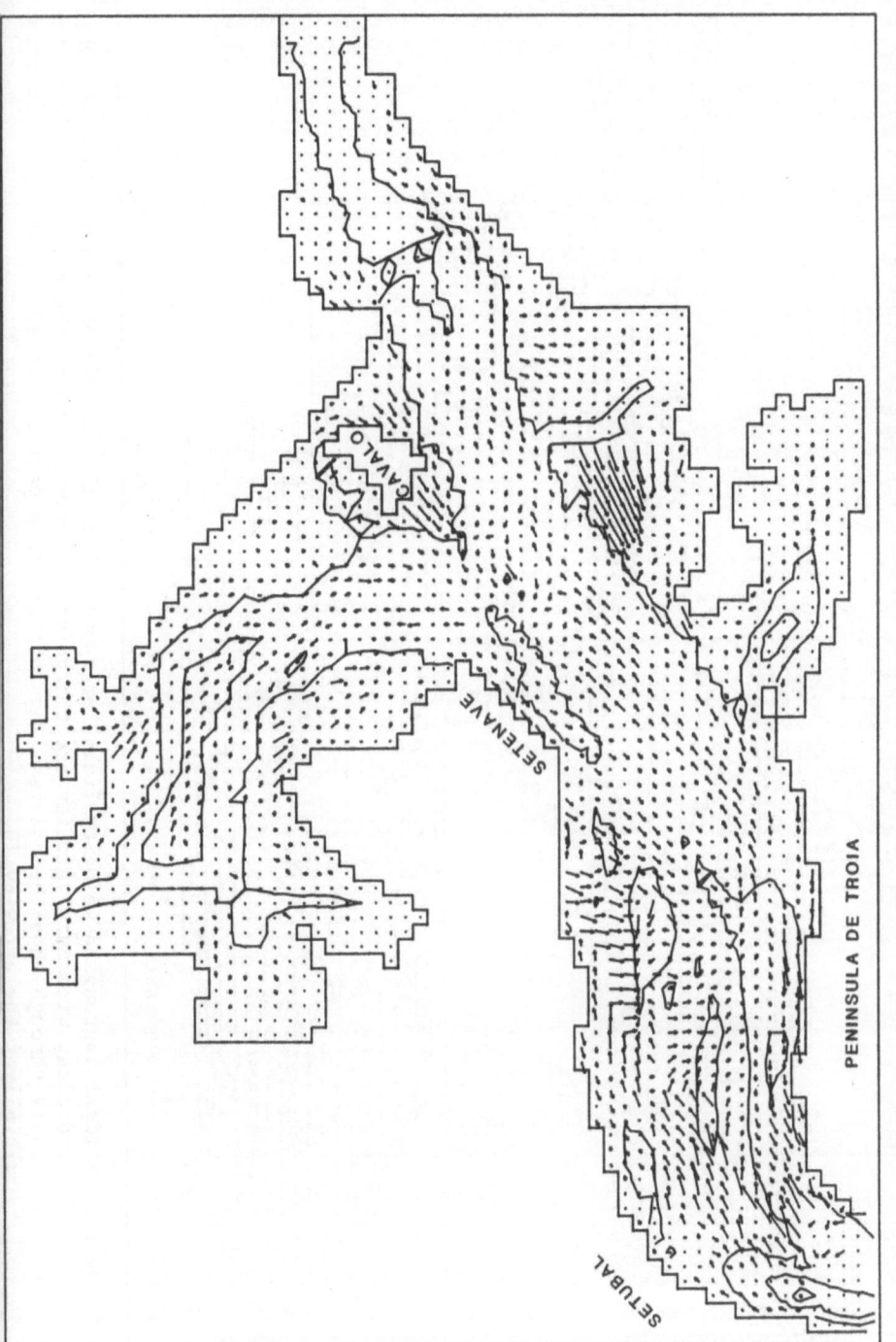

Figure 5 : residual lagrangian velocity field for $t_o = 6H$ after the high water. Points, near the open boundaries, with zero velocity are the points where was located the water that left the estuary through those boundaries. The location of a particle after a tidal cycle is found multiplying the arrow by 10.

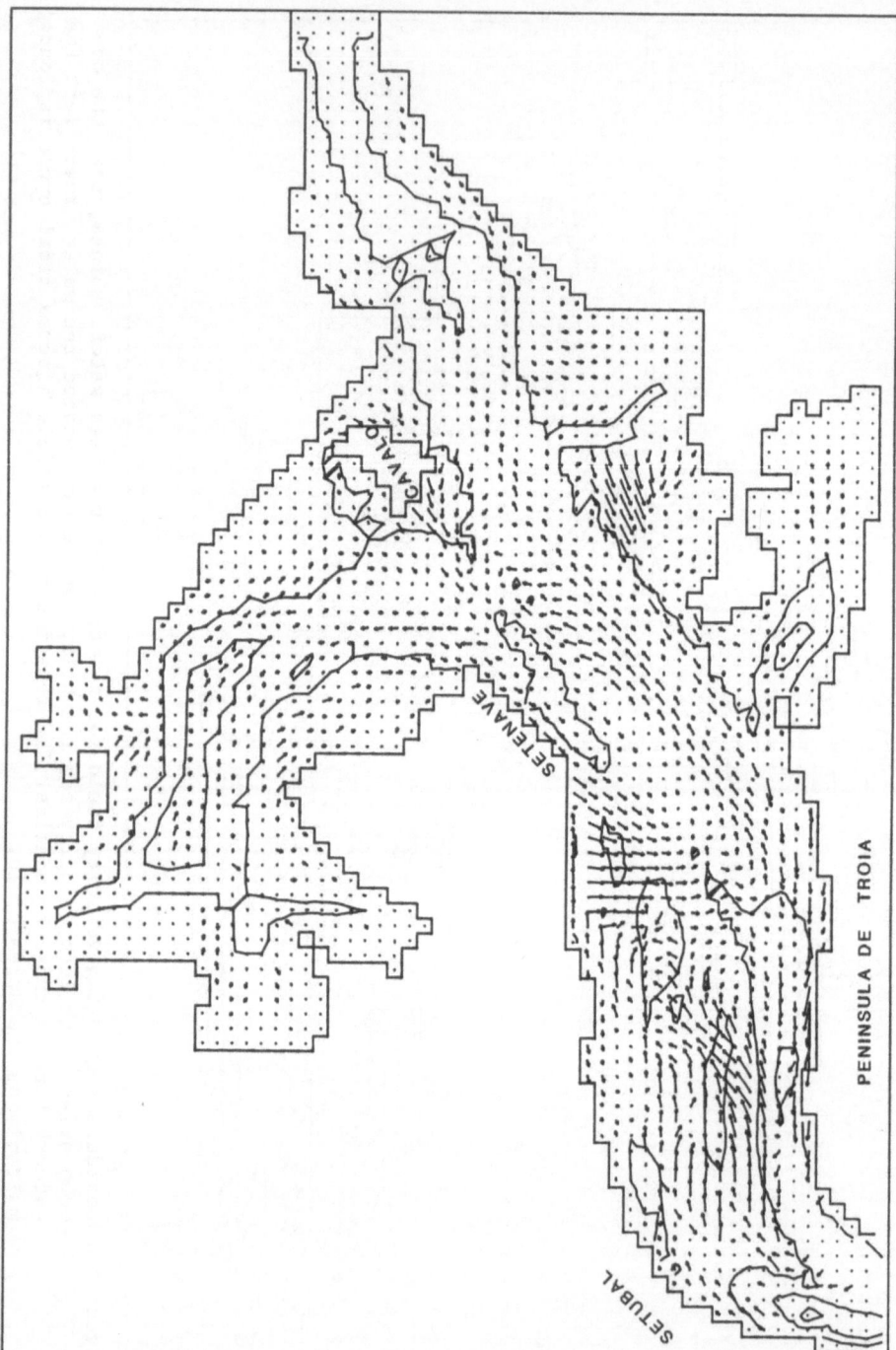

Figure 6 : residual lagrangian velocity field for $t_0 = 9H$ after the high water. Points, near the open boundaries, with zero velocity are the points where was located the water that left the estuary through those boundaries. The location of a particle after a tidal cycle is found multipliyng the arrow by 10.

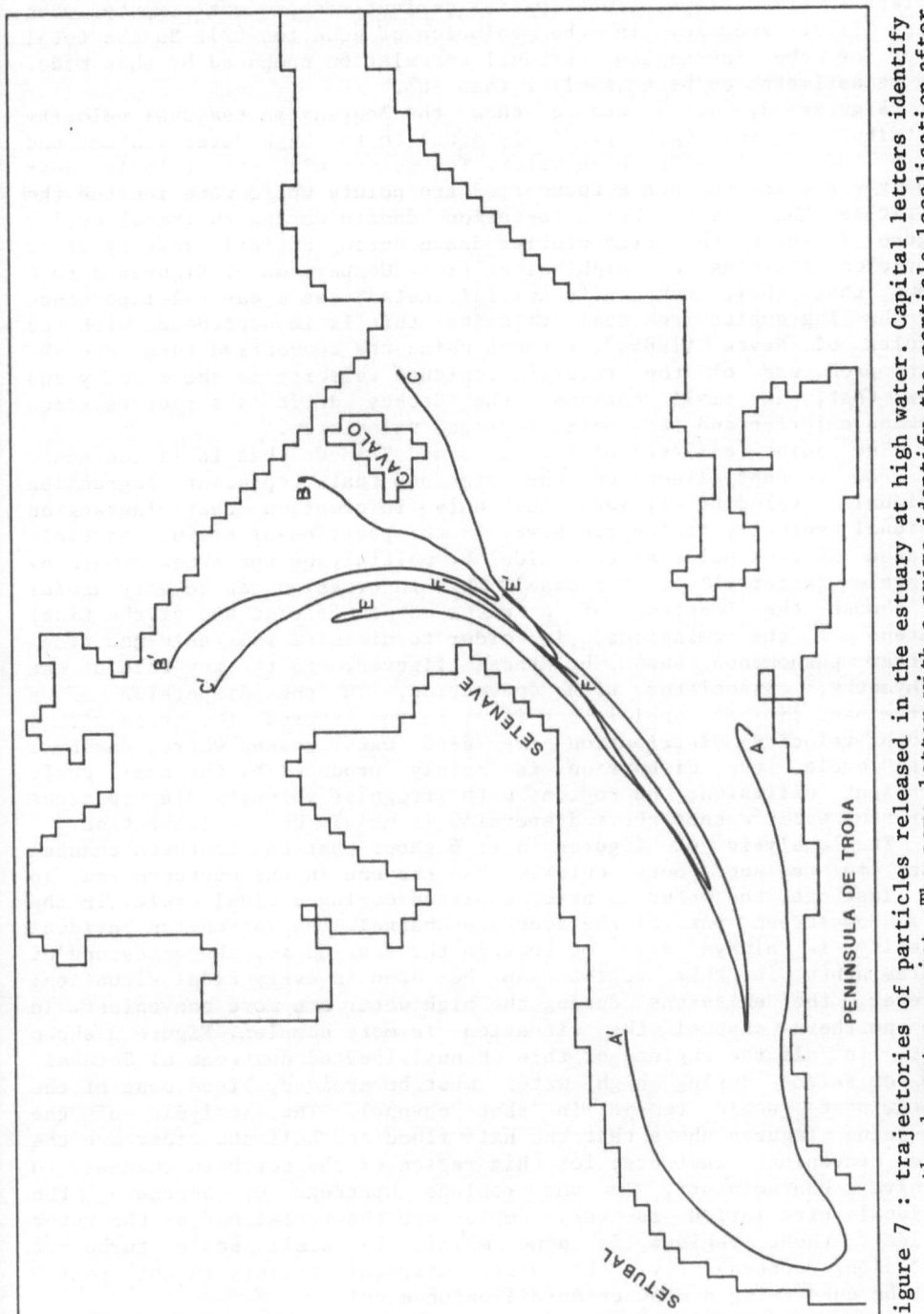

Figure 7 : trajectories of particles released in the estuary at high water. Capital letters identify the departure point. The same letter with a prime identify the particle localisation after a tidal cycle.

different time steps, since spatial derivatives are not computed when using first accuracy in the solution of equation (2). So the total error of the lagrangian residual circulation computed by this model can be estimated as being smaller than 10%.

Figures 3, 4, 5 and 6 show the lagrangian residual velocity distributions for t_0 respectively equal to the high water instant and 3, 6 and 9 hours after high water. The points with zero velocity near the river's and the sea's boundaries are points where were located the particles that left the integration domain during the tidal cycle. Figure 7 shows the trajectories drawn during a tidal cycle by fluid particles starting at high water time. Comparison of figures 3 to 6 shows that their movement's initial instant has a capital importance to the lagrangian residual velocity; this is in accordance with the results of Neves (1985a), through which the convective terms are the main producers of the eulerian residual velocity in the estuary and show that, in small bassins, the Stokes drift is a poor relation between eulerian and lagrangian residual velocities.

The joint analysis of figures 3 and 7 shows that it is non sense to draw current lines on the figures that represent lagrangian residual velocity fields. The only information that lagrangian residual velocity fields can give, is the position of a fluid particle located at each point after a tidal by multiplying the arrow length by a scale factor (10 in this case). This information can be very useful to choose the location of a waste water effluent and of the tidal instant of the emissions, in order to minimise its residence time. Another phenomenon shown by those figures, is the key role of the bathymetry, associated with convection, in the dispersion of a contaminant in an oscillatory flow; in the figures, the areas with a smooth velocity distribution represent water masses where, during a tidal cycle, the dispersion is mainly produced by the small scale turbulent diffusion; the regions with irregular velocity distributions represent water masses where dispersion is mainly due to convection.

The analysis of figures 3 to 6 shows that the southern channel water is replaced more quickly than the one in the northern one. In the last one, the water is never replaced during a tidal cycle. In the half downstream part of the southern channel, the lagrangian residual velocity is always directed towards the sea and so, the emissions of contaminants in this region can be done in every tidal situation; however, the emissions during the high water are more convenient. In the northern channel the situation is more complex. Figure 3 shows that, in all the regions of this channel located upstream of Setubal, the emissions during high water must be avoided, since most of the contaminant would remain in that channel. The analysis of the remaining figures shows that the half flood and half ebb tides are the more convenient instants, for this region of the northern channel, to receive contaminants. In the regions upstream of Setenave, the residual circulation is very complex and the replacemnt of the water inside those regions is done mainly by small scale turbulent diffusion; consequently, the waste disposal analysis in this region must be done using a convection-diffusion model.

4. CONCLUSIONS

A model for the computation of the lagrangian residual velocity and its application to the Sado estuary was presented. The dependance of the results on the instant at which the fluid particle started to be followed showed to be very important and, therefore, that the Stokes drift doesn't represent accuratly the difference between the eulerian and the lagrangian residual velocities.

The results of the lagrangian residual model showed to be very useful to choose the location and the tidal instant for the emission of pollutants inside the estuary, in regions where the dispersion of pollutants is mainly due to the convection associated to the bathymetry.

ACKNOWLEDGEMENTS:

The author is indebted to Prof. Nihoul and to Dr. Ronday for their valuable advice during the course of this work and to Prof. Ribeiro for his comments on the manuscript. Finantial support from Serviço de Estudos do Ambiente, Lisboa, Portugal, is also greatfully acknowledge.

REFERENCES

AMBAR, I.S.A., FIÚZA, A.F.G., SOUSA, F.M., LOURENÇO, I.R. (1980) "General circulation in the lower Sado river estuary under drought conditions". Proceedings of Actual Problems of Oceanography in Portugal, Edited by JNICT, Lisboa, Portugal.

CHENG, R.T. and CASULLI, V. (1982) "On lagrangian residual currents with applications in South San Francisco Bay, California". Water Res. Research, 18, 6, 1652-1662.

DOOLEY, H.D. (1974) "A comparison of drogue currentmeter measurements in shallow waters". Rapports et procès - verbaux des réunions. Conseil Permanent International pour l'Exploration de la Mer, 167, 225-230.

LONGUET-HIGGINS, M.S. (1969) "On the transport of mass by time-varying ocean currents". Deep-Sea Research, 16, 431-447.

MULDER, R. (1983) "Eulerian and lagrangian analysis of velocity in the Southern North Sea". North Sea Dynamics. Ed. by J. Sunderman and W. Lenz, Springer-Verlag.

NEVES, R.J.J. (1985a) "Bidimensional model for residual circulation in coastral zones. Application to the Sado estuary". To be published in Annales Geophysicae.

NEVES, R.J.J. (1985b) "A semi-implicit scheme for the solution of the shallow water equations. Application to the Sado estuary". To be published in the Proceedings of <u>Journées Océanographyques Belges</u> helded in Brussel in march 1985.

NIHOUL, J.C.J. (1982) <u>Hydrodynamic models of shallow continental seas. Application to the North Sea</u>. Etienne RIGA, Editeur.

RIBEIRO, M.M.C. and NEVES, R.J.J. (1982) <u>Caracterização hidrográfica do estuário do Sado</u>. (2 vol.) published by Serviço de Estudos do Ambiente, Lisboa, Portugal.

ROACHE, P.J. (1972) "On artificial viscosity". <u>J. Comp. Phys.</u>, <u>10</u>, 169-184.

SOBRAL, J.T. (1977) <u>Estuário do Sado. Observação de correntes de marés</u>. Repport of Instituto Hidrográfico, Lisboa, Portugal.

WOLLAST, R. (1978a) <u>Rio Sado. Campagne de mesures de juillet 1978</u>. Rapport technique. Published by Secretaria de Estado do Ambiente, Lisboa, Portugal.

WOLLAST, R. (1978b) <u>Rio Sado. Campagne de mesures de décembre 1978</u>. Rapport technique. Published by Secretaria de Estado do Ambiente, Lisboa, Portugal.

WOLLAST, R. (1979) <u>Rio Sado. Campagne de mesures de avril 1979</u>. Rapport technique. Published by Secretaria de Estado do Ambiente, Lisboa, Portugal.

DISPOSAL OF SEWAGE IN DISPERSIVE AND NON-DISPERSIVE AREAS:CONTRASTING
CASE HISTORIES IN BRITISH COASTAL WATERS.

T.H. Pearson

Dunstaffnage Marine Research Laboratory
P.O. Box No 3, Oban,Argyll, Scotland

ABSTRACT. The effects of sewage sludge disposal on three contrasting
areas round the British coastline are compared. In two, relatively
shallow areas (20-30 m), where dumped material is widely dispersed,
there is some evidence of localised accumulation of material occurring
patchily, but extensively, in the surrounding sediments. In one of
these areas summer plankton blooms cause localised eutrophic effects.
In the third, deeper, accumulating, area (70 m), the effects of the
dumped material are confined to a 100 km^2 around the disposal ground
and planktonic eutrophication has not disturbed the system. The
contrasting evidence from these areas is used to suggest that
disposal of sewage sludge to non-dispersive areas overlaid by a
relatively deep, well mixed water column might be a preferred policy
option.

1.   INTRODUCTION

        Disposal of sewage sludge at sea creates two broad problems in
hazard assessment:- a) the effects of localised over-enrichment and b)
the potential toxicity or pathogenicity of associated material. Carbon
and nutrient enrichment in hydrodynamically active open sea areas is
unlikely to pose more than temporary nuisance problems, as nutrient
inputs are rapidly incorporated into marine ecosystems, which are well
adapted to metabolize very large carbon inputs (iv). Conversely in
enclosed, shallow,or hydrodynamically inactive systems,severe but
localised problems may be created (xx). Appreciation of this dichotomy
has lead to the assumption on which much dumping at sea policy is based,
i.e. that disposal into dispersive (hydrodynamically active) areas is
preferable (the 'dispersal' rather than 'containment' option).

G. Kullenberg (ed.), The Role of the Oceans as a Waste Disposal Option, 577–595.
© *1986 by D. Reidel Publishing Company.*

Consideration of the problems posed by potential toxicity,or pathogenicity,of the sludge may call into question such assumptions, however. Disposal to any area, irrespective of hydrodynamic characteristics, does assume the eventual safe dilution and ultimate degradation, sequestration, depuration and/or fixation and burial of all potentially hazardous materials in the sludge. Current knowledge of the depositional pathways and ultimate fate of most potentially hazardous toxins and pathogens is limited and patchy for most areas and/or systems where dumping is currently taking place, and thus the potentially hazardous effect of such material is difficult to assess. Such assessment is particularly difficult if the material is widely dispersed, thus there is an argument for opting for containment rather than dispersal, particularly where the degree of depuration of hazardous materials following dispersal is not known. In order to obtain some comparative information, which may illustrate the relative merits of containment or dispersion as sewage sludge disposal options, the documented experiences of the effects of sewage sludge disposal to contrasting British coastal areas have been examined . In 1977 over eleven million tonnes of wet sludge were licensed for dumping at sea from the UK each year (ii) to approximately fifteen designated areas round the British coasts. About 75% of this material is dumped at only 3 sites i.e. 45% or 5 x 10^6 wet tonnes to the Thames Estuary over the Barrow Deep, 16% or 1.8 x 10^6 wet tonnes to Liverpool Bay and 14% or 1.5 x 10^6 wet tonnes to the Firth of Clyde on the Garroch Head ground (Fig 1). These three major dumping sites present interesting contrasts in hydrodynamic, sedimentological and biological characteristics.

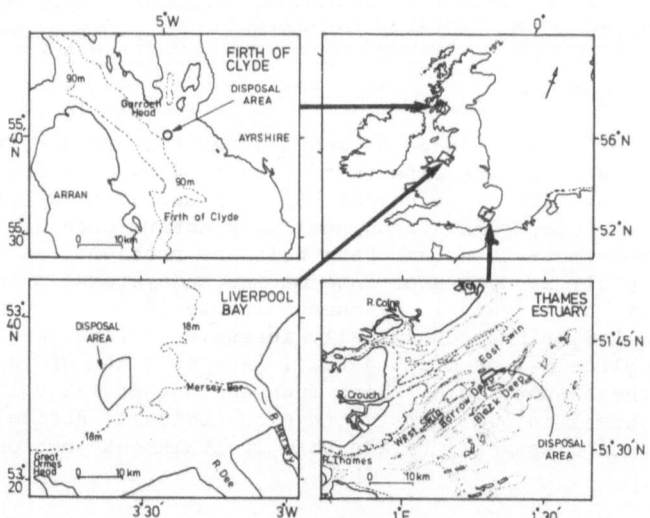

Figure 1. Location of the disposal grounds discussed. a) The Garroch Head disposal ground, Firth of Clyde; b) The Liverpool Bay disposal ground; c) The Barrow Deep disposal ground, Thames Estuary.

The Barrow Deep in the Thames Estuary and the Liverpool Bay site may be crudely defined as highly dispersive sites, whereas the Garroch Head site can be categorised as non-dispersive in its general hydrodynamic characteristics. Monitoring surveys of the prevailing environmental conditions of the distribution of sludge over the areas, and of its quantifiable environmental effects, have been carried out in all three areas for a number of years. The Thames estuary site is described in (xi) and (xxi). The Liverpool Bay ground is described in the reports of the Liverpool Bay Working Group, 1979, 1984 (iii) and (iv) and the Garroch Head site studies are reported in (iv) (v) and (vi). The comparisons made in this paper are based on information drawn from these sources unless additional references are given. Numerous other earlier published and unpublished reports on these sites are available, the majority of which are referred to in the sources mentioned above and are therefore not listed here. Brief descriptions will be presented here, of some comparative physical, chemical and biological characteristics of the three sites followed, by some interpretations and conclusions based on a comparative treatment of this data.

2.   THE GARROCH HEAD DISPOSAL SITE, FIRTH OF CLYDE

     The designated disposal grounds at Garroch Head cover an area some 6 km^2 in extent having a bottom depth of 70-80 m in the centre of the Firth, equidistant from the Arran and Ayrshire coasts (Fig. 1a).

2.1.   Physico-chemical characteristics of the sediments

     The sediments in the area are predominantly fine silty-clays with a background carbon content of between two and three percent. This is elevated in the centre of the disposal grounds to between 7 and 12%, with an area of some 10 km^2 around the centre showing higher than normal values (Fig. 2a). This high carbon area in the centre of the dumping grounds coincides with reduced sedimentary conditions (negative sedimentary redox potentials) and elevated values of a range of metal elements and other anthropogenic material. Fig. 2b,c, illustrates the distribution of high lead (Pb) and copper (Cu) values in the surface sediments which show typical distribution patterns. Thus maximum levels (300 ppm, i.e. up to 10 x background levels for Cu and and 5 x background for Pb) occur in the centre of the dumping grounds, but attenuate rapidly as distance from the centre increases. Levels in excess of 100 p.p.m. are found over an area of some 7 km^2 in the centre of the grounds. Background levels (Pb 43-86 ppm; Cu 7-39 ppm: (ix) recur within 3 kilometers of the centre of the dumping ground, with the exception of the area to the north of the grounds where intermittently high levels have been recorded over an area of some 15 km^2 in the vicinity of a former dumping area 2-3 km south of Garroch Head. Examination of gravity core sediment samples from the area indicates that a 15 cm deep sludge deposit exists in the centre of the grounds, but no separate sludge deposit is detectable beyond 1.5 km from the centre. Water conditions in the vicinity of the grounds have been

described in some detail (v). Residual currents over the disposal
area, driven by tidal streams, are generally low (< 10 cm s^{-1}).

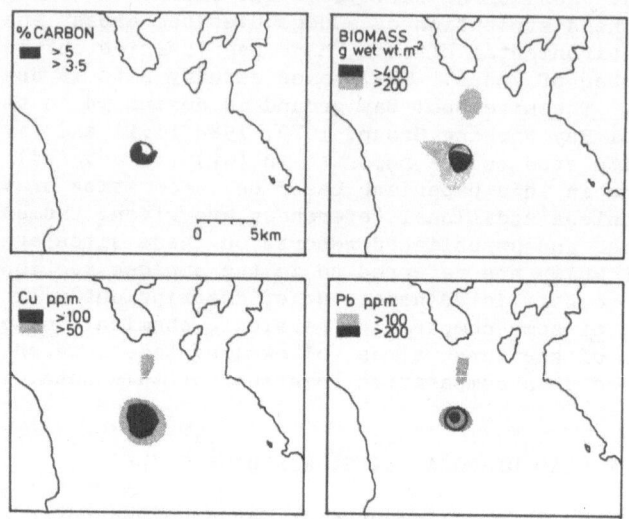

Figure 2.   Concentrations of Carbon, Lead (Pb) and Copper (Cu) in
the surface sediments of the Garroch Head grounds, and distribution of
macrofaunal biomass over the area. a) Carbon (%),   b) Pb (p.p.m.),
c) Cu (p.p.m.),   d) Biomass (g m^{-2}).

Additional wind driven flows cause relatively rapid renewal of the
bottom water at variable current speeds of between 5 and 15 cm s^{-1}
generally in a SE direction, but both speed and direction are
variable. The water immediately above the sediment surface remains
fully oxygenated at all times throughout the area.

2.2.   Distribution of benthic fauna

     The distribution of benthic fauna over the area has been examined
in some detail. In general very high macrofaunal biomass levels are
found within an area roughly coincidental with the area of elevated
sedimentary carbon and metal levels (Fig. 2d). Infaunal population
distributions over the area conform closely to the theoretical pattern
of successional change along a gradient of increasing sedimentary
organic enrichment described (vii). Thus along a twelve kilometer
E/W transect centred over the dumping area a characteristic pattern of
population change is observed (Fig. 3). In the centre of the grounds
species numbers are low but biomass and abundance are high. One
kilometer from the centre species numbers begin to increase whilst
biomass and abundance decline. A secondary biomass peak is found
between one and two kilometers from the centre associated with the

increasing variety of species present. Species numbers are highest
some four kilometers from the centre before declining towards the outer
ends of the transect,where all three parameters decline to the
relatively low levels characteristic of the normal communities of the
area. There has been very little variation in this pattern of
distribution over the past six years.

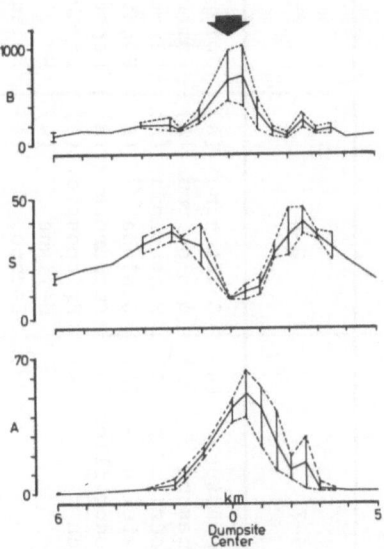

Figure 3.   Change in the principle faunal statistics along a transect
across the Garroch Head grounds (S, Total number of species present
m^{-2}; A, total density of animals x 10^3 m^{-2};   B, total biomass,
g. wet wt. m^{-2}). Mean and range of values obtained over five years
shown.

An examination of the successional change in species along the
enrichment gradient provides a more detailed illustration of the effects
of sludge input on the benthic fauna of the area. The centre of the
grounds is occupied by very large numbers of small capitellid and
spionid polychaete and nematode worms. A number of species which are
either absent,or present in very low numbers in the centre of the
grounds,appear on the edge of the central area,where they reach very
high numbers. These include capitellid and cirratulid polychaetes and
small bivalve molluscs. Beyond the central areas,towards either end of
the transect,this group in turn declines in numbers and a further
group, characteristic of the unpolluted areas beyond the influence of
the sludge dumping, becomes dominant. This includes a greater variety
of organisms with a number of different phyla being represented.
Polychaetes still predominate but molluscs, echinoderms and crustaceans
are all important. The characteristic taxa in this succession along
the gradient of organic enrichment are listed in Table 1.

Table 1. Characteristic taxa along the organic enrichment gradient over the Garroch Head sludge disposal grounds.

Distance from centre of disposal area (km)	< 1	1-2	3-5	> 5
Carbon content of surface sediments (%)	< 7	5-7	3-4	2-3
Characteristic Taxa (N, Nematoda; A, Annelida; M, Mollusca; E, Echinodermata)	Tubificoides (A) Capitella (A) Scolelepis (A) Pontonema (N)	Mediomastus (A) Notomastus (A) Ctenodrilidae (A) Cirratulus (A) Protodorvillea (A) Eumida (A)	Chaetozone (A) Diplocirrus (A) Pectenaria (A) Goniada (A) Ampharete (A) Prionospio (A) Melinna (A) Eteone (A) Corbula (M) Mysella (M) Abra (M) Thyasira (M)	Spiophanes (A) Lumbrinereis (A) Glycera (A) Rhodine (A) Nephtys (A) Lipobranchius (A) Nucula (M) Solenogastres (M) Amphiura (E) Brissopsis (E)

2.3.   Metal concentrations in fauna.

Samples of fin fish and shellfish taken from the vicinity of the
dumping ground have been analysed for metal levels (i). Concentrations
in a range of fish (muscle tissue) were in general low
(e.g. Pb and Cu 0.5-1 ppm dry weight of Cod tissue) but occasional
individuals were found to have higher levels of Cu and Zn in the liver
(e.g. Cu 4-70 ppm dry weight in Dab). Analysis of a number of mollusc
and crustacean species showed no consistently high values, but a wide
range of Zinc concentrations was found in <u>Buccinum undatum</u> collected
from the dumping ground area (38-866 ppm dry wt. of tissue). In
general metal levels in both fin fish and shellfish taken from the
area are similar to the levels recorded in the same species in other
coastal areas of the British Isles (i, x).

2.4.   Eutrophication effects.

Detailed assessment of phytoplankton populations in the vicinity of the
dumping area have not been made, but there is no evidence to suggest
that bloom conditions have occurred as a result of nutrient additions
from the dumped sludge.

2.5.   Accumulation or dissipation. Evidence and assumptions.

The Garroch Head disposal grounds represent a text book example of
an accumulating ground in so far as sedimentary biology and chemistry
are concerned. Steep gradients of effect in both biological and
chemical factors are measurable, orientated to maxima in the centre of
the grounds. Background levels recur at a radius of 5-6 kms from
the centre of the grounds, thus the total area of effect is confined to
some 95 km^2. It is assumed that effects beyond this area are
negligible. Although occasional elevated metal levels have been
recorded in individuals of some invertebrate species taken from the
vicinity of the dumping grounds, no evidence has been found of
consistent increases in metal levels in such species. It is assumed
that the elevated levels of potentially hazardous materials recorded
in the centre of the grounds represent the accumulated concentrations
of such materials in the dumped material and that significant
transference beyond the contaminated area does not occur. This
assumption is as yet unproven.

3.   THE LIVERPOOL BAY DISPOSAL SITE

The designated disposal grounds in Liverpool Bay lie some 33 km
west of the mouth of the river Mersey and cover an area of about 60 km^2
in an average water depth of 27 m (Fig. 1b).

3.1.   Physico-chemical characteristics   of the sediments.

Overall Liverpool Bay is broad and shallow, with a gently sloping bed. Superimposed on these broad scale features are local complexities, the main features of which are a series of ridges and troughs aligned mainly east-west, which form the base for trains of sand banks and linear sand waves. Local mud patches tend to accumulate somewhat ephemerally in the troughs of these waves and banks. Strong tidal currents and wind driven turbulence cause large scale sedimentary disturbance and translocation. Surface organic carbon (upper 25 mm) is high in muddy areas to the north and east of the dumping grounds in the path of the Mersey outflow. Organic matter content of the sediments (sand fraction) tends to vary in an irregular manner over the area as a whole (Fig. 4a), but is generally less than 1%. Concentrations of heavy metals in the sediments are somewhat elevated over the dumping grounds and in some areas to the south and west of the Mersey Channel, as exemplified by the distribution of copper and lead levels illustrated in Fig. 4b,c. Variability is high however and apparently dependent on local topography (xii and xiii). Metals appear to be largely associated with organic material and as this, in turn, is highest in the finer sediments, the high metal measurements are probably associated with the localised mud patches mentioned above. A detailed assessment of the distribution of dumped sludge and its effect on the benthos of the area distinguished five zones where sludge may influence the sediment and benthos of the bay (viii). These are 1) the dumping ground, subject to the immediate impact but, being current swept, not subject to long term deposition; 2) eastward dispersal tracks, generally the troughs of the east-west sand ridges, through which a relatively continuous movement of material may take place; 3) intermediate ephemeral deposition areas, where material may temporarily accumulate in erratic fluid mud patches or layers; and two distant accumulating zones where material may reside semipermanently, 4) inshore muddy sand pockets and 5) distant permanent deposits in offshore mud or estuarine marshes. Sludge particles are only temporarily resident in the first 3 zones which encompass the primary survey areas in the bay, thus it is not surprising that no measurable accumulations of sludge material are found in the sediments.

3.2.   Distribution of benthic fauna.

The benthos over an areas some 600 km^2 in south Liverpool Bay, centred on the dumping ground area, has been monitored on an annual basis since 1970. Since the area consists of current swept, relatively mobile, heterogeneous sediments the benthic fauna is both heterogeneous and often impermanent. Nevertheless, four major faunal grouping have been identified over the area which are principally associated with sediment type and distance from shore. Sub-groups have, in turn, been identified within each of the major groups, each of which is characterised by a different set of prominent species. Thus in the group associated with offshore clean sand with shell and gravel, five sub-groups have been defined whose distinction may be dependent on the amount of mud present in these generally coarse sediments.

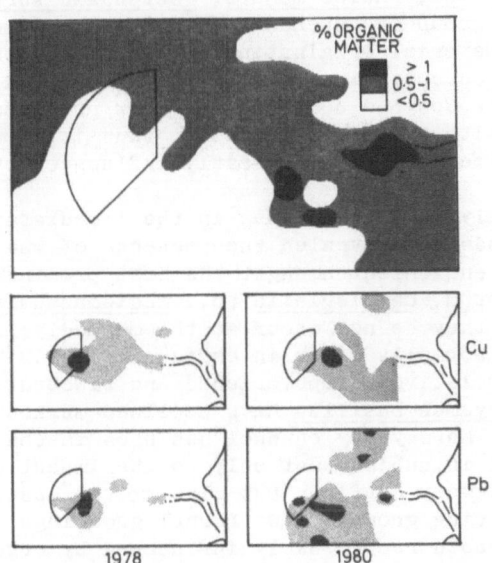

Figure 4. Concentrations of carbon, copper and lead in the surface
sediments in the vicinity of the Liverpool Bay grounds. (Data
from iv and xiii.)
a) Percent organic matter associated with the fraction of the sediments.
b) Copper (Cu). Light stippling; concentrations greater than 50 p.p.m.
   Dark stippling; concentrations greater than 100 p.p.m.
c) Lead (Pb). Light stippling; concentrations greater than 200 p.p.m.
   Dark stippling; concentrations greater than 500 p.p.m.

The group occupies the slight submarine ridges, particularly in the
southern half of the bay. Three of the sub-groups contain prominent
species which are commonly associated with organically enriched
sediments (see list of characteristic taxa from the Garroch Head site
and (vii)) e.g. Scoloplos armiger, Scalibregma inflatum, Chaetozone
setosa and may thus be influenced by the deposition of sludge material.
A second group associated with offshore mixed muddy sand has been
subdivided into four sub-groups which are again probably distinguished
by their affinity to differing amounts of fine particulate matter in
different areas. All the groups contain some prominent species which
are known to be associated with organically enriched sediments e.g.
S. inflatum, Cerianthus lloydi, Petenaria koreni and thus may be
influenced by the sewage deposition, The third major group again has
four sub-groups and occupies inshore muddy sands which occur in
restricted pockets all round Liverpool Bay. These pockets have
comparatively high faunal biomass, but are subject to frequent storm

disturbance (vi). Most of the prominent species in all four sub-groups
are well known as dominants in organically enriched sediments (e.g.
<u>Mysella bidentata</u>, <u>Abra alba</u>, <u>Pholoe minuta</u>, <u>Pectenaria koreni</u>) and it is
likely that this faunal group is dependent on sludge derived material
in addition to organic material originating in the Mersey outflow. The
fourth identified group occupies inshore shallow clean sand areas, is
not divisible into sub-groups and contains very few prominent species
that can be associated with organic enrichment, thus it is unlikely
that these types of sediment have been greatly influenced by sludge
deposition.

     A more detailed analysis of the fauna in the immediate vicinity of
the sludge dumping grounds has revealed the presence of many species
associated with organic enrichment amongst the most prominent species
(e.g. <u>Scalibregina inflatum</u>, <u>Caulleriella</u> sp., <u>Mediomastus fragilis</u>,
<u>Cerianthus lloydi</u>), but they do not occur at the markedly high
population levels which normally occur in enriched areas. The general
impression is one of a relatively impoverished and fluctuating fauna.
Echinoderms ,e.g. <u>Echinocyamus pusillus</u>,have declined markedly in recent
years. The fauna of the Mersey Bay channel has been intensively
examined, since the area is subject not only to the impact of
contamination from the Mersey outflow, but also to the eastward drift
of material from the dumping ground. Two faunal groupings can be
identified in the area, both recognisably influenced by organic
enrichment, with the south slope of the Mersey channel more markedly
affected. Ephemeral sludge deposition takes place in this area, with
residence times long enough to affect the fauna, which shows
impoverishment both in species numbers and abundance.

3.3.   Metal concentrations in fauna.

     Samples of finfish and shellfish taken from the vicinity of the
dumping grounds have been analysed for metal concentrations.
Concentrations in a range of fish (muscle tissue) are in general low
and did not change appreciably during the mid-seventies monitoring
period (Zinc, Copper). Mercury and lead levels were found to have
declined during this period, following a reduction in these metals in the
dumped sludge. Analysis of several types of molluscs and crustaceans
showed some evidence of increasing concentrations; e.g. of Cadmium,
Zinc and Copper in <u>Chlamys operculata</u> muscle tissue, and Mercury, Lead
and Copper in <u>Eupagurus bernhardus</u>, but levels were relatively low and
very variable.

3.4.   Eutrophication effects.

     Extensive and dense blooms of <u>Phaeocystis</u> occur both inshore and
offshore in the Bay during May and June every year. Such blooms
follow the spring diatom bloom and can be both toxic and create a
beach amenity nuisance. They are sustained by the nutrient
enrichment of the Bay waters, but the exact sources of this
enrichment, and its possible link with the sludge dumping, are not
known. Extensive research into the processes leading to bloom

development is being undertaken. (iv)

3.5.   Accumulation or dissipation:evidence and assumptions.

The Liverpool Bay disposal site is essentially dispersive. The dumping area is hydrodynamically energetic and the discharged material has only a short residence time in the immediate vicinity of the grounds. Nevertheless the chemical and biological evidence suggests that the dumped material influences benthic sediments over a considerable area surrounding the grounds, and that local pockets of accumulation do occur, particularly inshore in the direction of the Mersey channel. Evidence from metal analysis suggests that there may be accumulation of toxic material in some predatory species in the area, but the available information is relatively sparse and inconclusive. The occurrence of denser algal blooms over the area in the early summer can not be directly attributed to sludge dumping, but such dumping must inevitably contribute nutrients to the bay and thus exacerbate the general eutrophic trends in the area.

4.   THE BARROW DEEP DISPOSAL SITE IN THE THAMES ESTUARY

The Barrow Deep disposal site is situated in the outer Thames estuary some 15 km from the Essex coast and occupies an area 9 km in extent (Fig. 1c). Extensive surveys have been carried out throughout the middle and outer estuary in 1972 (xxi) and again in 1976-77 (xi).The information presented here is drawn primarily from the latter source, but comparison of the two surveys allows an assessment of changes over the intervening period.

4.1.   Physico-chemical characteristics of the sediments.

The outer estuary area is characterised by a series of parallel channels and shallow sand banks running north-east to south-west, intersected by numerous cross-channels. Strong tidal currents (up to 1.4 m s^{-1}) run through the whole area and the pattern of water movement is complex. Being generally shallow (< 20 m) the estuary is also subject to considerable wind driven turbulence and disturbance by storms,particularly in its outer areas. Dumped sludge is rapidly dispersed and spread throughout the Barrow Deep and into the adjacent West Deep and Middle Swin on the first tidal cycle. Sediments throughout this area contain particles originating from the sludge and show increased concentrations of organic carbon (Fig. 5a) and of heavy metals (Fig. 5b,c). The metal concentrations are associated with the finer sediment fractions and the highest concentrations are found in the vicinity of the dumping grounds where levels reach five to eight times background concentration (e.g. Pb, 300 p.p.m.. Cu, 110 p.p.m.). In general the sediments in the area are very mobile and it is thought that considerable secondary dispersion and reburial of sludge material takes place, with the finest particles being finally lost in suspension from the estuary via residual currents. However it is thought that

there are some areas of accumulation to the north-east of the
Barrow Deep, and in the Middle Deep on the northern side of the East
Barrow sands.

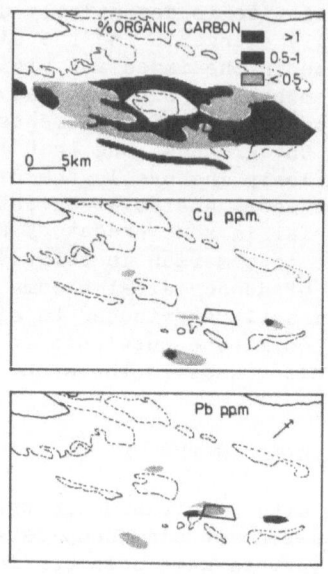

Figure 5.   Concentrations of carbon, copper and lead in the surface
sediments in the vicinity of the Barrow Deep disposal ground in the
Thames Estuary in 1976.   (Data from (xi)).
a) Percent organic carbon in the < 90μm fraction of the sediments.
   sediments.
b) Copper (Cu) concentrations in the < 90 μm fraction of the sediments.
   Light stippling: concentrations greater than 50 p.p.m.
   Dark stippling; concentrations greater than 100 p.p.m.
c) Head (Pb) concentrations in the < 90 μm fraction of the sediments.
   Light stippling; concentrations greater than 100 p.p.m.
   Dark stippling; concentrations greater than 200 p.p.m.

4.2.   Distribution of the benthic fauna.

     In 1972 the benthic fauna was sampled at 73 stations covering
an area of 780 km^2 extending from the sunk light vessel in the
north-east to Thameshaven in the south-west. A 1977 survey covered a
a smaller area (340 km^2) centred on the Barrow and Middle Deeps, more
intensively (171 stations).   There are quite marked differences in the
results of the two surveys.   In 1972 nine faunal groupings were
identified,whose distribution appeared to be governed principally by
sediment type and seabed mobility. Despite the coarse nature of the
sediments throughout most of the area total,density of animals was

relatively high and the fauna was dominated by polychaetes. Many of
the faunal groups were dominated by taxa   commonly associated with
organically enriched sediments e.g. <u>Mysella</u>, <u>Polydora</u>, <u>Heteromastus</u>.
Such groups were particularly associated with sediments near the Barrow
Deep dumping ground and the disused Black Deep dumping ground, where
faunal densities were particularly high. Two of the groups (B, E) were
dominated by polychaetes, some of which are known to be associated with
enriched sediments, but their overall densities are low. These
impoverished populations tended to be scattered througout the area. The
distribution of the 1972 faunal associations which can be described in
the above terms as either enriched or impoverished is illustrated in
Fig. 6.

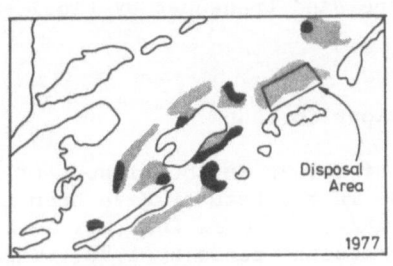

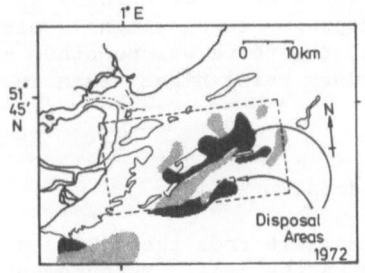

Figure 6.   Distribution of faunal associations over the survey areas
             in 1972 and 1977.
1972.   Stippling; associations where the fauna was impoverished.
         Hatching; associations where the fauna was enriched.
1977.   Light stippling; associations where the fauna was impoverished
         but contained species associated with organically enriched
         sediments.
         Dark shading; associations where the fauna was highly
         impoverished.

     The 1977 survey identified seventeen faunal associations,
reflecting the great heterogeneity of sediments and conditions in the
area, but only eight of these appeared at more than 10 stations. In
general faunal densities were very low throughout the area compared to

the 1972 survey and many of the species which had dominated areas in
the previous survey were now absent or present only in very low
numbers. This difference is ascribed to natural fluctuations arising
from sedimentary instability (xi). As in 1972 a number of the commoner
associations contain dominant species often found in high numbers in
organically enriched sediments,although in 1977 their densities were
low. These associations tend to be found in the immediate area of
the dumping ground and in areas of the Barrow and Middle Deeps and of
the East and West Swins,where deposition of sludge material is thought
to take place. Two of the identified associations (5,6) were
particularly impoverished in both species numbers and in faunal
densities and these tended to occur in areas of particular sedimentary
erosion and instability,e.g. to the south of the East Barrow Sands,and
in the deeper areas where sand transport by tidal currents is known to
take place.

4.3.   Metal concentrations in fauna.

        The concentration of metals in the tissues of finfish and
shellfish taken from the Thames Estuary have been reported (x). The
elevated levels of mercury in species from the area had fallen to
levels typical of other coastal zones by 1980. This fall paralleled a
progressive reduction in the amounts of Mercury in the discharged
sludge. Cadmium levels appeared to be slightly elevated in some shrimp
(Crangon crangon) samples, but there was no other evidence of metal
concentrations in the estuary being higher than in inshore waters
generally.

4.4.   Eutrophication effects.

        No information is available from the 1972 and 1976/77 surveys of
phytoplankton populations in the area. Measurements of inorganic
nitrogen compounds in the water column of the outer estuary were made
in 1978 which suggested that dumping has only a transient effect
locally on the concentrations of ammonia and nitrates. These findings
supported the earlier estimates,which suggested that dumping
contributed only 9% of the total nitrogen inputs to the estuary as a
whole. The shallow water column and strong currents ensure continuous
mixing throughout the area and stratification does not occur, thus
localised nutrient concentrations leading to bloom conditions in
particular areas are less likely in this system.

4.5.   Accumulation or dissipation:evidence and assumptions.

        It is apparent that the sludge disposal site in the Thames Estuary
is the most highly dispersive of the three being considered. Strong
water and sediment movements in the area ensure that sludge material is
rapidly dispersed throughout the estuary and indeed the finer material

may be lost to the southern North Sea beyond. The 1976/77 survey
evidence does suggest, however, that current dumping rates have
"exceeded the dispersive capacity of the area, resulting in readily
identifiable areas where organic matter and metals have accumulated".
(xi). Moreover the marked faunal impoverishment found in the Barrow
and Middle deep areas in 1977 when compared with the 1972 survey may,
in part, be due to the overall effect of sludge dumping rather than
be caused by sedimentary instability as suggested. Thus the commonest
associations contain a mixture of prominent species, some of which are
characteristic of coarse mobile sediments, and some of organically
enriched areas. This type of diversity was also evident in the 1972
survey, but densities then were much greater. It seems unlikely that
sedimentary mobility has changed greatly enough in the intervening
period to result in the observed impoverishment.

        Eutrophication is not considered to be a general problem in the
area, although various areas in the southern North Sea (but
principally along the eastern shores) are subject to summer plankton
blooms of considerable intensity (viii). There is little evidence for
the transfer of metals to the higher trophic levels in the system,
despite the occasional high sedimentary levels in some areas, though
the available data are fragmentary and relatively inconclusive.

5.   DISCUSSION

The Garroch Head ground is the least, and the Barrow Deep ground the
most, dispersive of the three areas discussed. Simple descriptive
scaling of this type serves little useful purpose, however, as the
hydrodynamic and sedimentological characteristics of each area are very
different. On the Garroch Head ground strong chemical and biological
gradients of effect exist in the sediments, radiating from the highly
affected areas in the centre of the grounds. In this area, although
nutrient enrichment of the water column undoubtedly takes place, there
is no evidence that it unduly intensifies summer plankton blooms. In
Liverpool Bay no strong sedimentary gradients can be linked to the
effects of dumped sludge but localised accumulations of sludge material
occur in pockets in various parts of the bay, more particularly in
shallow inshore areas. Generalised eutrophication of the bay as a
whole, resulting in strong summer plankton blooms, may be a partial
consequence of sludge enrichment, and certainly increases the organic
loading to the benthic ecosystem. This may well exacerbate local
deoxygenation events in the benthic boundary layer, which, combined
with ephemeral sludge desposition, would contribute to the observed
impoverishment of the benthic fauna. Thus it is suggested that sludge
dumping in dispersive areas might result in a sparse but diverse fauna
since the complexity of topography and mobility of sediments
inevitably results in a patchy distribution both of initial fauna and
of ephemeral concentrations of sludge (viii). The latter, rather
than providing a rich food source by _in situ_ degradation, would merely
cause localised catastrophic disturbance before being swept away. The

result would be a sparse and very variable fauna containing species
either resistant to anoxia and/or relatively mobile. This appears to be
exactly the mix of dominant species which is prevalent over large parts
of Liverpool Bay. Sludge dumping in the Thames Estuary may be having
a similar effect, since similar generalised impoverishment appears to
have occurred in the benthic communities of the area. Again no strong
gradients of effect can be detected in the sediments, but localised
'hot spots' of sludge-derived material occur throughout the area.
Intense summer plankton blooms have not been reported as of frequent
occurrence in the Thames estuary thus additional benthic enrichment
from eutrophication effects is probably not a further stress in this
area. However despite the highly dispersive characteristics of the
Thames Estuary system the input levels of 5×10^{6} tonnes yr^{-1} are
apparently sufficient to cause localised overloading. A general
assessment of the relative merits of the confinement and dispersion
options for disposal of waste at sea has been provided (vii). This
suggests that dispersion may be preferable for wastes of limited life
time, but that for waste of potentially permanent toxicity confinement,
possibly following by capping, may be the best choice. Unfortunately
sewage sludge, although consisting principally of highly degradable
organic material, also contains significant amounts of potentially
hazardous chemical and biological material. The choice is therefore
not straightforward. However, the evidence from the three sites
briefly reviewed here gives some support for a compromise option which
may be worth serious consideration. In both the highly dispersive
sites the wide dissemination of material resulted in the accumulation
of pockets of sludge derived material at considerable distances from
the original discharge points (far-field effects). Occasionally
elevated metal levels were recprded at such 'hot spots' and it is possible
that other toxins could also collect in such patches. The hazard
potential of such local patches is unknown, and very difficult to
assess. At the non-dispersive site high levels of metals (and other
potentially hazardous material) were confined to a small central area.
Because of the deep and relatively well mixed water column over the
site a very high level of benthic degradative activity is maintained
without anoxia occurring in the water overlying the sediment. The
organic component of the sludge fuels a productive and dense benthic
community which attenuates gradually into normal communities
as distance from the dumping area increases. Despite the high input
levels there is no impoverishment of benthic activity in the area. The
final fate of the hazardous materials in the sludge is not known for
certain however. It is assumed that the bulk is inactivated, buried,
or degraded in the central area, but this has not been conclusively
demonstrated. The possibility that there may be some transfer
outwith the area through food webs cannot be dismissed. However, given
the regular gradients of effect radiating from the central area, the
analysis of possible dissemination pathways and the monitoring of loss
rates and eventual sinks is practicable. Moreover further confinement
by capping is an additional option on a concentrated ground.
        It may be suggested then, that, drawing from the experience gained
in assessing the effects of sludge disposal on these three contrasting

areas, the type of ground represented by the Garroch Head area provides the best mix of characteristics for a disposal area i.e. a well mixed water column where stratification is never prolonged,over a conservative (relatively immobile) sedimentary environment. The former characteristic is essential for the rapid degradation of the organic component of sludge, the latter for the confinement of the more toxic or pathogenic elements. It is obvious that the amount of material to be dumped must also be considered in relation to such characteristics. Although experience in the Clyde suggests that 10^6 tonnes yr^{-1} are absorbed by that benthic ecosystem, it is difficult to predict the effect of five times that amount on such a system. Moreover should containment on that type of ground be considered, then extensive assessment should be made of the eventual fate of all toxins and pathogens in the dumped material. Ignorance in this respect is the weakest link in assessing the possible options for marine disposal.

REFERENCES

(i)   Davies, I.M. 1981.   Survey of trace elements in fish and shellfish landed at Scottish ports 1975-76. DAFS Scottish Fisheries Research Report. No. 19. 28 pp.

(ii)   DoE/NWC, 1979a.   Report of the Sub-committee on the Disposal of Sewage Sludge to sea 1975-78.(Standing Technical Committee report 18).   Department of the Environment and National Water Council, London 66 pp.

(iii)   DoE/NWC, 1979b.   Sewage Sludge Disposal in Liverpool Bay. Reseach into Effects 1975 to 1977. Part 1 - General. Standing Technical Committee report 16).   Department of the Environment and National Water Council, London. 44 pp.

(iv)   DoE/WTD, 1984.   Sewage Sludge Disposal in Liverpool Bay. Research into Effects 1975 to 1977. Part 2 - Appendices. Water Technical Division, Department of the Environment, London. 194 pp.

(v)   Dooley, H.D. 1979.   Factors influencing water movements in the Firth of Clyde.   Estuarine and Coastal Marine Science, 9, pp. 631-641.

(vi)   Eagle, R.A. 1975.   Natural fluctuations in a soft bottom benthic community.   Journal of the Marine Biological Association of the United Kingdom, 55, pp. 865-878.

(vii)   GESAMP, 1982.   Scientific Criteria for the Selection of Waste Disposal Sites at Sea.   IMCO/FAO/UNESCO/WMO/WHO/IAEA/UN/UNEP Joint Group of Experts on the Scientific Aspects of Marine Pollution (GESAMP).   Reports and Studies No. 16, 60 pp.

(vii)   Gillbright,M. 1983.   Einen "red tide" in der sudlichen Nordsee and ihre Beziehungen zur Umwelt. Helgolander Meeresunters., 36, pp. 393-426.

(ix)   Halcrow, W., MacKay, D.W., and Thornton, I. 1973.   The distribution of trace metals and fauna in the Firth of Clyde in relation to the disposal of sewage sludge. Journal of the Marine Biological Association of the United

Kingdom, 53, pp. 721-739.

(x)   Murray, A.J. and Norton, H.G. 1982.   The field assessment of
        effects of dumping wastes at sea: 10 Analysis of chemical
        residues in fish and shellfish from selected coastal
        regions around England and Wales. Fisheries Research
        Technical Report, MAFF Directorate of Fisheries Research,
        Lowestoft, No. 69, 42 pp.

(xi)   Norton, M.G., Eagle, R.A., Nunny, R.S., Rolfe, M.S.,
        Hardiman, P.A. and Hampson, B.L. 1981.   The field
        assessment of the effects of dumping wastes at sea:
        8 Sewage Sludge dumping in the outer Thames Estuary.
        Fisheries Research Technical Report, MAFF Directorate of
        Fisheries Research, Lowestoft, No. 62, 62 pp.

(xii)   Norton, M.G., Jones, P.G.W., Franklin, A. and Rowlatt, S.M.
        1984. Water quality studies around the sewage sludge
        dumping site in Liverpool Bay. Estuarine, Coastal and
        Shelf Science, 19, pp. 53-67.

(xiii) Norton, M.G., Rowlatt, S.M. and Nunny, R.S. 1984.   Sewage
        sludge dumping and contamination of Liverpool Bay
        sediments. Estuarine, Coastal and Shelf Science, 19,
        69-87.

(xiv)   Pearson, T.H. 1985.   The benthic ecology of an accumulating
        sludge disposal ground. Proceedings of the Fourth
        International Ocean Disposal Symposium, Plymouth, England,
        April 1983.

(xv)   Pearson, T.H. and Blackstock, J. 1983.   Selection of
        indicator species: a coordinated ecological and biochemical
        approach to the assessment of pollution. Oceanologica
        Acta. Proceedings 17th European Marine Biology
        Symposium, Brest, France, 27 September-1 October 1982,
        147-151.

(xvi)   Pearson, T.H., Gray, J.S. and Johannessen, P.J. 1983.
        Objective selection of sensitive species indicative of
        pollution-induced change in benthic communities. 2.
        Data analyses. Marine Ecology - Progress Series, 12,
        pp. 237-255.

(xvii) Pearson, T.H. and Rosenberg, R. 1978.   Macrobenthic
        succession in relation to organic enrichment and pollution
        of the marine environment. Oceanography and Marine Biology:
        Annual Review, 16, pp. 229-311.

(xviii) Rees, E.I.S. and Walker, A.J.M. 1984.   Macrobenthos and
        community monitoring studies across the dumping ground.
        Appendix F in Sewage sludge disposal in Liverpool Bay.
        Research into effects 1975 to 1977 Part 2 - Appendices
        DOE/WTD, pp. 113-163.

(xix)   Rohatagi, H. and Chen, K.Y. 1975.   Transport of trace metals by
        suspended particulates on mixing with sea water. Journal of
        the Water Pollution Control Federation, 47, 2298-2316.

(xx)   Rosenberg, R. (in press).   Marine eutrophication, a future
        coastal nuisance. Marine Pollution, Bulletin.

(xxi)   Talbot, J.W., Harvey, B.R., Eagle, R.A. and Rolfe, M.S. 1982.
        The field assessment of effects of dumping wastes at sea:
        9. Dispersal and effects on benthos of sewage sludge dumped
        in the Thames Estuary. Fisheries Research Technical
        Report. MAFF Directorate of Fisheries Research,
        Lowestoft, No. 63, 42 pp.

MOBILITY OF POLLUTANTS IN DREDGED MATERIALS -
IMPLICATIONS FOR SELECTING DISPOSAL OPTIONS

U. Förstner, W. Ahlf, W. Calmano, M. Kersten
Division of Environmental Engineering
Technical University Hamburg-Harburg
P.O. Box 90 14 03
D-2100 Hamburg 90
West Germany

ABSTRACT. Chemical speciation and bioassay data assist in establishing
criteria for the alternatives "marine vs. land disposal" of contaminated
dredged sediments. While lowering of pH and increasing E_h-values are
mainly affecting mobility of trace metals after land deposition, the
interactions with salt ions are particularly important for the disposal
of dredged materials in the marine environment

1.   INTRODUCTION

Sediments are increasingly recognized as both a carrier and a possible
source of contaminants in aquatic systems; further, that there are
potential biological effects from polluted solid materials disposed on
agricultural land. Metals, for example, are not necessarily fixed permanently by sediment, but may be recycled via biological and chemical
agents, both within the sedimentary compartment and the water column.
    This is especially valid for "dredged materials". As shipping
demands minimal water-action or current within the harbour basins, this
means that optimal conditions have been created for the sedimentation
of river or sea-borne material. In order to keep these ports and channels accessible to (marine) shipping, this material has to be removed
regularly by dredging (Van Driel et al., 1984). Material dredged from
waterways and harbors of the United States amounts approx. 250 million
m³ annually (Windom, 1976). In the river mouths to the southern coast
of the North Sea, approx. 20 million m³ have to be dredged from Rhine/
Meuse (Rotterdam Hr.) and approx. 10 million m³ from the rivers Scheldt
(Antwerpen), Weser (Bremen) and Elbe (Hamburg) (d'Angremond et al.,1978).
The total quantity of sediment which is dredged in The Netherlands,
Belgium and West Germany amounts to about twelve times the total suspended matter supply from the Rhine (Van Driel et al., 1984). The possibilities of disposal of these enormous quantities of material are
severely limited because of the pollution present in the dredged material.
    The major disposal alternatives are subaqueous (open-water) dispo-
sal, application to intertidal sites, and upland deposition; these cate-

G. Kullenberg (ed.), The Role of the Oceans as a Waste Disposal Option, 597–615.

gories differ primarily in the biological population exposed to the contaminated sediments, oxidation-reduction conditions, and transport processes potentially capable of removing contaminants from dredged material at the disposal site (Gambrell et al., 1978).

In discussions of environmental problems associated with contaminated sediments it is now widely accepted that the ecological significance of pollutant inputs is determined by the specific form and reactivity of the compound rather than by its rate of accumulation. This effect typically refers to adsorbed metal pollutants which may become partly mobilized in the aquatic milieu by changes in pH- and redox conditions, by increased salinities or concentrations of organic chelators. While the first two factors typically relate to upland disposal of dredged materials, the effect of salinity is more important for the deposition of freshwater sediments in intertidal and coastal marine environments.

2. METHODS

Pollutant transfer to organisms predominantly takes place via dissolved species. However, with regard to the selection of disposal options for dredged materials, the study of the water phase only would not be fully satisfying. While most of the actual situation would be reflected in these data, the potentialities of future adverse effects as well as the possible measures for reducing such hazards cannot be predicted. Two major concepts of assessing the environmental implications of pollutants for different disposal alternatives are: (1) Chemical characterization of critical phases, including elutriate test methods and sequential leaching techniques, and (2) bioassays, in which organisms are used to detect or measure the presence or effect of one or more substances or conditions (EPA, 1977).

2.1. Chemical Characterization

2.1.1. Elutriate test. To estimate short-term chemical transformations, the interrelations between solid phases and water has been increasingly subjected to laboratory experimentation. The advantage of such experiments is that especially important parameters can be directly observed and particularly unfavourable conditions simulated. The Army Corps of Engineers and the US Environmental Protection Agency have developed an elutriate test that is designed to detect any significant release of chemical contaminants in dredged material. This test involves the mixing of one volume of the dredged sediment with four volumes of the disposal site water for a 30-min shaking period. If the soluble chemical constituent in the water exceeds 1.5 times the ambient concentration in the disposal site water, special conditions will govern the disposal of the dredged material (Lee and Plumb, 1974).

2.1.2. Estimation of buffer capacity. On a regional scale acid precipitation is probably the most important single factor affecting metal mobility in surface waters, particularly for Al, Cd, Mn, and Zn. Acidity of surface waters imposes problems in all aspects of metal enrichment,

ranging from the toxification of drinking water to problems concerning
the growth and reproduction of aquatic organisms, the increased leaching
of nutrients from the soil and the ensuing reduction of soil fertility,
the increased availability and toxicity of metals, and finally to the
undesirable acceleration of Hg methylation in sediments (Fagerström and
Jernelöv, 1972; see section 3.3.2.). During land disposal of anoxic
sludges, acid solutions can result from the oxidation of sulfidic compounds; kinetics of these transformations are typically affected by
bacteria (Singer and Stumm, 1970). (The capability of certain bacteria
- e.g., Thiobacillus thiooxidans and T. ferrooxidans - to oxidize sulfur
and ferrous iron, while decreasing the pH-values from 4-5 to approximately pH 2 can be utilized for enhancing dissolution of metals from dredged materials; Calmano et al., 1983.)

For many metal examples, a linear relationship has been found between pH-values and dissolved metal concentrations. Therefore, with regard to the mobility of metals, the buffer capacity of the sediment
sample is of prime importance. It is shown in Figure 1 with 'titration
curves' for suspensions of dredged material from Neckar River and Hamburg Harbor (Elbe River) that content of calcium carbonate is the predominant factor affecting buffer capacity. For quantifying these properties and for better comparison of samples it is proposed to use the
term ' Δ pH', which is characterized by the difference of pH-values of
10-percent sludge suspensions in distilled water (pH_0) and in 0.1 N
sulfuric acid after 1 h shaking time (pH_x). Three categories of ' Δ pH'-
values can be established , ranging from $\Delta pH < 2$ (strongly buffered),
ΔpH 2-4 (intermediate) to $\Delta pH > 4$ (poorly buffered). While the sediment
from Neckar River belongs to the first category, the dredged sludges
from Hamburg Harbour exhibit very low buffer capacity, as evidenced by
the immediate lowering of pH-values upon addition of sulfuric acid.

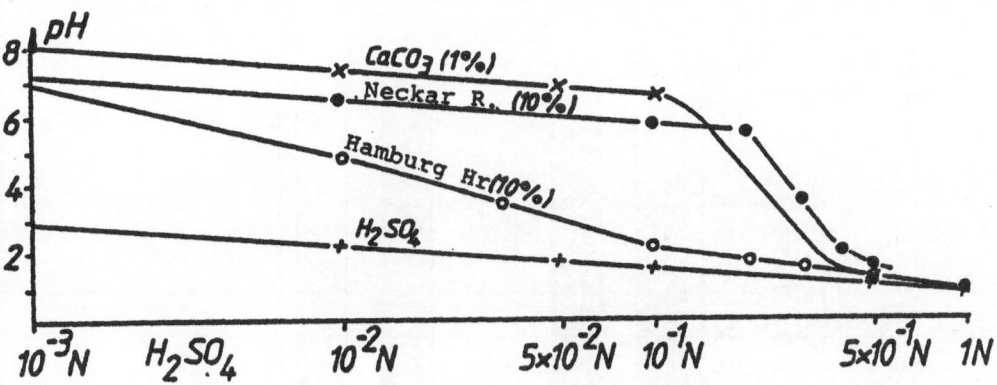

Figure 1. pH-changes of suspensions of calcium carbonate and river
sediments from additions of sulfuric acid in different concentrations
(after Förstner et al., 1985).

2.1.3. <u>Sequential chemical extraction</u>. Since adsorption of pollutants
onto particles is a primary factor in determining the transport, deposition, reactivity, and potential toxicity of these materials, analytical methods should be related to the chemistry of the particle's surface
and/or to the metal species highly enriched on the surface. Apart from
direct analysis on the particle's surface (Keyser et al., 1978) solvent
leaching techniques can provide information on the behavior of metal
pollutants under typical environmental conditions. Common single reagent
leaching tests, e.g., U.S. EPA, ASTM, IAEA, and ICES use either distilled water or dilute acetic acid (Theis and Padgett, 1983). In connection
with the problems arising from the disposal of contaminated dredged
materials and for classifying dredged substances, e.g. for ocean dumping,
upland disposal or storage in confined water bodies (Gambrell et al.,
1983), sequential extraction procedures have been developed which include the succesive leaching of metals from interstitial waters, and
from ion exchangeable, carbonatic, easily reducible, moderately reducible, organic and residual sediment fractions (Engler et al., 1974). One
of the more widely used techniques, which are still posing many pitfalls
(Calmano and Förstner, 1983) and being problematic for the assessment
of elements' "bioavailability" (Luoma, 1983) has been described by
Tessier et al. (1979).

    Because of the inherent instability of anoxic and especially inter-
mediate "post-oxic" (see below) metal forms it is clear that sample preparation is distinctly affecting the fraction of associations such as
exchangeable and easily reducible phases in a sediment sample. Previous
studies in our laboratory on the effect of various drying procedures on
the results of sequential leaching experiments for trace and minor elements have demonstrated that freeze-drying of anoxic sediments leads
to characteristic changes, as shown for iron and cadmium in Figure 2:

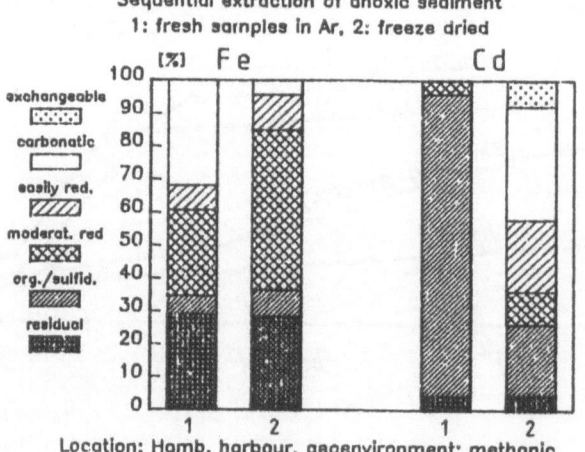

Figure 2. Sample drying as an important methodological error of the
sequential extraction procedure (Kersten et al., 1985).

Cadmium sulfidic associations are depleted in favour of exchangeable, carbonatic and easily reducible forms; this would clearly involve a weakening of the bonding strength of cadmium in the oxidized sediment, whereas transformations from carbonate to oxidic phases, e.g. for iron, would affect stronger fixation of metals in the sediment sample follo- wing oxidation. Thus all handling of fresh samples, including experi- mental performance prior to chemical analysis, has to be accomplished in a glove box under Ar-atmosphere.

2.2. Sediment Bioassay Techniques

Relatively few bioassays have been conducted to monitor the effects of contaminated sediment. Two general types of sediment bioassays may be carried out: One type is concerned with water column effects and the other type with the effects of the solid phase, which is ingested by many species of aquatic organisms. It is of importance that there are different vectors of metal uptake for aquatic organisms. Predicting, however, an effect of a pollutant in an ecosystem requires knowledge of ecological processes such as competition and predator-prey inter- actions. And we have to take into consideration than, for example, the sediment-water distribution observed in a closed laboratory experiment may not be the same as the distribution in open systems in nature. (The application of various bioassays to assess the potential pollution ari- sing from dredging activities, as required by legislative mandate in the U.S., has been reviewed by Engler, 1980; an overview of sediment-asso- ciated contaminants and their bioassessment is given by Munawar et al., 1984.)

An integrated approach correlating the bioassay data of a 'two- chamber exchange system' to the results from a six-step chemical extrac- tion sequence was designed by Ahlf et al. (1984). Here, an apparatus was used which·separates an algal population and suspended non-living solids by a 0.45 µm pore diameter membrane. Initial results indicate that there is an increase in the trace metal uptake by A. bibraianus in the sediment-water system in comparison to the water only control system. The uptake of nickel and cadmium by the algae indicate a significant correlation to the adsorbed/exchangeable phase as determined from the extraction studies. The two chamber system can be exposed to typical environmental factors, e.g. to an increase in salinity for studying the interactions in an estuary or during moulded subaqueous disposal. Here, it is shown that with increasing salinity the sorption of manga- nese by the test organisms decreased, whereas the availability of copper was increasing.

Interpretation of algal bioassays and application of toxicity data to natural sediments and waters is complicated because of growth-limi- ting factor interaction. The reverse relationship between algal growth and algal copper content corresponds to the fact, that copper is highly toxic to most aquatic plants. The relationship of toxicity data from 96-h sediment bioassay with algae indicated that in such short-term experiments chlorophyll content is probably the most sensitive response indicator.

3. RESULTS AND DISCUSSION: MOBILITY OF POLLUTANTS

3.1. Effects during Dredging Operations

Physical, chemical and biological effects of dredging operations have
been reviewed by Sly (1977), Gambrell et al. (1977), Sweeney (1975) and
Hebert and Schwartz (1983). It has been found that physical effects are
undoubtedly significant at a local level; however, the scale of events,
e.g. in Great Lakes, remains small in comparison to sediment suspension
resulting from wind-wave action (Sly, 1977). Some changes in dredged
material pH and redox potential may occur in the interstitial water components and solids when mixed with oxygenated surface water for extended
periods (Gambrell et al., 1977; Hebert and Schwartz, 1983). The oxidation of reduced material was found to cause approximately 12% of the
oxygen depletion in the central basin of Lake Erie (Burns and Ross,1972).
The effects of contaminant's release are as yet not fully understood;
however, with respect to the mobility of toxic metals, redox effects
seem to be more relevant than pH-changes during dredging (Weber et al.,
1982). Biological factors in respect to dredging operations include species transformations of toxic metals, e.g. by methylation (Saxena and
Howard, 1977), biodegradation, e.g. of PCBs (Clark et al., 1979), and
wide range of processes affecting "bioconcentration" of contaminants,
either by direct exchange of chemicals from suspended particulates to
gill or tissue surfaces or by stripping of chemicals from particulates
as they pass through digestive tracts. While there are many indications
that dredging has immediate impact on the benthic organisms (without
long-term deleterious changes; Sweeney, 1975), only the study of Seeyle
et al. (1982) as yet found direct effects on other components of the
ecosystem, i.e. accumulation of PCBs, zinc and mercury in fish. It has
been stressed, therefore, that before dredging decisions can be made
the safety of the ecosystem must be ensured with carfully designed and
monitored pilot studies (Hebert and Schwartz, 1983).

3.2. Effects of Redox Variations (mainly Land Disposal)

Decomposition of organic matter, which is mediated by microorganisms,
generally follows a definite succession in sediments depending upon the
nature of the oxidizing agent: The successive events are generally oxygen consumption (respiration), nitrate reduction, sulfate reduction, and
methane formation. On the basis of typical mineral associations Berner
(1981) has proposed a scheme for the recognition of ancient diagenetic
facies: MnO_2-type minerals are indicators for oxic conditions. Post-oxic
(weakly reducing) conditions, involving successively the reducting of
nitrate, manganese and iron, are characterized by the absence of sulfide
minerals. Sulfidic and methanic (both strongly reducing) environments
typically contain either newly formed or pre-existing sulfidic iron
minerals. Methane, siderite (Fe-carbonate) and Fe-phosphate form more
readily in non-marine sediments than in marine deposits, due to an
initially lower sulfate concentration in the water (Emerson and Widmer,
1978).

3.2.1. Early diagenetic reactions. Diagenesis of dredged sediments was studied experimentally in the field by Kerdijk et al. (in Salomons and Förstner, 1984, p. 86-90), where large pits (80 x 30 x 6 m) were dug in the Rhine Estuary below the water table and filled with material from freshwater, brackish and seawater environments. While the SO_4-content in pore water is continuously diminished by the formation of H_2S, the concentration of iron in pore water reaches a maximum (after 40-50 days of the one-year-experiment; after 70-100 days for the marine sediment), resulting from the reduction from iron(III) to iron(II); subsequently, there is precipitation as iron sulfide (mainly marine environment) or iron carbonate (fresh and brackish water). Similar developments have been found for copper in the freshwater and brackish sediments and for cadmium in the marine material.

With regard to pH-evolution (as an important variable for the solubility of trace metals) four different situations can be distinguished (Salomons et al., 1984): In the presence of either sulfate or reducible iron, pH-values are stabilized near pH 7 (6.75-7.25), whereas in the absence of both constituents (methanogenesis as the only macro-chemical oxidation-reduction reaction) the pH drops when the buffer substances such as calcite or earlier formed siderite are consumed. It has been stressed by Salomons et al. (1984) that in the marine environment all reducible iron is transformed to iron sulfide and also a large portion of the degradable organic matter is used up in this process.

3.2.2. Metal sulfide solubilities. Generally the solubility of metal sulfides is several orders of magnitude lower than the solubility of the respective oxide compounds. However, it is important to know whether metal concentrations are determined by precipitation-dissolution reactions or by adsorption processes (Salomons et al., 1984): With precipitation-dissolution processes the concentration will not depend on the total metal content, and an increased input will not affect the concentration in the pore waters; on the other hand, when adsorption is the main process for metal binding, the increased input will cause an increase in metal concentrations in the pore water. Comparison of calculated and measured equilibrium concentrations, pore water/sediment concentrations as well as adsorption experiments with and without sulfides suggest the predominance of solid metal sulfides in anoxic sediments (Salomons et al., 1984).

At increasing sulfide concentrations transition metals (Mn, Fe, Co, Ni) are increasingly forming solid metal sulfides, whereas mobility of "class B metals" (Cu, Ag, Zn, Cd, and Pb) can be significantly enhanced by the formation of strong complexes (Jacobs and Emerson, 1982). Pore water data of Boulegue et al. (1982) from organic-rich salt marsh sediments along the Delaware Estuary suggest, that at least 90% of total dissolved copper is strongly associated with organo-sulfur complexes, which apparently render it non-responsive to changes in the speciation of sulfur and the corresponding variations of the inorganic copper complexes. As the organo-sulfur molecules are more resistant to oxidation than inorganic sulfur molecules, complexes of Cu^+, Ag^+, Hg^+, Hg^{2+} and Cd^{2+} may be effectively transferred across the redox boundary via pore waters.

3.2.3. <u>Oxidation of anoxic sediments</u>. Under oxidizing conditions the
controlling solids may change gradually from metallic sulfides to carbonates, oxyhydroxides, oxides, or silicates, thus changing the solubility of trace metals. Field evidence for changing cadmium mobilities
was reported by Holmes et al. (1974) from Corpus Christi Bay harbour:
During the summer when the harbour water is stagnant cadmium is precipitated as CdS at the sediment/water interface; in the winter months,
however, the increased flow of oxygen-rich water into the bay result in
the release of some of the precipitated metal. The pH/redox dependencies
of exchangeable cadmium were studied experimentally by Gambrell et al.
(1977) with Mississippi River sediment: It is suggested that considerable cadmium release to relatively mobile forms may occur as cadmium-
contaminated sediment is transported from a near-neutral pH, reducing
environment to a moderately acid, oxidizing environment. Under these
conditions, cadmium levels of subsurface drainage water from upland disposal of dredged materials may be increased, and cadmium availability
to plants growing on the material enhanced.

Changes in cadmium forms have been determined by Hoeppel et al.
(1978) at dredged material land containment areas. Comparison of five
influent and effluent samples (Table I) indicates, that cadmium concentrations increased significantly in the carbonate phases, presumably as
a direct result of transformation from organic/sulfide phases present
in the influent slurries (see also Khalid, 1980).

TABLE I   Cadmium forms in solids from confined land disposal

Chemical Fraction	Percent of Total Cadmium Content	
	Influent	Effluent
Exchangeable [a]	21.0 %	18.0 %
Carbonate [b]	21.4 %	56.7 %
Easily reducible[c]	9.2 %	11.8 %
Remaining phases	49.3 %	13.5 %

[a] ammonium acetate [b] 1 M acetic acid extractable
[c] 0.1 M hydroxylamine hydrochloride in 0.01 M nitric acid

Apart from typical effects during land disposal it has been shown by
Kersten and Kerner (1985) from chemical extraction experiments (see
Fig. 2) that biological aeration can mobilize critical metals in deeper
parts of anoxic marsh sediments. Aquatic macrophytes can provide oxygen
for their soil rhizosphere, thus maintaining oxic to post-oxic environments up to several tenths of centimeters deep in the sediment. Investigations on the effects of redox changes from alternating flooding/drai-
ning of intertidal sediments were performed on a freshwater tidal flat
near Hamburg harbour (Kersten et al., 1985): The higher percentage of
labile Cd-forms in the oxic layers is accompanied by a marked depletion
in the total contents of the metal. We assume that approx. 30-50% of the
original cadmium content is extracted by a process of "oxidative pumping"
through tidal action, and can be exported from the estuary in dissolved
form.

3.3. Effects of Salinity (mainly Marine Disposal)

Land-derived particulate matter entering the estuarine and coastal marine environment can be affected by changes in the pH, chlorinity, cation concentration, and redox processes. Metal concentration and species in the water phase is typically influenced by the turbidity maximum and the formation of new particulate matter (Salomons, 1980); the removal of river-borne iron and in some estuaries of manganese at low salinity is well established (Duinker, 1980). The controversy regarding the interpretation of the seaward decrease in particulate metal concentrations - mobilization (De Groot, 1973) or mixing (Müller and Förstner, 1975) - is practically resolved from "either...or" to "not only...but also". Release of trace metals from particulate matter has been reported from several estuaries, and has partly been interpreted by the intensive breakdown of organic matter, whereafter the released trace metals become complexed with chloride and/or ligands from the decomposing organic matter (see review of Förstner, 1984). It has been suggested by Millward and Moore (1981) that the major cations, magnesium and calcium, are partially co-adsorbed and competition from these species for adsorption sites increases with increasing salinity.

Investigations on the behaviour of strong synthetic chelator nitrilotriacetic acid (NTA), which is used in some countries already in detergents for substituting polyphosphates, show characteristic differences for organically bound metals at the freshwater/seawater interface (Förstner et al., 1984): Combination of NTA and salinity causes lower sorption rates for cadmium than does either one of both factors, whereas for copper higher salt contents are clearly counteracting the mobilizing effect of increasing NTA-concentrations. This can be explained by the effect of calcium and magnesium concentrations, which are displacing heavy metals in their organic complexes; the 'free' metal ions may either hydrolize to become increasingly adsorbed onto the particulate matter (example: Cu), or may form chloro-complexes, which are partly kept in solution (example: Cd).

TABLE II Species distribution in seawater for Cu (Millero, 1980), Pb and Cd (Nürnberg, 1983) and Zn (Bernhard et al.,1975)

Metal Species	Copper	Lead	Zinc	Cadmium
Me^{2+}	3 %	1.8%	39.0%	1.9%
$MeCl^+$	1 %	8.6%	15.8%	29.1%
$MeCl_2/Me(OH)Cl$	0.6%	12.6%	10.0%	37.2%
$MeCl_3^-/MeCl_4^{2-}$	–	–	3.8%	31.0%
$MeSO_4$	0.4%	–	15.8%	–
$MeOH^+$	2.8%	30 %	0.6%	–
$Me(OH)_2$	1 %	0.5%	0.1%	0.8%
$MeCO_3$	80 %	43.0%	13.0%	–
$Me(CO_3)_2^{2-}$	–	3.7%	1.8%	–
Me-humate	11 %	–	–	–

3.3.1. Cadmium. The examples in Table II, where speciation data (model calculations) of various authors are compiled for trace elements in seawater, indicate typical associations of Cu with carbonate and humate complexes, elevated contents of hydroxo- and carbonato-species for Pb, of ionic species for Zn, whereas Cd is mainly associated with chlorospecies. Computations of equilibrium models by Sibley and Morgan (1975) showed that solid $CdCO_3$ was an important species in freshwater, while cadmium in the seawater was dissolved and was present as a chloride complex; maximum dissolved concentration of cadmium occurred at seawater/freshwater ratio of 1:1, the point where solid $CdCO_3$ dissolved. Since the negative charge on river-borne suspended colloids is not reversed by adsorption of seawater cations (Hunter, 1983) it can be expected that at least the negatively charged complexes (e.g., $CdCl_3^-$) are preferentially kept in solution.

Field investigations by Ahlf (1983) and Calmano et al. (1984) on longitudinal sections of the Elbe and Weser estuaries in northern Germany indicate characteristic mobilization of cadmium - different from other trace metals studied - at the salinity gradient. Investigations of Helz et al. (1975) on the fate of trace metals from wastewater effluents in the Chesapeake Bay estuary showed a decrease in the Cd/Zn ratio in the sediments in the seaward direction, thus confirming suggested remobilization effects from water data. It should be mentioned that another area of cadmium release from particulate matter is the seasurface microlayer (Lion and Leckie, 1981; see also Crecelius, 1980). Laboratory experiments with soils and freshwater suspended matter by Hahne and Kroontje (1973) and Van der Weijden et al. (1977) revealed much higher remobilization of cadmium than of other trace elements by application of saline water. Results of experiments on anoxic Rhine sediments by Salomons et al. (1982) are shown in Table III:

TABLE III   Metal mobilization from anoxic Rhine sediments by treatment with freshwater and seawater (Salomons et al., 1982)

	Zn	Cu	Ni	Cd	Pb
River Water	-0.8%*	0.9%	-2.0%*	1.0%	0 %
Seawater	2.2%	2.0%	2.5%	49 %	0.1%

* negative values indicate adsorption of metals from solution

Whereas in freshwater no remobilization was observed for all metals studied (in some cases even an additional adsorption from the water phase), treatment with seawater affects approx. 50% of the cadmium concentration to be released into the water. Similar effects were observed by Rohatgi and Chen (1975) at investigations on the release of trace elements from waste effluents on mixing with seawater, where up to 95% of cadmium was released from the suspended solids to the oxygenated seawater.

Two processes seem to be effective (Rohatgi and Chen, 1975):
(i) Oxidation either of organic particulates containing trace metals,

or oxidation of metal sulfides and the surface desorption of trace metals caused by a high dilution ratio (initial, rapid process), and (ii) complexation of trace metals to form soluble complexes of inorganic ligands such as Cl^-, and organic ligands, possible resulting from the oxidation of organic particulates. The slow oxidation of reduced materials possibly involves a change of controlling solids from very insoluble sulfides to higher solubility of carbonates (Khalid, 1980). Recent data of B. Prause (unpublished thesis Technical University Clausthal) suggest that the reaction kinetics - mainly with respect to the initial mobilization of metals from solids - are controlled by microbial activity; this would explain the findings of Salomons et al. (1982), that maximum mobilization of cadmium from anoxic sediments occurs only after six weeks of suspended interaction with seawater. Subsequent differentiations, however, are mainly influenced by thermodynamic factors,e.g. stability of chlorocomplexes, hydrolysis or readsorption to suspended solids.

3.3.2. Methylmercury. Methylation of inorganic mercury involves the non-enzymic reaction of mercuric mercury ions with methyl carboning compounds (e.g., Vitamin B_{12}), originating from bacterial synthesis. The monomethylmercury compound is avidly accumulated by fish and shellfish, whereas the dimethyl compound, having low solubility and high volatility, tends to vaporize from the water phase to the atmosphere where it may be subjected to photolytic decomposition (Clarkson et al., 1984). Data of Craig and Moreton (1984) and Wood and Wang (1983) suggest that S^{2-}-catalyzed disproportionation to volatile $(CH_3)_2Hg$ and insoluble HgS is reducing CH_3Hg^+-concentrations at higher sulfide levels, thus explaining the findings of particularly low concentrations of methyl mercury in natural sediment from estuarine and marine environments (Fagerström and Jernelöv 1972; Craig and Moreton, 1983).

The effect of salinity on the formation and stability of methylmercury has been studied by Blum and Bartha (1980) and Compeau and Bartha (1983); the following findings have significant practical implications: (i) Conversion of organic mercury to methyl mercury in anaerobic sediments is negatively correlated with salinity. As an explanation the theory was advanced that sulfide, derived from sea salt sulfate by microbial reduction, interferes with Hg^{2+} methylation by forming highly insoluble HgS (K = 10^{-53}). (ii) Bicarbonate, as a major component of seawater, noticably slows the methylation of Hg^{2+} under both aerobic and anaerobic conditions. In the presence of bicarbonate, other sea salts anions have no significant influence on the methylation of Hg^{2+}. The authors speculate that other conditions being equal Hg^{2+} will be more available for methylation in "soft" than in "hard" (bicarbonate-rich) freshwater systems. (iii) Elevated sulfide concentrations, as well as all other seawater anions had no effects on the stability of pre-existing monomethylmercury chloride (CH_3HgCl). This coincides with the findings of Craig and Moreton (1984) that preformed methyl mercury complexes may resist attack by sulfide species. However, if free methyl mercury is released from these complexes, as biological metabolites for example, then the higher the sulfide concentration in the sediment the greater the likelihood of reaction between methyl mercury and sulfide occurring, with subsequent release of mercury (Craig and Moreton, 1984).

4. CONCLUSIONS AND OUTLOOK

Significant release of pollutants from dredged sediments can originate from the following processes (Lu et al., 1978):

(1) Diffusion from the interstitial water

(2) Oxidation of reduced metallic sulfide solids, which are generall highly insoluble, to more soluble oxidized solids

(3) Formation of soluble metal complexes due to an increase of metal ligands in the soluble phase (such as the high levels of chloride and dissolved organic carbon in the influent samples)

(4) Ion exchange

(5) Oxidation and decomposition of organic compounds, mediated by microbial activity

(6) Desorption from clay minerals or other solid species

As shown in the foregoing chapters and in more detail by Khalid (1980) cadmium seems to be the critical element in most cases: With respect to the short-term impacts, metal release into adjacent receiving waters (land disposal) or at the contact with oxygenated surface water (open water disposal) can be expected; in the latter case, interactions with oxygenated, higher saline waters, e.g. in the estuarine mixing zones, would probably have the strongest effect on cadmium releases. While already Davey (1976) warned against the disposal of contaminants in productive estuaries, it is definitely stated by Khalid (1980) that the disposal of cadmium-rich sediments in ecologically productive, high-energy nearshore, estuarine, and inlet zones should be avoided.

Regarding the long-term impacts, concern has been expressed for uptake of contaminants from polluted sludge materials by aquatic and benthic organisms which may result in bioaccumulation in the food web (open water disposal) and for effects of contaminants' leaching into the groundwater from upland disposal sites. Laboratory lysimeter experiments and actual field studies by Mang et al. (1978) to determine the potential adverse impact of land-disposed sludges on groundwater suggest that cadmium, for example, released from the dredged material was attenuated by the underlying soils, presumably by sulfide precipitation under reduced soil conditions and adsorption by clay minerals, sesquioxides and humic materials. Scavenging effects of Fe/Mn oxyhydroxides may explain discrepancies, which have been found by Lu and Chen (1977) between experimental data and calculated equilibrium concentrations: Cd, Cu, Ni, and Pb are far below, Fe and Mn far above the equilibrium data. However, the movement of metals is influenced by several unknown factors; thermodynamic and kinetic influences make transport phenomena difficult to explain and predict (Khalid, 1980).

While contaminated material should be covered with a thick layer of clean soil to avoid excessive metal uptake by plants colonizing the disposal site (Gambrell et al., 1978), mechanical and chemical stabilization is providing additional security in long-term management plans:

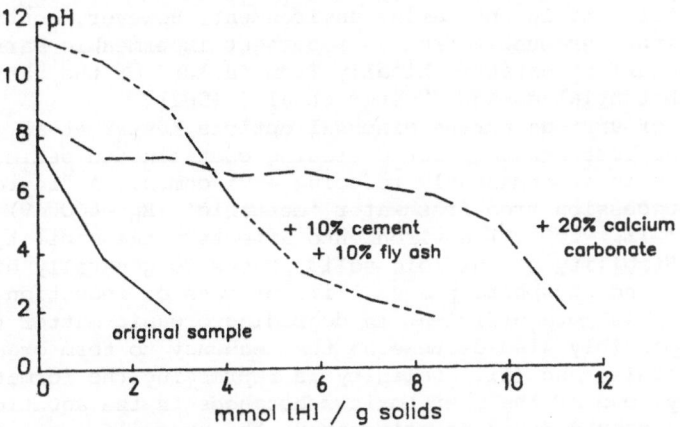

Figure 3. Effect of calcium carbonate and cement/fly ash additives on chemical stabilization of dredged sediment from Hamburg harbour (evaluation of 'buffer capacity' see section 2.1.2.)

Figure 3 shows 'acid titration curves' for Hamburg harbour mud without and after addition of limestone and cement/fly ash stabilizers. Best results are attained with calcium carbonate, since the pH-conditions are not changed significantly upon addition of $CaCO_3$. On the other hand it can be expected that both low and high pH-values will have unfavourable effects on the mobility of heavy metals.

Conventional wastewater treatment coagulants may not be practical because of problems with pH and volumes of the material required. However, several synthetic polymers have shown great promise in removing suspended particulates while having little effect on other chemical or physical properties of the settled solids (Wang and Chen, 1977; Jones et al., 1978; Gambrell et al., 1978).

Containment vs. Dispersion

Enclosure studies in the Marine Ecosystem Research Laboratory (MERL) suggest that trace metals are released from dispersed sediments by the oxidation of sulfides and organic matter; mobilization rates of Cu and Pb are higher from contaminated sediments than from less polluted ones (Hunt and Smith, 1983). Extrapolation from data on spiked sediments indicates that background levels of Cu and Pb will be reached after 44 and 400 years, respectively. Other findings with respect to metal mobilization include that after oxidation of the surface sediment the ecosystem is rapidly recovering and that an oxidized, bioturbated surface layer constitutes an efficient barrier against the transfer of most trace metals from below into the overlying water (Hunt and Smith, 1983). These data demonstrate the problematic effect of dispersing polluted sediments and suggest that containment is generally the more appropriate option for disposal of waste materials.

With respect to various containment strategies it has been argued that upland containment could provide a more controlled management than, for example, containment in the marine environment. However, contaminants released either gradually from an imperfect impermeable barrier (also to groundwater) or catastrophically from failure of the barrier could produce substantial damage (Kester et al., 1982).

In a review of various marine disposal options Kester et al.(1982) suggested that the best strategy for disposing contaminated sediments is to isolate them in a permanently reducing environment. A "regression" of biochemical succession from freshwater 'methanic' (E_h -400 mV) to marine 'sulfidic' (E_h -200 mV) environments affects metal mobility in three ways: (i) Stability of sulfidic solid phases is generally higher than of carbonate and phosphate phases, (ii) process of reduction of sulfate to sulfide is more efficient in degrading organic matter than methanogenesis, possibly also decreasing the tendency to form organic complexes with metals, and (iii) salinity is repressing the formation of methyl mercury, one of the most toxic substances in the aquatic milieu. Disposal in capped mound deposits above the prevailing sea-floor, disposal in subaqueous depressions, and capping deposits in depressions provide procedures for contaminated sediment; in some instances it may be worthwhile to excavate a depression for the disposal site of contaminated sediment that can be capped with clean sediment (Kester et al., 1982).

References

Ahlf, W.: The River Elbe: Behaviour of Cd and Zn during estuarine mixing. Environ. Technol. Letts. 4, 405-410 (1983)

Ahlf, W., Calmano, W., Förstner, U.: An algal assay method for determination of bioaccumulation and toxicity of sediment-bound heavy metals. In: P.G. Sly (ed.) Proc. 3rd Symp. Sediment and Water Interactions, Geneva, Aug. 1984. Springer-Verlag New York 1985

d'Angremond, K., Brakel, J., Hoekstra, A.J., Kleinbloesem, W.C.H., Nederlof, L., De Nekker, J.: Assessment of certain European dredging practices and dredged material containment and reclamation methods. Dredged Material Research Program, U.S. Army Engineer Waterways Experiment Station, Vicksburg, Miss., Technical Report D-78-58, Dec. 1978.

Berner, R.A.: A new geochemical classification of sedimentary environments. J. Sediment. Petrol. 51, 359-365 (1981)

Bernhard, M., Goldberg, E.D., Piro, A.: Zinc in seawater - an overview 1975. In: E.D. Goldberg (ed.) The Nature of Seawater. Dahlem Konferenzen, Physical and Chemical Sciences Report 1, p. 43-68 (1975).

Blum, J.E., Bartha, R.: Effect of salinity on methylation of mercury. Bull. Environ. Contam. Toxicol. 25, 404-408 (1980).

Boulegue, J., Lord, C.J., Church, T.M.: Sulfur speciation and associated trace metals (Fe, Cu) in the pore waters of Great Marsh, Delaware. Geochim. Cosmochim. Acta 46, 453-464 (1982).

Burns, N.M., Ross, C.: Project Hypo - discussion of findings. In: N.M. Burns, C. Ross (eds.) Project Hypo. CCIW Paper 6, 120-126 (1972).

Calmano, W., Förstner, U.: Chemical extration of heavy metals in polluted
    river sediments in Central Europe. Sci. Total Environ. 28, 77-90
    (1983)

Calmano, W., Ahlf, W., Förstner, U.: Heavy metal removal from contamina-
    ted sludges with dissolved sulfur dioxide in combination with bacte-
    rial leaching. Proc. Intern. Conf. Heavy Metals in the Environment,
    Heidelberg, Sept. 6-9, 1983, pp. 952-955, CEP Consult. Edingburgh

Calmano, W., Wellershaus, S., Liebsch, H.: The Weser Estuary: A study on
    the heavy metal behaviour under hydrographic and water quality con-
    ditions. Veröff. Inst. Meeresforschung Bremerhaven (1984, in press)

Clark, R.R., Chian, E.S.K., Griffin, R.A.: Degradation of polychlorinated
    biphenyls by mixed microbial cultures. Appl. Environ. Microbiol. 37,
    680-685 (1979).

Clarkson, T.W., Hamada, R., Amin-Zaki, L.: Mercury. In: J.O. Nriagu (ed.)
    Changing Metal Cycles and Human Health.Dahlem Konferenzen, Life Sci.
    Res. Report 28, 285-309. Springer-Verlag Berlin 1984.

Compeau, G., Bartha, R.: Effects of sea salt anions on the formation and
    stability of methylmercury. Bull. Environ. Contam. Toxicol. 31, 486-
    493 (1983).

Craig, P.J., Moreton, P.A.: Total mercury, methyl mercury and sulphide in
    River Carron sediments. Mar. Pollut. Bull. 14, 408-411 (1983)

Craig, P.J., Moreton, P.A.: The role of sulphide in the formation of di-
    methyl mercury in river and estuary sediments. Mar. Pollut. Bull.
    15, 406-408 (1984).

Crecelius, E.A.: The solubility of coal fly ash and marine aerosols in
    seawater. Mar. Chem. 8, 245-250 (1980).

Davey, E.W.: Trace metals in the oceans: problem or not? In: Water Qual-
    ity Criteria Research of the U.S. Environmental Protection Agency.
    U.S. EPA-600/3-76-079, pp. 13-22, Corvallis, Or. 1976.

De Groot, A.J.: Occurrence and behavior of heavy metals in river deltas
    with special reference to the rivers Rhine and Ems. In: E.D.Goldberg
    (ed.) North Sea Science. M.I.T. Press, Cambridge, Mass. p. 308-325
    (1973).

Duinker, J.C.: Suspended matter in estuaries. Adsorption and desorption
    processes. In: E. Olausson, I. Cato (eds.) Chemistry and Biogeochemi-
    stry of Estuaries. Wiley, Chichester/England, pp. 121-153 (1980).

Emerson, S., Widmer, G.: Early diagenesis in anaerobic lake-sediment.
    Thermodynamic and kinetic factors controlling formation of iron
    phosphate. Geochim. Cosmochim. Acta 42, 1307-1316 (1978).

Engler, R.M.: Prediction of pollution potential through geochemical and
    biological procedures: Development of regulatory guidelines and
    criteria for the discharge of dredged fill materials. In: R.A. Baker
    (ed.) Contaminants and Sediments. Ann Arbor Sci. Publ., Vol. 1, pp.
    143-169 (1980).

Engler, R.M., Brannon, J.M., Bigham, G., Rose, J.:'A practical selective
    extraction procedure for sediment characterization'.168th Meeting
    Amer. Chem. Soc., Atlantic City, N.Y., 17 p. (1974).

Environmental Protection Agency/Corps of Engineers (U.S.A.): 'Ecological
    Evaluation of Proposed Discharge of Dreged Material into Ocean
    Water'. U.S. Army Engineer Waterways Experiment Station, Vicksburg,
    Miss. 1977.

Fagerström, T., Jernelöv, A.: Aspects of the quantitative ecology of
    mercury. Water Res. 6, 1193-1202 (1972).

Förstner, U.: Effect of salinity on the metal sorption onto organic par-
    ticulate matter. In: R.W.P.M. Laane, W.J. Wolff (eds.) The Role of
    Organic Matter in the Wadden Sea. Neth. Inst. Sea Res. Publ. Ser.
    No. 10/84, pp. 195-209 (1984)

Förstner, U., Ahlf, W., Calmano, W., Sellhorn, C.: Metal interactions
    with organic solids in estuarine waters - experiments on the combi-
    ned effects of salinity and organic chelators. Proc. Intern. Conf.
    'Environmental Contamination', London, July 10-13, p. 567-572 (1984)

Förstner, U., Ahlf, W., Calmano, W., Kersten, M., Salomons, W.: Mobility
    of heavy metals in dredged harbor sediments. In: P.G. Sly (ed.)
    Proc. 3rd Symp. Sediment and Water Interactions, Geneva, August 1984
    Springer-Verlag New York (1985, in press).

Gambrell, R.P., Khalid, R.A., Patrick, W.H. jr.: 'Disposal Alternatives
    for Contaminated Dredged Material as a Management Tool to Minimize
    Environmental Effects'. U.S. Army Engineer Waterways Experiment Stn.
    Technical Report DS-78-8, Corps of Engineers, Vicksburg, Miss. 1978.

Gambrell, R.P., Reddy, C.N., Khalid, R.A.: Characterization of trace and
    toxic materials in sediments of a lake being restored. J. Water
    Pollut. Control Fed. 55, 1201-1213 (1983).

Gambrell, R.P., Khalid, R.A., Verloo, M.G., Patrick, W.H.jr.: Transforma-
    tion of heavy metals and plant nutrient in dredged sediments as
    affected by oxidation and reduction potential and pH. II. Materials
    and methods/results and discussion. U.S. Army Corps of Engineers,
    Dredged Material Research Program, Report D-77-4, 309 p., Vicksburg,
    Miss. 1977.

Hahne, H.C.H., Kroontje, W.: Significance of pH and chloride concentra-
    tion on behavior of heavy metal pollutants: mercury(II), cadmium(II),
    zinc(II), and lead(II). J. Environ. Qual. 2, 444-450 (1973).

Hebert, P., Schwartz, S.: Great Lakes Dredging in an Ecosystem Perspective
    - Lake Erie. Report submitted to the Dredging Subcommittee of the
    Water Quality Board, June 1983, 73 p.

Helz, G.R., Huggett, R.J., Hill, J.M.: Behavior of Mn, Fe, Cu, Zn, Cd,
    and Pb discharged from a wastewater treatment plant into an estua-
    rine environment. Water Res. 9, 631-636 (1975)

Hoeppel, R.E., Meyers, T.E., Engler, R.M.: Physical and chemical charac-
    terization of dredged material influents and effluents in confined

land disposal areas. U.S. Army Engineer Waterways Experiment Stn.,
Technical Report D-78-24, Corps of Engineers, Vicksburg/Miss. 1978.

Holmes, C.W., Slade, E.A., McLerran, C.J.: Migration and redistribution
of zinc and cadmium in marine estuarine systems. Environ. Sci. Technol. 8, 255-259 (1974).

Hunt, C.D., Smith, D.L.: Remobilization of metals from polluted marine
sediments. Can. J. Fish. Aquat. Sci. 40, 132-142 (1983).

Hunter, K.A.: On the estuarine mixing of dissolved substances in relation
to colloid stability and surface properties. Geochim. Cosmochim.
Acta 47, 467-473 (1983).

Jacobs, L., Emerson, S.: Trace metal solubility in an anoxic fjord. Earth
Planet. Sci. Lett. 60, 237-252 (1982).

Jones, R.H., Williams, R.R., Moore, T.K.: Development and Application of
Design and Operation Procedures for Coagulation of Dredged Material
Slurry and Containment Area Effluent. Technical Report D-78-54, U.S.
Army Engineer Waterways Experiment Stn., Vicksburg, Miss. 1978.

Kersten, M., Kerner, M.: Transformations of heavy metals and plant nutrients in intertidal flat sediment profiles of the Elbe Estuary as
affected by E_h and tidal cycle. Abstr. Intern. Conf. Heavy Metals
in the Environment, Athens/Greece, Sept. 1985.

Kersten, M., Förstner, U., Kerner, M.: Effect of tidal action on mineral/
water reactions and chemical forms of metals in a sediment core from
the Elbe River estuary. Preprint Ext. Abstr., Div. Environ. Chem.,
Amer. Chem. Soc. Meeting at Chicago, Ill., Sept. 1985 (submitted).

Kester, D.R., Ketchum, B.H., Duedall, I.W., Park, P.K. (eds.): Wastes in
the Ocean. Vol. 2: Dredged-Material Disposal in the Ocean. Wiley,
New York, 299 p. (1983).

Keyser, T.R., Natusch, D.F.S., Evans, C.A. jr., Linton, R.W.: Characterizing the surface of environmental particles. Environ. Sci. Technol.
12, 768-773 (1978).

Khalid, R.A.: Chemical mobility of cadmium in sediment-water systems. In:
J.O. Nriagu (ed.) Cadmium in the Environment. Vol. 1: Ecological
Cycling, pp. 257-304. Wiley, New York (1980).

Lee, G.F., Plumb, R.H.: Literature review on research study for the development of dredged material disposal criteria. U.S. Army Corps of
Engineers, DMRP, Report D-74-1, 145 p., Vicksburg, Miss. 1974.

Lion, L.W., Leckie, J.O.: The biochemistry of the air-sea interface.
Annu. Rev. Earth Planet. Sci. 9, 449-486 (1981).

Lu, J.C.S., Chen, K.Y.: Migration of trace metals in interfaces of seawater and polluted surficial sediments. Environ. Sci. Technol. 11,
174-181 (1977).

Lu, J.C.S., Eichenberger, B., Chen, K.Y.: Characterization of confined
disposal area influent and effluent particulate and petroleum fractions. U.S. Army Engineer Waterways Experiment Stn., Corps of Engineers, Technical Report D-78-16, Vicksburg, Miss. 1978.

Luoma, S.M.: Bioavailability of trace metals to aquatic organisms. A
    review. Sci. Total Environ. 28, 1-22 (1983)

Mang, J.L., Lu, J.C.S., Lofy, R.J., Stearns, R.P.: A study of leachate
    from dredged material in upland areas and/or in productive uses.
    U.S. Army Engineer Waterways Experiment Stn., Contract Report
    D-78-20. Corps of Engineers, Vicksburg, Miss. 1978.

Millero, F.J.: Chemical speciation of ionic components in estuarine
    systems. In: J.M. Martin, J.D. Burton, D. Eisma (eds.) River Inputs
    to Ocean Systems. UNEP/SCOR, Rome, p. 116-131 (1980).

Millward, G.E., Moore, R.M.: The adsorption of Cu, Mn, and Zn by iron
    oxyhydrate in model estuarine solutions. Water Res. 16, 981-985
    (1982).

Müller, G., Förstner, U.: Heavy metals in sediments of the Rhine and
    Elbe estuaries: Mobilization or mixing effect? Environ. Geol. 1,
    33-39 (1975).

Munawar, M., Thomas, R.L., Shear, H., McKee, P., Mudroch, A.: An Over-
    view of Sediment-Associated Contaminants and their Bioassessment.
    Canadian Technical Report of Fisheries and Aquatic Sciences, No.
    1253, 136 p.(1984).

Nürnberg, H.W.: Voltammetric studies on trace metal speciation in natu-
    ral waters. Part II: Application and conclusions for chemical ocea-
    nography and chemical limnology. In: G.G. Leppard (ed.) Trace Ele-
    ment Speciation in Surface Waters and its Ecological Implications.
    Plenum Press, New York, p. 211-230 (1983).

Rohatgi, N., Chen, K.Y.: Transport of trace metals by suspended particu-
    lates on mixing with seawater. J. Water Pollut. Control Fed. 47,
    2298-2316 (1975).

Salomons, W.: Adsorption processes and hydrodynamic conditions in estua-
    ries. Environ. Technol. Letts. 1, 506-517 (1980).

Salomons, W., Förstner, U.: Metals in the Hydrocycle. Springer-Verlag,
    Berlin, 349 p. (1984)

Salomons, W., Van Driel, W., Kerdijk, H., Boxma, R.: Help! Holland is
    plated by the Rhine (environmental problems associated with conta-
    minated sediments). Effect of Waste Disposal on Groundwater. Proc.
    Exeter Symp. IHAS, 139, 255-269 (1982).

Salomons, W., de Rooy, N., Kerdijk, H., Bril, J.: Sediments as a source
    of contaminants. In: R.L. Thomas (ed.) The Ecological Effects of
    In-Situ Sediment Contaminants. Proc. Workshop held at Aberystwyth/
    Wales, August 19-24, 1984. National Oceanic and Atmospheric Admini-
    stration, U.S. Department of Commerce (1985, in press).

Saxena, J., Howard, P.H.: Environmental transformation of alkylated and
    inorganic forms of certain metals. Adv. Appl. Microbiol. 21, 185-
    226 (1977).

Seeyle, J.G., Hesselberg, R.J., Mac, M.J.: Accumulation by fish of con-
    taminants released from dredged sediments. Environ. Sci. Technol.
    16, 459-464 (1982).

Sibley, T.H., Morgan, J.J.: Equilibrium speciation of trace metals in freshwater:seawater mixtures. Intern. Conf. Heavy Metals in the Environment, Toronto, pp. 319-340 (1975).

Singer, P.C., Stumm, W.: Acidic mine drainage: the rate-determining step. Science 167, 1121-1123 (1970).

Sly, P.G.: Some influences of dredging in the Great Lakes. In: H.L. Golterman (ed.) Interactions between Sediments and Fresh Water. Proc. Intern. Symp. Amsterdam, Sept. 6-10, 1976, pp. 435-445. Junk/ The Hague and PUDOC/Wageningen, The Netherlands (1977).

Sweeney, R. et al.: Impacts of the deposition of dredged spoils on Lake Erie sediment quality and associated biota. J. Great Lakes Res. 1, 162-170 (1975).

Tessier, A., Campbell, P.G.C., Bisson, M.: Sequential extraction procedure for the speciation of trace metals in particulates. Anal. Chem. 51, 844-851 (1979).

Theis, T.L., Padgett, L.E.: Factors affecting the release of trace metals . from municipal sludge ashes. J. Water Pollut. Control Fed. 55, 1271-1279 (1983).

Van der Weijden, C.H., Arnoldus, M.J.H., Meurs, C.T.: Desorption of metals from suspended material in the Rhine estuary. Neth. J. Sea Res. 11, 130-145 (1977).

Van Driel, W., Kerdijk, H.N., Salomons, W.: Use and disposal of contami- nated dredged material. Land + Water International, 53, 13-18 (1984)

Wang, C.-C., Chen, K.Y.: Laboratory Study of Chemical Coagulation as a Means of Treatment for Dredged Material. Technical Report D-77-39, U.S. Army Engineer Waterways Experiment Stn., Vicksburg, Miss. 1977.

Weber, W.J., Posner, J.C., Snitz, F.L.: Water Quality Impacts of Dredging Operations. Draft. U.S. Environmental Protection Agency. Grant No. R-801112. Grosse Ile Laboratory, Grosse Ile, Mi. (1982).

Windom, H.L.: Environmental aspects of dredging in the coastal zone. CRC Critical Reviews Environmental Control 5, 91-109 (1976).

Wood, J.M., Wang, H.-K.: Microbial resistance to heavy metals. Environ. Sci. Technol. 17, 582A-590A (1983).

MEASURING THE EFFECTS OF POLLUTION AT THE CELLULAR AND ORGANISM LEVEL

B.L. Bayne
Institute for Marine Environmental Research
Prospect Place, The Hoe
Plymouth PL1 3DH, U.K.

ABSTRACT. Responses by individual organisms to changes in the environment, including pollutants, can be measured as physiological and/or biochemical events. The former may lack specificity between stimulus (pollutant) and response but they relate directly to the fitness of the individual, so lending credibility to their use as general indices of the effects of pollution; the latter provide specificity and increased sensitivity but may be difficult to relate directly to traits of phenotypic fitness. Together, such response indices can be useful in assessments of environmental impact and current toxicological research provides a cogent, integrated framework for quantifying pollution damage. However, extrapolations from such studies to predictions of population damage are hampered by lack of information on the relationships between individual response and the population parameters of recruitment and mortality. Nevertheless, models of organism response, when linked to simulations of the physicochemical features of the environment, can provide useful tools for environmental impact assessment.

1. INTRODUCTION.

Organisms express suites of responses (biochemical, physiological, behavioural) to the biotic and abiotic features of the environment which, taken together, facilitate the survival and reproduction of the individual. It is commonplace to consider these responses as fundamentally adaptive, having evolved, by and large, so as to maximise individual fitness. However, phenotypic traits are not perfectly tailored to all potential environmental conditions. In a variable environment, an asymmetry exists between the observed expressions of the phenotype and the environmental demands on the organism; to measure this asymmetry is to be able to assess the current fitness of the individual and therefore to evaluate the impact of the environmental change.

Within this heirarchy of organismal traits are differences in specificity between stimulus and response. Recent research into the toxicology of marine invertebrates aims, in part, to characterise this specificity and to quantify general features of the stimulus/response relationship.

G. Kullenberg (ed.), The Role of the Oceans as a Waste Disposal Option, 617–634.
© *1986 by D. Reidel Publishing Company.*

Such studies provide an effective means for evaluating the impacts of waste disposal in the sea, meeting the criteria of sensitivity and specificity for short-term ("early-warning") assessments. However, measurements of these biochemical responses are at some distance, conceptually, from readily agreed features of phenotypic fitness and are therefore most convincing as indices of biological impact when linked to measures of organismal response, such as growth and reproduction. Together, biochemical and physiological determinations can provide a cogent statement of the biological effects of contamination on the organism; some of the processes measured also relate to possible population consequences, particularly recruitment. Nevertheless, it is difficult to extrapolate from the individual to the population in the absence of better understanding of likely compensatory mechanisms at the population level.

There is now a large and rapidly growing literature on the application of a variety of biological measurements to assess the effects on individuals of environmental contamination and waste discharge. Many of these studies are couched in terms of the "stress response", which is a concept derived from human medicine (Selye, 1973) but one that has proved widely applicable to fish (Mazeaud and Maseaud, 1981; Pickering, 1981) and to marine invertebrates (Bayne et al., 1984). This paper does not attempt a review of this material. Rather, I will briefly discuss some current research on the biochemical and physiological responses of marine molluscs, to illustrate the integrated nature of the toxicological (or "stress") syndrome in these animals, and suggest some of the possible consequences and applications of these studies in models designed to evaluate environmental impact. Although such studies may, by some criteria, have limited ecological scope, being confined to assessing effects on individual ("sentinel") organisms, they have a role to play in environmental management and are needed to complement the broader evaluations of community and ecosystem response.

2. BIOACCUMULATION.

A thorough analysis of the biological impact of contaminants must include consideration of the processes of bioaccumulation, which requires, in turn, knowledge of the chemical form (or "speciation") that the contaminant takes in the environment. Predictive modelling of biological impact requires statements relating chemical speciation to biological availability. An example concerns the Group B ("soft acid") elements which include the trace metals of most concern in pollution studies. The biological activity of these metals is dominated by their ligand binding characteristics. In the environment, concentrations of free metal ions are generally low. The dominant neutral species (e.g. $CdCl_2$ and $HgCl_2$) have been shown to penetrate lipid bilayer membranes and isolated epithelia some 10^3 and 10^7 times faster, respectively, than the free metal ion, due in part to their lipid solubility (Simkiss, 1983) and to their transport via phospholipid complexes within the cell membranes (George and Viarengo, 1984). Although trace metal speciation and bio-availability of inorganic moeities are reasonably well understood (Turner, 1985) there is less certainty regarding organo-metal

complexes, some of which may render the metals more available to the
organism by facilitating transfer from the bulk fluid phase to the cell
membrane, whereas others, which are not readily adsorbed on surfaces,
may reduce biological uptake by competitive binding to the metal.

For organic contaminants simple models based on octanol-water
partition coefficients have some utility (Gosset et al., 1983), but here
also complications arise when chemical form and biological uptake are
considered under natural conditions. The bio-availability of hydrocarbons is much modified in the presence of dissolved organic material,
for reasons not yet fully understood. The binding to particulates is
equally important since for some particles the organic molecules may be
occluded within the particulate matrix, rendering them relatively
unavailable to many organisms (F. Mantoura, pers. comm.). For other
particles the binding may be to freely available surfaces. Given the
diversity of organic compounds increasingly disposed to sea, a fundamental analysis is called for, relating features of molecular structure
and thermodynamic activity to environmental speciation. In the meantime
there will inevitably be some reliance on empirical measurements of
biological body burden to provide information on the availability of
contaminants to organisms.

3. CELLULAR AND BIOCHEMICAL RESPONSES.

3.1. Interactions with cytosolic components.

Trace metals are highly reactive within the cell, forming complexes with
both the substrates and products of enzymatic reactions, as well as with
the enzymes themselves. Immediate impacts may include the displacement
of naturally-occurring metal ions from the active sites of metalo-
enzymes, binding to deactivating sites, or the forcing of distortion on
the tertiary structure of the enzyme molecules. Enzyme activity will be
supressed as a result. However, such effects are not always discernible
in vivo, since denaturation of the enzyme may enhance the synthesis of
enzyme-protein so maintaining total enzymic activity (Jackim, 1974).
There are no simple relationships between in vitro and in vivo estima-
tions of metal effects on enzymic activity.

Livingstone (1984) has suggested that one test of the relevance of
observed enzyme effects concerns the extent of consequent alteration to
metabolic flux in vivo. The role of altered enzyme activities in
modulating metabolic flux is controversial, but where rate-limiting
enzymes are shown to be affected there can be some confidence that a
significant biochemical response has been invoked. However, the primary
virtue of biochemical tests lies in their stimulus/response specificity,
and analyses of biochemical processes that simply signal a general
response by the organism to an environmental challenge, without adding
to the information to be gained from physiological measures (see below)
does not seem worthwhile unless information is also provided on aspects
of phenotypic fitness; this criterion is seldom met.

One biochemical process that reflects a high specificity for trace
metal contamination is the synthesis and fate of metallothioneins.

These are low-molecular weight proteins with a high affinity for metal
cations (Roesijadi, 1980). Their normal function may be in the control
of Zn and Cu metabolism. On exposure to contaminating trace metals,
such as Cd, Hg and excess Cu, Zn is displaced from the thionein molecule
which then binds the metals, rendering them less toxic to the cell. If
the concentration of metals in the cell saturates the available pool of
metallothioneins, the synthesis of new thioneins may be stimulated,
either at the nuclear level or by enhanced translation of mRNA at the
ribosomes. If the influx of metals into the cell exceeds this enhanced
capacity for buffering by metallothioneins, the excess metal is then
available to interact with cellular components, with toxic results.

Another biochemical process that expresses a high specificity for a
particular type of contaminant is the series of reactions that metabol-
ise organic compounds, usually referred to as the mixed function oxidase
(MFO) enzyme system (Lee, 1981). The metabolism is in two phases: The
first phase involves the oxidation of the organic compound by reactions
catalysed by cytochrome P-450. The second phase involves further conversion and conjugation and the overall result is to transform the
organic molecule into more polar metabolites which might be excreted
more readily than the parent compounds. A feature of this system, as
with metallothioneins, concerns the increased synthesis ("induction")
of certain components following exposure of the organism to the appropriate contaminants. The response of the organism can be measured in
terms of overall activity within the detoxication system.

Recent studies by Livingstone et al. (1985) demonstrate the validity
of these measurements as indices of the effects of hydrocarbons on
mussels and on the gastropod mollusc Littorina littorea. Experiments
were carried out on animals exposed over 4-16 months to diesel oil (25
and 100 ppb) under natural environmental conditions in mesocosms at
Solbergstrand on the Oslo fjord. Cytological studies demonstrated that
animals in the control mesocosm did not suffer any decline in condition
relative to native animals in the fjord. Cytochromes P-450 and b5 and
NADPH-neotetrazolium and NADPH-cytochrome c reductase activities were
elevated in the digestive gland microsomes of exposed animals and were
highest in animals exposed to 100 ppb. Short-term exposures (8 days at
25 ppb) also resulted in elevated P-450 levels and, in these and in the
longer-term exposed animals, a recovery period of 8 days in clean seawater was sufficient to return values to control levels. Lysosomal
stability (see below) was considerably reduced in all exposed animals
(and recovered rapidly when transferred to clean seawater) and gametogenesis was seriously impaired at both 25 and 100 ppb exposures. Whole-
tissue concentrations of hydrocarbons reflected environmental levels
e.g. 2,3-dimethylnaphthalene equivalents of 0.3, 6.0, and 12.5 $\mu g.g^{-1}$ in
control, 25 and 100 ppb snails, respectively.

3.2. Interactions with cellular organelles.

The lysosomal system of the cell is involved in protein turnover, acting
as sites of hydrolytic enzyme activity and macro-molecular catabolism.
Metallothioneins, along with other proteins, are degraded within the
lysosome and their associated metals may be deposited, accumulating as

the lysosome ages into a residual body. The lysosomes can therefore act
as sites for metal accumulation and storage serving a secondary detoxication system. Residual bodies may eventually be excreted via exocy-
tosis in the kidney. Trace metals within the lysosomes are also now
known to stimulate lipid peroxidation and the formation of lipofuscin,
which then acts as a further sink for metal ions (George, 1982). As a
result of these various processes, however, the normal functioning of
the lysosomes may be disrupted; the lysosomal membranes become destabil-
ised (Viarengo et al., 1984) leading to leakage of hydrolytic enzymes
into the cell cytosol with consequent autolytic damage to the cell.

Lysosomal stability may also be disturbed more directly by organic,
lipid-soluble, compounds or by the highly reactive intermediates in the
MFO-mediated transformation of xenobiotics. It has therefore proved
possible to use measures of lysosomal membrane stability as a sensitive
and rapidly responsive index of generalised cellular injury in molluscs
(Moore and Lowe, 1984).

Other cellular organelles have been shown to be responsive to
pollutant exposure, such as the endoplasmic reticulum (the site of P-450
enzyme activity) and the mitochondria (e.g. as the result of lipid
peroxidation stimulated, as in the lysosome, by exposure to metals). An
example concerns genetic damage conferred directly on the chromosome by
nuclear/cytoplasmic exchange. Dixon (1982) has described increased
frequencies of chromosomal abnormalities in the embryos of the mussel,
Mytilus edulis, exposed to high concentrations of aromatic hydrocarbons
and has used techniques of sister chromatid exchange to quantify cytogenetic damage in laboratory experiments. In an extension to these
studies Dixon et al. (1985) have demonstrated a link between induced MFO
activity and cytogenetic damage in larval and adult mussels.

3.3. Cellular pathology.

Evidence of damage to cell organelles and to the nuclear material
provide pointers to causative links between biochemical responses to
pollutants and pathological symptoms within tissues. Such links have
been widely sought in recent years (Dethlefson, 1984; Malins et al.,
1984). One example concerning a bivalve mollusc has been described by
Lowe et al. (1981). Exposure of mussels to the water-accommodated
fraction of North Sea crude oil (30 µg total hydrocarbons 1^{-1}) resulted
in thinning and apparent atrophy in the cells of the digestive gland
(see Table I). These pathological symptoms were related to increases in
lysosomal volume associated with membrane destabilisation and accelerated lysosomal fusion and vacuolation; there was evidence of increased
free activity of hydrolytic enzymes within the cyoplasm of the cells.
The results of these and more recent studies (M.N. Moore and D.M. Lowe,
pers. comm.) indicate a convincing link from induced activity in the MFO
system (measured as NADPH-neotetrazolium reductase) to sub-cellular
damage to the lysosomal membranes and to pathological changes within cells
with concomitant impairment to digestive efficiency.

3.4. Discussion.

This chain of cause and effect from molecular events, through cellular
responses to pathological tissue damage represents an integrated toxicological response by the organisms and provides a practical means for
the assessment of pollutant impact at the cellular level. There is
evidence that these responses can provide information on the specific
classes of contaminant causing damage. The time-scale of response at
this level is rapid (e.g. enhanced metallothionein production, induced
cytochrome P-450 production, and lysosomal destabilisation within hours
to a few days of exposure). Many of the processes involved are
reversible and they are also very sensitive, measurable at contaminant
concentrations typical of some industrialised estuarine and coastal
sites (Table I).

TABLE I.   Cytochemical response of _Littorina littorea_ to hydrocarbons in
the field and in the laboratory (from Moore et al., 1982). NTR =
NADPH-dependent neo-tetrazolium reductase; * = P<0.05; ** = P<0.01;
Lysosomal latency in mins; Body concentrations in µg per g of wet tissue.

1: Field Data (Shetland Sites):

Sample	Lysosomal latency	Rel. activity of NTR	Concentrations in tissues Phenanthrene	Naphthalene
Ronas V.	24	100	0.11	0.11
Jetty 4	9**	131**	0.37 - 0.41	1.03 - 1.22

2: Laboratory Data (Phenanthrene at 400 µg l^{-1}):

Treatment	Lysosomal latency	Rel. activity of NTR
Control(5,13d)	22	100
Phenan.(5d)	3 - 7**	111 - 113*
Phenan.(5d) then clean sw(8d)	20	96

A general characteristic of these responses concerns the ubiquity
of response thresholds. Many of the processes can be viewed as having
evolved to deal with natural levels of potentially toxic compounds. The
processes may be capable of enhanced activity when levels of contaminants increase. It is when the maximum capacities of these detoxification processes are breached that toxic damage results. The organism
possesses its own "assimilative capacity" and pollution can be said to
occur only when this capacity proves insufficient to deal with the
influx of toxic material.

However, measures of these processes, even including consideration
of these threshold phenomena, are often inadequate in demonstrating a
detrimental effect on components of physiological fitness such as the
growth, reproduction and survival of the whole organism. Although
correlations can be drawn between cellular events and such fitness

traits, the specificity and sensitivity of the biochemical techniques are most effective when complemented by physiological determinations.

4. PHYSIOLOGICAL RESPONSES.

The role of physiological determinations of growth in assessing pollution effects on sessile invertebrates has recently been reviewed by Bayne (1984), Bayne and Widdows (1985) and Widdows (1984). The organisms are viewed as responding to changes in the environment, including changes in pollutant levels,by modulating physiological processes (e.g. rates of feeding and of metabolism) so as to maintain a constant scope for growth. When the pollutant level rises above a threshold value the homeostatic capacity of the animal is overloaded and new physiological balances are established, which represent a decline in functional efficiency and a reduction in fitness. Measures such as these are often considered part of the "tertiary effects of stress" (see Wedemeyer et al., 1984) and are closest to providing pointers to possible consequences at higher levels of biological organisation. These features may be analysed via the energy and/or nutrient budget of the organism. In its simplest formulation the scope for growth is calculated as the difference between the energy absorbed from the diet and the energy equivalents of the metabolic demand and of excretory losses. A decline in the scope for growth, which signals a "stressed" condition, has some parallels with Odum's (1967) concept of stress acting on ecosystems to divert energy away from doing useful work to increased expenditure on maintenance and repair.

An example from an experiment in which Mytilus edulis were exposed for 140 days to low levels of oil hydrocarbons in a flow-through aquarium is shown in Table II.

TABLE II. Responses of Mytilus edulis to the water-accommodated fraction of crude oil (30 µg total hydro-carbons 1^{-1}). Data from Widdows et al., (1982)

Animal Response	Experimental value relative to Control		
	Day 33	100:	140:
Feeding rate	0.83	0.79	0.75
Metabolic rate	1.00	1.39	1.41
Excretion rate	1.05	1.70	1.47
Scope for Growth	0.69	0.50	0.24
Lysosomal latency	0.58	0.14	0.07
Lysosomal volume	1.57	2.71	n.d

Rates of feeding were reduced, metabolic and excretion rates increased and the scope for growth declined, to 69, 50 and 24% of control values on days 33, 100 and 140 respectively. Table II also shows results of some lysosomal measurements. Statistically significant correlations between scope for growth and lysosomal function are a regular feature of these experiments.

4.1. Energy allocations to growth and reproduction.

Physiological determinations of the scope for growth in this basic form
do not provide information on the allocation of net available energy
between somatic growth (designated Pg) and reproduction (Pr). The
pattern of this allocation is age- and size-dependent. A convenient
measure of the proportional allocation betwsen Pg and Pr is provided by
the notion of reproductive effort:

$$\text{Reproductive effort} + \text{Pr}/(\text{Pr}+\text{Pg})$$
$$(\text{Pr}+\text{Pg}) = \text{Absorbed ration(A)} - \text{Respiratory losses(R)}.$$

Figure 1 illustrates the weight-dependence of this physiological
variable in <u>Mytilus</u>. The net energy available, or scope for growth,
increases to a maximum, in this example, at 1g dry flesh weight, before
declining due to the disproportionate increase in R relative to A.
Gamete production increases isometrically with weight in the adult

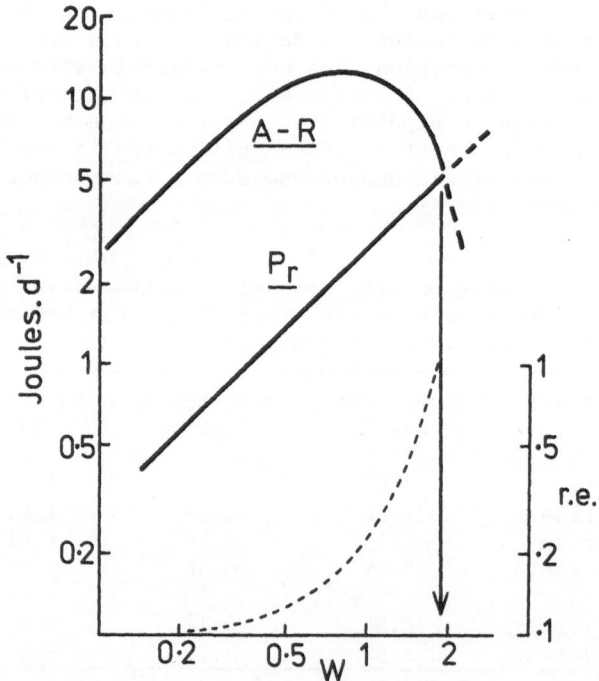

<u>Figure 1</u>. The scope for growth (A - R) plotted against body weight (W;
grams dry flesh), together with gamete production (Pr). The reproductive effort (r.e) is also plotted as a dashed line and the predicted
maximum attainable body size is identified by an arrow.

animal; the values for (A-R) and Pr intercept at a size equivalent to
the mean maximum body weight. Reproductive effort (the dashed line in
Fig. 1) increases exponentially with body weight, reaching a value of 1
at maximum body size, when all available energy is allocated to
reproduction.

Experiments have been carried out with mussels to determine the
effects of environmental changes on these processes. With increasing
stress (using the word "stress" to signify a range of environmental
stimuli which, by exceeding a threshold value, disturb normal animal
function) the total net energy available for growth and reproduction may
be reduced (see Table II). Pr may also be affected, with reductions in
fecundity accompanied, under severe stress, by the production of eggs of
reduced weight and impaired viability (Bayne et al., 1982). In naturally
occurring field populations considerable differences in fecundity and
reproductive effort have been observed (Table III).

TABLE III. Age-related fecundity and reproductive effort
in two populations of Mytilus edulis (from Bayne et al.,
1985)

Population	Age class	Fecundity eggs. 10^6	Reproductive effort
Lynher	4	1.04	0.24
	6	1.65	0.50
	8	2.09	0.69
Cattewater	4	0.16	0.03
	6	0.34	0.14
	8	0.52	0.33

These findings suggest that aspects of reproductive energetics in
this species are vulnerable to environmental changes and that traits
relevant to the fitness of the individual and to the maintenance of the
population may be used to assess the extent of sub-lethal, but nevertheless damaging, effects of pollution. In guiding the extrapolation from
effects on individuals to the likely consequences for the population,
can the available data be used to predict age-related reproductive
behaviour under different degrees of environmental stress? A simple
simulation model was set up to calculate life-time egg production by
individual mussels at different combinations of reproductive effort and
scope for growth. Relationships between the physiological variables and
body weight take the form shown in Fig. 1; different schedules for age-
related weight are simulated within limits of observed growth in natural
populations and the allocation pattern of resources to gamete production
that yields the maximum life-time fecundity is evaluated as the "optimal"
reproductive effort.

An example of the results from this simulation is shown in Figure 2.
The prediction is that the reproductive effort will rise under mild
conditions of stress (equivalent to 10-20% reduction in the energy

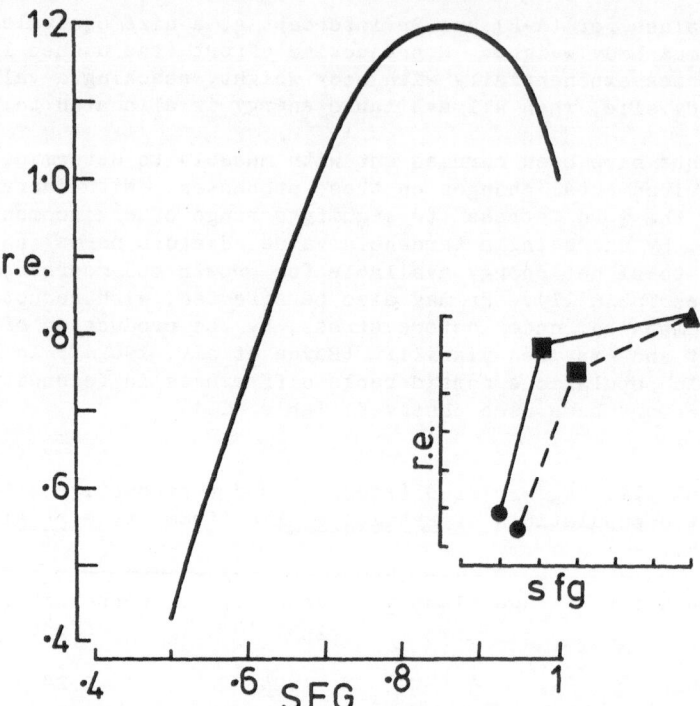

<u>Figure 2</u>. Reproductive effort (r.e) as a function of the scope for
growth (SFG), from a simulation as described in the text. <u>Inset</u>:
Relative values for r.e and SFG in three populations of mussels (Lynher,
triangle; Mothecombe, squares; Cattewater, circles) in two age classes
7 and 8 years.

available for growth) then fall steeply with further decline in the
scope for growth. As an inset to Fig. 2 some values for relative scope
for growth and relative reproductive effort in three populations and two
age classes are plotted; these results suggest that reproductive effort
may be disproportionately suppressed in populations (e.g. the Cattewater)
in which individual growth rate is depressed. We have other evidence
from laboratory experiments (R. Pipe, pers. comm.) that reproductive
effort in mussels increases as a result of short-term starvation.

    Although these relationships incorporate age-related features of the
physiology and suggest responses that reduce net energy flow, they do not
measure the likelihood of mortality and therefore do not extrapolate to
aspects of population response. The concept of reproductive value,
however, combines age-related fecundity and mortality schedules to
discriminate between current reproductive effort and a predicted future
investment in gametes (Williams, 1966). Residual reproductive value
(RRV) increases with age in mussels, reaching a maximum within the year
classes that express the maximal rates of growth and then declining
sharply with age. Mussels from stressed populations show a marked
depression in RRV and a shift in the timing of maximum RRV towards

younger age classes. In addition, within such populations there is some
evidence of an amplification in the fitness measures as they expand from
the individual to the population level (Bayne and Widdows, 1985).
Further research is needed to establish how these relationships might be
incorporated into population models of pollution effects.

5. POPULATION EFFECTS.

Given the observed effects of various contaminants on the physiological
energetics of growth, referred to above, it is not surprising that known
pollution events have provided evidence of reduced rates of growth in
natural populations. This has been documented particularly well for the
soft-shell clam, Mya arenaria. Following a spill of crude oil in
Chedabucto Bay, Nova Scotia, rates of growth of this clam were reduced
(Thomas, 1973, 1978), consistent with measured reductions in carbon flux
in clams from the oiled sediments (Gillifillan and Vandermeulen, 1978).
Due to the continued presence of the spilled oil in the sediments these
effects persisted for at least six years and, in addition to growth
effects on individuals, there has been reduced recruitment and increased
mortality of clams at the polluted site. Dow (1975, 1978) and Appeldoorn
(1981) have also reported disturbance to growth in Mya from various sites
subjected to pollution and evidence of recovery to normal growth at sites
where a former pollution has ceased.
      However, populations possess characteristic properties of compensa-
tion for environmental disturbance and many density-dependent processes
exist to limit the degree of population damage following effects on
individuals (Goodyear, 1980). Uncertainty over fundamental features of
intra-specific competition, of the stock/recruitment relationship, of
the force of inter-specific checks and balances within the community,
and of the "natural", background, variability in population size renders
the interpretation of population changes in response to any particular
environmental event a formidable task. On the other hand, the use of
population models to relate individual response to possible population
consequences (in effect, to calibrate physiological effects against
likely population response) holds some promise. Of the various approaches
possible, models based on the concepts of bio-energetics are recommended
where age-specific data and detailed physiological measurements are
available (Vaughan et al., 1984). Such models relate changes in individual body size to the balances between absorbed ration (A), respiration
(R) and reproductive output (G) and are extended to the population by
accounting for recruitment and mortality according to certain assumptions. The physiological features can be set dependent upon natural
environmental factors as well as the known effects of contaminants.
      Such a model is described for Mytilus edulis by Verhagen (1983) and
Harris et al. (1983/1984) based on formulations by Bayne et al. (1976).
      The Absorbed ration (A) is calculated according to:

$$A = s \, p \, c \, E \, q(a) \, f W^{0.6}$$

where s is the proportion of time the mussel is immersed, p is the particulate load, c is the proportion of organic matter in p, E is absorption efficiency, q(a) is the effect of a contaminant and $fW^{0.6}$ is the feeding rate of the mussel (litres cleared of particulates per day). An upper ceiling is set on the amount of material ingested per litre of water filtered by the expression $k(1)W^{0.2}$.

The Respiration rate (R) is calculated as:

$$R = q(r) \ W^{0.7} \ (r1 + r2)$$

where r1 represents standard (ration-independent) and r2 routine (ration-dependent) respiration and q(r) is the effect of the contaminant. Seasonal changes in routine respiration are represented as a sine-curve function reflecting ambient temperatures and the endogenous gametogenic cycle.

Reproductive output (G) is calculated as a function of the scope for growth (A - R), so incorporating contaminant effects as the cumulative impact on the animal's net energy acquisition:

$$G = k(2) \ (1 - e^{-0.05 \ (A - R - m)})$$

where k(2) is maximum G and m is fitted for the net energy value at which gamete production ceases.

This model has proved successful in reproducing observed rates of growth in a natural population of mussels and has been incorporated into an environmental quality simulation (see following section) and into a model of the population dynamics of mussels in an enclosed saline lake in Holland (Verhagen and Bayne, unpublished). The latter study included an analysis, by simulation, of the various factors likely to limit, recruitment and control mortality. Recruitment is modelled as being successful only if the food concentration immediately above the mussels exceeds a certain value; this therefore incorporates the concepts of potential food limitation for suspension feeders (Wildish and Kristmanson, 1979) and of intra-specific space and trophic competition as density-dependent population processes. Mortality is modelled as a function of density-dependent space limitation (i.e. the overall availability of suitable substrates for growth) and of storms (as a density-independent feature). In both this and the environmental quality model the simulation of growth of the mussels is linked to hydrodynamic features of the habitat; this linkage is essential if information on the biological effects of pollution is to be effective in facilitating environmental management.

6. AN ENVIRONMENTAL QUALITY MODEL.

Simulation modelling affords the opportunity of coupling three essential features of any general assessment of biological response to environmental contamination viz. the hydrodynamics of the region of concern; the chemical speciation, within the receiving waters, of the particular contaminants, and the implications this has for biological availability; and the biological response itself. Unless these three components are formulated together, the information gain from separate studies of the physics, chemistry and biology of the problem will be limited.

The model by Harris et al. (1983/1984, and subsequent unpublished improvements) of the dispersal and biological effects of contaminants in the Tamar estuary, south-west England, provides an example. A one-dimensional advection-diffusion simulation provides the basis for modelling the fluxes of dissolved contaminants. Suspended particles are modelled as marine, estuarine and fresh-water components, depending on their origin and behaviour within the estuary. Importantly, the presence of a turbidity-maximum near the limit of saline intrusion is simulated on the basis of mechanisms described by Uncles et al. (1985); such a turbidity-maximum is a common feature of this type of partly-mixed estuary and is of central importance in affecting the behaviour of chemical constituents that enter via river flows or as direct inputs to the body of the estuary (Morris, 1984).

Two chemical speciation sub-models have been set up, for hydro-carbons and for cadmium. Hydrocarbons are considered to partition between dissolved and particle adsorbed phases, according to fundamental features of their chemistry and as a function of the concentrations of particulate organic carbon and the octanol-water partition coefficient. Dissolved hydrocarbons are lost from the estuary by three independent, additive processes; volatilisation, photo-oxidation and bacterial degra-dation. Particle-bound hydrocarbons follow the seasonal cycle of estuarine particle flux, including seasonally variable exchanges, which are driven by river-flows, between sedimented and suspended states.

The uptake of hydrocarbons into mussels is determined from octanol-water partition coefficients and uptake characteristics observed in laboratory experiments. The effects on the mussels are simulated using the model briefly described in the preceding section. Figures 3 and 4 illustrate some results from hypothetical inputs of one kilogram of naphthalene per day to the top kilometre of the estuary in summer and winter. In the summer, tissue concentrations of naphthalene are predic-ted to be lower and less variable than in winter, due to lower suspended particle concentrations at the estuary mouth. Rates of growth are lower in the winter as the result of higher contaminant loading. Simulation results of this type provide evidence of likely biological impact (at biochemical, physiological and population levels) on various temporal (including tidal, seasonal and annual cycles) and spatial scales, essential information for any objective management of such a system.

7. CONCLUSIONS.

The research briefly reviewed in this paper demonstrates a linked suite of responses, within bivalve and gastropod molluscs, to pollution by hydrocarbons and by trace metals; the responses include effects on the natural cellular protection mechanisms (metallothioneins and the MFO system), sub-cellular organelles (including damage to the genetic material), on cellular and tissue pathology and, at the level of the whole organism, on the physiological processes of feeding, assimilation, respiration and growth. When these responses are coupled, through

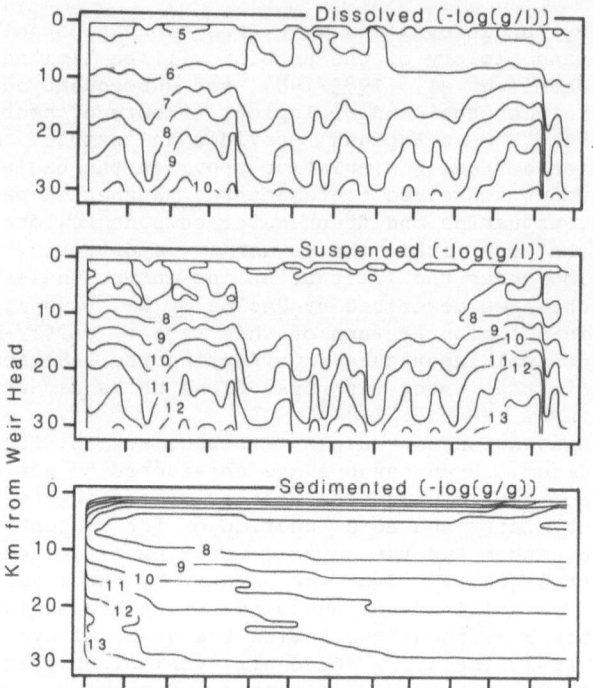

Figure 3. The distribution of naphthalene (dissolved, particulate, and sedimented) as resulting from a constant input of 1kg day^{-1} to the top kilometre of an estuary, in summer; results from a simulation model, from Harris et al. (1983/1984).

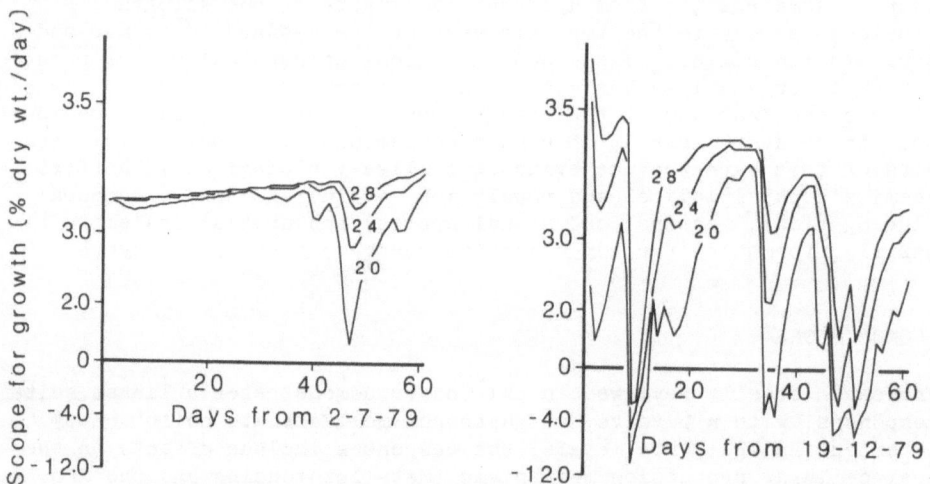

Figure 4. The scope for growth of mussels, in summer and winter, within three sections of an estuary (20, 24 and 28 km from the head) subjected to a constant input, at the head, of 1kg naphthalene day^{-1}; results from a simulation model (Harris et al., 1983/84).

knowledge of the mechanisms of bio-accumulation, to the physical and chemical features of the environment, in the form of linked simulation models, it is possible to provide a powerful tool for use in managing the discharge of wastes to coastal and estuarine environments. Modelling of this kind affords the opportunity to quantify the assimilative capacity of the waste-receiving system at many levels, from the thresholds of toxicity within organisms to the fine balance, within estuaries, between contaminant accumulation within sediments and the flux of particle-adsorbed material to sea (the "sink or drain" distinction).

The levels of biological response considered here do not directly include the population or the community, although it is proving possible to make some predictions of population effect from the behaviour of individuals (and, for cases involving species dominants within communities, to assess some likely ecosystem impacts through changes to nutrient dynamics and occupation of habitat space). Models of population dynamics, set up not to give detailed simulations of population change, but to investigate potential population responses, may be useful in calibrating individual effects given known contaminant loadings. However, an exclusive concentration on the use of sentinel species in environmental assessment is not recommended. Information is also required on community response to disturbance and the much wider issues of social cost (and benefit) cannot be ignored by the biologist. Nevertheless, the sensitivity, specificity and rapid response time of the measures discussed in this paper suggest that they will have a continuing and important role to play in analyses of the environmental impacts of waste disposal at sea.

8. ACKNOWLEDGEMENTS.

Much of the research discussed in this paper is due to my colleagues at IMER, particularly the Toxicology, Estuaries Ecology and Systems Analysis Groups and I thank them for their help and enthusiasm. IMER is a component laboratory of the Natural Environment Research Council (UK). The Institute continues to benefit from commissioned funding from the Department of the Environment (UK).

9. REFERENCES.

Appeldoorn, R.S. 1981. Responses of soft-shell clam (Mya arenaria) growth to onset and abatement of pollution. J. Shellfish Res. 1, 41-49.

Bayne, B.L. 1984. Responses to environmental stress: Tolerance, resistance and adaptation. In: Marine Biology of Polar Regions and Effects of Stress on Marine Organisms (eds: J.S. Gray and M.E. Christiansen). Wiley and Sons, Chichester. pp. 331-349.

Bayne, B.L. and Widdows, J. 1985. Strategies for increasing the biological effects of pollution: cellular and organism levels. In: Proceedings of Second International Mussel Watch Conference. In press.

Bayne, B.L., Widdows, J. and Thompson, R.J. 1976. Physiological
    integrations. In: Marine Mussels (ed. B.L. Bayne).
    Cambridge University Press, Cambridge. pp. 261-292.
Bayne, B.L., Widdows, J., Moore, M.N., Salkeld, P., Worrall, C.M. and
    Donkin, P. 1982. Some ecological consequences of the
    physiological and biochemical consequences of petroleum
    compounds on marine molluscs. Phil. Trans. Roy. Soc. Lond. B.
    297, 219-239.
Bayne, B.L., Salkeld, P.N. and Worrall, C.M. 1983. Reproductive
    effort and value in different populations of the marine
    mussel, Mytilus edulis. Oecologia (Berl.), 59, 18-26.
Bayne, B.L., Brown, D.A., Burns, K., Dixon, D.R., Ivanovici, A.,
    Livingstone, D.R., Lowe, D.M., Moore, M.N., Stebbing, A.R.D.
    and Widdows, J. 1984. The effects of stress and pollution on
    marine animals. Praeger, New York. pp. 384.
Dethlefson, V. 1984. Diseases in North Sea fishes. Helgo. Meeres-
    unters. 37, 353-374.
Dixon, D.R. 1982. Aneuploidy in mussel embryos (Mytilus edulis)
    originating from a polluted dock. Mar. Biol. Letts., 3,
    155-161.
Dixon, D.R., Jones, I.M. and Harrison, F.L. 1985. Cytogenetic evidence
    of inducible processes linked with xenobiotic metabolism in
    adult and larval Mytilus edulis. Sci. Total Environ. In
    press.
Dow, R.L. 1975. Reduced growth and survival of clams transplanted to
    an oil spill site. Mar. Poll. Bull. 6, 124-125.
Dow, R.L. 1978. Size-selective mortalities of clams in an oil spill
    site. Mar. Poll. Bull. 9, 45-48.
George, S.G. 1982. Subcellular accumulation and detoxication of metals
    in aquatic animals. In: Physiological mechanisms of marine
    pollutant toxicity (eds: F.J. Vernberg et. al.). Academic
    Press, New York. pp. 3-52.
George, S.G. and Viarengo, A. 1984. An integration of current know-
    ledge of the uptake, metabolism and intracellular control of
    heavy metals in mussels. In: Sixth Symposium on Pollution and
    Physiology of Marine organisms, 1983, Connecticut, USA. In
    press.
Gillfillan, E.S. and Vandermeulen, J.H. 1978. Alteration in growth and
    physiology of soft-shell clams, Mya arenaria, chronically
    oiled with Bunker C from Chedabucto Bay, Nova Scotia, 1970-
    1976. J. Fish. Res. Board Can. 35, 630-636.
Goodyear, C.P. 1980. Compensation in fish populations. In: Biological
    monitoring of fish (eds. C.H. Hocutt and J.R. Stauffer, Jr.)
    Lexington Books, Lexington. pp. 253-280.
Gosset, R.W., Brown, D.A. and Young, D.R. 1983. Predicting the bio-
    accumulation of organic compounds in marine organisms using
    octanol water partition coefficients. Mar Poll. Bull., 14,
    387-392.
Harris, J.R.W., Bale, A.J., Bayne, B.L., Mantoura, R.F.C., Morris, A.W.,
    Nelson, L.A., Radford, P.J., Uncles, R.J., Weston, S.A. and
    Widdows, J. 1983/1984. A preliminary model of the dispersal

and biological effect of toxins in the Tamar estuary, England. Ecol. Model., <u>22</u>, 253-284.

Jacklin, E. 1974. Enzyme responses to metals in fish. In: <u>Pollution and Physiology of Marine Organisms</u> (eds. F.G. Vernberg and W.B. Vernberg). Academic Press, New York. pp. 59-65.

Lee, R.F. 1981. Mixed function oxygenases (MFO) in marine invertebrates. <u>Mar. Biol. Letts</u>. <u>2</u>, 87-105.

Livingstone, D.R. 1984. Biochemical differences in field populations of the common mussel <u>Mytilus edulis</u> L. exposed to hydrocarbons: some considerations of biochemical monitoring. In: <u>Toxins, Drugs and Pollutants in Marine Animals</u> (eds: L. Bolis, J. Zadunaisky and R. Gilles). Springer-Verlag, Berlin, pp. 161-175.

Livingstone, D.R., Moore, M.N., Lowe, D.M., Nasci, C. and Farrar, S.V. 1985. Responses of the cyochrome P-450 monooxygenase system to diesel oil in the common mussel, <u>Mytilus edulis</u> L. and the periwinkle, <u>Littorina littorea</u>. <u>J. Aq. Toxicol</u>. In press.

Lowe, D.M., Moore, M.N. and Clarke, K.R. 1981. Effects of oil on digestive cells in mussels: quantitative alterations in cellular and lysosomal structure. <u>Aquat. Toxicol</u>. <u>1</u>, 213-226.

Malins, D.C., McCain, B.B., Myers, M.S., Brown, D.W., Sparks, A.K., Morado, J.F. and Hodgins, H.O. 1984. Toxic chemicals and abnormalities in fish and shell-fish from urban bays of Puget Sound. <u>Mar. Environ. Res</u>. <u>14</u>, 527-528.

Mazeaud, M.M. and Mazeaud, F. 1981. Adrenergic responses to stress in fish. In: <u>Stress and Fish</u>. (ed: A.D. Pickering) Academic Press, London. pp. 49-75.

Moore, M.N. and Lowe, D.M. 1984. Cytological and cytochemical measurements. In: <u>The Effects of Stress and Pollution on Marine Animals</u>. (eds: B.L. Bayne <u>et. al</u>.) Praeger, New York. pp. 46-74.

Moore, M.N., Pipe, R.K. and Farrar, S.V. 1982. Lysosomal and microsomal responses to environmental factors in <u>Littorina littorea</u> from Sullom Voe. <u>Mar. Poll. Bull</u>. <u>13</u>, 340-345.

Morris, A.W. (Ed.) 1983. <u>Practical procedures for estuarine studies</u>. Natural Environment Research Council, London. pp. 262.

Odum, H.T. 1967. Work circuits and system stress. In: <u>Symposium on Primary Production and Mineral Cycling in Natural Ecosystems</u>. (ed: H.E. Young) University of Marine Press, Orono.

Pickering, A.D. 1981. Introduction: The concept of biological stress. In: <u>Stress and Fish</u> (ed: A.D. Pickering). Academic Press, London. pp. 1-9.

Roesijadi, G. 1980. The significance of low molecular weight, Metallothionien-like proteins in marine invertebrates: current status. <u>Mar. Environ. Res</u>. <u>4</u>, 167-181.

Selye, H. 1973. The evolution of the stress concept. <u>Amer. Sci</u>. <u>61</u>, 692-699.

Simkiss, K. 1983. Lipid solubility of heavy metals in saline solutions. <u>J. mar. biol. Ass. U.K</u>., <u>63</u>, 1-7.

Thomas, M.L.H. 1973. Effects of Bunker C oil on intertidal and lagoonal biota in Chedabucto Bay, Nova Scotia. <u>J. Fish Res</u>.

Board Can., 30, 83-90.

Thomas, M.L.H. 1978. Comparison of oiled and unoiled intertidal
communities in Chedabucto Bay, Nova Scotia. J. Fish. Res.
Board Can., 35, 707-716.

Turner, D.R. 1985. Relationships between biological availability and
chemical measurements. In: Metal ions in Biological Systems,
Vol. 18. (ed: H. Sigel). Marcel Dekker, New York. pp.
137-164.

Uncles, R.J., Elliott, R.C.A. and Weston, S.A. 1985. Observed fluxes
of water, salt and suspended sediment in a partly mixed
estuary. Est. Coast and Shelf Sci. In press.

Vaughan, D.S., Yoshiyama, R.M., Breck, J.E. and De Angelis, D.L. 1984.
Modelling approaches for assessing the effects of stress on
fish populations. In: Contaminant Effects on Fisheries.
(eds: J.W. Cairns, P.V. Hodson and J.O. Nriagn). Wiley-
Interscience, New York. pp. 259-278.

Verhagen, J.H.G. 1983. A distribution and population model of the
mussel Mytilus edulis in Lake Grevelingen. In: Analysis of
Ecological Systems: State-of-the-Art in Ecological Modelling.
(eds: W.K. Lauenroth, G.V. Skogerboe and M. Flug). J. Wiley,
New York. pp. 373-383.

Viarengo, A., Pertica, M., Mancinelli, G., Orunesu, M., Zanicchi, G.,
Moore, M.N. and Pipe, R.K. 1984. Possible role of lysosomes
in the detoxication of copper in the digestive gland cells of
metal-exposed mussels. Mar. Environ. Res., 14, 469-470.

Wedemeyer, G.E., McLeay, D.J. and Goodyear, C.P. 1984. Assessing the
tolerance of fish and fish populations to environmental stress:
the problems and methods of monitoring. In: Contaminant
Effects on Fishes. (eds: V.W. Cairns, P.V. Hudson and J.O.
Nriagn). Wiley-Interscience, New York. pp. 164-195.

Widdows, J. 1984. Field measurements of the biological impacts of
pollutants. In: Assimilative Capacity of the Oceans for Man's
Wastes. SCOPE/ICSU, Academia Sinica, Taipei. pp. 111-129.

Widdows, J., Bakke, T., Bayne, B.L., Donkin, P., Livingstone, D.R.,
Lowe, D.M., Moore, M.N., Evans, S.V. and Moore, S.L. 1982.
Responses of Mytilus edulis on exposure to the water-
accommodated fraction of North Sea Oil. Mar. Biol., 67,
15-31.

Wildish, D.J. and Kristmanson, D.D. 1979. Tidal energy and sub-
littoral macrobenthic animals in estuaries. J. Fish Res.
Board Can., 36, 1197-1206.

Williams, G.L. 1966. Natural selection, the cost of reproduction and a
refinement of Lack's principle. Am. Natur., 100, 687-690.

BIOLOGICAL INDICES OF CHANGES INCLUDING PRIMARY PRODUCTION

J.S. Gray
Section of Marine Zoology and Marine Chemistry
Department of Biology
University of Oslo
P.O. Box 1064 Blindern
0316 Oslo 3
Norway

INTRODUCTION

Monitoring of pollution induced changes on populations and communities can be loosely divided into effects on functional attributes and effects on structural attributes. By functional attributes I mean the types, kinds and rates of flow of energy and materials through the system whereas structural attributes are the numbers of individuals, species and related characteristics such as biomass and how they vary through time. Clearly there are no hard and fast boundaries between functional and structural attributes since biomass can be regarded as both a functional and structural attribute. However, there are different strategies involved in applying monitoring programmes to functional and structural attributes and it is these that I wish to consider.

FUNCTIONAL ATTRIBUTES

Bayne (this volume) has already considered the physiological approaches applied to individuals and in particular to the integration of physiological characteristics into the measurement of scope for growth. Many physiological processes in organisms vary seasonally with changes in temperature. Bayne and his coworkers (Bayne & Widdows 1981, Widdows et al. 1981) have shown that in Mytilus edulis gametogenesis is a highly important process since energy is used preferentially for gamete production. In M. edulis gametogenesis occurs in winter when scope for growth is minimal leading to the well-known seasonal growth check. Gametogenesis in bivalves does not, however, always occur in winter. In Mercenaria mercenaria in N. Carolina Peterson et al. (1983) have shown that gametogenesis occurs in summer and the growth check previously assumed to be a winter-ring in this species is in fact laid down in summer. Thus functional attributes of individuals have to be measured regularly over the seasonal cycle to encompass temperature variations and detailed knowledge must be obtained of gametogenetic cycles. A high frequency of measurements must also be used for

G. Kullenberg (ed.), The Role of the Oceans as a Waste Disposal Option, 635–646.

estimating another functional attribute which is widely used in
pollution monitoring, primary production in the water column.

Usually primary production measurements are made at weekly
intervals over the whole year in order to obtain an annual figure. Yet
frequently constraints are put on monitoring programmes to limit the
number of sampling times. Table 1 shows data from Wulff (1982) from the
Baltic Sea. The funding authorities decided to reduce the sampling
programme to 4 time intervals and the consequences analysed by Wulff in
terms of deviations in estimates are dramatic and show that with only 4
sampling occasions primary production is not worth measuring. Even if
the sampling frequency is high problems arise since there are
considerable variations in production rates from year to year and place
to place. Fig. 1 shows data from Aertebjerg et al. (1981) for a station
in the Kattegat used in a pollution monitoring programme. The year to
year variations are high. The timing of the spring bloom is
particularly important, as it can vary by weeks from year to year at
the same station and thus fixed sampling times may miss this period. In
temperate areas 50% of the annual production can occur during the
spring bloom so it is essential to cover the bloom accurately. This
will entail intensive sampling over a short time period which is often
both impractical in terms of ship-time scheduling and impossible in
terms of funding.

The rationale for sampling primary production was that it was
widely believed that all the phytoplankton was consumed by the
secondary production and subsequently the benthos obtained its food
from faeces sedimenting from secondary producers (Steele 1974). In
other words the whole system was dependant on primary production.
Recent data has shown that this idea must be radically altered. It is
now known that in temperate areas over half of the annual primary
production sinks directly to the sediment. Table 2 shows some typical
data. If one considers only the spring bloom the amount settling is
greater and figures of up to 70% have been obtained (Smetacek 1982 from
Kiel Bay and Wassmann 1984 from Norwegian fjords). The reason for the
large amounts settling is that the spring phytoplankton bloom is poorly
coupled to secondary production in temperate areas, secondary
production being sometimes weeks later than the primary bloom. In polar
regions the coupling between phytoplankton and zooplankton blooms is
even less tight so that an even higher percentage of the primary bloom
sinks to the sea bed. Polychaetes have been found at 2000 m depth in
the Barents Sea with stomachs full of green settled phytoplankton cells
(Oug pers. comm.), so settlement is fast, (see also Honjo 1982 for a
review). In the tropics on the other hand coupling between
phytoplankton production and secondary production is tight so that
little material settles. These new findings show that the earlier
rationale for monitoring primary production, that it controlled
secondary production in both the water column and in the benthos were
tightly linked to primary production must be reassessed.

Furthermore, primary production by phytoplankton as measured by
the ^{14}C method does not represent the potential food resource for
secondary producers. The role of bacteria and other microorganisms has
recently been the subject of much research. The traditional idea that

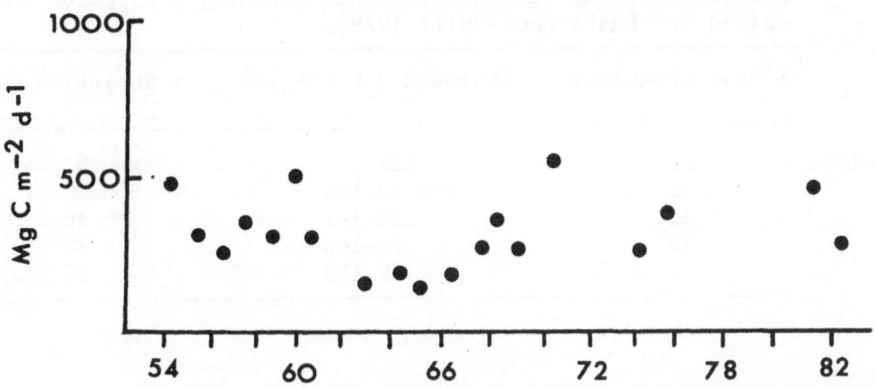

Fig. 1.   Long-term primary production data for the Kattegat, (from Aertebjerg et al. 1981).

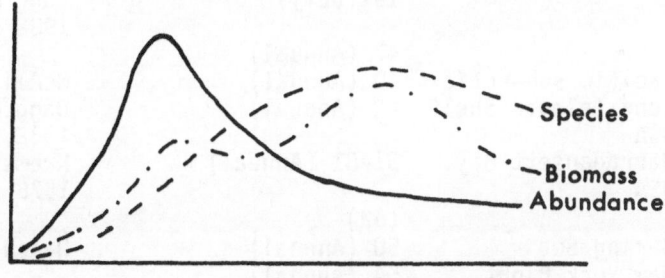

Fig. 2.   Generalised model of effects of organic enrichment on benthic communities of soft-sediments: the SAB curve (Species, Abundance, Biomass) of Pearson & Rosenberg 1978.

TABLE 1. effect of reduced sampling frequency on the precision of the estimate of annual primary production in the Baltic Sea (data from Wulff 1979).

Sample frequency y-1	Estimate g C m-2 y-1	% Deviation
52	127	0
26	120-138	10
18	128-141	10
13	93-145	30
6	69-183	50

TABLE 2. Grazing efficiency of zooplankton on phytoplankton.

Area	% of primary production grazed	Reference
Taguchi Bay, Japan (eutrophic)	0.1-1 (Spring bloom)	Taguchi & Fukuchi 1978
Hudson River Plume, USA	7 (October)	Malone & Chervin 1979
	10 (April)	
	26 (July)	
Georges Bank, USA	6.4 (April)	Riley & Bumpus 1946
	19 (July)	Dagg & Turner 1982
	47 (Annual)	
Pacific subarctic	40 (Annual)	McAllister 19
Long Islanad Shelf, USA	42 (Annual)	Dagg & Turner 1982
Narragansett Bay, USA	31-81 (Annual)	Kremer 6 Nixon 1978
	(62)	
Bering Sea	50 (Annual)	Walsh 1981
New York Bight	54 (Annual)	-"-
Kiel Bay	31-41 (Annual)	Smetacek, 1980
Norwegian fjords	43-80 (Annual)	Wassmann, 1984

bacteria remineralised large organic molecules such as settling faeces and thereby released nutrients for phytoplankton growth is not now felt to be correct. Bacteria are in fact important competitors with phytoplankton for available nutrients. Remineralisation occurs at higher levels in the trophic web when the flagellates which consume bacteria are in turn consumed by small microzooplankton. This has been called the bacterial loop (see Azam et al. 1984 for a review). Traditional ^{14}C methods of measuring primary production do not give estimates of bacterial production and recent data suggests bacterial production may be equally as important as the phytoplankton (Azam et al. 1984). In addition the use of the ^{14}C method has been called into question with some proponents (Gieskes et al. 1979, 1984) claiming that the traditional method (Steeman-Nielsen 1952) underestimates primary production by up to 50%. However, using alternative methods such as sensitive oxygen consumption and carbon dioxide production measurements (Williams et al. 1979) has shown that the traditional ^{14}C method is reasonably accurate.

In summary the usefulness of primary production measurements in monitoring programmes can be seriously questioned both in terms of the reliability of estimates in relation to sampling frequency and the fact that they neglect the important role of microorganisms in the water column.

STRUCTURAL ATTRIBUTES
The most commonly used structural attributes of communities are number of species, abundance of individuals and biomass, (the SAB curves of Pearson and Rosenberg 1978), diversity which integrates number of individuals and the number of species, dominance patterns and the distribution of individuals among species. The use of these attributes in a pollution monitoring context has been extensively reviewed by Sheehan (1984, a, b). There seems little point, therefore, in repeating Sheehan's arguments. Along heavy pollution gradients of which organic matter is perhaps the commonest marine pollutant there are usually clear trends moving towards the source of pollution. The number of species declines, abundance and biomass increase and then decline approaching the pollution source (fig. 2 illustrates these changes). Diversity declines towards the source and dominance usually increases. Simple transects from the source often indicate clear trends where the pollution load is heavy and acute.

Where the gradients are less obvious often multivariate techniques are needed to analyse the complex data sets. Multidimensional scaling is a widely used modern technique recommended by Field et al. (1982). Most multivariate techniques depend on species absences, since they use similarity indices based merely on presence and absence data such as the Bray-Curtis index (Bray and Curtis 1957). Yet absence of species indicates that pollution effects are relatively severe. The ideal goal would be to find techniques that record significant changes in abundance patterns along pollution gradients. Two techniques have received much recent attention the so-called log-normal distribution of individuals among species of Gray and Mirza (1978) and dominance diversity methods revived by Shaw, Platt and Lambshead (1984).

The log-normal distribution in its original form (Gray and Mirza 1978) has been criticised by Platt et al. (1983) as being confusing to interpret and often misapplied. Clearly any method can be misapplied if users do not follow the constraints laid down in the original method description. Certainly there have been naive applications of the method, which has also often been used on inappropriate data sets. Platt et al.'s criticisms were rebutted by Gray (1983) who showed that the data claimed by Shaw et al. to be hard to interpret showed a clear trend using the method described in the original paper. However, the method has been developed further and I agree with Shaw et al. that the term <u>log-normal</u> is perhaps inappropriate as there appear to be not one log-normal distribution of individuals among species within any given data set from unpolluted areas, but usually three or more (Ugland and Gray, 1983). Using this new finding Gray and Pearson (1983), and Pearson, Gray and Johannessen (1983) were able to show that simple plots of the distribution of individuals among species showed clearly different patterns when approaching pollutant sources. The changing patterns resulted from changes in abundance of not only the common species (much used in dominance analyses) but also in moderately common species. Using the moderately common species alone, Gray and Pearson (1983) were able to objectively identify groups of species which changed abundance patterns along the gradiente. This method was applied to a range of different pollution types in northern Europe and the suite of species identified was similar but not identical over the areas studied.

    In conclusion a simple plot of individuals among species rather than trying to fit a mathematical model (the log-normal) to the data gives clear changes in pattern which can also indicate changes in abundance of sensitive species along the gradient. From this analysis a sampling programme can later be concerned only with selected species instead of the whole community.

    Dominance diversity methods have recently been advocated as useful in a pollution monitoring context (Shaw et al. 1984). That dominance increases with increased organic loading has been known for many years. Shaw et al. introduce a method called 'k' dominance or Lorenzen curves. Fig. 3 shows that the polluted station is less diverse, but the slopes of the plots are similar showing that the sole difference is in the intercept which is determined by the dominance of the first ranked species. Analyses of other plots presented by Shaw et al. show this to be the general case. 'k' dominance curves, therefore, present little biological information and cannot be compared to the information available from simple plots of individuals among species using Gray and Pearson's (1983) method. Furthermore, in unpolluted assemblages from benthic sediments dominance in fact increases with increasing diversity (Birch 1981). This is in marked contrast to most other assemblages. Without further study doubts must be raised as to the wisdom in applying dominance diversity methods when the relationship between diversity and dominance is in dispute.

    All the above structural attributes of assemblages rather than populations. Clearly once sensitive species have been identified thant monitoring can be concentrated on abundance changes in these species.

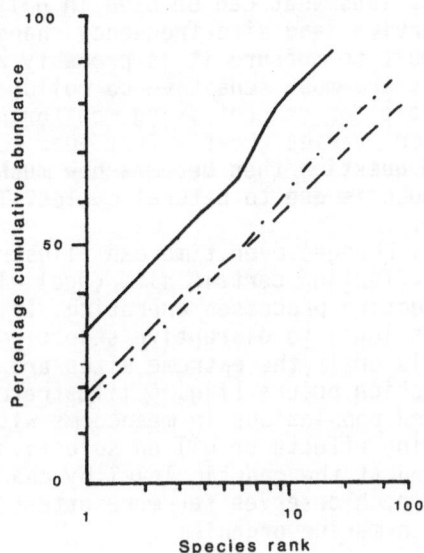

Fig. 3. Generalised model of effects of pollutants on size distribution of marine organisms showing two possible results of selective processes.

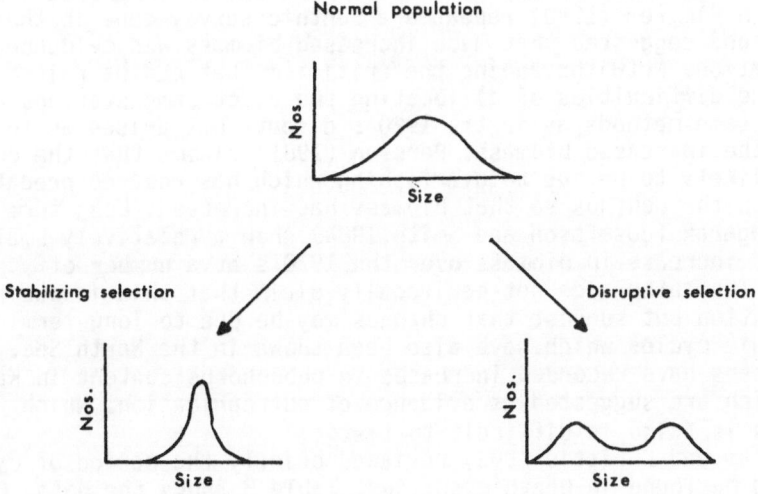

Fig. 4. 'k' dominance curve of benthic nematodes from the Clyde dumping ground. Solid line - polluted site broken lines - control sites, (from Platt et al. 1984).

Changes can be recorded along gradients or from one area over time.
Other characteristics of populations that can be used in pollution
monitoring are recruitment, survival and size-frequency changes.
Although recruitment is difficult to measure it is probably at this
life-history stage that species are most sensitive to pollutants. Lewis
(1976) has given a full rationale for concentrating monitoring on
recruitment. Recruitment however, varies greatly from year to year
(Buchanan et al. 1978) and the question then becomes how much of this
variation is natural and how much is due to natural cycles? This will
be covered in the next section.

Analyses of size-frequency changes over time can illustrate how
pollutants are differentially affecting certain size (age) classes.
This can lead to different selective processes operating. If the mean
size is most affected than this leads to disruptive selection of the
smallest and largest individuals or if the extreme sizes are most
affected than stabilizing selection occurs (fig. 4 illustrates these
processes). By using semi-closed populations in mesocosms with marked
individuals we have been studying effects of oil on selective processes
both at the population level and at the genetic level by changes in
enzyme polymorphisms. This approach deserves far more attention in
studying effects of pollution on marine organisms.

TIME SERIES ANALYSES IN BIOLOGICAL MONITORING

One of the major problems that bedevils biological monitoring is
interpreting time series analyses. Are the changes in numbers, species
biomass or whatever over time due to effects of pollution or are the
changes observed part of long-term natural cycles? Considerable debate
has been going on concerning claimed eutrophication in the Baltic Sea.
Cederwall & Elmgren (1980) repeated a benthic survey done in the 1920's
by Hessle and suggested that much increased biomass was evidence of
eutrophication. Notwithstanding the criticism that can be raised
against the difficulties of a) locating the exact same stations and b)
using the same methods as in the 1920's dispute has arisen as to the
cause of the increased biomass. Persson (1981) claims that the changes
are more likely to be due to overfishing which has reduced predation
pressure on the benthos so that biomass has increased. Long-term data
on the Skagerak (Josefsson and Smith 1984) show a relatively small but
consistent increase in biomass over the 1970's at a number of
stations. The authors do not equivocally claim that this is due to
eutrophciation but suggest that changes may be due to long-term
hydrographic cycles which have also been shown in the North Sea. Yet
other workers have recorded increases in phosphorus content in Kattegat
waters which are suggested as evidence of eutrophication. Which
hypothesis is false is difficult to test.

I (Gray and Christie 1983) reviewed briefly the period of cycles
that could be found in benthic species. Table 3 shows the data. Few
pollution monitoring surveys cover a decade or more and, therefore, one
cannot examine the likely cycles that can be expected in north
temperate latitudes. The analyses of time-series data becomes
particularly problematical because the cycles are superimposed upon one

Table 3. Hydrographic cycles with examples of species showing
similar cycles, (see Gray & Christie, 1984 for detailed
references).

Cycle (yrs)	Biological data	
3- 4	Zooplankton, N. Atlantic	Canuella, Tachidius, Paronychocamptus, Belgium
6- 7	Zooplankton, N. Atlantic	Pontoporeia, Baltic; Ciona, Skagerrak; Bluecrab catch; Lanice, Abra, Scalibregma, Pectinaria, Liverpool Bay; Amphiura, Brittany
10-12	Zooplankton, N. Atlantic	Harmothoe, Echiurus, N.Sea; Paronychocamptus, Belgium; Blue crab catch; Shrimp catch, Irish Sea; Laminaria, Scotland
18-20	Blue crab catch	
100	Macroalgae, France	

another and do not give smooth sine curves that the naive expect. One
can perhaps hope that suitable biological material will be found that
has records over long time periods such as tree ring data. Recently
Isdale (1984) has discovered growth rings in the coral Porites sp.which
using X-ray tecnhiques can be measured automatically and show records
over 400 years and with some specimens probably can measure growth
conditions over 2000 years! Within the growth record are clear effects
of El Nino and the southern oscillation and fluorescent bands have been
found corresponding to rainfall which can be quantified to indicate the
size of the one hundred year storm. This data together with that from
the Bristle Cone Pine links biological and geological time scales.
Similar data which correlated with hydrographic conditions of the North
Sea and North Atlantic would give us the necessary background for
assessing effects of pollutants over time. Lacking such data the only
viable approach is spatial scale monitoring comparing a polluted site
with non-polluted control areas. Time-series monitoring is unlikely in
my opinion to provide an adequate data base to give definite answers on
effects of pollutants on marine ecosystems.

REFERENCES

Aertebjerg, G., Jacobsen, T., Gargas, E., and Buch, E., 1981. The Belt
       Project: Evaluation of the physical, chemical and biological
       measurements. The National Agency of Environmental Protection,
       Denmark.
Azam, F., Fenchel, T., Field, J.G., Gray, J.S., Meyer-Reil, L.A. and
       Thingstad, F., 1984. 'The ecological role of water-column bacteria
       in the sea. Mar. Ecol. Progr. Ser.
Bayne, B.L., 1985. 'An overview of biological effects techniques
       applied to marine organisms.' (This volume).
Bayne, B.L. and Widdows, J., 1978. 'The physiological ecology of two
       populations of Mytilus edulis L.' Oecologia (Berl.), 37, 137-162.
Birch, D.W., 1981. 'Dominance in marine ecosystems.' Am.
       Nat., 118, 262-274.
Bray, J.R. and Curtis, J.T., 1957. 'An ordination of the upland forest
       communities of southern Wisconsin.' Ecol. Monogr., 27, 325-349.
Buchanan, J.B., Sheader, M. and Kingston, P.R., 1978. 'Sources of
       variability in the benthic macrofauna off the South Northumberland
       coast, 1971-6.' J. mar. Biol. Ass. U.K., 58, 191-210.
Cederwall, H. and Elmgren, R., 1980. 'Biomass increase of benthic
       macrofauna demonstrates eutrophication of the Baltic Sea.' Ophelia
       Suppl., 1, 287-304.
Dagg, M.J. and Turner, J.T., 1982. 'The impact of copepod grazing on
       the phytoplankton of Georges Bank and the New York Bight.' Can.
       J. Fish. Aquat. Sci., 39, 979-990.
Field, J.G., Clarke, K.R. and Warwick, R.M., 1982. 'A practical
       strategy for analysing multispecies distribution patterns.' Mar.
       Ecol. Progr. Ser., 8, 37-52.
Gieskes, W.W.C., Kraay, G.W. and Baars, M.A., 1979. 'Current ^{14}C
       methods for measuring primary production: gross underestimates in
       oceanic waters. Neth. J. Sea Res., 13, 58-78.

Gieskes, W.W.C. and Kraay, G.W., 1984. 'State of the art in primary production measurements. In Fasham, M. (ed.) Flows of Energy and Materials in Marine Ecosystems. Plenum Press, New York, pp 171-190.

Gray, J.S. and Christie, H., 1983. 'Predicting long-term changes in marine benthic communities.' Mar. Ecol. Progr. Ser., 13, 87-94.

Gray, J.S. and Mirza, F.G., 1979. 'A possible method for the detection of pollution-induced disturbance on marine benthic communities.' Mar. Poll. Bull., 10, 142-146.

Gray, J.S. and Pearson, T.H., 1982. 'Objective selection of sensitive species indicative of pollution-induced change in marine benthic communities. 1. Comparative methodology.' Mar. Ecol. Progr. Ser., 9, 111-119.

Honjo, S., 1982. 'Seasonality and interaction of biogenic and lithogenic particulate flux at the Panama Basin.' Science, N.Y., 218, 883-884.

Isdale, P. (1984). Fluorescent bands in massive corals record centuries of coastal rainfall. Nature, Lond., 310, 578-579.

Josefsson, A.B. and Smith, S., 1984. 'Färändringar av benthos-biomassa i Skagerak-Kattegat under 1970-talet; ett resultat av slumpen, klimat-förändringar eller eutrofiering?' Medd. Havfiskelab. Lysekil, 292, 111-121.

Kremer, J.W. and Nixon, S.W., 1978. A Coastal Marine Ecosystem. Springer-Verlag, Berlin. 217 pp.

Lewis, J.R., 1976. 'Long-term ecological surveillance: practgical realities in the rocky littoral.' Oceanogr. Mar. Biol. Ann. Rev., 14, 371-390.

Lewis, J.R. and Bowman, R.S., 1975. 'Local habitat induced variability in the populatioan dynamics of Patella vulgata L.' J. Exp. Mar. Biol. Ecol., 17, 165-204.

Lambshead, P.J.D., Platt, H.M. and Shaw, K.M., 1983. 'The detection of differences among assemblages of marine benthic species based on an assessment of dominance and diversity.' J. Nat. Hist., 17, 859-874.

Malone, T.C. and Chervin, M.B., 1979. 'The production and fate of phytoplankton size fractions in the plume of the Hudson River, New York Bight.' Limnol. Oceanogr., 24, 683-696.

McAllister, C.D., 1972. 'Estimates of transfer of primary production to secondary production at Ocean station P.' In A.Y. Takenouti (ed.) Biological Oceanography of the North Pacific. Idemitsu Shoten, Tokyo pp 575-579.

Pearson, T.H., Gray, J.S. and Johannessen, P.J., 1983. 'Objective selection of sensitive species indicative of pollution-induced change in benthic communities. 2. Data analyses.' Mar. Ecol. Progr. Ser., 12, 237-255.

Pearson, T.H. and Rosenberg, R., 1978. 'Macrobenthic succession in relation to organic enrichment and pollution of the marine environment.' Oceanogr. Mar. Biol. Ann. Rev., 16, 229-311.

Persson, L.-E., 1981. 'Were macrobenthic changes induced by thinning out of flatfish stocks in the Baltic proper?' Ophelia, 20, 137-152.

Peterson, C.H., Duncan, P.B., Summerson, H.C. and Safritt, G.W. Jr.,
    1983. 'A mark-recapture test of annual periodicity of internal
    growth band deposition in shells of hard clams Mercenaria
    mercenaria from a southeastern population.' Fish. Bull.
Platt, H.M., Shaw, K.M. and Lambshead, P.J.D., 1984. 'Nematode species
    abundance patterns and their use in the detection of environmental
    perturbations.' Hydrobiologia, 118, 59-66.
Riley, G.A. and Bumpus, D.F., 1946. 'Phytoplankton-zooplankton
    relationships on Georges Bank.' J. Mar. Res., 6, 33-46.
Sheehan, P.J., 1984(a). 'Effects on individuals and populations.' In
    Sheehan, P.J., Miller, D.R., Butler, G.C. and Bourdeau, PH.
    (eds.). Effects of Pollutants at the Ecosystem Level. SCOPE. John
    Wiley, Chichester, pp 23-50.
Sheehan, P.J., 1984(b). 'Effects on community and ecosystem structure
    and dynamics.' In Sheehan, P.J., Miller, D.R., Butler, G.C. and
    Bourdeau, Ph. (eds.). Effects of Pollutants at the Ecosystem
    Level. SCOPE. John Wiley, Chichester, pp 51-100.
Smetacek, V., 1980. 'Annual cycle of sedimentation in relation to
    plankton ecology in Western Kiel Bight.' Ophelia,
    Suppl., 1, 65-76.
Steele, J.H., 1974. The Structure of Marine Ecosystems. Harvard,
    Cambridge Mass. 128 pp.
Steeman-Nielsen, E., 1952. 'The use of radioactive carbon (^{14}C) for
    measuring organic production in the sea.' J. Cons. Inst. Explor.
    Mer,, 18, 117-140.
Taguchi, S. and Fukuchi, M., 1975. 'Filtration rates of zooplankton
    community during spring bloom in Akkeshi Bay.' J. Exp. Mar. Biol.
    Ecol., 19, 145-164.
Ugland, K.I. and Gray, J.S., 1982. 'Lognormal distributions and the
    concept of community equilibrium.' Oikos, 39, 171-178.
Walsh, J.J., 1981. 'Shelf-sea ecosystems.' In A.R. Longhurst (ed.)
    Analysis of Marine Ecosystems. Academic Press, New York. pp
    159-196.
Wassmann, P., 1984. 'Sedimentation of organic material in West
    Norwegian fjords.' Dr. Scient. thesis, Univ. of Bergen, Norway.
    238 pp.
Widdows, J., Bayne, B.L., Donkin, P., Livingstone, D.R., Lowe, D.M.,
    Moore, M.N. and Sallkeld, P.N., 1981. 'Measurement of the
    responses of mussels to environmental stress and pollution in
    Sullom Voe: a baseline study.' Proc. R. Soc. Edinb., 80b, 323-338.
Williams, P.J. Le B., Raine, R.C.T. and Bryan, J.R., 1979. 'Agreement
    between the ^{14}C and oxygen methods of measuring phytoplankton
    production: reassessment of the photosynthetic quotient.'
    Oceanol. Acta, 2, 411-416.
Wulff, F., 1979. 'The effects of sampling frequency on estimates of the
    annual pelagic primary production in the Baltic.' In H. Hytteborn
    (ed.) The Use of Ecological Variables in Environmental
    Monitoring. The Swedish Environment Protection Board, Report PM
    1151. pp 147.150.

PHYSIOLOGICAL AND CELLULAR RESPONSES OF ANIMALS TO ENVIRONMENTAL STRESS - CASE STUDIES

T. J. Lack
Water Research Centre
Medmenham Laboratory
Henley Road, Medmenham
PO Box 16, Marlow
Buckinghamshire, SL7 2HD
UK

J Widdows
Natural Environment Research
 Council
Institute for Marine
 Environmental Research
The Hoe, Plymouth, PL1 3DH
UK

ABSTRACT. This review is concerned with the sensitive and sublethal responses shown at the individual and cellular levels by animals exposed in the field, either naturally or by transplanting, to various levels of environmental contamination. The relationships between contaminant concentrations in the tissues and the biological responses are given particular emphasis.

Results from eight reported case-studies have been reviewed. Sites range from open coastal waters, bays and estuaries with contaminants arising from the petroleum industry, waste disposal operations and more general anthropogenic inputs. Responses measured are physiological; (Scope for Growth) and biochemical (stability of lysosomal membranes, presence of metal-protein compounds and activity of blood-cell NADPH - neotetrazolium reductase).

On the evidence of the field studies there appears to be a good relationship between the biological responses and tissue concentrations of contaminants, particularly the aromatic hydrocarbons. There is little direct field evidence of the toxic effects of metals related to the tissue concentrations but present chemical analysis of tissues may not be appropriate for the detection of such a relationship. Recommendations are made as to the future deployment of field and laboratory tests in order that impact on the environment can be assessed on a common scale of values.

1. INTRODUCTION

Numerous and varied biological responses have been recommended as potential techniques for quantifying and monitoring the environmental impact of marine waste disposal operations and these have been recently reviewed (Bayne - this meeting, Bayne et al. (1985) and McIntyre and Pearce, 1980). Physiological and biochemical responses to contaminants in the environment are most likely to be correlated with concentrations in the tissues rather than in the external medium so this review is limited to the sensitive and sublethal responses shown, at the individual and cellular levels, by animals exposed in the field, either

G. Kullenberg (ed.), The Role of the Oceans as a Waste Disposal Option, 647–665.

by their natural occurrence or by transplanting, to a gradient of
environmental contamination.

Case studies reported so far have involved different approaches.
Firstly, the physiological responses (Scope for Growth (SFG), Widdows
1985) of native populations (usually Mytilus spp.) have been measured
using a mobile laboratory under ambient field conditions. Secondly, SFG
of transplanted animals has been measured under ambient field conditions
and thirdly, SFG of transplanted animals has been measured in a
laboratory-based seawater facility under standard conditions of
temperature, salinity and food availability approximating to the field
situation.

Biochemical responses (e.g. lysosomal stability, neotetrazolium
reductase activity and presence of metal-binding proteins) have been
carried out on native (Mytilus spp. and Littorina littorea) and
transplanted animals (Moore 1985).

The techniques have been applied at sites ranging from open coastal
waters to enclosed bays and estuaries and sources of contamination
include activities of the petroleum industry, sewage and sewage sludge
waste disposal operations and more general anthropogenic inputs.

Where the objective is a comparison of degree of stress or fitness
at different sites where animals are exposed to different stressors or
pollutants, then the transplanting of animals from a clean population to
the study site is to be recommended. Animals of similar body size from
the same population and in similar reproductive condition and with a
similar environmental history can be placed at selected sites along a
pollution gradient removing any reliance on the sporadic distribution of
naturally occurring populations of the sentinel organism.

2.   CASE STUDIES

2.1. Casco Bay, Maine, USA

An early use of SFG to measure the long-term effects of an oil spill was
reported by Gilfillan et al (1977). Two years after an oil spill in
Casco Bay, Maine, the authors recorded a significantly reduced carbon
flux (equivalent to SFG) in clams (Mya arenaria) from contaminated
sites. Carbon flux was found to be negatively correlated with aromatic
hydrocarbon concentrations in the body tissues, (figure 1).

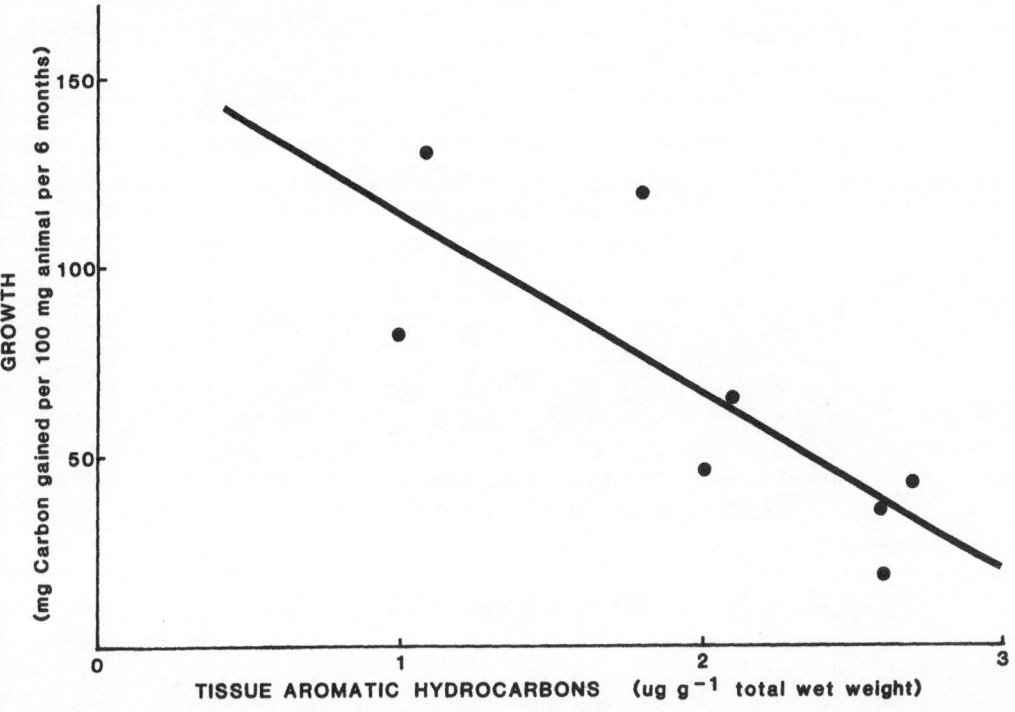

Figure 1.   Relationship between total carbon flux by <u>Mya arenaria</u>
(100 mg) and the tissue aromatic hydrocarbon
concentration.   (From Gilfillan et al, 1977.)

2.2. Narragansett Bay, Rhode Island, USA

An example of a field programme involving the transplantation of mussels
along a pollution gradient is described by Widdows et al. (1981 a).
Mussels were transplanted to four sites along an established pollution
gradient in Narragansett Bay and after 1 month the SFG under standard
laboratory conditions was measured.   Body tissues were subsequently
analysed for metals and hydrocarbons.   There was found to be a
significant decline in SFG associated with increasing concentrations of
contaminants (hydrocarbons and some metals).   Natural environmental
stressors (temperature, dissolved oxygen, salinity and seston
concentration) were relatively constant at the transplant sites.
Consequently, there was good evidence that the reduced SFG in the
northern (landward) end of Narragansett Bay was caused by anthropogenic
stressors (figure 2).

a) Location of Pollution Gradient

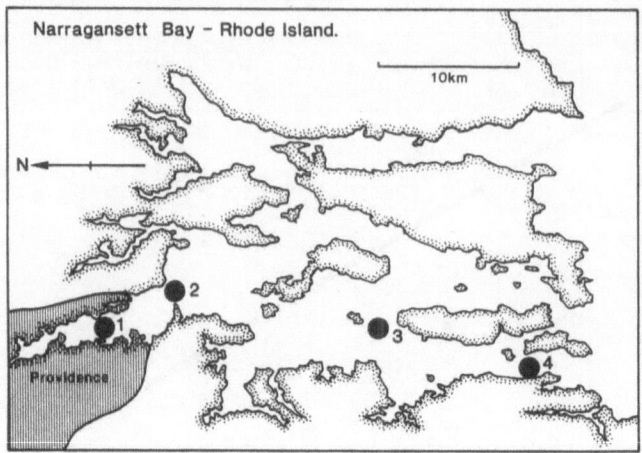

b) Environmental Contamination (Bioaccumulation of Petroleum
    Hydrocarbons and Metals by Transplanted Mussels).

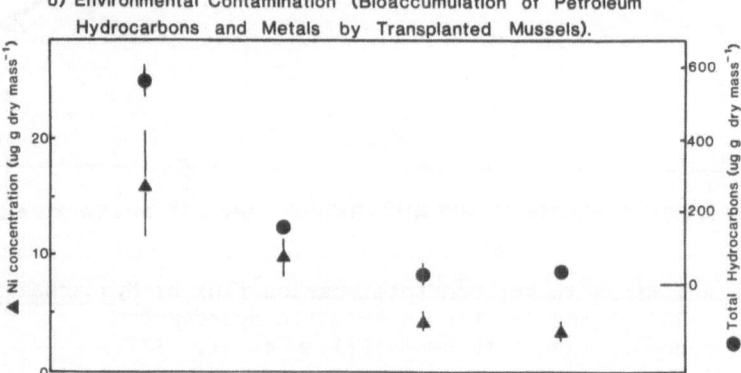

c) Biological Impact of Pollution (Decline in Growth of
    Transplanted Mussels).

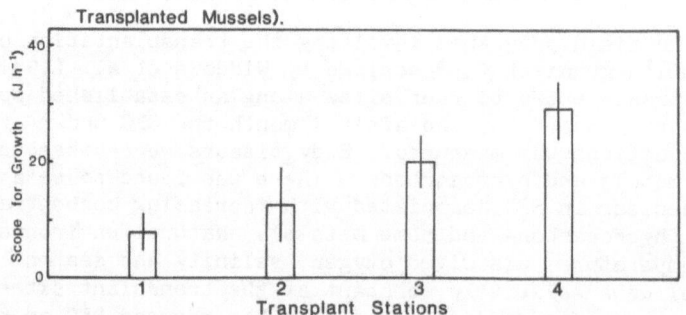

Figure 2.   A. Transplant stations in Narragansett Bay. B. Tissue
            concentrations of nickel and total hydrocarbons (µg/g
            dry mass) in transplanted mussels. C. Scope for
            Growth (Joules/hour) of transplanted mussels.

Figure 3.   Sites for resident mussel collections (Mytilus edulis)
from Tomales and San Francisco Bays.   (From Sevareid
and Ichikawa, 1983.)

2.3. San Francisco Bay, California, USA

A more recent example of the use of native mussels for quantifying
sublethal effects of environmental contaminants is reported by Sevareid
and Ichikawa (1983). Mussels (Mytilus edulis) were collected from a
clean area (Tomales Bay) and 5 sites in south San Francisco Bay, the
sites being selected in relation to a gradient of heavy metal and trace
organic pollution. Scope for Growth declined markedly along the
pollution gradient (figure 3). There was a high correlation between
increasing tissue concentrations of some metals (e.g. Cr, Cu, Hg, Al, Zn
and Ag) and decreasing SFG. The synthetic organic compounds, chlordane
and dieldrin were similarly correlated (figure 4). The San Francisco
Bay sites are arranged in order from north to south, an order that also
represents increasing ambient levels of metals and organics.

2.4. San Diego Bay, California, USA

Mussels (M. edulis) were transplanted from Tomales Bay to 4 sites in
San Diego Bay (Martin 1985) and left for 2 months (figure 5). A stress
gradient was revealed where SFG was greatly reduced and this was
correlated with PCB bioaccumulation. At two of the transplant sites SFG
was zero or negative indicating a high degree of stress. Laboratory
bioassays showed that exposure to PCB over a 6 week period produced
tissue concentrations over 2000 ng/g (dry mass) and a significant
reduction in SFG after 2 weeks falling to around 0 Joules/hour in the
animals exposed to the highest concentrations (8 ng/l) for 6 weeks. In
the field studies, transplanted animals with zero or negative SFGs had
tissue concentrations between approximately 1000-7000 ng/g (dry mass).

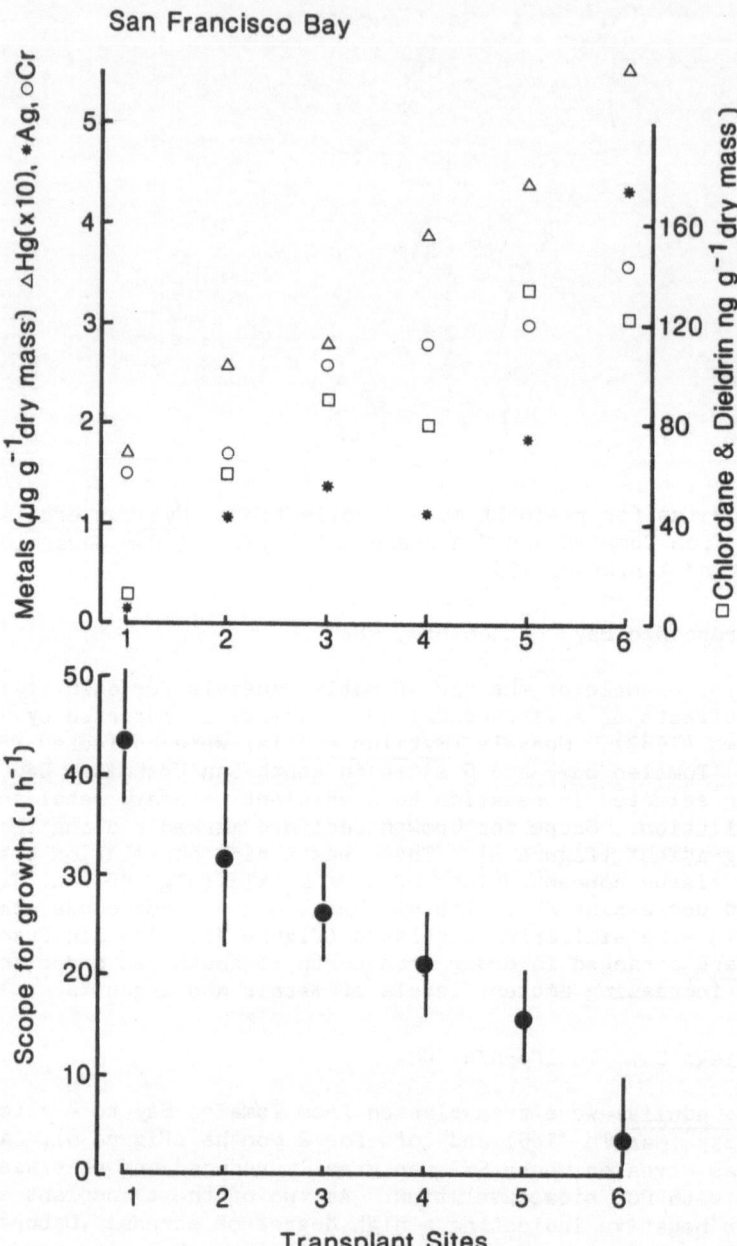

Figure 4.   Tissue concentrations of metals (µg/g dry mass) and
             organics (chlordane + dieldrin ng/g dry mass) of
             mussels collected from Tomales and San Francisco Bays.

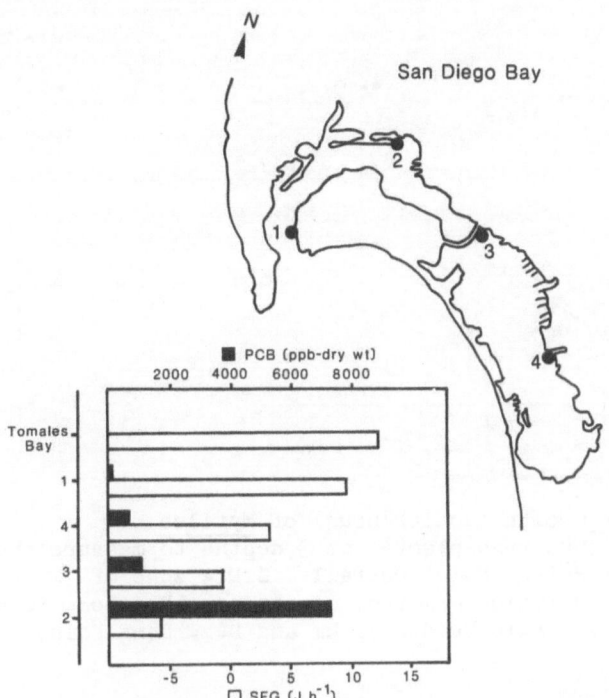

Figure 5.   Tissue concentrations of PCB_{1254} (ng/g dry mass) and
Scope for Growth (Joules/hour) in <u>Mytilus edulis</u>
transplanted in San Diego Bay for 2 months. (From
Martin 1985).

2.5. Whites Point Outfall, Los Angeles, California, USA

Mussels (<u>M. californianus</u>) were transplanted from Bodega Head (near
Tomales Bay) to measure the effect of Los Angeles County Sanitation
District's Whites Point Outfall. This discharges about 21 m^3/sec of
primary sewage effluent, the largest discharge in California (Martin
1985). The effluent has a large contribution from industrial wastes and
some of the highest concentrations of trace metals and synthetic
organics in any discharge into Californian marine waters. The diffuser
system in 85 m of water is located 2.5 km offshore of Whites Point,
Palos Verdes Peninsula. Mussels were exposed for 3 months at 10, 25 and
50 m depths at; the zone of initial dilution 0 km from the diffuser;
Long Point 7.4 km; Palo Verdes 12 km and Dana Point 63 km. Lowest SFG
was measured in mussels transplanted in the immediate vicinity of the
discharge (figure 6). Values were around zero or negative around the
discharge but by 12 km mussels exposed at 10 m had significantly
improved. The only chemical determinand reported was tissue
concentrations of silver which may be considered to act as a marker for
a variety of other contaminants and lowest SFGs were recorded in animals
having higher body levels of silver.

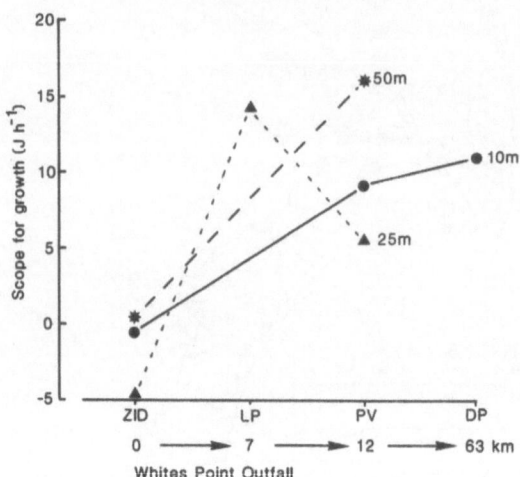

Figure 6.   Scope for Growth (Joules/hour) of <u>Mytilus</u>
            <u>californianus</u> transplanted at 3 depths to measure the
            effect of Whites Point Outfall. ZID = zone of
            immediate dilution OKm from discharge; LP = Long Point
            7.4 km; PV = Palo Verdes 12 km and DP = Dana Point
            63 km.

2.6. Plymouth Sewage Sludge Disposal Ground, UK

The feasibility of using sentinel organisms to monitor the environmental
impact of sewage sludge disposal operations at sea was carried out off
Plymouth during summer 1984 (Lack and Johnson 1985). Mussels
(<u>Mytilus edulis</u>) were taken from a clean site and exposed near the
surface and near the bed for approximately 60 days in baskets along a
suspected pollution gradient passing through a sludge disposal ground
(figure 7). A complicating factor in the assessment of these results
was that 'trickle spawning' occurred during the exposure period.
However, because the results have value in a comparative sense, the data
presented in Lack and Johnson (1985) have been converted to relative SFG
with Site 7 lower animals being ascribed the controls and all other data
being expressed as percentages of the control (figure 8). According to
SFG all the animals were stressed except those at Site 1 (the nearshore
site) and the upper (near surface) caged animals were more stressed than
the lower. The most stressful location for upper animals was Site 5, on
the edge of the disposal ground and Site 2 for the lower animals.
Lysosomal destabilisation represents a general cellular stress response
which has been shown to be applicable over a wide range of environmental
conditions. The labilisation period of latent lysosomal enzymes in the
digestive cells of mussels can range from 25-2 minutes; the higher
figure being representative of unstressed animals (Moore 1980).
Cytochemically determined activity of blood-cell NADPH - neotetrazolium
reductase (NTR) which is believed to be a component of the microsomal
detoxication system (Moore 1980, 1985) is used to quantify responses to

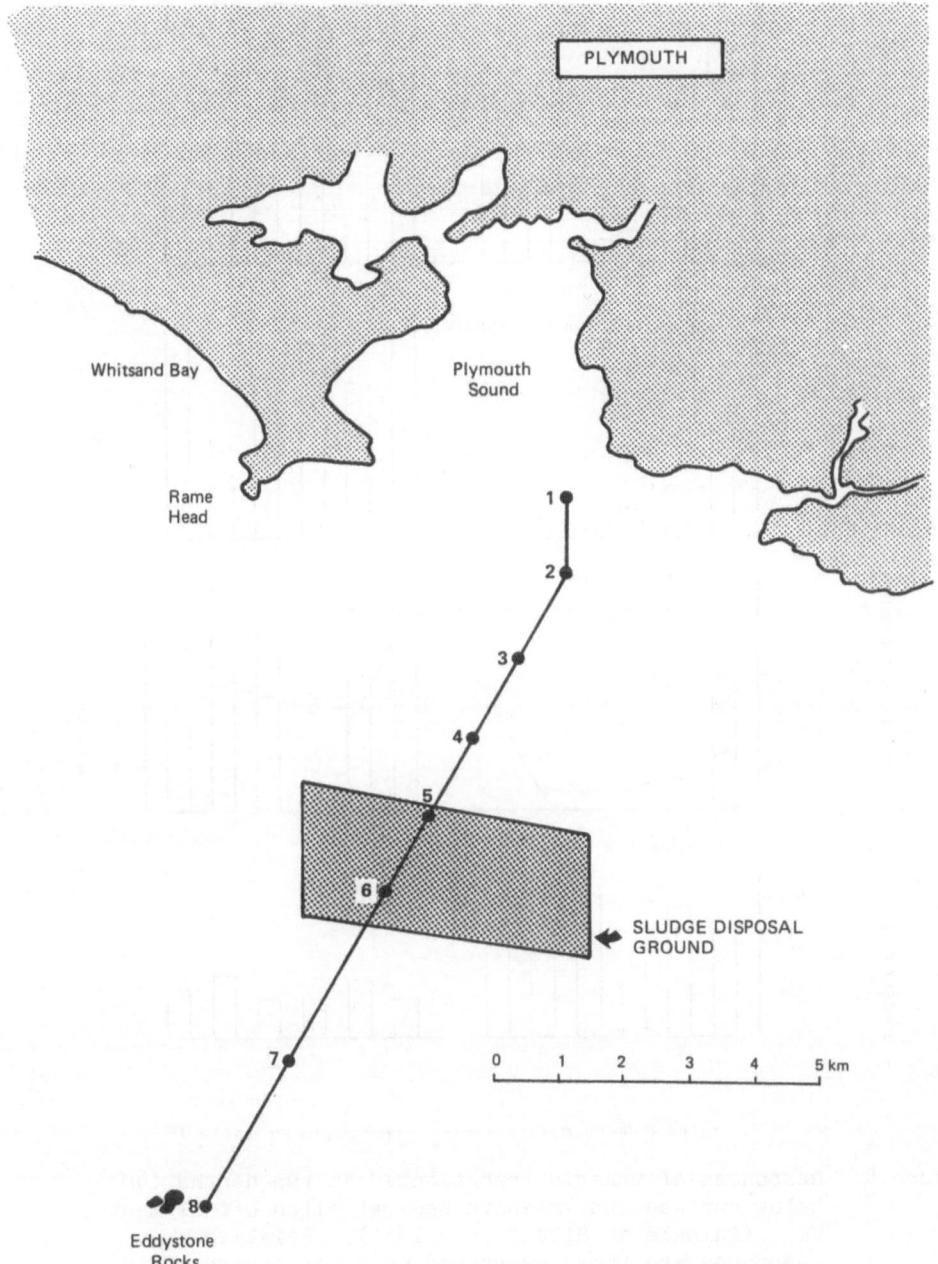

Figure 7. Location of mussel transplant sites (1-7) off
Plymouth, UK in relation to the sludge disposal
ground. (Site 8 is a clean water bioassay site).

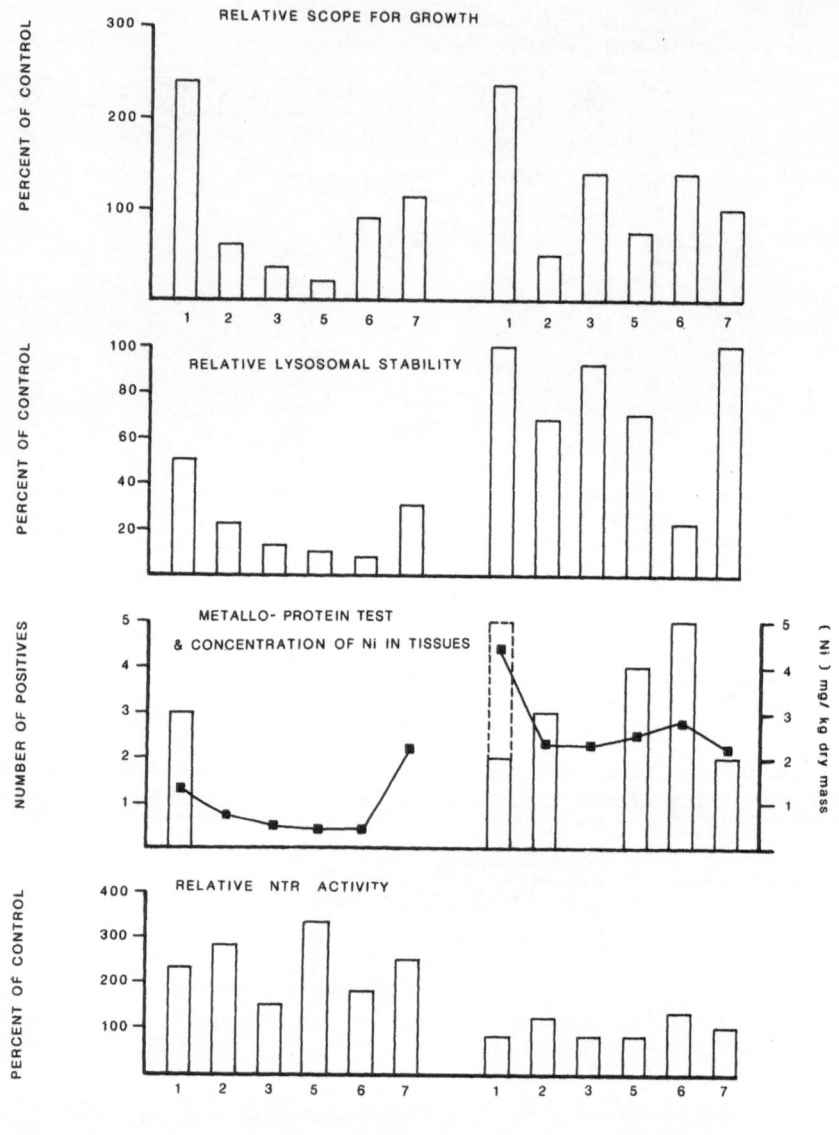

Figure 8.   Responses of mussels transplanted at two depths (4m
             below surface and 4m above bed) at sites off Plymouth,
             UK.   (Animals at Site 4 were lost).   Relative
             responses are those expressed as a percentage of the
             response of the control animals; Site 7 lower.   The
             concentration of Nickel (mg/kg dry body mass) is shown
             as being typical of the pattern of concentrations of
             other metals (Pb, Zn, Cu and Cr).

polynuclear aromatic hydrocarbons. According to the Lysosomal Stability test (also converted here to values relative to the control) all the upper animals were stressed particularly those (Sites 5 and 6) most exposed to the finely particulate sewage sludge solids. Of the lower animals only those in the centre of the disposal ground (Site 6) were stressed.

The lower animals showed evidence that the metal detoxication system had been induced, particularly at Sites 5 and 6 within the disposal ground. The upper animals showed no such evidence (except at Site 1) and the concentrations of Nickel in the exposed animals to some extent reflects the different responses (figure 8).

Although no analyses were made of organic contaminant concentrations in the tissues of the exposed mussels there is some evidence that the stress in the upper animals measured by SFG and Lysosomal Stability could have been caused by such contaminants as the upper animals had elevated levels of relative activity of Neotetrazolium Reductase (NTR); part of the organic xenobiotic detoxication system. Certainly, there was no evidence that metal concentrations in the tissues of the upper animals could be responsible.

2.7. Sullom Voe North Sea Oil Terminal in Shetland, UK

A field study in which the biological impact of pollution has been assessed by measuring scope for growth of native M. edulis and Littorina littorea in association with lysosomal and microsomal responses has been carried out at sites around the Sullom Voe Oil Terminal (figure 9). The base-line study was reported by Widdows et al. (1981 b).

Mussels taken from Gluss Voe, a clean reference site, had SFGs between 15.1 and 22.8 at various times between September 1981 and July 1983, the range probably reflecting a natural seasonal variability (Table I).

Tissue concentrations of 2-ring aromatic hydrocarbons were low (0.04-0.05 µg/g wet mass). Mussels from Scatsta Voe, due south of the oil terminal and potentially vulnerable to spillages, showed somewhat higher hydrocarbon concentrations in the tissues but a barely detectable decrease in SFG which probably has little or no environmental significance. Mussels sampled from Mavis Grind, a 'contaminated' site where there is oily run-off from previous contamination of a quarry, had three times the tissue hydrocarbon concentration of those at Scatsta Voe but barely significantly lower SFGs. Samples taken from the oil jetty in August 1984 had 17 times the tissue hydrocarbon concentration of the control site and the SFG was found to be about half of the control. These variations in the fitness of native mussels are mirrored by cellular and biochemical changes in winkles (Littorina littorea) sampled at other sites in Sullom Voe (Table II).

Table I

Scope for growth (S.F.G.; J g^{-1} h^{-1}) and 2 ring aromatic hydrocarbon concentrations (A.H.C.; μg g^{-1} wet mass) in the tissues of Mytilus edulis collected from sites in the vicinity of the Sullom Voe North Sea Oil Terminal, Shetland. (From Widdows et al. 1983)

SAMPLE SITE	SEPT 1981	JULY 1982		JULY 1983		AUG 1984	
	S.F.G.	S.F.G.	A.H.C.	S.F.G.	A.H.C.	S.F.G.	A.H.C.
Gluss Voe (reference)	15.1	21.1	0.04	22.8	0.05	21.1	0.08
Scatsta Voe	13.9	17.1	0.67	17.0	0.28	-	-
Mavis Grind	10.8	-	-	14.2	0.86	-	-
Tanker Jetty 3	-	-	-	-	-	10.4	1.37

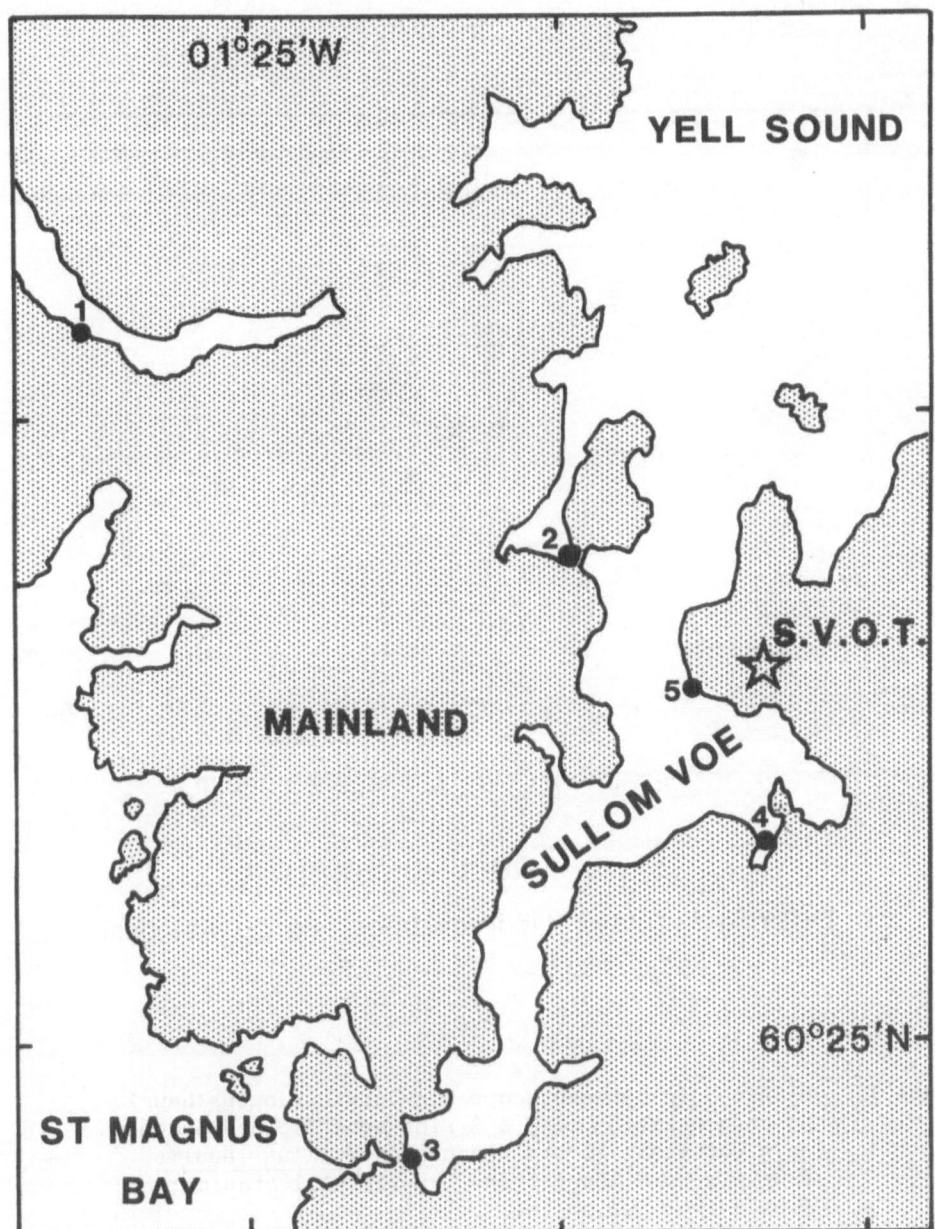

Figure 9. Location of sampling sites, Sullom Voe, Shetland;
(1) Ronas Voe, (2) Gluss Voe, (3) Mavis Grind,
(4) Scatsta Voe and (5) Tanker Jetty.

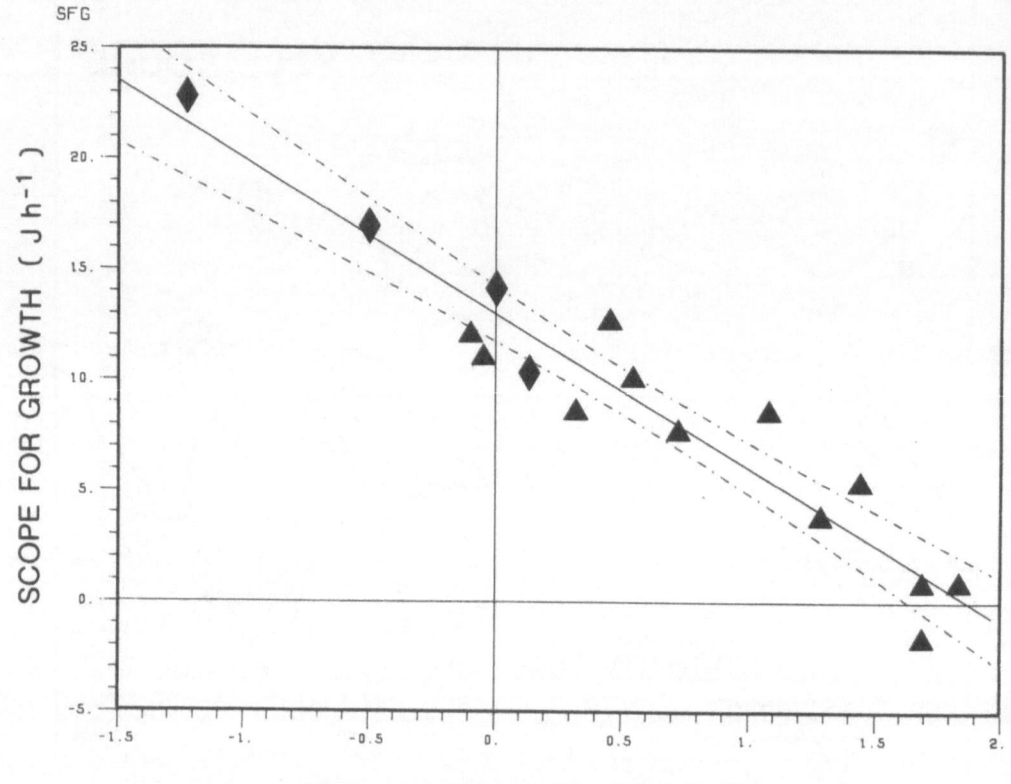

Figure 10. Relationship between Scope for Growth (Joules/hour)
and concentration of 2 & 3 ring aromatic hydrocarbors
(μg/g wet mass) in the tissues of <u>Mytilus edulis</u>.
▲ Data from Solbergstrand Experimental Station;
◆ Data from Sullom Voe.

Table II

Cellular stress responses (Lysosomal stability and neotetrazolium reductase – N.T.R.) expressed as % of reference site and 2 ring aromatic hydrocarbon concentrations (A.H.C.; µg g^{-1} wet mass) in the tissues of Littorina littorea collected from sites in the vicinity of the Sullom Voe North Sea Oil Terminal Shetland. (From Moore et al. 1982; Widdows et al. 1983.)

SAMPLE SITE	SEPT 1981		JULY 1982			JULY 1983
	Lysosomal Stability	N.T.R.	Lysosomal Stability	N.T.R.	A.H.C.	A.H.C.
Ronas Voe (reference)	100%	100%	100%	100%	0.11	0.09
Scatsta Voe	47%	128%	75%	106%	–	0.20
Mavis Grind	53%	114%	25%	148%	–	0.72
Tanker Jetty 4	15%	130%	37%	130%	0.91	0.45

Biological effects data not available for 1983 due to spawning.

The latency of lysosomal enzymes, aryl sulphatase in 1981
and β-glucaronidase in 1982, were measured cytochemically (Moore et al.,
1982). In both years lysosomal stability was lower at the
'contaminated' sites and there was an increase in tissue hydrocarbon
concentration similar to that found in Mytilus. Increase in NTR
activity relative to the control site showed a similar pattern to the
lysosomal stability. There appeared to be relatively little difference
between lysosomal stability and NTR activity in winkles sampled from
Mavis Grind or at the tanker jetty in the oil terminal. The sensitivity
of the physiological and biochemical techniques employed at Sullom Voe
and the other field case-studies, particularly with regard to
hydrocarbon contamination in the tissues, is well demonstrated by
referring to a particular experiment that has been carried out at the
Solbergstrand Marine Research Station.

2.8. Solbergstrand Marine Research Station, Oslo Fjord, Norway

Although this review is restricted to field case-studies it is relevant
to include some results from an enclosed experimental facility at the
Solbergstrand Marine Research Station (Widdows et al. 1985, Moore
unpublished data). Two indoor concrete basins with a total water volume
of 320 m^3 and established rocky shore communities (including
Mytilus edulis) were chronically exposed to low (30 μg/l) and high
(130 μg/l) concentrations of diesel oil for 8 months. Scope for Growth,
lysosomal stability (latency of β-hexosaminidase) and NTR activity have
been measured on mussels exposed in the experimental tanks and compared
with clean controls. Tissue concentrations of 2+3 ring aromatic
hydrocarbons were also measured, (Table III).
Hydrocarbon concentration and SFG in the control animals, kept in Oslo
Fjord water was very similar to those found at Mavis Grind and the
Tanker jetties at Sullom Voe. In the low oil exposures, SFG was halved,
lysosomal stability significantly reduced and NTR activity increased.
Mussels exposed to the high levels of oil had very low SFG, low
lysosomal stability and NTR activity was nearly double the controls.
The hydrocarbon concentration in the tissues was approximately 50 times
that of the controls (and the Sullom Voe contaminated samples). Further
data derived from oil depuration and recovery experiments at
Solbergstrand are shown together with the Sullom Voe data (figure 10)
demonstrating the relationship between SFG and hydrocarbon concentration
in the tissues over a wide range.

DISCUSSION & CONCLUSIONS

In preparing this review we have found it difficult to make inter-study
comparisons of relationships between biological effects measurements and
tissue concentrations of organic contaminants partly because of the lack
of compatibility in the chemical data probably arising from the use of
different analytical methods in different laboratories. A notable
exception is the good agreement between the results obtained in the
'mesocosm' experiments at Solbergstrand and the field studies at

Table III

Effects of chronic diesel oil exposure (8 months) on the scope for growth (S.F.G), lysosomal stability (minutes), neotetrazolium reductase (N.T.R.) and 2+3 ring aromatic hydrocatbon concentrations (A.H.C.; μg g^{-1} wet mass) in the tissues of <u>Mytilus edulis</u>. (Solbergstrand experimental station).

	S.F.G. Joules/Hour	Lysosomal Stability	N.T.R.	A.H.C.
CONTROL (5.6μg 1^{-1})	10.8	25	100%	0.9
LOW OIL (30μg 1^{-1})	5.3	5	138%	28
HIGH OIL (130μg 1^{-1})	1.8	2	184%	49

Sullom Voe because there is consistency both in the biological and chemical methodology.

On the basis of the field studies so far carried out there does appear to be a good relationship between physiological and biochemical responses and tissue concentrations of contaminants, particularly the aromatic hydrocarbons.

There is some evidence that the effects of environmental contaminants are additive which suggests that in many cases (particularly for monitoring rather than research) preliminary chemical analysis could be simplified in order to achieve a total body burden rather than a detailed breakdown of individual organic contaminants. There is, however, a need for standardisation of analytical procedures that are cost-effective.

Current research is concerned with the use of structure-activity relationships to describe and predict the sublethal toxic effects of organic pollutants on marine animals. The ultimate objective is to provide a toxicological interpretation of tissue residue data.

There is little direct field evidence of the toxic effects of
metals related to the tissue concentration. Results reported so far
have been concerned with total concentration in the tissues. It is
likely that metals exist in two basic fractions; the detoxified or
sequestered component and the toxic component probably free in the
cytosol. Total metal concentration may not therefore be directly
related to biological effects because of the protection afforded by the
detoxication systems. Therefore it would appear necessary for chemical
monitoring (and chemical analysis of tissues) to take account of this
when trying to determine the impact of the release of metals to the
environment.

The general stress responses (SFG and lysosomal stability) appear
to have value in quantifying the intensity and extent (spatial and
temporal) of environmental impact over a wide range of inputs. The more
specific stress responses (e.g. induction of metallo-proteins and NTR)
require further and wider deployment in the field to assess their
validity although NTR activity does appear to be a valuable quantifier
of stress caused by exposure to hydrocarbons.

For the present, field studies need to be complemented by
controlled laboratory experiments involving chronic exposure of test
organisms to potential toxicants in order to calibrate the general and
specific dose-response relationships; to examine the relative toxicities
of different contaminants and to identify causative agents. An
important advantage of this approach to biological effects monitoring is
that the techniques are applicable to laboratory, 'mesocosm'
(e.g. Solbergstrand) and field studies.

Ultimately, it should therefore be possible to establish the
relative toxicity of different classes of contaminants released into the
marine environment by waste disposal operations and other human
activities (including relatively uncontrollable sources) in order that
the impact on the environment can be assessed on a common scale of
values.

REFERENCES

Bayne, B.L., Brown, D.A., Burns, K., Dixon, D.R., Ivanovici, A.
Livingstone, D.R., Lowe, D.M., Moore, M.N., Stebbing, A.R.D. &
Widdows, J. (1985). The effects of stress and pollution on marine
animals. Praeger Scientific, New York. pp384.

Gilfillan, E.S., Mayo, D.W., Page, D.S., Donovan, D. & Hanson, S.
(1977). Effects of varying concentrations of petroleum hydrocarbons in
sediments on carbon flux in Mya arenaria. In: Physiological Responses
of Marine Biota to Pollutants (Vernberg F.J., Calabrese, A.,
Thurberg, F.P. and Vernberg, W.B., eds.), 299-314, Academic Press.
New York.

Lack, T.J. & Johnson, D. (1985). Assessment of the biological effects
of sewage sludge at a licensed site off Plymouth. Marine Pollution
Bulletin, 16,(4), 147-152.

Martin, M. (1985). State mussel watch: Toxics surveillance in California. Marine Pollution Bulletin, 16,(4), 140-146.

McIntyre, A.D. & Pearce, J.B. (1980). Biological effects of marine pollution and the problems of monitoring. (Proceedings of an ICES Workshop at Beaufort, North Carolina 1979). Rapp. P.-v. Réun. Cons. int. Explor. Mer, 179, pp346.

Moore, M.N. (1980). Cytochemical determination of cellular responses to environmental stressors in marine organisms. In: Biological effects of marine pollution and the problems of monitoring (McIntyre, A.D. & Pearce, J.B. eds.) Rapp. P.-v. Réun. Cons. int. Explor. Mer, 179, 7-15.

Moore, M.N. (1985). Cellular responses to pollutants. Marine Pollution Bulletin, 16,(4), 134-139.

Moore, M.N., Pipe, R.K. & Farrar, S. (1982). Lysosomal and microsomal responses to environmental factors in Littorina littorea from Sullom Voe. Marine Pollution Bulletin, 13, 340-345.

Sevareid, R. & Ichikawa, G. (1983). Physiological stress (scope for growth) of mussels in San Francisco Bay. In: Waste Disposal In the Oceans: Minimizing Impact, Maximizing Benefits. (Soule, D.F. & Walsh, D. eds.). 152-170. Westview Press. Boulder. Colorado.

Widdows, J. (1985). Physiological responses to pollution. Marine Pollution Bulletin, 16,(4), 129-134.

Widdows, J., Phelps, D.K. & Galloway, W. (1981 a). Measurement of physiological condition of mussels transplanted along a pollution gradient in Narragansett Bay. Mar. Environ. Res. 4, 181-194.

Widdows, J., Bayne, B.L., Donkin, P., Livingstone, D.R., Lowe, D.M., Moore, M.N. & Salkeld, P.N. (1981 b). Measurement of the responses of mussels to environmental stress and pollution in Sullom Voe: a base-line study. Proc. Roy. Soc. Edinb. 80B, 323-338.

Widdows, J., Cleary, J.I., Dixon D.R., Donkin, P., Livingstone, D.R., Lowe, D.M., Moore, M.N., Pipe, R.K., Salkeld, P.N., and Worrall C.M. (1983). Sublethal biological effects monitoring in the region of Sullom Voe, Shetland. 1983 SOTEAG Report. pp34.

Widdows, J., Donkin, P. & Evans, S.V. (1985). Recovery of Mytilus edulis L. from chronic oil exposure. Marine Environ. Research (in press).

BIOLOGICAL EFFECTS STUDIES AT VARIOUS LEVELS ALONG THE U.S. PACIFIC COAST

Alan J. Mearns
Pacific Office
Ocean Assessments Division
National Oceanic and Atmospheric Administration
7600 Sand Point Way NE
Seattle, Washington 98115
U.S.A.

ABSTRACT. Recent biological effects research and monitoring activities along the U.S. Pacific Coast are reviewed from both disciplinary and decision-making points of view. Innovations and applications are being developed at all levels of biological organization from community and population studies to new measurements of the biochemical fate of pollutants. There are clear needs to publish the results of longterm monitoring studies, develop more quantitative approaches for using biological effects studies in decision-making and to increase inter-laboratory comparability of results.

1. INTRODUCTION

Biological effects studies are field or laboratory investigations that attempt to quantify changes in the performance, structure or health of organisms exposed to excessive amounts of alledged hazardous materials. Changes in performance and health may be measured at various levels of organization from communities and populations to chromosomes and biochemical processes.

This report reviews some of the biological effects studies being used to examine marine pollution and waste management problems along the U.S. west coast. The review is intended to familiarize the Atlantic reader with "What's going on out there in the Pacific?" It is quite simply a literature survey, not an exhaustive synthesis, and is merely intended to guide the reader toward more information. However, it is also intended as first step toward a long-needed regional synthesis and has been organized by several criteria needed for placing biological effects studies in the context of monitoring and waste-management decision-making, factors important to those who wish to see wider use of biological effects measurements in marine pollution assessment (1).

1.1 Why are Biological Effects Studies Important?

Accepting the GESAMP definition that pollution is a damaging excess of materials (2), then biological effects studies can provide the ultimate

667

G. Kullenberg (ed.), The Role of the Oceans as a Waste Disposal Option, 667–690.
© *1986 by D. Reidel Publishing Company.*

measure of pollution: damage to marine organisms. Indeed, they provide
a mechanism for distinguishing between contamination or other
environmental alterations, which may or may not be of consequence, and
the actual or potential effects of that contamination or alteration,
which may be of great consequence.

1.2 Why a One-Coast Summary?

The U.S. coastal states of Washington, Oregon and California, together
with the neighboring Canadian Province of British Columbia and the
Mexican State of Baja Califoria del Norte, border on a common resource,
the Northeast Pacific Ocean. The combined human population of over 30
million draw upon a common legacy of fishery stocks of the Japanese and
California Currents, share common mineral resources of the shelf,
freely exchange commerce and have similar, very young and
regionally-distinctive histories of population growth and industrial
development. Until quite recently, waste disposal has been considered a
proper and planned use of the ocean. Deep water is close to shore, even
in the inland sea of Puget Sound, and has been relied upon as a
cost-effective depository for sewage, industrial wastes and disposal of
dredged sediments. Indeed, the ability to discharge wastes directly
into deep water has helped reduce waste loading and contamination
nearshore and in harbors, lakes and estuaries.
    Some of the region's scientists have long recognized the need for
regional management of coastal resources and regional exchange of ideas
and information. In fisheries and oceanography, that exchange has been
underway for decades. In the area of marine environmental quality and
marine waste disposal management, that exchange is just begining but
has not yet resulted in regional syntheses of information. However,
there is now a new recognition that three of the region's most
populated embayments, Puget Sound, San Francisco Bay and the Southern
California Bight, share a common legacy of contamination by chlorinated
hydrocarbons, petroleum hydrocarbons, trace metals and pathogens. The
public is demanding new management actions to identify hazards
associated with this contamination and to "clean up" hotspots. Since
the localities share common resources, what is learned in one area is
applicable to the others.

1.2 Why Should NATO Nations be Concerned with Pacific Coast Studies?

The simple answer is that Pacific scientists, like their counterparts
elsewhere, are making significant contributions to the general body of
knowledge about marine pollution and how to measure it. By doing so at
specific contaminated localities these scientists are also contributing
to a body of case-history information that may help others identify and
solve similar waste management problems elsewhere. But equally
compelling reasons for sharing this knowledge have to do with the
natural similarities between the two coastal areas.
    For example, the shores of both Europe and North America are
exposed to the climates of eastern boundary current regimes and thus
experience similar variances in climatology and oceanography that

differ from those of western boundary current regimes such as the east coasts of North America and Asia. And, while there are clearly taxonomic similarities between the western and eastern Atlantic there are also some critical taxonomic similarities between Europe and Pacific North America. For example, the flatfish genera Glyptocephalus, Hippoglossoides, Hippoglossus, Limanda, Microstomus and Platichthys, and the cod-like fish genera Merluccius and Gadus are not only represented on both coasts but are also demonstrably affected by marine pollution or are otherwise subjects of marine pollution investigations. This overlap of congeneric species may make the results of genetic, biochemical and physiological and ecological effects measurements much more transferable than they would be otherwise between any other two temperate coastal regions of the world.

2. METHODS

Five criteria were used to select biological effects studies for inclusion in this review: 1) availability of reports to non-resident readers, 2) relevancy to marine pollution assessment, monitoring and management, 3) use of non-traditional approaches, 4) new and unusual applications to assessments and management, and 5) quality control.

The results of a study were considered available if they appeared in the published and/or peer-reviewed scientific literature. The criteria of relevancy required that the study was done along demonstrated contamination or physical gradients, either at sea or in the laboratory. An untradtional approach was defined as a) the succesful use of new methods, or b) the use of techniques resulting in improved resolution using traditional methods. Since there is a tradition of NOT using biological effects studies in waste management decision making, I considered any biological effects study that was demonstrably and quantitatively used in a management decision action as an unusual application. Finally, to meet the criteria of quality control I attempted to seek out examples of inter-laboratory experiments or other inter-agency interactions that resulted in giving credence to wider application of the method or results.

The survey was conducted by simply reviewing existing regional literature and interviewing colleagues. Particular attention was focused on files and reports of the Ocean Assessments Division (OAD) of the National Oceanic and Atmospheric Administration (NOAA). OAD has conducted and sponsored a wide spectrum of biological effects research, monitoring and development activities throughout the United States including within contaminated areas such as Puget Sound, Washington and San Francisco Bay and the Southern California Bight in California. Results of internal and sponsored work appear in a variety of bibliographies, final reports, technical memoranda, and technical reports as well as in peer-reviewed journals. Summaries examined included regional reports (3,4), activity reports (5) and disciplinary reports (6,7). Other summaries included waste disposal case histories (8,9,10,11) specific resource impact summaries (12) and annual reports of the Southern California Coastal Water Research Project.

I chose to organize the information according to various levels of
biological organization starting with populations and communities and
concluding with hydrocarbon toxification and detoxification mechanisms.

3. POPULATIONS AND COMMUNITIES

An ultimate concern of the public is insuring that waste disposal
activities do not affect the integrity of fishery and wildlife stocks
or permanently damage marine communities. The scientific problem is how
to measure populations, how to identify pollution-related alterations
of populations, and how to predict future impacts at the population and
community level.

In terms of human resources and monitoring budgets most of the
biological effects studies along the U.S. Pacific Coast are focused on
assessments of the abundance of populations and on the abundance,
diversity and structure of marine communities. Dozens of studies are
conducted at existing, recovering and proposed disposal and discharge
sites. Study methods involve both direct capture of organisms as well
as non-destructive sampling techniques coupled with a variety of
statistical manipulations.

3.1 Patterns and Trends from Algae, Invertebrate and Fish Surveys

Pollution-related gradients in effects at the community and population
level have been clearly defined near several sewage outfalls along the
U.S. Pacific coast. Gradients range in extent from several tens of
meters to more than 10km from sources and generally conform to the
Pearson and Rosenberg (13) model for effects of organic enrichment
(ie., increasing biomass and abundance and decreasing diversity
approaching the zone of highest contamination). These kinds of
biological effects gradients are demonstrable for intertidal macroalgae
and macrofauna inshore of discharges in Southern California (14,15,16)
and Puget Sound (17) and for benthic infauna around deep-water (60 to
100m) sewage disposal sites at Palos Verdes (18,19,20) in Santa Monica
Bay (8,9,21) and off the coasts of San Diego and Orange counties in
Southern California (8,9). Large-scale gradients in benthic community
structure have also been identified from synoptic region-wide benthic
surveys in Southern California (9,22) and Puget Sound (3,4).

Long-term monitoring studies conducted near outfalls have
demonstrated waste-related changes in the relative abundances of
bottomfish offshore of Orange County (23) and Los Angeles County (24)
and Seattle (25). Nichols (26) has recorded large-amplitude changes in
the abundances of of common benthic species in a 10-year study of
intertidal mudflat benthos in San Francisco Bay and in a 20-year study
of deep basin (200m) benthos in Puget Sound. Stochastic processes such
as annual variations in freshwater runoff may explain some of the
year-to-year variations; the long-term increase in of biomass in Puget
Sound may also be due to increasing emmissions from a nearby sewage
outfall.

The benthic infauna and and bottomfish studies along the Pacific
Coast continue to be done using traditional tools, namely bottom grab
samplers (eg., 0.1m^2 Van Veen) and small (7.2m headrope length) otter
trawls. Advances are not so much in improvements in collecting methods
but in the use of more sophisticated multivariate statistical
techniques (27) and indicies (28), commensurate required improvements
in the correct and complete taxonomic identification of species,
attention to additional measures for evaluating population trends (such
as complete size- or length-frequency analyses, 29) and increasing
adherence to standard or regionally consistent sampling methods (30)
and taxonomic protocols (31).

3.2 Use of Remote and Non-Destructive Sampling Techniques

Non-destructive population and community biological effects studies are
gaining increasing attention on this coast. Wilson et al (32) have
perfected aerial infra-red photography overflight techniques for
monitoring the recovery of giant kelp (Macrocystis pyrifera) canopies
following pollution abatement and restoration activity around the Los
Angeles County outfalls at Palos Verdes. Larson and DiMartini (33)
developed improved and calibrated visual belt transects with movie
cameras for measuring trends in subtidal fish communities near a major
cooling-water discharge. At another power plant site, Stephens et al
(34) succesfully monitored longterm trends in subtidal fish communities
using direct visual census techniques. Direct visual censusing has also
been succesfully used to measure pollution-related gradients, and
abatement-related recovery, of subtidal macroalgae, macroinvertebrate
and fish communities near the Los Angeles County outfalls (35,36).

In deep water, Moore and Mearns (37) and Allen et al (38) mapped
the distribution of bottomfish and macroinvertebrates around the
Hyperion Treatment Plant outfalls in Santa Monica Bay using remote
35-mm cameras and photographing from a submersible. The surveys
demonstrated that trawls and grabs were seriously underestimating the
extent of gorgonian, sea pen and giant brittle-star communities, the
size of fish and the abundance of sablefish, sharks and hagfish
surrounding deepwater sewage outfalls. Underwater television systems
(39) helped identify the occurrence of planktivorous pelagic fish in
deep-water sludge deposits (author, unpublished data), and have now
been used to survey deepwater populations of large clams in Puget Sound
areas closed to fishing because of proximity to sewage outfalls (D.
Jamison, Washington Department of Natural Resources, personal
communication). Bottom rig fishing (40) also indicated that trawls were
not capturing numerous large benthic fish around the Southern
California outfalls.

Direct visual censusing was also succesfully employed to document
improved fledgling success of Californa brown pelican (Pelecanus
occidentalis) colonies following abatement of DDT discharges into the
Southern California Bight (41). Indeed, marine bird and mammal censuses
are major activities of resource agencies in California, Oregon,
Washington and Alaska and the techniques are constantly being refined
(eg., 42); however, no one has published results demonstrating adverse

effects, at the population level, related to pollution gradients. Such
an activity is being completed for populations of Harbor seals (Phoca
vitulina) in Puget Sound (E. Long, OAD/NOAA, Seattle, personal
communication).

Attempts have been made to use acoustic methods for identifying
and quantifying abundances of nektonic animals near waste disposal
sites. Acoustic surveys helped identify propensity of various fish to
become entrained by a power plant intake (43). Slight gradients in
nekton densities south of a Seattle sewage outfall were recorded using
acoustic techniques by English and Thorne (44) and Piper and Dmohowski
(45) have proposed, and subsequently tested, a submerged towed vehicle
for mapping densities of specific kinds of plankton and nekton around
deepwater waste discharges in Southern California.

3.3 Ecological Processes and Predictive Capabilities

Additional studies have been conducted to develop better understanding
of factors affecting recruitment and development of populations in
polluted areas. Benthic-pelagic coupling is a subject of great interest
along the U.S. Pacific coast. Cross et al (46) and Becker (47) examined
food habits of benthic fish and concluded that pollutant-related
availability of prey was a major factor regulating abundances of
specific species. Despite a fin erosion epidemic, young flatfish appear
to suffer less mortality around an outfall site than at control sites
(24). Bottomfish around Southern California sewage outfalls clearly
attain much of their growth from sewage-derived carbon and nitrogen
(48) while trophic relations among fish and other nektonic organisms
can be defined by their cesium content (49). Using both prey-item
information and morphology, Allen (50) developed a model of fish
community structure that can be used to predict types and sizes of
bottomfish expected around impacted sites such as sewage outfalls.

There is also new empirical information available for predicting
something about the rates of development of pollution impacts on
invertebrate communities. Schoener (51) was able to relate decreased
colonization rates and equilibrium abundances of fouling organisms to
chemical pollution gradients in Puget Sound. Likewise, Anderson et al
(52) and Vanderhorst et al (53) used experimentally oiled sediments to
quantify colonization and recovery rates in intertidal benthic communities. Likewise, our ability to predict the scale of impacts may be
close at hand. For example, the area of seafloor occupied by impacted
benthic communities can be partially predicted from information about
waste solids mass emmision rates (10). In addition, Hendricks (54)
found good agreement between Infaunal Tropic Indicies (ITI) computed
from a predictive model of flux of organic material from a sewage
outfall and actual ITI values obtained at the outfall site.

The only direct U.S. Pacific Coast laboratory experiments I am
aware of that focus on the effects of pollutants on community structure
are the fresh-watrer microcosm studies of Dr. Frieda Taub; for example,
Harrass and Taub (55) found that juvenile fish predation on zooplankton
lead to seriously under-estimating the effects of algicidal toxicants
on community standing crops; the results suggest abundance data alone

may be misleading when interpreting the effects of pollutants on marine populations and communities; production may be the more important, and sensitive, response.

4. ACUTE AND CHRONIC TOXICITY

Dozens of species of marine invertebrates have been used by U.S. Pacific Coast investigators to determine the acute (up to 4 day) and chronic (up to 28 day) toxicity of contaminants. The activity of conducting these bioassays and toxicity tests probably rivals field surveys in terms of human energy and pollution assessment budgets. Most work to date involves testing discrete chemicals or chemical mixtures and the data resulting from these tests has been extensively used in establishing marine water quality criteria (56). In addition, there is increasing use of similar bioassays to determine the toxicity of complex effluents, recieving waters, sediments and dredge spoils, and sea surface micro-layer material. Finally, there is increasing activity to define species sensitivities to various contaminants and to determine inter-laboratory variability of toxicity tests.

4.1 Invertebrate Toxicty Tests

Polychaete toxicity tests are common tools in many U.S. Pacific Coast laboratories, largely as a result of developmental work by Dr. D.J. Reish and his associates at Long Beach, California. Several methodologies appear in Standard Methods (57). Similarly, Brinkhurst et al (58) has developed and applied a method for using marine oligochaete for testing the toxicity of chemicals and environmental samples.

Toxicity tests using echinoderm and mollusk larvae are well established on the U.S. Pacific Coast (59-63). Embryo mortality is a principal endpoint. Reduction of recieving water toxicity, as determined by decade-long monitoring using oyster larvae, was the principal factor demonstrating benefits to Puget Sound of advanced treatment of pulp mill wastes; recieving water toxicity is no longer a regionwide problem here (60).

A benthic community (annelid, mollusks, crustacean, echinoderm) toxicity testing system proposed by Swartz et al (64) was subsequently modified to focus on a sensitive infaunal amphipod, <u>Rhepoxynius abronius</u> (65) and the method has already been used to define sediment toxicity gradients over large areas of the Puget Sound sea floor (summarized in 7) and elsewhere in the U.S. In addition, the species has been subject to detailed ecological and behavioral studies that confirm its sensitivity to sewage sludge and sediment metals (66-68). Additional experiments and biochemical investigations confirm the sensitivity of <u>R. abronius</u>, relative to other invertebrates, to sediment toxicants in general (65,69) and to toxic organic chemicals in particular (70, 71). And the method proposed by Swartz et al (65) appears to have reasonably low inter-laboratory variability (68).

Other amphipods as well as mysids have been proposed for effluent toxicity testing as described in Linfield et al (72) and in Standard Methods (57).

4.2 Fish Toxicity Tests

There are still relatively few fish toxicity test systems available. The northern anchovy, _Engraulis mordax_, together with the white croaker, _Genyonemus lineatus_, have been proposed as principal species for effluent toxicity testing (72). Other U.S. Pacific Coast fishes that have been used in acute and chronic chemical toxicity testing include striped bass, _Morone saxatilis_ (73,74) California halibut (_Paralichthys californicus_) and English sole (_Parophrys vetulus_; as cited in 72). Chapman et al (75) used the stickleback, _Gasterosteus aculeatus_, a euryhaline fish, to assay the toxicity of marine sediments, but it was quite insensitive relative to other organisms examined.

4.3 Microbial Toxicity Tests

There is considerable interest among U.S. Pacific coast agencies in using a MICROTOX (trade name) _Photobacterium_ system (as cited in 72) as a rapid (5 min) toxicity screening tool for chemicals, sediments and receiving waters; the organism's endpoint, inhibition of luminescence, has a sensitivity to various toxicants comparable to acute mortalities in higher organisms (see 72 for proposed application).

5.0 REPRODUCTIVE EFFECTS

Reproduction is here defined as the suite of processes from development of gametes to successful fertilization. Most pollution-related reproductive-effects reseach along the Pacific Coast has been based on whole- or partial-life cycle bioassays using marine seaweeds or invertebrates exposed to individual chemicals, effluents and marine sediments and waters. A recent and detailed summary of numerous tests including reproductive effects has been published in connection with a review of marine bioassays for effluent toxicity testing (72).

5.1 Seaweeds

Considerable attention has been focused on marine seaweeds such as giant kelp since they are extremely valued resources along the Pacific Coast. Gametophyte bioassays have been attempted and proposed as methods for evaluating impacts of pollutants and waste disposal on marine plant reproduction. Glass slides are innoculated with spores from ripe sporophylls and the resulting male and female gametophytes allowed to develop and produce eggs and sperm. Upon fertilization, successful embryos are then monitored for their ability to settle and grow into new young sporophytes. Tests have been done with copper (as cited in 72) and DDT (C. Barilotti, KELCO Co., pers. comm.) and in situ outplanting near power plant discharges (76). In addition, trace element requirements and toxicities have been described for giant kelp (77).

5.2 Invertebrates

Many species of marine polychaetes have been examined for their use in reproduction bioassays (eg., 78, 79). An example is the use of the nest-building nereid, Neanthes arenaceodentata, to determine reproductive thresholds for highly-toxic hexavalent chromium and non-toxic trivalent chromium (Oshida et al, 1982 as reviewed in 80). Reproduction in the polychaete Capitella capitata has also been used to help identify the presence of toxic substances in sediments from Puget Sound and adjacent areas (81).

Successful partial life cycle tests also include methods for examining fertilization of oysters and echinoderms exposed to specific chemicals, diluted effluents and recieving water and sediment samples (59-63); the tests are usually extended to include examination of short-term survival and occurrence of developmental abnormalities. Bay et al (82) developed a method for rapidly quantifying the relative abundance of fertilized and developing echinoderm larve using the specific pigment, echinochrome.

Complete or partial life-cycle tests including the examination of reproductive succes· have not been completely developed for U.S. Pacific Coast crustaceans or fishes, but several methods have been proposed based on successes elsewhere with mysids and amphipods (72). Northern anchovies may be the "most culturable" California marine fish as a result of methods developed by the National Marine Fisheries Service.

There have apparently been few efforts to determine reproductive effects of pollution on invertebrates at sea. Siegel and Wenner (83) have identified what appears to be a pollution related gradient in reproductive damage to sand crab (Emerita analoga) in California.

5.3 Fish

Despite emphasis on laboratory experiments, there are field observations suggesting effects of pollutants on reproduction in wild fish populations. Egg condition in striped bass (Morone saxililis) from the San Francisco Bay-Delta region was significantly poorer than in fish from Coos River (Bay), Oregon, and the Hudson River, New York; within the San Francisco Bay-Delta populations, young prespawning females had altered egg maturation rates and resorption associated with petrochemical concentrations (84). Low-moleucular weight petrochemicals, mainly toluene, occurred in highest concentration in spent females (84). Cross et al (85) observed high rates (28 - 49% of fish examined) of atresia (resorbtion of more than 10% of oocytes) in two species of nearshore bottomfish near Southern California outfalls and only slightly lower rates away from the outfalls; all fish gonads

had high levels of chlorinated hydrocarbons. Spies et al (86) cbserved
a strong correlation between fertilization success and hepatic mixed
function oxidase (MFO) activity in starry flounder (Platichthys
stellatus) from sites in San Francisco Bay. Hose et al (87) identified
effects of a polycyclic aromatic hydrocarbon exposed adult flatfish on
subsequent development of larvae. NOAA's Ocean Assessments Division is
sponsoring new work to determine 1) what connection, if any, there is
between liver disease and reproduction in flatfish (U. Varanasi, NMFS,
Seattle, personal communication, 2) the severity of reproductive damage
in Southern California nearshore fishes (J. Cross, Southern California
Coastal Water Research Project and J. E. Hose, Occidental College, Los
Angeles, personal communication), and 3) the extent to which contaminants in the sea surface microlayer (surface slicks, as described by
88) may inhibit fertilization and development of flatfish eggs and
embryos (J. Hardy, Batelle, personal communication).

6. SYSTEMIC EFFECTS: GROWTH AND RESPIRATION

Growth and respiration are recieving increased focus as systemic
indicators of distress in marine organisms exposed to pollutants while
diseases, including both internal and external malformations, have been
a major focus of U.S. Pacific coast biological effects studies for
nearly 3 decades.

6.1 Growth Rates

Growth and growth rates of whole organisms have been examined in caged
mussels deployed near an outfall in San Francisco Bay, and flatfish
living near and away from Southern California's sewage outfalls. Using
simple but careful measurements, Roth et al (89) has been able to
measure significant gradients in shell growth of mussels (Mytilus
californianus) deployed for several weeks near a San Francisco Bay
outfall. Fabrikant (19) confirmed strong relationships between
sediement organic loading near the Los Angeles County outfalls and
exceptional size in a deposit feeding bivalve, Parvilucina
tenuisculpta. Harris and Mearns (90) observed that recently-settled
juvenile Dover sole (Microstomus pacificus) grew faster during their
first year near the Los Angeles County outfalls than at other sites.;
growth rates beyond the first year were similar among the sites.
However, by age 6, all the Southern California Dover sole were smaller
than their counterparts in northern California, Oregon, and
Washgington. Since Southern California is near the southern limit of
the species range, it is possible the entire local poipulation is under
a number of stresses not imposed on populations to the north. Growth of
striped bass is materially affected in San Francisco Bay relative to
other estuarine areas (84).

6.2 Organ-Somatic Indicies

Relationships between the size of organs and the total size of fish
(i.e., organ-somatic ratios) may be useful indicators of pollution.
Liver weight-body weight ratios (liver somatic index, LSI) increase
with proximity to sewage outfalls (24, 91), with increases in tissue
levels of chlorinated hydrocarbons (91) and with increasing esposure
time to contaminated sediments by recently-settled post-larval flatfish
(24).

6.3 Scope For Growth and Respiration

Martin and his colleagues have successfully identified
pollutant-related gradients in scope for growth (SFG) of mussels
deployed in several areas of California including along the shoreline
of San Francisco Bay (92, 93), near an outfall in San Francisco Bay
(94) and near outfalls in Southern California (M. Martin, Calif. Fish
and Game, pers. comm.). In a now long-forgotten study, a suite of
measurements similar to those used in SFG were used to quantify stress
to salmon migrating through the contaminated Duwamish estuary near
Seattle, Washington (95); the study is noteworthy because it was done
aboard a specially constructed floating marine physiology laboratory,
an experimental platform recently advocated for continued in situ
marine pollution investigations in San Francisco Bay (84).

Anomalies in respiration of oligochaetes and rainbow trout gonad
cells have been used to document the presence of obnoxious materials in
sediments from contamination gradients in Puget Sound (69). Oxygen
consumption and food intake of sea urchins which dominate the deepwater
macrobenthic community of the southern California continental slope are
two variables being used in a model to predict effects of deepwater
sludge disposal (96). Pamatmat (97) measures metabolic effects of
sediment pollution directly using thermal properties.

7. SYSTEMIC EFFECTS: DISEASES AND ABNORMALITIES

There is now an extensive body of knowledge concerning the occurrence
and distribution of pollution-related diseases in fish and shellfish
from Pacific Coast sites near Los Angeles and in San Francisco Bay,
Puget Sound and coastal areas of Vancouver Island British Columbia.
From a historical perspective, initial concern was raised by Young (98)
about possible relationships between pollution and externally visible
lesions, fin rot and tumors in fish from Los Angeles Harbor and Santa
Monica Bay. Since then there have been numerous investigations of
external diseases in Southern California bottom fish (91, 99, 100), in
San Francisco Bay striped bass (73, 84), and in Puget Sound bottomfish
(91, 101, 102). Since 1977, however, effort and interest have shifted
to the examination of lesions in internal organs (103-108).

7.1 External Lesions and Disorders

The oft-cited and wide-spread skin tumor disease of young pleuronectid
flatfish and cod-like fishes of the northeast Pacific (eg, 109) has
clearly been shown to be associated with an apparently-infectious
amoeba, at least in cod-like fishes (110). Some investigators still do
not rule out pollution as a factor in increased prevalence at some
sites (111). In contrast to other diseases, noted below, the skin tumor
disease is clearly a disorder of young fish.

The more serious external disorders that can clearly be related to
pollution gradients include fin erosion in flatfish at several outfall
sites in Southern California and in Seattle's Duwamish River (91) and
the ulcer-like parasite-associated wounds in striped bass from San
Francisco Bay (84). Tissue metals and DDT concentrations can be ruled
out as primary causes of the fin erosion syndrome in general; but
tissue levels of PCB's cannot be ruled out as a causative factor (91).
Unlike skin tumors, the fin erosion disease is not narrowly size or
age specific, occurring rapidly in recently-settled post-larval fishes
at a Southern California outfall site but also occurring at high
frequencies in larger older fish (24). Further, Whipple (84) describes
relationships between petrochemicals, mainly lower molecular weight
hydrocarbons and PCB's, parasitism, and ulcer-like wounds in San
Francisco Bay stripped bass; these disorders and associated hydrocarbon
concentrations, occur at higher prevalences and concentrations,
respectively, in San Francisco Bay striped bass than in fish from the
relatively uncontaminated Coos River, Oregon, or the PCB-contaminated
Hudson River, New York (84).

7.2 Internal Idiopathic Lesions

U.S. Pacific Coast researchers are now recognizing two general classes
of internal lesions in organs of fish (1) hepatomas and related
pre-neoplastic and neoplastic lesions that may be caused by
carcinogens, and (2) apparently non-neoplastic idiopathic (of unknown
cause) disorders such as increased melanin macrophage centers (91) and
many others (105, 112). The group of cancer-like disorders clearly
increase with increasing age in Puget Sound English sole (105, 112 and
Scott Becker, Tetra Tech Inc., Bellevue, Washington, personal
communication) so prevalences must be normalized to age or size to
compare sites for spatial and temporal trends (Scott Becker,
Tetra-Tech, Inc., Bellevue, Washington, and Lucia Susani, OAD/NOAA,
Rockville, Maryland, personal communication). Similar cancer-like
lesions are produced in trout fed high doses of carcinogens (113).
Discovery of these cancerlike lesions in fish along both the Atlantic
and Pacific coasts has sparked new national interest in using fish to
detect carcinogens in the marine and estuarine environment. The search
for these disorders is now part of NOAA's new Status and Trends Program
(114).

Correlative and regression analyses from field data suggest that
aromatic hydrocarbons may be significant factors in causing the
cancer-like disorders (105), although other chemicals cannot be

entirely ruled out. New studies in San Francisco Bay (R. Spies, pers. comm.) and in Puget Sound are attempting to focus on relationships between pollutant levels, detoxification mechanisms, tissue damage and reproductive damage.

7.3 Deformities and Asymmetry

It has been nearly 15 years since Valentine observed a large-scale (1000 km) gradient in bone deformities and asymmetry in sand bass (Paralabrax nebulifer) related to proximity to Los Angeles and to tisssue levels of total chlorinated hydrocarbons (115, 116). Considering that there hyas been a major decline of chlorinated hydrocarbon inputs into the region it may be be intriguing to repeat this study.

7.4 Consequences and Causes

Despite the great amount of recent work on fish diseases on the west coast, there remain two areas in need of further investigation: relation between disease and total health of fish stocks and (2) experimental induction and thus confirmation of the diseases under controlled conditions. The studies by Sherwood (91), Whipple (84) and Landolt et al (112) did focus on collective measures of total fish health including growth, parasitism, pollutant exposure and disease. A factor not yet tested, but of general interest, is dietary imbalances imposed on benthic fish by pollution-induced alterations of benthic food organisms (eg, elimination of crustaceans). With respect to induction of disease, Sherwood (91) conducted fish-sediment exposure experiments that lead to the induction of fin erosion disease in dover sole exposed to contaminated sediments from the Palos Verdes shelf; the process required 13-months of continuous contact, a period that seems inexorably long; the study also suggested that of the pollutants examined, PCB's could not be ruled out as causal factors (others could). On the other hand Hendricks et al (113) induced hepatomas in trout within a few months of exposure to carcinogens but PCB's through food actually reduced the prevalence.

8. BIOCHEMICAL EFFECTS:TOXIFICATION/DETOXIFICATION

U.S. Pacific Coast researchers are contributing greatly to an increasing body of knowledge about marine toxicological processes, knowledge that may have great influence on future sewage and industrial waste management, biological effects monitoring and identifying the causes of pollution-related diseasses in marine fishes. Below, I focus on work related to toxic organic chemicals; similar intensive studies are underway to understand metal detoxification mechanisms.

PCB's, DDT's and polycyclic aromatic hydrocarbons (PAH's) are ubiquitous contaminants in U.S. Pacific Coast urbanized estuaries and embayments; monocyclic aromatic hydrocarbons (MAH's) have also been identified as ubiquitous in San Francisco Bay fishes. Many of the

metabolic byproducts of the first three chemical groups are mutagenic
and carcinogenic. A major problem is to identify the extent to which
valued marine resource species are at risk from the mutagenic and
carcinogenic consequences of these and other organic chemicals.
Accordingly, there have been numerous studies focusing in various
species on different aspects of the problem including measurements of:
1) the enzyme systems that mediate production of metabolites; 2) the
metabolites themselves; 3) the possible protective or detoxifying
systems; and, 4) the cellular and chromosomal damage presumably
inflicted by excessive concentrations or production of metabolic
products.

A key to the toxification/detoxification process is the extent to
which marine organisms are cabable of metabolizing parent organic
chemicals. The bothid flatfish species Citharichthys (two species) and
Platicthys stellatus, the pleuronectid flatfish species Parophrys
vetulus, the scorpaenid Scorpaena guttata and the sciaenid Genyonemus
lineatus all appear to be active producers of potentially self-damaging
metabolites as reflected in either increased hepatic mix function
oxidase (MFO) activity, as in Citharichthys and Platichthys (85, 116),
or actual presence of specific metabolites and free radicals as in
Parophrys (70, 71, 108, 118 and 119) and Genyonymus (120). MFO
activity in the sanddabs (Cithaichthys spp.) was increased, relative to
baseline conditions, in fish collected near a natural oil seep as well
as in others collected along known pollution gradients (near major
sewage outfalls, 117); this observation supports the assumption that
these kinds of metabolic mechanisms are not only natural in fish but
are in fact activated under natural conditions. Of equal importance is
the finding that some fish tissues, such as blue gill sunfish gonad (a
freshwater species used in some fish bioassays) and some invertebrates,
such as mussels (Mytilus), clams (Macoma) and polychaetes (Neanthes) have
low MFO activity or are apparently nearly incapable of metabolizing
PAH's (71). The implication is that such organisms are not at risk
from mutagenic or carcinogenic damage from PAH's. Conversely, these
organisms may be useful tool for monitoring the presence of the parent
compounds in the marine environment (ie., Mussel Watch) whereas the
absence of the parent chemicals in fish does not mean that they have
not been exposed to, or accumulated them. In between these extremes are
crustaceans which may vary widely in their ability to metabolize these
organic toxicants. For example, the amphipod Rhepoxynius abronius
(recommended by Swartz et al, 65, as a sensitive bioassay organism) is
an active metabolizer of PAH's whereas another amphipod species is not
(70, 71). Bile metabolites appear to be useful in monitoring exposure
in active organisms (119), a tool now being used in NOAA's National
Status and Trends Program (114).

9.0 DISCUSSION AND CONCLUSION

McIntyre (1) attempted to identify why biological effects methods are
not widely used in marine pollution assessment and decision making
activities. It is obvious from the review just presented that even on

one coast there are a great number of biological effects studies underway or completed. I close by offering some comments supporting McIntyre's contrast between existence of methods and application.

The biological effects studies reported here generally fit at least four of the five criteria I adopted to help identify information I percieved should be relevant to management. But many did not and thus were not included. This does not mean the missing studies are not important, only that some action is necessary to make them so.

A most serious deficiency in this report is an underrepresentation of long-term field monitoring conducted by various local and state government agencies and industries. Some of the monitoring programs go back in time nearly three decades and include obsrvations at the community, population or systemic effects levels of organization. The data are available to scientists aware of them, but the results are not published or citable in the traditional literature. They should not be ignored since they may provide our only link with the past and may well be telling us the extent to which things are getting better or worse.

A second serious deficiency is lack of direct use of the results of biological effects studies in waste management decision-making. This is not to say that the results are not used indirectly; indeed, it is the author's opinion that most of the results of the studies reported here have been brought to the attention of authorities and have been used successfuly to draw attention to incipient problems. But, they have not been used in the sense that an engineer would use hydrographic measurements to locate and design an ocean outfall; they have not been used to set quantitative biological effects criteria or goals. I don't necessarily advocate that they should be but merely make the observation and suggest a challenge (ie, sediment quality criteria?).

A third serious deficiency appears to be the lack of quality assured investigations sufficient to support the wider use of biological effects method (ie, by more than one investigator). Important exceptions were mention, including the taxonomic intercalibration activities of the Southern California Association of Marine Invertebrate taxonomists (SCAMIT, 31) and several direct interlaboratory experiments. Otherwise, there is a major void.

On the other hand, I believe a great many of the U.S. Pacific Coast biological effects studies are attempting untraditional approaches and increasing resolution of identifying biological damage at all levels of biological organization. Some predictive relationships are close at hand. Coupling this innovation and increasing level of detection with renewed attention to identifying cause-effect relationships, testing preditions, resurrection of historical data and explicit quality assurance efforts , will bring U.S. Pacific Coast biological effects studies squarely into the decision-making arena.

10.0 ACKNOWLEDGEMENTS

I thank all those who gave freely of their time in the libraries and on the telephone to direct me toward sources of information. I thank Karen Conlan and Gerry Arbios for assisting with the manuscript.

11.0 LITERATURE CITED

1. McIntyre, A.D. 1984, 'What happened to biological effects
   monitoring?' Mar. Poll. Bull. 15, pp. 391-2.
2. Cole, H.A. 1977, 'What to protect - a matter of definition.' Mar.
   Poll. Bull. 8, pp. 1-2.
3. Long, E.R. 1982, 'An assessment of marine pollution in Puget Sound.'
   Mar. Poll. Bull. 13, pp. 380-3.
4. Long, E.R. 1982, 'Commencement Bay: Resource-use conflicts at a
   marine superfund site.' Conf. Record, Oceans 82, IEEE,
   Washington, D.C., pp. 1086 - 91.
5. Matta, M.B. 1985, 'A directory of selected research supported by
   NOAA's Ocean Assessments Division.' NOAA Rpt NOS OMA 1, Nat. Ocean.
   Atm. Admin., Rockville, Md., pp. 1 - 147.
6. Chapman, P.M. and Long, E.R. 1983, 'The use of bioassays as part of
   a comprehensive approach to marine pollution assessment.' Mar. Poll.
   Bull. 14, pp. 81-84.
7. Long, E.R. 1984, 'Sediment bioassays: a summary of their use in
   Puget Sound.' Seattle Project Office Spec. Publ., Ocean Asses. Div.,
   Nat. Ocean. Atm. Admin., Seattle, Wa. pp. 1 - 30.
8. Mearns, A.J. and Young, D.R. 1983, 'Characteristics and effects of
   municipal wastewater discharges to the Southern California Bight, A
   case study.' In E.P. Myers (ed), Ocean disposal of municipal
   wastewater: Impacts on the coastal environment, Vol 2 MITSA-83-33,
   Mass. Inst. Tech, Cabbridge, Ma pp. 763-819.
9. Bascom, W. 1982, 'The effects of waste disposal on the coastal
   waters of Southern California.' Env. Sci.Tech. 16, pp. 226A - 236A.
10. Mearns, A.J. and O'Connor, T.P. 1984, 'Biological effects versus
    pollutant inputs: the scale of things.' In H. H. White(ed), Concepts
    in marine pollution measurements, UM-SG-TS-03, Univ. Maryland,
    College Park, pp. 693 - 721.
11. Soule, D.F and Walsh, D. (eds) 1983, Waste disposal in the oceans:
    minimizing impact, maximizing benefit. Westview Press, Boulder, Co.
12. Bascom, W. (ed) 1983. The effects of waste disposal on kelp
    communities, So. Calif. Coast. Wat. Res. Proj., Long Beach. 328 pp.
13. Pearson, T.H. and Rosenberg, R. 1978, 'Macro benthic succession in
    relation to organic enrichment and pollution of the marine
    environment.' Oceanog. and Marine Biology Ann. Rev. 16, pp. 229-311.
14. Harris, L. 1980, 'Changes in intertidal algae at Palos Verdes.' In
    W. Bascom (ed), Coast. Wat. Res. Proj. Bien. Rpt., 1979 - 1980. So.
    Cal. Coast. Wat. res. Proj., Long Beach, Ca. pp. 35 - 75.
15. Littler, M.M. and Murray, S.N. 1978, 'Influence of domestic wastes
    on energetic pathways in rocky intertidal communities.' J. Appl.
    Ecol. 15, 581-595.
16. Dorsey, J.H., Green, K.D., and Rowe, R.C. 1983, 'Effects of sewage
    disposal on the polychaetous annelids at San Clemente Island,
    California.' In D.F. Soule and D. Walsh (eds), Waste disposal in the
    oceans: minimizing impacts, maximizing benefit Westview Press,
    Boulder, Co. pp. 209 -245.
17. Thom, R.M., Armstrong, J.W., Staude, C.P., Chew, K.K. and Smith,
    R.E. 1977, 'The impact of sewage on the benthic marine flora of the

Seattle area.' In K. Chew (ed), Studies of intertidal biota at five
Seattle beaches. Puget Sound Interim Studies Rpt., Municipality of
Metropolitan Seattle, pp. 35-58.
18. Stull, J.K., Haydock, C.I., and Montagne, D.E. 1985, 'Effects of
Listriolobus peloides (echiura) on coastal shelf benthic communities
and sediment modified by a major California wastewater discharge.'
Est. Coast. Shelf Sci. In press.
19. Fabrikant, R. 1984, 'The effect of sewage effluet on the population
density and size of the clam Parvilucina tenuisculpta.' Mar. Poll.
Bull. 15, pp. 249-253.
20. Swartz, R.C., Schults, D.W., Ditsworth, G.R., DeBen, W.A., and
Cole, F.A. 1985, 'Sediment toxicity, contamination, and
macrobenthic communities near a large sewage outfall.' In T. P.
Boyle, (ed), Validation and predictability of Laboratory Methods for
Assessing the Fate and Effects of Contaminants in Aquatic
Ecosystems, ASTM STP 865, Amer. So. Test. Mat., Philadelphia, Pa.
pp. 152-175
21. Hyperion Treatment Plant 1984, 'Santa Monica Bay monitoring study
annual report, 1983.' Hyperion Treatment Plant Rpt., City of Los
Angeles, Playa del Rey, Ca.
22. Word, J.Q. and Mearns, A.J. 1979, '60-meter control survey off
Southern California.' TM 229, So. Calif. Coast. Wat. Res. Proj.,
Long Beach. pp. 1-58.
23. Mearns, A.J. 1979, 'Abundance, compostion and recruitment of
nearshore fish assemblages on the Southern California mainland
shelf.' Calif.Coop. Ocean. Fish. Invest.Rpt. 20, pp. 111-119.
24. Sherwood, M.J. and Mearns, A.J. 1981, 'Fate of post-larval
bottom fishes in a highly urbanized coastal zone.' Rapp. P.-v. Reun.
Cons. int. Explor. Mer. 178, pp. 104-111.
25. Miller, B.J., McCain, B.B., Wingert, R.C., Borton, S.F., Pierce,
K.V., and Griggs, D.T. 1977, 'Ecological and disease studies of
demersal fishes in Puget Sound near METRO-operated sewage treatment
plants and in the Duwamish River.' Puget Sound Interim Studies inal
Rpt., Municipality of Metropolitan Seattle, FRI-UW-7721, Seattle,
Wa.
26. Nichols, F.H. 1985, 'Abundance fluctuations among benthic
invertebrates in two Pacific estuaries.' Estuaries 8(2A), pp.
136-44.
27. Boesch, D.F. 1977, 'Application of numerical classification in
ecological investigations of water pollution.' EPA Ecol. Res. Ser.,
EPA 600/3-77-033. 113 pp.
28. Word, J.Q. 1979, 'The infaunal trophic index.' In W. Bascom (ed),
Coast. Wat. Res. Proj. Ann. Rpt. 1978. So. Calif. Coast. Wat. Res.
Proj., Long Beach, Ca. pp. 19-39.
29. Mearns, A.J., Allen, M.J., Moore, M.D., and Sherwood, M.J. 1980,
'Distribution, abundance and recruitment of soft-bottom rockfishes
(Scorpaeidae:Sebastes) on the Southern California mainland shelf.'
Calif. Coop. Ocean. Fish. Invest. Rpts. 21, pp. 180-190.
30. Bascom, W. 1982, 'A regional ocean monitoring program.' In W.
Bascom (ed), Coast. Wat. Res. Proj. Bien. Rpt. 1981-1982. So.
Calif. Coast. Wat. Res. Proj., Long Beach, pp. 81-82.

31. Martin, A. and Mearns, A.J. 1983.   'Region-wide taxonomic
    inter-calibration program underway in Californis.' Coast. Ocean
    Poll. Asses. News. 2, p.31.
32. Wilson, K.C., Haaker, P.L., and Hanan, D.A. 1978, 'Kelp
    restoration in Southern California.' In J. Krauss (ed), The marine
    plant biomass of the Pacific Northwest coast. Oregon State Univ.
    Press, Corvallis. pp. 183-202.
33. Larson, R.J. and DiMartini, E.E. 1984, 'Abundance and vertical
    distribution of fishes in a cobble-bottom kelp forest off San
    Onofre, California.' Fish. Bull.(U.S.) 82, pp. 37-53.
34. Stephens, J.S. Jr., Morris, P.A. and Westphal, W. 1983, 'Assessing
    the effects of a coastal steam generating station on fishes
    occupying its recieving waters.' In D.F. Soule and D. Walsh (eds),
    Waste disposal in the oceans: minimizing impact, maximizing benefit.
    Westview Press, Boulder, Colo. pp. 194-208.
35. Grigg, R.W. 1978, 'Long term changes in rocky bottom communities
    off Palos Verdes.' In W. Bascom (ed), Coast. Wat. res. Proj Ann.
    Rpt.1978, So. Calif. Coast. wat. res. Proj., Long Beach, Ca. pp.
    157-184.
36. Meistrell, J.C. and Montagne, D.E. 1983, 'Waste disposal in
    Southern California and its effects in the rocky subtidal
    habitat.' In W. Bascom (ed) , The effects of waste disposal on
    kelp bed communities. So. Calif.Coast. Wat. Res. Proj., Long
    Beach, pp. 84-102.
37. Moore, M.D. and Mearns, A.J. 1980, 'Photographic survey of benthic
    fish and invertebrate communities in Santa Monica Bay.' In W. Bascom
    (ed), Coast. Wat. Res. Proj.Bien Rpt. 1979-1980. So. Calif. Coast.
    Wat. Res. Proj., Long Beach, pp. 143-147.
38. Allen, M.J., Pecorelli, H. and Word, J.Q. 1976, 'Marine organisms
    around outfall pipes in Santa Monica Bay.' J. Wat. Poll. Cont. Fed.
    48, pp 1881-93.
39. Bascom, W. 1976, 'An underwater television system.' In W. Bascom
    (ed), Coast. Wat. Res. Proj. Ann. Rpt. 1976. So. Calif. Coast. Wat.
    Res. Proj., Long Beach. pp. 171-174.
40. Allen, M.J., Isaacs, J.B. and Voglin,'R.M. 1975, 'Hook-and-line
    survey of demersal fishes in Santa Monica Bay.' TM 222, So. Calif.
    Coast. Wat. Res. Proj., Long Beach, 23 pp.
41. Anderson, D.V., Jurek, R.M., and Keith, J.O. 1977, 'The status of
    brown pelicans at Anacapa Island in 1975.' Calif. Fish. Game 63, pp.
    4-10.
42. Berkson, J.M. and DeMaster, D.P. 1985, 'Use of pup counts in
    indexing population changes in pinnipeds.'Can. J. Fish. Aquat. Sci.
    42, pp. 873-79.
43. Thorne, R.E., Thomas, G.L., Acker, W.C., and Johnson, L. 1979,
    'Two applications of hydroacoustic techniques to the study of fish
    behavior around coastal power generating stations.' Univ. Wash. Sea
    Grant Tech. Rpt. 79-2. Seattle, Washington. 26 pp.
44. English, T.S. and R.E. Thorne 1977, ' Acoustic and net surveys of
    fishes and zooplankton.' Puget Sound Interim Studies Final Rept.,
    Municipality of Metropolitan Seattle, M77-17, 64 pp.
45. Piper, R.E. and Dmohowski, J.A. 1983, 'The application of

acoustics in marine monitoring.' In D.F. Soule and D. Walsh (eds), Waste disposal in the oceans: minimizing risk, maximizing benefit. Westview Press, Boulder, Colo. pp. 76-89.

46. Cross, J.N., Roney, J. and Kleppel, G.S. 1985. 'Fish food habits along a pollution gradient.' Calif. Fish. Game 71, pp. 28-39.

47. Becker, D.S. and Chew, K.K. 1984, 'Fish-benthos coupling in seage enriched marine environments.' Final Report, Grant NA80RAD00050, Nat. Ocean. Atm. Admin., Ocean Asses. Div., Rockville, Md. 78 pp.

48. Rau, G.H., Sweeney, R.E., Kaplan, I.R., Mearns, A.J., and Young, D.R. 1981, 'Differences in animal 13C, 15N and D abundance between a polluted and an unpolluted coastal site: likely indicators of sewage uptake by a marine food web.' Est. Coast. Shelf Sci. 13, pp 701-7.

49. Young, D.R. 1984, 'Methods of evaluating pollutant biomagnification in marine ecosystems.' In H.H. White (ed), Concepts in marine pollution measurements, Univ. Maryland Sea Grant, College Park, Md., pp. 261-278.

50. Allen, M.J. 1982, 'Functional structure of soft-bottom fish communities of the Southern California shelf.' PhD Dissertation, Univ. of Calif., San Diego. 577 pp.

51. Schoener, A. 1983, 'Colonization rates and processes as an index of pollution severity.' NOAA Tech Mem. OMPA-27, Nat. Ocean. Atm. Adm., Ocean Asses. Div., Rockville, Md. 37 pp.

52. Anderson, J.W., Riley, R.G., and Bean, R. M. 1978, 'Recruitment of benthic animals as a function of petroleum hydrocarbon concentrations in the sediment'. Jour. Fish Res. Bd. Can. 35, pp.76-90.

53. Vanderhorst, J.R., Blaylock, J.W., Wilkinson, P., Wilkinson, M., and Fellingham, G. 1981, 'Effects of experimental oiling on recovery of Strait of Juan de Fuca intertidal habitats.' DOC/EPA Interagency Energy/Environment R&D Prog. Rpt. EPA-600/7-81-008. U.S. Env. Prot. Agen., Washington D.C. 129 pp.

54. Hendricks, T.J. 1983, 'Numerical model of sediment quality near an ocean outfall.' Final Rpt, NOAA Cont. NA80RAD00041, Nat. Ocean. Atm. Admin., Ocean Asses. Div., Rockville, Md.

55. Harrass, M.C. and Taub, F.B. 1985, 'Effects of small fish predation on microcosm community bioassays.' In R.C. Cardwell, R. Purdy and C. Bahner (eds), Aquatic toxicology and Hazard Assessment: Seventh Symposium, ASTM STP 854, Amer. Soc. Test. Mat., Philadelphia, Pa. pp. 117-133.

56. Klapow, L.A. and Lewis, R.H. 1979, 'Analysis of toxicity data for California marine water quality standards.' J. Wat. Poll. Cont. Fed. 51, pp. 2054-70.

57. APHA 1976, Standard Methods for the Examination of water and wastewater, 14th Edition, Amer. Publ. Health. Assn., Washington D.C., 1193 pp.

58. Brinkhurst, R.O., Chapman, P.M., and Farrell, M.A. 1983, 'A comparative study of respiration in some aquatic oligochaetes in relation to environmental stress.' Int. Revue Ges. Hydrobiol. 68, pp. 683-9.

59. Woelke, C.E. 1972, 'Development of a receiving water quality bioassay criterion based on the 48-hour Pacific oyster (Crassostrea

gigas) embryo.' Wash. Dep. Fish. Tech. Rpt. 9, 93 pp.
60. Cardwell, R.D., Woelke, C.E., Carr, M.I., and Sanborn, E.W. 1977,
    'Evaluation of the efficacy of sulfite pulp mill pollution abatement
    using oyster larvae.' In F.L. Mayer and J.L. Hamelink (eds), Aquatic
    Toxicology and Hazard Evaluation ASTM STP 634, Amer. Soc. Test.
    Mat., Philadelphia, Pa., pp. 291-295.
61. Dinnel, P.A., Stober, Q.J., Crumley, S.C., and Nakatani, R.E.
    1982, ' Development of a sperm cell toxicity test for marine
    waters.' In J.G. Pearson, R.B. Foster and W.E. Bishop (eds),
    Aquatic Toxicology and Hazard Assessment: Fifth Conference, ASTM
    STP 766, Amer. Soc.
    Test. Mat., Philadelphia, pp. 82-98.
62. Dinnel, P.A., Stober, Q.J., Link, J.M., Letourneau, M.W., Roberts,
    W.E., Felton, S.P., and Nakatani, R.E. 1983, 'Methodology and
    validation of a sperm cell toxicity test for testing toxic
    substances in marine waters.' Final Rpt. Grant R/TOX-1, U.S.E.P.A.,
    FRI-UW-8306, Fish. Res. Inst., Univ. Wash., Seattle. 208 pp.
63. Oshida, P.S., Goochey, T.K., and Mearns, A.J. 1981, 'Effects of
    municipal wastewater on fertilization, survival and development of
    the sea urchin, Strongylocentrotus purpuratus. In F.J. Vernberg, A.
    Calbrese, F.T.P. Thurberg and W.B. Vernberg (eds), Biological
    monitoring ofMarine Pollutants, Academic Press, Ney York, pp.
    389-402.
64. Swartz, R.C., DeBen, W.A., and Cole, F.A. 1979, 'A bioassay for the
    toxicity of sediment to marine macrobenthos.' J. Wat. Poll. Cont.
    Fed. 51, pp. 944-50.
65. Swartz, R.C., DeBen, W.A., Jones, J.K.P., Lamberson, J.O. and Cole,
    F.A. 1985, 'Phoxocephalid amphipod bioassay for marine sediment
    toxicity.' In R.D. Cardwell, R. Purdy and R.C. Bahner (eds), Aquatic
    Toxicology and Hazard Assessment: Seventh Symposium, ASTM STP 854,
    Amer. Soc. Test. Mat., Philadelphia, Pa. pp. 284-306.
66. Swartz, R.C., Schults, D.W., Ditsworth, G.R., and DeBen, W.A.
    1984, 'Toxicity of sewage sludge to Rhepoxynius abronius, a marine
    benthic amphipod.' Env. Cont. Toxicol., 13, pp 207-16.
67. Oakden, J.M., Oliver, J.S., and Flegal, A.R. 1984, 'Behavioral
    responses of a phoxocephalid amphipod to organic enrichment and
    trace metal in sediment.' Mar. Ecol.- Prog. Ser., 14, pp. 253-7.
68. Mearns, A.J. 1985, 'Interlaboratory comparison of a sediment
    toxicity test.' In S.M. Woods (ed), Report on Ocean Dumping, R&D
    Pacific Region, Dep. Fish. Oceans, 1983-84, Can. Cont. Rpt, Hydro.
    Ocean. Sci. No. 20, Inst Ocean Sci., Sydney, B.C., Can. pp. 31-36.
69. Chapman, P.M., Dexter, R.N., Morgan, J., Fink, R., and Mitchell, D.
    1984, 'Survey of biological effects of toxicants upon Puget Sound
    Biota -III, Tests in Everett Harbor, Samish and Bellingham Bays.'
    NOAA Tech. Mem. NOS OMS 2, Nat. Ocean. Atm. Admin., Rockville, Md.
    48 pp.
70. Reichert, W.L., Le Eberhart, B.-T., and Varanasi, U. 1985,
    'Exposure of two species of deposit-feeding amphipods to
    sediment-associated 3H benzo(a)pyrene: uptake, metabolism and
    covalent binding to tissue macromolecules.' Aquat. Toxicol. 6, pp.
    45-56.

71. Varanasi, U., Reichert, W.L., Stein, J.E., Brown, D.W., and Sanborn, H.R. 1985, 'Bioavailability and biotransformation of aromatic hydrocarbons in benthic organisms exposed to sediments from an urban estuary.' Env. Sci. Tech., In Press.

72. Linfield, J.T., Martin. M., and Norton, J. 1985, 'Bioassay species selection and recommended protocols: First Progress Report.' Calif. State. Wat. Res. Cont. Bd., Sacramento, 113 pp.

73. Whipple, J.A., Eldridge, M.B. and Benville, P., Jr. 1981. 'An ecological perspective of the effects of monocyclic aromatic hydrocarbons on fishes.' In F.J. Vernberg, A. Calabrese, F.P. Thurberg and W.B. Vernberg (eds), Biological Monitoring of Marine Pollutants, Academic Press, New York, pp. 483-551.

74. Benville, P.E., Jr., Whipple, J.A., and Eldridge, M.B. 1985. 'Acute toxicity of seven alicyclic hexanes to striped bass and bay shrimp.' Calif. Fish Game, In Press.

75. Chapman, P.M., Dexter, R.N., Kocan, R.M., and Long, E.R. 1985, 'An overview of biological effects testing in Puget Sound, Washington: methods, results and implications.' In R.D. Cardwell, R. Purdy and R.C. Bahner (eds), Aquatic Toxicology and Hazard Assessment: Seventh Symposium, ASTM STP 854, Amer. Soc. Test. Mat., Philadelphia, Pa., pp. 344-63.

76. Dean, T.A. and Deysher, L.E. 1983, ' The effects of suspended solids and thermal discharges on kelp.' In W. Bascom (ed), The Effects of Waste Disposal on Kelp Communities, So. Calif. Coast. Wat. Res. Proj., Long Beach, pp. 114-35.

77. Kuwabara, J.S. 1983, 'Effects of trace metals and natural organics on algae.' In W. ascom (ed) The Effects of Waste Disposal on Kelp Communities, So. Calif. Coast. Wat. Res. Proj., Long Beach, pp. 136-46.

78. Reish, D.J., Piltz, F., Martin, J.N., and Word, J.Q. 1976, 'The effect of heavy metals on laboratory populations of two species of polychaetes with comparisons to the water quality conditions and standards in southern California marine waters.' Water Res. 10, pp. 299-302.

79. Reish, D.J. 1980, 'Use of polychaetous annelids as test organisms for marine bioassay experiment.' n A.L. Buikema and J. Cairns (eds), Aquatic Invertebrate Bioassays, ASTM STP 715, Amer. Soc. Test. Mat., Philadelphia, Pa., pp. 140-54.

80. Mearns, A.J. 1985. 'Biological implications of the management of waste materials: the importance of integrating measures of exposure, uptake and effects.' In R.D. Cardwell, R. Purdy and R.C. Bahner (eds), Aquatic Toxicology and Hazard Assessment:Seventh Symposium, ASTM STP 854, Amer. Soc. Test. Mat., Philadelphia, Pa. pp. 335-43.

81. Chapman, P.M., Munday, D.R., Morgan, J., Fink, R., Kocan, R.M., Landolt, M.L., and Dexter, R.N. 1983, ' Survey of Biological effects of toxicants upon Puget Sound Biota. II. Tests of reproductive impairment.' NOAA Tech. Rept. NOS 102 OMS 1, Nat. Ocean. Atm. Admin., Rockville, Md., 58 pp.

82. Bay, S.M., Jenkins, K.D., and Oshida, P.S. 1982, 'A new bioassay based on echinochrome pigment sysnthesis.' In W. Bascom (ed) Coast. Wat. Res. Proj. Bien. Rpt. 1981-1982, So. Calif. Coast. Wat. Res.

Proj., Long Beach, pp. 205-215.

83. Siegel, P.R. and Wenner, A.M. 1984, 'Abnormal reproduction of the sand crab, Emerita analoga in the vicinity of a nuclear generating station in Southern California.' Mar. Biol., 80, pp. 341-5.

84. Whipple, J.A. 1984, 'The impact of estuarine degredation and chronic pollution on populations of anadromous striped bass (Morone saxatilis) in the San Francisco Bay-Delta, California: A summary for managers and regulators.' Admin. Rpt. T-84-01, Southwest Fish. Cent., Nat. Mar. Fish. Serv., Tiburon, Ca. 47 pp.

85. Cross, J.N., Raco, V.E., and Diehl, D.W. 1984, 'Fish reproduction around outfalls.' In W. Bascom (ed), Coast. Wat. Res. Proj. Bien. Rpt. 1983-1984, So. Calif. Coast. Wat. Res. Proj., Long Beach, pp. 211 -227.

86. Spies, R.B., Rice, D.W., Jr., and Ireland, R.R. 1984, 'Preliminary studies of growth, reproduction and activity of hepatic mixed-function oxidase in Platichthys stellatus.' Mar. Env. Res., 14, pp. 426-8.

87. Hose, J.E., Hannah, J.B., Landolt, M.L., Miller, B.S., Felton, S.P., and Iwaoka, W.T. 1981, 'Uptake of benzo(a)pyrene by gonadal tissue of flatfish and its effects on subsequent egg development.' J. Toxicol. Environ. Health, 7, pp. 991-1000.

88. Hardy, J.T., Apts, C.W., Crecilius, E.A., and Bloom, N.S. 1985, 'Sea-surface microlayer metals enrichments in an urban and rural bay.' Est. Coast. Shelf Sci., 20, pp. 299-312.

89. Roth, J.C., Williams, R.L., Horne, A.J., Smith, D.W., and Commins, M.L. 1984, 'Dilution-field bioassay for local effects monitoring of wastewater discharge into San Francisco Bay. II. A demonstration study of toxicity based on the growth and condition of caged mussels (Mytilus edulis).' Univ. Calif./San.Env. Eng. Res. Lab. Rpt. 84-1, 76 pp.

90. Harris, L. and Mearns, A.J. 1975, 'Age, length and weight relationships in Southern California populations of Dover sole.' Tech. Mem. 219, So. Calif. Coast. Wat. Res. Proj., Long Beach, 17 pp.90. Sherwood, M.J., 1980, 'Fin erosion, liver condition and trace contaminant exposures in fishes from three coastal regions'. In A. F. Mayer (ed) Ecological Stress in the New York Bight: Science and Management. Est. Res. Found., Columbia, S. C.

91. Sherwood, M.J. 1980, ' Fin erosion, liver condition and trace contaminant exposures in fishes from three coastal regions.' In G.F. Mayer (ed), Ecological Stress and the New York Bight: Science and Management. Est. Res. Fed., Columbia, S.C.

92. Sevareid, R. and Ichikawa, G. 1983, 'Physiological stress (scope for growth) of mussels in San Francisco Bay.' In D.F. Soule and D. Walsh (eds) Waste Disposal in the Oceans: Minimizing Impact, Maximizing Benefits, Westview Press, Boulder, Co., pp. 152-70.

93. Martin, M., Ichikawa, G. and Goetz, J. 1984, 'San Francisco Bay Aquatic Habitat Program: local effects monitoring demonstration project using mussels.' Rpt. Calif. Fish Game, Calif. Fish Game, Monterey, Ca., 51 pp.

94. Martin, M., Ichikawa, G., Goetyl, J. de los Reyos, M., and Stephenson, M. 1984, 'Relationships between physiological stress and

trace toxic substances in the Bay mussel, Mytilus edulis, from San Francisco Bay, California.' Mar. Env. Res. 11, pp. 1-20

95. Smith, L.S., Cardwell, R.D., Mearns, A.J., Newcomb, T.W., and Watters, K.D., Jr. 1972, 'Physiological changes experienced by Pacific salmon migrating through a polluted urban estuary.' In M. Ruvio (ed), Marine Pollution and Sea Life, Fishing News (Books) Ltd., London. pp. 322-25.

96. Thompson, B., Laughlin, J.D., and Tuskada, D.T. 1984, 'Ingestion and oxygen consumption by slope echinoids.' In W. Bascom (ed), Coast. Wat. Res. Proj. Bien. Rpt. 1983-1984, So. Calif. Coast. wat. Res. Proj., Long Beach, pp. 93-107.

97. Pamatmat, M.M. 1983, 'Metabolism of a burrowing polychaete: precaution needed when measuring toxic effects.' Mar. Poll. Bull.13, pp. 364-6.

98. Young, P.H. 1964, 'Some effects of sewer effluent on marine life.' Calif. Fish Game, 50, pp. 33-41.

99. Mearns, A.J. and Sherwood, M.J. 1977, 'Distribution of neoplasms and other diseases in marine fishes relative to the discharge of waste water.' N.Y. Acad. Sci. 298:210-24.

100. Sherwood, M.J. and Mearns, A.J. 1977, 'Environmental significance of fin erosion in Southern California demersal fishes.' N.Y. Acad Sci., 298, pp. 177-189.

101. McArn, G.E., Chuinard, R.E., Miller, B.S., Brooks, R.E. and Wellings, S.R. 1968. Pathology of skin tumors found on English sole and starry flounder from Puget Sound, Washington. J. Nat. Can. Inst., 41, pp. 229-42.

102. Wellings, S.R., Alpers, C.E., McCain, B.B. and Miller, B.S. 1976, ' Fin erosion disease of starry flounder (Platichthys stellatus) and English sole (Parophrys vetulus) in the estuary of the Duwamish River, Seattle, Washington. J. Fish. Res. Bd. Can., 33, pp.2577-86.

103. Hawkes, J.W., 1980, 'The effects of xenobiotics on fish tissues: morphology studies.' Comp. Pharm. Aquat. Sci. Fed. Proc., 39, pp. 3230-36.

104. Meyers, T.R. and Hendricks, J.D. 1982, 'A summary of tissue lesions in aquatic animals induced by controlled exposures to environmental contaminants, chemotherapeutic agents, and potential carcinogens.' Mar. Fish. Rev.(U.S.), 44(12), pp. 1-17.

105. Malins, D.C., McCain, B.B., Brown, D.W., Chan, S-L., Myers, M.S., Landahl, J.T., Prohaska, P.G., Friedman, A.J., Rhodes, L.D., Burrows, D.G., Gronlund, W.D., and Hodgins, H.O. 1984, 'Chemical pollutants in sediments and diseases of bottom-dwelling fish in Puget Sound, Washington.' Env. Sci. Tech., 18, pp. 705-13.

106. Perkins, E.M., Brown, D.A., and Jenkins, K.D. 1982. Contaminants in white croakers Genyonemus lineatus (Ayres, 1855) from the Southern California Bight: III. Histopathology. In W.B. Vernberg, A. Calabrese, F.P. Thurberg and F.J. Vernberg (eds) Physiological Mechanisms of Marine Pollutant Toxicity, Academic Press, New York, pp. 215-231.

107. Rosenthal, K.D., Brown, D.A., Cross, J.N., Perkins, E.M., and Gossett, R.W. 1984, 'Histological condition of fish livers.' In W. Bascom (ed) Coast. Wat. Res. Proj. Bien. Rpt. 1983-1984, So. Calif.

Coast. Wat. Res. Proj., Long Beach, pp. 229–45.

108. Malins, D.C., Myers, M.S., and Roubal, W.T. 1984, 'Organic free radicals associated with idiopathic liver lesions of English sole (Parophrys vetulus) from polluted marine environemnts.' Env. Sci. Tech. 17, pp. 679–85.

109. Stich, H.F., Acton, A.B., Oishi, K., Yamazaki, F., Harada, T., Hibino, T., and Moser, H.G. 1977, 'Systematic collaborative studies on neoplasms in marine animals as related to the environment.' N. Y. Acad. Sci. 298, pp. 374–388.

110. Dawe, C.J., Bagshaw, J., and Poore, C.M. 1979, 'Amebic pseudotumors in pseudobranchs of Pacific cod (Gadus macrocephalus). Proc. Amer. Assoc. Cancer Res. 20, p. 245.

111. Cross, J.N. 1984, 'Tumors in fish collected on the Palos Verdes shelf.' In W. Bascom (ed) Coast. Wat. Res. Proj. Bien. Rpt. 1983-1984., So. Calif. Coast. Wat. Res. Proj., Long Beach, pp. 81–91.

112. Landolt, M.L., Powell, D.B., and Kocan, R.M. 1984. Renton Sewage Treatment Plant Project: Seahurst baseline study. VII. Fish Health.FRI-UW-8413, Fish. Res. Inst., Univ. Wash. Seattle, 160 pp.

113. Hendricks, J.D., Pitnam, P.T., and Sinnhuber, R.O. 1980, 'Null effects of dietary Arochlor 1254 on hepatocellular carcinoma incidence in rainbow trout (Salmo gairdneri) exposed to aflatoxin B, as embryos. J. Env. Path. Toxicol., 4, pp. 9–16.

114. Cantillo, A.Y., Calder, J.A., Long, E.R., and Chapman, P. 1984, "A new emphasis on coastal and estuarine environmental quality assessment.' Oceans '84 Conf. Rec.: Industry, Government, Education, Designs for the future. IEEE, pp.302–308.

115. Valentine, D.W., and Bridges, K.W. 1969, 'High incidence of deformities in the serranid fish, Paralabrax nebulifer from Southern California. Copeia 3, pp. 637–8.

116. Valentine, D.W., Soule, M.E. and Samollow, P. 1973, 'Asymmetry analysis in fishes: a possible statistical indicator of environmental stress.' Fish. Bull. (U.S.), 71, pp. 357–370.

117. Spies, R.B., Felton, J.J., and Dillard, L. 1982, 'Hepatic mixed-function oxidases in California flatfishes are increased in contaminated environments and by oil and PCB ingestion.' Mar. Biol., 70, pp. 117–27.

118. Varanasi, U. and Malins, D.C. 1977, 'Metabolism of petroleum hydrocarbons: accumulation and biotransformation in marine organisms.' In D.C. Malins (ed), Effects of Petroleum on Arctic and Subarctic Marine Environments and Organisms. III. Biological Effects. Academic Press, New York, pp. 175–270.

119. Krahn, M.M., Myers, M.S., Burrows, D.G., and Malins, D.C., 1984, 'Determination of metabolites of xenobiotics in the bile of fish from polluted waterways.' Xenobiotica 14, pp. 633–46.

120. Brown, D.A., Gossett, R.W., and Jenkins, K.D. 1982, 'Contaminants in white croaker Genyonemus lineatus (Ayres, 1855) from the Southern California Bight: II. Chlorinated hydrocarbon detoxification/toxification'. In W.B. Vernberg, A. Calabrese, F.P. Thurberg and F.J. Vernberg (eds) Physiological Mechanisms of Marine Pollutant Toxicity, Academic Press, New York, pp. 197–213.

SOCIAL ASPECTS OF WASTE DISPOSAL

R.B. Clark
Department of Zoology
The University
Newcastle upon Tyne
U.K.

ABSTRACT. There is generally sufficient information to allow the
prescription of waste disposal strategies that will cause least
environmental damage, but what may be technically the preferred option
must be balanced against financial cost and public reaction. Marine
disposal of wastes is increasingly viewed with public disfavour,
though the public is rarely informed about the financial and environmental costs of the alternative disposal routes. The final decision
between disposal options for various wastes is essentially political,
not scientific. The general public deserves to be more comprehensively
informed about the various options for disposal of a waste.

THE DILEMMA

Human society inevitably produces wastes that have to be disposed of
somewhere, somehow, and industrialised, urbanized societies produce a
large volume and great variety of wastes, many of them noxious. In
several countries, disposal options for wastes are rapidly disappearing
as they become unacceptable in the public mind.

The dilemma this poses for waste disposal authorities arises even
over such unemotional material as sewage sludge. In Britain, the Northwest Water Authority is responsible for disposing of the wastes from
Liverpool, Manchester and the surrounding industrial area. Some of the
sewage and industrial waste is discharged, untreated, to the estuary of
the river Mersey and the sea, but much of it is given secondary treatment, generating large quantities of sewage sludge. There is little
land available for landfill or the agricultural disposal of sludge
although these routes are used, and over 1.7 million wet tonnes of
sludge per year are dumped in Liverpool Bay. There is evidence of
accummulation, oxygen depletion of bottom waters and impoverishment of
the benthic fauna at the dump site. The Mersey, on which Liverpool
stands is probably the most heavily polluted estuary in Britain and
additional sewage treatment plants to remedy this are under construction
or planned. They will generate increased volumes of sludge for disposal.

G. Kullenberg (ed.), The Role of the Oceans as a Waste Disposal Option, 691–699.

Sea dumping of sewage sludge, particularly when it is damaging as in this case, is increasingly viewed with disfavour. Were sea disposal in Liverpool Bay to be banned, since land disposal sites are already used almost to capacity, the only remaining option would be incineration. For this, it is estimated that three incinerators would be required by the end of the century, each capable of handling 60,000 tonnes of dry sludge per year. But when the Authority recently attempted to build a 25,000 tonne per year incinerator, public opposition prevented it from doing so.

This example can no doubt be multiplied many times in most countries, but when the wastes have emotional overtones - radioactive wastes are a prime example - public opposition to all disposal options becomes even more vociferous. Indeed, not only is there violent opposition to the disposal options but even investigations of the feasibility of alternative strategies is strenuously resisted. A geological survey in Britain to identify sites that might be suitable for deep storage or disposal of radioactive wastes was abandoned in part because of local alarm at each site that was investigated. A current search for suitable sites for shallow disposal of low level radioactive waste is already attracting opposition. Research into the sub-seabed disposal of radioactive wastes is at present halted because of a prohibition by the seamen's trade union, although no radioactive material is involved in the investigation.

THE OPPOSITION

The opposition to waste disposal options has several strands.

At one end of the spectrum there are those who are not much concerned about the natural environment or remote health hazards, but will object to the siting of sewage treatment plant, incinerator or rubbish tip in their own immediate neighbourhood. This is the NIMBY ('not in my backyard') syndrome. Since this reaction can be expected in any area where the local population is politically active, few areas in a densely populated country can be used for waste disposal without attracting local opposition.

But more sparsely populated areas which might be used in an attempt to avoid this opposition are often designated as 'wilderness areas', 'national parks' or 'areas of outstanding natural beauty', or they are productive agricultural land. If such areas are used for waste disposal, opposition can be guaranteed from those concerned about conservation of the natural environment and the protest will be from town dwellers as much as, or more than from those living in the affected areas.

More comprehensive and more considered opposition comes from the Environmentalist of Green movement. It favours the 'small is beautiful' philosophy and in its more radical forms disapproves of modern technology in all its aspects, and sees the solution to all environmental problems in a return to small, self-sustaining rural communities. A less radical and more persuasive view accepts that we cannot dismantle western urban society and all adopt a simple life on the land. Instead, it emphasizes the need for a low and non-waste technology, recycling and, in the context of the present discussion, solving waste disposal problems by severely reducing the volume of waste that is generated.

Excessive packaging and manufactured goods with a life measured in months rather than a lifetime are offensive to older members of the population who found it hard to adjust to the 'throwaway'economy. Some policies of the Environmentalists strike a responsive note in them, but it may be doubted if the majority of the population is yet ready for the radical change in life-style that a full realization of the Environmentalists' objectives would entail. That has not prevented the Environmentalist movement making political progress: no political party can claim that all its supporters wholeheartedly accept every element in its manifesto, nor even its main outlines. At least in the short term, one or two popular and dramatic issues are sufficient to win widespread support and scientific considerations have then only a very small role.

Further clouding of scientific issues is caused by the moral overtones that sometimes enter the debate. In an extreme form there is the question of whether humans have superior rights over other organisms living on the planet. Though I am not sure what that proposition means, it surfaces in one form or another from time to time and is capable of attracting a favourable response when waste disposal threatens a rare species (often one previously known only to a handful of taxonomists) or excludes a popularly favoured species from a particular area. Another moral issue concerns the source of a waste. The same waste is not always equal. Radioactive waste from hospitals is acceptable, that from power generation or industry much less so. Radioactive waste from military installations is scarcely acceptable at all.

Environmentalists of all shades, whether of the NIMBY or Green variety, have realized, like many other groups, that their objectives are most likely to be achieved through large or small scale political influence. Public opinion has to be mobilized and the most effective way of doing this is by exploitation of the news media. The media, and particularly television, have a notoriously short attention span. Issues have to be simplified and dramatized, preferably visually, in order to attract the attention of television. The visual dramatization may take the form of mass meetings or demonstrations and does not stop short of illegal actions and violence. Violent protest by aggrieved minorities against decisions that displease them are now commonplace, whether it is by farmers, football fans, truck drivers or animal welfare supporters. The causes are so various that it is hardly surprising that waste disposal should attract its share of violent protest.

WASTE DISPOSAL OPTIONS

Unlike some issues that attract protest, problems caused by the disposal of wastes cannot be avoided by postponing a decision or preserving the status quo: the wastes continue to be generated every

day and have to be dealt with somehow. Although one disposal route
may attract opposition, there is no guarantee that alternative options
will prove less damaging or less unpopular. There are usually no easy
solutions.

It is to take an objective view of the alternative options for
disposing of various kinds of wastes that the strategy of 'the Best
Practicable Environmental Option' (BPEO) has been strongly advocated
in Britain. Implicit in this approach is the view that waste treatment
and disposal, or even recycling, is rarely possible without causing
some environmental impact or creating some risk to human health. The
objective is therefore to select the disposal option that has the
least environmental impact and the lowest health risk. This requires
a comprehensive examination of waste disposal strategy that takes into
account all the potentially polluting substances in a waste, their
pathways, disposal routes and effects.

This admirably rational approach has a number of qualifications
and implications that may be unwelcome but have to be faced.

Many wastes are extremely complex mixtures and their exact
constitution is unknown. Often there is sufficient accumulated
knowledge and experience of their more important components to predict
their pathways, fates and effects when discharged into different
environments to make the assessment required for a BPEO study. But
there is always the risk of unpleasant surprises, particularly with
the development of new industrial processes and new technologies.
Unsuspected or new constitutents of a waste, harmless in one environment may prove very damaging in another. As with any other strategy
for deciding waste disposal options, BPEO cannot provide a definitive
and permanent solution but must be subject to constant review. Today's
preferred solution may be a time-bomb for tomorrow. Monitoring of
effects is essential so that if a waste disposal option has unforeseen
consequences, they can be detected and evaluated, and the disposal
strategy changed if necessary.

Next, in making a forward projection of the impact of a waste
disposal strategy, considerable reliance has to be placed on modelling.
Ecological modelling has now reached a sophisticated level. Modellers
can speak to other modellers and make a critical assessment of each
other's work, but other scientists, still less the concerned general
public, have no contact with the logic of the analysis and must take
the conclusions on trust. They will not. It is often possible to
make a few calculations on the back of an envelope to show the effect
of a waste disposal option. That is probably as good as the most
sophisticated model and has the advantage that it is understandable
to the layman. It will not satisfy the professional modeller, however,
and when experts disagree, the layman naturally distrusts all
professional advice. A scientific analysis of the relative merits of
alternative waste disposal option is therefore faced with the problem
of deciding at what level the analysis should be conducted.

Models can be general or site-specific. In searching for the
BPEO, should we compare the environmental and health impact of discharging low and intermediate level radioactive waste at sea with those
of land disposal, or compare sea disposal at the north east Atlantic

dump site with disposal at specific sites in France or Britain? The
general public is naturally concerned about specific sites, but there
is an almost infinite number of those, and it is not practicable to
examine all the conceivable disposal options in a BPEO study. Yet
without site specificity the exercise lacks reality.

If we are to seek the least damaging environmental solution to
waste disposal, it follows that all disposal routes should be considered
including sea disposal. If landfill sites are acceptable on land, why
not their equivalent in the sea where, at least, they are not visually
offensive? But for some reason which is not clear to this marine
biologist, the sea is often given a special status and regarded as
sacrosanct. Since most coastal communities and seafarers have a long
history of throwing their wastes into the sea, this attitude is not one
shared by those most intimately connected with the marine environment.
It is true that with the greatly increased discharges to the sea,
localized impoverishment of this environment can be and has been caused,
but that is no different from what happens on land. The BPEO approach
implies that there has to be a trade-off between one environmental cost
and alternative environmental costs : the land versus freshwaters versus
the air versus the sea.

All waste disposal options have their costs, financial as well as
environmental. The environmental lobby rarely takes much account of
the former, but it is questionable if environmental protection has
higher priority in the public mind than medical services, education,
policing, street lighting or a dozen other things. All are necessary,
but some balance has to be struck between these calls on the national
economy, and environmental protection cannot be given overriding
priority but must take its place along with the rest.

While an examination of the Best Practicable Environmental Option
provides an objective scientific assessment of the consequences of the
various strategies for waste disposal, it does not avoid the need for
political decisions. What proportion of national expenditure, private
and public, should be devoted to waste disposal compared with other
activities? What weighting should be given to damage in one environment compared with damage in another? If some hazard to human health
is unavoidable, is a large risk to a few preferable to a small risk to
many? These decisions are the responsibility of politicians whose role
it is to assess where the balance of public opinion and of the public
good lies.

Although I have dwelt on one particular approach to deciding
between alternative waste disposal strategies - the BPEO which is
currently advocated in Britain - I have used it only as an example.
The problems it reveals do not differ from the scientific questions
raised in any other approach to the problems of waste disposal.

THE SEA

The sea has a special status in the public mind. It cannot be quantified and is not consistently applied, but it is a factor to be reckoned
with. Degradation of marine environments, as of any other, follows if

they are overloaded with wastes, but these are very localized effects
and the great capacity of the sea to absorb wastes without detectable
effects is quite evident. Nevertheless, widely publicized claims that
the seas are dying or have been irreparably damaged by pollution,
although obviously untrue, are believed.

If marine environments are to be given a special status, different
criteria for the management of waste disposal have to be employed for
the sea than for other environments. It is a reflection of this that
a widely favoured approach to waste disposal to the sea should limit
discharges to the receiving capacity of the local area. Assimilative
capacity is a concept taken over from the management of waste disposal
into freshwater where it originally related to the disposal of organic
wastes that are subject to bacterial degradation. In the sea it relates
to all wastes, degradable or not, and is the expression of the objective of balancing inputs to the diluting and dispersive capacity of the
receiving waters so that no measurable change is brought about by disposing of wastes into them. Chemical analytical techniques have now
become so sensitive, however, that unusual and persistent compounds or
elements can be detected at incredibly low concentrations far from
their source. This testifies to the success of the dispersion and
dilution objective, but only adds fuel to the concern about the widespread contamination of the sea resulting from the marine disposal of
wastes. It says nothing, however, about pollution which by all
accepted definitions implies that the contamination has some damaging
effect.

Caesium-137, included in the discharges from the Sellafield nuclear
reprocessing plant on the northwest coast of England can be detected in
waters around Spitzbergen and southern Greenland. The local level of
radioactivity is not detectably increased, but it is naturally suggested
by the anti-nuclear lobby that the United Kingdom is exporting its
wastes to the detriment of others who are not themselves beneficiaries
of the British nuclear energy programme. The contamination - at a
microscopic level - cannot be disputed, but the detriment is far from
established and exceedingly unlikely.

Environmentalists argue that the oceans are part of the 'global
commons', areas beyond territorial boundaries which should be enjoyed
by all and not exploited by one country to its own benefit, but to the
detriment of other users. This argument has been applied to the ocean
dumping of wastes even if the detriment to others is more imaginary
than real and, given the sensitivity of modern chemical analytical
techniques, strikes at the heart of the dilution and dispersion
solution to waste disposal problems. Indeed, since discharges of
persistent materials to the atmosphere and freshwaters inevitable
result in some inputs to the sea through rain-out, fall-out or run-off,
most forms of waste disposal can come under attack if this philosophy
is persued to its logical conclusion.

SOLUTIONS

It would be easy to dismiss opposition to waste disposal options as

irrational or stemming from fear based on ignorance. There is some
truth in this, but it is by no means the whole, nor even the most
important part of the problem.

To the extent that it is true, the solution lies in educating the
public to understand and appreciate the basis for a scientific judgement of the best strategies for disposing of various kinds of waste.
That will not be easy for a number of reasons. It will be argued that
this is merely an attempt to brain-wash the general public into accepting the options of experts without question. Unfortunately, 'experts'
have a bad image : the tragedies at Minamata, Seveso, Bhopal, the
disastrous consequences of the use of Thalidomide, the environmental
impact of the widespread use of organochlorine pesticides, the accident
at Three-Mile Island are all laid at their door, and experts are no
longer trusted, any more than are governments. Large scale industrial
accidents are not new, but they are now regarded as the fault of modern
science and technology and in the aftermath of each disaster, the view
gains ground that such risks should be avoided, not by improved
technical safeguards but by abandoning the technology.

The consequences of this are not usually appreciated. During the
late 1960s there was great public concern in Europe and North America
about the effect of DDT on wildlife, particularly on falcons and hawks
and other predatory birds. In 1970, the World Health Organization was
sufficiently alarmed at the possibility that the manufacture of DDT
would cease, that it mounted a campaign to explain that its very
successful programme to eradicate malaria and other insect-borne
disease in the tropics would be halted if DDT were withdrawn before
an equally effective, safe and cheap alternative was available.

For reasons that I have already discussed, pollution issues are
presented to the public in simplistic terms. There is little awareness
that environmental improvement in one area may lead to an increased
human health risk elsewhere, that improvement in one environment may
involve deterioration in another. Financial costs of environmental
improvement are rarely considered: the cost of alternative waste
disposal options can be estimated with accuracy, but evaluating the
environmental benefits in cash terms is difficult and, although economists claim to be able to do so, their efforts have not proved acceptable to the public at large. Finally, the public perception of risks
bears little relation to their statistical reality: the hazard of
exposure to radioactivity is regarded much more seriously than the
hazards of road traffic accidents although the statistical evidence
points in entirely the opposite direction. One hazard is acceptable,
the other not.

Experience suggest that schoolmasters and professors in classrooms
and lecture halls will not suddenly produce a generation of rational
beings who will view environmental matters in a detached, dispassionate
way. Nevertheless, even if formal education will not transform the
situation, that is no argument for failing to try and give a more
balanced and sophisticated view of human interactions with the natural
environment than is usually the case at present.

Government, too, has a crucial role to play, not only over
immediate issues but in a longer-term educational process. Government,

by regulation, determines how and where wastes are disposed. In doing
this, it must weigh the scientific and economic aspects of each case,
take a broad view of national rather than simply local benefits and
disbenefits, and have regard to its international obligations. Above
all, it is sensitive to the political consequences of its decisions.
If it is convinced that one course of action is scientifically and
economically the best option, it may still have to win public support
and understanding for it. Unfortunately, such is the polarization of
views that almost anything government spokesmen say will often be
regarded as mere propaganda and disbelieved, and even scientific
evidence dismissed. There have been sufficient examples of this that
it is understandable that government should resort to secrecy while
options are being explored and to issue a diktat when a final decision
is reached. This simply feeds the suspicion that government has something to hide and brings science into even greater disrepute. Painful
and inconvenient though it may be, this unhappy situation will not
change unless official bodies are frank and open at all stages in the
decision process.

The press and television, at least in Britain, also fail to play
the educational role in environmental matters that they might. Although
environmental matters are of almost constant interest to the press, it
is remarkable how little informed comment appears in the media.
Politicians tend to portray the world in black and white terms, but
there is no shortage of serious, well-informed commentary about
politics in the media, exploring the complexities of a situation and
the available options together with the advantages and disadvantages
of each. If environmental and pollution problems received the same
dispassionate analysis that is given to political issues, the thinking
section of the public might better appreciate the complexities of
waste disposal and that sometimes there are no easy options.

None of this addresses the more fundamental question posed by the
Green lobby, that even when waste disposal options are discussed
openly,the debate is concerned only with superficial matters and the
central issues are ignored. So, a public enquiry into the siting of
a nucleau power-station had as its hidden core a debate about the
desirability of persisting with nuclear power generation, a public
enquiry into the route of a new motorway raises fundamental issues
about transport strategy in general. It may well be true that an
overwhelming majority of the population does not favour a low and non-
waste technology or a non-materialist society, but these are guiding
motives of an important section of the most articulate environmentalists. In the absence of an open discussion of those issues, the
environmentalist lobby is free to attack developments which are
contrary to their long-term strategy and mobilize public opinion on
narrow issues.

It can be argued with some, though not total justification, that
social attitudes to marine pollution reflect an unduly pessimistic
view of the resilience of marine environments and because it is less
visible, the sea is regarded as a more mysterious environment than the
land or freshwaters. Marine ecological processes may be less well
understood than those in other environments, but we are by no means

ignorant of them. There is abundant evidence that we are well able to
manage our use of the marine environment and its resources in a rational
way if there is the political will to do so.

Of course, the sea does not respect national boundaries and its
management demands the international cooperation that has been growing
in recent years. The need for international agreement may add to the
difficulty of achieving a rational exploitation of the marine resource,
but does not affect the scientific issues. This applies as much to
disposing of wastes into the sea as to taking fish out of it.

Governments now show a greater awareness of environmental issues
than formerly. If in their sensitivity to public fears about the
health of the marine environment, governments give it a preferred
status over other environments and do not take advantage of its capacity
to receive wastes, this will be at greater cost to other environments
and often at greater financial cost to the detriment of other public
activities. If this be the final decision, so be it, but before that
stage is reached the public at large should be better informed about
the full implications of all the options for waste disposal, not given
selective objections and equally selective answers to them. I do not
underestimate the difficulty of popularizing complex issues, but it is
a task that needs to be undertaken.

SUMMARY OF THE WORKSHOP

G. Kullenberg
University of Copenhagen
Department of Physical Oceanography
Haraldsgade 6
DK-2200 Copenhagen N

1. DEFINITION OF THE PROBLEM

The Workshop originated from discussions on how to choose between
options available for waste disposal. It was considered that criteria
for selection between options would be of several kinds, and that the
relative values attached to the criteria would vary depending upon
e.g. the area, the waste, the state of development, and the economy.
Generally, division of the environment into separate compartments should
be avoided, since the best option can only be selected when all aspects
of the environment are considered as a whole. Again, what is regarded
as the best option will vary according to the particular values attached
to the various parts of the environment. In some cases it is impossible
or inappropriate to attach any price value at all.

Here, the best option is understood as one being selected upon a
scientific basis, using scientific criteria. These should be objective
and ought to be universally applicable. However, they may not provide
a unique solution or they may provide a range of uncertainty as it may
not be possible or justifiable to arrive at one single critical number.
The scientific basis itself may also be hampered by lack of knowledge
regarding certain processes and by the necessity of making assumptions.
These, then, should of course be based on scientific insight. A certain
amount of judgment will influence most cases.

The basic idea of the workshop was to attempt to define a
scientific basis for the selection of options by using a series of case
studies covering different regional characteristics and different waste
categories.

Focus was put on the water sphere.

Throughout the world natural terrestrial waters are used either
deliberately or unwittingly for reception of wastes. It is well
recognized that such use prejudices all other uses that society wishes
to make of these waters, for example as drinking water sources, for
irrigation or for industrial supply, and as habitats for aquatic
organisms. Lowland rivers and estuaries bordered by population areas
have inevitably been used for the reception of wastes, sometimes to
such a degree that serious pollution has occurred. In most countries

G. Kullenberg (ed.), The Role of the Oceans as a Waste Disposal Option, 701–717.
© 1986 by D. Reidel Publishing Company.

it is now recognized that the assimilative capacity of these waters to
receive waste is limited. Serious attempts are being made to control
the pollution, by various means, such as more extensive purification
of waste waters, arranging for discharge at sites offering greater
dispersion, dumping of sludge in controlled areas, and land reclamation
use of waste material. Because of pressure on terrestrial water
resources, on the financial resources needed to preserve them, and on
the shortage of land for siting of treatment plants, it has been
natural for coastal countries to discharge their wastes in the sea.
Furthermore, because the sea offeres potentially the largest available
dilution and the opportunity to take wastes far away from human
habitation, it has also been natural to regard the sea as a potentially
suitable receptor for particularly hazardous wastes. It is recognized
that the sea has a limited capacity to assimilate wastes, and in order
to evaluate the case for such uses of the sea it is necessary to consider
the options, including disposal on land, to terrestrial waters, and to
the atmosphere.

The problem, then, is to define the scientific basis for selection
of, and between, options and for the role of the oceans as a waste
disposal option. An attempt to achieve this goal was made through
consideration of a series of cases which dealt with similar waste
categories, but in different areas of economic and environmental
conditions. In this way, it was argued, a review would be obtained
showing how the scientific, and to some extent the socio-economic facts,
have influenced the choice of options.

As a common basis it may be considered that the goal of waste
management is to store or to dispose of the waste while maintaining
life quality and life supporting functions of the environment. The
scientific, technological, and economic criteria for rational handling
of waste are being developed, usually on site specific and waste
specific bases. Special attention is directed to the handling of toxic
wastes which constitute about ten per cent of the total, and include
metals, readioactive species, organic materials such as intermediates
in chemical production, and fossil fuel residues. Disposal of such
wastes on land can threaten critical resources, such as surface and
groundwaters. Can the oceans be considered as an effective disposal
option? To what extent is the choice between different opitons in
different communities the consequence of different technical and
scientific problems or of economic legal or political systems? The
study of alternative solutions can elucidate each choice.

2. STRUCTURE OF THE WORKSHOP

In order to cover the rather different aspects of the subject, a
series of different sessions were composed with four presentations,
on the average, each followed by discussion. The Workshop started
with a series of papers in which the problem of waste management and
waste disposal was dealt with in a perspective manner. These and the
associated discussions set the scene for the subsequent sessions.

The arguments behind the selection of substances and associated

cases together with the session structure were developed by the
organizing committee, as follows.

The disposal of sewage is a universal problem and formed one of the
central themes of the meeting. Apart from dumping at sea there is a
range of options, including incineration, disposal on land, and a
range of possible treatments before disposal. "Clean" sewage, i.e.
without industrial wastes, represents largely an eutrophication problem,
but the presence of metals and organochlorines raises the question of
toxicity. These matters were examined in the context of case histories
which illustrated the practice in different countries, the options
considered, and the problems encountered at different sites. In
addition, specific attention was given to the disposal of nutrients into
marine and freshwater environments, and the effects of these disposals
were discussed considerably, the eutrophication problem presently being
a serious one in many areas.

The improper management of wastes containing toxic metals has
resulted in epidemic poisoning (methyl mercury entering marine systems
and cadmium entering terrestrial systems). Surveillance of these metals
and others such as chromium, tin, and arsenic in industrial and domestic
wastes is essential for an assessment of the relative abilities of
marine and terrestrial sites to accomodate them most reasonably.
Overviews were made on selected metal containing wastes and several
case studies were discussed where marine, freshwater and land disposal
and subsequent contamination had also occurred.

Comparisons were also made through the examination of practices of
dealing with some toxic organic wastes, in respect to marine and freshwater environments.

There will most probably be continuing pressure to put low- and
medium-level radioactive wastes into the oceans as a consequence of
their continued production in nuclear energy facilities. Past
experiences with the discharge of low-level wastes into the Irish Sea
from Windscale and into the English Channel from Cherbourg, have
emphasized that with appropriate controls, impacts upon public health
can be made negligible. Still, additional inputs must be managed
effectively to maintain the present record. Many other countries in
the northern hemisphere use land disposal for low- and medium-level
wastes. Comparisons between sea disposal and land disposal were made
and experiences examined.

On land or in the sea, pollutants may be confined to specific areas
or disposed of as widely as possible. The aim of the latter is to take
concentrations to low levels; the problem is to define the distribution
and, specially, to estimate the impact over large heterogeneous and
variable environments. Acid rain on land and endemic hydrocarbon
contamination at sea are examples where both difficulties occur. The
alternative, confinement in a small region by natural or artificial
means can imply obvious impact on local areas and, consequently, a
fairly clear cut definition of the areal extent and of the severity of
the disturbance. Land fill and sewage sludge dumping in quiescent sea
water are examples. Dispersion characteristics are of a major concern
in relation to the determination of an assimilative capacity and in
disposal site selection criteria. Because of the form (solid, liquid)

of pollutants, an exact choice is not always available, but it would appear that containment is favored on land and dispersal at sea. Is this always the correct approach?

Through reviewing different cases an attempt was made to bring out the relative advantages, when options were available, to be considered in the context of monitoring and management.

The immediate impact from waste disposal in the sea, apart from physical smothering, is from the input of chemicals. Adequate detection and quantification of these in the water and sediments is therefore an essential part in assessing biological effects. In addition, the determination of residues in organisms (e.g. by means of the mussel water approach) offers a means of integrating chemical contamination in time and space and provide an index of pollution. A complete assessment requires that the chemical residues can be linked to effects on organisms.

Can effects be demonstrated on populations, communities or ecosystems? Such effects are regarded as the potentially most critical evidence in evaluating pollutant disposal, particularly for widespread or longer term contamination. Yet, for these cases such as plankton blooms or fish kills, incontroversial evidence is rarely, if ever, available. Only for local severe contamination is community change, usually combined with chemical data, sufficiently clear cut to be useful to management. These inadequacies occur both in terrestrial and aquatic systems. The underlying problems are the great natural variability, especially at larger space and time-scales, and our present inability to relate these changes unequivocally to causes in the physical or chemical environment. Given the present emphasis on these ecosystem changes, it is essential that we examine critically the basis for their use as general pollution indices at sea or on land in the expectation of ecosystem monitoring as a management tool. Various examples were examined with the aim of producing criteria for the use of different techniques.

The public perspective and understanding of the waste production, management, and disposal problem is a very important factor influencing the decisions made in selecting different options. The marine environment appears to be a particularly sensitive issue, probably due to concern that contamination and damage will be irreversible and the fact that a few inflict harm on our common heritage. The necessity of conveying to the general public in an appropriate way information on scientific findings and explaining the implications of these findings on the waste problem was discussed on several occasions and was the subject of one paper.

One aim of the discussions was to define the absolutely crucial information required for choosing between options. It was, however, not the aim to formulate recommendations, and it was not possible during a workshop of this kind to formulate a set of guidelines.

3. SUMMARY OF DISCUSSIONS

3.1 Perspective papers: Chapters 1 - 4

The validity of the contention that there was no widespread damage
to the sea was questioned with reference to evidence of change in
phytoplankton populations, reduction in primary productivity in the
North Sea, and greater prevalence fo fish diseases. However, it was
argued that to the extent that such problems were shown validly to be
the result of pollution, something could be done about them. An
important issue was whether this was best done by application of uniform
emission standards as practiced by some countries, or by measures
tailored to the local circumstances. Reference was made to the disposal
of drilling muds (freed of diesel oil), or petroleum hydrocarbons, and
of heavy metals, which may not be as serious as originally conceived.
The question of identification of future problems, e.g. toxaphene,
low molecular weight chlorinated hydrocarbons, was raised and it
appeared to be widely thought among the audience that all one could do
was to draw up lists and assign priorities to relevant investigations,
using informed judgment. The question was raised of using more
application of economics in the assessment of options and greater use of
financial incentives in promoting solutions of problems. The approach
using the best practicable environmental options was argued to provide
a rational economic solution.

Examples were given where effects did reverse rapidly (e.g. in the
Thames, and after major oil spills), following remedial action. However,
it was also argued that because of the very long residence times in the
oceans (of the order of 1000 years) it would inevitiably take a very
long time to reduce the concentration of a conservative pollutant if it
has been allowed to build up to levels which were ultimately found to be
unacceptable. This was particularly true of sediments whose retention
time would tend to be longer than that of the water. On the other hand,
it was felt that in relation to the residence times of the oceans, very
few substances could be regarded as truly conservative and that even
half-lives of several decades were probably quite low in this context.

The difficulty of setting agreements as to whom should pay for
clean-up when international waters were found to be excessively
contaminated was noted, but it was argued that the problems often had
identifiable local sources that could be dealt with by those responsible.
In the particular case of dumping, if there were adverse local effects
then there was the option of shifting the dumping ground. A pragmatic
approach was to devote the financial resources that society was willing
to spend to where they could do the most good in the short-term. It
was noted that with the exception of studies of radioactivity there was
a dearth of substantial investigations into the condition of the open
oceans, so it was perhaps unwise to argue too strongly that there had
been no adverse effects. This prompted the question as to whether an
open ocean baseline study would be worth mounting. Debating this
question speakers argued that such a study would be very expensive and
it would be difficult to persuade decision-makers faced with obvious
short-term problems to allocate the necessary funds. However, it was
recognized that these local problems might be the source of more
widespread ones and there appeared room for a compromise, in which some
resources would be allocated to monitoring the open oceans. It was
noted that many regulatory agencies were seeking to develop general

policies that could be expressed as "single numbers" pertaining to all situations. Single numbers might be satisfactory for say the surface or lower   layers of the open oceans which individually might be regarded as roughly   homogeneous, but could be counterproductive for local situations, which tended to be highly site-specific. The answer to this, perhaps, was to produce guidelines which would define the circumstances in which generalisations were possible on the one hand, and where they would not be helpful, on the otherhand.

Finally, the issue was raised as to whether it was better to concentrate and store or dilute and disperse in achieving the protection required. The latter was often easier but it led to contamination of the whole marine environment, to acceptable levels if it was done correctly but to widely dangerous ones if it was not so done.

Around the question of TBT and the dumping of sludge, concern was expressed that the public were confused about the issues because scientists could not agree about the severity of the problems. One of the dilemmas referred to was that even where no effects were discernible over quite some time, some people were still not convinced that there was not a problem just around the corner. However, it was noted that one had to be clear about what were realistic time-scales for judging whether those were real effects or not. The point was made that people expected to be assured that risks were zero. In all honesty, it was not possible to do this. Instead scientists should express risks in probabilistic terms and put the probabilities into perspective with other risks that were habitually regarded as quite acceptable. The point was also made that people's judgment of the importance of a risk depended on the extent to which they were personally exposed to it, and personally responsible. The difficulty of human perceptions or foresight being limited within the scale of human life span was mentioned, and it was suggested that this limitation could be overcome by the development and use of models.

The validity of the contention that the space and time-scales over which observations had to be taken to detect effects must be increased as the variability of the natural system increased and the severity of the effect looked for decreased, appeared to be accepted at least in a qualitative sense. However, not everyone was prepared to accept that the distinction between internally dominated terrestrial phenomena (such as the defoliating effects of bud worms on trees), and externally dominated effects of pollution on the marine environment was a reflection of a general difference between the terrestrial and marine environments. Several speakers contended that there was in reality not a great deal of difference between environments in relation to waste disposal. There was a suggestion that if age-structured models could be developed for analyzing population dynamics in the ocean some of the difficulties in interpretation induced by variability would be reduced. However, it was noted that more data would be required to develop models and at the moment progress was distinctly data-limited, especially having regard to the long time-scales over which data had to be collected to establish the more subtle effects. One suggestion made was that perhaps data-acquisition could be facilitated by the use of high flying aircraft, which would provide usefully intermediate coverage between satellites and boats.

Speakers asked whether, because the time-scale of effects was so long, monitoring as currently practiced was of little value. It was suggested that might be the case for some types of subtle effects in areas where pollutants were dispersed over long distances, but in other circumstances, for examples in quiescent areas, the majority of impact could possibly be adeuqately assessed by relatively short-term studies. It was pointed out that quite long-term records were now available for some water bodies on quality characteristics and primary production. So total pessimism about shortage of data would not be justified.

It was suggested that money devoted to long-term monitoring might be better reallocated to fundamental, though problem-orientated studies of processes with a view to the production of better deterministic models.

Many in the audience appeared to feel that small-scale short-term effects could be effectively quantified but it was agreed that identification of the larger-scale, longer-term effects was more difficult. The possible value of models was recognized in predicting future events though the difficulties of producing and particularly verifying such models was emphasized. If useful long-term information was to be obtained from models it would probably have to be expressed in terms of statistical probabilities. It was felt that, in relation to the state of the art in modelling, the basic equations incorporated in current models were well established and understood, but to cope with the complexities of marine systems greater complexity had to be introduced into the models.

There was general agreement that, taking account of the present understanding of the make up of the world's oceans and their value to man as a resource, relative to the value and availability of water or other precious resources on land, disposal of waste in the deep ocean (depth > 1000 m) should be considered as a valid option on purely rational grounds. It was agreed that this should not be regarded as an excuse not to avoid waste production. Much can already be achieved and more undoubtedly should be sought by way of reducing the arisings of highly toxic or otherwise hazardous wastes. Some waste will, however, always have to be disposed of. Some will be relatively innocuous, others more harmful and sea disposal can often be regarded as the safest option from a standpoint of Man's well-being.

Perversely our understanding of processes in the more productive shelf areas is perhaps less adequate, especially when their greater value in terms of productivity and food production is considered. Accordingly, it was agreed that the extent to which these areas can continue to be used for waste disposal will depend upon extending our predictive ability. Nevertheless, it is apparent that our understanding is such that there is ample cause to believe some wastes can be disposed of in shelf waters without damaging the resources of value to Man.

There was general agreement that the oceans cannot be regarded as a single entity and that some areas are more valuable than others. It was suggested that it would be extremely difficult to envisage serious despoilation of the marine environment on anything other than limited geographical scales. As an example, it was suggested that if all the

the radioactive wastes produced were dumped into the deep ocean the
levels of radioactivity in ocean waters after dispersion would only be
doubled. However, it was accepted that such statements could lead to
overestimates of the capacity of the marine environment to absorb wastes
without serious negative effects, because distribution is certainly not
instantaneous and some areas and interests are more valuable and
vulnerable than others.

3.2 Engineering aspects: Chapters 5-8

     The subsequent four chapters deal with technical and engineering
aspects of sewage sludge disposal together with options of disposal and
impacts of such disposals. Topics raised during the discussions
included:

- Many countries will not have the technology or the finance to engage
  in successful treatment schemes. In such cases a long sea outfall
  for untreated sewage may be the best options and the cheapest.

- The question of pathogens is clearly of importance. At the purely
  technical levels it was pointed out that good models exist which
  allow bacterial numbers to be forecast with an accuracy of x 5. This
  is acceptable since we may start with 50 million bacteria per 100 ml
  and aim to end with a few hundred per 100 ml. However, concern was
  expressed that the models dealt in terms of coliforms, while the
  microorganisms which do most harm could be viruses. Secondary treatment does not kill microorganisms, but chlorination does.

- The recycling techniques, such as reverse osmosis, could remove the
  taste of water - this could be restored, but at a cost; low molecular
  weight organics can sometimes be a problem but there are techniques
  for removing them; a range of techniques are available so that most
  problems arising in obtaining the desired sludge quality can be dealt
  with, but it would be unnecessarily expensive to instal them unless
  they really were needed, - each case should be tackled on its merits;
  for deep outfalls it was not necessary to go for secondary treatment,
  but in the USA, the EPA insists on this, although a waiver can be
  granted.

- Incineration of sewage sludge should be regarded as a treatment rather
  than a disposal option, the remains being a metal - rich flyash.

- In the context of sewage sludge disposal, our knowledge of the
  receiving capacity of soils is probably deficient.

- We should at least recognize as an option the possibility of
  destroying limited areas of sea bed by giving them over as disposal
  grounds.

- In calculating the costs of disposal options, we should try to
  recognize hidden costs or benefits. Thus the costs of sea dumping

should include the cost of the effects, and in costing diposal on
agricultural land, we should substract the cost of the fertilizer not
used.

- Sewage sludge should not be put on land unless we are sure we can
control it. Before deciding on any option for sludge disposal, we
should ask if it is possible to separate the harmful components from
the rest.

3.3 Case studies: Chapters 9 - 21

Following the engineering discussions a series of case studies from
different regions were analyzed, covering freshwater environments, parts
of the Mediterranean, the US coast, and the North and Baltic Seas. The
regional or local case studies gave examples of disposal of wastes
leading to the receiving capacity of the locality being exceeded in
relation to oxygen, and that species compositions have changed. The
effects of inputs of organic material and nutrients were clearly
demonstrated. An increase of nitrogen was shown to occur in ground and
marine water, being caused, at least in part, by the increasing use of
fertilizers over the last 2-3 decades. It was also demonstrated that
appropriate actions, such as control and change of technology could
reduce the waste inputs very significantly. This was the case for large
point sources. The difficulties of correctly determining and reducing
diffusive inputs were demonstrated for several regions, in particular as
regards nutrients. The economic consequences of imposing severe
purification or treatment requirements were discussed and it was pointed
out that in many areas restrictions could only be gradually imposed.
        An overview of nutrient levels and primary production in several
marine and freshwater regions related input rates to production, and
indicated critical values generating a transition from oligotrophic to
eutrophic and to hypertrophic conditions.
        Several attempts to model distributions of contaminants were
presented and the necessity of including all relevant processes and
interactions in the model was shown. Physical dispersion alone cannot
account for the distributions of inputs from land-based sources and
rivers.
        A paper on fates and pathways of nutrients, acids, trace metals and
radionuclides in lakes and wetlands drew on some twenty years of
experimental studies in a group of freshwater lakes in Canada, which
contrasted the behaviour of whole limnic systems under different
artificially induced physioco-chemical conditions. The discussion
centered on the utility of extrapolating the data to other situations.
The results of these experiments were thought to have been particularly
helpful in assessing the impact of phosphorous on the Great Lakes
ecosystem and to have influenced policy makers in that situation. It was
suggested that the results might be helpful in assessing the
consequences of nutrient enrichment in the Baltic, but only if
particular experiments were designed with that end in mind. The
discussion concluded with a consideration of the general differences
between nutrient cycling in freshwater and marine systems.

A contribution on the relationship between nutrient input and primary production across a range of ecosystems, provoked a discussion on the relationship between the release of nitrogen and phosphorous from sediments, and their rate of uptake by organisms in the water column.

An ecosystem behaviour model of the Southern North Sea had been designed to provide a water quality assessment of the North Sea. It was a conservative model representing water transport and particulate loading throughout the Southern North Sea. It was to be used to assess the concentrations of particular pollutants for different areas, to define their origins and to calculate risk indices for different areas. Several participants pointed out that a thorough validation of the model had not been carried out and several weaknesses in the input data were highlighted. It was felt that the failure to include any sedimentary and ecological processes in the model seriously weakened its utility and raised serious doubts about the applicability of its predictions. It was pointed out that physio-chemical and ecological processes were more important than residual circulation in the Southern North Sea, yet the model's predictions were based only on tidal flow calculations. The calculations of risk indices and critical levels of particular pollutants, and the mapping of these over the area from model predictions, was thought to be misleading and it was suggested that such exercises should not be carried out before a thorough revision of the model had taken place. This should include the addition of processes and a thorough validation. Nevertheless, some participants thought that the attempt was a useful beginning and it was suggested that the various groups currently constructing North Sea models should come together to pool their ideas and data and reach a common agreement on the type of North Sea model needed.

A paper on copper distributions in two contrasting Portuguese lakes aroused considerable interest. The contrasting oxic and anoxic regimes in the two areas created very different conditions for the speciation of copper, and these were described in some detail. There was some discussion on the solubility of copper under natural conditions and use of stability constants in calculating complexation.

A common set of criteria evaluating ecological damage and human health impacts may be devloped for land-based and ocean-based disposal options. A matrix was presented of evaluation parameters and ecosystem types, summarizing present knowledge of ecosystem responses to disturbance and concerns about human health impacts. This created much discussion, demonstrating once again that model formulations, of analytical, numerical or conceptual format, tend to generate very fruitful exchanges of views.

It was generally agreed that our present knowledge was sufficient to allow formulations of these kind of models, provided that they were used to highlight our areas and ranges of uncertaintity. This can be developed as a predictive tool.

Several presentations were made regarding the North Sea summarizing experiences in England, Scotland, Holland, and the Federal Republic of Germany. These presentations dealt with waste disposal, inputs from river runoff, and dumping in various areas. Comparisons with alternative disposal sites on land and at sea were made. In some cases it was clear that marine disposal was the preferable option. In many situations, the North Sea can be used as an example in relation to waste disposal. It may be compared with the Great Lakes on the freshwater side, as regards the role of atmospheric inputs and the rate of river inputs in some parts of the area, as may the Baltic Sea.

3.4 Radioactive waste disposal: Chapters 22 - 25

The following four presentations dealt with low-level disposals from point sources into coastal waters, modelling for prediction concerning open sea disposal, modelling for comparison of land and sea disposal, evaluations of various options in relation to high level waste disposal, and considerations of burial of radioactive waste in the deep sea clay (this presentation is not included in this proceedings, reference is made to Hollister and Smedes (1984)). From the very lively discussion associated with these presentations the following views emerged.

3.4.1 Low-level release. It was acknowledged that the C^{14} release from the Sellafield reprocessing plant is now traceable over long distances. However, the proportion reaching Man can be shown to be less than 1/1000 of 1% of that discharged. Levels of exposure to all radionuclides released from Sellafield, the U.K. and other populations, have always been well below the level of safety defined by ICRP and all are decreasing. Nevertheless, there is a definable increased risk of mortality as a result of radiation exposure which, in terms of the U.K. population, amounts to some five deaths, relative to approximately 5×10^5 per year from other causes. It was noted that the most sensitive species was assumed to be Man. Experiments with fish or shellfish exposed to various forms of radioactivity show it is possible to detect effects but indicate that these would be trivial at the levels of exposure which are safe to Man.

3.4.2 NEA dumpsite. There was general agreement that the models used to assess the NEA dumpsite showed that past dumping at that site, and even continued dumping at 10 times the previous rates, would have no adverse effect on Man or the marine environment, whether at a species or individual animal level. In answer to a number of questions, it was pointed out that the models used involve very few estimates or assumptions. Most values used were measured or derived from proven analogies. There was some discussion as to whether monitoring at the dumpsite should be conducted, despite the fact that both logistics and the model indicated that no changes or radioactivity would be detectable. It was acknowledged that some water samples had been collected which appeared to contain levels of radioactivity above those normally expected. However, it was pointed out that the differences were so small that they could be due to counting errors. Finally, it was pointed out that the maximum individual dose to Man as a result of dumping at the NEA site was shown by the model to be well within 0.01% of the variation in exposure to natural background radiation on land.

3.4.3 Comparisons of land and sea. It was suggested that the comparisons of land or sea disposal routes had not taken account of possible accidents to the dumping vessel which might lead to premature release. However, it was noted that the standards for containment of radioactive materials transported by ships are such that accidental loss would not occur. It was noted that most of the hypothetical exposure scenarios in the land disposal models, and certainly all those which use was likely to be designated acceptable, involved lower exposures than those which arise from accepted practices in uranium ore mining and handling procedures.

It was pointed out that the assumptions made about land containment were rather conservative and that leakage would occur rather earlier than had been assumed. Finally, it was noted that the main reason the collective doses turn out to be comparable whether land or sea disposal is used, is because the major route of exposure in each case was through C^{14}, all of which are likely to enter the carbon cycle.

3.4.4 <u>Geological</u>. Time-scales of subductions might be higher than half-life time of many of the wastes. Subductions may presently be oversold due to inherent problems, and substances introduced in such zones might be turbated during going down. One of the options discussed in Sweden is encapsulating the waste in copper, but it was felt that copper is not always an ideal metal because it could be oxidized in some places while it stays uncorroded in others. Techniques are developed where phosphate is used as an absorbant of radium, which could be kept away from leaking into groundwater. Groundwater and soil needs the strongest protection. The least stable thing is earth, and the best predictability exists for stability of the sea floor, and mechanics of the sea floor are better known than the mechanics of the earth.

Difficulties exist in persuading the public and politicians that geological options could be the best. It was important to mention that time to wait exists for disposal of high energy waste. For low-level wastes it is important to know whether long-lived material is present or not. If the material contains short-lived wastes at low concentrations, it is another problem. The more important differentiation is small and high volumes of wastes.

It was agreed that despite the fact that it is technically feasible to bury radioactive (or any other) forms of wastes in deep sea clay in a way which will minimize possible subsequent exposure to Man, some uncertainties or identifiable risks will remain. As a consequence, the public and decision-makers are likely to remain unconvinced. A similar argument applies to the proposed disposal at sea of nuclear submarines from which the reactors and most of the radioactivity have been removed. Attempts must be made to bridge the communication gap. It was agreed that the type of discussion/workshop run by the Keystone Center might assist in this matter, but several attendants expressed reservations as to whether or not all parties were prepared to be convinced away from their positions. Concern was also voiced as to whether a consensus, especially a consensus to disagree, really helped bridge the communication gap or not.

The theme running throughout the discussion was that when disposal of wastes at sea is raised, whether the waste is radioactive or not, the requirements seem to be guarantees that neither marine organisms nor Man will be affected in any way. This is far more than is required for any other forms of waste disposal where damage and risk are both accepted without any real attempt to predict their scale, even though in most cases the damage to Man's interests can be shown to be substantially greater than those which will arise through disposal to the sea.

3.5 Containment versus dispersal: Chapters 26 - 29

The subsequent section of five papers (one not submitted for publication) addressed the question of containment versus dispersal of the waste material, which may be specified as follows:
(i)   Is guaranteed containment of waste possible?
(ii) Is this a reasonable option when considering costs involved?
(iii) Which chemicals can and cannot be contained, what are the biological factors?

One factor emphasized was the differences between sea and land disposal as far as the dispersive conditions are concerned. Several models were presented, showing that great progress has been made over the last 10 years in our understanding of mixing and circulation in shelf seas and coastal zones. A plea was made for continued support of such research, for one thing so that we would be ready scientifically when the need for greater use of the coastal zone would inevitably arise.

The question was raised of the influence of episodic events such as storms on the models presented. It was pointed out that the diffusion equation used with the assumption of random walk works well. The model assumes random motion on the underlying advection processes and storms can be modelled, it was claimed, although it was suggested that storms were far from random.

Questions were raised on the decay terms in the equations used and on the pattern of particle settlement applied. It was pointed out that the system modelled is an open one flushing to infinity. The main sink was the continental shelf, but some material went off the edge. Due to storms there was no permanent settlement of sludge. Due to bioturbation processes geologists have great difficulty in defining accurately where sinks are located. Resuspension was included in the model and particles perform a mixed motion of settlement and resuspension.

In relation to dump sites, the question of whether there ultimately could be a build up of nutrients far remote from the dispersion site was raised. The author felt it was unlikely that this could occur except in some local areas of say Liverpool Bay. It was suggested that the discharging of sludge was a "no win situation" since in non-dispersive areas it was claimed chemicals build up whereas from dispersive sites the build up would occur at remote sink sites, neither case therefore being acceptable. However, after dispersion concentrations must be appreciably lower that at discharge so the remote sink danger of accumulation was probably slight. In reply to the question

on accumulating versus dispersive sites the author suggested that one of
the key advantages of non-dispersive sites was the fact that the benthic
communities remove carbon, accumulate organochlorines and heavy metals
and remove them from the system. In such sites the benthic organisms
are adapted to these functions and deal with chemicals, whereas in
dispersive sites, e.g. The Thames Dumping Ground, there is evidence that
the communities are not adapted to the discharge because of the
dispersive nature of the sites and as a consequence show signs of damage.

In discussion of remobilization and interactions it was asked if it
was possible to model such a system given the complexity of the chemical
reactions involved. It was felt that at this stage realistic models
could not be made and small-scale models based on pilot experiments were
needed. The techniques outlined gave reasonable results for the bulk
metals but not for details of speciation.

In reply to a question on diagenesis in manganese, Dr. Förstner
said that he was in favor of placing dredge spoils in dyked containment
systems. A concern expressed was that containment is not 100% and that
it relied solely on diagenetic processes for detoxification.

Following the presentation on mustard gas depositions in the Baltic
Sea, reference was made to disposal of gas and bombs in deep areas of
the mid-Atlantic by sinking vessels. This was no longer an option
following international conventions. It was felt that the mustard gas
story had important lessons since methods thought 25 years ago to be
satisfactory were clearly not acceptable, yet the same methods are still
being used today.

In discussing the application of the estuarine circulation model
it was asked if the model could be used to predict particle flow because
metal discharge upstream was a problem. The author said this was quite
possible. The model had been validated against current meter data and
against fine particles and chemicals associated with these particles.

In summing up the session the chairman suggested that the meeting
had not thoroughly appraised containment as an option, but we could
conclude that containment could never be 100%. Ultimately it will be
cost that dictates whether or not this option is in fact used.

3.6 Biological effects of pollutants: Chapters 30 - 33

The last sequence of presentations dealt with the question of
biological effects of pollutants, how to measure or detect these and
how to relate chemical monitoring to biological effects, which after
all are the essential factors to control.

Following the presentation of various techniques and tests, the
discussion centered around the specificity of the physiological tests
proposed by the author. It was noted that most of the enzyme reactions
tested were not specific and hence there would always be a need to carry
out biochemical analysis to determine the cause of the observed effects,
and that different species behave differently. The significance of the
physiological response was also questioned. The author replied that
when the threshold for response was overtaken, then toxicity was
observed. There was a general consensus that chemical analysis of test
organisms would always be necessary. Further discussions of limitations

of physiological tests suggested that a) natural environmental stress
could not be ignored; b) responses were not specific ; c) back-up
analytical work was required to determine environmental changes; and
d) other types of monitoring should be carried out to support physiological tests. Despite the limitations there was some evidence that
physiological response tests are beginning to be used by regulatory
authorities e.g. at the Shetland Oil Terminal and by the U.S. EPA in
California, where mussel watch type sutdies were used. There was not
thought to be much to be gained from experience with physiological
testing of terrestrial animals.

    After presentation of a paper showing the strengths and weaknesses
of benthic monitoring as a tool for determining the effects of
anthropogenic input, the discussion centered on two issues. There was
a consensus that temporal differences could not be used as a basis for
determining changes, and that emphasis should be placed on spatial
comparisons although there were difficulties finding the "right"
Control area. The second part of the discussion was related to the
significance of environmental disturbance. It is becoming easy for the
scientist to show biological changes, but when should action be taken?
There was a consensus that it was the responsibility of the scientist
to show changes, and the administrator/politician on behalf of society
to decide when action should be taken. Some thought the scientist
should have greater influence in the decision-making process.

    After hearing of the scale of US west coast biological monitoring
programmes, it was suggested that the results of the data should be
more widely known. On the question of intercalibration of techniques
there was some meausre of uniformity between those doing tests (80%
agreement) but others were subject to wide variations. More needs to be
done to improve reproducibility. Despite this, some well-established
techniques are suitable for wider use, including third world countries.

    In the session, a wide range of biological assessment methods were
discussed, and although a number of them were successful in determining
changes, the limitations of the techniques, the complexity of the
systems, and the uncertainties of interpretation are the major reasons
for their limited use, when compared with chemical/physical methods.
It is clear that in the foreseeable future, chemical monitoring
supported by selected biological monitoring techniques will continue to
be the basis for monitoring. There was no discussion on biological
modelling as a basis for making assessments, indicating the current
weakness of this art.

3.7 Public information: Chapter 34

    The final formal presentation addressed the question of how to
inform the general public, objectively and understandably, about practical
scientific results on waste treatment and disposal on land, at sea, and
to the atmosphere. This general question was frequently posed during
the discussions in view of the necessity of explaining the available
options and their implications to the population, who clearly influence
the decision process. It was the general feeling that open information
was required and that the uncertainties should be brought out, rather

than suggesting that single values and definite answers could be given
in all cases. Information through the press, through radio, open
university activities, and through television, in the form of a series
of films and lectures should be used and explored more than presently is
the case.

4. GENERAL COMMENT

During the last decade, the concern for the environment has changed.
It is now generally recognized that development, economic stability, and
growth can occur together with an appropriate concern for and protection
of the environment. At the same time it is more and more apparent that
the environment must be treated as a whole and should not be divided
into separate compartments. The importance of the interactions and the
cycling of elements between earth-atmosphere-ocean is becoming evident.
Similarly, the severe limitations of freshwater and soil and the
vulnerability of these components are giving rise to serious concern
in many regions.

Society will continue to produce wastes and means of disposal must
be made available, although recirculation and elimination of some
waste components is being gradually achieved. New substances and new
processes are, however, introduced, often generating new problems. The
limitations of the environment in receiving substances without giving
rise to unacceptable changes are also exemplified on relatively large
scales, where it was thought earlier that the environmental capacity
would not be surpassed. Exapmles are groundwater contamination,
eutrophication in shelf sea areas, and release of organic material in
coastal zones. Experiences with other contamination types (e.g. DDT,
PCB and Hg) also show, however, that trends of increasing contamination
can be reversed by proper actions of control and limitations of use.

All the experiences show the necessity of segregation and careful
use of the different components of the environment as recipients of
different types of wasts. It is necessary to consider the
characteristics of the different environmental components and the
interactions together with the properties of the wastes in order to
arrive at solutions which make optimal use of the environment.
Important tools in arriving at the best solutions are models of various
kinds, which are conceptual as well as operational mathematical
formulations. These also help focus and identify need for further study.
Inherently it is necessary to make use of several different areas of
expertise to disseminate the problems. Interdisciplinary groups need
to work together.

Wastes produced by different societies and processes are different.
Likewise, environmental characteristics are different, e.g. the Baltic,
the North Sea, and the European Mediterranean. These differences
should be taken into account when criteria and standards are developed.
Technological developments and possibilities of maintenance of
installations can also differ between different regions of the world.
This must also be considered when options of waste treatment and
disposal are investigated.

The scientific and technical investigations produce a variety of optional solutions to a given problem, with ranges of uncertainties and interpretations of the findings, and perhaps suggestions for a decision. The decision to select one of the options must, however, be made by the decision- and policy-makers. The establishment of communication between these different groups, as well as between the different scientific and technical disciplines, seems to be an important task. It is also necessary to create better communication between the science-technology-economy aspect and the general public.

Finally, this meeting also showed the need for meetings of this kind, perhaps with better coverage of all science and technology disciplines and also including economists.

REFERENCE

Hollister, C.D. and H.W. Smedes 1984. Selecting sites for radioactive waste repositories. In: Hazardous Waste Management, In Whose Backyard, pp. 63-97, M. Harthill (ed.), The American Association for the Advancement of Science, Westview Press, Boulder, Colorado.